Lecture Notes in Electrical Engineering

Volume 97

Wensong Hu (Ed.)

Electronics and Signal Processing

Selected Papers from the 2011 International Conference on Electric and Electronics (EEIC 2011) in Nanchang, China on June 20–22, 2011, Volume 1

Wensong Hu,
Nanchang University
Xuefu Avenue 999,
New Honggutan Zone,
Nanchang, Jiangxi,
China
E-mail: hwsncu@yahoo.com.cn

ISBN 978-3-642-21696-1   e-ISBN 978-3-642-21697-8

DOI 10.1007/978-3-642-21697-8

Lecture Notes in Electrical Engineering   ISSN 1876-1100

Library of Congress Control Number: 2011929653

© 2011 Springer-Verlag Berlin Heidelberg

This work is subject to copyright. All rights are reserved, whether the whole or part of the material is concerned, specifically the rights of translation, reprinting, reuse of illustrations, recitation, broadcasting, reproduction on microfilm or in any other way, and storage in data banks. Duplication of this publication or parts thereof is permitted only under the provisions of the German Copyright Law of September 9, 1965, in its current version, and permission for use must always be obtained from Springer. Violations are liable to prosecution under the German Copyright Law.

The use of general descriptive names, registered names, trademarks, etc. in this publication does not imply, even in the absence of a specific statement, that such names are exempt from the relevant protective laws and regulations and therefore free for general use.

Typeset & Cover Design: Scientific Publishing Services Pvt. Ltd., Chennai, India.

Printed on acid-free paper

9 8 7 6 5 4 3 2 1

springer.com

EEIC 2011 Preface

The present book includes extended and revised versions of a set of selected papers from the International Conference on Electric and Electronics (EEIC 2011), held on June 20-22 , 2011, which is jointly organized by Nanchang University, Springer, and IEEE IAS Nanchang Chapter.

The goal of EEIC 2011 is to bring together the researchers from academia and industry as well as practitioners to share ideas, problems and solutions relating to the multifaceted aspects of Electric and Electronics.

Being crucial for the development of Electric and Electronics, our conference encompasses a large number of research topics and applications: from Circuits and Systems to Computers and Information Technology; from Communication Systems to Signal Processing and other related topics are included in the scope of this conference. In order to ensure high-quality of our international conference, we have high-quality reviewing course, our reviewing experts are from home and abroad and low-quality papers have been refused. All accepted papers will be published by Lecture Notes in Electrical Engineering (Springer).

EEIC 2011 is sponsored by Nanchang University, China. Nanchang University is a comprehensive university which characterized by "Penetration of Arts, Science, Engineering and Medicine subjects, Combination of studying, research and production". It is one of the national "211" Project key universities that jointly constructed by the People's Government of Jiangxi Province and the Ministry of Education. It is also an important base of talents cultivation, scientific researching and transferring of the researching accomplishment into practical use for both Jiangxi Province and the country.

Welcome to Nanchang, China. Nanchang is a beautiful city with the Gan River, the mother river of local people, traversing through the whole city. Water is her soul or in other words water carries all her beauty. Lakes and rivers in or around Nanchang bring a special kind of charm to the city. Nanchang is honored as 'a green pearl in the southern part of China' thanks to its clear water, fresh air and great inner city virescence. Long and splendid history endows Nanchang with many cultural relics, among which the Tengwang Pavilion is the most famous. It is no exaggeration to say that Tengwang Pavilion is the pride of all the locals in Nanchang. Many men of letters left their handwritings here which tremendously enhance its classical charm.

Noting can be done without the help of the program chairs, organization staff, and the members of the program committees. Thank you.

EEIC 2011 will be the most comprehensive Conference focused on the various aspects of advances in Electric and Electronics. Our Conference provides a chance for academic and industry professionals to discuss recent progress in the area of Electric and Electronics. We are confident that the conference program will give you detailed insight into the new trends, and we are looking forward to meeting you at this world-class event in Nanchang.

Contents

Research on Novel Fuzzy and Intelligent Resource
Management in Grid Computing 1
FuFang Li, Guowen Xie, DeYu Qi, Fei Luo, DongQing Xie

Wireless Sensor Network System for the Real-Time Health
Monitoring 9
Chen Li-Wan, Chen Qiang, Li Hong-Bin

Study of Automatic Construction and Adjustment of Social
Accounting Matrix (SAM) 15
Jun Wang, Lan-Juan Liu, Ji-Hui Shi

Research on Testing Methods of I-V Characteristics of Solar
Photovoltaic Cell Array 23
Yunhai Hou, Ershuai Li, Shihua Sun

The Performance Comparison of ANFIS and
Hammerstein-Wiener Models for BLDC Motors 29
Abbas Nemati, Mohammadreza Faieghi

Application of Real-Time Image Mosaics for Video Processing
System 39
Wang Lutao, Zhang Lei

Estimate the All Vanishing Points from a Single Image 47
Yongyan Yu, Zhijian Wang, Yuansheng Lou

Fault Diagnosis of Three Level Inverter Based on Improved
Neural Networks 55
Wang Wu, Wang Hong-Ling, Bai Zheng-Min

The Application for Diagnose of Ginseng Disease and Insect
Pests-Es Based on Case-Based Reasoning 63
Zhang Li-Juan, Li Dong-Ming, Chen Gui-Fen, Chen Hang

Identification of Pipeline Circuit Design 71
Jiatian Zhang, Jianming Ma, Zhengguo Yan

Vehicle Velocity Calculation in Accident Based on Morphology
of Broken Vehicle Windshield Glass 77
Jie Zhang, Chuanjiao Sun, Hongyun Chen

A Low Power Wide Linear Range OTA and Its Application in
OTA-C Filter 85
An Li, Chunhua Wang, XiaoRong Guo, Jingru Sun

Research of Self-calibration Location Algorithm for ZigBee
Based on PSO-RSSI 91
Chengbo Yu, Yimeng Zhang, Jin Zhang, Yuxuan Liu

A New Method for Bad Data Identification of Integrated
Power System in Warship Based on Fuzzy ISODATA
Clustering Analysis 101
Lei Wu, Li Xia, Yong Shan

Analysis of Fingerprint Performance among Left-Handed and
Right-Handed People 109
Nasrul Humaimi Mahmood, Akram Gasmelseed

The Design of Reconfigurable Remote Monitor and Control
System Based on CAN Bus 117
Wang Qing

Adaptive Friction Compensation of Robot Manipulator 127
Wang Sanxiu, Jiang Shengtao

An Example of Cloud Business Application Development
Using the Continuous Quantity Identifier of the Cellular Data
System 135
Toshio Kodama, Tosiyasu L. Kunii, Yoichi Seki

Research on a Novel Piezoelectric Linear Motor Driven by
Three Langevin Type Vibrators 145
He Honglin, Chen Wenjun, Long YuFan

Reduction of Production Costs in Machining the Wood Panels
MDF Using Predictive Maintenance Techniques 153
José Augusto Coeve Florino, Leonimer Flávio de Melo

Minimum Step Control System Design 159
Youmu Zhang, Dongyun Luo

Handling Power System Alarm Cascade Using a Multi-level
Flow Model... 165
*Taotao Ma, Jinxing Xiao, Jingqi Xu, Chuangxin Guo, Bin Yu,
Shaohua Zhu*

Design and Implement for Information System Intelligent
Monitoring System of Power Grid 177
Lei Huang, Taotao Ma, Jufang Li

Design of Controlling System of Higher Layer Elevator 185
Yushui Huang, Wenbin Peng, Chong Liu

A New Denosing Method in Seismic Data Processing Based
on Wiener Filter.. 191
Wang Shi-Wei

Trotting Gait Planning and Implementation for a Little
Quadruped Robot 195
Bin Li, Yibin Li, Xuewen Rong, Jian Meng

Optimal Reducing the Solutions of Support Vector Machines
Based on Particle Swam Optimization 203
Jih Pin Yeh, Chiang Ming Chiang

Improved Relaxation Algorithm for Passive Sensor Data
Association 215
Cheng Ouyang, Hong-Bing Ji, Jin-Long Yang

A Novel Integrated Cmos Uhf Rfid Reader Transceiver
Front-End Design... 227
Chao Yuan, Chunhua Wang, Jingru Sun

Hammerstein-Wiener Model Predictive Control of Continuous
Stirred Tank Reactor 235
Man Hong, Shao Cheng

The Relativity Analysis between Summer Loads and Climatic
Condition of Ningbo... 243
Yiou Shao, Xin Yu, Huanda Lu

Application of the Modified Fuzzy Integral Method in
Evaluation of Physical Education Informationization in High
School 251
Tongtong Liu, Ting Li

Research on the IT Governance Framework of Sports
Products Industry 257
Baozeng Qu, Jinfeng Li

An Efficient Incremental Updating Algorithm for Knowledge
Reduction in Information System 263
Changsheng Zhang, Jing Ruan

Principle of Radial Suspension Force Generation and Control
System Simulation for a Bearingless Brushless DC Motor 273
Leigang Chen, Xiaodong Sun, Huangqiu Zhu

Based on Regularity District Scheduling of Resources for
Embedded Systems 281
Min Rao, Jun Xie, Di Cai, Minhua Wu

ICAIS: Improved Contention Aware Input Selection
Technique to Increase Routing Efficiency for
Network-On-Chip 289
Ebrahim Behrouzian Nejad, Ahmad Khademzadeh, Kambiz Badie, Amir Masoud Rahmani, Mohammad Behrouzian Nejad, Ahmad Zadeali

A New Method of Discriminating ECG Signals Based on
Chaotic Dynamic Parameters 299
Canyan Zhu, Aiming Ji, Lijun Zhang, Lingfeng Mao

Research on Iterative Multiuser Detection Algorithm Based
on PPIC 307
Canyan Zhu, Lijun Zhang, Aiming Ji, Yiming Wang

Research of Aero-Engine Robust Fault-Tolerant Control
Based on Linear Matrix Inequality Approach 315
Xue-Min Yang, Lin-Feng Gou, Qiang Shen

Research on m Value in Fuzzy c-Means Algorithm Used for
Underwater Optical Image Segmentation 325
Shilong Wang, Yuru Xu, Lei Wan

A Novel Method on PLL Control 335
Youhui Xie, Yanming Zhou

Analysis and Design of Mixer with DC-Offset Cancellation 343
Jianjun Song, Shuai Lei, Heming Zhang, Yong Jiang

UWB Low-Voltage High-Linearity CMOS Mixer in Zero-IF
Receiver 349
Huiyong Hu, Shuai Lei, Heming Zhang, Rongxi Xuan, Bin Shu, Jianjun Song, Qiankun Liu

Design and Simulation of Reserved Frame-Slotted Aloha
Anti-collision Algorithm in Internet of Things 355
Jun-Chao Zhang, Jun-Jie Chen

A Hop to Hop Controlled Hierarchical Multicast Congestion
Control Mechanism 363
Jun-Chao Zhang, Rong-Xiang Zhao, Jun-Jie Chen

A Moving Mirror Driving System of FT-IR Spectrometer for
Atmospheric Analysis.. 371
*Sheng Li, Yujun Zhang, Minguang Gao, Liang Xu, XiuLi Wei,
JingJing Tong, Ling Jin, Siyang Cheng*

Research on Intelligent Cold-Bend Forming Theory and
Fitting Method of Actual Deformation Data 379
Guochang Li

Research on Cold-Belt Deformable Characteristic Eigenfinite
Strip Filter and Characteristic Agent Model 387
Guochang Li

TPMS Design and Application Based on MLX91801 395
*Song Depeng, Wan Xiaofeng, Yang Yang, Gan Xueren, Lin Xiaoe,
Shen Qiang*

Research on Control Law of Ramjet Control System Based on
PID Algorithm 403
Duan Xiaolong, Mao Genwang, Xu Zhongjie, Wu Baoyuan

Layout Design and Optimization of the High-Energy Pulse
Flash Lamp Thermal Excitation Source 411
*Zhang Wei, Wang Guo-wei, Yang Zheng-wei, Song Yuan-jia,
Jin Guo-feng, Zhu Lu*

Effective Cross-Kerr Effect in the N-Type Four-Level Atom 421
Liqiang Wang, Xiang'an Yan

The Design and Implementation of Groundwater Monitoring
and Information Management System 427
Yujie Zhang, Yuanyuan Zhang

Design of Hydrological Telemetry System Based on Embedded
Web Server 435
Yujie Zhang, Yuanyuan Zhang, Sale Xi, Jitao Jia

Passive Bistatic Radar Target Location Method and
Observability Analysis 443
Cai-Sheng Zhang, Xiao-Ming Tang, You He, Jia-Hui Ding

Analysis of Coherent Integration Loss Due to FSE in PBR 453
Cai-Sheng Zhang, Xiao-Ming Tang, You He, Jia-Hui Ding

Research on Double Orthogonal Multi-wavelet Transform Based Blind Equalizer 461
Han Yingge, Li Baokun, Guo Yecai

A Novel Approach for Host General Risk Assessment Based on Logical Recursion 469
Xiao-Song Zhang, Jiong Zheng, Hua Li

A Time Difference Method Pipeline Ultrasonic Flowmeter 477
Shoucheng Ding, Limei Xiao, Guici Yuan

A Study of Improvements on the Performances of Load Flow Calculation in Newton Method 485
Xiangjing Su, Xianlin Liu, Ziqi Wang, Yuanshan Guo

A Design Approach of Model-Based Optimal Fault Diagnosis ... 493
Qiang Shen, Lin-Feng Gou, Xue-Min Yang

Dynamic Adaptive Terminal Sliding Mode Control for DC-DC Converter 503
Liping Fan, Yazhou Yu

Adoption Behavior of Digital Services: An Empirical Study Based on Mobile Communication Networks 509
Bing Lan, Xuecheng Yang

Review of Modern Speech Synthesis 517
Cheng Xian-Yi, Pan Yan

Design of the Stepper Motor Controller 525
Bo Qu, Hong Lin

Design of Sensorless Permanent Magnet Synchronous Motor Control System 533
Bo Qu, Hong Lin

The Design of Synchronous Module in AM-OLED Display Controller 541
Junwei-Ma, Feng Ran

A Survey of Computer Systems for People with Physical Disabilities 547
Xinyu Duan, Guowei Gao

A Monopole Scoop-Shape Antenna for 2.4GHz RFID Applications 553
Mu-Chun Wang, Hsin-Chia Yang

EEIC 2011 Organization

Honor Chairs

Prof. Chin-Chen Chang   Feng Chia University, Taiwan
Prof. Jun Wang   Chinese University of Hong Kong, HongKong

Scholarship Committee Chairs

Chin-Chen Chang   Feng Chia University, Taiwan
Jun Wang   Chinese University of Hong Kong, HongKong

Scholarship Committee Co-chairs

Zhi Hu   IEEE IAS Nanchang Chapter, China
Min Zhu   IEEE IAS Nanchang Chapter, China

Organizing Co-chairs

Jian Lee   Hubei Normal University, China
Wensong Hu   Nanchang University, China

Program Committee Chairs

Honghua Tan   Wuhan Institute of Technology, China

Publication Chairs

Wensong Hu   Nanchang University, China
Zhu Min   Nanchang University, China
Xiaofeng Wan   Nanchang University, China
Ming Ma   NUS ACM Chapter, Singapore

Parasitic Effect Degrading Cascode LNA Circuits with $0.18\mu m$
CMOS Process for 2.4GHz RFID Applications 561
Mu-Chun Wang, Hsin-Chia Yang, Ren-Hau Yang

Minimization of Cascade Low-Noise Amplifier with 0.18 μm
CMOS Process for 2.4 GHz RFID Applications................. 571
Mu-Chun Wang, Hsin-Chia Yang, Yi-Jhen Li

Simulation and Analysis of Galileo System...................... 579
Wei Zhang, Weibing Zhu, Huijie Zhang

High-Power Controllable Voltage Quality Disturbance
Generator 587
Gu Ren, Xiao Xiangning

High-Power Controllable Current Quality Disturbance
Generator 593
Gu Ren, Xiao Xiangning

Nonlinear Control of the Doubly Fed Induction Generator by
Input-Output Linearizing Strategy............................... 601
Guodong Chen, Luhua Zhang, Xu Cai, Wei Zhang, Chengqiang Yin

Passive Location Method for Fixed Single Station Based on
Azimuth and Frequency Difference Measuring 609
Tao Yu

Online Predicting GPCR Functional Family Based on
Multi-features Fusion 617
Wang Pu, Xiao Xuan

Information Architecture Based on Topic Maps.................. 625
Guangzheng Li

Implementation of 2D-DCT Based on FPGA with Verilog
HDL... 633
Yunqing Ye, Shuying Cheng

Impact of Grounding Performance of AMN to Conducted
Disturbance Measurement 641
Xian Zhang, Qingdong Zou, Zhongyuan Zhou, Liaolan Wu

Optimization of PID Controller Parameters Based on
Improved Chaotic Algorithm 649
Hongru Li, Yong Zhu

Generalizations of the Second Mean Value Theorem for
Integrals... 657
Chen Hui-Ru, Shang Chan-Juan

Research into Progressiveness of Intermediate Point and
Convergence Rate of the Second Mean Value Theorem for
Integrals 663
Chen Hui-Ru, He Chun-Ling

Flooding-Based Resource Locating in Peer-to-Peer Networks ... 671
Jin Bo

A Kind of Service Discovery Method Based on Petroleum
Engineering Semantic Ontology 679
Lixin Ren, Li Ren

Study of Coal Gas Outburst Prediction Based on FNN 687
Yanli Chai

Study of Coal and Gas Prediction Based on Improvement of
PSO and ANN 695
Yanli Chai

A New Web-Based Method for Petroleum Media Seamless
Migration 703
Jiaxu Liu, Yuanzheng Wang, Haoyang Zhu

The Summary of Low-Voltage Power Distribution Switch
Fault Arc Detection and Protection Technology 711
Aihua Dong, Qiongfang Yu, Yangmei Dong, Liang Li

Multi-disciplinary Modeling and Robust Control of Electric
Power Steering Systems 719
Ailian Zhang, Shujian Chen, Huipeng Chen

A Novel Analog Circuit Design and Test of a Triangular
Membership Function 727
Weiwei Shan, Yinchao Lu, Huafang Sun, Junyin Liu

Novel Electronically Tunable Mixed-Mode Biquad Filter 735
Sajai Vir Singh, Sudhanshu Maheshwari, Durg Singh Chauhan

Localization Algorithm Based on RSSI Error Analysis 743
Liu SunDong, Chen SanFeng, Chen WanMing, Tang Fei

Moving Target Detection Based on Image Block
Reconstruction 749
Guo Sen, Liao Jiang

Explore Pharmacosystematics Based on Wavelet Denoising
and OPLS/O2PLS-DA 755
Zhuo Wang, Jianfeng Xu, Ran Hu

Contents

The Study on Pharmacosystematics Based on
WOSC-PLS-DA 763
Zhuo Wang, Jianfeng Xu, Ran Hu

Research on Sleep-Monitoring Platform Based on Embedded
System 771
Kaisheng Zhang, MingXing Gao

Study on Sleep Quality Control Based on Embedded 777
Kaisheng Zhang, Zhen Li

Study on Embedded Sleep-Monitoring Alarm System 785
Kaisheng Zhang, Wenbo Ma

CUDA Framework for Turbulence Flame Simulation 791
Wei Wei, Yanqiong Huang

Multi-agent Crowd Collision and Path Following Simulation 797
Wei Wei, Yanqiong Huang

Study on a New Type Protective Method of Arc Grounding
in Power System 803
Wenjin Dai, Cunjian Tian

The Development of Economizing-Energy Illumination
System 811
Wenjin Dai, Lingmin Liang

The Membrane Activity's Design and Evolution about acE
Service Flow 817
Xiaona Xia, Baoxiang Cao, Jiguo Yu

Design of Intelligent Manufacturing Resource Planning III
under Computer Integrated Manufacturing System 825
Baoan Hu, Min Wu

The Design of Improved Duffing Chaotic Circuit Used for
High-Frequency Weak Signal Detection 831
Wenjing Hu, Zhizhen Liu, Zhihui Li

A Method of Setting Radio Beacon's Frequency Based on
MCU 839
Yongliang Zhang, Jun Xue, Wenhua Zhao, Xinbing Fang

Study on Modeling and Simulation of UPFC 847
Tingjian Zhong, Zunnan Min, Raobin, Huangwei

The Research Design of the Pressure-Tight Palette for Oil
Paintings 859
Ruilin Lin

Product Innovative Development of the Automatic Protection
Facility Product for Pool Drain Covers 865
Ruilin Lin

A Dynamic Optimization Modeling for Anode Baking Flame
Path Temperature 871
Xiao Bin Li, Leilei Cui, Naijie Xia, Jianhua Wang, Haiyan Sun

An Improved Ant Colony System for Assembly Sequence
Planning Based on Connector Concept 881
Hwai-En Tseng

Study on Hysteresis Current Control and Its Applications in
Power Electronics 889
Ping Qian, Yong Zhang

Design of Pulse Charger for Lead-Acid Battery 897
Ping Qian, Maopai Guo

Hot-Line XLPE Cable Insulation Monitoring Based on Quick
Positive and Negative DC Superposition Method 903
Jianbao Liu, Zhonglin Yang, Liangfen Xiao, Leping Bu, Xinzhi Wang

Study on the Propagation Characteristics of Electromagnetic
Waves in Horizontally Inhomogeneous Environment 913
Lujun Wang, Yanyi Yuan, Mingyong Zhu

Multimedia Sevice Innovation Design of Graffiti in Modern
Commercial Design 921
Xiaoyan Wang, Zhanxi Zhao, Yantao Zhong

Using Bypass Coupling Substrate Integrated Circular Cavity
(SICC) Resonator to Improve SIW Filter Upper Stopband
Performance 929
Boren Zheng, Zhiqin Zhao, Youxin Lv

Lossless Robust Data Hiding Scheme Based on Histogram
Shifting 937
Qun-Ting Yang, Tie-Gang Gao, Li Fan

Open Fault Diagnose for SPWM Inverter Based on Wavelet
Packet Decomposition 945
Zhonglin Yang, Jianbao Liu, Hua Ouyang

The Method for Determination the Freshness of Pork Based
on Edgedetection 953
Tianhua Chen, Suxia Xing, Jingxian Li

Remote Monitoring and Control System of Solar Street
Lamps Based on ZigBee Wireless Sensor Network and GPRS ... 959
Lian Yongsheng, Lin Peijie, Cheng Shuying

Non-line of Sight Error Mitigation in UWB Ranging Systems
Using Information Fusion 969
Xiangyuan Jiang, Huanshui Zhang

Ensemble Modeling Difficult-to-Measure Process Variables
Based the PLS-LSSVM Algorithm and Information Entropy 977
Jian Tang, Li-jie ZHao, Shao-wei Liu, Dong Yan

E-Commerce Capabilities and Firm Performance: A Empirical
Test Based on China Firm Data 985
Kuang Zhijun

The Development of Experimental Apparatus for
Measurement of Magnetostrictive Coefficient 993
Lincai Gao, Bao-Jin Peng, Fei Xu, Jia-Qi Hong, Wenhao Zhang, Xiao-Dong Li, Cui-Ping Qian

A New Driving Topology for DC Motor to Suppress Kickback
Voltage 1001
Jiang Sun, Bo Zhang, Haishi Wang, Ke Xiao, Peng Ye

The Reduction of Facial Feature Based on Granular
Computing 1015
Runxin He, Nian He

Control Methods and Simulations of Micro-grid 1023
He-Jin Liu, Ke-Jun Li, Ying Sun, Zhen-Yu Zou, Yue Ma, Lin Niu

Harmonic Analysis of the Interconnection of Wind Farm 1031
Jidong Wang, Xuhao Du, Guodong Li, Guanqing Yang

Simulation of Voltage Sag Detection Method Based on DQ
Transformation 1039
Jidong Wang, Kun Liu, Guanqing Yang

Simulation of Shunt Active Power Filter Using Modified
SVPWM Control Method 1047
Jidong Wang, Guanqing Yang, Kun Liu

Author Index 1055

Research on Novel Fuzzy and Intelligent Resource Management in Grid Computing

FuFang Li[1,2], Guowen Xie[3], DeYu Qi[4], Fei Luo[1], and DongQing Xie[2]

[1] College of Automation Science and Engineering, South China University of technology, Guangzhou 510640, China
[2] School of Computer Science & Educational Software, GuangZhou University, Guangzhou 510006, China
[3] School of Life Science, GuangZhou University, Guangzhou 510006, China
[4] School of Computer Science and Engineering, South China University of Technology, Guangzhou 510006, China
lifuf@mail.csu.edu.cn, xgw168@163.com, deyuqi@gmail.com, aufeiluo@scut.edu.cn

Abstract. Resource management and scheduling are the most important problems in grid computing. This paper tries to present a novel Agent & Small-world theory based Grid resource management model (A_Swt_Grm) and related Optimized Scheduling Algorithm of Grid Resource based on Fuzzy Clustering (OSA_GR_FC) for the model. Being constructed by agent technology and small-world theory, the model of A_Swt_Grm has good intelligence, self-adjustment, efficiency, and high performance. The proposed algorithm can subtly schedule appropriate resource to accurately satisfy the user's multi-QoS needs of resource, while effectively avoid assigning most powerful resources to the user which largely exceeds the user's needs. Experimental results show that the proposed model and algorithm works better than other similar algorithms.

Keywords: Grid Resource Management; Agents; Small-World Network Theory; Fuzzy Clustering.

1 Introduction and Related Work

Resource management and scheduling is one of the most important and key problem in the field of grid research. To improve the efficiency of grid resource management, numerous works have been done by researchers in the area. By using cooperative game to construct a math model, paper [1] presented out a grid resource assignment model and related scheduling algorithm based on Nash-balance theory, which soundly reduce the waiting time of tasks. Enlightened from Sufferage algorithm, paper [2] proposed a grid resource mapping algorithm of QoS-Sufferage. To satisfy the user's multi-QoS requests, the algorithm of QoS-Sufferage firstly introduces a synthetic matrix for grid resources, and then maps the grid resources to appropriate tasks. With combination of DLT theory and LCM method, Shah [3] introduced a novel scheduling strategy and algorithm for hybrid grid resources, which largely improved the performance of scheduling. Soni [4] put forward a constraint-based job and resource scheduling algorithm, which reduced the processing time, processing cost and enhance the resource utilization in comparison to similar algorithms.

Agent technology has provided a new way in solving the problem of grid resource management and task scheduling which needs more cooperation, and has been widely used in the area[5~7]. Paper [7] addressed an agent-based grid resource management model and relevant scheduling algorithm (named HGSSA) by integrating Genetic Algorithm (GA) and Simulated Annealing Algorithm (SA). The algorithm of HGSSA possesses the advantages of both GA and SA, but the algorithm itself is too complicated which affects the performance of the algorithm. Recent research has proved that many real networks such as WWW and Internet possess typical character of small-world network [8]. Organized according to the small-world theory, grid resources in the grid resource management system are more reachable to each other, which would largely accelerate the searching speed while scheduling grid resources.

The above research works have achieved better results, but how to effectively manage and schedule the hybrid and fast changing grid resources in real time has not yet been resolved satisfactorily far away. Aiming at the shortage in real-time performance and efficiency of existing grid resource scheduling methods, a novel grid resource management model and relevant optimized scheduling algorithm based on Agents, Small-World theory and fuzzy clustering thoughts is put forward in this paper. The proposed model and algorithm can meet multiple QoS needs of the tasks at the same time, and also possesses sound intelligence, self-adaptability, robustness and high performance and efficiency.

2 Agent and Small-World Theory Based Grid Resource Management Model

The proposed Agent & Small-world theory based Grid resource management model (A_Swt_Grm) is a of distributes layered model, and is shown in Figure 1. Figure 1 shows that the whole grid computing system is composed of n Autonomous Areas (AA), and each AA includes m, p, …, q grid resources respectively.

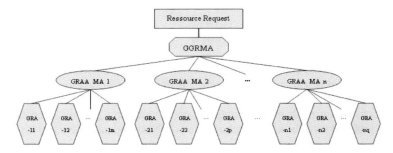

Fig. 1. The grid resource management model of A_Swt_Grm

As is shown in Figure 1, the proposed resource management model is logically divided into 3 layers: (1) The grid resource virtualization layer. This layer virtualizes various grid resources deployed in grid nodes as grid service resources with specific functions. Grid Resource Agents (GRAs) are introduced to present theses virtualized grid service resources. On the one hand, the GRAs are responsible for collecting real-time base information of grid nodes, such as CPU frequency, memory size, network width, network connection degree, etc; on the other hand, they are also in charge of gathering up real-time state, function description, calling parameter and its' format, etc of the grid service. (2) The autonomy-area grid resource management layer. All GRAs try to consult with each other, and thus to join and organize a certain number of Grid Resource Autonomy-Areas (GRAAs), according to small-world network theory. In each GRAA, a CNAA (Core Node of Autonomy-Area), with a GRAA_MA (Grid Resource Autonomy-Area Management Agent) deployed in, is selected out to manage and schedule the grid service resources within this autonomy area. (3) The global grid resource management layer. In this layer, with the guidance of small-world network theory, a GGRMN (Global Grid resource Management Node) should be chosen out from all nodes in the whole grid system through negotiations between GRAA_MAs of the whole system. In the node of GGRMN, a GGRMA (Global Grid Resource Management Agent) is deployed, which is responsible for managing and scheduling the grid resources in the whole grid system. Based on agent technology and small-world theory, the model of A_Swt_Grm can effectively increase the speed of finding out specific grid resources, and hence soundly improve the performance and efficiency of the grid resource scheduling system.

3 Optimized Scheduling Algorithm of Grid Resource Based on Fuzzy Clustering (OSA_GR_FC) in A_Swt_Grm

To satisfy user's multi-QoS needs, this paper introduces an Optimized Scheduling Algorithm of Grid Resource based on Fuzzy Clustering (OSA_GR_FC) in A_Swt_Grm. The main idea of the algorithm of OSA_GR_FC is as follows: Firstly, we try to respectively build Grid Resource Vectors (GRVs) from grid resources and Requested Grid Resource Vectors (R-GRVs) with same format; Secondly, we put together GRVs and R-GRVs to construct hybrid vectors, and next we do hybrid fuzzy clustering on the hybrid vectors. According to the meaning of fuzzy clustering [9], the hybrid vectors would be divided into a certain number of groups, and similarity degree between the GRV(s) and R-GRV(s) in the same group would be high, which shows that in the same group, the GRV(s) would exactly match the R-GRV(s). At last, we schedule the grid resource represented by a GRV to the user represented by a R-GRV, which the very GRV and R-GRV are in the same group. By this way, we precisely schedule the appropriate resource to the user that subtly satisfy his requests of multi-QoS, while reserving the resources whose power is greatly exceeds the needs of the current task for future use. To reduce the computation complexity of clustering, we do hybrid fuzzy clustering in both autonomy-area grid resource management layer and global grid resource management layer, which the performance and efficiency of the proposed algorithm are largely increased. The algorithm of OSA_GR_FC is described in more detail as below.

(1) Activate all GRAs deployed in every grid nodes, and all GRAs try to obtain information to build GRVs;

(2) By negotiation and according to small-world theory, all GRAs try to organize a certain number of GRAAs, and select out a CNAA for each GRAA. Then activate all GRAA_MA in every CNAA, and all GRAA_MA begin to collect and arrange the GRVs of the autonomous area;
(3) All GRAA_MAs try to build global grid resource management system according to small-world theory, and the GGRMN should be chosen for the whole system simultaneously. Then, activate GGRMA in GGRMN at the same time.
(4) GGRMA accepts the resource request from the users, and builds R-GRV(s) according to the user's requests;
(5) GGRMA submit R-GRV(s) to every GRAA_MA; Then, GRAA_MAs do hybrid clustering in their own autonomous area; At last, GRAA_MAs return candidate grid resource of its' area to GGRMA;
(6) GGRMA collects all candidate grid resources from all autonomous areas, and GGRMA do hybrid clustering again;
(7) GGRMA pick up the clustering sub set of S with R-GRV(s) included in;
(8) GGRMA randomly selects a GRV from sub set S, and then schedule the grid resource related with the GRV to the user;
(9) End.

4 Experimental Results and Evolution

To investigate and verify the proposed model and algorithm, we do simulation experiment on Netlogo, a multi-agent simulation platform. To carry out the experiment, we design and implement three kinds of Agents: GRA, GRAA_MA and GGRMA. We also design and implement an Agent to act as real grid resource during the process of simulation experiment. The actual experiment scheme is shown in table 1. As is shown in table 1, we do simulation experiment under 8 kinds of experimental conditions. In each condition, the simulation experiment is repeated 20 times, and the average of the results is taken as the final result. For the convenience of comparison, we also do same experiment by using similar scheduling algorithm of QoS-Suf (of reference [2]) and HGSSA (of reference [7]) under the same conditions. Experimental results are shown in Figure 2, 3 and 4. Figure 2, 3 and 4 show the makespan, average waiting time and failure ratio of resource scheduling respectively.

Table 1. Eight condition of resources and related requests

No.	1	2	3	4	5	6	7	8
Number of resources	10	25	40	55	70	85	100	115
Number of requests	60	150	350	550	800	1100	1500	2000

As is shown in Figure 2, 3 and 4, the proposed model and algorithm of OSA_GR_FC in this paper, is better than other two similar algorithms of QoS-Suf and HGSSA. On one hand, the proposed model and algorithm are designed based on agent technology and small-world theory, which makes it be more intelligent and

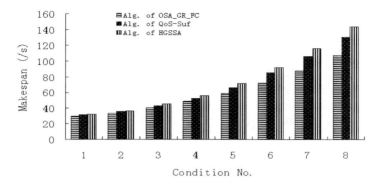

Fig. 2. Makespan of three algorithms

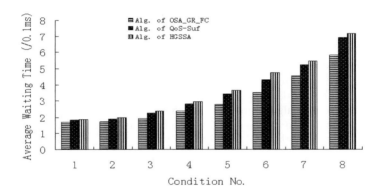

Fig. 3. Average waiting time

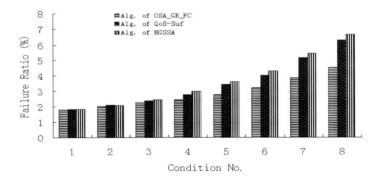

Fig. 4. Failure ratio of scheduling

self-adaptive. By using small-world theory to construct grid resource autonomous area, resources can be scheduled more quickly, and thus better performance and higher efficiency have been achieved. On the other hand, by applying fuzzy clustering method to conduct resources scheduling, the proposed algorithm can exactly assign appropriate resources to the tasks which subtly meet the task's multi-QoS needs, while saving most powerful resources for future use. By this way, this algorithm can effectively avoid unreasonable scheduling, and failure ratio of scheduling can also be reduced recognizably, and consequently, the resources in the whole system would be more fairly used. Furthermore, seeing from Fig. 2,3 and 4, we can find out that, when the number of resources and tasks becoming more bigger, the algorithm of OSA_GR_FC presents more significant advantages in the respects of makespan, average waiting time and failure ratio.

5 Summaries and Future Work

This paper presents a novel Agent & Small-world theory based Grid resource management model (A_Swt_Grm) and relevant optimized resource scheduling algorithm (OSA_GR_FC). The proposed model and algorithm can effectively satisfy user's multi-QoS requests of grid resources, with sound intelligence, self- adaptability, robustness and high performance and efficiency. Simulation experiments show that our model and algorithm works better on makespan, average waiting time and failure ratio compared with similar algorithms. In the future, we'll try to design and implement knowledgebase and learn mechanism for Agents, so as to further improve the intelligence and self- adaptability. We will also do more work on making the proposed model and algorithm be applied to practical engineering projects.

Acknowledgements

Thanks for the self-giving helps of colleagues from School of Computer Science and Technology and College of Automation Science and Engineering, South China University of technology. Their idea and suggestion help us to finish the work more smoothly and efficiently. The paper has been co-financed by GuangZhou Municipal High School Science Fund (under grant No. 10A009), National Natural Science Foundation of China (under grant No. (60903165, 60774032, 30970191), and Foundation of comprehensive Strategic Cooperation between the Guangdong Province and the Chinese Academy of Sciences (under Grant No 2009B091300069).

References

1. Fang, J., Xu, T., Wang, X.: Grid Resource Management Based on Nash Equilibrium. Journal of Beijing University of Technology 36(10), 1418–1422 (2010)
2. Hua, R., Fu, Y., Du. Research, Y.: on QoS-guaranteed Resource Mapping Policy in Service Grid. Computer Engineering 36(15), 43–45, 48 (2010)

3. Shah, S.N.M., Bin Mahmood, A.K., Oxley, A.: Hybrid Resource Allocation Method for Grid Computing. In: 2010 Second International Conference on Computer Research and Development (ICCRD 2011), Kuala Lumpur, pp. 426–431 (2010)
4. Soni, V.K., Sharma, R., Mishra, M.K., Das, S.: Constraint-Based Job and Resource scheduling in Grid Computing. In: 2010 3rd IEEE International Conference on Computer Science and Information Technology (ICCSIT 2010), Chengdu, pp. 334–337 (July 2010)
5. Kesselman, C.: Applications of Intelligent Agent Technology to The Grid. In: Proceedings of IAT 2004: IEEE/WIC/ACM 2004 International Conference on Intelligent Agent Technology, pp. xxv+569 (2004)
6. Li, F., Xie, D.-Q., Qi, D., et al.: Research on Agent-based grid resource management. Computer Engineering and Applications 45(10), 30–33 (2009)
7. Zeng, Z., Shu, W.: Grid resource management model based on mobile agent. Computer Engineering and Applications 44(21), 138–141 (2008)
8. Albert, R., Barabási, A.-L.: Statistical mechanics of complex networks. Rev. Mod. Phys. 74, 47–97 (2002)
9. Ruspini, E.H.: New experimental results in fuzzy clustering. Information Science 18(2), 273–284 (1973)

Wireless Sensor Network System for the Real-Time Health Monitoring

Chen Li-Wan, Chen Qiang, and Li Hong-Bin

College of Electronic and Information Engineering, Chongqing Three Gorges University,
404100, Chongqing China
clw164@126.com

Abstract. Based on the conventional physical parameters measurement of human body as an example, this paper has put forward a scheme of real-time health monitoring on wireless sensor networks. The scheme has been integrated into most detection sensors of human physiological parameters, And human physiological parameters for data acquisition, and the measured data is transmitted to the monitoring center through wireless sensor network the monitoring center, and these physiological parameters processed, the human body health evaluated, and information immediate feedbacked for prevention and treatmen.

Keywords: wireless sensor networks; heath monitoring; zigbee; data acquisition.

1 Introduction

Common physiological parameters of human include body temperature, blood pressure, heart rate, oxygen, etc. Through the monitoring of these indicators we can roughly understand the body's physical health and illness. By medicine subjects, we know that the body's normal average temperature is between 36-37 ℃ (Armpit), the body's normal blood pressure is systolic blood pressure, less than 120, in special circumstances 120-139 is normal, diastolic blood pressure less than 80, in special circumstances 80-90 is normal, blood oxygen content greater than or equal to 95 Normal, pulse between 60-100 times per minute is normal. The traditional method is to check the patient to the hospital or specialized clinic by a doctor using fixed diagnostic medical equipment, the biggest drawback of the method is inconvenient for patients, not flexible, especially for elderly patients, disabled patients and remote Mountain patients, and because of fatigue and back and forth to the hospital to see patients with adverse reactions and other complex equipment of the psychological pressure, making the diagnosis not matching the data with the real situation may lead to wrong diagnosis. In this paper it designs a new type of wireless network monitoring systems, using a variety of micro-sensors automatically collect patient physiological indicators, the use of multi-channel high-frequency wireless data transmission, a doctor's examination and consultation, the biggest advantage of the approach is that eliminates patients' running around between home and hospital; patients accept checks in the free space and mind of peace, measure accurately; enhance the hospital's modern information management and increase efficiency; especially for rural patients in underserved[1] areas and chronic diseases and aging patient care. The wireless sensor network is to remote health care to provide a more convenient, fast and accurate means of technology.

2   System Design

This paper presents a new type of wireless sensor network health monitoring system, which includes a wireless sensor node and destination node-specific form of the sensor monitoring network, the gateway to external connections, external transmission network, base stations, external data processing networks (Internet, etc.) , Remote medical monitoring center, a variety of users as shown in Figure 1.

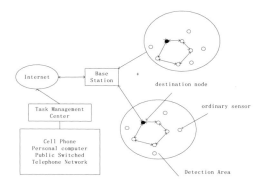

Fig. 1. Wireless sensor network health monitoring system structure

In the network, to detect the body temperature, blood pressure, pulse, oxygen and other special micro-sensor nodes installed in the patients who need to check the corresponding parts of a large number of sensor networks through the organization constitutes a form of patient monitoring area, the sensor node is need to monitor the physiological data collection, data obtained after treatment by simple transmission of the node to the destination node, the access node at the same time is a wireless sensor network and external network communications gateway nodes, gateway nodes through a single-hop links or a series of wireless network access nodes of the transmission network, the data sent from the monitoring area to provide remote access and data processing base stations, base station is to tansport data management center through the external network (Internet or satellite communication networks), and finally, the use of the application of the collected data are analyzed by the remote data center professional doctors diagnosis, by a variety of ways through the remote data transfer communications to the patient feedback, make suggestions and advices , for example by the RS-232, it is connected to the computer display, publish information by the GSM mobile phone, by the modem using a fixed telephone answering message, a list by means of the printer to print directly to the user, another user and remote data management center can also be through an external network, and the destination node interact with the

---

[1] Fund Project: Wanzhou District Science and Technology Project (2010-2).

sensor node to the destination node which can query and control instructions issued, and return to the target sensor node receiving the information.

Protocol for wireless sensor networks, using Zigbee technology, is a new close, low complexity, low power, low cost automation and wireless control for wireless network technology, and it defines the relevant set of IEEE802.15.4 net, security and application software, technical standards, in thousands of tiny sensors to achieve communication between the co-ordination. These sensors requires very little energy to relay the data adopted by radio waves from one sensor to another sensor, and the communication efficiency is very high. Compared with various existing wireless communication technology, ZigBee technology will be the lowest power consumption and cost of technology. ZigBee is the starting point to achieve a low-cost wireless networks easy to implement, while its low power consumption allows battery-powered products for 6 months to several years of work time. The low data rate and communication range of minor characters, determine the ZigBee technology which is suitable for carrying data traffic in small business. ZigBee can be easily achieved through a variety of networking devices. As a global standard, ZigBee is for the realization of ubiquitous network to create the conditions. The system meets the requirements Zigbee Alliance, while the standard is open, their products do not pay royalties.

3 Hardware Design

The nodes in the local network are divided into two kinds:

1) General sensor nodes
The node has functions of real-time collection of temperature, blood pressure, pulse and blood oxygen indicator, and by radio communication, data are transported to the destination node as shown in Figure 2.

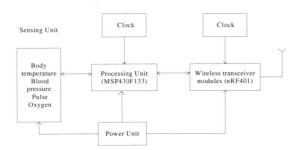

Fig. 2. Structure of General sensor nodes

The sensor nodes include: sensors, data processing and storage of parts, RF communication and power supply parts. Temperature sensor uses GE's ZTP135S-R infrared temperature sensor measurement module; pressure sensor is produced in Taiwan TaiDoc blood pressure measurement module, oxygen and pulse sensor is produced by BCI blood pressure pulse measurement module in Bei Rui Company. Data processing and memory are produced by low-power MSP430 MCU in TI Company, the sensor is mainly used for storing temporary data collected, as part of the wireless data transmission providing job status control lines and two-way serial data transmission lines, eliminate redundant data to reduce network load and transmission package of wireless radio transmission and verification, the answering remote base station queries for data forwarding and storage; regional nodes route maintenance; node energy management, rational standby mode set to save energy has been extended nodes life. RF radio frequency communications by Nordic company's low-power wireless transceiver nRF401 chip that uses a strong anti-jamming power FSK modulation, with frequency stability and reliability, are suitable for ISM band. RF transceiver meets IEEE802.15.4/Zigbee standards, and is the ideal small wireless network transceiver module.

2) Purpose sensor nodes
The node has aggregated data from the local network, sent queries to the network or the distribution of tasks, and been as a gateway or relay, via wireless link to connect to the remote management center and users to exchange information with the outside world as shown in Figure 3.

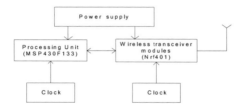

Fig. 3. Structure of Purpose sensor nodes

4 Software Design Process

4.1 Local Network Software

The software is divided into four parts: signal acquisition, signal processing, signal transmission, signal reception and emission. The program running flow of common sensor nodes shown in Figure 4. The program flow of purpose sensor is shown in Figure 5.

Signal acquisition process is to complete two major functions: (1) drive the work of the human body monitoring sensors; (2) collect data by AD of MSP430 MCU. This part of the code program is running in the MSP430F133 chip.

The main function of the signal processing is the filtering, and this part of the code program is running in the MSP430F133 chip.

The main function of the signal transfer process is handled well by the MSP430F133 chip with ZigBee wireless data communication protocol stack through transfer to another wireless sensor node, this part of the code in the kernel 8051 of RF chip nRF401.

Signal receiving program has two main functions: (1) to receive data and other sensor nodes sending data through the serial port microcontroller, this part of the code running in the kernel 8051 of CC2430 RF chip; (2) micro-controller data processing, data storage, this part of the code running in the microcontroller.

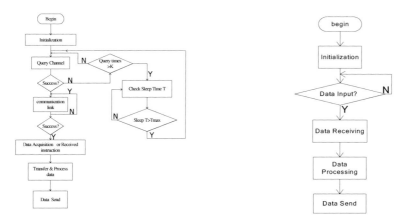

Fig. 4. Program flow of General sensor nodes   **Fig. 5.** Program flow of Purpose sensor nodes

When ordinary sensor node and the destination node are set up, we must rely on ZigBee protocol stack to achieve point to point communication. This design uses the 1.4.2 version of the ZigBee protocol stack API function to implement.

4.2 Remote Mission Control Center

Software management of remote task management center is developed by the C + +, using Microsoft Access2002 version of the database. When the indicators need to query a node, and simply enter the corresponding node number, we can see the corresponding node test parameters and the acquisition time, if without manual collection, we can set a certain time interval collection, the system will focus on collecting and maintaining data. At the same time after diagnosis by medical professionals, the diagnostic conclusions by SMS or fixed phone delivery to patients and the provision of consulting services.

5 Conclusion

The design of wireless sensor networks is used in the field of medical care with a tentative application of the practice, it achieves a purpose of human monitoring sensor nodes and a data transfer between nodes. The innovation is to use several special micro-sensors to monitor, through the network, sensor nodes collect information on human physiological indicators, and through remote task management center a variety of physiological parameters are on process of the diagnosis or consultation. It will not only help patients in developed areas access to health services, but patients in remote mountainous areas is conducive to obtaining the necessary medical services. In future work, it will further develop the software and hardware of system to improve stability and reliability of system, with a special need to develop appropriate software management platform.

References

1. Bin, S., Ye, X.: MSP430-based wireless sensor network design. Microcontroller and Embedded System Applications (7), 5–7 (2006)
2. Zheng, G., Li, J.-D., Zhou, Z.-L.: Wireless sensor network MAC layer protocol in the status quo. Automation Technology 34(3), 305–316 (2008)
3. Chao, S., Ting, D.T., WANG, R.-c., et al.: Wireless sensor network hardware platform for the research and design. Electronic Engineers, 48–51 (2005)
4. Sun, L., Li, J., Yu, C., Zhu, H.: Wireless sensor networks, pp. 8–30. Tsinghua University Press, Beijing (2005)
5. Hong, T., Jing, X., Lu, Y., Lun, T.: Principles and applications of wireless sensor networks, pp. 1–160. Posts & Telecom Press, Beijing (2010)
6. Chen, L.: Wireless sensor network technology and applications, pp. 1–25. Publishing House of Electronics Industry, Beijing (2009)
7. Szwczyk, R., et al.: An analysis of a large scale habitat monitoring applicaion. Proc. SenSys, 214–226 (2004)
8. Ji, X., Zhao, B., Peng, M. (translation): Wireless sensors and components: network, design and application, pp. 188–224. Machine Press (2008)

Study of Automatic Construction and Adjustment of Social Accounting Matrix (SAM)

Jun Wang[1], Lan-Juan Liu[2], and Ji-Hui Shi[2]

[1] Zhejiang Water Conservancy and Hydropower College, Hangzhou 310018, P.R. China
[2] School of Information Mangagement & Engineering, Shanghai University of Finance & Economics, Shanghai 200433, P.R. China

Abstract. SAM is a most important data platform of many economic analysis methods, but great deal time and energy are immersed in SAM construction and adjustment, the problem that we try to solve in this paper by aggregate-disaggregate rules, automatic calculation program, and summary information iteration in Excel so that model constructers and analysts can concentrate themselves on such professional works as model research, shock study, results analysis, and so on.

Keywords: Social Accounting Matrix (SAM), Input-Output (IO) Table, Aggregate-disaggregate Rule.

1 Introduction

Social Accounting Matrix (SAM) is the main data foundation of Computable General Equilibrium (CGE) analysis[1], SAM multiplier analysis[2, 3], shock path analysis and many other macroeconomic analysis. It is an extension of input-output (IO) table, because it contains not only the relationship among production sectors and the information of primary distribution of income, but also the redistribution information and the relationship between production activities and factors and household [4].

Almost each specific SAM is a special one. Different analysis purposes require different accounts and different accounts need different data collection and processing. No one could reckon on that former construction works of SAM could help to reduce any jobs of a new SAM construction. You have to calculate cells one by one from IO table or any other statistical tables and fill them carefully in your SAM sheet. And if you are fortunately proficient in formulas in EXCEL or something else to help you, you will work more efficiently. But you have to write hundreds of thousands of formulas in your SAM and it is much a hard job to check one by one to make sure every one is right. Therefore, plenty of time is wasted while constructing a project oriented SAM in counting rows and columns, adding from cell to cell, and examining formula to formula, especially for constructing a micro SAM. After hundreds of thousands of cell calculations, no one can ensure that there has no error definitely.

In order to save analysts' time and energy to data collection, model research, shock study, and results analysis, firstly, we create a VBScript Macro program in Excel so as to take advantage of Excel to calculate cells and then, everyone can use it like any embedded function of excel to fill cells of SAM automatically. Secondly, finished SAM can work as a template to renew an old SAM or construct a new SAM in only several minutes with only the adjustment of aggregate-disaggregate rule.

2 SAM Construction

We often treat the column account of SAM as expenditure account, which means that all cells in this column are expenditures from this account, and treat the row account as income account, which means that all cells in this row are incomes to this account. Each cell of the SAM shows the payment from the column account to the row account. Thus, for each account, total revenue (row total) must equal total expenditure (column total) of the same account. If someone wants to analyze something more disaggregated or more detailed than what this SAM can provide, accounts of the SAM can be divided to show more detail information. For example, if detailed shocks of factors are required to be analyzed, the account F can be divided into Capital, Land and Labor with every cell related to F account divided into three cells for Capital, Land and Labor accordingly.

To construct such a SAM, we must seek many official statistical books or reports to find authoritative data as many as possible to fill hundreds of thousands of cells. Usually, we will get most information from IO table if the accounts in SAM are corresponding to those in IO table, which is often issued every 5 years. There are several types of IO table with different accounts in China, such as 22 segments, 33 segments, 42 segments, 122 segments, 135 segments, and so on. The more disaggregated the IO table is, the more information it can provide, and the more flexible the future analysis can be.

2.1 Aggregate-Disaggregate Rule of Accounts

As we all know, there are more and more statistical books with digital version and mainly in Excel format, which facilitate us read, find and calculate information easily. Since IO table is the main source of SAM, automatic construction is mainly based on IO table.

IO table and SAM are both formed in square matrix style (IO table can be regarded as square with the fourth quadrant missing) and they are organized in rows and columns as income or expenditure. Usually, accounts in SAM are the same to but often much less and more aggregated than those in IO table, which means one account in SAM is aggregated from one or several disaggregated accounts in IO table. We can call this relationship between SAM and IO table aggregate-disaggregate rule, a-d rule for short, which is the main fruits of former account analysis. Therefore the value of a cell in SAM often means the summation of all cross cells of row disaggregated accounts and column disaggregated accounts in IO table.

For example shown in figure 1, if the row account of the cell in SAM is aggregated from 27 accounts in IO table and the column account is aggregated from 5 accounts in IO table, the value of the cross cell of the aggregated row account and column account in SAM is summed by all those crossed cells of disaggregated row and column accounts, scattered in four cross zones circled with red line in IO table. As we know, to sum such a complicated zone would write a long "sum" string with great care to make it accurate in Excel and can't use the powerful function of Excel, drag-to-fill.

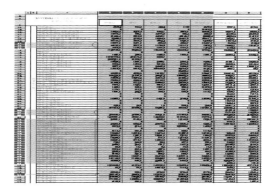

Fig. 1. Summation Zone in IO Table of a Cell in SAM[1]

2.2 Idea of Automatic Calculation

The a-d rule help us much to construct the SAM. But the long formulas are so fragile that any change in the a-d rule makes all related formulas invalidity. We can record the content of a-d rule in one cell somewhere of Excel, which can clearly show what the row account or column account of SAM contains in IO table. Then the address of the cell recording a-d information is the key clue to calculate. We can try to encapsulate a function to access the addresses to gain enough a-d information in order to combine address strings of all crossed disaggregated cells in IO according to the surjection operation and sum then to generate the value of the cell in SAM. If we record all the a-d rules in order of what their corresponding accounts in SAM are, their values can be generated through drag-to-fill easily according to the a-d rule, with the help of the encapsulated function.

We have finally successfully programmed such a function in Excel named "SumF" with a-d information as parameters to sum all the disaggregated crossed cells in IO table. For example, after account analysis and organization, we record the a-d rules in column "A" and row 1. The value of cell "B2" is summed from all those disaggregated crossed cells of column accounts shown in "B1" and row accounts shown in "A2" in IO table, which means that the "B2" in SAM is an aggregated account of row accounts in IO table from row 37 to row 39, row 65 to row 67, row 69 to row 99 and row 92, and column accounts in IO table from column P to column AI. Suppose the IO table is named "IO" and available in the Excel file, we only need to write a very easy formula with "SumF" to generate values without any other operation in SAM from IO table shown in figure 2.

---

[1] We only want t o illustrate that there have many cells to be added, in continuous or discrete zones, not the right contents of what should be added. So we compressed our picture.

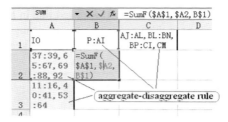

Fig. 2. Call Formation of User-defined Function

2.3 Algorithm of Automatic Calculation

IO file information, a-d rule of row account and a-d rule of column account must be provided to the function and it will find all the cross cells in IO table and sum them together. It is much simple to combine IO file information to our summation string, so our emphasis is mainly on how to locate those cross cells and sum them up. We have two ways to sum cells in Excel, Using the standard summation function "SUM" to sum continuous square zone or adding cells one by one with "+" symbol no matter they are continuous or scattered. We adopt the first way to avoid too long summation string.

Given the aggregated row account of a cell in SAM have "M" discontinuous accounts in IO table and the column account have "N" discontinuous accounts, we have totally "M*N" cross zone to sum with "SUM" function. The a-d rule information is recorded as string in Excel, so it must be divided into readable column or row IDs and then combine them according to the surjection operation. It is well known that the beginning cell and the terminal cell by row number and column id must be specified with colon connection while using "SUM" function to sum this continuous zone and comma symbol is used to separate multiple continuous zones. We need to use nested loop structure to find and combine all the "M*N" cross zones between rows and columns, form the long "SUM" string, and evaluate the string to generate the summation value.

One account in SAM, no matter row account or column account, may aggregate from one or several accounts in IO table, which will be represented with none or several commas and one or several colons. As we all know, the colon symbol is used to connect the start cell and the end cell of a rectangle zone and the comma symbol is used to connect multiple calculation zones. So we must to find all the colons and commas from the a-d string in SAM to get readable row IDs or column IDs in order to locate corresponding row accounts or column accounts in IO table clearly. Then, we cross them to define all cross zones.

Maybe commas or colons will both appear in the a-d string, maybe only one symbol of them appears, and maybe none. But we need to find commas in the string firstly because comma means tow zones to calculate and colon means the start or end of a whole zone. So, while defining the cross zones, it is mostly important to extract a substring divided by a comma, not a colon. We can calculate the cross zone between row accounts and column accounts without any comma in only one summation phrase.

For example, if the row account is "3, 5" and the column account is "C: E", there must have two summations with formula "SUM (3C:3E) +SUM (5C:5E)" in Excel. Even if one can use "SUM (3C:3E, 5C:5E)", it could not change the essence of two summations. But if the row account is "3:5", we get a big continuous zone from "3C" to "5E" with only one summation in essence and we can use "SUM (3C: 5E)" to calculate.

After the above process, we get a substring without comma for one summation operation. But we still have to make sure that if there has any colon which will change the summation string. Given single row ID is represented by the character of "x" or "y" and single column ID is represented by the character of "A" or "B", there have totally four possibilities to be solved listed as follows:

- There is no colon in both row substring and column substring, which means that there is only one row like "x" and one column like "A". So the summation string is "SUM (Ax)", e.g. "SUM (F2)".
- There is only one colon in row substring which means that the row account in SAM is composed of several continuous rows in IO table like "x: y". So the summation string is "SUM (Ax: Ay)" e.g. "SUM (F2:F8)".
- Similar to the case of one colon in column substring like "A: B", the summation string is "SUM (Ax: Bx)" e.g. "SUM (F2:G2)".
- There are two colons which means that both the row account and the column account in SAM are composed of several continuous rows or columns like "x: y" and "A: B". So the summation string is "SUM (Ax: By)" e.g. "SUM (F2:G8)".

We can program and tested directly in Excel with Visual Basic Script and save it as .xla file to ensure that the function is valid in any computer installed MS Excel 2003 with only a simple setting[2]. Some embedded string functions are used in our program listed as follows:

- Instr(s, str, substr). Return the fist appearance position of substring "substr" in the father string "str" from the starting position "s".
- Len(str). Return the length of the string, "str".
- Trim(str). Remove the front or back blank spaces of the string, "str".
- Mid(str, i, j). Return the substring with "j" characters from the "i-th" character in the string, "str".
- Left(str, i). Return a substring with "i" characters counting from left to right of the string, "str".

---

[2] One may search on Internet or Microsoft Excel help about how to setting in MS Excel 2003 to load user defined macros.

The main algorithm of this whole process can be described as follows:

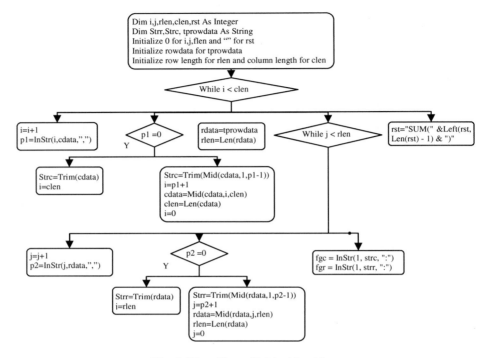

Fig. 3. Flow Chart of Main Algorithm

We only need to evaluate the result summation string in the Visual Basic function to get the final value of what cell we want to calculate in SAM, no matter it is aggregated from only several accounts or a great deal of accounts in IO table. By the way, there is a length limitation for the evaluation function in Excel that string length that is longer than 250 is not permitted. It is not a big problem to programmers and they just need to take a little time to handle it.

2.4 Summary Information Citation

IO table is not the only data source of SAM, and we still have some cells valued based on some other yearbooks to form a complete SAM. But account organization rules are often different so that we seldom get information in accord with IO table and SAM accounts directly. We can usually only get summary information which means we have to divide it into several cells based on some changeless rules relatively to match accounts in SAM. In order to keep all those cells consistent all the time, citation is a good idea and strong recommended. That is putting the summary information in some cell of the Excel and citing it whenever it is required. Once the summary information is changed for different years or different regions, all values of those cells that cite the summary information will change accordingly automatically.

3 Work as a Template

While finishing firstly analysis of accounts organization of SAM, one need not to sum cell by cell, row by row, and column by column dazedly to fill the SAM again and again, but only need to record a-d rules somewhere properly in Excel, write a simple "SumF" function with proper parameters, drag-to-fill all applicable cells, and write a few formulas citing summary information, no matter how complex it is. With the help of automatic calculation function as figure 1 shows, all you need to do to construct a new SAM are just those several simple steps. It is very import to ensure drag-to-fill function valid that put a-d rules in order of accounts order in SAM. Since it can be regarded as the same to functions embedded in Excel, we can just drag it or copy it to cover all those similar cells to form SAM quickly without any additional works shown as figure 4.

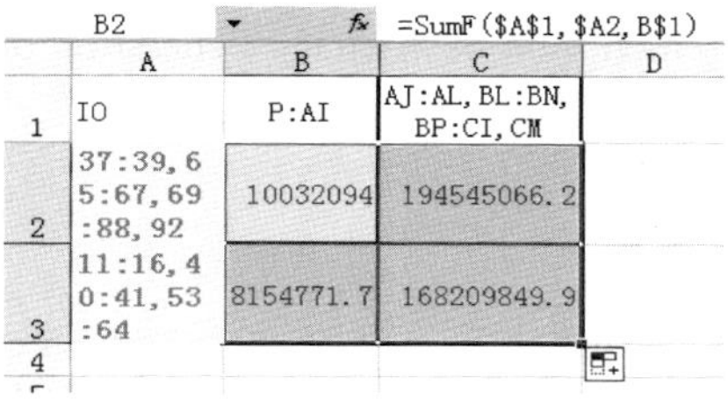

	A	B	C	D
1	IO	P:AI	AJ:AL, BL:BN, BP:CI, CM	
2	37:39, 65:67, 69:88, 92	10032094	194545066.2	
3	11:16, 40:41, 53:64	8154771.7	168209849.9	
4				

B2 fx =SumF(A1, $A2, B$1)

Fig. 4. Drag Results of Function

It is very usual that former account organization requires some adjustment to suit the objective better with the progress of project research. Such adjustment on account organization may be a disaster through writing long and boring "sum" string cell by cell. In fact, we only need to adjust the a-d rules or summation information a little, and all values in SAM will be adjusted automatically according to the new setting with the help of our automatic calculation function. Even if the SAM has to be reconstructed for another topic, there is nothing serious. Just replace the IO table and summary information if necessary and change the a-d rules according to project target, and a new SAM will be constructed immediately with out any additional operations. Therefore, we can save our finished SAM structure and make it work as a template. After all, the structure of SAM is relatively stable and main variable things have been expressed in a-d rules and summation information.

4 Conclusions

It is totally a big, complicated, time consuming, and fallible job to construct and adjust a project oriented SAM without methods and works focused in this paper. But such a manual job should not occupy much of model constructers and analysts, because they should concentrate mainly on such brain works as problem analysis, model selection, methods research, results explanation and so on.

Massive works of calculating values of cells from different and scattered accounts of IO table seem no rule to follow while constructing a SAM. However, if we conclude the a-d rule between SAM and IO table, record it in Excel as only a string, compose a function to read the information and calculate the cell's value, and cite summary information with changeless formulas relatively, all things become so easy. All things need to do is just writing a very simply formula and dragging the formula properly to construct or adjust a SAM instantly. The program in VBScript can be copy to any computer and loaded in their MS Excel 2003 or other versions to work as convenient as any embedded function in Excel so that model constructers and analysts can concentrate mainly on what they should.

Acknowledgement

Our works are sponsored by Major Program of National Social Fund of China (Grant No. 08&ZD047) and research fund of Zhejiang Federation of Humanities and Social Sciences Circles (Grant No. 2010N73) of China.

References

1. Kehoe, T.J.: Social Accounting Matrices and Applied General Equilibrium Models (1996)
2. Breisinger, C., Thurlow, M., Thurlow, J.: Social Accounting Matrices and Multiplier Analysis: An Introduction with Exercises. In: Food Security in Practice, p. 42. International Food Policy Research Institute, Washington, D.C (2009)
3. Sanz, M.T.R., Perdiz, J.V.: SAM multipliers and inequality measurement. Applied Economics Letter 10(7), 4 (2003)
4. Qiwen, W., L.S., Ying, G.: Principle, Methods and Application of Social Accounting Matrix. Tsinghua University Press, Beijing (2008)
5. Devarajan, S., Go, D.S., Lewis, J.D., Robinson, S., Sinko, P.: Simple General Equilibrium Modeling, p. 30 (1997)
6. Essama-Nssah, B., Building and Running General Equilibrium Models in EViews. World Bank Policy Research Working Paper, 2004(3197)
7. Zhao, Y., Wang, J.F.: CGE Model and Its Application in Economic Analysis. China Economic Publishing House, Beijing (2008)
8. Wang, Z., X.J., Zhu, Y., Wu, J., Zhu, Y.: CGE Techonologies of Policy Simulation and Analysis of Economy Development. Science Publishing House, Beijing (2010)
9. Lofgren, H., Harris, R.L., Robinson, S.: A Standard Computable General Equilibrium Model in GAMS. Microcomputers in Policy Research 5 (2002)

Research on Testing Methods of I-V Characteristics of Solar Photovoltaic Cell Array

Yunhai Hou[*], Ershuai Li, and Shihua Sun

School of Electric & Electric Engineering, Changchun University of Technology,
130012 Changchun, China
houyunhai@mail.ccut.edu.cn, {liershuai,sunshihua100}@126.com

Abstract. By testing the I-V characteristics of the solar photovoltaic cell array and referencing the experimental data, it can effectively evaluate the PV power plant control and design standards. In order to get the accurate test to the characteristics of solar photovoltaic cell array data, test its I-V characteristics, we use the dynamic capacitance charging test method, according to the characteristics of capacitors, take the dynamic capacitance as a solar PV array variable load, through the current and voltage sampling on the whole process of charging the capacitor for the photovoltaic cell array, then get I-V curve of PV array.

Keywords: Solar photovoltaic cell array, I-V characteristics, Dynamic capacitance, Sampling.

1 Introduction

Solar energy is recognized as one of the most promising new energy sources. There are three main ways to use solar energy directly: photovoltaic conversion, photochemical conversion and thermal conversion. Photovoltaic conversion called solar power is one of the very important part [1].Based on the principle of photovoltaic effect, the PV conversions energy into electricity by using solar photovoltaic cells solar.

Since energy absorption efficiency of solar photovoltaic cells is impacted serious by the external conditions, so it can use the testing method of characteristics of solar photovoltaic cells to study the effect of external conditions on the solar photovoltaic cell output characteristics, then improve the efficiency of solar cells.

2 Model and the Electrical Characteristics

Solar photovoltaic system consists of an array of solar photovoltaic cells, power conditioners, batteries (not according to the conditions), the load, the control protection devices and other accessories. The energy of the system is solar, and solar photovoltaic cells consisted of semiconductor devices is the core of the system [2].

---

[*] Corresponding author.

Combination of solar photovoltaic cell array, its efficiency remained unchanged, there is a corresponding growth curves (the characteristics of a single cell or component remain unchanged).So when studying the characteristics of an array of solar photovoltaic cells, it should begin to analyze the characteristics of solar photovoltaic cells monomers.

For example, silicon-based solar photovoltaic cells, the ideal form and the actual form of its equivalent circuit are shown in Fig. 1 (a), (b) below:

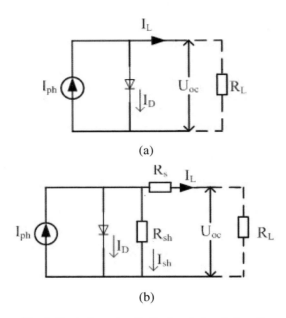

Fig. 1. Equivalent circuit of solar photovoltaic cells.

In which, I_{ph} is the photocurrent, I_D is the dark current, I_L is the load current of solar photovoltaic cell output, U_{oc} is open circuit voltage of the battery. R_s and R_{sh} are the inherent resistance of silicon-based solar cells,which is the battery internal resistance.As an ideal solar photovoltaic cells, due to a small series resistance R_s, a large parallel R_{sh}, so to a calculation of ideal circuit, they can be ignored, resulting in an ideal equivalent circuit only equivalent to a constant current source I_{ph} in parallel current (Fig. 1 (a) above).

From the above definition, it lists the equations of the variables in the equivalent circuit of solar photovoltaic cells as follows:

$$I_D = I_0 \left(\exp \frac{qU_D}{AkT} - 1 \right) \tag{1}$$

$$I_L = I_{ph} - I_D - \frac{U_D}{R_{sh}} = I_{ph} - I_0\left[\exp\left(\frac{q(U_{oc}+I_L R_s)}{AkT}\right)-1\right] - \frac{U_D}{R_{sh}} \quad (2)$$

$$I_{sc} = I_0\left[\exp\frac{qU_{oc}}{AkT} - 1\right] \quad (3)$$

$$U = \frac{AkT}{q}\ln\left(\frac{I_{sc}}{I_0} + 1\right) \quad (4)$$

In which, I_0 is equivalent to the P-N junction diode reverse saturation current of equivalent diode within solar photovoltaic cells, I_{SC} is short-circuit current of the battery, U_D is the voltage of equivalent diode, q is the electron charge, k is the Boltzmann constant, T is the absolute temperature, A is the curve constant of the P-N junction.

In low light conditions, because $I_{ph} \ll I_0$, so $U_{oc} = \frac{AkT}{q} \cdot \frac{I_{ph}}{I_0}$;while in strong light conditions, because $I_{ph} \gg I_0$, so $U_{oc} = \frac{AkT}{q}\ln\frac{I_{ph}}{I_0}$.

Thus, when the sun is weak, the open circuit voltage of the silicon-based solar cell changes linearly with the intensity of the light,when the sun is too strong, then the light intensity changes with logarithmic.The open circuit voltage of silicon-based solar cells is generally between 0.5-0.58V.

In the ideal form(set $R_s \to 0; R_{sh} \to \infty$), the equivalent circuit equations is

$$I_L = I_{ph} - I_D - \frac{U_D}{R_{sh}} \approx I_{PH} - I_D \quad (5)$$

It can be seen, if we put the battery technology parameters such as I_{sc}, U_{oc}, R_{sh} into the mathematical model of solar photovoltaic cells, then get the I-V characteristic curve of solar photovoltaic cells [3].

3   I-V Characteristic

According to the formula (1)-(5), it obtains voltage-current curve of solar photovoltaic cells that is volt-ampere characteristic curve (in Fig. 2), it shows the relationship between output voltage and output current of the solar photovoltaic battery in a certain strength and temperature conditions of sunlight, referred to the voltage characteristic [4].

Fig. 2 shows the output current of solar photovoltaic cells maintain approximately constant in most of the work current range, near the open circuit voltage, the electric current declines greatly. The node of the output I-V characteristic curve of solar

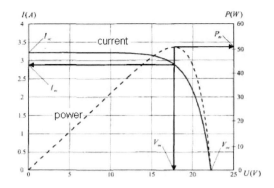

Fig. 2. The I-V characteristic curve of solar photovoltaic cell array.

photovoltaic cell and current-axis is short-circuit current I_{sc} and voltage-axis is the open circuit voltage U_{oc} [5].

4 Dynamic Capacitance Charging Test Method

The basic principles of testing characteristics of solar photovoltaic cell array by charging dynamic capacitance is shown in Fig. 3.

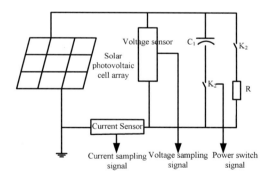

Fig. 3. Test schematic of characteristics of solar cell array.

The test method of dynamic capacitance charging is based on the characteristics of capacitors, the capacitance as a variable load is connected to the solar PV array output, when the photovoltaic cell arrays charge to the capacitor, continuous to sampling circuit voltage and current in the period of charging, by descriptiing the sampling points, then gets the I-V curve of the solar photovoltaic cell array in the test environment [6].

In this paper, with arithmetic mean filter, it reduces the impact of interference and improves the smoothness of the curve. Through the arithmetic mean filter, the signal to noise ratio improved $\sqrt{N}$ times, and the filtering means filtering on the signal level depends entirely on the N. According to the actual situation of the system, select N = 10 as a sample number [7].

5 Simulation and Analysis of Test Results

Based on mathematical models, in the MATLAB simulation environment, it uses simulink simulation tool to build test generic simulation model of the characteristics of a solar photovoltaic cell array, according to the model, it obtains the voltage and current curves of the solar photovoltaic cell array under the variation of light and temperature, as shown in Fig. 4-5.

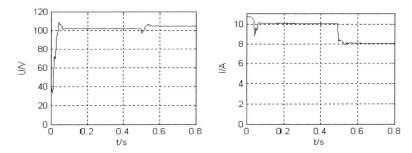

Fig. 4. The output voltage and current waveforms of solar panel when the light intensity changes.

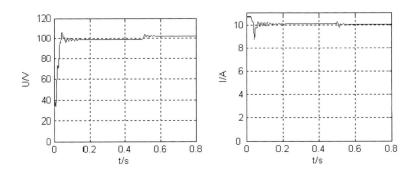

Fig. 5. The output voltage and current waveforms of solar panel when the temperature intensity changes.

Through the measure and simulation results of the characteristics of solar photovoltaic cells array, it shows that: the dynamic capacitance charging has high testing speed and high precision, the characteristics of solar photovoltaic cell array can be directly displayed in the form of a curve, the test results are intuitive, it can satisfy the needs of engineering applications greatly.

References

1. Wen, D.X.: Research and Evaluation on solar array's output chsraCteristics: [Master thesis]. Hefei: Hefei University of Technology (2007)
2. Liu, Y., Wan, f., Wang, D., et al.: Adaptive Fuzzy paste algorithm of MPPT of photovoltaic systems, pp. 657–661. Solar (2008)
3. Zhao, Z., Liu, J., et al.: Solar photovoltaic power generation and its application, pp. 27–29. Science Press, Beijing (2005)
4. Wang, Y.: Study on MPPT control method on PV generation system: [Master thesis]. North China Electric Power University, Baoding (2007)
5. Zhang, G., Di, X., Su, J., et al.: Based on the dynamic capacitance charging photovoltaic array I-V tester. Electronic Measurement and Instrument (2009)
6. Qu, X.: Research on tester of photovoltaic array I-V characteristic curve: [Master thesis]. Hefei University of Technology, Hefei (2007)
7. Zhou, D., Zhao, Z., et al.: Analysis of characteristics of the solar photovoltaic cell array based on simulation model. Tsinghua University (Natural Science) 47, 7 (2007)

The Performance Comparison of ANFIS and Hammerstein-Wiener Models for BLDC Motors

Abbas Nemati[1] and Mohammadreza Faieghi[2]

[1] Department of Electrical Engineering, Miyaneh Branch,
Islamic Azad University, Miyaneh, Iran
[2] Young Researchers Club, Miyaneh Branch,
Islamic Azad University, Miyaneh, Iran
nemati@m-iau.ac.ir

Abstract. Emerging technologies such as Fuzzy Logic (FL) or Neural Network (NN) have received wide attention in the field of system identification. These techniques have been claimed to yield excellent results for some applications. In this paper, one of the emerging identification techniques i.e. Adaptive Neuro-Fuzzy Inference System (ANFIS) which utilizes both advantages of FL and NN is compared with the well-known Hammerstein-Wiener models for identification of Brushless DC (BLDC) Motor. Numerous simulation studies carried out to provide a comprehensive comparison between two aforementioned techniques. Simple structure with high accuracy and fulfilling robustness is desired. Numerical simulations show that ANFIS outperforms Hammerstein-Wiener models.

Keywords: BLDC motor drive; system identification; ANFIS; Hammerstein-Wiener model.

1 Introduction

Brushless DC (BLDC) motor is a well-known DC motor receiving a surge of interest in many industrial applications due to its high torque density, high efficiency and small size [1]. However, it requires more complicated drive in the comparison with brushed DC motors. Also, it is a typical example of highly coupled nonlinear systems and requires complex control methods to control its position, velocity and torque.

To design a proper controller for the motor at constant load applications an exact mathematical model of the motor is required. The mathematical model of the motor [2] could be a useful tool to aim the procedure of controller design. However, this requires an exact knowledge of the motor parameters such as stator resistance, stator inductance, torque constant and flux linkage. These parameters are completely dependent to the way of usage and environmental conditions.

System identification is especially useful for modeling systems that it cannot easily model from first principles or specifications, such as engine subsystems, thermo fluid processes, and electromechanical systems. It also helps us simplify detailed first-principle models, such as finite-element models of structures and flight dynamics models, by fitting simpler models to their simulated responses. Using system identification in designation of a control system could ease the design procedure and make it more efficient. There are many approaches proposed in the literature to provide a powerful framework for identification of systems. However there always exists a question, which method is the most powerful? Although it cannot be claimed that a method is the most efficient in general, we want to investigate the answer of this question in a special case of study. The goal of this paper is analyzing two common different approaches for system identification of BLDC motor: ANFIS and Hammerstein-Wiener models.

ANFIS was introduced by Jyh-Shing Roger Jang in 1993 which can serve as a basis for constructing a set of fuzzy if-then rules with appropriate membership functions to generate the stipulated input-output pairs [3, 4]. By using different learning algorithms, the ANFIS can construct an input-output mapping based on both human knowledge (in the form of fuzzy if-then rules) and stipulated input-output data pairs [4], so it can be used for system identification as an efficient tool [5]-[8].

Block-oriented nonlinear models that consist of linear dynamic subsystem and memory-less nonlinear static functions such as Wiener, Hammerstein, and Hammerstein-Wiener models [9] have been often used to describe the nonlinear dynamics of many chemical, electrical, and biological processes [10]. As is well known these models contain one or two nonlinear static blocks and a liner block. Different nonlinear estimators have been used in nonlinear blocks such as polynomials [9], Multilayer Perceptron (MLP) [9, 10], Orthogonal Wavelet Neural Networks [12], Radial Basis Function Networks (RBFN) [13], and ANFIS [14] to mention few.

In this paper, first we will develop several ANFIS structures with different attributes to approximate the BLDC motor characteristics. In the second part, we will tackle this problem with Hammerstein-Wiener models. Based on numerous simulations we will find the best Hammerstein-Wiener structure. Finally, the most efficient structures of ANFIS and Hammerstein-Wiener models will be compared.

The rest of this paper is organized as follows: Theoretical background is given in section 2. Section 3 presents ANFIS identification of BLDC motor. In section 4 illumination of identification of BLDC motor based on Hammerstein-Wiener model is presented. In section 5 comparisons between ANFIS and Hammerstein-Wiener is illustrated. Finally we conclude the paper in section 6.

2   BLDC Motor Drive

The three-phase BLDC motor is operated in a two-phase-on fashion, i.e. the two phases that produce the highest torque are energized while the third phase is off. Which two phases are energized depends on the rotor position. The signals from the position sensors produce a three digit number that changes every $60°$ (electrical degrees) as shown in Fig.1 (H1, H2, H3). These signals are used for current commutation.

Current commutation is done by a six-step inverter as shown in Fig.2. The switches presented in Fig.2 are IGBT, but MOSFETs are more common. Table 1 shows the switching sequence [15].

For the purpose of controlling BLDC motors, a common approach is based on appropriate adjusting the input DC voltage of the inverter. Hence, it worth nothing to investigate the dynamic behavior of the motor while input voltage of the inverter fluctuates. Therefore, we concentrate on identification of nonlinear relationship of motor speed and input voltage of the inverter.

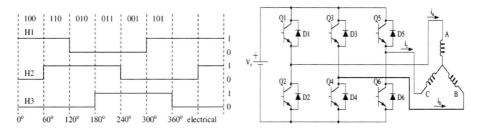

Fig. 1. Position sensors signals according

Fig. 2. Simplified scheme of BLDC motor to electrical degreedrive

Table 1. Switching sequence

Switching interval	Seq. number	Pos. sensors H1	H2	H3	Switch closed		Phase Current A	B	C
0° − 60°	0	1	0	0	Q1	Q4	+	−	off
60° − 120°	1	1	1	0	Q1	Q6	+	off	−
120° − 180°	2	0	1	0	Q3	Q6	off	+	−
180° − 240°	3	0	1	1	Q3	Q2	−	+	off
240° − 300°	4	0	0	1	Q5	Q2	−	off	+
300° − 360°	5	1	0	1	Q5	Q4	off	−	+

3 ANFIS Identification

The first step in identification of BLDC motor is generating an appropriate data set. We have applied a white noise as input voltage to the motor drive and measured the rotor angular velocity every 1 ms. Fig. 3 depicts generated data set includes 2000 input-output pairs.

Having constructed the data set, second step is to determine the most suitable combination of regressions to be considered as ANFIS inputs for the purpose of identification. We have used exhaustive search technique [5] to find the best combination of regressions. In this method 36 different ANFIS structures with different inputs had been generated and their training errors were compared. Fig. 4 shows the training error of these ANFIS structures. As it shown, combination of $y(k-2), y(k-1)$ and $u(k-1)$ has the minimum training error and is the most adequate to be used for identification.

In order to analyze ANFIS properties in system identification, several ANFIS structures based on different data clustering algorithms, training procedure and membership functions have been generated and compared.

We have used three different data clustering algorithms: Fuzzy C-Means (FCM), subtractive clustering and gird partitioning [22]. FCM was used to generate ANFIS with 3, 5 and 8 membership functions. Subtractive clustering was used with different cluster radius: 0.3, 0.5 and 0.8. and grid partitioning was used to generate ANFIS with 2 and 3 membership functions.

Remark 1.3 Using grid partitioning, yields more complicated structures, therefore in this case just 2 and 3 membership functions is considered.

Error Back propagation (B.P) and hybrid learning [4] are known as two common training algorithms in training ANFIS. Hence, we have employed both methods to evaluate their performance. For simplicity, we have used these training algorithms in one epoch with 1000 data pairs for the purpose of estimating and 1000 data pairs for validating. An adaptive learning procedure is carried out, that is we have set the decrease/increase speed of the learning rate by 0.5 factors.

Table 2 summarizes simulation results. Properties of 16 different ANFIS structures and their training errors are illustrated. Comparing the attributes of all structures shown in Table 2, dispose us to yield some interesting results.

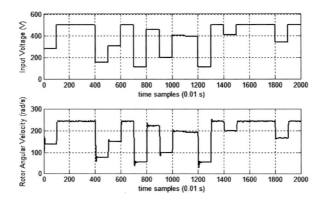

Fig. 3. Data set for ANFIS identification

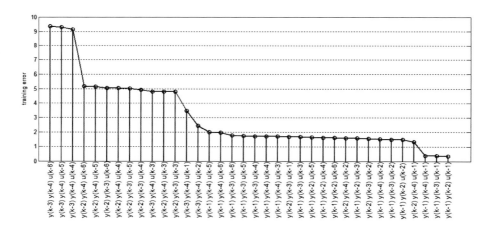

Fig. 4. 36 different ANFIS models training error

Among all the generated structures, one can conclude that ANFIS#14 has the best performance. Its structure is depicted in Fig.5 and Fig. 6 shows the ANFIS#14 and BLDC motor response to step input.

Remark 2.3. However some structures have very small training errors, they have large validating errors, due to over fitting phenomenon; there should be some corrections in the train procedure of these structures.

Remark 3.3. In the comparison between training algorithms it is clear that the hybrid learning operates better.

Remark 4.3. Building structures by using grid partition make the structure more complex. In some cases this complexity is not desirable and makes the ANFIS implementation hard. In addition, complex structure requires more computational time.

Table 2. Comparison of 16 different ANFIS structures to find the best for system identification

No. of ANFIS	Data Clustering Method	No. of Membership Functions	Training Algorithm	No. of Rules	Training Error	Validating Error
1	FCM	3	B.P	3	1.2050	1.1221
2	FCM	3	hybrid	3	1.1401	1.0527
3	FCM	5	B.P	5	1.2050	1.1221
4	FCM	5	hybrid	5	1.0565	3.3748
5	FCM	8	B.P	8	1.2050	1.1221
6	FCM	8	hybrid	8	0.9810	58.8974
7	Subtractive	4	B.P	4	1.1716	0.9305
8	Subtractive	4	hybrid	4	1.1189	1.1972
9	Subtractive	3	B.P	3	1.1313	1.0368
10	Subtractive	3	hybrid	3	1.1246	1.0555
11	Subtractive	2	B.P	2	1.1957	0.9846
12	Subtractive	2	hybrid	2	1.1757	1.0634
13	Grid Partition	2	B.P	8	185.5609	209.3311
14	Grid Partition	2	hybrid	8	0.6663	0.7542
15	Grid Partition	3	B.P	27	185.5609	209.3311
16	Grid Partition	3	hybrid	27	0.3561	31.4502

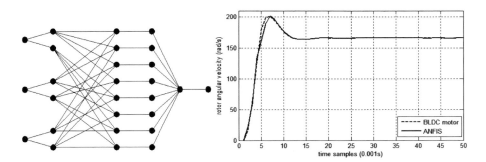

Fig. 5. Structure of ANFIS#14   **Fig. 6.** BLDC motor and ANFIS#14 step response

4 Hammerstein-Wiener Model Identification

The data set generated in the previous section has been employed in order to Hammerstein-Wiener model identification of BLDC motor. In accordance to Hammerstein-Wiener models principles, here there is no need to build regression of inputs or outputs. 16 different structures have been constructed. Their differences are based on the number of nonlinear blocks (Hammerstein models or Wiener models), the order of the linear block and the nonlinear estimators that are used in the nonlinear blocks. We have used WNN and one-dimension polynomial nonlinear estimators. The choice of these estimators is deliberate as there is a significant volume of research on these particular estimators.

Simulation results are presented in Table 3. Structures that have unit gain in the second nonlinear block are Hammerstein models and structures which have unit gain in the first nonlinear bock are Wiener models. Considering the simulation results, one can conclude following remarks.

Table 3. Comparison of different Hammerstein-Wiener structures

No. of the structure	First nonlinear block	Second nonlinear block	No. of linear block zeros	No. of linear block poles	Percentage of fitting
1	WNN	Unit Gain	1	2	95.39
2	WNN	Unit Gain	2	3	97.69
3	Polynomial	Unit Gain	1	2	94.79
4	Polynomial	Unit Gain	2	3	97.44
5	Unit Gain	WNN	1	2	94.7
6	Unit Gain	WNN	2	3	97.31
7	Unit Gain	Polynomial	1	2	93.9
8	Unit Gain	Polynomial	2	3	50.49
9	Polynomial	Polynomial	1	2	94.78
10	Polynomial	Polynomial	2	3	97.63
11	WNN	WNN	1	2	95.51
12	WNN	WNN	2	3	97.76
13	WNN	Polynomial	1	2	94.73
14	WNN	Polynomial	2	3	97.50
15	Polynomial	WNN	1	2	95.36
16	Polynomial	WNN	2	3	97.69

Remark 1.4. Generally, higher order linear blocks yield greater accuracy.

Remark 2.4. It is obvious that WNN operates provide more accuracy than polynomials.

Remark 3.4. Hammerstein models outperform Wiener models. However, as expected Hammerstein-Wiener models offer the best performance at the cost of more complexity.

Remark 4.4. Hammerstein-Wiener model#12 can be considered as is the most powerful structure. It consists of two WNN as nonlinear estimators and a linear block with 2 zeros and 3 poles.

In order to evaluate the effects of increasing complexity in linear blocks and nonlinear estimators, other case of simulations have been carried out for structure #12 due to its superiority in table 3. We have increased the number of neurons in the output layer WNNs to 7 and 10. Moreover we have considered higher-order linear blocks as shown in table 4.

As expected the increase in the number of neurons in nonlinear blocks and also, increasing the order linear block provide better performance.

5 Comparison of ANFIS and Hammerstein-Wiener Models

In previous sections, several structures are implemented to deal with identification of BLDC motor. This section devotes to compare the performance of the most accurate structures generated in the last sections.

We have considered ANFIS#14 and Hammerstein-Wiener model#27 as the most adequate models of BLDC motor. In Fig. 7 the step response of these structures are compared with the actual motor response. One can clearly observe that in transient state, both models have some negligible inaccuracies, especially in the overshoot, but ANFIS#14 offers more accurate approximation in the comparison with Hammerstein-Wiener mode#27. In the steady-state, it is clear that both models can fit the actual motor adequately.

Overall, the consideration of the performance of both models shows that ANFIS can provide better performance for system identification. Not only, ANFIS offers more accuracy, but also have simpler structure.

Table 4. Hammerstein-Wiener models with differences in WNN and linear block

No. of the structure	No. of linear block zeros	No. of linear block poles	No. of neurons	Percentage of fitting
17	2	3	7	97.82
18	2	3	10	97.77
19	3	4	5	97.82
20	3	4	7	97.80
21	3	4	10	97.77
22	4	5	5	97.78
23	4	5	7	97.75
24	4	5	10	97.87
25	5	6	5	97.78
26	5	6	7	97.83
27	5	6	10	97.92

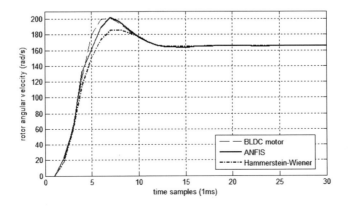

Fig. 7. BLDC motor, ANFIS and Hammerstein-Wiener step response

6 Conclusion

In this paper, with reference to BLDC motor the performance of neuro-fuzzy and Hammerstein-Wiener models were compared. To find out that which models can have better performance of system modeling, three steps employed.

In the first step 16 different ANFIS structures have been generated and their accuracy was compared. This comparison yields that hybrid learning, clearly works better than error back propagation in order to train the ANFIS. Also, in order to use data clustering algorithms to generate ANFIS structures there is no much different between Fuzzy C-Means and subtractive clustering, however if there is no limitation in implementation procedure, grid partition can have better results, but it might achieve structures more complex. Finally an ANFIS with 3 input, 8 rules and 2 membership function for each input is selected.

In the second step 27 different Hammerstein-Wiener models were generated and compared. The comparison between the results show that Hammerstein models have better accuracy than Wiener models, but using Hammerstein-Wiener structures yields better accuracy. Also, using higher-order linear blocks increase the modeling accuracy. In order to determine which nonlinear estimator can be selected as the best one for the nonlinear blocks WNN and one-dimension polynomials have been used appears that WNN is better selection however it depends on the order of the estimator. It is possible that higher order polynomials have better performance than WNN that is not considered in this paper. At last a Hammerstein-Wiener model with 5 zeros and 6 poles in the linear block and nonlinear blocks with WNN containing 10 neurons in output layer is selected.

In the third step, the best models of ANFIS and Hammerstein-Wiener models were compared to achieve the goal of this paper: Which models have better performance? Simulation studies show that in the case of accuracy ANFIS model is more accurate in transient state but in the steady state both models operate well. In the case of robustness there is no much difference between these models and both of them have well robustness. The answer of the above question now is determined. Its answer is: ANFIS. But there are many other questions, such as: How about using ANFIS as nonlinear estimators in Hammerstein-Wiener models? or How about the Fuzzy Wavelet Neural Networks? and so on. These are questions that can be further studied.

Acknowledgment

The authors would like to express their gratitude to the Islamic Azad University, Miyaneh Branch for their financial support and assistance; also this paper is extracted from the research project with title of "*Survey of 3-phases induction & BLDC motors and presentation of the new procedure for decreasing THD of 3-phases induction motors*", 2011.

References

1. Xia, C., Guo, P., Shi, T., Wang, M.: Speed Control of Brushless DC Motor Using Genetic Algorithm Based Fuzzy Controller. In: Clerk Maxwell, J. (ed.) Proceeding of the 2004 International Conference on Intelligent Mechatronics and Automation, Chengdu, China, 3rd edn. A Treatise on Electricity and Magnetism, vol. 2, pp. 68–73. Clarendon, Oxford (1892)

2. Chiasson, J.: Modeling and High Performance Control of Electrical Machinery. IEEE Press Series on Power Engineering. John Wiley and Sons, Chichester (2004)
3. Jang, R.: Adaptive-Network based Fuzzy Inference Systems. IEEE Transactions on Neural Networks 3(5), 714–723 (1992)
4. Jang, R.: Neuro-Fuzzy Modeling for Dynamic System Identification. In: Young, M. (ed.) Soft Computing in Intelligent Systems and Information Processing. The Technical Writer's Handbook. University Science, Mill Valley (1989)
5. Jang, R.: Neuro-Fuzzy Modeling for Dynamic System Identification. In: Soft Computing in Intelligent Systems and Information Processing (1996)
6. Jang, R., Sun, C.T.: Neuro-Fuzzy Modeling and Control. Proceedings of the IEEE (March 1995)
7. Faieghi, M.R., Azimi, S.M.: Design an Optimized PID Controller for Brushless DC Motor by Using PSO and Based on NARMAX Identified Model with ANFIS. In: IEEE 12th International Conference on System and Modeling, United Kingdom (May 2010)
8. Buragohain, M., Mahanta, C.: ANFIS Modeling of Nonlinear System Based on Vfold Technique. In: IEEE International Conference on Industrial Technology, ICIT (2006)
9. Janczak, A.: Identification of Nonlinear Systems Using Neural Networks and Polynomial Models. Springer, Heidelberg (2005)
10. Amralahi, M.H., Azimi, S.M., Sarem, Y.N., Poshtan, J.: Nonlinear Model Identification for Synchronous Machine. In: 6th International Conference on Electrical Engineering/Electronics, Computer, Telecommunications and Information Technology (2009)
11. Sung, S.W., Je, C.H., Lee, J., Lee, D.H.: Improved System Identification Method for Hammerstein-Wiener Processes. Korean Journal Chemical Engineering 25(4), 631–636 (2008)
12. Fang, Y., Chow, T.W.S.: Orthogonal Wavelet Neural Networks Applying to Identification of Wiener Model. IEEE Transactions on Circuits and Systems—I: Fundamental Theory and Applications 47(4) (April 2000)
13. Hachino, T., Deguchi, K., Takata, H.: Identification of Hammerstein Model Using Radial Basis Function Networks and Genetic Algorithm. In: IEEE 5th Asian Control Conference (2004)
14. Jiat, L., Chiut, M.S., Get, S.S.: Neuro-Fuzzy System Based Identification Method for Hammerstein Processes. In: 5th Asian Control Conference. IEEE, Los Alamitos (2004)
15. Baldursson, S.: BLDC Motor Modeling and Control – A MATLAB®/Simulink® Implementation, Institutionen För Energy Och Miljö, Master Thesis (May 2005)
16. Sugeno, M., Kang, G.T.: Structure Identification of Fuzzy Model. Fuzzy Sets and Systems 28, 15–33 (1988)
17. Takagi, T., Sugeno, M.: Fuzzy Identification of Systems and its Application to Modeling and Control. IEEE Transactions on Systems, Man and Cybernetics 15, 116–132 (1985)
18. Xia, C.-L., Xiu, J.: Sensorless control of switched reluctance motor based on ANFIS. In: King, I., Wang, J., Chan, L.-W., Wang, D. (eds.) ICONIP 2006. LNCS, vol. 4234, pp. 645–653. Springer, Heidelberg (2006)
19. Mathworks, MATLAB, system identification toolbox (2009)
20. Lin, W., Zhang, H., Liu, P.X.: A New Identification Method for Hammerstein Model Based on PSO. In: IEEE International Conference on Mechatronics and Automation (June 2006)
21. Dhaouadi, R., Al-Assaf, Y., Hassouneh, W.: A Self Tuning PID Controller Using Wavelet Networks. In: IEEE Power Electronics Specialists Conference (2008)
22. Mathworks, MATLAB, Fuzzy Logic Toolbox (2009)

Application of Real-Time Image Mosaics for Video Processing System

Wang Lutao[1] and Zhang Lei[2]

[1] University of Electronic Science and Technology of China
Chengdu, China
`wltuestc@163.com`
[2] Space Star Technology Co., Ltd
Beijing, China
`zhanglei004171@163.com`

Abstract. This paper presents the application of the image mosaics on video processing system and describes an image mosaics solution based on scale invariant feature transform (SIFT) algorithm to get the wider view. The SIFT algorithm is a stable and fast image feature extraction techniques, and it can solve the matching problem with translation, rotation and affine distortion. The result shows that SIFT algorithm is an effective image mosaics method in video processing system.

Keywords: image mosaics, SIFT, image registration, image fusion.

1 Introduction

Video processing system [1] is the key technology of computer vision and artificial intelligent. Recently, video image mosaic is a new arising high-tech application field in video processing system. With the rapid growth of video display requirements, the video scene becomes wider and wider. The view of the camera is limited, so that it is important to find a way to get the full view. Accompanied by the development of the technology of the image processing, image mosaics can be used to resolve this problem. As the structure shown in Fig.1, this paper focuses on the step of image mosaics to blend the multiple video images into a single larger output image.

The most important task in video image mosaics is to get the mosaic imaging in real time. In order to solve this problem, there are two factors to be concerned. Firstly, the video images should be got in real-time. Secondly, the video images should be stitched in real-time. It means that a fast image mosaics algorithm must be used. At present, a lot of researchers have proposed many different matching algorithms, such as the algorithm based on area, histogram and so on. Each algorithm has its advantages and application situation. Scale Invariant Feature Transform (SIFT) algorithm is adopted for video image mosaics in this paper, which can get the stable features fast. SIFT is based on the feature transformation of scale-invariant. This paper presents a complete solution based on SIFT algorithm for video image mosaics application.

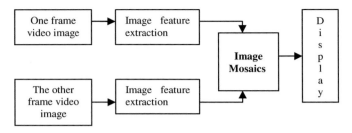

Fig. 1. The structure of video image mosaics processing

2 Overview of the System

The video image mosaics processing system contains image acquisition, image feature extraction, image registration and image fusion. Image acquisition is the precondition of the video image mosaics. The image mosaics methods and results are different from different image acquisitions. There are main three methods to acquire the video image. The first one is to use two fixed video cameras to capture the image. Make sure the visual angle of the two cameras have the overlap area. The second method is to move one video camera along the horizontal line to capture the whole observation scene. In the third method, the video images are captured from the rotary video camera. The rotary angle should be controlled to make the adjacent images have overlap area.

The image registration is the most difficult and important procedure in image mosaics. Before image registration, the image feature points should be extracted effectively. In this paper, we study the SIFT algorithm, and use it to extract the image SIFT feature points. The SIFT algorithm will be described in Section 3. After the correct matching points in image registration, the images can be blended into one image.

3 Sift Algorithm

The SIFT algorithm proposed by David. G. Lowe [2] is an approach for detecting and extracting local feature points from an image. The SIFT algorithm belongs to the feature-based method, and its advantages are invariant to objects scaling and rotation, and partially invariant to change in illumination. The SIFT feature point descriptor is a 128-dimension vector based on a 4 x 4 array neighborhood centered at a feature point extracted from the scale-space. The SIFT algorithm can be summarized into five major steps, as follows [3]:

One: Scale-space extreme detection. The first step is to compute all scales and image locations. The scale space of an image is defined as a function, $L(x, y, \sigma)$, which is produced from the convolution of a variable-scale Gaussian kernel, $G(x, y, \sigma)$, with an input image, $I(x, y)$. An efficient approach is to use scale-space extreme in the difference-of-Gaussian (DOG) function convolved with the image. DOG function is computed from the difference of two nearby scales separated, shown as below,

$$\begin{aligned} D(x,y,\sigma) &= (G(x,y,k\sigma) - G(x,y,\sigma)) * I(x,y) \\ &= L(x,y,k\sigma) - L(x,y,\sigma) \end{aligned} \qquad (1)$$

where * is the convolution operation in $I(x, y)$, and $G(x,y,\sigma) = \frac{1}{2\pi\sigma^2}e^{-(x^2+y^2)/2\sigma^2}$, k is the multiplicative factor.

The candidate key points are selected if they are extreme compared to local neighbors in the current and the adjacent scales.

Two: Keypoint localization. In DOG image, each pixel is compared to its 8 neighbors in the same scale and 9x2 neighbors in above and down scales. Keypoints are selected based on measures of their stability. Then remove the low contrast points and edge responses.

Three: Orientation assignment. In this step, each keypoint is assigned an orientation to make the descriptor invariant to rotation. The keypoint orientation is calculated from the gradient orientation histogram in the neighborhood of the keypoint. For the Gauss smoothed image $L(x, y)$ at scale k_s, the gradient magnitude $m(x, y)$ and orientation $\theta(x, y)$ is computed by:

$$m(x,y)=\sqrt{(L(x+1,y)-L(x-1,y))^2+(L(x,y+1)-L(x,y-1))^2} \quad (2)$$

$$\theta(x,y)=\tan^{-1}((L(x,y+1)-L(x,y-1))/(L(x+1,y)-L(x-1,y))) \quad (3)$$

The peaks in the histogram correspond to the main orientation, and other local peak within 80% of the highest peak is used also to create a keypoint with that orientation.

Four: Keypoint descriptor. Once a keypoint location, orientation, scale are obtained, the next step is to create the feature descriptor. A keypoint descriptor is created by computing the gradient magnitude and orientation at each image sample point in a region around the keypoint location. It is computed based on a 4x4 array of histograms with 8 orientation bins in each. This leads to a SIFT feature descriptor with 4x4x8=128 elements.

Five: The matching of feature points. The best candidate match for each keypoint is found by identifying its nearest neighbor in the database of keypoints. Euclidean distance between each keypoint descriptor is used to search each keypoint's closest two feature points. The ratio of distance between the closest neighbor and the second closest neighbor is computed. If the ratio of distances is less than the threshold, the keypoint and its closest neighbor are kept as a pair of matching points.

Although most pairs of matching point will be correctly matched, a small number of them have bad match. Therefore, it would be useful to have a way to discard features that do not have any good match. If $I_1(x_1, y_1)$ and $I'_1(x'_1, y'_1)$, $I_2(x_2, y_2)$ and $I'_2(x'_2, y'_2)$ are two pairs of matching points, vector $D_1=I_1-I'_1$ and $D_2=I_2-I'_2$ should have the same characters. So we can set a threshold which is equal to the average of all x-axes or y-axes value of the matching points. Each difference of matching point is compared to the threshold, and the greater one will be deleted.

4 Image Mosaics

It is the first step to extract the SIFT feature points for image mosaics. Then the image transformation model between the two adjacent images should be built for the image fusion. The next problem is to compute the transformation model parameters.

4.1 Image Transformation Model

There are many kinds of image transformation models, such as: translation transformation, rotation transformation, rigid transformation, affine transformation, projective transformation and so on. The selection of the image transformation model is based on the camera motion. The projective is suitable for the camera translation, rotation and zoom. The rigid and affine transformation are the special case of the projective transformation. In order to get the image mosaics method more useful, the projective transformation is chose. The relationship between the overlapping images can be described by a homography plane. The projective transformation coordinates between $I(x, y)$ and $I'(x', y')$ can be computed using:

$$\begin{bmatrix} x \\ y \\ 1 \end{bmatrix} = \begin{bmatrix} h_0, h_1, h_2 \\ h_3, h_4, h_5 \\ h_6, h_7, 1 \end{bmatrix} \times \begin{bmatrix} x' \\ y' \\ 1 \end{bmatrix} = H \bullet \begin{bmatrix} x' \\ y' \\ 1 \end{bmatrix} \qquad (4)$$

H matrix is the homography between the two images. There are 8 parameters to compute in the matrix. Many methods can be used to resolve this problem. One of them is RANSAC [4] algorithm.

4.2 Image Fusion

Image fusion is to blend the two overlapped images into one bigger image. If the images are stitched directly, the blending image will contain obvious boundaries due to the luminance difference. There are many image fusion methods which can blend two images smoothly. One of the effective methods is the weighted average method [5]. It can be implemented to stitch images gap less, and it is easy to use. $I(x, y)$, the value of the pixel in the overlapping image is the weighted average value of the corresponding pixel in the two images. It can be calculated by:

$$I(x, y) = I_1(x, y) \times W_1 + I_2(x, y) \times W_2 \qquad (5)$$

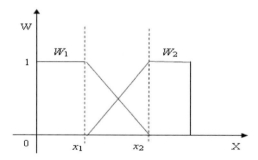

Fig. 2. The weighted average image fusion method

The weighted average image fusion method is shown in Fig. 2. W1 is the weighed coefficient, which can be computed by the x1 and x2. (x1, x2 are the abscissa of the overlapping area).

$$W_1 = x_i / (x_2 - x_1) \quad , \qquad W_2 = 1 - W_1 \tag{5}$$

5 Experimental Results

In this experiment, the results of applying our video image mosaics system are presented. The SIFT algorithm described above has been implemented using Visual C++ 6.0 on a computer with 2.5 GHz processor. A number of real image sequences have been used to test our method. The video images are captured by the video camera rotating on a fixed axis.

(a)

(b)

Fig. 3. (a) and (b) are two frame images of the initial image sequences

Fig. 4. The SIFT feature points of (a) and (b) two images

Fig. 5. The image mosaics result of (a) and (b) two images

The matching targets are the video image sequences, and it is a time-consuming task to deal with all the images. So, the matching image frame should be sampled from all the image frames according to the system run-time and the result vision demands. Fig. 3 just shows two frame images of the video image sequences. The SIFT feature points extracted by our method are shown in Fig. 4, and the image mosaics result of the two frame images is shown in Fig. 5.

6 Conclusions

This paper focuses the work on the image mosaics technology in the video processing system. The most important task in image mosaics is to get the stable image features and describe the transformation between the two images. The SIFT algorithm is used to extract the stable and effective image feature points in this paper. These features are reasonable invariant to scaling, translation and rotation. The corresponding feature points are matched by Euclidean distance. RANSAC algorithm is used to compute the projective transformation parameters. Finally, the images are stitched smoothly by using the weighted average method.

The performance of the proposed method has been demonstrated by blending the images. The mosaic results prove that our method is practical and effective.

Acknowledgment

I would like to express my gratitude to my wonderful Pro. Jin, who has been giving my knowledgeable guidance through the work of this project and this paper.

References

1. Irani, M., Anandan, P., Hsu, S.: Mosaic based representations of video sequences and their applications. In: ICCV 1995, pp. 605–611 (1995)
2. Lowe, D.G.: Distinctive Image Features from Scale-Invariant Keypoints. International Journal of Computer Vision 60(2) (November 2004)
3. Yang, P., Mao, Z.: Development of Electronic Panoramic Mirror Based on Image Mosaics. In: SOPO (2010)
4. Fisehier, M.A., Bolles, R.C.: Random Sample Consenuss:A Paradigm for model fitting wih applications to image analysis and automated cartogrpahy. Communication Association Machine 24(6), 381–395 (1981)
5. Irani, M., Anandan, P.: Video indexing based on mosaic representations. Proceedings of the IEEE 86(5), 905–921 (1998)

Estimate the All Vanishing Points from a Single Image

Yongyan Yu[1,2], Zhijian Wang[2], and Yuansheng Lou[2]

[1] School of Computer Engineering, Huaiyin Institute of Technology, 223003 Huaian, China
[2] School of Computer and Information Engineering, Hohai University,
210098 Nanjing, China
`shanshan_yyy@163.com`

Abstract. Vanishing point is the important precondition for camera self-calibration from a single image. Previously proposed solutions, either relies on voting in the Gaussian sphere space, or iteratively time and again base on Maximal Likelihood Estimator. All these methods expend lots of time, have large error and low level of efficiency. Hence, a new scheme will be presented with a recently proposed algorithm called J-Linkage, in which each edge is represented with the characteristic function of its preference set and vanishing points are revealed as clusters in this conceptual space. First, it estimates all possible vanishing point, and then refines them by Expectation Maximization. Finally, two experiments show that algorithm reduces the number of variables, error measures are done in the image, a consistency measure between a vanishing point and an edge of the image can be computed in closed-form. So that it has a low computation and high precision.

Keywords: preference set, consistency measure, Jaccard distance, clustering, refine.

1 Introduction

Under perspective projection, parallel lines in 3-D space project to converging lines in the image plane. The common point of intersection, perhaps at infinity, which is called the vanishing point (VP). has became very consequence in Computer Vision. To calibrate accurately the VPs would be the most important precondition for camera calibration, pose estimation, autonomous navigation and 3-D reconstruction, etc. Detection of vanishing points, which is of little help in natural outdoor scenes, becomes of prime importance in the man-made environment where regular block looking structures or parallel alignments (streets, pavements, railroad) abound. For that reason, most efforts have been put into algorithms for their automatic detection in images, several methods have been presented to address these problems.

These methods can be divided into two categories, the first one depends on Hough Transform.Since Barnard [1], detection was performed on a quantized Gaussian sphere using a Hough transform, until this was shown to lead to spurious vanishing points. Subsequently, Evelyne [2] proposed a method to locate three vanishing points on an image based on two cascaded Hough transforms, Andrea [3] advanced that the estimation of the coordinates of the vanishing point can be retrieved directly on the Hough Transformation space or polar plane. However these makes the approach ill-suited for accurately estimating the true vanishing point location. If a low variance estimate is required, the histogram must be partitioned very finely, negating some of the computational benefits of the approach.The second one is based on the statistical.For example, Magee [4] compute intersections of pairs line segments directly, using cross product operations.Vanishing points are detected as clusters of intersection points on the sphere. Since computed points are maintained and distances compared uniformly over the sphere, vanishing points can be estimated with greater accuracy. The drawback is that examining all pairs of line segments yields a complexity of $O(n^2)$. Andre [5] developed a new detection algorithm that relies on the Helmoltz principle, both at the line detection and line grouping stages. This leads to a vanishing point detector with a low false alarms rate and a high precision level, which does not rely on any a priori information on the image or calibration parameters, and does not require any parameter tuning.Hadas's approach detects entire pencils based on a property of 1D affine-similarity between parallel cross-sections of a pencil[6].One typical approach uses the RANSAC principle: intersecting randomly selected couples of lines and choosing those that are consistent with a large group of lines[7]. It is a pity that those methods have enormous numeration and not fit mult-model estimation.

Thereby, we present a novel algorithm with a recently proposed algorithm called J-Linkage. First, it estimates all possible vanishing point, and then refines them by Expectation Maximization, finally identifies the three vanishing points corresponding to the Manhattan directions.

2 Estimate Presumptive VPs Using J-Linkage

We assume that a set of edges which is denoted as $E_{edge} = \{e_n \mid n=1...N\}$, have been detected from a image in pixel. The nth edge of the E_{edge} is given by e_n. The end points of e_n are given by e_n^1, e_n^2, and their centroid is given by $\overline{e}_n$. The line passing by the $\overline{e}_n$ is given by $l_n = (l_n^1, l_n^2, l_n^3)^T$. A especial example is show in Fig. 1.

Since a image possibly consists of many VPs,then VP estimation belong to typical multiple model estimation.The problem of simultaneous multiple model estimation has became a hotspot in computer vision. At present, the most popular methods for resolving above issue are multi-RANSAC and RHT [8,9].multi-RANSAC method is effective and very robust, but the number of models is user specified, and this is not acceptable in some applications.On the other hand,RHT does not need to know the

Fig. 1. The Structure of a Edge

number of models beforehand. However, RHT suffers from the typical shortcomings of Hough Transform methods, such as limited accuracy and low computational efficiency. Ultimately, the choice and the discretization of the parameter space turn out to be crucial.

Recently, Toldo and Fusiello [10] presented an algorithm called J-Linkage that overcome the weakness of multi-RANSAC and RHT. The kernel idea about J-Linkage is that each edge would be represented by its Preference Set(PS) which models a edge has given consensus to, and then cluster the edges based on the Jaccard distance of PSs. The models are built using minimal sample sets similar to RANSAC. Our approach for detecting VPs relies on the J-Linkage algorithm. The following are basicl steps.

2.1 Create the Minimal Sample Sets

The method starts with random sampling: M model hypothesis are generated by drawing M minimal sets of data edges necessary to estimate the VP, called Minimal Sample Sets (MSS), marked as s_m. The first step is to randomly create a MSS consist of one edge pair $\{e_1, e_2\}$. Those lines through the edges of that MSS are l_1 and l_2 respectively, which could determine a candidate VP given by $\mathbf{v}_m = l_1 \times l_2$. Thereout, we obtain a set of MSS $S = \{s_m \mid m = 1...M\}$.

2.2 Construct the Preference Sets

The following step is to build a matrix $\mathbf{P}_{N \times M}$, called preference matrix(PS). Each row of $\mathbf{P}$ is associated with a edge e_n of E_{edge}, and each column of that matrix corresponds to a hypothesis $\mathbf{v}_m$, i.e. which models it prefers. The value of a element p_{nm}, i.e. the entry of matrix $\mathbf{P}$, is made by

$$p_{nm} = \begin{cases} 1 &, C(\mathbf{v}_m, e_n) \leq \Box t \\ 0 &, C(\mathbf{v}_m, e_n) > \Box t \end{cases} \quad (1)$$

where $C(\mathbf{v}_m, e_n)$ is a consistency measure function, use to measure the consistency between hypothesized vanishing point and edge. $\Box t$ is a premeditated consensus threshold, we used $\Box t = 2$ pixel.

In Fig.2, we gained a line by traversing $\mathbf{v}_m$ and $\overline{e}_n$, called as $\hat{l} = \overline{e}_n \times \mathbf{v}_m = [\overline{e}_n]_\times \mathbf{v}_m$. The orthogonal distances of the end points to $\hat{l}$ are given by d_1 and d_2, respectively. Apparently, $d_1 = d_2$.

We find the one minimizing the maximal distance for $\hat{l}$ to end points e_n^1, e_n^2. It is given in closed-form by

$$C(\mathbf{v}_m, e_n) = \max_{i=1,2} \mathrm{dist}(e_n^i, \hat{l}) = \mathrm{dist}(e_n^1, \hat{l}) \quad (2)$$

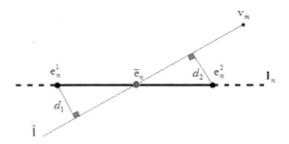

Fig. 2. The principle of Consistency Measure

If $C(\mathbf{v}_m, e_n) \leq \Delta t$, then show that edge e_n is consistency with $\mathbf{v}_m$. Thus, we define Consensus Set (CS) for vanishing point $\mathbf{v}_m$, i.e. a set of edges which are associate to the entries with true value in m^{th} column of $\mathbf{P}$, given by $CS_m = \{e_n\}$. The other way round, those candidate VPs relating to the entries with true value of nth row can form a Preference Set (PS) about e_n, given by $PS_n = \{v_m\}$. It's means that each row indicates which models a points has given consensus to, i.e., which models it prefers.

2.3 Calculate Jaccard Distance

Subsequently, it is needed to calculate the Jaccard distance between two PS, i.e. e_i^{PS} and e_j^{PS} of edges e_i, e_j. It is given by

$$d_{Jaccard}(e_i^{PS}, e_j^{PS}) = \frac{|e_i^{PS} \cup e_j^{PS}| - |e_i^{PS} \cap e_j^{PS}|}{|e_i^{PS} \cup e_j^{PS}|} \tag{3}$$

The Jaccard distance measures the degree of overlap of the two sets and ranges from 0 (identical sets) to 1 (disjoint sets). Over here, if e_i^{PS} is superpose to e_j^{PS}, then $d_{Jaccard} = 0$. If e_i^{PS} and e_j^{PS} are detached completely, $d_{Jaccard} = 1$. Finally we have a set of Jaccard distance, $\mathbf{D}_{Jaccard} = \{d_{Jaccard}^i \mid i = 1 \ldots N(N-1)/2\}$.

2.4 J-Linkage Clustering

VPs will be extracted by agglomerative clustering of edges in the conceptual space, where each edge is represented by its preference set. Our agglomerative clustering algorithm proceeds in a bottom-up manner: Starting from all singletons, each sweep of the algorithm merges the two clusters with the smallest Jaccard distance.

Step 1. To create a set of atom-cluster, given by $C = \{C_i \mid i = 1, 2, ..., N\}$, and then each atom-cluster consists of a edge, i.e. $C_i = \{e_i\}$.

Step 2. Let us to find the entry $d_{Jaccard}^k$ corresponding to the smallest Jaccard distance from $D_{Jaccard}$. When $d_{Jaccard}^k < 1$, we can detect the two atom-clusters C_i, C_j (i<j) that was determined by $d_{Jaccard}^k$. Next, Replace C_i cluster with the union of the two original ones, $C_i = C_i \cup C_j$, and eliminate C_j form C.

Step 3. We defines the PS of a cluster of edges as the intersection of the preference sets of its members. Thus we can estimate a PS of the New C_i.

Step 4. Update the $D_{Jaccard}$. At first, we will calculate renewedly Jaccard distance between New C_i and other clusters of C. Second, those entries associated to C_j will be cleaned out from $D_{Jaccard}$.

Step 5. Repeat from step 2 while the smallest Jaccard distance is lower than 1.

The result of clustering using J-Linkage is that all edges was classified into finite several clusters, formed a set of cluster $S' = \{s_m' \mid m = 1...M'\}$. Each cluster has a imaginabale vanishing point which lies on the PS of all edges in this cluster.

2.5 Estimate Vanishing Point Renewedly

Once clusters of edges are formed, a vanishing point can be computed for each of them. If there are more than two edges in a cluster, we must find a effectual method to estimate vanishing point. Aim at this case of more than two edges, we describe the function $V(s_m')$ that computes a vanishing point v_m based on a set of edges s_m', given by

$$V(s_m') = \arg\min_{v_m} \sum_{\varepsilon_j \in s_m'} \text{dist}^2(\hat{l}_n, e_n^1) = \arg\min_{v_m} \sum_{\varepsilon_j \in s_m'} \text{dist}^2([\overline{e}_n]_\times v_m, e_n^1) \quad (4)$$

According to (4), it shown that a point minimizing the summation of perpendicularity distance from constraints line $\hat{l}_n$ to endpoint of per edge e_n in s_m' maybe the vanishing point. Apparently, this scheme doesn't rely on extra parameters $\hat{l}_n$, only relates with the edge, therefore would reduce the number of variables。

3 Refine Vanishing Point Using EM

The most frequent mistake made by the J-Linkage algorithm is to divide a cluster of edges of a vanishing point into two groups. However, this is easily corrected by refining the solution using Expectation Maximization(EM).The EM method performs both classification and estimation tasks by alternating between finding the best classification given the current estimates (the E-step) such as VP v_j', and finding the best estimates given a classification (the M-step). EM is guaranteed to converge on the optimal solution given a fixed number of mixtures and a reasonable initialization.

3.1 E Step

Given an estimate of a VP v_j' and its associated variance σ_j^2, we can compute the probability that a given edge e_n belongs to this VP. Here we use a weighted zero-mean Gaussian model

$$P(e_n | v_j') = \frac{1}{\sqrt{2\pi}\sigma_j} \exp\left(\frac{-\theta_{nj}^2}{2\sigma_j^2}\right) \qquad (5)$$

Where, $\theta_{nj} = \sin^{-1}(e_n \cdot v_j')$, it denotes the angular deviation of edge from the plane.

From these conditional probabilities, we use Bayesian arguments to derive the reverse conditionals,

$$P(e_n) = \sum_{j=1}^{M'} P(e_n | v_j') P(v_j') \qquad (6)$$

$$P(v_j' | e_n) = \frac{P(e_n | v_j') P(v_j')}{P(e_n)} \qquad (7)$$

where $P(v_j')$ is the proportion of those edges belonging to v_j' as $P(v_j') = K_j / N$, called priori probability.

The probabilities $P(v_j' | e_n)$ give each e_n a likelihood of belonging to each v_j', providing a weighting mechanism for subsequent fitting steps.

3.2 M Step

Given a set of $P(v_j' | e_n)$ for each edge and each VP, we estimate the variable quantities σ_j^2, $P(v_j')$ and v_j'. The estimation of prior probabilities and variance is straightforward:

$$P(v_j') \approx \frac{1}{N} \sum_{n=1}^{N} P(v_j' | e_n) \qquad (8)$$

$$\sigma_j^2 = \frac{\sum_{n=1}^{N} \theta_{nj}^2 P(v_j' | e_n)}{N P(v_j')} \qquad (9)$$

Since $\theta_{nj} \approx x_n \cdot d_j$ for small deviations from the plane, the VP can be estimated using the weighted linear least-squares formulation

$$\min_{v_j} \left\| W_j^{n \times n} A_j^{n \times 3} v_j' \right\|^2 \tag{10}$$

Where W_j is a diagonal matrix containing the weights $P(v_j' | e_n)$. A_j is a matrix whose rows are the edge e_n, assuming $\text{rank}(A_j) \geq 2$.

3.3 Validation

Thanks to that a real image has a lots of useless edges possibly, thus a validation step is performed to verify that the VP estimated by EM are statistically significant. To be considered significant, each VP must meet several criteria:

- Edge count metric $K_j > \mu_e - \gamma_e$
- Variance metric $-\log(\sigma_j^2) > \mu_v - \gamma_v$

where (μ_e, γ_e) and (μ_v, γ_v) are the respective mean and standard deviation of the edge count and variance metrics.

If there is no change between the current set of VPs and the output of the validation step, the process terminates. Otherwise, EM is performed on the validated VPs and the process repeats.

4 Experiments

In the interest of validating accuracy and efficiency of our idea, we will take a test about indoor and outdoor image respectively. Each image has three orthogonal vanishing points at least, and much possibly non-orthogonal vanishing points. Our main goal is to detect all vanishing points, without knowing the camera internal parameters. For the J-Linkage algorithms, we used $\Delta t = 2$ pixels and M = 200 in all our tests. The results are shown in Fig.3.

(a) Outdoor image   (b) Indoor image

Fig. 3. The Estimation of Vanishing Points

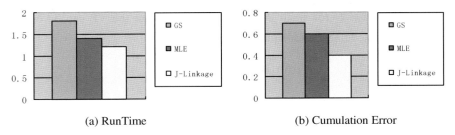

Fig. 4. Compare the three methods in RunTime and Cumulation Error

Fig.4 shows that results of comparing with Gaussian sphere(GS), Maximal Likelihood Estimator (MLE) and ours based on J-Linkage.

5 Conclusion

The detection of vanishing points in images is a typical problem of multi-model estimation. The provided new algorithm based on J-linkage avoids clustering on the Gaussian sphere, but also does not require prior specification of the number of vanishing points, nor it necessitate manual parameters tuning. Our method demonstrated its effectiveness in comparison with existing algorithms.

References

1. Barnard, S.T.: Interpreting perspective images. Artificial Intelligence 21(4), 435–462 (1983)
2. Lutton, E., Mahe, H., Lopez-Krahe, J.: Contribution to the Determination of Vanishing Points Using Hough Transform. IEEE Transactions on Pattern Analysis and Machine Intelligence 16(4), 430–438 (1994)
3. Matessi, A., Lombardi, L.: Vanishing point detection in the hough transform space. In: Amestoy, P.R., Berger, P., Daydé, M., Duff, I.S., Frayssé, V., Giraud, L., Ruiz, D. (eds.) Euro-Par 1999. LNCS, vol. 1685, pp. 987–994. Springer, Heidelberg (1999)
4. Magee, M.J., Aggarwal, J.K.: Determining Vanishing Points from Perspective Images. CVGIP 26, 256–267 (1984)
5. Almansa, A., Desolneux, A., Vamech, S.: Vanishing Point Detection without Any A Priori Information. IEEE Transactions on Pattern Analysis and Machine Intelligence 25(4), 502–507 (2003)
6. Kogan, H., Maurer, R., Keshet, R.: Vanishing Points Estimation by Self-Similarity. IEEE Transactions on Pattern Analysis and Machine Intelligence 32(9), 755–761 (2009)
7. Aguilera, D.G., Lahoz, J.G., Codes, J.F.: A new method for vanishing points detection in 3d reconstruction from a single view. In: Proceedings of the ISPRS Commission V (2005)
8. Vidal, R., Ma, Y., Sastry, S.: Generalized principal component analysis (GPCA). IEEE Transactions on Pattern Analysis and Machine Intelligence 27(12), 1945–1959 (2005)
9. Zuliani, M., Kenney, C.S., Manjunath, B.S.: The multiransac algorithm and its application to detect planar homographies.In International Conference on Image Processing (2005)
10. Toldo, R., Fusiello, A.: Robust multiple structures estimation with J-linkage. In: Forsyth, D., Torr, P., Zisserman, A. (eds.) ECCV 2008, Part I. LNCS, vol. 5302, pp. 537–547. Springer, Heidelberg (2008)

Fault Diagnosis of Three Level Inverter Based on Improved Neural Networks

Wang Wu, Wang Hong-Ling, and Bai Zheng-Min

School of Electrical and information Engineering, Xuchang University, Xuchang, China
`jhwlz@tom.com`

Abstract. Inverter and related system were widely used in power electrical system and motor drive system to enhance the reliability and efficiency, the faults with various types and difficult to isolate with traditional techniques, so a new method based on neural network was presented. The neural-point clamped three level invert systems were analyzed and fault features were created by harmonica spectral analysis. The neural network was designed with algorithm programmed, with the fault diagnosis as inputs of neural network, by neural networks adaptive self-learning and take the outputs as judgment of fault types and then the faults occurred in inverter system was isolated . The new technique proposed in this paper was a theoretical foundation for invert system fault diagnosis and practical application for motor driver system, the simulation shows this method is effective and can be widely into inverted fault diagnosis system and relevant fault diagnosis system.

Keywords: neural networks; inverter; fault diagnosis; power electronics; simulation.

1 Introduction

With the rapid development of power electronics technology, power electronic devices have been applied widely in industry area; it becomes an urgent matter to diagnose power electronic devices in real time. Playing roles as controllers or power supplies, the stability of power electronic devices make the assurance of continuous production. Severe fault of power electronic devices can even paralyze the whole system. Therefore, in order to improve reliability and maintainability of power electronic system, researches on fault diagnosis of power electronic circuits is essential. With the comprehensive application of inverter in high voltage and large power field, its fault problem becomes very distinct, the protection circuit of inverter and fault diagnosis algorithm are very useful to improve life and efficiency of device and to reduce fault rate [1]. For the character of nonlinear, inverter can't use accurate mathematical model for fault diagnosis, a novel algorithm based on neural network was presented to detect and diagnose in inverter. Artificial neural networks (ANN) are simplified mathematical models inspired by the biological structure and functioning of the brain, imitating the human brain's physical structure and with its powerful parallel computing and the ability to think, artificial neural network is very suitable for the equipment fault diagnosis, the equipment failure diagnosis using of neural network can greatly enhance the accuracy [2].

In this paper, three level inverter system of neural-point clamp inverter was analyzed and the typical faults were proposed, with $d-q$ transform of three output voltage signal and by Fourier transform, magnitude and phase of the fault signal harmonics are obtained as neural networks inputs. Also the improved neural networks algorithm was proposed to determine the hidden layer neurons of neural networks, the simulation result shows this is a good method for inverter fault diagnosis.

2 Inverter and Typical Fault Analysis

2.1 Inverter Structural and Fault Feature

A typical main circuit topology of neural-point clamped three-level inverter system as shown in Fig. 1, the inverter DC bus voltage was U_d, two series capacities divide equally the input voltage and each capacity's voltage is $U_d/2$, each phase was composed of four switching power electronics devices and two clamp diodes, because the circumstance of each phase is basically same, the paper takes the single-phase circuit as a study object. When switches Sa1 and Sa2 turn on at the same time, output voltage is $U_d/2$. When Sa2 and sa3 turn on ate the same time, output voltage is 0. When sa3 and sa4 turn on ate the same time, output voltage is $-U_d/2$ [3].

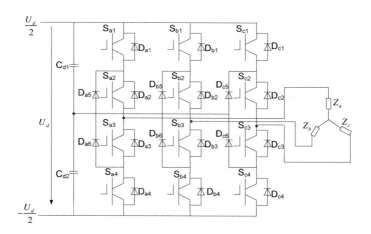

Fig. 1. Main Circuit topology of Three-level inverter

With the principle of the circuit, the relation between output voltage and main switch states as shown in Table 1.

The inverter system may have faults mode as:

(1) Base driven fault of each switching device (F_1);
(2) short-circuit fault of each switching device (F_2);
(3) Intermittent breaker failure of each switching device (F_3)
(4) Meanwhile break failure of switching device at same phase (F_4);
(5) Short-circuit fault of switching device at same phase (F_5);
(6) break failure of switching device at cross bridge arm (F_6);
(7) Passive components failure (F_7);

Table 1. The relation between output voltage and main switch states

Switch state	$S_{x,1}$	$S_{x,2}$	$S_{x,3}$	$S_{x,4}$	Output voltage
P	ON	ON	OFF	OFF	$U_d/2$
O	OFF	ON	ON	OFF	0
N	OFF	OFF	ON	ON	$-U_d/2$

2.2 Fault Features and Harmnoic Spectrum Analysis

Three level inverters with complicated fault circumstance, here we suppose the fault was occurred on two elements at most, so the fault can be divided into 9 types, the typical 5 types are as following:

Type1: normal condition, without failure in any power electronics device;

Type2: one power electronics device failure, which with 12parts, we can described as $S_{ij}, i = a, b, c; j = 1, 2, 3, 4$

Type3: two power electronics device failure at same bridge arm, which also with 12 parts, we can described as: S_{ij} and $S_{ik}, i = a, b, c; j = 1, 2; k = 3, 4$

Type4: two power electronics device failure at cross arm of upper bridge, which with 12 parts, we can describe as: S_{ij} and $S_{kl}, i, k = a, b, c; i \neq k; j, l = 1, 2$

Type5: two power electronics device failure at cross arm of under bridge, which with 12 parts, we can describe as: S_{ij} and $S_{kl}, i, k = a, b, c; i \neq k; j, l = 3, 4$

Also with other four types and we take it as non typical faults and not mentioned in this paper.

Three phase voltage output was monitored and with $d-q$ transform.

$$\begin{bmatrix} U_d \\ U_q \end{bmatrix} = \sqrt{2/3} \begin{bmatrix} 1 & -\frac{1}{2} & -\frac{1}{2} \\ 0 & \sqrt{3}/2 & -\sqrt{3}/2 \end{bmatrix} \begin{bmatrix} U_a \\ U_b \\ U_c \end{bmatrix} \qquad (1)$$

The fault features above mentioned was often surpassing the required range of neural networks, it may be difficult for NN's training, so the data was normalized before NN training. Then was analyze with Fourier Transform, and the amplitude and phase of harmonics was extracted, and the DC component, base amplitude, base phase and second harmonics phase was selected as feature parameters for neural networks input.

3 Neural Networks Model for Fault Diagnosis

The essence of BP algorithm is a nonlinear optimization problem with gradient algorithm and iterative operation to resolve weight value. Correlation pruning algorithm was used on the basis of correlation between hidden layer nodes, which probed and discussed by many scholars. Sietsma analyze the correlation of hidden layer nodes and delete nodes which with high correlation [4]. Castellano delete nodes which with minimum output energy for all patterns [5], Song Qing-kun etc proposed structural optimization of BP neural networks with correlation pruning algorithm [6], Rao-hong proposed circular self-configing algorithm for neural networks structure optimization [7]. Here hidden layer nodes merge method introduced.

3.1 Correlation Analyze for Hidden Layer Nodes

Assume $\{V_{ip}\}$ and $\{V_{jp}\}$ is the output sequence of node i and node j in same hidden layer, the average value are $\overline{V}_i$ and $\overline{V}_j$ respectively, so the correlation coefficient of $\{V_{ip}\}$ and $\{V_{jp}\}$ can be given [8]:

$$r_{ij} = \frac{\sum_p V_{ip} V_{jp} - n\overline{V}_i \overline{V}_j}{(\sum_p (V_{ip} - \overline{V}_i)^2)^{\frac{1}{2}} (\sum_p (V_{jp} - \overline{V}_j)^2)^{\frac{1}{2}}} \qquad (2)$$

The standard deviation of nodes output can be given:

$$s^2_i = \sum_p (V_{ip} - \overline{V}_i)^2 \qquad (3)$$

The relation between two nodes can be classified three types: maybe with high positive correlation, high negative correlation, these two types can be defined as high correlation and the nodes can be merged. Maybe low correlation without merge or delete any node. Based on the correlation of hidden layer nodes, the redundant nodes in networks can be deleted or merged.

3.2 Algorithm for Merging

The algorithm can be realized with the following steps [9]:

Step1: initialization. Define networks structure, initial weights, learning coefficient, momentum coefficient, the combined time error ε_1, destination error ε_2, merge threshold θ_1 and θ_2, make learning number with $k=1$.

Step2: BP neural network training. Amend weight of neural networks with delta learning rules, computing training error and output sequence of hidden layer simultaneously.

Step3: determine merge or not. If current error $E > \varepsilon_1$, step into next learning with $k = k+1$, then goto step2. if the conditions satisfied with $\varepsilon_2 < E < \varepsilon_1$, then goto step4. and if $E < \varepsilon_2$ then learning stopped. The combine time can be selected as:

$$\frac{1}{2}\sum_p \sum_k (O_{kp} - T_{kp})^2 \leq \varepsilon_1 \tag{4}$$

Where T_{kp} and O_{kp} are $k-th$ desired output and actual output of networks, combined time error ε_1 slightly larger than destination error ε_2.

Step4: computing standard deviation and correlation coefficient with (3) (4).
Step5: merge hidden nodes of high correlation, when the correlation coefficient and standard deviation of $\{V_{ip}\}$ and $\{V_{jp}\}$ satisfied with (6), two nodes merged.

$$|r_{ij}| > \theta_1 \text{ And } s_i^2 > \theta_2, s_j^2 > \theta_2 \tag{5}$$

Where θ_1 and θ_2 were threshold for merging.
With linear regression given as (7) (8) (9):

$$V_j = aV_i + b \tag{6}$$

$$a = \frac{\frac{1}{n}\sum_{p=1}^n V_{ip}V_{jp} - \overline{V}_i\overline{V}_j}{\frac{1}{n}\sum_{p=1}^n V^2_{ip} - \overline{V}^2_{ij}} \tag{7}$$

$$b = \overline{V}_j - \overline{V}_i \tag{8}$$

The input of next layer node can derive:

$$(\omega_{ki} + a\omega_{kj})V_i + (\omega_{kb} + b\omega_{kj})*1 + \sum_{l \neq i,j} \omega_{kl}V_l \tag{9}$$

So the node j deleted and set:

$$\omega_{ki} \rightarrow \omega_{ki} + a\omega_{kj} \tag{10}$$

$$\omega_{kb} \rightarrow \omega_{kb} + b\omega_{kj} \tag{11}$$

Step6: offset nodes merging. If nodes of hidden layers satisfied with $s_i^2 < \theta_2$, merging with offset nodes and step into next learning with $k = k+1$, goto step2.

When the standard deviation of output sequence smaller, output can be replaced by average value, make $V_i = \overline{V}_i$, the input of next layer nodes derived:

$$(\omega_{kb} + \overline{V}_i\omega_{ki})*1 + \sum_{l \neq i,j} \omega_{kl}V_l \tag{12}$$

So the node i deleted and set:

$$\omega_{kb} \rightarrow \omega_{kb} + \overline{V}_i\omega_{ki} \tag{13}$$

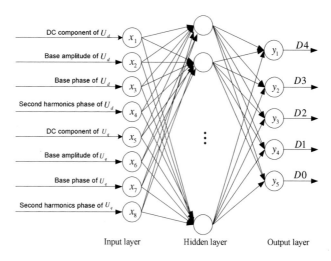

Fig. 2. The structure of neural networks

3.3 Learning Samples for Neural Networks

Neural networks with 8 input neurons and 5 output neurons, hidden layer neurons number was determined by correlation pruning algorithm, the structure of neural networks as shown in Fig. 2. The fault types were denoted by code of NN's output $D4D3D2D1D0$ which binary code was for classify fault type and training with the 8 inputs which were DC component, base amplitude, base phase, second harmonics phase of U_d and U_q respectively.

4 Simulations and Conclusions

The convergence curve of normal BP neural networks and BP neural networks with correlation pruning algorithm was shown in Fig. 3.

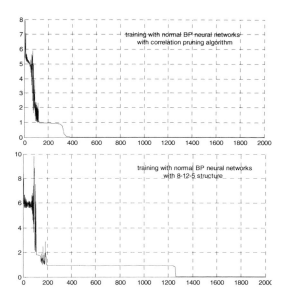

Fig. 3. Comparation of normal BP and improved BP neural networks

The hidden layer was hard to determine, here we take correlation pruning algorithm and simulation under MATLAB software [10], [11], the hidden layer neuron was 24. with normal BP neural networks, we take the structure of BP neural network with 8-12-5 and with learning rate is 0.2 and inertia coefficient 0.8, it will take about 1300 epochs for NN's training and the fault recognition rate is 92.83%, but with correlation pruning algorithm as proposed in this paper, only 380 epochs taken for NN's training and the fault recognition rate is 97.48. the simulation shows the neural network can reflect the nonlinear relation of fault features and fault type, by simulation also we can see that the method presented in this paper can analyze the correlation of NN and determine hidden layer neurons automatically, the improved neural networks with higher training speed and high precision, with it's generalize ability and the fault recognition rater is higher than normal BP neural networks. Faced with the inverter with large fault types which were hard to determine, this new method was effective and the fault features can get with output binary code, this method can realize fault diagnosis and applied into practical applications.

Acknowledgments

It is a project supported by natural science research in office of education, HeNan Province. With the Grant NO: (2011B470005) and Grant NO: (2011B510016).

References

1. Khomfoi, S., Tolbert, L.M.: Fault Diagnosis system for a multilevel inverters using a neural network. In: IEEE Industrial Electronics Conference, vol. 4(5), pp. 2188–2193
2. Wang, F.: Sine-triangle versus space-vector modulation for three level PWM voltage source inverters. IEEE Transactons on Industry Applications 38(2), 500–506
3. Kastha, D., Bose, B.K.: Investigation of fault modes of voltage mode inverter system for induction motor. IEEE Trans. Ind. Applicat. 30, 1028–1037
4. Sietsma, J., Dow, R.J.F.: Neural net pruning – why and how? In: EEEE ICNN, pp.325–333 (1988)
5. Castellano, G., Fanelli, A.M., Pelillo, M.: An iterative pruning algorithm for feedforward neural networks. IEEE Trans. Neural Networks 8(3), 519–531 (1997)
6. Song, Q.-k., Hao, M.: Sturctural Optimization of BP Neural Network Based on Correlation Pruning Algorithm. Control Theory and Applications (25), 4–6 (2006)
7. Rao, H., Fu, M.-f., Chen, L.: Structural Optimization for Neural Network Based on circular Self-configuring Algorithm. Computer Engineering and Design (29), 411–417 (2008)
8. Kun, W.H.: Theory and Mehod for Neural Networks Structure Design, pp. 131–138. National Defense Industrial Press, Beijing (2005)
9. Qiao, J.-F., Han, H.-G.: Optimal structure design for RBFNN structure. Acta Automatica Sinica 6(6), 865–872 (2010)
10. Liu, J.K.: MATLAB Simulation for Sliding Mode Control, vol. 10, pp. 237–279 (2005)
11. Liu, J., Sun, F.: A novel dynamic terminal sliding mode control of uncertain nonlinear systems. Journal of Control Theory and Applications 5(2), 189–193 (2007)

The Application for Diagnose of Ginseng Disease and Insect Pests-Es Based on Case-Based Reasoning

Zhang Li-Juan[1,*], Li Dong-Ming[2,**], Chen Gui-Fen[2], and Chen Hang[2]

[1] College of Computer Science and Engineering, Changchun University of Technology, YanAnRoad 2055,Changchun, Jilin Province, China
ldm0214@163.com
[2] School of Information Technology, Jilin Agriculture University, XinCheng Road 2888, Changchun, Jilin Province, China
guifchen@163.com

Abstract. This article addresses Ginseng disease and insect pests diagnostic expert system. The aim of the work presented here is twofold. First, we diagnose the origin of disease; Second, how to prevention and cure phase. On the basis of this, in the light of diagnostic Ginseng the origin of disease and insect pests, which is the model of the case, we use CBR technology for computer-based problem solvers and relevant algorithm. This paper presents the two machine learning techniques incorporated are case-based reasoning and genetic algorithms(GA).These algorithms look for improving the results obtained by human experts and the previous statistical model.

Keywords: Case Based Reasoning(CBR), Ginseng disease and insect pests, Genetic Algorithms(GA), Expert System(ES), K-Nearest Neighbor(K-NN).

1 Introduction

The purpose of this article is twofold: first, we diagnose the origin of disease, in the other words, diagnostic what disease, what virus or insect gives rise to the disease; Second, we find the solution about how to prevention and cure disease. when solving the first problem, we use the two machine learning techniques incorporated are Case Based Reasoning and Genetic Algorithms, in order to find the origin of disease .From human experts conclusion and previous result statistical model, the incorporation of CBR and GA ,which improves the accuracy of diagnostic the origin of disease .The second problem solving method, we give first place to Rule-Based Reasoning(RBR),thus use Case Based Reasoning as subordinating. According to the origin of disease describe, we can find previous disease and insect pests how to prevention and cure method from the knowledge base. If the origin of disease is not exist in knowledge base, according to all features of it finding the best prevention and cure method

---

[*] The research work is supported by youth foundation project for Science and Technology development of Jilin Province(201101093).
[**] The research work is supported by youth research and development project of Jilin Province(20100155).

in case base .The whole system can be considered as totality structure shown in Fig.1.This article will mainly introduce the application in the CBR of diagnostic "the origin of disease" system.

The paper is structured as follows: section 2 presents a brief description of Case-Based Reasoning, section 3 describes the key segment of CBR and disposing of scheme in "diagnostic the origin of disease" system, section 4 tests the system and summaries the conclusions.

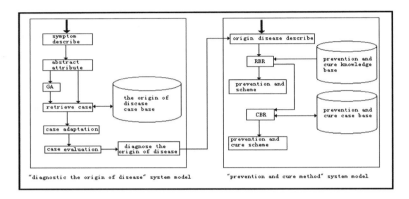

Fig. 1. The totality structure of Ginseng disease and insect pests diagnostic expert system.

2 An Overview of Case-Based Reasoning

Case Based Reasoning (CBR) system is a particular type of intelligence and ability reasoning system which has a development from the surface of machine imitating to the deep layer of machine thought [4]. The goal of CBR is infer a solution for a current problem description in a special domain from solutions of a family of previously solved problems, the case base or case memory, in consequence, getting the conclusion of the current problem .Therefore, it is the deep layer researching of CBR that can help for discussing human beings solving problems and learning new knowledge methods; Helpful for resolving to getting knowledge difficult and inference fragile in expert system; Helpful for not complete known domain ,according to previous experience making some hypothesis and inference then help to face current problems.

2.1 The Totality Idea of CBR

The core idea of CBR is that revise the solutions of a family of previously solved problems, apply to current problem similar to previous problems[1,3].The main impetus of CBR research comes from building a type of human consider and cognitive model, then establish computer system which can more effective, stronger resolve current world problems.

2.2 The Core Step of CBR

Basically, the CBR core steps are:1.abstract case feature ,load basic case for case base;2.retrieving past cases that resemble the current problem;3.adapting past solutions to the current situation[5] ,then applying these adapted solutions and evaluating the results;4.updating the case base[3] .Basically CBR system makes inferences using analogy to obtain similar experience for solving problems .Similarity measurements between pairs of features play a central role in CBR. In essence when developing a CBR system, determining useful case features that are able to differentiate one case from others must be resolved firstly [2]. Furthermore, the weighting values used to determine the relevance of each selected feature has to be assigned before proceeding with the case matching process .To provide an alternative solution this paper presents a genetic algorithm(GA)-based approach to automatically construct the weights by learning the historical data.

3 The Key Segment of CBR and Disposing of Scheme in "Diagnostic the Origin of Disease" System

In "diagnostic the origin of disease "system, to be solved mainly problems, there are three content ,as follows:1 define the pattern of storage cases;2 retrieval case ;3 adapt solutions and evaluate the results.

3.1 Case Base Building of "Diagnostic the Origin of Disease" System

Whether the whole system can accuracy play, decided by the pattern of storage cases ,abstract of the case features and describe of basic case .In"diagnostic the origin of disease "system,we use the pattern of data base to save cases,which make retrieval case and update case more convenient and more speedy.According to main features of the origin of disease ,we conclude the feature of case as follows: the position of flaring up the disease ,the color of the disease ,the shape of disease ,the diameter of disease ,the time of flaring up the disease ,local temperature ,local humidity , the age of flaring up disease,the type of tooth on insect pest ,whether infect[8,9] .The structure of data base which use to save cases defined as :features+"diagnostic the origin of disease",case base structure is shown in table 1.Because the process of reasoning in applying system is on the basis of basic cases,the basic cases will be described more accurate and precise ,meanwhile ,the basic cases should not be more similar to each other ,which prevent to deduce more similar cases in the course of retrieveing ,which probably mislead the conclusion of system.

3.2 Achieve Case Retrieval about "Diagnostic the Origin of Disease" System

Case retrieval and match are important segment about case based reasoning ,to some extent which decide the quality of an expert system whether well or not .In CBR system, there are three widely applying retrieval algorithms: K-nearest neighbor tactics ,inducing inference tactics and knowledge guiding tactics .In this system, many of feature-value of the origin of disease is some part of normal direct quantization values, therefore, the system use K-nearest neighbor tactics to retrieval cases[2,7].

Table 1. Express case base of "diagnostic the origin of disease" system

Field name	Data type	Content	Value range
Position of disease	Enum	Root,stalk,leaf,fruit,blossom	
Color of disease	Number	0-255	255
Shape of disease	Character	Triangle,circular shape, Rectangle,irregular shape	
Diameter of disease	Number	1-50	50
Time of flaring up the disease	Date	1.1—12.31	12.31
Local temperature	Number	-35---+35	70
Local humidity	Number	10%--90%	80
Age of Ginseng disease	Boolean	T: annual , F: perennial	
Type of tooth on insect pest	Character		
Whether infect	Boolean	T: ture , F: false	
diagnostic the origin of disease	Character	Conclusion	

At present, K-nearest neighbor tactics is widely applying to CBR domain as an effective matching method for measuring similarity between two events . In the following formula a typical numerical function (Eq(1))of K-NN is shown:

$$NN(n,o) = \frac{\sum_{i=1}^{n}(w_i * sim(f_i^n, f_i^o))}{\sum_{i=1}^{n} w_i} \qquad (1)$$

Where w_i is the weight of the ith feature, f_i^n is the value of the ith feature for the input case , f_i^o is the value of the ith feature for the retrieved case, $sim()$ is the similarity function for f_i^n and f_i^o.

Similarity function can be expressed as the following equation(Eq(2)):

$$sim(f_i^n, f_i^o) = 1 - \left|\frac{f_i^n - f_i^o}{R_i}\right| \qquad (2)$$

Where R_i is the longest distance between two extreme values for the ith feature.

The implicit meaning of nearest-neighbor matching is that each feature of a case is a dimension in the search space .A new case can be added into the same space according to its feature values and the relative importance of the features. Most of times weighting values are determined using human judgment, and therefore the retrieved solutions can not always be guaranteed. To overcome this shortcoming in the traditional case retrieval process, this study presents a GA approach to support the determination of the most appropriate weighting values for each case feature.

GA is an optimization technique inspired by biological evolution. Its procedure can improve the search results by constantly trying various possible solutions with the reproduction operations and mixing the elements of the superior solutions. To determine a set of optimum weighting values, the search space is usually quite huge. This is because the search process must consider countless combinations of variant possible weighting values for each of the feature against all of the cases stored in the case base [6].

To solve a problem, the GA randomly generates a set of solutions for the first generation. Each solution is called a chromosome that is usually in the form of a binary string. According to a fitness function, a fitness value is assigned to each solution. The fitness values of these initial solutions may be poor. However, the fitness values will rise as better solutions survive in the next generation [7]. A new generation is produced through the following three basic operations.

Selection. Solutions with higher fitness values will be reproduced with a higher probability. Solutions with lower fitness value will be eliminated with a higher probability.

Crossover. Crossover is applied to each random mating pair of solutions.

Mutation. With a very small mutation rate Pm , a mutation occurs to arbitrarily change a solution that may result in a much higher fitness value.

The basic principle of GA can be described as followings:

    choose an initial population
    determine the fitness of each individual
    perform selection
    repeat
        perform crossover
        perform mutation
        determine the fitness of each individual
        perform selection
  until some stopping criterion applies

3.3 The Evaluation Process

The similarity value(OSD) is the key determinant for assessing the similarity between the input case and the old case .For each the similarity process batch executed with a specific input case $Case_j$,then n OSDs are produced since n old cases have been compared with the input $Case_j$ the higher the OSD ,the more likely the retrieved old case matches the input case.

Usually there may exist several old cases that are inferred to be similar (either exact or near Identical) to the input case. In other words, the solution of each old case could be proposed as a solution for a certain training case .To determine which cases whose outcome feature can be adapted as the outcome feature for the input case ,this research proposes that for the majority of the outcome feature among the top 10% OSDs in those old cases are used to represent the final solution .The derived final expected outcome feature is denoted as O'_j as opposed to the real outcome feature O_j .Though testing we know that the more chance O'_j is deemed as equal to O_j , the high probability that the appropriate weighting values can be produced. Therefore, the evaluation function is expressed as the following equation[7]:

$$Y = \sum_{j=1}^{p} y_j \qquad (3)$$

Where p is the total number of training cases; y_j is the matched result between the expected outcome and the real outcome, if $O'_j = O_j$, then y_j is 1; otherwise y_j is 0.

4 Test the System and Summay the Experiment Results

In Ginseng disease and insect pests diagnostic expert system the proposed GA-based CBR technology is through some training(GA),apply to practice which shows significant results, and improve the accuracy of inference .For example, input cases as following:

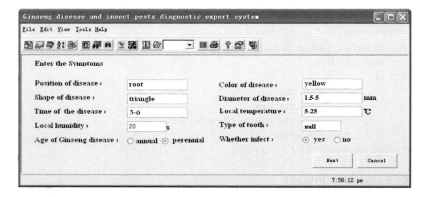

Fig. 2. Ginseng insect pests diagnostic expert system

By using above input, in case base it can be found only one case whose OSD higher 10%, which the origin of disease fungus is Ginseng Sclerotinia SP., therefore inference result is completely identical with practice.

When we input cases model too complicated, it is not easy to find input case results. This research adopted GA-CBR approach in order to further explore that are most likely and most unlikely situation, sometimes it also needs man-made conclusion to help system for going step further analysis, which enable the system more effective and more accurate about learning and updating case base [5,6].

5 Concluding Remarks

Ginseng disease and insect pests expert system belong to a small-scale CBR expert system, according to the finite scope of case base (the finite sort of Ginseng disease and insect pests).It is only aimed at Ginseng disease and insect pests aspect, on the basis of this we can use CBR and GA and so on core technologies modify the other aspects of Ginseng expert system, certainly this need go step further explorations and research on CBR and GA technology.

References

1. Yu, J.: Product Packing Design System With Case-based Reasoning. Computer Applications, 39–40 (2003)
2. Zhang, N., Gan, R.: Complex Case Retrieval Based on Intelligent Clustering. Computer Science, 75–78 (2002)
3. Zheng, Q., Wang, J., Wu, X., Cai, Q.: The Working Principle, Application and Current Study of CBR. Intelligent Control and Automation, 242–245 (2000)
4. Cai, Z., Xu, G.: Artificial Intelligence: Principles & Application, 2nd edn. Tsinghua University, China (1998)
5. McSheey, D.: Demand-Driven Discovery Adaptation Knowledge. In: IJCAI 1999, pp. 222–227 (1999)
6. Hullermeier, E.: Toward a Probabilistion Formalization of Case-based Inference. In: IJCAI 1999, pp. 248–253 (1999)
7. Chiu, C.: A case-based customer classification approach for direct marketing. Expert systems with Applications (2002)
8. Putnam, M.L., Meltchell, J.: Phytopathology 854 (1984)
9. Bai, R., Wang, Z.: Plant Pathology Learned Journal 79 (1990)

Identification of Pipeline Circuit Design

Jiatian Zhang, Jianming Ma, and Zhengguo Yan

The Key Laboratory of Photoelectricity Gas & Oil Logging and Detecting Ministry of Education,
Xi'an Shiyou University, Shaanxi Xi'an 710065, China
mjiopq@163.com

Abstract. In order to identify oil pipeline, this paper puts a kind of pipeline circuit for recognition, and presents the principle diagram of system, and analyses the circuit in detail. This circuit can identify the single pipeline and many pipelines connecting each other at the terminal. The experiment's results show that the scheme is practicable.

Keywords: Pipeline; Identify; Testing.

1 Introduction

With the development of science and technology, hand-held device more and more applied to industry, because analog circuits have a large size, so the integrated circuits become peoples' choosing firstly. The paper adopts some integrated chips to design a hand-held device that can identify pipeline quickly. When knocking the pipeline at the knocking point, then we can get the testing results by the equipment at the testing point. As shown in figure 1. The testing results alarm by light and sound, so we use it conveniently. This equipment is used to identify pipelines quickly where it is underground and the midway is not visible. So it could help people to get the messages about the pipeline, then you can do something for it.

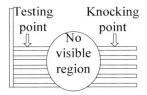

Fig. 1. Connected piping at terminal

2 System Overall Scheme

This equipment consists by four kinds of chip in the circuit mainly, they are amplifier, comparator, 555 timer and trigger. System structure diagram shows by figure 2.

For the situation that pipelines connected at the terminal, no matter which pipeline you knocked, the equipment could receive signals in the testing point. thus, needing to

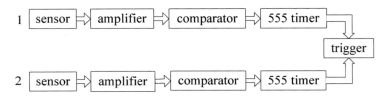

Fig. 2. Structure diagram of system

judge direction of signal sources. The circuit has two channels, and they have same structure, but there is some difference time in received signals. Firstly, the shock signals are received by vibration sensor, after that the signals are enlarged to volt level, then the signals are processed by comparator, and become to a low level and square-wave of narrowband. The low-level enter the timer and output a high-level of delay, putting the two signals that they are from the two channels into D foot and CP foot of flip-flop finally. Then we can judge the direction of signal sources by flip-flop. If the signals enter equipment directly, we call it positive direction, or we call it reversed direction. Then, the positive direction alarmed by light and sound, showing us that it's the right one; or if it alarmed by light only, showing us that it's the wrong one.

This equipment can apply to single pipeline too, because there is no reversed direction, so it would receive signals from the positive direction only. In the other words, it could identify the pipeline. This paper will discuss the pipelines connected at the terminal mainly.

3 System Hardware Design

3.1 Amplifier

The AD623 offers superior user flexibility by allowing single gain set resistor programming, and conforming to the 8-lead industry standard pin out configuration. With no external resistor, the AD623 is configured for unity gain ($G = 1$) and with an external resistor, the AD623 can be programmed for gains up to 1,000. The AD623 holds errors to a minimum by providing superior AC CMRR that increases with increasing gain.

The differential output is

$$V_O = [1 + 100K\Omega / R_G]V_C \qquad (1)$$

The AD623's gain is resistor programmed by R_G. Note that for $G = 1$, the R_G terminals are unconnected ($R_G = \infty$). For any arbitrary gain, R_G can be calculated by using the formula

$$R_G = 100K\Omega /(G-1) \qquad (2)$$

3.2 Comparator

LM211 is a single high-speed voltage comparators. This device is designed to operate from a wide range of power supply voltages, including ±15V supplies for operational amplifiers and 5V supplies for logic systems. The output levels are compatible with most TTL and MOS circuits.

Single threshold voltage comparator despite circuit is simple, high sensitivity, but its anti-interference ability is poor. Improving the anti-jamming ability of a plan is to use hysteresit comparator. It is generally known, no linear relationship about the output voltage V_O and input voltage V_i. According to the different output voltage V_O (V_{OH} or V_{OL}) value, the higher threshold voltage V_{T+} and the lower threshold voltage V_{T-} are classified as

$$V_{T+} = \frac{R_1 V_{REF}}{R_1 + R_2} + \frac{R_2 V_{OH}}{R_1 + R_2} \tag{3}$$

$$V_{T-} = \frac{R_1 V_{REF}}{R_1 + R_2} + \frac{R_2 V_{OL}}{R_1 + R_2} \tag{4}$$

Threshold width or back for poor voltage

$$\Delta V_T = V_{T+} - V_{T-} = \frac{R_2 (V_{OH} - V_{OL})}{R_1 + R_2} \tag{5}$$

Threshold voltage of hysteresit comparator changes by the output voltage V_O. Its sensitivity is lower, but anti-interference ability is improved greatly. This paper selects LM211 comparator connected with reversed hysteresit comparator.

3.3 Timer

The Maxim ICM7555 is respectively single and dual general purpose RC timers capable of generating accurate time delays or frequencies. The ICM7555 has a Low cost and reliable performance, need external several resistor, capacitor only, can achieve multivibrator, monostable trigger and Schmitt toggle circuit.

The ICM7555 provides two benchmark voltage VCC/3 and 2VCC/3. The monostable trigger used to the timer circuits, changing the RC value can get different duration. In other words, $T = R*C$. If $C = 1\mu F$ and $R = 330K$, then $T = 330ms$. This paper selects ICM7555 timer connected with monostable trigger circuit.

3.4 Trigger

The 74HC175 has four edge-triggered, D-type flip-flops with individual D inputs and both Q and $\bar{Q}$ outputs. The common clock (CP) and master reset ($\overline{MR}$) inputs load and reset (clear) all flip-flops simultaneously. All Q_n-lead outputs will be forced low independently of clock or data inputs by a voltage low-level on the $\overline{MR}$ input. The device is useful for applications where both the true and complement outputs are required and the clock and master reset are common to all storage elements. This paper selects 74HC175 trigger, using the rising edge of CP to trigger.

4 Signal Processing Analysis

4.1 Signal Amplifier Circuit

The AD623 maximum may enlarge 1,000 times, we choose to enter 100mv at here, and select $R_G = 2K\Omega$, because

$$V_O = [1 + 100K\Omega / R_G]V_C \qquad (6)$$

Then we could get $V_O = 51V_C$, in the other words, $V_O = 5.1V$, figure 3 is the waveform about input voltage and output voltage.

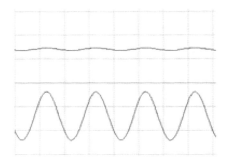

Fig. 3. Amplifier input/output voltage waveform

4.2 Hysteresis Comparison Circuit

Selecting $R_1 = 5K\Omega$, $R_2 = 1K\Omega$, $V_{REF} = 3V$, and $V_{OH} = 5.1V$, because we select single-supply at here, so $V_{OL} = 0V$, then we get $V_{T+} = 4.75V$, $V_{T-} = 0.5V$, and $\Delta V_T = 4.25V$. Because the 555 timer is effective when it receives a low-level, so need to input a signal as narrow and low-level pulse. When the input voltage is greater than V_{T+}, then getting a low level, and when the input is less than V_{T-}, then getting a high level. So we get a narrow and low-level pulse signal, as shown in figure 4 below.

Identification of Pipeline Circuit Design 75

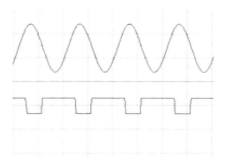

Fig. 4. Comparator input/output voltage waveform

4.3 Timer Circuit

The 555 timer is triggered by input low-level pulse. When the low-level coming, the timer output the high level. It continued some time, here is $T = R*C = 330ms$, then output low-level. When input a high-level, the light-emitting diodes will be lighted at the buffer circuit, for judging whether receives knocking signals. Figure 5 is the input and output voltage waveform of timer.

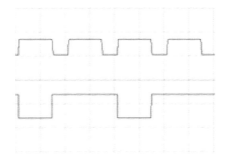

Fig. 5. The timer input/output voltage waveform.

4.4 Trigger Circuit

Trigger received two pulse signals from different channels, one input D-lead, and the other input CP-lead. It will appear four states at the moment. (1) if D=0, CP↓, then Q=0; (2) if D=1, CP↓, then Q=0; (3) if D=0, CP↑, then Q=0; (4) if D=1, CP↑, then Q=1. The Q-lead is output. In the other words, when CP-lead jump to high-level from low-level, the Q-lead accord with D-lead. Figure 6 is the two channels'pulse of input signals. If the first channel belong D-lead, the second channel belong CP-lead, then the trigger output a high-level, the buzzer will alarm from buffer circuit. It means the knocking pipeline is the testing pipeline at the moment.

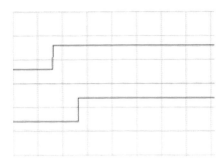

Fig. 6. Trigger input pulses of two channels

5 Summary

The system applied commonly integrating chips for designing. It is low cost, and easy to carry. It could identify pipeline about the midway is not visible, for instance conduit, tubing, tracheal etc. Through testing, this system is stable, reliable, high sensitivity and judgment timely, so it can satisfy the customers.

Note: the paper is funded by the key discipline construction project of Measurement technology and Instrument in Shaanxi province.

References

1. Zhang, J., Zhao, J.: The Performance and Applications of Instrument Amplifier AD623. Instrumentation Technology, 5 (2002)
2. Kang, H.: Foundamentals of Electronic Technique. Higher Education Press, Beijing (2003)
3. Wang, Y.: Digital Circuit Logic Design. Higher Education Press, Beijing (2004)

Vehicle Velocity Calculation in Accident Based on Morphology of Broken Vehicle Windshield Glass

Jie Zhang, Chuanjiao Sun, and Hongyun Chen

Research Institute of Highway, Ministry of Transport of China
Beijing, China
{zhang.jie,cj.sun,hy.chen}@rioh.cn

Abstract. The dynamic approach has been widely applied to measure the velocity of accident vehicles, while it is difficult to perform sometimes due to the shortage of parameters. In this paper, the algorithm is proposed to measure the velocity of the accident vehicle based on morphology of broken vehicle windshield glass. A relationship model between the velocity of accident vehicle and objects' parameters which include structure, quality and speed before collision was established by image processing technology. Five different types of collision objects, in different velocities are selected to simulate the collision process. The results show that, the concentration area of broken vehicle windshield glass is decided by the velocity and quality of the collision objects, the morphology of broken vehicle windshield glass is related to the velocity, quality and structure of the collision objects. Good idea is given to reconstruct the velocity of the vehicle velocity based on the morphology of broken glass.

Keywords: morphology of broken vehicle windshield glass; image processing; image enhancement, noise filtering; velocity; quality.

1 Introduction

Overspeed is the major factor which causes road traffic accidents in China. Because of the lack of the identification of the velocity of accident vehicle before collision, it is hard to make cognizance on the responsibilities of an accident by either traffic police department or court. Although the dynamic approach has been widely applied now and the velocity of the accident vehicle can be calculated by comprehensive analysis on the vehicle and trace on site, in many cases, it is still difficult to achieve a reasonable and accurate result by the dynamic approach since some parameters are uncertain and the trace on road is hardly to retain, measure accurately or being repeated, either.

In the process of traffic accident, vehicle windshield glass is easily rupture or broken because of other objects' impact and will show different morphologies. These morphologies hint important details about the status of traffic accident. It is an important aspect in road traffic accident reconstruction that how to reverse the shape of collision objects, force-bearing point and velocity before the collision.

According to the existing literature search, few research results can be found about the morphology of broken vehicle windshield glass, especially lack of the research results which is combined with the image processing technology. But these research results are very important in analysis of the whole accident process. Nowadays, the research about the relationship between the broken windshield glass and the velocity of vehicle mainly focuses on the following aspects: (1) Study on theoretical model of glass splinter throw distance for calculating automobile collision velocity [1-4]. (2) Model of vehicle velocity calculation based on deflection of windshield [5].

In this paper, through physical experiments and analyzing the broken glass by image processing technology, the model of vehicle velocity calculation based on the morphology of broken vehicle windshield glass is established. Good idea is given to study the morphology of broken glass and this research also plays important role in researching road traffic accident reconstruction.

2 Technical Route

The flow diagram of our algorithm contains three steps. Step 1 implements the experiment of the objects collision with the vehicle windshield glass. Step 2 processes the image of the broken vehicle windshield glass. Step 3 analyzes the experiment data. Fig. 1 shows the flow diagram of our algorithm.

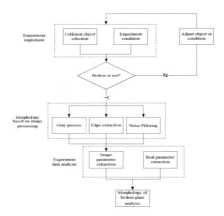

Fig. 1. Flow diagram of our algorithm

3 Experiment Implement

In the experiment, Jetta's windshield glass is selected. Collision object is brick, coconut, steel ball, concrete block, grapefruit. Firstly, the type of collision object is decided. Secondly, the quality and height of collision object is determined. Thirdly, the steel ball makes freely falling body motion and collides with the windshield glass. Specific experimental conditions are shown in Table 1:

Table 1. Information Table of Experiment Implement

No	Collision Object	Quality(kg)	Height(m)	Velocity(km/h)
01	Brick	1.48	8	45
02	Brick	1.48	9	48
03	Brick	1.48	9	48
04	Brick	1.48	5.3	37
05	Brick	1.48	3.5	30
06	Coconut	2.2	8	45
07	Steel ball	0.18	8	45
08	Steel ball	0.18	9	48
09	Steel ball	0.045	9	48
10	Steel ball	0.18	5.3	37
11	Steel ball	0.18	3.5	30
12	Steel ball	0.18	4.5	34
13	Concrete block	0.36	9	48
14	Concrete block	0.36	9	48
15	Concrete block	0.36	5.3	37
16	Grapefruit	1.5	8	45
17	Grapefruit	2.5	8	45
18	Grapefruit	2.5	5.3	37
19	Grapefruit	2.5	3.5	30

4 Morphology Based on Image Processing

After classifying and summarizing the experimental results, there are eight types of morphologies, shown as Figure 2~Figure 9.

Fig. 2 shows that, center of the broken glass is hole-shaped trace and the morphology of the broken glass is intensive arc trace and radial trace. Fig. 3 shows that, center of the broken glass is rectangular-shaped trace and the morphology of the broken glass is intensive arc trace and radial trace. Fig. 4 shows that, center of the broken glass is strip-shaped trace and the morphology of the broken glass is intensive arc trace and radial trace. Fig. 5 shows that, center of the broken glass is curved surface-shaped trace and the morphology of the broken glass is sparse arc trace and radial trace. Fig. 6 shows that, the glass has no broken center and the morphology of the broken glass is bar-shaped trace. Fig. 7 shows that, center of the broken glass is hole-shaped trace and the morphology of the broken glass is radial trace. Fig. 8 shows that, center of the broken glass is hole-shaped trace and the morphology of the broken glass is sparse arc trace and radial trace. Fig. 9 shows that, center of the broken glass is curved surface-shaped trace and the morphology of the broken glass is radial trace and bar-shaped trace.

Fig. 2.   Fig. 3.   Fig. 4.

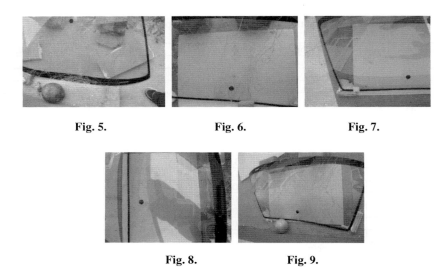

Fig. 5.   Fig. 6.   Fig. 7.

Fig. 8.   Fig. 9.

4.1 Color Image Gray Processing

Color image gray processing defines the process that color image turns into gray image. There are many gray processing methods, but the following three methods are frequently used: maximum value method, mean value method, weighted mean value method. In this paper, weighted mean value method is selected.

According to the importance of the three primary colors or other indicators, R, G, B is given different weights, namely:

$$R = G = B = (\omega_r R + \omega_g G + \omega_b B)/(\omega_r + \omega_g + \omega_b) \quad (1)$$

where, ω_r, ω_g, ω_b is respectively the weight of R, G, B.

If $\omega_r = 0.299$, $\omega_g = 0.587$, $\omega_b = 0.114$, then

$$R = G = B = 0.299R + 0.587G + 0.114B \quad (2)$$

At the moment, the value of R, G, B is equal to the gray value of the point. The gray value calculation formula is most commonly used.

After processing the image by (2), a gray image is obtained, shown as Fig. 10.

Fig. 10. Image of broken glass by gray processing

4.2 Region of Interest Extraction

Because the image is captured in the experiment scene, each image contains a large number of surrounding information. Most of the information is unrelated to velocity measuring. We use grid method to extract region of interest in order to improve the processing speed and accuracy of the whole algorithm.

The image is divided into small grids by the grid method, and then under the rule of region of interest (ROI), the rectangular window is extracted. Figure 3 is the image which is divided by a 100 × 100 pixel grid. The size of grid will directly affect the processing speed. When the image resolution is high, the grid should set a little larger. In this way, the number of following feature points will not be too large and the processing speed can be improved.

Through analyzing the divided frames, in this case, the ROI starts at Row 7~16, Column 13~25. The extracted ROI shows in Fig.12.

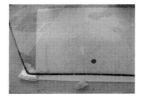

Fig. 11. Image divided by grid   **Fig. 12.** Extracted ROI

4.3 Edge Detection

In this paper, Otsu [6] algorithm which is a global binarization algorithm is used to segmentation the traces of broken glass. Otsu algorithm is a typical global threshold binarization algorithm. Global threshold binarization algorithm is to select a threshold to divide the value of image into two categories and Otsu algorithm is used to find the threshold which makes the variance within the class be the minimum value. The relationship shows as following:

$$\sigma_B^2(k^*) = \max_{1 \le k < L} \sigma_B^2(k) \tag{3}$$

where, k^* is the best threshold, $\sigma_B^2(k^*)$ is the maximum between-class variance. L is the gray value of the image. $\sigma_B^2(k)$ is the between-class variance when the value of the threshold is k. After processing the image by Otsu algorithm, an image is obtained, which is shown as Fig. 13.

4.4 Noise Filtering

After analyzing the image of broken glass, it can be seen in the image that there is a certain noise. The noise has a great influence on the following process. In this paper, the median filtering algorithm [7] is selected to remove the image noise. Considering the image size and other factors, the filtering operator W is [5, 5]. The image after noise filtering is shown as Fig. 14.

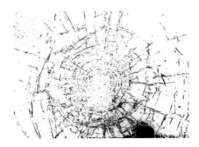 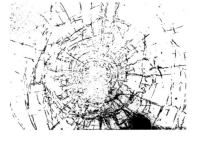

Fig. 13. Image by edge extraction   **Fig. 14.** Image by filtering and noise reduction

5 Analysis of Experiment Results

5.1 Image Parameter Extraction

The image is divided by a pixel grid. The distance in the image field which is from the broken center to the feature point is L_{img} pixels. The distance in the real field which is from the broken center to the feature point is L_{real} centimeters. So we can obtain relationship of distance between in image and in actual field $L_{img} \sim L_{real}$.

The distance in the image field which is from the broken center to the top of the arc trace is L_{img_1} pixels. The distance in the real field which is from the broken center to the feature point L_{real_1} is calculated in (4).

$$L_{real\_1} = L_{real} / L_{img} \times L_{img\_1} \quad (4)$$

The distance which is from the broken center to the right of the arc trace is L_{img_2} pixels in the image field. The distance which is from the broken center to the right of the arc trace L_{real_2} is calculated in (5).

$$L_{real\_2} = L_{real} / L_{img} \times L_{img\_2} \quad (5)$$

The area of the broken trace can be calculated as following.

$$Area = k \times L_{real\_1} \times L_{real\_2} \quad (6)$$

5.2 Experiment Results Summary

After processing the image of the broken glass, the experiment results are shown in Table 2:

Table 2. Information table of experimental results

No.	Colliding part	Broken center	Trace of broken glass	Area of broken glass
01	Vertex of brick	Hole-shaped	Intensive arc trace and radial trace	264.33
02	Plane of brick	Rectangular-shaped	Intensive arc trace and radial trace	2551.76
03	Edge of brick	Bar-shaped	Intensive arc trace and radial trace	1995.25
04	Vertex of brick	Hole-shaped	Intensive arc trace and radial trace	289.53
05	Plane of brick	No broken center	Bar-shaped trace	706.86
06	Coconut	Curved surface-shaped	Sparse arc trace and radial trace	4536.46
07	Steel ball	Hole-shaped	Intensive arc trace and radial trace	201.06
08	Steel ball	Hole-shaped	Intensive arc trace and radial trace	304.01
09	Steel ball	Hole-shaped	Radial trace	3.80
10	Steel ball	Hole-shaped	Sparse arc trace and radial trace	3.14
11	Steel ball	No broken center	No trace	0
12	Steel ball	Hole-shaped	Radial trace	2.54
13	Concrete block	No broken center	Bar-shaped trace	1983.16
14	Concrete block	Hole-shaped	Intensive arc trace and radial trace	1809.55
15	Concrete block	Hole-shaped	Intensive arc trace and radial trace	615.75
16	Grapefruit	Curved surface-shaped	Radial trace and bar-shaped trace	921.56
17	Grapefruit	Curved surface-shaped	Radial trace and bar-shaped trace	1350.49
18	Grapefruit	Curved surface-shaped	Radial trace and bar-shaped trace	706.86
19	Grapefruit	Curved surface-shaped	Radial trace and bar-shaped trace	481.50

After analyzing the experiment results, two results can be concluded:

(1) The broken morphology and the broken center of vehicle windshield glass is decided by the velocity, structure and quality of the collision objects:

- If the structure of object is spherical body and the diameter of spherical body is a little small, broken center is hole-shaped. With the increase of the diameter of spherical body, the morphology is from no trace, radial trace, sparse arc trace and radial trace to intensive arc trace and radial trace.
- If the structure of object is spherical body and the diameter of spherical body is a little big, broken center is curved surface-shaped and the morphology is radial trace and bar-shaped trace.
- If the structure of object is cuboid, when different parts of cuboid collide with the windshield glass, different morphologies will appear. If the colliding parts are vetex, plane, edge of the cuboid, the broken center is bar-shaped, rectangular-shaped, hole-shaped respectively. The morphology is intensive arc trace and radial trace.

- If the structure of object is irregular, the broken center is decided by the colliding part's structure. The morphology is intensive arc trace and radial trace.
(2) The concentration area of broken vehicle windshield glass is decided by the velocity and quality of the collision object, the heavier the collision object, the faster the velocity, then the larger the area.

6 Conclusion

In this paper, three different types of ball, in five different speeds are selected to simulate the collision process, and image processing technology is used to deal with the broken glass. The relationship between the morphology and the quality and the speed is given. Good idea is given to reconstruct the speed and quality of the collision objective based on the morphology of broken glass. However, because the number of samples is a little fewer, we can only study the relationship of the three objects qualitatively. In the future research work, with the number of samples increasing, we can study the relationship of the three objects quantitatively.

References

1. Emori, I.: Automobile Accident Engineering. People's Education Press, Beijing (1987)
2. Xu, H.-g., Ren, Y., Wang, L.-f., Lin, Q.-f.: Modeling of Generalized Motion Distance for Throwing Object on Traffic Accident. Journal of Jilin University of Technology (3), 10–14 (2002)
3. Du, X.-y., Du, X.-q., Li, Y.-j.: Road traffic accident scene deposition tutorial. Chinese People's Public Security University Press, Beijing (2005)
4. Xu, H.-g., Gao, Y.-l., Chen, L.-f.: Study on Theoretical Model of Glass Splinter Throw Distance for Calculating Automobile Collision Velocity. Automotive Engineering (4), 246–251 (1995)
5. Xu, J., Li, Y.-b.: Model of Vehicle Velocity Calculation in Vehicle-pedestrian Accident Based on Deflection of Windshield. Journal of Mechanical Engineering (July 2009)
6. Otsu, N.: A threshold selection method from gray-level histograms. IEEE Transactions on Systems, Man and Cybernetics (January 1979)
7. Zhang, Y.-j.: Image Engineering, 2nd edn. Tsinghua University Press, Beijing (2007)

A Low Power Wide Linear Range OTA and Its Application in OTA-C Filter

An Li, Chunhua Wang, XiaoRong Guo, and Jingru Sun

School of Computer and Communication
Hunan University
Changsha, Hunan Province, P.R. China
wonder139@126.com, wch1227164@sina.com, jt_guoxr@hnu.cn,
sjrbird@163.com

Abstract. A fully-differential symmetric CMOS operational transconductance amplifier (OTA) with negative current feedback is presented in this brief. The source degeneration topology in previous researches is modified to achieve higher linearity in the proposed circuit. Simulation results show that the IIP3 is 24.80dBm and the input voltage swing is up to 1 V. The power consumption of the proposed OTA is less than 14μw. A third order low-pass OTA-C filter implemented with the presented OTA is verified in the end.

Keywords: CMOS, OTA, OTA-C filter, negative current feedback.

1 Introduction

Operational transconductance amplifiers (OTAs) are important building blocks for analog circuit and systems, OTA-C topology that is a good choice for realizing continuous-time filter, which has better performance in frequency response and tuning capability, but suffers from poor linearity. Thus designing an OTA with good linearity tends to be a constraint and challenging task.

The linearity of OTA strongly depends on its input voltage swing. Good linearity means that the OTA have a flat Gm curve for all input range. There are three basic OTA linearization techniques in [1]-[10]: i) attenuation through floating gate MOS transistors; ii) cross-coupling; iii) source degeneration. In [1]~[3], Floating gate MOS transistors are used in the input stage of the OTA, which act as source followers along with a degeneration resistor. In [4]~[6], linearization is achieved by cross-coupling multiple differential pairs, properly scaling W/L ratios and tail currents to cancel the third-order harmonic distortion.

The main drawbacks of technique i) and ii) are excess power consumptions and transconductance loss. It seems that a better technique for linearity improvement in OTA is source degeneration because of its independence to work region of transistors and less power consumption. In [8], both enhanced adaptive biasing and passive resistor source degeneration techniques are combined to achieve superior linearity at higher frequency. In [9], MOS transistors in the triode region are used to replace the degeneration resistor. In [10], a modified source degeneration circuit is presented that uses negative feedback to maintain the linearity and a MOS transistor working in the linear region is added to regulate the transconductance. In the previous research [8]-[10], source degeneration which can significantly improve linearity also has several defects such as insufficient linear range and weak stability. Therefore, a modified OTA with source degeneration circuit feathers a wide linear range and low power consumption is proposed in this paper.

2 Proposed OTA Circuit

Fig 1 shows the circuit of OTA with conventional source degeneration. The linearity of the conventional source degeneration circuit would be degraded since the degenerated resistor would be small when achieving large transconductance. Voltage feedback topology [11] is a solution to enhance the linearity of traditional source degeneration structure. Unfortunately, the loop gain is highly degraded at high speed, and thus the improved linearity is not sufficient. Therefore, the negative current feedback circuit feathers wider linear range would be proposed.

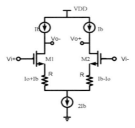

Fig. 1. Conventional source degeneration circuit

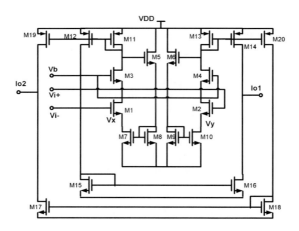

Fig. 2. The proposed OTA circuit

Fig. 2 shows the proposed source degenerated OTA. M1~M4 are cascade transistors used as input stage which can ensure stability. M5~ M10 are added to form a negative current feedback to enhance the linearity. M11~ M16 are current mirrors which used to generate positive output current i_{o1} and M17~ M20 are utilized to produce negative output current i_{o2}. The proposed circuit operates as follows:

At first, an increasing input voltage Vi+ is given, but the voltage at node V_x does not follow this variation. This will make the voltage between the gate and the source of transistor M1 to increase, the drain current of transistor M1 increases and so does the drain current of transistor M3 when Vb is given a proper value, the voltage between the drain and the source of transistor M3 also increases. Thus the voltage across the drain and the source of the transistor M11 also increases. For this reason, the gate voltage of transistor M5 decreases. Then, the gate-to-source value of transistor M5 decreases and so does the drain current of transistor M5. The decreased current will mirror through transistor M7. Therefore, the drain current of M1 will be pulled down, and the voltage at source of will be pushed up. Therefore, the negative current feedback circuit would enforce the voltage at node V_x to follow the variation of the input voltage.

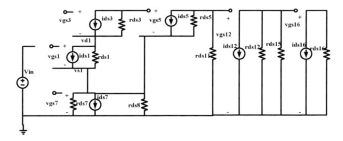

Fig. 3. The small signal model of the proposed OTA circuit

Intuitive small signal analysis [12] is adopted to carry out quantitative analysis. The small signal circuit of the proposed OTA is shown in Fig. 3. Based on the model, the relationship of the current and the voltage in the proposed OTA can be expressed as follow:

$$g_{m7} \times v_{g8} + v_{s1}/r_{ds7} = i_{ds1} \quad (1)$$

$$i_{ds1} = g_{m1} \times (v_{i+} - v_{s1}) \quad (2)$$

$$v_{g11} = g_{m1} \times (v_{i+} - v_{s1})/g_{m11} \quad (3)$$

$$v_{g8} \times g_{m8} = g_{m5} \times (v_{g11} - v_{g8}) \quad (4)$$

$$i_{ds12} = v_{g11} \times g_{m12} \quad (5)$$

Where g_{m1}, g_{m5}, g_{m7}, g_{m8}, g_{m11}, g_{m12} are the transconductance of the transistor M1, M5, M7, M8, M11, M12, the transconductance can be expressed as (6):

$$g_m = \frac{\partial i_D}{\partial v_{GS}}\bigg|_{V_{DS} \geq V_{GS}-V_{TH}} = \mu_n C_{ox} \frac{W}{L}(V_{GS}-V_{TH}) \qquad (6)$$

Where i_D is the drain current, v_{GS} is the voltage between gate and source, μ_n is the electron mobility, c_{ox} is the gate–oxide capacitance per-unit area, W is the gate width of the transistor, L is the gate length of the transistor, v_{th} is threshold voltage.

Based on the small signal, the current i_{ds16} can be expressed as equation (9): In the output node, i_{o1} can be gotten by Kirchhoff's law.

$$i_{o1} = i_{ds14} - i_{ds16} = \frac{g_{m16}}{g_{m15}+g_{m12}} \frac{g_{m1}g_{m7}g_{m12}(g_{m5}+g_{m8})}{g_{m11}(g_{m5}+g_{m8})(g_{m7}+g_{m11})-g_{m1}g_{m5}g_{m7}} v_{i+} - \frac{g_{m2}g_{m10}g_{m14}(g_{m6}+g_{m9})}{g_{m13}(g_{m6}+g_{m9})(g_{m10}+g_{m13})-g_{m2}g_{m6}g_{m10}} v_{i-} \qquad (11)$$

Due to the symmetry structure, i_{o2} can be gotten similarly.

$$i_{o2} = i_{ds15} - i_{ds17} = \frac{g_{m17}}{g_{m18}+g_{m16}} \frac{g_{m2}g_{m10}g_{m16}(g_{m6}+g_{m9})}{g_{m13}(g_{m6}+g_{m9})(g_{m10}+g_{m13})-g_{m2}g_{m6}g_{m10}} v_{i-} - \frac{g_{m1}g_{m7}g_{m12}(g_{m5}+g_{m8})}{g_{m11}(g_{m5}+g_{m8})(g_{m7}+g_{m11})-g_{m1}g_{m5}g_{m7}} v_{i+} \qquad (12)$$

Based on the definition of the OTA, $i_{o1} = -i_{o2} = G_m(v_{i+}-v_{i-})$, thus

$$G_m = \frac{g_{m12}}{g_{m15}+g_{m12}} \frac{g_{m1}g_{m7}g_{m12}(g_{m5}+g_{m8})}{g_{m11}(g_{m5}+g_{m8})(g_{m7}+g_{m11})-g_{m1}g_{m5}g_{m7}} \qquad (14)$$

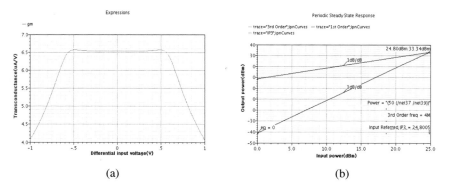

(a)   (b)

Fig. 4(a). Transconductance vs. Differential input voltage; **(b).** IIP3 plot of the proposed transconductor

The simulation is made using CADENCE software. Fig. 4(a) shows the transconductance of the OTA. We can found the proposed circuit has good flatness performance and wide linear range. Fig. 4(b) shows the PSS performance of the OTA, the measurement of the IIP3 is up to 24.80dBm, which is very large. Table 1 shows the main specifications of proposed OTA compared with some similar works referenced in Section 1. From table 1, it can be seen that the proposed circuit has very large linear range and very low power consumption.

Table 1. Specifications of proposed OTA compared with similar works

Referenc	This work	[1]	[7]	[13]	[14]
Process(μm)	0.18	0.18	0.13	0.13	1.2
Supply volage(V)	1.5	0.9	1.2	1.2	1.5
Linear range(V)	1.0	0.5	0.4	0.3	0.7
power(W)	13.6μ	456μ	20.8μ	1.1m	27.86μ
IIP3(dBm)	24.80	-	7.6	7	-

3 The Third Order Filter Structure

In order to verify the proposed OTA, a 3rd low-pass Gm-C filter using the proposed OTA is shown in Fig. 5. The circuit includes seven OTAs and eight capacitors. The filter is also simulated using CADENCE software. Fig. 6(a) shows the frequency characteristics of the filter. From Fig.6(a), it is clear that the filter has very sharp trailing edge. The IIP3 performance is shown in Fig. 6(b). From Fig.6(b), it is clear that the IIP3 is equal to 13.51dBm which is very large. Then the power consumption of the filter is less than 2mw (1.6mw).

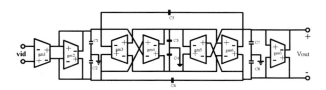

Fig. 5. The structure of the Gm-C filter

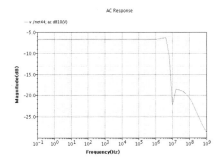

Fig. 6(a). Frequency response of the filter

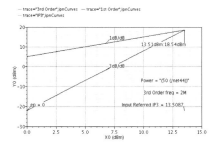

Fig. 6(b). IIP3 plot of the filter

4 Conclusion

In this paper, based on the analysis of previous design of OTA, the comparative advantage of source degeneration is higher linearity and lower power consumption. A modified OTA with negative current feedback is proposed and simulation result shows that the proposed OTA has wide linear range and low power consumption. A third-order low-pass Gm-C filter implemented with the presented OTA is shown. Simulation indicates that the filter could operate well at the required speed and achieve excellent performance.

References

1. Zare-Hoseini, H., Kale, I., Morling, C.S.: High linear transconductance topology using floating transistors. Electronics Letters 42, 151–159 (2006)
2. Carvajal, R.G., Galan, J., Torralba, A.: A very linear OTA with V-I conversion based on quasi-floating MOS resistor. In: 10th IEEE International Symposium on Circuits and Systems, New Orleans, pp. 473–476 (2007)
3. Rezaei, F., Azhari, S.J.: Ultra low-voltage rail-to-rail input/output stage OTA with high linearity and its application in a gm-c filter. In: 2010 11th International Symposium on Quality Electronic Design, San Jose, pp. 231–235 (2010)
4. Almazan, S.P.R., de Leon, M.T.G.: A 3rd order Butterworth Gm-C Filter for WiMAX Receivers in a 90nm CMOS Process. In: 2010 12th International Conference on Computer Modeling and Simulation, Cambridge, pp. 625–630 (2010)
5. Gao, Z., Wang, J., Lai, F.: Wideband Reconfigurable CMOS Gm-C Filter for Wireless Application. In: 16th IEEE International Conference on Electronics, Circuits, and Systems, Yasmine Hammamet, pp. 179–182 (2009)
6. Shaker, M.O., Mahmoud, S.A., Soliman, A.M.: NEW CMOS fully-differential transconductor and application to a fully differential gm-c filter. ETRL Journal 28, 175–179 (2006)
7. Mubarak, M., Navajo, M., Silva-Martinez, J.: Attenuation Predistortion Linearization of CMOS OTAs with digital correction of process variations in OTA-C filter applications. IEEE Journal of Solid-State Circuits 45, 351–365 (2010)
8. Chen, J., Sánchez-Sinencio, E., Silva-Martinez, J.: Frequency Dependent harmonic distortion analysis of a linearized cross coupled CMOS OTA and its application for OTA-C filters. IEEE Transactions on Circuits and Systems 53, 499–509 (2006)
9. El Mourabit, A., Sbaa, M.H., Alaoui-Ismaili, Z.: A CMOS transconductor with highly linear range. In: 14th IEEE International Conference on Electronics, Circuits and Systems, Marrakech, pp. 1131–1134 (2007)
10. Lo, T.-Y., Kuo, C.-L., Hung, C.-C.: A 250MHz gm-c filter using negative current feedback OTAs. In: 16th IEEE International Conference on Electronics, Circuits, and Systems, Yasmine Hammamet, pp. 367–370 (2009)
11. Hori, S., Maeda, T., Matsuno, N.: Low Power Widely Tunable Gm-C Filter with an Adaptive DC-blocking, Triode-biased MOSFET Transconductor. In: Proceeding of the 30th European Solid-State Circuits Conference, ESSCIRC 2004, pp. 99–102 (2004)
12. Allen, P.E., Holberg, D.R.: CMOS Analog Integrated Circuits Design. McGraw-Hill, New York (2007)
13. Kavala, A.: Kondekar P. N: A Low voltage low power linear pseudo differential OTA for UHF Applications. In: IEEE International Workshop on Antenna Technology, Santa Monica, p. 1 (2009)
14. Veeravalli, A., Sánchez-Sinencio, E., Silva-Martínez, J.: A CMOS Transconductance Amplifier Architecture With Wide Tuning Range for very low frequency applications. IEEE Journal of Solid-State Circuits 37, 776–781 (2002)

Research of Self-calibration Location Algorithm for ZigBee Based on PSO-RSSI

Chengbo Yu, Yimeng Zhang, Jin Zhang, and Yuxuan Liu

Research Institute of Remote Test and Control,
Chongqing University of Technology, Chongqing, China, 400054

Abstract. With WSN widely used in all areas of society, nodes self-calibration localization seems more and more important, it becomes one of the key technique of WSN. For nodes that palced in practical application evironment, signal strength is always influenced by the environment, this paper presents a location algorithm based on error correction. We use the PSO-RSSI which based on Particle Swarm Optimization (PSO) to optimize a number of LQI(which exist some deviation)which unknow node received from sink node. Then transform from LQI to RSSI to get distance. The result of the experiment shows that this algorithm can improve the precision of location and it has general application significance.

Keywords: WSN, localization, Particle Swarm Optimization(PSO), PSO-RSSI.

1 Introduction

In WSN, location information are essential for monitoring sensor networks, the location of the incident or the location of nodes which access to information is the important information contained in monitoring information. The monitoring which has no location information is always meaningless. Sensor nodes must be made clear the location of their own; therefore, location is the important support technology of WSN. Currently, GPS(global position system) is the location system which is most widely and maturely used. GPS has high precision positioning, real time, anti-jamming. But GPS node user always has high energy consumption, large volume, high cost and require fixed infrastructure etc., Which makes it does not apply to low-cost self-organizing sensor networks. Therefore, this approach is not feasible in many applications.[1]

Currently, as the methods of localization, the self-localization of WSN has two types of Range-based and Range-free. Where typical Range-based localization algorithm include TOA, TDOA, AOA, RSSI etc..[2] The localization based on RSSI has known the transmitting signal strength of transmitter node, according to the signal received by the receiving node, calculate the transmission loss of the signal, use theoretical and empirical model to make the transmission loss into a distance and then use the algorithm which has known is used to calculate the location of the node. This method doesn't need the information of distance and angle, only use the connectivity information among nodes to locate. Based upon localization algorithm of RSSI, a method of optimal location algorithm from calibrate the angle of RSSI values is proposed, which reduce the ranging error effectively by using the maximum likilihood method to caculate the node position[3]. While the method that the signal strength value to use unknow node receives from beacon node is used to convert into the distance, multiple RSSI values are collected and then the distance d is obtained by using formula after averaging. However, in practical evironment, the influence on RSSI values by environmental factors is large, and there exist a big errors by averaging to obtain the RSSI value. In order to reduce the errors that caused by RSSI values, this paper propose using Particle Swarm Optimization (PSO) to calibrate the RSSI values that collect, then make node localization to reduce the errors.

2 Algorithm Model

2.1 Wireless Propagation Path Loss Model

The models of common propagation path are the following: free space propagation model, log-distance path loss model, Hata model, log-distance distribution etc.. The model of free space propagation model and log-distance distribution is used in this paper.[4]

Free space propagation model is:

$$Loss = 32.4 + 10n\lg(d) + 10n\lg(f) \qquad (1)$$

Where d is the distance to beacon node(km). f is frequence(MHz), n is the factor of path loss.

Many disturbing factors exist in the practicle environment. There have changes between wireless transmission path loss and theoretical value. So log-distance distribution is introduced.

$$PL(d)[dBm] = PL(d_0)[dBm] - 10n\log(d/d_0) - \begin{cases} nW \times WAF, nW < C \\ C \times WAF, nW \geq C \end{cases} \qquad (2)$$

This model can be used to calculate the path loss when nodes received the information from beacon nodes. Where $P(d)$ is the signal strength that base station received from user nodes; $P(d_0)$ is the signal strength that base station received from reference point d_0 and we assuming that the signal strength that all the nodes transmit are the same. n is the scale factor between path length and path loss, depending on the structure and building materials used; d_0 is the distance between reference nodes and base station; d is the distance between nodes that need to calculate and base station; nW is the the number of walls that between nodes and base station; C is the threshold value that signal can pass through the wall.

From Eq. (1) and (2), we obtain conclusion that the signal strength that unknown nodes recevied from beacon; that is,

$$RSSI = P + G - PL(d) \qquad (3)$$

Where P is transmit power; G is antenna gain.

2.2 Particle Swarm Optimization

PSO algorithm is based on swarm, according to the fitness of the environment to move individual into the good regions. It isn't use as evolution operator to individual, but every individual is regarded as a particle(point), that has no volume in the D dimension search space and flies in the search space with a certain speed adjusted dynamically according to the flying experience of itself and companion. The i particle is $Z_i = (z_{i1}, z_{i2},..., z_{iD})$, the best position where the i particle experienced(has the best fitness) denoted as $P_i = (p_{i1}, p_{i2},..., p_{iD})$, called pbest. The best position that all the particles in the swarm experienced called gbest. The speed of particle i is $V_i = (v_{i1}, v_{i2},..., v_{iD})$. For each generation, its S d-dimension($1 \leq d \leq D$) changing according the equation following:

$$v_{id}^{k+1} = \omega v_{id}^k + c_1 r_1 (p_{id} - z_{id}^k) + c_2 r_2 (p_{gd} - z_{id}^k) \quad (4)$$

$$z_{id}^{k+1} = z_{id}^k + v_{id}^{k+1} \quad (5)$$

Where ω is the inertia weight, and played as the roles of tradeoff between local optimum and the ablity of global optimum. $i = 1, 2,...m$, $d = 1, 2,...,D$, k is the number of iterations. r_1 and r_2 are the random numbers between[0,1],and the two numbers are used to maintain the diversity of swarm. c_1 and c_2 are learning factor and also called acceleration factor, which make particles have the ablities of self-summary and learning to the excellent individual of the swarm, so to close to the best point of itself in the history and the best point of inside of the swarm in the history. The two parameters acted as convergence are not very big. However, if the two parameters are adjusted appropriately, the distress of local minimum can be reduced and the speed of convergence can be accelerated also. The second part of Eq.(4) is cognition part, which means particle on their own learning, and the third part is social part, which means cooperation between particles. The Eq. (4) is the particle speed based on its previous iteration, and update speed changed according to its current location and the distance between the best experience of itself and the best experience of the swarm, and then the particle flies to the new location according to the Eq. (5).[5]

While the inertia weight is smaller than 0.8, and if PSO can find the global optimum, all the paricles tend to bring together quickly. If the optimum solution is in the inside of the initial search space, PSO would find the global optimum easily; otherwise it wouldn't find the global optimum. While the inertia weight is biger than 1.2, PSO would be similar with a global search method, which needs much iteration to reach the global optimum, and has more difficulty to find the global optimum. While the inertia weight is moderate, PSO would have more chance to find the global optimum. According to the analysis above, that inertia weight is not set as a fixed value, but set as a function which linearly decreasing with time, the function of inertia weight usually be in the form

$$\omega = \omega_{max} - \frac{\omega_{max} - \omega_{min}}{k_{max}} \quad (6)$$

Where ω_{max} is the initial weight, ω_{min} is the final weight, k_{max} is the maximum number of iterations, k is the current number of iterations[6].

Figure 1 is the flow chart of PSO:

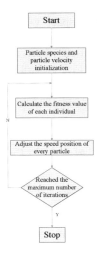

Fig. 1. The flow chart of PSO

3 The Caculation and Simulation of Algorithm

3.1 The Caculation of Fitness Value

In PSO, individual extreme and global extreme are dicided by the fitness of every particle[7]. The range of LQI is 0~255, and the greater the LQI is, the better the signal strength is. In a LQI sample space, the individual sample collected fluctuated generally in the vicinity of a value which occured more furquently in a probability. According to this analysis, if a LQI is small, a worse fitness value would be obtained, so that it will not select as a extremum of iterative calculation. If litter probability of LQI values occurred is larger than greater probability of the LQI values occurred, its initial fitness value is greater, and according to the change of the location during iteraton, the LQI would be replaced also by the particles which have better position, and wouldn't affect the normal evolution of particle swarm.

According to the analyzing above, the normalized function to obtain the ideal fitness function is used:

$$fitness(l_i) = 256 \times \frac{l_i - l_{min}}{l_{max} - l_{min}} \tag{7}$$

Where l_i is a sample of sample space of LQI which collected. l_{min} is the minimum of the sample space of LQI which collected. l_{max} is the maximum of the sample space of LQI which collected.

3.2 The Caculation Process of PSO-RSSI Algorithm

The flow chart of PSO-RSSI is shown in figure 2, its caculation process is as follows:
(1) Use a beacon node to measure unknown nodes' LQI, collecting 100 values, caculate the fitness of every particle, and set the LQI of particle which has the greatest fitness as the initial gbest, and other particles set individual extremum pbest themselves as their initial location.
(2) By iteration, select the particle which has the greatest fitness as global extremum gbestk, and every particle selecting individual extremum pbestk according to the flight records itself.
(3) Put the value of gbestk, pbestk and z_i into the Eq. (4) and Eq. (5), update the location and speed of itself, regain the new fitness value after every particle iterated accord to the Eq. (7), go to step (2).
(4) After a suitable number of iterations, get the global extremism LQI, exit loop.
(5) According to the Eq. (1), Eq. (2), Eq. (3), obtained distance d.

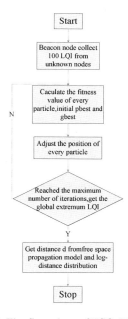

Fig. 2. The flow chart of PSO-RSSI

3.3 Measure the Distance and Simulation

While experiment, the model of JN5139-Z01-M02 which produced by Jennic Company is used as wireless sensor node to collect the data, and the position to do the experience is a courtyard outdoor. Using serial port debugging tool to record 100 LQI which unknown node received from beacon node, and then using PSO to process these 100 values, according to the formula, the LQI converts into RSSI, finally the Eq. (1),(2),(3) are used to get the distance d. Using 10-dimensional matrix, that is the porpulation is 10, the number of particle swarm is 10, the maximum number of iterations k_{max} =200, initial inertia weight ω_{max} =1.2, final inertia weight ω_{min} =0.9, acceleration factor $c_1 = c_2$ =2. Taking an unknown node as a example, using serial port debugging tool to get 100 LQI shown as 10-demensional matrix express as:

$$\begin{bmatrix} 96 & 78 & 90 & 78 & 90 & 90 & 78 & 90 & 78 & 84 \\ 90 & 90 & 90 & 90 & 72 & 90 & 90 & 90 & 90 & 90 \\ 90 & 102 & 78 & 90 & 90 & 90 & 78 & 90 & 90 & 90 \\ 72 & 78 & 90 & 96 & 90 & 90 & 90 & 90 & 90 & 78 \\ 78 & 90 & 90 & 90 & 90 & 90 & 90 & 96 & 90 & 90 \\ 90 & 78 & 90 & 90 & 90 & 84 & 90 & 90 & 90 & 90 \\ 90 & 90 & 78 & 90 & 90 & 90 & 90 & 90 & 72 & 78 \\ 84 & 78 & 90 & 90 & 90 & 90 & 78 & 96 & 90 & 90 \\ 90 & 90 & 90 & 90 & 90 & 78 & 90 & 90 & 102 & 72 \\ 78 & 72 & 90 & 90 & 78 & 90 & 78 & 78 & 90 & 78 \end{bmatrix}$$

The global extremum is 90.169 by MATLAB simulating. The conversion formula of wireless radio model of LQI and RSSI which based on JN5139 is :

$$u16RSSI = [\frac{(\frac{u8LQI \times 880000000}{2^8 - 1})}{10000000}] - 98 \qquad (8)$$

Putting the global extremum of LQI 90.169 into Eq.(7), RSSI value is -66.88dBm, using Eq. (1),(2)and(3) to get the corresponding distance d, among which, transmitting frequenc f =2.4GHz. In the environment of courtyard, the path loss $PL(d_0)$ =-34.2dBm at the reference point d_0 =1m, path loss factor n =2.8, WAF depends on the layout and materials of location enviroment, and usually WAF =3.1.Getting the distance d =14.79m by caculating. The actual distance is 14.3m, and error is only 0.49m.

The normal averaging method is used, and LQI=87. According to Eq. (8), LQI convert into RSSI, then get the distance is 16.069m, error is greater than the algorithm of PSO-RSSI. Figure 3 shows that by comparing the LQI using the algorithm of

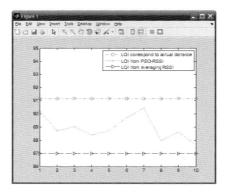

Fig. 3. The comparison figure of LQI which using two kind of algorithm to get

PSO-RSSI to get with the LQI using the method of averaging RSSI to get and LQI corresponding to the actual distance, it can be seen that get LQI which using the algorithm of PSO-RSSI to get is in the vicinity of actual LQI, and the value is better than the value that averaging LQI.

3.4 Maximum Likelihood Method to Locate[8]

Maximum likelihood estimation is shown as figure 4, the coordinates of 1,2,3,n nodes are known as $(x_1, y_1), (x_2, y_2), (x_3, y_3), ..., (x_n, y_n)$, and the distance to node D are $d_1, d_2, d_3, ..., d_n$. Assuming the coordinates of the node D is (x, y), So:

$$\begin{cases} (x_1 - x)^2 + (y_1 - y)^2 = d_1^2 \\ \vdots \\ (x_n - x)^2 + (y_n - y)^2 = d_n^2 \end{cases} \quad (9)$$

Starting from the first equation minus the last equation, respectively:

$$\begin{cases} x_1^2 - x_n^2 - 2(x_1 - x_n)x + y_1^2 - y_n^2 - 2(y_1 - y_n)y = d_1^2 - d_n^2 \\ \vdots \\ x_{n-1}^2 - x_n^2 - 2(x_{n-1} - x_n)x + y_{n-1}^2 - y_n^2 - 2(y_{n-1} - y_n)y = d_{n-1}^2 - d_n^2 \end{cases} \quad (10)$$

The linear equation of Eq. (9) is:

$$AX = b \quad (11)$$

Which:

$$A = \begin{bmatrix} 2(x_1 - x_n) & 2(y_1 - y_n) \\ \vdots & \vdots \\ 2(x_{n-1} - x_n) & 2(y_{n-1} - y_n) \end{bmatrix} \quad b = \begin{bmatrix} x_1^2 - x_n^2 + y_1^2 - y_n^2 + d_n^2 - d_1^2 \\ \vdots \\ x_{n-1}^2 - x_n^2 + y_{n-1}^2 - y_n^2 + d_n^2 - d_{n-1}^2 \end{bmatrix} \quad X = \begin{bmatrix} x \\ y \end{bmatrix} \quad (12)$$

Using standard minimum mean square error, we get the coordinates of node D; that is

$$\hat{X} = (A^T A)^{-1} A^T b \quad (13)$$

The node localization can be achieved by the maximum likelihood method. Using Matlab simulation is shown as figure 4 and figure 5. In figures 4 and 5, rhombus icon is actual node location, red star icon is theoretical location of the node which measured by algorithm. We can know that using the method of averaging RSSI to location, the distance between the theoretical coordinate points and the actual coordinate points is further in figure 4, and using PSO-RSSI algorithm to location, the distance between the theoretical coordinate points and the actual coordinate points is closer in figure 5. From figure 4 and figure 5 can be known that the precision which using PSO-RSSI algorithm to locate is higher than using the method of averaging the RSSI.

 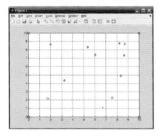

Fig. 4. The simulation of averaging RSSI to locate   **Fig. 5.** The simulation of PSO-RSSI to locate

4   Conclusion

Simulation result shows that the PSO-RSSI self-calibration location algorithm can locate the node more precisely than the traditional method which averaging RSSI. In addition, this self-calibration location algorithm is less demanding on the hardware, and meet the requirements of WSN of low cost and low consumption which is a practical optimal location. In the future, it can use some algorithm to solve the problem which PSO would emerge the partial optimal to improve the performance of location better.

Acknowledgment

This research was sponsored by Project of ChongQing Economy and Information Technology Commission(Yu Economy and Information Technology[2010]No.9) and Project of ChongQing Jiulongpo District Technology Committee(Jiulongpo District Technology Committee[2009]No.52).

References

1. Sun, L., Li, J., Chen, Y., et al.: Wireless Sensor Networks, pp. 5–15. Tsinghua University Press, Beijing (2005)
2. He, Y., Li, H.: Self-calibration Location Algorithm for Wireless Sensor Network Based on RSSI. Radio Engineering 40(3), 7–9 (2010)
3. Song, N., Kwak, B., Song, J., et al.: Enhancement of IEEE 802.11 distributed coordination function with exponential increase exponential decrease backoff algorithm. In: IEEE VTC 2003, pp. 2775–2778 (Spring 2003)
4. Tang, L., Gao, B.: Research on Node Location in Wireless Sensor Network. Computer and Modernization 2, 19–21 (2010)

5. Xing, M., Li, L.: Application of particle swarm optimization to positioning for wireless sensor networks. Computer Engineering and Applications 45(32), 72–74 (2009)
6. Ji, Z., Liao, H., Wu, Q.: Particle Swarm Optimilizaion and applications, pp. 16–21. Science Press, Beijing (2009)
7. Deng, J., Varshney, P.K., Haas, Z.J.: A new backoff algorithm for the IEEE 802.11 distributed coordination function. In: Proc. CNDS 2004 (January 2004)
8. Ledeczi, A., Kiss, G., Feher, B., et al.: Acoustic source localization fusing sparse direction of arrival estimates. Intelligent Solutions in Embedded Systems (6), 1–13 (2006)

A New Method for Bad Data Identification of Integrated Power System in Warship Based on Fuzzy ISODATA Clustering Analysis

Lei Wu, Li Xia, and Yong Shan

Institute of Electrical and Information Engineering,
Naval University of Engineering
Wuhan, China
wulei840127@163.com

Abstract. Bad date detection and identification of integrated power system in warship is the basis for Establishment of reliable data-base and implement of ship energy management. Due to the problems(pollution and submerge of residual error) of conventional method applied to integrated power system in warship, a new method for bad date detection and identification is used. Based on fuzzy ISODATA clustering analysis, Weight of the objective function is improved on. Estimation in advance with residual error and improved fuzzy ISODATA clustering analysis are combined to carry out bad date identification. Emulation of bad date identification is carried out to simulate power system in big warship. Emulation results show that the method can detect and identify bad date effectively, and overcome the phenomenon of pollution and submerge of residual error.

Keywords: power system in warship; bad data; detection and identification; fuzzy clustering analysis.

1 Introduction

As the relatively simple structure of traditional warship power system and small amount of data , the issue of bad data detection and identification is not considered. There will be many new features in large modern integrated power system in warship, for example: a larger scale, AC-DC co-exist, asymmetry of each phase, multi-harmonic and electromagnetic interference, great impact on power grid in the process of propulsion motor starting, braking and speed, so bad data is more likely to occur.

Bad data detection and identification is a very important part of state estimation of integrated power system in warship. The purpose is to detect and handle bad data appearing in sampling and improve the reliability of state estimation, ultimately establish a reliable database[1].

Current home and abroad method for Bad data detection and identification include residual search algorithm[2], Non- quadratic criteria method, zero residual method, estimation identification method and so on[3]. These methods both use hypothesis test of probability and statistics, then determine a threshold at a certain confidence level, finally load "either-or" binary logic judgment of measurement data. There may be the phenomenon of pollution and submerge of residual error which led to the missed and false of bad data[4-5]. The common drawback of methods described in paper[6-7] is the phenomenon of pollution and submerge of residual error. There will often occur error identification in the case of multiple bad data. In recent years, although domestic and foreign scholars have made many useful improvements[8-9], the number of iterations increase even converge.

With the emergence of new theories, various new methods[10-12] are used in bad data identification. Paper[13] use right measurement in typical conditions as backpropagation neural network training samples to detect and identify bad data correctly when real-time monitoring. Paper[14] structure a neural network based on (GMDH), and use regular information as input variables to detect and identify bad data. But the shortcomings of neural networks is that the representation of training samples will directly affect the detection and identification results. Based on the non- quadratic criterion identification method, how to eliminate false positives and missing, the specific methods to get a more ideal state estimates results through the intervention are introduced by Paper[15]. Paper[16] summarizes the home and abroad research results on state estimation theory is applied in the power system.

There is the possibility of miscalculation and missing when traditional method is applied to bad data identification of integrated power system in warship. So it is difficult to overcome pollution and submerge of residual error. A new method is used in this paper. Measurement data is classify according to standard residual size using the concept of membership degree. Estimation in advance with residual error and improved fuzzy clustering analysis are combined to carry out bad date identification.

2   Fuzzy ISODATA Clustering

2.1   Fuzzy Classification

Fuzzy classification is proposed based on the concept of fuzzy sets. Samples classified by objects ollection belong to such a class with a certain degree of membership, So the classification matrix corresponding to each classification results is a fuzzy matrix R.

$$R = \begin{bmatrix} \mu_{11} & \mu_{12} & \cdots & \mu_{1n} \\ \mu_{21} & \mu_{22} & \cdots & \mu_{2n} \\ \vdots & \vdots & \vdots & \vdots \\ \mu_{c1} & \mu_{c2} & \mu_{c3} & \mu_{cn} \end{bmatrix} \qquad (1)$$

2.2   Fuzzy ISODATA Cluster Analysis

Fuzzy clustering method can be broadly divided into three categories. In this paper, the soft classification of space-based Fuzzy Clustering is researched. Such as ISODATA(Iterative Self-Organizing Data Analysis Technique A). In the cluster analysis, suppose the set of classified objects is:

$$X = \{x_1, x_2, \cdots, x_n\} \qquad (2)$$

Each of which sample x_i has m characteristics index matrix

$$\begin{bmatrix} x_{11} & x_{12} & \cdots & x_{1m} \\ x_{21} & x_{22} & \cdots & x_{2m} \\ \vdots & \vdots & \vdots & \vdots \\ x_{n1} & x_{n2} & x_{n3} & x_{nm} \end{bmatrix} \qquad (3)$$

We divide sample X set into c categories $(2 \leq c \leq n)$, suppose clustercenter vectoras follow, in order to obtain an optimal fuzzy classification, we must choose one of the best fuzzy classification from fuzzy classification space in accordance with the clustering criteria.

$$V = \begin{bmatrix} V_1 \\ V_2 \\ \vdots \\ V_c \end{bmatrix} = \begin{bmatrix} v_{11} & v_{12} & \cdots & v_{1m} \\ v_{21} & v_{22} & \cdots & v_{2m} \\ \vdots & \vdots & \vdots & \vdots \\ v_{c1} & v_{c2} & v_{c3} & v_{cm} \end{bmatrix} \qquad (4)$$

Fuzzy ISODATA clustering criterion is to achieveminimumobjective function.

$$J(R,V) = \sum_{i=1}^{n} \sum_{h=1}^{c} (\mu_{hi})^q \| x_i - V_h \|^2 \qquad (5)$$

In which, V_h express cluster center of category h, $\| x_i - V_h \|$ express distance from sample x_i to cluster center V_h, q is Parametergreater than 0. In order to change the relative membership degree of flexibility, general values q as 2, Too much value will cause information distortion.

2.3 Fuzzy ISODATA Clustering Criteria

Generally speaking, it si too hard to solve the objective function extremum, but when $q \geq 1, x_i \neq V_h$, we can implement iterative calculation through the following way, and the computing process is convergent. The steps are as follows:

(1) Select c, $(2 \leq c \leq n)$, take an initial fuzzy classification matrix $R^{(0)} \in M_{fc}$, Iteration step by step, $l = 0, 1, 2, \cdots$

(2) For $R^{(l)}$, calculate cluster center vector

$$V^{(l)} = \left(V_1^{(l)}, V_2^{(l)}, \cdots, V_c^{(l)} \right)^T \qquad (6)$$

$$V_h^{(l)} = \frac{\sum_{i=1}^{n} (\mu_{hi}^{(l)})^q x_i}{\sum_{i=1}^{n} (\mu_{hi}^{(l)})} \qquad (7)$$

(3) Amend fuzzy classification matrix $R^{(l)}$

$$\mu_{hi}^{(l+1)} = \frac{1}{\sum_{j=1}^{c} \left(\frac{\|x_i - V_h^{(l)}\|}{\|x_i - V_j^{(l)}\|} \right)} \qquad (8)$$

(4) Compare $R^{(l)}$ to $R^{(l+1)}$, to Given $\xi > 0$,

$$\max \left| \mu_{hi}^{(l+1)} - \mu_{hi}^{(l)} \right| \leq \xi \qquad (9)$$

then $R^{(l+1)}$ and $V^{(l)}$ are Solutions, stop iteration. Else: $l = l+1$, back to step2, repeat calculating, we will get the optimal fuzzy classification.

The optimal classification appied of algorithm above is relative to the initialclassification matrix $R^{(0)}$, Matrix norm, ξ and optimal classification under the conditions of m.

3 Improved Bad Data Identification Method Based on Fuzzy ISODATA Clustering

3.1 Improvement to Fuzzy ISODATA Clustering

The above objective function has not consider the effect to the classification results from different weights of each index. In fact, their effect is different, so above classification may lead to unreasonable results. Then change the objective function to:

$$J(R,V) = \sum_{i=1}^{n} \sum_{h=1}^{c} (\mu_{hi})^q \sum_{k=1}^{m} [\omega_k (x_{ik} - V_{hk})]^2 \qquad (10)$$

In which, ω_k ——weights of characteristic index of k.

This shows: membership function expresse the level by which Characteristic index is subordinate to a classification. ω_k use membership under normal circumstances.

This option is feasible. In Practical situations, evry feature indicator has different effect to Classification, so Formula(10) use differet weight to different indicators. It is more realistic. But Traditional ISODATA method do not consider the impact on the classification from weights. Weights are considered equal. This may lead to inaccurate even unreliable classification result. So improved fuzzy ISODATA clustering in this paper is more rigorous theoretically, and clustering results is closer to reality.

3.2 Estimation in Advance with Residual Error

Estimation in advance with residual error means that the data from estimation is detected by standardized residuals. Then measurement data is divided into two parts According to detect results. We assume that the measurement data of which standardized residuals is less than a given value is reliable data, otherwise, the measurement data is suspicious data. Then suspicious data is deleted from the measurement data, and re-state estimation is carried out based on reliable data. We think that the estimation result is not contaminated from the error and closer to the true value. Estimated residual error can truly reflect the degree of deviation of suspicious data from the true value.

Fuzzy ISODATA clustering is carried out according to initial classification matrix form estimated residual error. The result would be better. Emulation results show that the method can detect and identify bad date effectively, and overcome the phenomenon of pollution and submerge of residual error.

Specifically as follows: set parameter 2.81 as the threshold of standard residual error method, when estimated residual error $r_N \geq 2.81$, take degree of membership that measurements Z_i belonging to bad data set A to a number more than 0.5 and less than 1, degree of membership that measurements Z_i belonging to positive data set B to a number more than 0 and less than 0.5, and meet the the following formula:

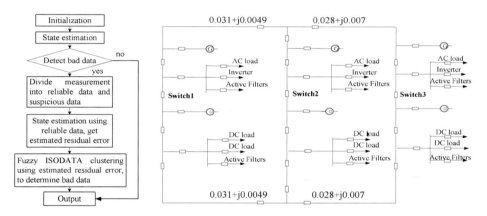

Fig. 1. The diagram of bad data identification

Fig. 2. Example of power system

$$\mu_A(z_i) + \mu_B(z_i) = 1 \qquad (11)$$

when estimated residual error $r_N \leq 2.81$, $0 < \mu_A(z_i) < 0.5, 0.5 < \mu_B(z_i) < 1$, the initial classification fuzzy matrix $R^{(0)}$ can be obtained as follows:

$$R^{(0)} = \begin{bmatrix} \mu_A(z_1) & \mu_A(z_2) & \cdots & \mu_A(z_h) \\ \mu_B(z_1) & \mu_B(z_2) & \cdots & \mu_B(z_h) \end{bmatrix} \qquad (12)$$

Fuzzy ISODATA clustering is carried out , then we can get general classification using judgment principle of the new sample and identfy bad data from Measurement data. Detection and identification process as shown in Figure 1.

4 Numerical Example and Analysis

Power system in warship which is independent system is different form land grid. Simulation is carried out in simulate in this paper, Trapezoidal power system in warship as shown in Figure 2. Six generating units can work independently, can also be combined to different working conditions. Switch 1 is taken off in this example, then it can be simplified into 4-bus system. Take 4-node as the reference node. Assume that the system has seven active measurement and nine reactive measurement, as shown in Table 1. Use standard residual error as feature vector, $m = 1$, when $r_N \geq 2.81$ $\mu_A(z_i) = 0.825$, $\mu_B(z_i) = 0.175$, otherwise, $\mu_A(z_i) = 0.175$, $\mu_B(z_i) = 0.825$, $q = 2$. Simulation as follows:

Table 1. Measurement Data

point	measurement items	value z	Weight
1	P1	0.476	0.58
2	P13	0.454	0.48
3	P23	-0.374	0.44
4	P3	0.509	0.48
5	P31	-0.443	0.37
6	P32	0.385	0.48
7	P34	0.539	0.59
8	Q1	0.115	0.5
9	Q13	0.121	0.49
10	Q23	0.095	0.48
11	Q3	-0.296	0.69
12	Q31	-0.113	0.58
13	Q32	-0.103	0.37
14	Q34	-0.098	0.58
15	V1	1.014	0.8
16	V3	1.099	0.8

(1) set single bad data point: In the 4-node system, assume bad data is P_1. Test results shown in Table 2:

Table 2: when standard residual error method is used, except standard residual error of P_1 is greater than the threshold value, standard residual error of P_{13} and P_{31} are greater than the threshold value, too. The phenomenon of residual error pollution occurs. Method used in this paper, only membership degree of P_1 (μ_A) is greater than 0.5, so P_1 is the bad data.

(2) set multiple bad data point: In the 4-node system, assume bad data are P_1, P_3. Test results shown in Table 3:

Table 3: when standard residual error method is used, except standard residual error of P_1 is greater than the threshold value, standard residual error of P_{13} and P_{31} are greater than the threshold value, too. The phenomenon of residual error pollution occurs. But standard residual error of P_3 is less than the threshold value, phenomenon of residual error submerge occurs. Method used in this paper, only membership degree of P_1 and P_3 are greater than 0.5, so P_1 and P_3 is the bad data.

Table 2. Detection results (bad data: P_1)

point	Standard residual method		Method in this paper	
	Standard residual	Threshold	μ_A	μ_B
P1	3.8	2.81	0.7684	0.2316
P13	2.9	2.81	0.0278	0.9722
P23	0.2	2.81	0.0195	0.9805
P3	-1.6	2.81	0.0398	0.9602
P31	3.5	2.81	0.0195	0.9805
P32	0.6	2.81	0.0637	0.9363
P34	0.1	2.81	0.3217	0.6783
Q1	0.3	2.81	0.0851	0.9149
Q13	-0.3	2.81	0.0857	0.9143
Q23	1.1	2.81	0.3949	0.6051
Q3	0.3	2.81	0.1298	0.8702
Q31	0.4	2.81	0.3771	0.6229
Q32	0.2	2.81	0.0621	0.9379
Q34	-0.5	2.81	0.3215	0.6785
V1	0.4	2.81	0.1905	0.8095
V3	0.1	2.81	0.0393	0.9607

Table 3. Detection results (bad data: P_1, P_3)

point	Standard residual method		Method in this paper	
	Standard residual	Threshold	μ_A	μ_B
P1	1.5	2.81	0.7525	0.2475
P13	3.1	2.81	0.0276	0.9724
P23	0.9	2.81	0.1131	0.8869
P3	0.7	2.81	0.6858	0.3142
P31	3.3	2.81	0.3313	0.6687
P32	1.1	2.81	0.0157	0.9843
P34	1.9	2.81	0.0288	0.9712
Q1	0.4	2.81	0.2751	0.7259
Q13	-1.8	2.81	0.0211	0.9789
Q23	1.2	2.81	0.3917	0.6083
Q3	0.4	2.81	0.0156	0.9844
Q31	0.7	2.81	0.3798	0.6202
Q32	0.2	2.81	0.0271	0.9729
Q34	-0.9	2.81	0.2137	0.7863
V1	0.8	2.81	0.0387	0.9613
V3	0.3	2.81	0.0618	0.9382

5 Conclusion

Weight of the objective function is improved in this paper based on fuzzy ISODATA clustering. Estimation in advance with residual error and improved fuzzy ISODATA clustering analysis are combined to carry out bad date identification. Emulation results show that the method can detect and identify bad date effectively, and overcome the phenomenon of pollution and submerge of residual error.

References

1. Yu, E.: State Estimation of Power System. Hydraulic and Electric Power Press, Beijing (1985)
2. Handschin, E., Schweppe, F.C., Kohals, J.: Bad date analysis for power system state estimation. IEEE Transactions on Power Apparatus and System 94, 329–336 (1975)
3. Sun, G., Wei, Z., Zhou, F.: The Application of ISODATA to Bad Data Detection and Identification Based on Genetic Algorithms. Proceedings of the CSEE 26(11), 162–166 (2006)
4. Li, B., Xue, Y., Gu, J.: Identification of Residuals Contamination Based on Signs of Measurement Resuduals. Automtion of Electric Power Systems 24(5), 5–8 (2000)
5. Zhang, X., Mao, Y., Zhu, J.: Detection and Identification of Multi-bad Data Using Graph Theory. Proceedings of the CSEE 17(1), 69–72 (1997)
6. Mili, L., Chenial, M.G., Rousseeuw, P.J.: Robust State Estimation of Electrical Power Systems. IEEE Transaction on Circuits and Systems(1): Fundamental Theory and Applications 41(5) (1994)
7. Celik, M.K., Ali, A.: A Robust WLAV State Estimator Using Transformations. IEEE Trans. on Power Systems 12(1) (1997)
8. Li, Z.: Identification of Bad Data of Electric Power System State Estimation. Journal of Qinghai University, Natural Science Edition 19(1), 49–51 (2001)
9. Liu, H., Cui, W.: The Detection and Identification Method of Bad Data Combined RN Detection and State Forecast. Proceedings of the EPSA 13(2), 39–43 (2001)
10. Yang, W., Hu, J., Wu, J.: The identification algorithm of bad data in power system based on GSA. Relay 33(22), 41–43 (2005)
11. Lu, Z., Zhang, Z.: Bad Data Identification Based on Measurement Replace and Standard Residual Detection. Automation of Electric Power Systems 31(13), 52–56 (2007)
12. Liu, J.: An Operation-mode-based Approach for Bad Data Detection and Identification. Bulletin of Science and Technology 19(5), 432–433 (2003)
13. Salehfar, H., Zhao, R.: A neural network pre-estimation filter for bad date detection and identification in power system state estimation. Electric Power System Reserch 34, 127–134 (1995)
14. Souza, J.C.S., Leite da Silva, A.M., Alves da Silva, A.P.: Information debugging in forecasting-Aided state estimation using a pattern analysis approach. In: 12th PSCC, Dresden, pp. 1214–1220 (1996)
15. Wei, Q., Wang, K., Han, X.: A Correction Method for Distributor When Bad Data Identification Mistake Happens. Journal of Northeast China Institute of Electric Power Engineering 23(1), 34–38 (2003)
16. Liu, L., Zhai, D., Jiang, X.: Current situation and development of the methods on bad-data detection and identification of power system. Power System Protection and Control 38(5), 143–146 (2010)

Analysis of Fingerprint Performance among Left-Handed and Right-Handed People

Nasrul Humaimi Mahmood and Akram Gasmelseed

Biomedical Instrumentation and Electronics Research Group
Faculty of Electrical Engineering
Universiti Teknologi Malaysia, 81310 UTM Johor Bahru,
Johor, Malaysia
nasrul@fke.utm.my

Abstract. Fingerprint is one of the main features in identifying a person and the recognition of fingerprint is much easier to compute and analyze compared to another recognition such as iris recognition and voice recognition. This paper investigates the fingerprint performance between the left-handed and right-handed people by determining which one of these two groups has the easier computation using MATLAB. The minutiae matching are used to determine the feature of fingerprint image. Minutiae are the main feature of fingerprint identification. By using MATLAB as a tool for calculating the minutiae compared to the manual calculation, the error is reduced. As a result, it can be concluded that the minutiae matching is the best way in identify and analyze the fingerprint performance and pre-processing image is needed to produce clear images as well as the ridges and furrow could be differentiated easily.

Keywords: Fingerprint, left-handed people, right-handed people, minutiae.

1 Introduction

Nowadays, identification of a person is really important in performing certain work or task. With the identification, almost all hard work got easy to do. All the identification of a person is connected with biometrics. The biometrics had been used to identify a person, although the person is physically damage or destroyed. Biometrics refers to the automatic identification or verification of an individual or a claimed identity by using certain physiological or behavioral traits associated with the person such as fingerprints, hand geometry, iris, retina, face, hand vein, facial thermo grams, signature, voiceprints and others [1,6].

Biometric indicators have an edge over traditional security in that these attributes cannot be easily stolen or shared. Among all the biometric indicators, fingerprint-based identification is the oldest and most popular method among all the biometric techniques being used today, which has been successfully used in numerous applications. The main reason for the popularity is of course its high level of reliability. This is because, everyone is known to possess a unique fingerprint, and its features remain invariant with age [7,5].

Fingerprint has been used as identifications for individuals since the late 19th centuries, and it has been discovered that every individual has different fingerprints even for identical twins. Fingerprints have the properties of distinctiveness or individuality, and the fingerprints of a particular person remain almost the same (persistence) over time. These properties make fingerprints suitable for biometric uses [1,2]. Fingerprints are actually the ridge and furrow patterns on the tip of finger. The characteristics to be extracted in a given fingerprint image can be divided into two main categories, global or high level features and local or low level features. Core and delta are the global features while ridge ending and bifurcation of the fingerprint ridge are the local features. Local features are commonly named minutiae [3,4,9]. All features are stored as principal component data.

It is important to have reliable personal identification due to growing importance of information technology. A biometric system is the best way to be the core of identification compares to verification. The most popular biometric is a fingerprint. Fingerprint is unique and differs from one person to another [7,8]. If there is a situation, which is a person lost his right finger, is there a possibility that the left finger is the same pattern of the right finger or is it the left finger's pattern is just the mirror of right finger. In this case, the software must be implemented to detect the pattern of fingerprint for both right hand and left hand also the similarity (if possible) of the pattern. Besides that, in this world, there are billion people will use fingerprint as their identification. Some of them are right-handed user and other left-handed user. However, are there single or large different between the left-handed and right-handed user? Hence, the software should also be implemented to analyse and detect the different between left-handed and right-handed people.

2 Methodology

Two representation forms for fingerprints separate the two approaches for fingerprint recognition. The first approach, which is minutiae-based [4,9], represents the fingerprint by its local features, like terminations and bifurcations. This approach has been intensively studied, also is the backbone of the current available fingerprint recognition products. The second approach, which uses image-based methods, tries to do matching based on the global features of a whole fingerprint image. It is an advanced and newly emerging method for fingerprint recognition. And it is useful to solve some intractable problems of the first approach.

Most of the existing methods used for fingerprint verification are based on local visible features called minutiae. Fingerprint preparation for minutiae extraction needs several complex steps. These time-consuming steps are fingerprint enhancement, directional filtering, segmentation, and thinning. They may erroneously introduce false minutiae and reject some real minutiae points. Therefore, some additional steps must be provided to alleviate these errors. In addition, it is hardly possible for low-quality images to extract minutiae points and a complementary matching method is needed for fingerprint verification. Hence, discriminatory information has been considered. The three stage approaches are used by researchers, which are pre-processing, minutiae extraction and post-processing stage.

To remove the false minutiae, histogram equalization is used. Histogram equalization expands the pixel value distribution of an image so it will increase the perceptional information. The "thinning" process is used to skeletonise the binary image by reducing all lines to a single pixel thickness. The approach iteratively deletes edge point pixels from the region until just the skeleton remains. Finally, the minutia match is performed. Minutia match algorithm determines whether the two minutiae sets are from the same finger or not. This includes the alignment stage and the match stage [9].

2.1 Pre-processing Using ImageJ Software

Before proceeding with MATLAB software, the pre-processing image is done to enhance the image clearer and easy to understand by MATLAB. The fingerprint images were taken by using fingerprint stamp pad and A4 paper. Thus, the images are not clear with a lot of noise. Normally, in the white and black colour image there is salt and pepper noise. This will occur if there is the dead pixel. The image will have dark pixels in bright regions and bright pixels in the dark regions. Thus, the image should be filtered to obtain the good quality of image. The images from A4 paper were scanned and were crop to get the desired region of interest (ROI) of the image. Finally, the images were saved in the different name according to the group which is left-handed people and right-handed people. The pre-processing of fingerprint images had successfully done using ImageJ software and the steps taken to do this pre-processing are:

Step 1: The image of PNG format is loaded in the ImageJ software.
Step 2: The image is duplicated to make a comparison between original image and modified image.
Step 3: The image is converted into 8 bit image. This is important as the fingerprint recognition software in MATLAB only accepted the image range between 0 to 255, while the original image is in RGB colour.
Step 4: The brightness and contrast of image such as the darker the dark region and brighter the bright region is adjusted.
Step 5: The image is filtered to reduce salt and pepper noise in the image. Besides that, the unsharp mask is used to enhance the image.

2.2 Fingerprint Recognition Using MATLAB

Figure 1 is a flowchart of the recognition process of fingerprint using the MATLAB. After the grayscale image is loaded, the histogram equalization is used to expand the pixel values of the image. A Fourier Transform method is connecting the falsely broken region and furrow. Then, the binarisation is applied to convert the grey scale image value from 0 to 255 to only white and black colour for further analysis. Direction and region of interest are selected to demonstrate the direction of the ridges on fingerprint as well as the interested region to be operated. Some noised are removed such as H break and spike by thinning process. The minutia of fingerprint, including the false and correct path is extracted, and then the real minutia is applied to remove spurious minutia. Finally, the data is saved for performance analysis between left-handed and right-handed people.

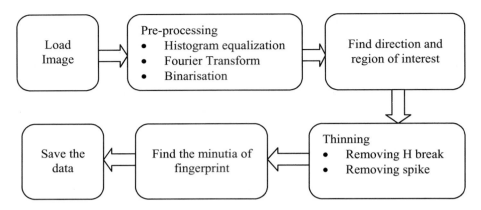

Fig. 1. Flowchart of fingerprint recognition process using MATLAB

2.3 Matching Process

The data from the fingerprint recognition process are compared to each other to find the similarities among them. In the Figure 2, the Template 1 and Template 2 are compared according to the region of interest (ROI), the direction of ridges and furrow, and the real minutia of a fingerprint. The template contains the feature of the fingerprint. These features determine the similarity or the different between one fingerprint to another fingerprint.

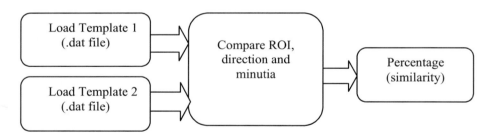

Fig. 2. Flowchart of the matching process

3 Result and Discussion

The fingerprint images are obtained from students, staff and lecturers around Universiti Teknologi Malaysia (UTM). The result of each operation is shown in Figure 3. This operation is needed to produce a good data for comparison between left-handed and right-handed people. The result of the recognition process shows the data of fingerprint in the form of ".dat file". The data is contained three parts of important features of fingerprint, which is the region of interest (ROI), the direction of ridges and furrow, and the real minutia of a fingerprint.

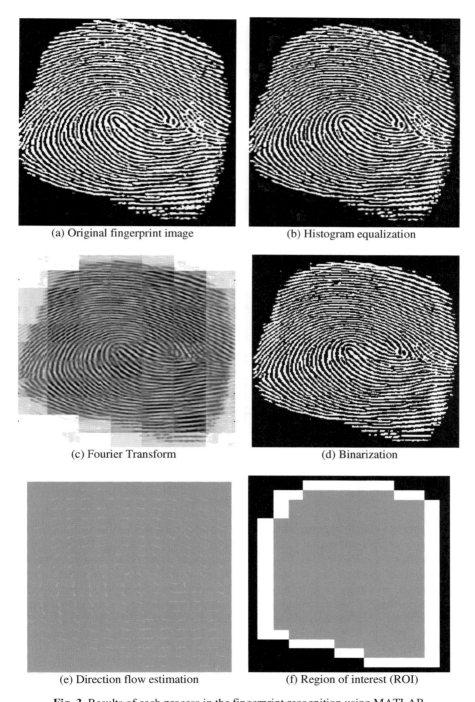

Fig. 3. Results of each process in the fingerprint recognition using MATLAB

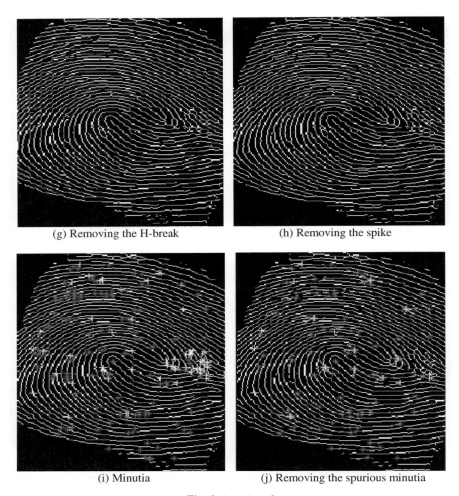

Fig. 3. (*continued*)

After performing histogram equalization, Fourier transform, binarisation, orientation of flow estimation, region of interest (ROI), thinning, removing H break, removing the spike and removing the spurious minutia, the image data is saved in the term of principal component. The comparison is based on this data. The result showed the percentage of similarity of any fingerprint is below than 90 percent. This proved the fingerprint image is not belonging to the same person. Besides that, the fingerprint recognition software work properly and follow the fact that every single person has different type and shape of fingerprint. The result also showed the comparison of minutia for each fingerprint give the best output as the minutia is the feature of any fingerprint of a person. From the data obtained from left-handed and right-handed people, the statistical analysis is performed. The results of statistical analysis are shown in Figure 4 which consisting results of right-handed people (with both right and left fingerprint) and left-handed people (with both right and left fingerprint).

Analysis of Fingerprint Performance among Left-Handed and Right-Handed People 115

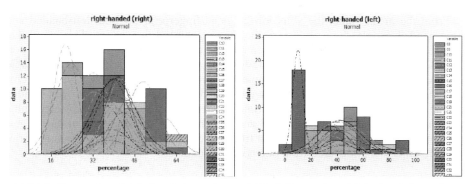

(a) Right fingerprint of right-handed people.   (b) Left fingerprint of right-handed people.

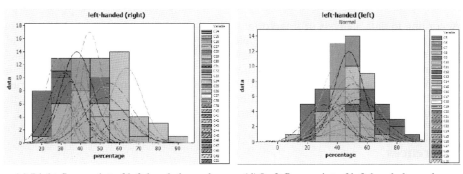

(c) Right fingerprint of left-handed people.   (d) Left fingerprint of left-handed people.

Fig. 4. Histogram plotted for different preferences of people.

According to the histogram in Figure 4, the value of each histogram for all graph are between 0 to 90 percent. Therefore the similarity between one fingerprint to another is not more than 90 percent. The histogram is also show the large distribution percentage value of histogram is between 40 to 60 percent. To ensure the data is sufficient to use, the mean value for each data is calculated. The calculated values are represented in Table 1.

Table 1. Total percentage and mean percentage of fingerprint

Categories	Left hand/fingerprint	Right hand/fingerprint
Left-handed people	Total percentage:19374.9	Total percentage: 19689.6
	Mean percentage: 44.54	Mean percentage: 45.26
Right-handed people	Total percentage:16829.6	Total percentage: 17835.0
	Mean percentage: 38.68	Mean percentage: 41.00

* Total percentage is the total value of similarity for each category.
\# Mean percentage is the average similarity value for each category.

From the mean value, the value of left hand/fingerprint for both left-handed and right-handed people show smaller value than right hand/fingerprint. This show that the left hand of a person is easy to differentiate compared to the right hand of a person. Besides that, from the mean value for both right-handed and left-handed people the mean value for right-handed people are smaller than left-handed people. This demonstrated right-handed person is easy to differentiate compare to left-handed people. Although the left-handed and right-handed people have slightly differed value of mean, but the value are between 35 and 46. This show the similarity of one fingerprint to another fingerprint is less than fifty percent. Thus, the value show fingerprint of a person is totally differed from one person to another.

4 Conclusion

According to data analysis of fingerprint image, it shows that the left hand of a person either left-handed or right-handed people have smaller value than right hand. This make the left hand fingerprint image is easy to compute using MATLAB software compare to the right hand of fingerprint image. Besides that, the average value of right-handed people is smaller than left-handed people. Thus, most of the fingerprints of right-handed people have many different features such as region of interest, the direction of ridges and furrow and minutia.

References

1. Sargur, N.S., Harish, S.: Comparison of ROC and Likelihood Decision Methods In Automatic Fingerprint Verification. International Journal of Pattern Recognition and Artificial Intelligence 22(3), 535–553 (2008)
2. Krawczyk, S.: User Authentication Using On-line Signature and Speech. Master thesis, Michigan State University, 6–9 (2005)
3. Ross, A., Govindarajan, R.: Feature Level Fusion Using Hand and FaceBiometrics. In: Proc. SPIE Conf. Biometric Tech. for Human Identification, USA, vol. II, 5779, pp. 196–204 (2005)
4. Liang, X.F., Asano, L.: Fingerprint Matching Using Minutiae Polygons. In: Japan Advanced Institute of Science and Technology, JAIST, pp. 1046–1049 (2006)
5. Musa, M.M., Rahman, S.A.: SAB.: Directional Image Construction Based On Wavelet Transform For Fingerprint Classification And Matching. In: International Conference on Computer Graphics, Imaging and Visualization (CGIV 2004), vol. 3, pp. 325–340 (2007)
6. Waymen, J., et al.: Biometric Systems Technology, Design and performance Evaluation, vol. 14, pp. 43–54. Springer, London (2008)
7. Maltoni, Davide, Maio, Jain, Prabhakar: Handbook of Fingerprint Recognition, vol. 3, pp. 725–732. Springer, Heidelberg (2009)
8. Wayman, J.L., Jain, A.K.: Performance Evaluation of Fingerprint Verification Systems. IEEE Transactions on Pattern Analysis Machine Intelligence 28(1), 3–18 (2006)
9. Alessandro, F., Kovacs-Vajna, Z.M., Alberto, L.: Fingerprint minutiae extraction from skeletonized binary images. Pattern Recognition 32(4), 877–889 (1999)

The Design of Reconfigurable Remote Monitor and Control System Based on CAN Bus

Wang Qing

Wuhan Digital Engineering Institute Wuhan, Hubei, China
wqpatric@263.net

Abstract. The paper describes the design and realization of a reconfigurable remote monitor and control system based on CAN bus. It illustrates the total structure of the system, the design of hardware and software of CAN node, embedded network server and remote monitor and control terminal. It also illustrates the CAN node network structural reconfiguration emphatically. The system designed with this method has the abilities such as flexible, efficient and autonomy.

Keywords: CAN bus; reconfigurable; ARM.

1 Overview

The network has become a social foundation information facility, which is the main method for information circulating. The remote monitor and control system is an important maintenance way according to the network, which has already become a main trend for industry control.

CAN bus is a kind of communication protocol developed to solve the data exchange among the numerous electronic controlled modules in modern cars. It has been widely used in automobile, industry, consumer electronics and so on because it admits the bus access of multi-master stations according to the priority and it adopts the non-destructive bus arbitration and it can complete the examination to error of the correspondence data and the distinction of the priority with the data length of 8 bytes at most, short in transmission, strong anti-interference ability and the correspondence speed may reach at most 1Mbit/s.

This article proposed a design of reconfigurable remote monitor and control system based on CAN bus. According to the industry scene's environment characteristic, it developed the gateway application based on the embedded network server to realize the monitoring to the apparatus working condition of the remote industry field. Simultaneously it introduced the design of software and hardware and reconfigurable of the network emphatically.

2 The System Structure

The remote monitor and control system based on Ethernet can be divides into reconfigurable CAN bus network, embedded network server and remote monitor and control

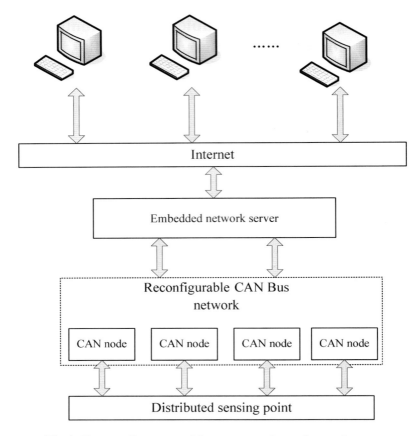

Fig. 1. The overall structure of the remote monitor and control system

terminal. The remote monitor and control terminal connects with the embedded network server through the Ethernet and monitors the field bus equipment, stores the monitor data to the database and determines the reasonable control according to the equipment information in the database. The embedded network server is responsible for receiving and dispatching information flow from the CAN bus and the monitor terminal connected with the Ethernet and realizing the correspondences of the two different kind of network through the protocol conversion. The network server judges the type and meaning of the acquired information by recognizing and describing the information between network equipments to determine the transmitting position. And there is no need to analyze the information in this way it has saved the network server's processing time greatly. It can also configure the network server automatically through the parameter provided by the monitor terminal and dispose any node in CAN bus automatically through the network server to realize the dynamic disposition of the network sensor and the restructuring function and dispose the number of the intelligent equipment node hooked on the system with flexibility, which reduces the cost of the manufacture and maintenance of the equipment network interface. When failure occurs, the network server may handle it in time or send warnings to the monitor terminals. The monitor terminal has been designed with image monitoring interface which makes it possible to know the situation in real time and make the decision. The embedded network server realizes the data correspondence between the CAN bus and Ethernet. In the mean time it enhances the corresponding efficiency and the reliability of the system. The overall structure of the remote monitor and control system shows as above.

3   The Function Realization of Each Module of the System

The specific realization of each module as follows:

(1) The CAN node design of reconfigurable CAN bus network.

Reconfigurable CAN bus network use the node structure. Each node use the CAN bus to correspond with the embedded network server, which realizes the interface connection and the information exchange between embedded network server and field bus. The CAN node as the network sensor monitors the overall operating condition of the field equipments. The hardware circuit is composed of LPC2292 microcontroller with ARM (with 2 standard CAN interfaces), the 2 CAN transceiver CTM8251D with buffer, A/D converter and anti-jamming circuit. This CAN node takes the ARM microprocessor as a core, through the sensors acquiring the analog signals such as power, temperature and humidity and so on, transforming into digital signal after A/D converter. In the end data is processed by the microprocessor. The ARM microprocessor loads the embedded operating system μCOSII. It realizes dynamic load of the operation function through the multi-task scheduling mechanism. The hardware architecture of the CAN node is shown in Figure 2.

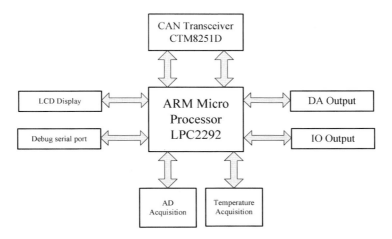

Fig. 2. The hardware structure of CAN node

(2)The embedded network server realizes the data transmission

This article has developed the gateway application based on the embedded network server and realized the seamless connection of CAN bus network and Ethernet. The hardware of the embedded network server includes: ARM9 core processor S3C2410A, 64 MB NOR Flash, 256MB RAM, the periphery expanded two CAN bus control unit SJA1000, CAN bus transceiver CTM8251D and Ethernet controller DM9000A. The software transplanted real-time core under the embedded Linux-2.4.18 operating system, through the development of the application to realize the transformation of the 2 kinds protocols achieve the data communication. The hardware structure of the embedded network is as shown in Figure 3.

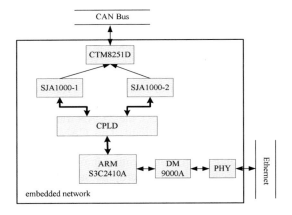

Fig. 3. The hardware structure of the embedded network

(3) Remote monitor terminal realizes the data storage and the access

The network database server was located at the remote monitor terminal which is mainly used to process the data, store the received data in and the data accessing. The design of the network database mainly includes the onstage user surface, the backstage database as well as the correspondence of the hardware equipments. The user may carry on the monitor in any place to the devices.

4 The Restructuring of CAN Bus Network Architecture

The CAN bus network usually adopts linear topology bus structure. All nodes link on the stem of twisted-pair line (or optical fiber) as branches, which use the multi-host ways and there is no owner-member relationship between the nodes but there is priority between them. In the application system, the configuration parameter usually has space attribute. Therefore, the CAN network stem takes on diversity along with different application systems. In the engineering application, the change of the structure status is usually a gradation process from a point and the whole surface. So, it should pay extra attention to the nodes whose structure state change is big and their peripheral nodes set too. In the CAN network, it can raise its priority and give a quick respond to the structure change through the information acquisition and localization of the disposal, the key technology of which is to realize the clustering of the priority of the reconfigurable. Reconfigurable is an efficient measure to ensure the reliability and the self-healing of the network. When the nodes or the communication links get an error or part of the network got security threaten, a network with reconfigurable ability can reconfigure in time and resume the normal communication and network service.

Self Organizing Feature Map (Self Organizing Feature Map, SFOM) is a processing system imitating nervous system of living creatures. It is competitively studying networks which can study automatically under no supervise. The structure of SFOM network is shown in figure 4. It is composed of input layer and competition layer. The neuron number of the input layer is N and the competition layer include $M = m \times n$ neurons, to make a two-dimensional array. The two layers was connected through two kinds of connection weight: One is the connection weight of neuron to the external input reaction; The other is the connection weight between the neuron the size of which controlled interaction between the neuron. For each inputted vector, its value was compared with weight coefficient and it aroused competitions between neurons. The neuron whose weight coefficient vector was the most familiar with the input vector was considered to have the strongest reaction to the input vector and was thought to be the winner. Then according to the rule "might makes right" as a neuron, it took the winning neuron as the centre. Those who were near to it took on exciting side feedback and those who were far away from it took on inhibitory side feedback. Through this side feedback course, SFOM network pressed the weight coefficient vector of the neurons in the surrounding area approaching to the input vector in self organization. In this way, it got the input vector with similar character together to achieve the classification which the cluster analysis processed from data to the attribute (the output of the network).

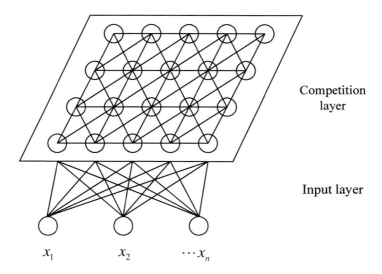

Fig. 4. SOFM structure

SOFM can do the clustering unsupervised and can map any dimension input mode into discrete graphics of one-dimensional or two-dimensional, keeping its topology unchanged. Considering that CAN network of this system has many nodes, its feature complexity, uncertainty in network rollout path and so on. It chose SOFM network to do the feature clustering to the priority of CAN network nodes.

Assume the distributed node was $x_i(i=1,2,\cdots,k)$, structure evaluation indicators was $f_j(j=1,2,\cdots,n)$, and the structure condition assessment Matrix was:

$$S = \begin{matrix} & \begin{matrix} f_1 & f_2 & \cdots & f_n \end{matrix} \\ \begin{matrix} x_1 \\ x_2 \\ \vdots \\ x_k \end{matrix} & \begin{bmatrix} s_{11} & s_{12} & \cdots & s_{1n} \\ s_{22} & s_{23} & \cdots & s_{2n} \\ \vdots & \vdots & \vdots & \vdots \\ s_{k1} & s_{k2} & \cdots & s_{kn} \end{bmatrix} \end{matrix}$$

Assume the contribution of the indicator f_j to the node was $w_j(j=1,2,\cdots,n)$, composite indicator was shown as

$$g_i(s_{ij}, w_j) = \sum_{j=1}^{n} s_{ij} \Delta w_j, \Delta w_j = w_j / \sum_{j=1}^{n} w_j$$

The number of structure condition set requiring extra attention was m. Cluster the condition Matrix S with SFOM, the algorithm was as follow:

① Set the connection weight vector: $W_j = (w_{j1}, w_{j2}, \cdots, w_{jn}), j = 1, 2, \cdots, m$, with a less assignment in initialization. t was the study time, initialization $t = 0$.

② Input mode: $S_i = (s_{i1}, s_{i2}, \cdots, s_{in}), i = 1, 2, \cdots, k$

③ Calculate the Euclidean distance:

$$d_{j^*} = \|R_i - W_j\| = \sqrt{\sum_{p=1}^{n}[r_{ip} - w_{jp}]^2}$$

$d_j = \min_j\{d_j\}$, the neuron j^* which d_j was corresponding was the winning unit.

④ Adjust j^* and the neuron weight value of its neighborhood L_{j^*}.

$w_{jp}(t+1) = w_{jp}(t) + \eta(t)[r_{ip}(t) - w_{jp}(t)]$

$\eta(t)$ was the plus item descend with time, $0 < \eta(t) < 1$.

⑤ update $\eta(t)$ and $L_{j^*}(t)$, read:

$$\eta(t) = \eta(0)(1-\frac{t}{T}) \quad L_{j^*}(t) = INT[L_{j^*}(0)(1-\frac{t}{T})]$$

T was the maximum number of learning, $INT(x)$ was the rounding sign

⑥ $t \approx t+1$, repeat the processes until $t = T$

Calculate the mean value of the m priority clustering composite indicators; arrange the priority according to the mean value.

Clustering algorithm was realized by the upper computer calculating according to the present state of the network node to get the newest network structure parameters and then assign them to each node. According to this algorithm, the nodes with similar structure state attribute character were classified to the same kind and had the same priority. In the mean time, it did the localization in the interior to the structure state information and the results were sent back to the upper computer. In this way, it realized the dynamic adjustment to the network structure and the real time process to the abrupt change of partial structure state.

5 System Software Design

System software design was focused on the embedded network sever, including 4 parts: Boot code U-boot, Embedded Linux-2.4.18 operating system core, File system and User application management software. The first three system-class software was generated by some necessary transplantation based on the source code provided by the authority, while user application management software was free to design completely. User application management software was composed of CAN interface communication program, frame information management program and protocol conversion program.

Frame information management program received the data and command from the monitor terminals and executed configure, add, alter, delete, maintain, read and write. Then it distinguished the network equipment information, processed and sent information to the monitor terminals.

Protocol conversion included two kinds: (1) Protocol conversion from CAN to TCP/IP. Because the gateway will interrupt when receiving the CAN message, it needed to converse the message to the Ethernet data packet when sending the CAN message to Ethernet. (2) Protocol conversion from TCP/IP to CAN. Because the gateway will query when receiving the Ethernet data packet, it also needed conversion when sending Ethernet data packet to the CAN bus. The flow chart was as follows.

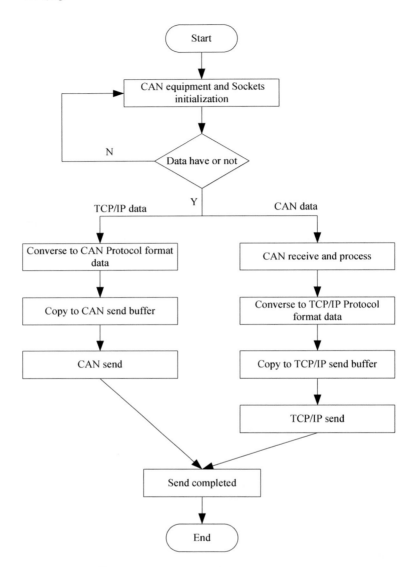

Fig. 5. Main flow chart of protocol conversion

6 Conclusion

The article showed the software and hardware design of reconfigurable remote monitor and control system based on CAN bus, introduced the structure of the system and the software and hardware of the CAN node, embedded network server and remote monitor control terminal and realized the reconfiguration of the CAN node network. The system achieved the data conversion between the CAN bus and the Ethernet and made the field intelligent devices, the devices and the control room a network system, which complied with the trend of intelligence, network, decentralization, reconfiguration of the control system.

References

1. Jaman, G., Hussain, S.: Structural Monitoring Using Wireless Sensors and Controller Area Network. In: Proc. of CNSR 2007, Fredericton, Canada (2007)
2. Zheng, W., Cui, R.-r., Lu, P.: Design of Reconfigurable SHM System Based on CAN Bus. Computer Engineering 36(15), 228–232 (2010)
3. Zhao, X.-j., Su, H.-x., Ren, M.-w.: Remote Monitor and Control System based on ARM9 and CAN bus. Computer Engineering 36(5), 231–233 (2010)
4. Ekiz, H., Powner, E.T., Kutlu, A.: Design and Implementation of a CAN/Ethernet Bridge. In: Proceedings of IEEE ISPAN 1996, IEEE Press, Beijing (2006)
5. Thomas, N., Mikael, N., Hans, A.: Real-time Server-based Communication with CAN. IEEE Trans. on Industrial Informatics 1(3), 192–201 (2005)
6. Kim, W.H.: Bandwidth Allocation Scheme in CAN Protocol. IEEE Trans. on Control Theory and Application 147(1), 37–44 (2000)

Adaptive Friction Compensation of Robot Manipulator

Wang Sanxiu[1,2] and Jiang Shengtao[1]

[1] Taizhou University, Taizhou, 318000, China
[2] College of Information Engineering, Zhejiang University of Technology, Hangzhou, 310032, China

Abstract. In this paper, an adaptive tracking control scheme is proposed for a friction model, which includes coulomb friction, viscous friction and stribeck effect. Parameters are adjusted on-line by the adaptive controller. The experiments have been implemented on one-link robot manipulator and demonstrate the validation of the proposed control scheme.

Keywords: Adaptive control; Friction compensation; Robot manipulator.

1 Introduction

Friction is a nonlinear phenomenon that causes performance degradation especially in low velocities when the friction terms dominate the dynamic system. Nonlinear friction is difficult to describe because friction characteristics change with various environmental factors, such as load variations, lubrication, temperature and the status of machine, may result in steady state errors, limit cycles and stick-slip motion. However, nonlinear friction is an unavoidable phenomenon frequently experienced in mechanical and robotic system. So compensating for friction has been one of the main research issues over the years. In applications, control design to compensate for friction is usually a problem of formulation of nonlinear friction model, identification of its parameters and the corresponding compensation scheme.

Numerous friction compensation schemes have been proposed by researchers. Traditional PD control algorithm is simple, but can not achieve satisfactory results because of steady-state error, and the high gain may cause system instability [Wu and Paul, 1980]. In order to achieve favorable performance, adaptive scheme has been proposed to compensate for nonlinear friction, which is based on on-line estimation of parameters of dynamic friction models. [Canudas de Wit and Lischinsky, 1997] present an adaptive friction compensation scheme with partially known dynamic friction model. In [Misovec and Annaswamy, 1999], an adaptive controller is proposed to handle static, Coulomb and viscous friction components as well as inertia and stribeck effects. An Adaptive friction compensation schemes for robot tracking control is proposed to handle parametric uncertainty with LuGre dynamic friction model [Tomei, 2000].

An accurate friction model that predicts the behavior of friction is required in all the above methods. However, modeling nonlinear friction effects is not so straightforward. Different friction models have been studied by numerous researchers in literatures. [Olsson, Astrom, et al 1997] has given a summary of some classical static and dynamic friction models. A single state elastoplastic friction models is proposed in [Dupont, Hayward, et al, 2002].

In this paper, we propose a nonlinear friction compensation algorithm based on adaptive controller for one-link robot manipulator. Due to the complexities and nonlinearities of friction parametrically modeling, the adaptive law is employed to adjust parameters. In this paper, we compare the proposed design with neural network control scheme and demonstrate the validation of the proposed adaptive control strategy.

This article is organized as follows: section 2 gives a brief summary of the friction models and describes the friction model considered in this paper. Section 3 proposes a nonlinear friction compensation technique using an adaptive control strategy. Section 4 provides the simulation results to demonstrate the validation of proposed method based on one-link robot manipulator. Finally, conclusions are presented in section 5.

2 Description of Friction Model

As pointed out that the presence of friction in robot and mechanical control system can give rise to undesired effects, such as steady-state error, limit cycling and stick-slip behavior. For eliminating or reducing the effects of the friction force, model-based friction compensation technique is an active control strategy. And it is believed that the friction modeling is more accurate, the tracking error is smaller. In this section, we will give a brief summary of some classical friction models, then, the friction model considered in this paper is mainly introduced.

Friction is usually modeled as a static map between velocity and friction force that depend on the sign of the velocity, which has static, coulomb and viscous friction components. As shown in figure1.

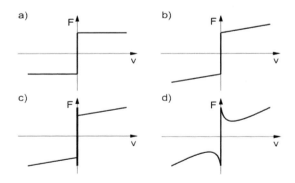

Fig. 1. Static friction models.
a) Coulomb friction model
b) Coulomb and viscous friction model
c) Stiction, coulomb and viscous friction model
d) Stribeck effect, static, coulomb, and viscous friction model

Coulomb friction is a constant opposing torque for nonzero velocities. For zero velocity, the stiction will oppose all motions as long as the torques is smaller in magnitude than the stiction torque. This model is represented in fig.1.a). The Coulomb friction model has often been used for friction compensation because of its simplicity. Viscous friction is the frictional force opposing the motion and is proportional to the velocity, which is often combined with Coulomb friction as shown in Fig.1.b). Stiction is short for static friction as opposed to dynamic friction. Static friction counteracts external forces below a certain level and thus keeps an object from moving. It is hence clear that friction at rest cannot be described as a function of only velocity [Ciliz and Tomizuka, 2007]. The friction components combine with stiction, coulomb and viscous friction, see Fig.1.c). When the stiction torque level is overcome, the friction toque decreases making a downward bend and then increases again proportional to the velocity. This downward bend sometimes is called the stribeck effect, as shown in Fig. 1.d).

In this article, A static friction model as a summation of the coulomb friction, viscous friction and the stribeck effect is employed to approximate friction parametrically, which can be describe as a nonlinear function of the velocity as Eq.1. The justification for using this friction model is provided by Armstrong's experimental results in [Armstrong, 1988].

$$F(\dot{\theta}) = [\omega_0 + \omega_1 e^{-\gamma_1 |\dot{\theta}|} + \omega_2 (1 - e^{-\gamma_2 |\dot{\theta}|})] \operatorname{sgn}(\dot{\theta}) \quad (1)$$

Where, F is the friction force, $\dot{\theta}$ is the angular velocity. ω_0 represent coulomb friction, ω_1 is static friction and ω_2 represent the viscous friction. The stribeck effect is modeled with an exponential second item. Thus, the complete friction model is characterized by five static parameters, $\omega_0, \omega_1, \omega_2$ and γ_1, γ_2.

3 Robust Adaptive Friction Compensation Scheme

3.1 Description of Robotic System

The system under consideration in this paper is one-link robot manipulator with friction. The robot is model as a single inertia I, subjected to a viscous friction torque $d\dot{\theta}$, a friction torque $f(\dot{\theta}, u)$, and a control input u. Then the dynamic model for the one-link robot can be described as follows

$$I\ddot{\theta} + (d + \delta_1)\dot{\theta} + \delta_0 \theta + mgl \cos \theta = u - f(\dot{\theta}, u) \quad (2)$$

In which, $I = \dfrac{4}{3} mg^2$, θ is the angular position, $mgl \cos \theta$ is gravity item. δ_0 is the elastic friction coefficient, δ_1 is the uncertainty of viscous friction.

If the movement plane of robot is parallel with the horizontal plane, then the gravity term can be ignored. And the model of uncertain one-link robot can be expressed by following second order differential equations

$$\ddot{\theta} + \alpha_1 \dot{\theta} + \alpha_0 \theta = \beta_0 u - \beta_1 f(\dot{\theta}, u) \tag{3}$$

Where, $\dot{\theta}$ is angular velocity, the parameters α_i, β_i $(i = 0,1)$ are positive bounded constant.

For the system (3), import the following steady reference model:

$$\ddot{\theta}_d + a_1 \dot{\theta}_d + a_0 \theta_d = br \tag{4}$$

In which, θ_d is the desired position trajectory, r is actuation input, a_i and b are positive constant.

According to the Eq.(3) and Eq.(4), It is easy to obtain the dynamic error equation as Eq.(5)

$$\ddot{e} + a_1 \dot{e} + a_0 e = br - \beta_0 u + \beta_1 f(\dot{\theta}, u) + (\alpha_1 - a_1)\dot{\theta} + (\alpha_0 - a_0)\theta \tag{5}$$

Where $e = \theta_d - \theta$ is the position error, $\dot{e} = \dot{\theta}_d - \dot{\theta}$ is the velocity error.

Let $x = [e, \dot{e}]^T$, then Eq. (5) can be expressed in the following state space equation:

$$\dot{x} = Ax - \begin{bmatrix} 0 \\ \beta_0 \end{bmatrix} u + \begin{bmatrix} 0 \\ \Delta \end{bmatrix} = Ax + Bu + D\Delta \tag{6}$$

Where, $A = \begin{bmatrix} 0 & 1 \\ -a_0 & -a_1 \end{bmatrix}$, $B = \begin{bmatrix} 0 \\ -\beta_0 \end{bmatrix}$, $D = \begin{bmatrix} 0 \\ 1 \end{bmatrix}$,

$\Delta = br + \beta_1 f(\dot{\theta}, u) + (\alpha_1 - a_1)\dot{\theta} + (\alpha_0 - a_0)\theta$

3.2 Adaptive Controller Design

As has been pointed out friction is a major source of loss performance in the robot control system. If friction is not properly compensated, it may course significant positioning errors in point-to-point control and large tracking errors at low velocities. So the key problem of the robot control is to find an effective control law such that the tracking error can be convergence to zero.

In this section, an adaptive nonlinear friction compensation control strategy is proposed.

Because A is a stable matrix. According to the Lyapunov stability theory, for an arbitrary given positive definite matrix Q, there exists a symmetric positive definite matrix P and satisfies the following relationship:

$$PA + A^T P = -Q \tag{7}$$

By use of the solution P, an auxiliary signal is defined as $\bar{e} = [0 \quad 1]Px$, and the adaptive control law is designed as follow

$$u = f(\overline{e},r)r + g(\overline{e},\theta)\theta + h(\overline{e},\dot{\theta})\dot{\theta} \quad (8)$$

Where, parameters $f(\overline{e},r)$, $g(\overline{e},\theta)$ and $h(\overline{e},\theta)$ are adaptive gain coefficient.

In application, the adaptive law of parameters are designed as follow

$$\begin{bmatrix} \dot{f}(\overline{e},r) \\ \dot{g}(\overline{e},\theta) \\ \dot{h}(\overline{e},\dot{\theta}) \end{bmatrix} = \begin{bmatrix} \lambda_0 & 0 & 0 \\ 0 & \lambda_1 & 0 \\ 0 & 0 & \lambda_2 \end{bmatrix} \overline{e} \begin{bmatrix} r \\ \theta \\ \dot{\theta} \end{bmatrix} \quad (9)$$

Where, λ_i (i=0,1,2) is the coefficient of the adaptive law.

4 Experiments

In this section, the experiment is implemented on a one-link robot manipulator. In order to test the validation and effectiveness of the proposed adaptive friction compensation control algorithm, performance of the proposed scheme will be compared with that of a neural network controller.

The dynamic model of the one-link robot system in the presence of friction is given by $\ddot{\theta} + \alpha_1\dot{\theta} + \alpha_0\theta = \beta_0 u - \beta_1 f(\dot{\theta},u)$. Reference model is chosen as $\ddot{\theta}_d + a_1\dot{\theta}_d + a_0\theta_d = br$. Where, $r = \sin(0.5t) + 0.5\sin(0.25t)$.

The friction model has discussed in section 2, and is given by

$$f(\dot{\theta}) = [\omega_0 + \omega_1 e^{-\gamma_1|\dot{\theta}|} + \omega_2(1 - e^{-\gamma_2|\dot{\theta}|})]\mathrm{sgn}(\dot{\theta})$$

All the parameters used in the experiments are chosen in table 1.

Table 1. Experiment parameters

parameter	value	parameter	value
α_0	0.1	ω_0	1.0Nm
α_1	0.3	ω_1	0.5Nm
β_0	$10 + \sin(\pi t)$	ω_2	0.3Nms/rad
β_1	$\sin(\pi t)$	γ_1	3
a_0	4	γ_2	2
a_1	5	λ_0	5
b	6	λ_1	10
ε	0.02	λ_2	20

First, the adaptive controller is run in the experiment. It is attempted to reduce the effect of friction by an adaptive controller. The simulation results show in figure2. The position tracking, control input have shown in figure (a), (b) respectively. In the fig.2.(a), blue curve represent reference trajectory, red curve is the actual output trajectory.

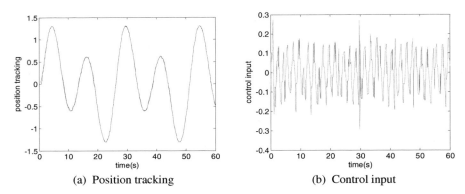

(a) Position tracking   (b) Control input

Fig. 2. Simulation results of adaptive controller

From Figure2, we can find the performance of proposed adaptive controller is favorable. Especially from the tracking error curve, which has shown in fig.3., it is clearly show that the tracking errors convergence fast. The maximum tracking error is only about 0.12.

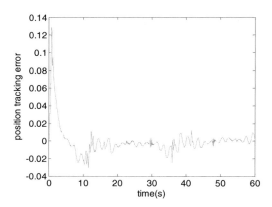

Fig. 3. Tracking error of adaptive controller

In order to further validate the superiority of the proposed control strategy, we compare it with the neural network control scheme. A RBF neural network, with m inputs, n outputs and r hidden or kernel units, can be expressed by $\varphi_i = f(\|x - c_i\|^2 / \sigma^2_i)$, $y = W\varphi(x)$ is utilized in this experiment. Where

$x \in R^m$ is the input, $\varphi(x) = [\varphi_1, \varphi_2, ..., \varphi_r]^T \in R^r$ is the output of the hidden layer, $y \in R^n$ is the output of the network, W is the weight matrix, while c_i and σ_i are the centre and width of the *ith* kernel unit respectively. And the Euclidean norm $\|\cdot\|$ is employed. The detailed neural network controller parameters used in this experiment are seen in reference literature [Feng G, 1995]. The tracking performance using neural network control strategy is shown in figure 4.

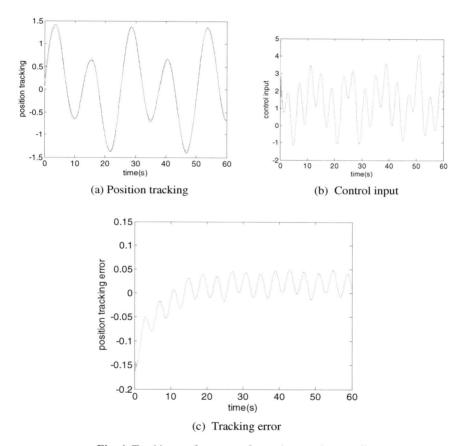

(a) Position tracking

(b) Control input

(c) Tracking error

Fig. 4. Tracking performance of neural network controller

According to the comparison of the adaptive control strategy and neural network controller, it has clearly shown that the tracking performance of the adaptive control is significantly better than that of the neural network control. The maximum tracking error of neural network controller reached 0.15, which is greater than 0.12 of the adaptive controller. Besides, average tracking error of neural network controller is also far more than the tracking error of the proposed method. Consequently, all these experiment results verify the effectiveness of the proposed algorithm.

5 Conclusion

This paper proposes an adaptive control strategy, which is very simple but effective in compensating for friction. The parameters are adjusted online. The proposed nonlinear friction compensation method is compared with neural network control strategy through tracking experiments on one-link robot manipulator. All these simulation results illustrate the superiority and effectiveness of the proposed scheme.

References

1. Wu, C.H., Paul, P.: Manipulator compliance based on joint torque control. In: Proceedings of 9th IEEE Conference on Decision and Control, Albuquerque, NM, USA, pp. 88–94 (1980)
2. Canudas de Wit, C., Lischinsky, P.: Adaptive friction compensation with partially known dynamic friction model. Int. J. Adapt. Contr. Signal Process. 11, 65–80 (1997)
3. Misovec, K.M., Annaswamy, A.M.: Friction compensation using adaptive nonlinear control with persistent excitation. Int. J. Control 72(5), 457–479 (1999)
4. Tomei, P.: Robust adaptive friction compensation for tracking control of robot manipulators. IEEE Trans. Automat. Contr. 45(6), 2164–2168 (2000)
5. Olsson, H., Astrom, K.J., Canudas de wit, C., et al.: Friction Models and Friction Compensation (1997)
6. Dupong, P., Hayward, V., Armstrong, B., Altpeter, J.: Single state elasto-plastic friction models. IEEE Trans. Automat. Contr. 47(5), 787–792 (2002)
7. Ciliz, M.K., Tomizuka, M.: Fricition modeling and compensation for motion control using hybrid neural network models. Engineering Applications of Artificial Intelligence 20, 898–911 (2007)
8. Armstrong, B.: Friction: Experimental determination, modeling and compensation. In: Proceeding of IEEE International Conference on Robotics and Automation, pp. 1422–1427 (1988)
9. Feng, G.: A compensating scheme for robot tracking based on neural networkds. Robotics and Autonomous Systems 15, 199–206 (1995)
10. Han, S.I., Lee, K.S.: Robust friction state observer and recurrent fuzzy neural network design for dynamic friction compensation with backstepping control. Mechatronics 20, 384–401 (2010)

An Example of Cloud Business Application Development Using the Continuous Quantity Identifier of the Cellular Data System

Toshio Kodama[1], Tosiyasu L. Kunii[2], and Yoichi Seki[3]

[1] Maeda Corporation, Advanced Computer Systems, Inc., 3-11-18 Iidabashi, Chiyoda-ku, Tokyo 102-0072 Japan
kodama@lab.acs-jp.com, kodama.ts@jcity.maeda.co.jp
[2] Morpho, Inc., Iidabashi First Tower 31F, 2-6-1 Koraku, Bunkyo-ku, Tokyo 112-0004 Japan
kunii@ieee.org, kunii@acm.org
[3] Software Consultant, 3-8-2 Hino-shi, Tokyo 191-0001 Japan
yseki@amber.plala.or.jp

Abstract. In the era of 'cloud' computing, cyberworlds as information worlds have grown rapidly and on an extremely large scale to encapsulate real world activities like finance, commerce, education and manufacturing in the form of e-financing, e-commerce, distance education and e-manufacturing despite the lack of a firm mathematical foundation. In software engineering, many software development methodologies, including the waterfall model and object-oriented model have been introduced. But these current methodologies cannot yet solve the combinatorial explosion problem. A new approach, different from conventional engineering approach, is needed. We have developed a data processing system called the Cellular Data System (CDS) as a new methodology, based on the Incrementally Modular Abstraction Hierarchy (IMAH) in the cellular model, which offers powerful mathematical background in data modeling. In this paper, we develop a continuous quantity identifier to deal with objects that can express continuous quantity on the presentation level of IMAH and integrated it into the data search function of CDS. By taking advantage of continuous quantity identifiers with the data search function of CDS, development of business application logic using continuous quantity becomes much simpler and significantly reduces development and maintenance costs of the system. In addition, we show the effectiveness of a continuous quantity identifier by taking up some examples of core logic development of a car sharing reservation system.

Keywords: cyberworlds, cellular model, formula expression, continuous quantity identifier, condition formula.

1 Introduction

The cyberworlds in cloud computing are distributed systems. Data and its dependencies are constantly changing within them. Cyberworlds are more complicated and fluid than any other previous worlds in human history and are constantly evolving.

Millions of people use Twitter or Facebook every minute through Web services on mobile phones which are one of main elements of cyberworlds. At the same time, user requirements for cyberworlds also change and become more complicated as cyberworlds change. If you analyze data using existing technology, you have to modify the schema design and application programs whenever schemas or user requirements for output change. That leads to combinatorial explosion, because user requirements, and their combinations and schemas must be specified clearly at the design stage. That is a fundamental problem, so we have to reconsider development from the data model level.

Is there a data model that can reflect the changes in schemas and user requirements in cyberworlds? We consider that Incrementally Modular Abstraction Hierarchy (IMAH) of the cellular model built by one of the authors (T. L. Kunii) to be the most suitable model. The IMAH can model the architecture and the changes of cyberworlds and real worlds from the homotopy level which is most general to the view level which is the most concrete preserving invariants while preventing combinatorial explosion [1][13]. From the viewpoint of IMAH, existing data models are positioned as special cases. For example, UML can model objects at levels below the presentation level, and in the relational data model, a relation is an object at the presentation level which extends a cellular space because it has necessary attributes in which a type is defined, while the processing between relations is based on the set theoretical level. In the object-oriented model, an object is also the object at the presentation level, which extends a cellular space, while the relation between Classes is the tree structure, which is a special case of a topological space. An Object in XML is considered a special case of a cellular space that extends a topological space, because an attribute and its value are expressed in the same tag format.

In our research, one of the authors (Y. Seki) proposed an algebraic system called Formula Expression as a development tool to realize the cellular model. Another (T. Kodama) has actually implemented CDS using Formula Expression [12]. In this paper, we have introduced the concept of a continuous quantity identifier into CDS. A continuous quantity formula is effective when a continuous quantity is dealt with in business application development. In addition, we have placed emphasis on practical use by taking up some examples. Firstly, we explain CDS and its main data search function briefly in Section 2, 3. Secondly, we design the properties of a continuous quantity by Formula Expression, and integrate them into the condition formula search function in Section 4. Next, we implement them in Section 5. We demonstrate the effectiveness of the continuous quantity identifier by developing a business application system, thereby abbreviating the process of designing and implementing most application programs in Section 6. The business application system is the core logic of a car sharing reservation system, where cars are reserved in accordance with customers' required times and car reservation schedules. Related works are mentioned in Section 7. Lastly, we conclude in Section 8.

2 The Cellular Data System (CDS)

2.1 Incrementally Modular Abstraction Hierarchy

The following list constitutes the Incrementally Modular Abstraction Hierarchy to be used for defining the architecture of cyberworlds and their modeling:

1. the homotopy (including fiber bundles) level
2. the set theoretical level
3. the topological space level
4. the adjunction space level
5. the cellular space level
6. the presentation (including geometry) level
7. the view (also called projection) level

In modeling cyberworlds in cyberspaces, we define general properties of cyberworlds at the higher level and add more specific properties step by step while climbing down the Incrementally Modular Abstraction Hierarchy. The properties defined at the homotopy level are invariants of continuous changes of functions. The properties that do not change by continuous modifications in time and space are expressed at this level. At the set theoretical level, the elements of a cyberspace are defined, and a collection of elements constitutes a set with logical operations. When we define a function in a cyberspace, we need domains that guarantee continuity, such that neighbors are mapped to a nearby place. Therefore, a topology is introduced into a cyberspace through the concept of neighborhood.

Cyberworlds are dynamic. Sometimes cyberspaces are attached to each other, an exclusive union of two cyberspaces where attached areas of two cyberspaces are equivalent. It may happen that an attached space is obtained. These attached spaces can be regarded as a set of equivalent spaces called a quotient space, which is another invariant. At the cellular structured space level, an inductive dimension is introduced into each cyberspace. At the presentation level, each space is represented in a form which may be imagined before designing the cyberworlds. At the view level, the cyberworlds are projected onto view screens.

2.2 The Definition of Formula Expression

Formula Expression in the alphabet is the result of finite times application of the following steps.

1. a $(a \in \Sigma)$ is Formula Expression
2. unit element ε is Formula Expression
3. zero element φ is Formula Expression
4. when r and s are Formula Expression, addition of r+s is also Formula Expression
5. when r and s are Formula Expression, multiplication of r×s is also Formula Expression
6. when r is Formula Expression, (r) is also Formula Expression
7. when r is Formula Expression, {r} is also Formula Expression

3 The Condition Formula Search of CDS

If users can specify search conditions, data searching will become more functional. Here, we introduce the function for specifying conditions defining a condition formula by Formula Expression into CDS. Let propositions P, Q be sets which include characters p, q respectively. The conjunction, disjunction and negation of them in logical operation are defined by Formula Expression as follows:

1) Conjunction
$$P \wedge Q = p \times q \,. \tag{1}$$

2) Disjunction
$$P \vee Q = p+q \,. \tag{2}$$

3) Negation
$$\neg P = !p \,. \tag{3}$$

A formula created from these is called a condition formula. Here "!" is a special factor which means negation. Recursivity by () in Formula Expression is supported so that the recursive search condition of a user is expressed by a condition formula.

Condition formula processing is processing that results in a disjoint union of terms that satisfy a condition formula from a formula. When condition formula processing is considered, the concept of a remainder of spaces is inevitable. The processing consists of two maps: a quotient acquisition map f that derives a term that includes a specified identifier and a remainder acquisition map g that derives a term that does not include a specified identifier.

If you assume x to be a formula and $p, !p, p+q, p \times q, !(p+q), !(p \times q)$ to be condition formulas, the images of (x, $p+q$), (x, $p \times q$), (x, $!(p+q)$), (x, $!(p \times q)$) by f, g are the following:

$$x = f(x, p) + g(x, !p) \text{ where } f(x, p) \cap g(x, !p) = \varphi. \tag{4}$$

$$f(x, p+q) = f(x, p) + f(g(x, p), q) \tag{5}$$

$$f(x, p \times q) = f(f(x, p), q) \tag{6}$$

$$f(x, !(p+q)) = g(g(x, p), q) \tag{7}$$

$$f(x, !(p \times q)) = g(f(f(x, p), q) \tag{8}$$

It is obvious that any complicated condition formula can be processed by the combinations of the above four correspondences.

4 A Continuous Quantity Identifier and Its Application to the Condition Formula Search of CDS

4.1 Definition

The continuous quantity identifier is defined as one case of identifiers in Formula Expression, and therefore it follows the general operation of Formula Expression. If it is assumed that r and s ($r<s$) are arbitrary numerical identifiers, a continuous quantity

identifier to express continuous quantity from r to s is defined $[r+s]$. If you assume t, u $(t<u)$ and v, w $(v<w)$ are also arbitrary numerical identifiers and a is an arbitrary letter factor, the continuous quantity identifier has the following properties:

$$[s+r] = \varphi \text{ (if } r<s\text{)} . \tag{1}$$

$$[r+r] = \varepsilon . \tag{2}$$

$$a\times[r+s] = [r+s]\times a . \tag{3}$$

$$\begin{aligned}[r+s]+[t+u] &= [r+u] \text{ (if } r\leq t<s\leq u) \\ &= [t+s] \text{ (if } t\leq r<u\leq s) \\ &= [r+s] \text{ (if } r\leq t<u\leq s) \\ &= [t+u] \text{ (if } t\leq r<s\leq u) \\ &= [r+s]+[t+u] \text{ (if } t\leq u\leq r<s \text{ or } r<s\leq t\leq u) .\end{aligned} \tag{4}$$

$$\begin{aligned}[r+s]\times[t+u] &= [t+s] \text{ (if } r\leq t<s\leq u) \\ &= [r+u] \text{ (if } t\leq r<u\leq s) \\ &= [t+u] \text{ (if } r\leq t<u\leq s) \\ &= [r+s] \text{ (if } t\leq r<s\leq u) \\ &= \varphi \text{ (if } t\leq u\leq r<s \text{ or } r<s\leq t\leq u) .\end{aligned} \tag{5}$$

Next, the quotient acquisition map f of condition formula processing is applied to the continuous quantity identifier according to the above definitions as follows:

$$\begin{aligned}f([r+s],[t+u]) &= [t+s] \text{ (if } r\leq t<s\leq u) \\ &= [r+u] \text{ (if } t\leq r<u\leq s) \\ &= [t+u] \text{ (if } r\leq t<u\leq s) \\ &= [r+s] \text{ (if } t\leq r<s\leq u) \\ &= \varphi \text{ (if } t\leq u\leq r<s \text{ or } r<s\leq t\leq u)\end{aligned} \tag{6}$$

$$\begin{aligned}f([r+s],![t+u]) &= [r+t] \text{ (if } r\leq t<s\leq u) \\ &= [u+s] \text{ (if } t\leq r<u\leq s) \\ &= [r+t]+[u+s] \text{ (if } r\leq t<u\leq s) \\ &= \varphi \text{ (if } t\leq r<s\leq u) \\ &= [r+s] \text{ (if } t\leq u\leq r<s \text{ or } r<s\leq t\leq u) .\end{aligned} \tag{7}$$

4.2 Applied Map g

The applied map g, which replaces a formula including continuous quantity identifier(s) with the remainder of a specified continuous quantity identifier using the above f, is designed as follows;

If you assume the entire set of formulas to be A, g: A , and arbitrary terms r, s, t, u follow these rules:

$$g([r+s],u+v) = g([r+s],u)+g([r+s],v) . \tag{8}$$

$$g([r+s],(u)) = (g([r+s],u)) . \tag{9}$$

$$g([r+s], a \times [p+q] \times b) = a \times f([r+s], ![p+q]) \times b . \quad (10)$$

A simple example of the map g is shown here.
$g([1+24], \text{Kodama}([12+14]+[16+18])) = \text{Kodama}([1+12]+[14+16]+[18+24])$

5 Implementation

This system is a Web application developed using JSP and Tomcat 5.2 as an application server. The client and the server are the same machine. (OS: Windows XP; CPU: Intel Core2 Duo, 3.00GHz; RAM: 3.23GB; HD: 240GB).

The quotient acquisition map f is the main function of condition formula processing. In this algorithm, the absolute position of the specified factor by the function of the language and the term including the factor are acquired first. Next, the nearest brackets of the term are acquired, and because the term becomes a factor, a recursive operation is performed. Details are abbreviated due to the restriction on the number of pages.

6 Case Study: A Car Sharing Reservation System

6.1 Outline

Car sharing is a model of car rental where people rent cars for short periods of time, often by the hour. We take up the example of development of core logic of a car sharing reservation system as a business application using CDS. In this case study, we assume that there are five customers; customer (A, B, C, D, E) and three shared cars; car (1, 2, 3), and that each reservation is to be arranged by adjusting customers' required times and car reservation schedules within a given period from 0 to 20.

Firstly, formulas for customer's requirement times and car reservation times are designed using a continuous quantity identifier and an operation for getting the formula for each's available time from them (6.2). Secondly, sample data of customers and cars are inputted and the operation is carried out according to the design. Thirdly, required data is outputted by the condition formula processing map (6.3).

6.2 Space Design

We design a formula for the space and the operation as follows:

1. The formula for a customer and their required times as a topological space

$$\text{Customer}(\Sigma customer\ id_i \times \Sigma [r_i + s_i]) . \quad (1)$$

$customer\ id_i$: a factor which identifies a customer
$[r_i + s_i]$: a continuous quantity identifier from r_i to s_i

2.   The formula for shared cars and their reserved times as a topological space

$$\text{Shared Car}(\Sigma car\ id_i \times \Sigma[p_i+q_i])\ . \tag{2}$$

$car\ id_i$: a factor which identifies a car
$[p_i+q_i]$: a continuous quantity identifier from p_i to q_i

3.   The operation for getting the formula for available times of shared cars

$$g([0+20], \text{Shared Car}(\Sigma car\ id_i \times \Sigma[p_i+q_i]))\ . \tag{3}$$

[0+20]: a continuous quantity identifier of the entire period (from 0 to 20)

6.3 Data Input/Output

If customer A requires a shared car from time 7 to 8 and from time 13 to 15, a term for customer A is created according to the space design (**1**) as below:

$$\text{Customer(customer A} \times ([7+8]+[13+15]))$$

And if customer B requires a shared car from time 4 to 5 and from time 15 to 17, a term for a customer B is created and added to the previous term as below:

$$\ldots + \text{Customer(customer B} \times ([4+5]+[15+17]))$$

Terms for customer C, customer D and customer E are also created as below and added to the previous formula:

$$\begin{aligned}&\text{Customer(customer A} \times ([7+8]+[13+15]) + \text{customer B} \times ([4+5]+[15+17]) + \text{custo}\\&\text{mer C} \times ([8+20]) + \text{customer D} \times ([6+10]) + \text{customer E} \times ([7+8]+[13+15]))\ .\end{aligned} \tag{4}$$

In the same way, a term for car reservations is created and added to the previous formula according to the space design (**2**) as below:

$$\text{Shared Car}(car1([0+4]+[10+14]) + car2([4+7]+[14+17]) + car2([7+10]+[12+14]+[17+20]))\ . \tag{5}$$

The following operation is performed to get the formula for times when cars are available from formula (**5**) according to the operation design (**3**), which is then added to the formula (**4**).

*formula (**4**)+g([0+20],formula (**5**))*
$=\text{Customer(customer A} \times ([1+3]+[16+18]) + \text{customer B} \times ([0+2]+[5+8]) + \text{customer C} \times ([4+7]+[11+13]) + \text{customer D} \times ([10+14]+[17+19]) + \text{customer E} \times ([2+6]+[12+15]+[17+20])) + \text{Shared Car}(car1([4+10]+[14+20]) + car2([0+4]+[7+14]+[17+20]) + car3([0+7]+[10+12]+[14+17]))\ .$ (6)

The resulting figure is shown in Fig 6.3.

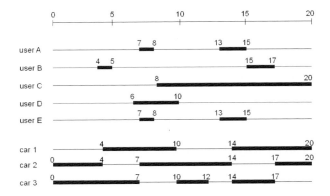

Fig. 1. Times that customers require a car and when cars are available

If you want to answer the question *"When and which cars can customer A use between times 10 and 20?"*, first you make the condition formula *"Customer(customer A)[10+20]"* from the question and get the image of formula (**6**) and the condition formula by the map f as below:

$$f(formula\ (\mathbf{6}), "Customer(customer\ A)[10+20]")$$
$$=Customer(customer\ A)[13+15]\ . \tag{7}$$

And then you get the image of formula (**6**) and the condition formula *"Shared Car[13+15]"* by the map f:

$$f(formula\ (\mathbf{6}), "Shared\ Car[13+15]")$$
$$=Shared\ Car(car2\times[13+14]+(car1+car3)[14+15])\ . \tag{8}$$

From the results, you can know that *customer A can use car2 between times 13 and 14 and can also use car1 or car3 between times 14 and 15.*

Next, if you want to answer the question *"Which customer wants to use car1 from time 8 to 10?"*, you get the image of formula (**6**) and the condition formula *"(Customer+car1)[8+10]"* by the map f:

$$f(formula\ (\mathbf{6}), "(Customer+car1)[8+10]")$$
$$=(Customer(customer\ C+customer\ D)+Shared\ Car(car1))[8+10]\ . \tag{9}$$

From the results, you can know that *customer C, D* want to use *car1 from time 8 to 10.*

Next, if you want to answer the question *"Which customer can use which car between times 8 and 10?"*, you get the image of formula (**6**) and the condition formula *"(Customer+Shared Car)[8+10]"* by the map f:

$$f(formula\ (\mathbf{6}), "(Customer+Shared\ car1)[8+10]")$$
$$=(Customer(customer\ C+customer\ D)+Shared\ Car(car1+car2))[8+10]\ . \tag{10}$$

From the results, you can know that *customer C and D can use car1 or car2 from time 8 to 10*.

6.4 Considerations

Condition formula processing has become more effective by integrating a continuous quantity identifier when a continuous quantity such as time, distance, temperature, etc. is dealt with.

If the existing method of business application development is used in this case study instead of the continuous quantity identifier function of CDS, complicated input/output programs would have to be developed according to needs, and the maintenance costs required to meet various and unexpected user requirements would be considerable.

7 Related Works

The distinctive features of our research are the application of the concept of topological process, which deals with a subset as an element, and that the cellular space extends the topological space, as seen in Section 2. Relational OWL as a method of data and schema representation is useful when representing the schema and data of a database [2][5], but it is limited to representation of an object that has attributes. Our method can represent both objects: one that has attributes as a cellular space and one that does not have them as a set or a topological space.

Many works applying other models to XML schema have been done. The motives of most of them are similar to ours. The approach in [8] aims at minimizing document revalidation in an XML schema evolution, based in part on the graph theory. The X-Entity model [9] is an extension of the Entity Relationship (ER) model and converts XML schema to a schema of the ER model. In the approach of [6], the conceptual and logical levels are represented using a standard UML class and the XML represents the physical level. XUML [10] is a conceptual model for XML schema, based on the UML2 standard. This application research concerning XML schema is needed because there are differences in the expression capability of the data model between XML and other models. On the other hand, objects and their relations in XML schema and the above models can be expressed consistently by CDS, which is based on the cellular model. That is because the tree structure, on which the XML model is based, and the graph structure [3][4][7], on which the UML and ER models are based, are special cases of a topological structure mathematically. Entity in the models can be expressed as the formula for a cellular space in CDS. Moreover, the relation between subsets cannot in general be expressed by XML.

Although CDS and the existing deductive database apparently look alike, the two are completely different. The deductive database [11] raises the expression capability of the relational database (RDB) by defining some rules. On the other hand, CDS is a new tool for data management, and has nothing to do with the RDB.

8 Conclusions

In this paper, we have designed and implemented a continuous quantity identifier and applied it to condition formula processing, which is the main data search function of CDS. A continuous quantity such as time, distance, etc. can be expressed as a factor in Formula Expression and integrated into business logic modeling in business application development. As a result, the cost of cloud application program development can be significantly reduced.

References

1. Kunii, T.L., Kunii, H.S.: A Cellular Model for Information Systems on the Web - Integrating Local and Global Information. In: Proc. of DANTE 1999, pp. 19–24. IEEE Computer Society Press, Los Alamitos (1999)
2. Antoniou, G., van Harmelen, F.: Web Ontology Language: OWL. In: Handbook on Ontologies, International Handbooks on Information Systems, Part 1, pp. 91–110. Springer, Heidelberg (2009)
3. Lukichev, S.: Improving the quality of rule-based applications using the declarative verification approach. International Journal of Knowledge Engineering and Data Mining 1(3), 254–272 (2011)
4. Embley, D.W.: Semantic priming in a cortical network model. Journal of Cognitive Neuroscience 21(12), 2300–2319 (2009)
5. Vysniauskas, E., Nemuraite, L.: Transforming Ontology Representation From Owl to Relational Data. Information Technology and Control 35, 333–343 (2006)
6. An, Y., Mylopoulos, J., Borgida, A.: Building semantic mappings from databases to ontologies. In: Proc. of AAAI 2006, pp. 1557–1560. AAAI Press, Menlo Park (2006)
7. Dolev, S., Schiller, E.M., Spirakis, P.G., Philippas, P.: Strategies for repeated games with subsystem takeovers implementable by deterministic and self-stabilising automata. International Journal of Autonomous and Adaptive Communications Systems 4(1), 4–38 (2011)
8. Mlynkova, I., Pokorny, J.: From Xml Schema To Object-Relational Database – An Xml Schema-Driven Mapping Algorithm. In: Proc. of IADIS International Conference WWW/Internet 2004, pp. 115–122. IADIS Press (2004)
9. Lósio, B.F., Salgado, A.C., do Rêgo GalvÐo, L.: Conceptual Modeling of XML Schemas. In: Proc. of WIDM 2003, pp. 102–105. ACM Press, New York (2003)
10. Mellor, S.J., Balcer, M.J.: Executable UML: A Foundation for Model Driven Architecture. Addison-Wesley, Reading (2002)
11. Arni, F., Ong, K., Tsur, S., Wang, H., Zaniolo, C.: The Deductive Database System LDL++. Theory and Practice of Logic Programming, pp. 61–94. Cambridge University Press, Cambridge (2003)
12. Kodama, T., Kunii, T.L., Seki, Y.: A New Method for Developing Business Applications: The Cellular Data System. In: Proc. of CW 2006, pp. 65–74. IEEE Computer Society Press, Los Alamitos (2006)
13. Ohmori, K., Kunii, T.L.: Designing and modeling cyberworlds using the incrementally modular abstraction hierarchy based on homotopy theory. The Visual Computer: International Journal of Computer Graphics 26(5), 297–309 (2010)

Research on a Novel Piezoelectric Linear Motor Driven by Three Langevin Type Vibrators

He Honglin[1,2], Chen Wenjun[1], and Long YuFan[1]

[1] Nanchang Hangkong Univ, Nanchang, 330063, P.R. China
[2] Zhejiang Univ., The State Key Lab. Of Fluid Power and Mechatronic Systems, Hangzhou, 310027, P.R. China

Abstract. A piezoelectric linear motor is proposed and designed. It employs the first order longitudinal vibration modes of the horizontal bar and vertical bars of the motor's H-shaped stator to drive the particles on the vertical bars moves in elliptical trajectory so as to push the slider moving. The principle of the motor is detailed. The driving cycle of the stator is detailed. The elliptical trajectory has been proved. A dynamical model used to determine the relationships between the working modes and the size of the stator is derived. A finite element analysis models being utlized to calculate the modal frequency and to optimize the sturcture size of the stator has been built with ANSYS software. The optimal structure of the motor is obtained. Three Langevin type vibrators using the d33 effect of piezoelectric ceramics have been designed for the stator, which are employed to excite the working modes of the horizontal bar and vertical bars relatively. It is concluded that the motor proposed in this paper is able to generate larger thrust , and it is of higher positioning accuracy.

Keywords: piezoelectric linear motor; H-shaped stator; design; optimization.

1 Introduction

With the development of the manufacturing, with addtions, the astronautics and aerospace, the requirements of linear actuators becomes increasingly more and higher [1-2]. Of all sorts linear actuators, the most widely used one is the linear motor which takes three main forms. The first form is the one that uses screw to convert the rotational electromagnetic motor's motion to linear motion. It takes the advantages of lower cost and simpler principle, but its accuracy and performance are worse. The second one is the linear electromagnetic motor whose performanc is much better than the first one, but its structure is complicated and it is difficult to be minimized. The third one is the piezoelectric linear motor (PLM), which characterizes of larger power density, fast response speed, diversity. The PLM is of wide application prospect in many fields[3-4] . Many scholars have studied the PLM and they have developed several prototypes[5]. However, according to the related reports, even in today both the theory and techniques of the PLM are not mature. There still exist many problems in PLM e.g. lower thrust, lower efficiency, etc., which influences the promotion of the motor greatly and makes it an important research aspect to improve the motors perfoamnce[8]. Meanwhile, the released PLMs are too fewer to meet the various demands from engineering, which makes it an imperative thing to develop new PLMs. Amied to enrich the PLM types, a motor based on longitudinal modes is proposed in this paper. As we know, several PLMs based on the longitudinal modes have been developed at home and abroad[6-8],and what distinguishes this motor from the others is that it is capable of producing larger thrust and it is of high positioning accuracy.

2   Working Principle of the Motor

As shown in Fig.1. the motor mainly consists of a stator, a slider, piezoelectric ceramics (PZT) stacks, The stator is used to push the slider moving. The PZT are used to excite the working modes of the motor. The ball bearings is employed to guide the slider moving and reduce the frictional resistance. The stator is composed of a horizontal bar (H-Bar) and two vertical bars (V-Bars), which makes the stator be H-shaped, as shown in Fig. 2. All the bars are designed to be Langevin type vibrators which make use of d33 effect of the PZT to excite working modes. There are two grooves on both ends of the H-Bar to be used as the slideways of the V-Bars. The 1st order longitudinal vibration modes of the H-Bar and V-Bars are selected as working modes of the motor. Once the working vibration phase difference between the H-Bar and V-Bar is equal to $\pi/2$, any particle on the V-Bars tracks elliptical trajectory.

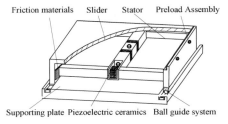 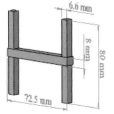

Fig. 1. Structure of the motor   **Fig. 2.** H-shaped stator

The drive principle of the motor is illustrated in Fig. 3. To excite working modal vibration, a sinusoidal signal $E\sin\omega t$ is applied to the H-Bar while the cosine signals $E_1\cos\omega t$ and $-E_1\cos\omega t$ are applied to the V-Bars. Assumed in the time of $t=0$, the ends of the right V-Bar are in the place being farest away from the bar's middle while the ends of the left V-Bar being nearest to its middle and the H-Bar's deformation is zero, it results the right bar to contact with the slider while the left bar not to, as shown in Fig. 3(a). The driving cycle of the motor is described as follows:

In the first quarter of the vibration cycle, the right bar shrinks, its ends move toward the bar's middle, the deformation of the H-Bar decreases from positive maximum to zero to make the right bar keep contacting with slider,while the left bar stretches to increases its deformation from negative maximum to zero to make it separate from the slider. In contrast, the H-Bar stretches to increase its deformation from zero to maximum, which results the right bar pushing the slider in a step of distance λ, as shown in Fig. 3(a).

In the second quarter of the vibration cycle, the right bar keeps on shrinking, its deformation decrease from zero to negative maximum, which results the right bar to separate from the slider. In contrast, the left bar keeps on stretching so as to increase its deformation from zero to maximum to make the left bar keeps on contacting with the slider. Meanwhile, the H-Bar shrinks to increase its deformation from negative maximum to zero, which enables the ends of the left bar to push the slider moving forwards in a new distance λ, as shown in Fig. 3(b).

In the 3rd quarter of the vibration cycle, the right bar stretches outwards to increase its deformation from negative maximum to zero gradually, which makes the right bar's ends separate from the slider. In contrast, the left bar shrinks to decrease its deformation from positive maximum to zero, which enables the left bar to contact with the slider. Meanwhile, the H-Bar keeps on shrinking to decreases its deformation from zero to negative maximum, which results the ends of the left bar pushing the slider moving in the third step of distance λ, as shown in Fig. 3(c).

In the fourth quarter of the vibration cycle, the right bar keeps on stretching to increases its deformation from zero to maximum, which makes the ends of the right bar contact with the slider. In contrast, the left bar keeps on shrinking to decreases its deformation from zero to negative maximum, which results that the left bar separates from the slider. Meanwhile, the H-Bar stretches to increase its deformation from negative maximum to zero, which enables the left bar to push the slider moving forward in the fourth step of distance λ, as shown in Fig. 3(d).

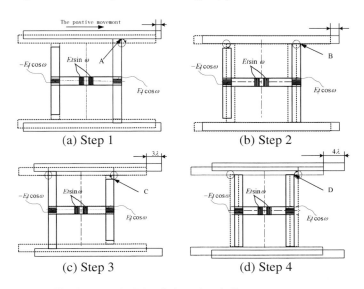

Fig. 3. The principle of piezoelectric linear motor

The stator repeats the drive cycle again and again to pushes the slider moving step by step. Once a cycle is completed, the slider moves in a distance of 4λ. If the phase leading and lagging between the vibration of H-Bar and V-Bar reverses, the direction of the elliptical trajectory reverses, which makes the slider to move backwards.

3 Proof of the Elliptical Trajectory

According to the vibration theory, the first order longitudinal vibration modes of the H-Bar and V-Bars which is of free boundary can be written as follows

$$U_H = A\cos(\pi x/l_h), U_V = B\cos(\pi y/l_v) \quad (1)$$

where A and B are the vibration amplitude of the H-Bar and V-Bar respectively, l_h and l_v are their length. The 1st longitudinal vibration of the H-Bar and V-Bar are

$$X = U_H \sin(\omega t + \alpha), Y = U_V \sin(\omega t + \beta) \quad (2)$$

Eliminating the time parameter in formula (1) and formula(2), yields

$$\frac{X^2}{U_H^2} - \frac{2XY}{U_H U_V}\cos(\beta - \alpha) + \frac{Y^2}{U_V^2} = \sin^2(\beta - \alpha) \quad (3)$$

When $\alpha - \beta = \pi/2$, it is obtained that

$$\frac{X^2}{U_H^2} + \frac{Y^2}{U_V^2} = 1 \quad (4)$$

Apparently, if the V-Bar's ends is determined as the drive surfaces of the stator, any particle on the surfaces will track in a elliptical trajectory.

4 Dynamics Analysis of the Stator

As described above, since the V-Bar can slide through the groove in the H-Bar, the H-Bar longitudinal vibration is not coupled to that of V-Bars. On that account, we divide the H-Bar into two parts and hence build a dynamical model for the H-Bar, as shown in Fig. 4. According to this model, the vibration of the H-Bar can be written as[9]

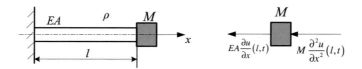

Fig. 4. Theoretical dynamical model of the H-Bar

$$U(x,t) = \left(A\sin\frac{\omega}{a}x + B\cos\frac{\omega}{a}x\right)\sin(\omega t + \alpha) \quad (5)$$

Since the left end of the bar is fixed, the bar's boundary condition takes the form of

$$U(0,t) = B = 0 \quad (6)$$

Substituting the formula (5) into equation (6), yields

$$U(x) = A\sin\left[(\omega/a)x\right] \quad (7)$$

The force exerting on the lumped mass is shown in Fig. 5, and the dynamical balance state of the mass can be written in form of

$$EA\frac{\partial u}{\partial x}(l,t) + M\frac{\partial^2 u}{\partial t^2}(l,t) = 0 \quad (8)$$

Substituting Eq. (7) into the Eq. (8), and considering that $a^2 = E/\rho$, yeilds

$$\frac{\rho Al}{M} = \frac{\omega l}{a}\tan\frac{\omega l}{a} \quad (9)$$

Assumed $\beta = \omega l/a$, the form of the above equation becomes

$$\frac{\rho Al}{\beta M} = \tan\frac{1}{\beta} \quad (10)$$

where E is the elastic modulus, A is the H-Bar's cross-sectional area, M is the V-Bar's mass, l is the half length of the H-Bar. Eq. (10) implied the relationships between the working modal frequency and the size of the stator. It can be solved with numerical method. Eq. (10) is used as the theoretical basis to determine the stators size,

5 Optimization of the Stator

According to the principle of the motor, to generate normal elliptical trajectories on ends of the V-Bar, it requires that the H-Bar and V-Bar be of same working modal frequency, that is the frequency consistency which is related to the mass configuration and size of the stator according to Eq. (10). In order to quantify the relationships, it demands to build a finite element model for the motor. To accomplish this, a 3D solid model of the stator is established with CATIA at first, and then the model is imported to the ANSYS so sa to generate the FEM model, which is used to carry out modal analysis. By varying the bars' size in the FEM model, a series of modal frequencies are calculated. As a consequence, the relationships between the modal frequency and the bar's size can be obtained, which is used to optimize the motor structure and size.

As the explicit relationship between the modal frequencies and the H-Bar's size is being calculated, the cross-sectional size of the H-Bar is selected as variable. Fig. 5(a) gives the caculation results. It could been seen that both the modal frequencies of the H-Bar and V-Bar increase as long as the H-Bar's section size increases, and when the size of H-Bar is 8cm, the modal frequencies of the V-Bars are approximately equal to that of the H-Bars. Similarly, when the explicit relationships between the modal frequency and V-Bar's size is being calculated, the V-Bar's cross-sectional size is selected as variable which varies its value from 6.2mm to 6.9 mm. Fig. 5(b) gives the

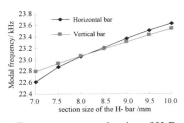
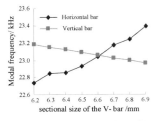

(a) Frequency versus the size of H-Bar (b) Frequency versus the size of V-Bar

Fig. 5. The relationships between modal frequency and the stator's size

relationships between the working modal frequencies and the V-Bar's cross-sectional size. It is can be seen with the increasing of the V-Bar's sectional size the H-Bar's modal frequency increases while the V-Bar's modal frequencies decreases, the section size of V-Bar is more sensitive to the modal frequency of the motor than that of the H-Bar. Based on the above caculations, an optimal structure for the stator is obtained, as shown in Fig. 2, whose modal frequency difference is 10 Hz, which is no more than 0.43% to the modal frequency.

6 Conclusions

A novel piezoelectric linear motor with new principle and structure and H-shaped stator is designed. It utilizes three 1st order longitudinal modes of the bars of the stator to make the particles on the ends of the V-Bar track elliptical trajectory. In order to excite the working modal vibration and increase the vibration amplitude to improve the thrust of the motor, all the bars in the stator are designed to be Langevin type vibrators which make use of the d33 effect of the piezoelectric ceramics directly. The forming of the elliptical trajectory is proved. The optimal structure of bars which can ensures the consistency of the modal frequencies well is calculated and given. It is concluded that this motor is capable of producing larger thrust and of higher position resolution, and it exists application potential in the fields such as precise driving and tiny servo system.

Acknowledgement

This work is supported by NSFC (50865009), Visiting Scholar Foundation of State Key Lab of Fluid Power and Control in Zhejiang Univ.(GZKF-201006) and the Science Foundation of Educational Department of Jiangxi Province(GJJ10024).

References

1. Zhao, C.S.: The technology and applications of the ultrasonic motors. Science Press, Beijing (2007)
2. Hemsel, T., Wallaschek, J.: Survey of the present state of the art of piezoelectric linear motors. Ultrasonics 38(3), 37–40 (2000)

3. Bai, Y.G.: Application research of ultrasonic motor used on precision positioning worktable. Micromotor 41, 85–86 (2008)
4. Janker, P., Claeyssen, F.: New actuators for aircraft and space application. In: Actuator 10th international Conference on New Actuators, Bremen, pp. 324–330 (2006)
5. Uchino, K., Cagatay, S., et al.: Micro Piezoelectric Ultrasonis. Electroceramics 13, 393–401 (2004)
6. Chen, W.S., Li, X., Xie, T.: Application of ultrasonic motor in space detection. Small & Special Electrical Machines 35(1), 42–45 (2007)
7. He, H.L., Huang, G.Q., Zhao, C.S.: Design of a powerful Linear stage driven by two Langevin type stator. China Mechanical Engineering 20(3), 354–357 (2009)
8. Shi, Y.L., Li, Y.B., Zhao, C.S.: Design and experiment of butterfly linear ultrasonic motor with two driving tips. Optics and Precision Engineering 16(12), 772–775 (2008)
9. Xie, G.M.: The Mechanics of Vibration. National Defence Industry Press, Beijing (2007)

Reduction of Production Costs in Machining the Wood Panels MDF Using Predictive Maintenance Techniques

José Augusto Coeve Florino[*] and Leonimer Flávio de Melo[**]

State University Of Londrina UEL, Brazil
jose.florino@usc.br,
leonimer@uel.br

Abstract. The electric motors suffer from efforts differently when subjected to cutting, trimming or finishing and are directly related to the material being machined and edging tool. Choosing the right tool for this operation depends on an expected result. The optimization between energy consumption and the state of the tool can be found to reduce operational costs of production, besides determining the exact time to make the set-up of worn tool. The reduction in operating costs is an item of sustainability that outlines the strategic positioning on companies to become competitive. From a practical way the problems of minimal operating cost and higher productivity will be dealt with in this paper with the moderns equipments of maintenance as power quality analyzer, thermal imager, profile projector and microscope. The result of this paper is the optimization of the cutting operation and energy consumption and optimum point of operation on a given machine.

Keywords: Reduced Operating Costs; Production Management; Machining Operation.

1 Introduction

With the increasing demand to production industries, tend to adopt competitive strategies and sustainability becomes a determining factor in the continued presence in strategic market.

Control of expenditures on raw materials and fixed costs of production are the factors that contribute significantly to the growth of the sales price of the products. It is a strategy of determining company to stay in business, because consumers seek low cost with high quality.

---

[*] Student of Master level in the Department of Electrical Engineering on State University of Londrina, Electrical Engineer, Mechanical Production Engineer and Coordinator the Course of Production Engineering at the University Sacred Heart - USC, Bauru, SP, Brazil.

[**] Electrical Engineer, Master and Ph.D. in Electrical Engineering and Professor of Electrical Engineering Department at State University of Londrina - UEL, Londrina, PR, Brazil.

One theory can be seen in Fig. 1 which shows the old and current views on the sale price of the product. Currently the market demands of processing industries reduce their fixed costs of production in order to maintain an acceptable profit margin [1].

Fig. 1. Theorem of the equations of the sale price of the products [1].

Assuming operating costs within the cost of implementing an automated system, the time of return, commonly called the payback will be in a shorter time. Thus enabling the manager has focus on new investments with a short time of payback.

The money significance in an operation that involves costs on financial decisions can be measured by techniques [7].

The cutting tools used in industry are part of a manufacturing process which adds value to a particular raw material, making a simple piece to wood into an object to use by the consumer market.

The Quick-Change Tool (QCT) may be a differential to increase the equipment availability [2].

For ensuring stability of the process should be avoided breakage of machinery or variations in production [3].

One of the factors that cause downtime of the production process is the tool wear; one should combine the stops to TRF with small interventions of maintenance [3].

If not, the production capacity will be compromised [4].

1.1 Objective

Search of the determining factor of excessive expenditure of energy on a particular machine.

Create a methodology to minimize costs, focused on low power consumption and lifetime of the correct tool, in order to production a leaner and more profitable.

This paper will help determining the need to stop to set-up, the parade is based on the analysis of motor behavior.

2 Technology Applications

The process of cutting operation was studied in sheets of MDF with a thickness of 15 mm of eucalyptus, to be produced pieces of sheets in the desired sizes.

The motor responsible for movement of the saw is triphasic induction type, with 7.5 hp potency, operating voltage of 220vac and rotation of 3600 rpm.

The saw which is being utilized in this process has 60 teeth, diameter of 250 mm and material HW (hard metal).

2.1 Labor Development

Observing the tools life in the productive sector, it is determined in an intuitive way, every 15 days it is replaced.

Cognitively process of analysis and decision making for determining the moment of effecter set-up the machine did not meet one or methodology a scientific basis for it.

Also is not taken into account the volume of production, because it oats: it may be cut too much in a period and not in another.

The Fig. 2(a) shows a defect classified as apparent quality by the production sector due to the large volume of production. Fig. 2(b) shows the pattern of cut desired.

This defect is due to wear of the cutting tool, specifically in the state of sharpening of their teeth.

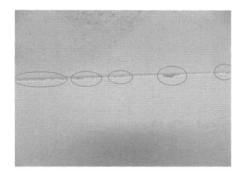

(a) Edge of a chipped MDF.   (b) Edge of a sheet of MDF in perfect condition after the cut.

Fig. 2. Edge of MDF

The apparent quality of the cut is compromised due to the morphological change of the cutting edges of the teeth of saws.

When the apparent qualities, of the final products are starting to be compromised, the line speed of production is decreased.

The problem raised in this activity is to decrease the sector's productive capacity and consequently a reduction in production.

2.2 Production Costs

Production costs related to sharpening existing tools and the acquisition of new tools for the machining of reconstituted wood panels of the type MDF were raised.

The cost of the tool HW is R$ 215.00 for sharpening your tooth is R$ 0.30 which corresponds to R$ 18.00 per sharpening. This tool supports up to 13 sharpening, after that it should be discarded, because its geometry is compromised due to withdrawal from the sharpening process.

3 Parameters Achieved

We measured various parameters such as voltage, current, power, temperature of the motor contacts, and temperature during the cutting.

The values obtained with the infrared thermal imaging camera were simply to determine if some discrepancy in the engine power to accuse an unusual change, a problem that has not happened.

In Fig.3 we can found an increase in current due to the increase of cut pieces.

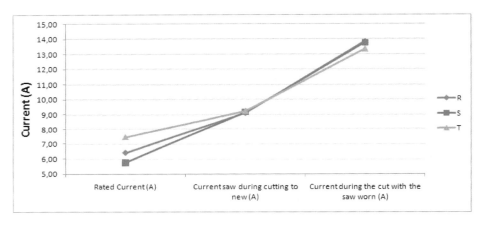

Fig. 3. Comparison the currents.

(a) Measuring the cutting edge after use.   (b) Measuring the cutting edge before use.

Fig. 4. Measuring the cutting edge.

This current increase is related to the morphological change of the status of the cutting edges of the tool, because it undergoes changes as the accumulation of pieces.

The angles were measured before and after and have not changed; the profile projector was used for this result.

The change that occurred was in the morphology of the tool cutting edge, as can be seen in Fig. 4, obtained through a microscope.

With increasing current, the power consumption also increased. It was determined using 1 hour per day and 20 days per month with the price of a kilowatt to R$ 0.36.

4 Conclusions

There is a moment that the current and potency began to rise, because of increased engine torque so that the tool will be attrition, that point is where the engine began to be asked more potency, behaving as a large consumer of electric power potential and exceeding the cost of edging tool.

Illustrated in Fig. 5 is the consumer price of electricity which was related variables results collected during the study.

In the same figure is related, future value being increased the amount of sharpening being graphically inverted, according to the consumption of electricity, so at the point where the two graphs intersect is the optimum point of operation.

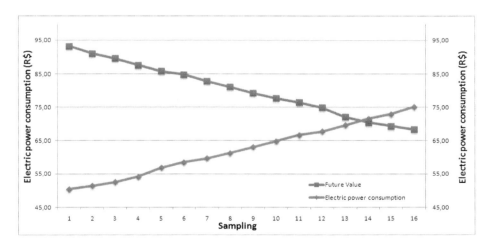

Fig. 5. The optimal point of operation.

The optimal point of operation is between samplings 13 and 14. In sample 13, the value of the operation has not yet reached the value of spending edging tool. As for the 14 sampling the value of the operation exceeded the price cutting edging tool.

In these situations were not submitted the apparent defects in cut pieces, but spending on electricity consumption justify replacement. What happens in this situation is a false perception that the process is more economical.

The study can be a didactic material based on the determination of the optimal point of production machinery manufacturing mechanical machining of materials.

Every machine or equipment must have its individual study, since it must be considered the particularities of each case.

With a great purpose, not only in electricity consumption, the apparent quality of the final product is no longer the only parameter for determining the replacement of worn tool.

The optimal point of operation was determined before it that the apparent quality of the products began to be compromised, so the taken of managerial decisions can be taken parameterized.

One way that can be adopted for determining the optimum point of operation and a machine or equipment is to install an ammeter in line for determining the time for a replacement tool.

The current engine operating nominally should be noted, with the new tool and then with the worn tool. Thus a comparison can be made and parameterized with the studies in this work.

References

1. Smith, T.F., Waterman, M.S.: Identification of Common Molecular Subsequences. J. Mol. Biol. 147, 195–197 (1981)
2. Slack, N., Chambers, S., Johnston, R.: Administração da Produção, Editora Atlas, Segunda Ediçãao, SP (2008)
3. Leão, S.R.R.D.C., dos Santos, M.J.: Aplicação da Troca Rápida de Ferramentas (TRF) em Intervenções de Manutenção Preventiva. Revista Produção Online 9, 1676–1901 (2009)
4. Ishikawa, K.: Controle de Qualidade Total à Maneira Japonesa; Total Quality Control in Japanese Manner, Editora Campus, SP (1993)
5. Ferraresi, D.: Fundamentos da Usinagem dos Materiais, Edgard Blücher, SP (1977)
6. Filho, J.M.: Instalações elétricas industriais, Livros Técnicos e Científicos, SP (2007)
7. dos Reis, L.B., Cunha, E.C.N.: Energia Elétrica e Sustentabilidade: Aspectos Tecnológicos, socioambientais e Legais, Manole Ltda, SP (2006)
8. Gitman, L.J.: Princípios de Administração Financeira: essencial, Bookman, SP (2006)

Minimum Step Control System Design

Youmu Zhang and Dongyun Luo

Jiujiang University, College of Electronic Engineering, Jiujiang, China
dz_zhangyoumu@jju.edu.cn, ldongyun2001@126.com

Abstract. Minimum step control system includes a pattern control system and no pattern control system, it belongs to one of the discrete control system, which aims to enter in the 3 typical digital controller is designed under the system and take the minimum and limited the number of end Control process, and do not in the sampling error of the moment, so take control of the minimum number of control system design in the design of this controller design is an important part of the design. In the digital control system which, in the complex plane Z domain to achieve a stable, non-error control can be achieved.

Keywords: Minimum film; control; ripple; non ripple.

1 Introduction

In the automatic control system, as long as the deviation of the time there is always hope in the shortest possible time, to eliminate bias, when the input has changed, the output follows the input changes, then the limited sampling time in which the control system to achieve balance. Typically, a digital control process is called a beat sampling period. Beat control is the smallest system in the typical input signal (such as accelerometers, signal step signal, speed signal, etc.) under the action of at least one sampling period after the system output steady-state error is zero[1-2] . Therefore, the minimum time step control is actually the least control. Beat control system design task is to design a digital controller, allowing the system to reach at least the time needed to stabilize and control the output of the sampling period can accurately track the value of the input signal, there is no static error. Between any two sampling period of the course requirements will not do. Beat control system design for minimum requirements are: the shortest settling time, the system tracks the input signal at least the required number of sampling period; closed-loop system must be stable; in the sampling point no static error, ie, a specific reference input signal, After steady state is reached, the system samples the input signal to achieve accurate tracking. Digital controller designed to be physically achievable. Figure 1 shows this design principle at least beat the system block diagram of a ripple[3-4].

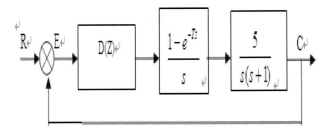

Fig. 1. Principle of at least beat the system block diagram of a ripple

2 Minimum Beat Control System Design

2.1 Minimum Beat Determination of Closed-Loop Pulse Transfer Function

First, the control system based on performance requirements and other constraints, structural system of the closed-loop pulse transfer function. Minimum design requirements for making the control system is a specific reference signal, the system reaches steady state, the system static error in the sampling point is zero. This constraint can be constructed according to the system pulse transfer function of error. A typical computer control system structure shown in Figure 2.

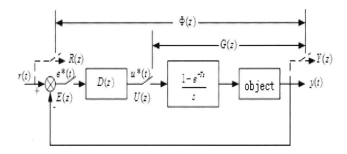

Fig. 2. The minimum step control block diagram of typical control

By the discrete control theory, at least beat the error control system pulse transfer function

$$\varphi_e(Z) = \frac{R(Z)}{E(Z)} = 1 - \phi(Z) \tag{1}$$

$$E(Z) = \varphi_e(Z) * R(Z) \tag{2}$$

Typical of the general control system has three input forms:

$$R(z) = \frac{1}{1-z^{-1}} \tag{3}$$

$$R(z) = \frac{Tz^{-1}}{(1-z^{-1})^{-2}} \tag{4}$$

$$R(z) = \frac{T^2 z^{-1}(1+z^{-1})}{2(1-z^{-1})^3} \tag{5}$$

Enter the above expression can be expressed as:

$$R(Z) = \frac{A(Z)}{(1-Z^{-1})^m} \tag{6}$$

2.2 Took Control of the Minimum Hardware Design of a Pattern

There is also overflow computing processing, when the computer control output over 00H-FFH (corresponding to the analog-5V-+5V), the computer outputs the corresponding extreme value 00H or FFH. After controlling the amount of each calculation, the computer output immediately, and will adopt various errors into the calculation of the output with the various operations for the delay, and finally as part of the next output control calculation [5]. So that when taken into the next error signal, it can reduce the computation times, thereby reducing the computer's pure delay time. Charged object has analog circuit simulation, because the circuit in the resistor, capacitor parameters have a certain error, it should be setting, can be setting the first order inertia, and then setting the integrator should be made both in series as close as possible given mass function of the mathematical model. Constitute at least making a ripple system shown in Figure 3, the A1, A2, A3, A4, A5, B1 and B5 cell formation. Sampling period T = 1S, A5 of the time constant of 1, A6 of the time constant of 1, K = 5. For observations in the experiment results, as long as running LCAACT program, select the appropriate computer control experiment under the menu item, then select start the experiment, the interface will pop up a virtual oscilloscope, click on Start will automatically load the appropriate source file, this time can use an ordinary oscilloscope can also use virtual oscilloscope (B3) cells of the CH1, CH2 measure hole measured waveforms, some of the virtual oscilloscope. Note: The observation, if the user presses the experimental board "RST" key, you must repeat the process in order to continue observation[6]. In general, the design for a typical input function obtained closed-loop pulse transfer function systems, for the number of lower input function, the system will be a large overshoot, system response time will increase, but the sampling time the error is zero; when the input function for a typical system design be used for closed-loop pulse transfer function of the number of high input function, the output will not fully track the input, resulting in a static error. For a given system, the pulse transfer function of controlled object G (z) contains a term, T is sampling period, due to design features at least beat the system is to require the fastest response, the system transforms the input signal is relatively poor ability to adapt, the output response only guarantee zero error on the sampling points, sampling points can not ensure that the error between the values of zero, that is, at least beat the system, the system output response in the presence ripple between sampling points. Because of the ripple generated in the zero-order hold input terminal, that is, by sampling the output of digital controller reach a relatively stable after the switch, and thus make the system output fluctuations between sampling points; if the input error E (k) is zero, the input pulse sequence to maintain a constant value, then the output C (t) will not arise between non-sampling ripple. Since G (z) contains items, the sampling cycle changes also have an impact on the output ripple, if T tends to infinitesimal, then the discrete system into continuous system, there will be no ripple, so a person accused of the same object G (s), the digital controller of the control algorithm D (z) design is also closely related with the sampling period. Since the control algorithm design is theoretical, because of the controlled object in the actual discrete parameters, making the G (s) and G (z) and the actual system error.

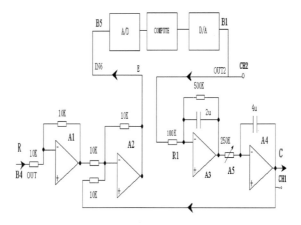

Fig. 3. Taken with the minimum hardware design pattern

2.3 Took Control of the Minimum Hardware Design without Pattern

Servo system for at least beat the adaptability of the input signal is poor, the output response can only guarantee the error of the sampling points 0, can not ensure that the error between the sampling points for 0. In other words, there is the ripple between sampling points. Output ripple not only caused the error, but also the executing agency of the drive power consumption, increased mechanical wear. Minimum beat ripple-free design, in addition to eliminating the ripple between sampling points, but also to some extent reduce the control energy, reducing the sensitivity of the parameters, but it is still designed for a particular input on other input adaptation is still not good. Constitute at least beat ripple-free system as shown, from A1, A2, A3, A4, A5, B1 and B5 cell formation. Sampling period T = 1S, A5 of the time constant, A6 of the time constant, K = 5. For observations in the experiment results, the available general scope, also can choose the test machine supporting virtual oscilloscope. If you use a virtual oscilloscope, just run LCAACT program, select the appropriate computer control experiment under the menu item, then select start the experiment, the interface will pop up a virtual oscilloscope, click on Start will automatically load the appropriate source files, you can use the test machine supporting the virtual oscilloscope (B3) cells CH1, CH2 porosimetry measurement waveform. Experimental specific usage instructions see chapter II, part of the virtual oscilloscope. Note: The observation, if the user presses the experimental board "RST" key, you must repeat the process in order to continue observation[7].

3 Experimental Results

At least beat the system, the system output response in the presence ripple between sampling points. Because of the ripple generated in the zero-order hold input terminal, that is, by sampling the output of digital controller reach a relatively stable after the switch, and thus make the system output fluctuations between sampling points in figure 4.

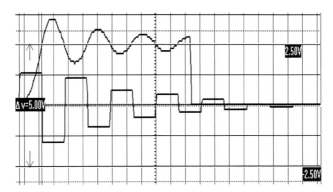

Fig. 4. The smallest grain control results beat

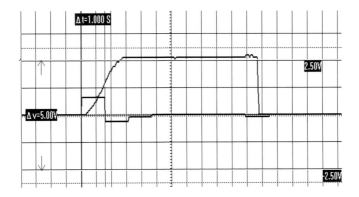

Fig. 5. The minimum step results without pattern control

Minimum beat ripple-free design, in addition to eliminating the ripple between sampling points, but also to some extent reduce the control energy, reducing the sensitivity of the parameters in Figure 5.

References

1. Li, H.: The computer control system, vol. (5), pp. 100–105. Machinery Industry Press (2007)
2. Clerk Maxwell, J.: In the marine and computer control system, vol. (6), pp. 71–82. Machinery Industry Press (2007)
3. Xie, J.-Y.: Micro computer control technology, vol. (9), pp. 76–85. National Defence Industry Press (2001)
4. Tang, W.: IPC particle flux in the pre-melting material ratio of. Anshan Iron and Steel Technology (3), 66–67 (2001)
5. Maxwell: Computer monitoring system based on single-chip digital filtering technique. Anhui Institute of Electronics and Information Technology (5), 100–102 (2006)
6. Yorozu, Y., Hirano, M., Oka, K., Tagawa, Y.: Electron spectroscopy studies on magneto-optical media and plastic substrate interface. IEEE Transl. J. Magn. 2, 740–741 (1987), Digests 9th Annual Conf. Magnetics, Japan, p. 301 (1982)
7. Wu, X.: The computer system of comprehensive monitoring treatment. Anhui University of Technology (Natural Science Edition) (3), 33–35 (2003)

Handling Power System Alarm Cascade Using a Multi-level Flow Model

Taotao Ma[1], Jinxing Xiao[1], Jingqi Xu[1], Chuangxin Guo[2], Bin Yu[2], and Shaohua Zhu[3]

[1] SMEPC Qingpu Power Supply Company, Qingpu District 201700, Shanghai, China
[2] College of Electrical Engineering, Zhejiang University, Hangzhou 310027, Zhejiang Province, China
[3] Jiaxing Electric Power Bureau, Jiaxing 314033, Zhejiang Province, China
mtt3220@163.com

Abstract. In power grids, a severe fault often leads to several consequent alarms which is called alarm cascade phenomenon. This paper presents a novel method to handle the alarm cascade generated on the energy manage system (EMS) computers of power grids. The method consists of a multi-level flow model (MFM) and heuristic rules, in which the MFM is used to present the relationships between different alarm events while the heuristic rules are employed to pretreat the alarm events, perform message synthesis and indicate the cause of alarm cascade. The method can overcome the disadvantages of the traditional rule-based alarm processing solutions based on limited human experience or need additional training work to generate rules. Also MFMs are geographic modeling languages which do not involve complex calculation. Furthermore the proposed method has been evaluated via the simulation studies which are undertaken based on 2-substation system, IEEE 9-bus, 14-bus and 39-bus system respectively.

Keywords: Alarm processing, Multi-level flow model, Heuristic rule.

1 Introduction

In a power grid, a severe and complex fault often leads to several consequential faults resulting in alarm events. Normally, alarms arrive irregularly, depending on the power system topology, alarm settings and current loads. Therefore it is difficult for operators to analyze fault developing states, and to take actions to alleviate the states or perform an effective recovery of the grid in time. An alarm cascade is the most difficult problem to handle, as it could appear in the situation when the correct alarm action is expected. So alarm processing is necessary to be used to filter false or unimportant alarm events generated on EMS computer and detect the originating events in alarm cascade.

First of all, the number of alarms occurring in a power system simultaneously when a fault occurs is a crucial issue to concern. Once the Hydro-Québec Regional Control Center reported the maximum number of alarms which could be triggered by several types of events as follows (Durocher, 1990):

- up to 150 alarms in 2 seconds for a transformer fault;
- up to 2000 alarms for a generation substation fault, the first 300 alarms being generated during the first 5 seconds;
- up to 20 alarms per seconds during a thunderstorm;
- up to 15000 alarms for each regional center during the first 5 seconds of a complete system collapse.

As a result, attempts have been made to find the way of reducing the number of alarms presented to power system operators. It has been suggested in (Chan, 1989) that the alarm filtering could be treated as a pattern recognition problem and that neural networks could be used to identify the cause of these alarm events. Also expert systems could be employed to handle alarm events as proposed a decade ago (Bijoch et al, 1991; Li et al, 1996). The first result, concerning alarm filtering and fault diagnosis, obtained using expert system techniques was reported in (Wollenberg, 1986). The development of an intelligent alarm filter was reported recently for Portuguese substation control centers using a knowledge-based method (Vale, 1997). In conventional expert systems, rules are established using the knowledge of power system experts. However one of the disadvantages for the rule-based expert system is that the provision of the comprehensive knowledge cannot be guaranteed. The model-based approach can be used to solve this problem. A multilevel flow model based approach was investigated to handle status change alarms (Roudén, 2008). In this work, a model that included power flow and logic calculation was used to establish the relationship between the continuous-time and status-change alarms. However, model based methods has their own weaknesses, such as they are difficult to model message synthesis procedure and give proper suggestions for power system operators.

In this paper, utilizing a multilevel flow mode and heuristic rules forms a hybrid model to solve the above problems. By using the model based approach, the completeness of the knowledge can be sure, while using the heuristic rules will ensure the flexibility. Message synthesis is performed by heuristic rules, while the MFM is used to obtain the relation between alarm events.

2 Basic Concept of MFM

Detailed description of MFM is presented in (Larsson, 1994; Larsson, 2002; Lind, 1990). MFM is a graphically represented modeling approach, in which the intentional properties of a complex system can be described.

The basic concept types of MFM are: goals, functions, and physical components. The purposes of the system and its subsystems are modeled with goals. The functions are the means by which the goals are obtained. The physical components are equipments of the system. In MFM, different types of relations, such as achieve relations, condition relations and realize relations, are used to describe the relationship among goals, functions and components.

Function is the important concept of MFM. MFM describes the functional structure of a system as a set of flow structures, which consist of connected flow functions. In MFM the flow structures include mass flows, energy flows and information flows. The function symbols of MFM are shown in Fig. 1.

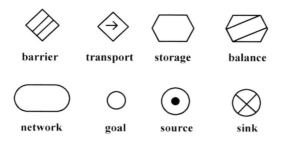

Fig. 1. The functions symbols of MFM model

The main flow function types are sources, transports, storages, balances, barriers, and sinks, and they describe either mass or energy flows. Observers, decision makers, and actors describe information flows. The manager function describes control systems. Each flow network can be connected to one or several goals via achieve relation, which means that functions in the network achieve the goal. A goal may be connected to one or several functions via condition relations, meaning that the goal is condition for function.

3 Construction of MFM for Power Grid

In Fig. 2, we see IEEE 9-bus system, with generators, bus, transmission line, and load.

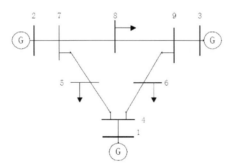

Fig. 2. The topology of IEEE-9 bus system

The goals of the system are described as:

- G1: Produce energy;
- G2: Deliver energy ;
- G3: Transfer energy;
- G4: Consume energy.
  The functions are divided into:
- F1: Produce energy;
- F2: Mix and Deliver energy;

- F3: Transfer energy;
- F4: Use energy.

The third types of objects are the physical components:
- C1: Generator;
- C2: Bus;
- C3: Transmission line;
- C4: Load.

These are the sets of goals, functions, and components. However, the relations between these objects are important. First, the goal G1 is superior to G2 and G3, i.e., the latter are sub goals of G1. Thus, there is a goal hierarchy, formed by goal sub-goal relations. These are also relations between goals, functions, and components. For example, the generator component is used to realize the function of produce energy. In Fig. 3, both the goal hierarchy and the means-end relations are shown in a graph.

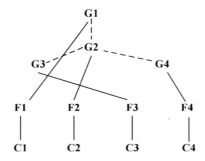

Fig. 3. The relationship of goal, function, and component in IEEE-9 bus system

In an MFM model, the goals, functions, and relations are represented in a graphical language. In Fig. 4, the MFM model of IEEE 9-bus system is presented.

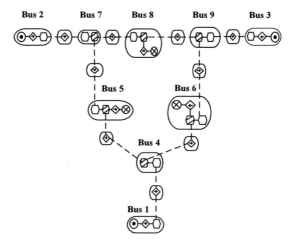

Fig. 4. The MFM model IEEE-9 bus system

The MFM model consists of some basic objects as mentioned above. Also Basic components, such as generators, buses transmission lines and etc, form a complex power grid. So the MFM model of a certain power grid can be automatically built from the existing diagram in EMS database, which represents the physical grid topology. The automatic model generation ideal is described shortly as:

The model can be obtained automatically by adding model fragments and connecting them. And new model fragments can be created, if new grid components are added. The information needed can be fetched from the grid topology database of EMS. Each time the grid and database is updated; the automatic generation process is triggered.

4 Consequence Propagation Rules of MFM Functions

Larsson (Larsson, 1992) introduced general consequence propagation rules for describing the causality between different connected flow functions. These rules apply to all systems modeled in MFM. The rules used in MFM model of power grids are as follows.

Rule 1: A source *low capacity* will force the connected transport to have a *low flow*.

Rule 2: A transport *low flow* may cause a storage connected at the inlet of the transport to have *high volume*, and a storage connected at the outlet to have a *low volume*. It may cause another transport connected in the same direction via a balance to have a *low flow*. If the balance has no other connections the same alarm will be forced.

Rule 3: A transport *high flow* may cause a connected source or sink to have a *low capacity*. It may cause a storage connected at the inlet of the transport to have a low volume, and a storage connected at the outlet to have a *high flow*. It may cause a transport connected in the same direction via a balance to have a *high flow*. If the balance has no other connections, the same alarm will be forced. It may cause another transport connected in the opposite direction via a balance to have a *low flow*.

Rule 4: A storage *low volume* may cause an outgoing connected transport to have a *low flow*.

Rule 5: A storage *high volume* may cause an incoming connected transport to have a *low flow*, and it may cause an outgoing connected transport to have a *high flow*.

Rule 6: A storage *leak* may cause the same storage to have a *low volume*.

Rule 7: A storage *fill* may cause the storage to have *high volume*.

Rule 8: A balance *leak* may cause a connected outgoing transport to have a *low flow*, and a connected incoming transport to have a *high volume*.

Rule 9: A balance *fill* may cause a connected incoming transport to have a *low flow* and a connected outgoing transport to have a *high flow*.

A sink *low capacity* will force the connected transport to have a *low flow*.

As show in Fig.5, the system produce two alarm events, *loflow*(transport) and *hivol*(storage). In Fig.5, we can see that the storage is in the outgoing direction of transport.

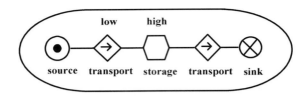

Fig. 5. Example of alarm events relation analysis by MFM judgment logic

According to Rule 2, it should be *loflow* (transport), and then the outlet storage produces a *lovol* (storage) alarm. With Rule 3, the result is that *hivol* (storage) causes the incoming transport *loflow* (transport) alarm which is the same as the actual alarm situation. So the *hivol* (storage) alarm is the primary alarm, and then the *loflow* (transport) alarm is the consequent alarm.

5   The Algorithm for Handling Power System Alarm Cascades

The algorithm for handling power system alarm cascades consist of alarm relation analyst methods and heuristic rules.

The alarm relation analyst method are based on the consequence propagation rules of MFM functions above, which is important part of MFM model language. And it can be applied in many industry areas for both known and unknown alarm states.

EMS is the main source of alarm events in control centers. Given a set of alarms, it is possible to decide which of the alarms that must be primary ones, and which ones that may be secondary. It is important to observe, however, that one cannot be certain that a fault is indeed secondary; there might be multiple faults. Thus, the method will differentiate between positively primary alarms, and alarms that may be either primary or secondary.

As soon as new alarm value is discovered in EMS, such as transmission lines is overload, the corresponding alarm of the corresponding alarm of the concerned flow function is set to an alarm value. Then all rules that can be applied to the new alarm are tried, in order to see if they match the EMS alarm situation. If so, the primary failed component and primary alarm event can be found out.

The alarm relation analyst method is described in detailed. When alarm events arrive in EMS, signing a high or low value for the corresponding functions of the MFM model, and then treat the signed function as the leaf node, use the tree searching method to traverse the tree. The traversed algorithm is as follows:

Step 1: Search downward for a leaf node

Step 2: If the leaf node is a function signed a alarm value (high or low), perform the consequence propagation down to the leaf node.

Step 3: If the root node is now considered a primary alarm for this path then perform the consequence propagation down to the leaf node.

Step 4: Search downward for a new leaf node (choose another alarm event) and repeat from 2 until all paths have been analyzed.

Using IEEE 9-node system above as an example, F1, F2, F3 ... Fn are short for the functions in Fig. 4. When transmission line 2-7 (between bus 2 and bus 7), has an overflow alarm, signing the corresponding transport function with a high value. In Fig. 6 treating this function (F4) as the leaf node, traverse the tree until the primary alarm is found out or the entire path has been analyzed.

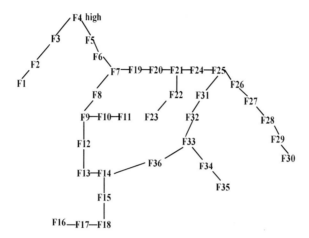

Fig. 6. The searched tree based on MFM judgment logic

The heuristic rules are used to pretreat the incoming alarms from EMS and synthesize alarm messages. These heuristic rules are described as follows:

Rule1: The alarms of EMS are classified as communication outage, EMS itself working state fault, digital values such as switch or breaker state change, and analog values such as current, voltage limit exceed. Only the alarm events concerned with power grid are fetched from EMS.

Rule2: Identify the cause of the alarm events by MFM.

Rule3: If the topology of a power grid is changed, the MFM should be updated.

Rule4: If the alarm event received is status change alarm of a switch or a breaker, then create a relation between the switch or breaker and the line where the switch or breaker is located.

Rule5: If the cause of the alarm cascade cannot be identified, then wait for more alarm events from EMS.

Rule6: If the time has passed for 10 seconds after the cause alarm is identified by multi-level flow model as in rule1, then treat the alarm events as the consequences of the cause alarm.

Rule7: If the time has passed for 10 seconds since the last alarm event is received from EMS, then synthesis process is undertaken for the alarm events stored in the alarm processor database.

6 Case Studies

Using power system analysis toolbox, PSAT, as software tool, different faults are simulated in IEEE 9-bus system, IEEE 14-bus system, and IEEE 39-bus system. In IEEE 9-bus system, when three phases to ground fault happened, it produces 6 to 11 alarm events related to location of lines.

For example at t = 0, three phases to ground fault occurred on bus 7, circuit breaker opens and removes the source of failure at t = 1.1s, circuit breaker 1 closes at t = 4s, during this procedure a series of alarm is producing, as in Table 1.

Table 1. Alarm events in IEEE 9-bus system while bus 7 three phase to ground fault occurs

No.	Time	Event
1	0.5	Bus 7 voltage exceeds the lower limit
2	0.65	Line 5-7 power flow over the upper limit
3	0.7	Line 7-8 power flow over the upper limit
4	1.1	Break 1 closed to open
5	0.9	Bus 5 voltage exceeds the lower limit
6	0.9	Bus 6voltage exceeds the lower limit
7	0.9	Bus 8 voltage exceeds the lower limit
8	0.9	Bus 1 reactive power exceeds the upper limit
9	1.1	Bus 2 reactive power exceeds the upper limit
10	1.1	Bus 3 reactive power exceeds the upper limit
11	4	Break 1 open to closed

Using the method discussed in this paper, bus 7 voltage exceeds the lower limit is the primary alarm, and three phase to ground fault on bus 7 is figured out.

Single phase to ground, phase to phase, and transmission line broken faults are simulated in different locations of IEEE 9-bus system. In each fault situation, alarm events are processed by the hybrid method presented here. 320 different faults simulations are performed, the method figures the primary alarm successfully for 317 times. In IEEE 14-bus system, 430 different faults simulations are performed, the method figures the primary alarm successfully for 428 times. In IEEE 39-bus system, 625 different faults simulations are performed, the method figures the primary alarm successfully for 623 times. The testing results for each system are in Table 2.

Table 2. Successful rate for alarm cascade handling

Test system	Successful rate
IEEE-9 bus	99.06%
IEEE-14 bus	99.53%
IEEE-39 bus	99.68%

This method can be also used inside substations. A transmission network and two substations system are show in Fig. 7.

Handling Power System Alarm Cascade Using a Multi-level Flow Model 173

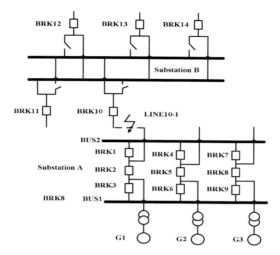

Fig. 7. The topology of two substation system

Different fault patterns are applied on the substations and network: protection relay missing operate; transmission line to ground transient fault; transmission line to ground faults; transmission line to ground fault plus primary protection relay refuse to act; transmission line to ground fault plus one breaker refuse to open; breaker missing operate. The MFM model of the transmission network and two substations system is presented in Fig. 8.

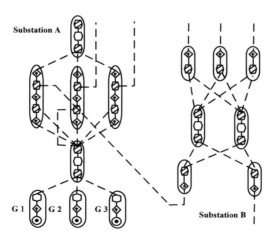

Fig. 8. The MFM model of two substation system

When an transmission line to ground fault happens on line 10-1, plus refuse to open fault on break 6. The alarm events generated are show in Table 3.

Table 3. Alarm events before process

No.	Time	Event
1	0.015	Relay 10 act
2	0.015	Break 3 closed to open
3	0.021	Break 9 closed to open
4	0.022	Relay 11 act
5	0.022	G2 disconnect

The hybrid method of MFM model and heuristic rules is applied to solve the problem. The result by primary alarm diagnosis is that at time 0.022 G2 disconnect; and the synthesis result is that Bus 1 fault and Break 6 refuse open. Other testing results also show the diagnosis method correctly detected the primary alarm and the synthesis result gives advice for fault diagnosis. The successful rate is about 99.33% (298/300).

7 Conclusions

Alarm processing is an very important issues of power grid monitoring and control, especially when alarm cascade happens in power grid. Lots of solutions have been suggested; however each of them has its own disadvantages, such as the problem of completeness of knowledge base and expert system. MFM has the advantages a relatively easy knowledge engineering task due to graphical and highly abstract nature. Research with the MFM area is maturing. But there is few method based on MFM being used in power grid industry practice.

In this work, how to build a MFM model of power grid is introduces carefully. For the topology diagram most EMS provided is in SVG format, by using proper XML parser, such as Xerces, an MFM model can be easily generated, also the alarm status of corresponding power system component can be fetched for EMS.

The method discussed in this work is based on consequence propagation rules of MFM functions and heuristic rules. The tree traversed algorithm also has impact on the performance of the method.

In the future work, developing a alarm process based on the method mentioned here for industry test is considering. Furthermore the tree traversed algorithm may be improved for better performance.

References

1. Durocher, D.: Langage: An expert system for alarm processing. In: Eleventh Biennial IEEE Workshop on Power Systems Control Centers, pp. 19–21 (1990)
2. Chan, E., Application, H.P.: of neural-network computing in intelligent alarm processing. In: Proc. Power Industry Computer Applications Conf., pp. 246–251 (1989)
3. Bijoch, R.W., Harris, H.S., Volkmann, T.L., et al.: Development and implementation of the NSP intelligent alarm processor. IEEE Transaction on Power System 6(2), 806–812 (1991)
4. Li, H.P., Malik, O.P.: An expert system shell for transmission network alarm processing. In: IEE Power System Control and Management Conference, pp. 143–147 (1996)

5. Wollenberg, B.F.: Feasibility study for an energy management system intelligent alarm processor. IEEE Trans. Power Systems 1(2), 241–247 (1986)
6. Vale, Z.A., Moura, A.M.: Sparse: An intelligent alarm processor and operator assistant. In: IEEE Expert AI in Power System, pp. 86–93 (1997)
7. Roudén, K., Kraftnän, S., Larsson, J.E.: Real-time Detection of originating events in large alarm cascades. In: CIGRE 2008, C2-104 (2008)
8. Larsson, J.E.: Diagnostic reasoning strategies for means-end models. Automatica 30(5), 775–787 (1994)
9. Larsson, J.E.: Diagnostic reasoning based on explicit means-end models: experiences and future prospects. Knowledge-Based Systems 15(2), 103–110 (2002)
10. Lind, M.: Representing goals and functions of complex systems-an introduction to multi-level flow modeling. Institute of Automatic Control Systems, Technical University of Denmark, Technical report, 90-D-38 (1990)
11. Larsson, J.E.: Knowledge-Based Methods for Control Systems, Doctor's thesis. Department of Automatic Control, Lund Institute of Technology, Lund, TFRT-1040 (1992)

Design and Implement for Information System Intelligent Monitoring System of Power Grid

Lei Huang, Taotao Ma, and Jufang Li

Qingpu Power Supply Company, SMEPC, Qingpu District 201700, Shanghai, China
mtt3220@163.com

Abstract. Information system intelligent monitoring system (ISIMS) can be used to monitoring and manage data which are separate in different information systems in power grid. With the help of ISIMS, the unique of data source can be ensured. The data and power grid topology graphics form the accuracy model of the power gird. Engineers can easily query the data once decentralize in various information systems in ISIMS. Furthermore the application functions of ISIMS can help to improve the monitoring and management capabilities of various power gird information systems.

Keywords: power system; information system; intelligent monitoring.

1 Introduction

More and more micro-computer based equipments are being used in power grid, such as micro-computer aided relay protections, fault recorders, and safety control devices. These information devices makes power grid running more automatically and ensure the safety, efficiency, and stability of power grid. However all these devices are running individually, the data are stored separately in different power grid monitoring and control systems, such as SCADA, EMS, GIS and MIS, and also the functions of these devices are seldom integrated. Furthermore as the power grid becomes larger, the types of these micro-computer aid devices are various, which increase the maintaining burden.

The operation conditions of power grid are various, especially when fault happens, the phenomena is quite complicated. Information devices play an important role in power grid operation and control. However in traditional management strategy, the working states of information devices sometimes are neglected. To confirm the exact working states, engineers need to go to field. Especially, there is no anti-misoperation system for the operation of information devices, such as hard strap and protection fuse of relay protections. When complicated work needs to be done, it may lead to misoperation. Cognitive theory indicates when a system become complex, human judge capability drops as shown in figure 1.

Besides, the logic experiment, daily operation of strap and using of relay protection works involves various departments of a power supply company. A closed loop management methods is urgently needed to ensure exactness of the procedure.

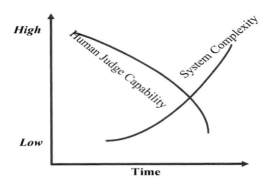

Fig. 1. Relation between system complexity and ability of human judgment (Cited in [1])

Information devices are footstone of smart grid. To overcome the problems mentioned above, this paper introduces the design and implement of an intelligent system for monitoring power grid information systems. With the help of this system, original work which is carried out by engineers is taken place by the system. The working efficiency and accuracy are highly improved.

2 System Design Strategy

The designs of information system now in use are oriented to electrical equipments which lead to equipments separate from power grid and the working states information of equipments separate from working information of power grid. As a result, many working flow of power grid operation and control are self-obturated. However, the power flow is companied by information flow. Various information systems describe power grid from different point of view. For example current of certain transmission line can be obtained in SCADA or relay protection systems. So each information system related to power grid operation and control should be treated as a whole.

Based on the idea above, information system intelligent monitoring system (ISIMS) associates power grid operation and control related information stored in different information systems together. In allusion to the characteristic of these information systems, such as variety is various, data is stored separately. The ISIMS design strategy is described as follows:

● Computer graphical user interface is used to achieve centralized information management, in order to solve spread state of information and data;
● By simulation and field data associated with the graphical topology, it will greatly improve data using efficiency;
● By carrying out application of electronic processes, it will enhanced controllability of the workflow;
● Through the system configuration diagram, simulation operations and the realization of abnormal early warning functions, it will enhance the reliability of information system operations and information timeliness.

ISIMS is based on computer network technology, database technology and graphics technology aimed to integrate multi-dimensional information stored in different power grid operation and control information system. ISIMS provides visual monitoring and decision-making tools for information system management.

3 Key Issues of ISIMS Implement

- synthesis of distributed graph model

Power gird is typical a distributed system. Different information systems have their own power grid model. Meantime these systems acquire filed data individually and the data are stored in different power grid monitoring and control systems, from EMS to MIS. As a result it forms various information islands. ISIMS should integrate these decentralized data to fulfill the intelligent monitoring function of these systems.

Graphics and the original data is the core of ISIMS. ISIMS is based on CIM[3-4] of IEC 61970[2]. CIS[5] and Scalable Vector Graphics[6] (SVG) are adopted as component interface specification and graphic standard. With these specifications, Internet browsing, operation and remote maintenance can be achieved. The SVG is development by the W3C organization as standard for vector graphics which is suitable for network applications. SVG is flexible, scalable. it can express a rich graphical content and achieve a strong interaction, reusability, and scalability.

Model is the foundation of power grid information integration. ISIMS adopts CIM which consists of public class, attribute, relation facts as model specification. Class and object are abstract of power grid and so that CIM can be used in various power grid information systems. CIM establishes logic data structure. Base on this, information exchange model of different information systems can be defined. Generally, CIM provides a whole logic view of power grid information which represents main objects of power grid by describing public class and attribute.

Based on CIM and SVG, ISIMS achieves synthesis of distributed graph model. By analyzing power grid mode separates in different systems, a whole picture view of power grid can be achieved. The sub-model used by ISIMS is fetched from various information systems. A synthesizing procedure takes places to form the entire model.

- Loose coupling software architecture.

Service oriented architecture[7] (SOA) is loose coupling to application integration and re-useable. A system based on SOA is flexible and extendable. Addition of new function has less impact on existed systems. SOA and IEC 61968 are used to build the software architecture of ISIMS to fulfill the demand of information integration. The principle of IEC 61968 is the same as SOA. IEC 61968 defines expression and semasiology of information exchange

- Across different platforms

The data is stored in different systems which are built on different operation systems and hard wares. The types of operation systems are windows NT4.0/2000/xp, Posix/Unix/Linux. Hardware includes work station, server and micro-computer. The main brands are COMPAQ、ALPHA、SUN、IBM and HP.

In order to monitor various information systems with different architectures, the characteristic of across different platforms and support of distributed environment[8] are required by ISIMS. ADAPTIVE Communication Environment[9] （ACE） middleware[10] is adopted as the development tools of under stratum. ACE shields the complexity of different hard wares and operation systems. And ACE provides flexibility for the develop work, meantime has less impact on the performance of original systems.

- Across regions

Information systems are divided into different safety regions from region I to region IV. ISIMS disposes data acquire nodes in region II and region III. These two nodes are in charge of fetching data from information systems on sides of physical insulating devices. The data in region I and II are gathered on data acquire node in region II. And the node sends the gathered data through the physical insulating devices to the main server of ISIMS.

4 Functions Design of ISIMS

The main functions of ISIMS are data inquire, operation management, repair management, report forms management, users management. All the functions share the some data and power gird model by object oriented method. The functions diagram is shown in figure 2.

Data inquire function includes: a. query of information system configuration diagram. Corresponding to topology diagram, on the configuration diagram, the relay type, change date and setting value of equipments, such as transmission lines and transformers can be found out. This inquire can be processed on power stations and its relative devices. b. providing fuse states information, such as types and reserve part.

Operation management function includes: a. hard/soft strap simulation diagram. Base on the state of field hard/soft strap, a corresponding hard/soft strap simulation diagram is generated. From the diagram, the relative relay protection of strap can be easily found out. Also the logic of relay, user guide of certain protection in different working condition can be referred. These shortcut improve the working efficiency; b. simulation of strap operation. By recording of operation of strap on the simulation diagram according to field data, it ensure the engineers grip the real-time condition and monitor the states of straps on line; c. fault alarming, such as power lose of automatic bus transfer device, breakdown of information systems

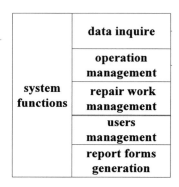

Fig. 2. The function design of ISIMS

Repair work management includes: combined equipment testing and annual cycle of condition-based maintenance conclusions, an annual maintenance plan for next year is generated. The module is used as status monitoring module for information systems. It will create history database files for the specific information devices which track the anti-fault measures and bugs. These database files together with relay protection inspection conclusions ensure the whole process management of information devices. In addition, professional inspection database is created for inspection usage.

User management function includes: based on the business processes and the nature of the user's posts, this module provides the appropriate user roles and permissions settings features.

Report forms generation is described as follows: custom reporting style, based on their needs, users can select the appropriate column field, and logical relations, the model will generate the required reports. And also this function supports export and print.

Power system is large, complex systems which consist of various devices, different technology and professional. Electrical equipments and information devices of power grid are closely linked and inseparable. Also the operation and management process power grid information devices involve lots of power system engineers from different departments of a power company.

To better reflect the actual power system integration, ISIMS will adopt a unified interface. Through power grid topology, the configuration diagram is as main line, the importance substation is as unit node. By this strategy, the ISIMS application functions are integrated into a unified application interface. On the one hand, this ensure the rational realization of various functions, on the other hand it reflects the variety of data, functional interrelation. Unified visual interface achieves more efficient function, in which the relationship between the synergy is clear.

The MMI hierarchy of ISIMS is shown in figure 3.

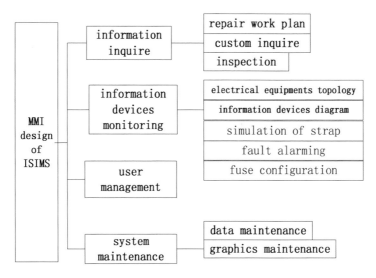

Fig. 3. The MMI hierarchy of ISIMS

5 System Architecture of ISIMS

Topology is the core of ISIMS. It is combined with equipment parameters, power grid states data, a whole picture view of power grid can be reached. Based on background calculation and logic judgment, the functions of ISIMS are realized.

ISIMS can be divided into 3 parts, as shown in figure 4.

a. foundation part, this part is in charge of communications management, security management, interface management, storage management, traffic management, security settings and etc, which is the basic part of ISIMS.

b. graphics and data part, this part includes topology management, equipment parameters management, relay protection setting, strap simulation, real time data collection. The data and model used by ISIMS are generated by it.

c. application part, this part includes the application functions of ISIMS, such as strap simulation, fuse management, repair work plan and etc.

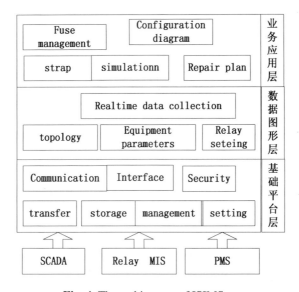

Fig. 4. The architecture of ISIMS

The server of ISIMS is deposited safety region III according to power system information security division regulations, the disposition diagram is shown in figure 5.

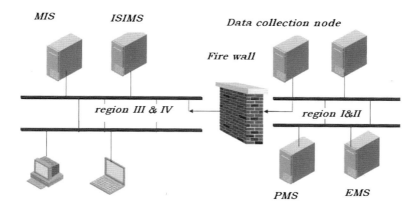

Fig. 5. The strategy of ISIMS disposition

6 Conclusions

Information system intelligent monitoring system (ISIMS) is in charge of monitoring various separated information systems now used in power grid by graphical interface. ISIMS reduces the management burden of information systems by data integration, simulation, repair work plan and fault alarming means. ISIMS includes 7 main models, such as state monitoring, relay protection management, maintenance status management, smart tips indication; reports form queries, system logs, system management. ISIMS improves the management level of power grid information systems.

References

1. Utility Consulting International. IntelliGrid Architecture: Power System Functions and Strategic Vision[R/OL], Frances Cleveland Presented to EdF/RTE on September 12, p. 5 (2005),
http://xanthus-consulting.com/Publications/IntelliGrid%20Architecture%20-%20Issues%20Stemming%20from%20IntelliGrid.pdf
2. Zhang, S.M., Huang, H.F.: Architecture of power dispatching automation system based on IEC 61970 standard. Automation of Electric Power Systems 26(10), 45–47 (2002)
3. Final Report, Common Information Model (CIM): CIM 10 Version [R] (November 2001)
4. Britton, J.P., Devos, A.N.: CIM-based standards and CIM evolution. IEEE Trans. Power System 20(2), 758–764 (2005)
5. Becker, D., Falk, H., Billerman, J., Mauser, S., Podmore, R., Schneberger, L.: Standards-based approach integrate utility applications. IEEE Comput. App. Power 14(4), 13–20 (2000)
6. SVG [R/OL], http://www.w3.org/TR/SVG

7. Zhu, J.: Web services provide the power to integrate. IEEE Power Energy 1(6), 40–49 (2003)
8. Tanenbaum, A.S., Steen, M.V.: Distributed Systems: Principles and Paradigms. Prentice-Hall, Upper Saddle River (2002)
9. ACE [R/OL], http://riverace.com/index.htm
10. Fu, Q.X.: Application level data duplication via middle-ware tonglink/q. Automation of Electric Power Systems 27(16), 45–47 (2003)

Design of Controlling System of Higher Layer Elevator

Yushui Huang, Wenbin Peng, and Chong Liu

Information Engineering School, Nanchang University
NanChang, 330031, China
luckingp@163.com

Abstract. Aiming at the disadvantage of the traditional relay control of elevator system, a higher layer elevator controlling system based on FX2N PLC is introduced, including the floor induction, calling display and other tasks. Practice has proved that the controlling system has good design, high reliability, convenient controlling and widely application.

Keywords: PLC, elevator, elevator controlling system.

1 Introduction

In recent years, with the development of science and technology, China's elevator production technology has developed rapidly. Some improvement in elevator designing successfully also design and process changes to product more new model elevator. Elevator mainly divided into two parts: mechanical system and control system. With the theory of automatic control and microelectronics technology development, the way of elevator drag and control model are changed greatly. Exchange control is the mainly development direction of elevator drag. Currently the elevator controlling system is mainly have two kinds of control ways[1][2]: Relay control system and PLC control system. Relay control system is to be being washed out gradually due to the high failure, poor reliability, flexibility control mode and high power consumption. PLC control system has been widely used in the elevator controlling system because of its high reliability of the convenient usage and maintenance, strong anti-interference, short design and commissioning period.

This paper use PLC to design elevator to require function change flexible, simple programming, low malfunction, low noise, easy maintenance, save energy, strong anti-jamming capability and control box cover of small area[3].

2 Elevator PLC Controlling System Input/Output Points

FX2N programmable controller is a new kind of programmable controller developed by Mitsubishi Company[4][5]. It is reliable, strong function, storage capacity and easy programming. Therefore, it can satisfy the elevator of electrical control system and series small PLC of FX2N can be applied to various automation system. The input/output tables are as follows[3]. (See Table 1, 2)

Table 1. System Input points.

Number	Name	Input	Number	Name	Input
0	One layer travel button	X11	8	Three layer inside the call button	X21
1	Two layer travel button	X12	9	Four layer inside the call button	X10
2	Three layer travel button	X13	10	One layer up call button	X1
3	Four layer travel button	X14	11	Two layer up call button	X2
4	Elevator open door button	X15	12	Two layer down call button	X3
5	Elevator close door button	X16	13	Three layer up call button	X4
6	One layer inside the call button	X7	14	Three layer down call button	X5
7	Two layer inside the call button	X20	15	Four layer down call button	X6
...	...	...	...	...	...

Table 2. System Output points.

Number	Name	Output	Number	Name	Output
0	One layer internal call display	Y1	10	Outside two layer down call display	Y32
1	Two layer internal call display	Y2	11	Outside three layer up call display	Y33
2	Three layer internal call display	Y3	12	Outside three layer down call display	Y34
3	Four layer internal call display	Y4	13	Outside four layer down call display	Y35
4	One layer display	Y41	14	Open door display	Y11
5	Two layer display	Y42	15	Close door display	Y12
6	Three layer display	Y43	16	Up display	Y13
7	Four layer display	Y44	17	Down display	Y14
8	Outside one layer up call display	Y30	18	Up-Stop display	Y15
9	Outside two layer up call display	Y31	19	Down-Stop display	Y16
...	...	...	...	...	...

3 User Programming Design

This paper use Mitsubishi FX2N programmable controller to complete the elevator car about the hall summoned instruction, floor position indicator to stop, the controlling of opened and closed door and other tasks.

3.1 Floor Induction and Display

Floor induction signal is an important signal in a circuit, because it involves many other controlling circuits. Such as enclosure instructions, vestibular summons, refers to the floor and so on. Generally, when elevator rises, we always make the enclosure bottom as a standard, when enclosure bottom into under layer as the starting signal of this layer, it also when enclosure bottom leave under layer as the ending signal of this layer. But when the elevator drops we use the top of the car signal as a standard. (See Fig. 1).

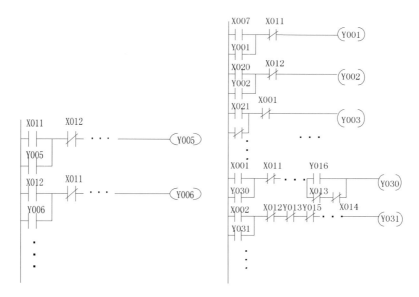

Fig. 1. Floor induction display   **Fig. 2.** Elevator of calling display

3.2 Elevator of Calling Display

Vestibular summons is a maintain links (call button is generally pulse signal).Besides the top layer and the root layer has a call-down or a call-up button[2][3], any other layers has two buttons. So it is important to consider how to eliminate the response about calling. (See Fig. 2).

3.3 Elevator Choose Direction and Display

Choose direction circuit controlled by two signals. One is choose layer in the car .The other is controlled by vestibular called. But the top and bottom is rather special, so it is to be treated separately. The direction divide into two kinds about rising and falling[3]. (See Fig.3).

3.4 Elevator Arrived and Stopped Display

Due to calling divide into two kinds about up-calling and down-calling.
   So when the calling direction is same as the elevator running direction the elevator can stop it[2][4]. (See Fig.4).

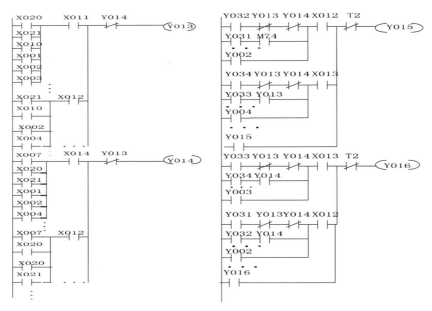

Fig. 3. Elevator choose direction display   **Fig. 4.** Elevator stopped display

3.5 Controlling of Elevator Opening and Closing

The elevator opened or closed door is divided into manual and automatic. In the process of automatic elevator opened or closed support manual. Lift up only in neither will not fall to the elevator up to stop, and whether if have dropped or stop by a response, the elevator will be closed after opening, when elevator stopped it need stops 2s and then continue to delay 5s closed.(see Fig. 5)

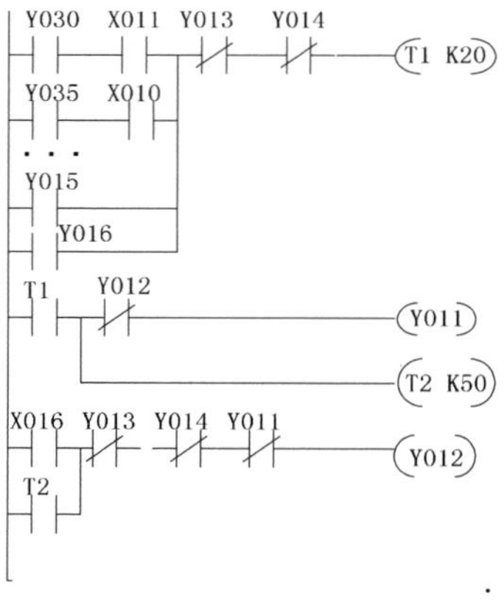

Fig. 5. Controlling system of elevator opening and closing

4 Conclusion

In higher layers of elevator controlling system, this paper use Mitsubishi FX2N programmable controller to design and complete the elevator car outside the hall summoned instruction, floor position indicator to stop, the controlling of opened and closed door and other tasks. After testing in a 20 layers apartment, the system operation stable and reliable for two years. So practice has proved that the PLC controlling of elevator has good design, testing and has great practical value.

References

1. Liao, C.: PLC application technology. Chongqing University Press, Chongqing (2006)
2. Xieg, W.: Programmable logical controller principle and application. China's Power Press, Beijing (2006)
3. He, H.: Application of PLC &Inverter in Improvement of Elevator System. The motor control and application 36(9), 53–55 (2009)
4. Ma, Y., Yao, C.: Elevator systems real-time controlling and software modeling by PLC. Control Engineering 16(6), 787–790 (2009)
5. Liao, C.: PLC programming and application of FX series. Mechanical Industry Press, Beijing (2005)

A New Denosing Method in Seismic Data Processing Based on Wiener Filter

Wang Shi-Wei[1,2]

[1] School of Mechanical and Electronic Engineering, Wuhan University of Technology, Wuhan, Hubei Province, China, 430070
[2] Information Engineering Department, Henan College of Finance &Taxation, Zhengzhou, Henan Province, China, 450002

Abstract. There are lots of random noises in seismic data. It is very important to remove random noises in seismic data processing. Fourier transform is the traditional method in signal analysis. In this paper, a novel method is put forward. Wiener filter method can effectively remove random noises. It gets over the limit of Fourier transform. By comparing the denoising results of Fourier transform method and Wiener filter method, Wiener filter method is prior to the traditional method.

Keywords: Image processing, Wiener filter, Denosing, Random noises.

1 Introduction

Computer image processing technology has been widely used in seismic data [1-2]. The computer image denosing processing technology can remove the noise from the seismic data by acting the seismic cross section as the digital image and using suitable mathematical method [3].

There are a lot of methods to remove the noise [4]. Traditional methods often increase the signal-to-noise ratio (SNR) by decreasing the resolution of the seismic data. Digital image processing technology can increase the SNR and keep the resolution of the seismic data, so it has been more widely used to denoise in seismic data.

There are lots of denoisng methods in digital image processing, and Wiener filter can adjust the parameters according to the part difference of the image. The method not only keeps the edge part but also the high frequency information. It can efficiently remove the noise.

2 Principle

There is a original signal (u) contaminated with the white noise (n). The observed signal is shown as the following equation:

$$f = u + n \qquad (1)$$

Wiener filter can find the estimated value ($\hat{f}$) when the statistic error function ($e^2 = E\{(f - \hat{f})^2\}$) is minimum. E is the operator of expectation value and f is the original image. The equation can be changed in frequency domain:

$$\hat{F}(u,v) = \left[\frac{1}{H(u,v)} \frac{|H(u,v)|^2}{|H(u,v)|^2 + S_\eta(u,v)/S_f(u,v)} \right] G(u,v) \quad (2)$$

Wiener filter improve the image quality by using the space self-adaption if the image is stable in local neighborhood. Every pixel of image is center (x), and th filtering window ($M \times N$) is stable. The estimated value of the mean and the variance of pixel (x) in filtering window can be obtained by weighted mean method:

$$\mu_f(x) = \frac{1}{MN} \sum_{y \in O(x)} f(y) \quad (3)$$

$$\sigma_f^2(x) = \frac{1}{MN} \sum_{y \in O(x)} -\mu_f^2(x) \quad (4)$$

Size of filtering window is very important using using the space self-adaption. If the filtering window is too small, the noise will influence very much on the estimated value of the mean and the variance. Size is usually chosen as 3×3, 5×5, or 7×7. The size, 5×5, is used to remove the noise of the seismic data.

3 Practical Application

Fig. 1 shows the theoretical seismic data, and is signal. Fig. 2 shows the result with the white noise. Fig. 3 shows the denoising result with Wiener filter. The results show that the noise can be pressured and the resolution of the signal improves. Fig. 4 shows the cross section of the practical seismic data record. Fig. 5 shows the denoising results with Wiener filter. The results show that the Wiener filter is prior to FFT denoing method. The resolution of the cross section improves.

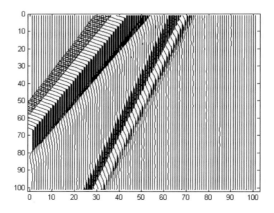

Fig. 1. Theoretical seismic data

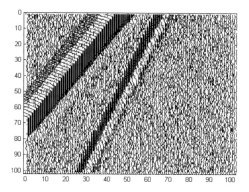

Fig. 2. Seismic data with the white noise

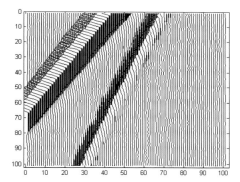

Fig. 3. Denoising signal with Wiener filter

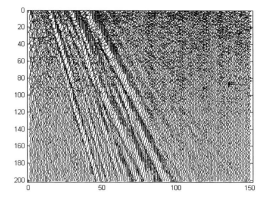

Fig. 4. Practical seismic data record

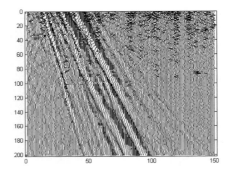

Fig. 5. Practical data with Wiener filter denoisng

4 Conclusion

The paper puts forward to the novel method to remove the noise from the seismic data based on Wiener filter of the digital image processing. Wiener filter can adjust the parameters according to the part difference of the image. The method not only keeps the edge part but also the high frequency information. It can efficiently remove the noise. Compared to the traditional denoising method, FFT, the results show that Wiener filter method is prior to FFT method.

References

1. Zhang, L.-l., Wu, J.-s., Wang, J.-s.: Application of image processing techniques to geophysics. Oil Geophysical Prospecting (3), 317–323 (2003)
2. Boore, D.M.: On pads and filters: processing strong motion data. Bull. Seismol. Soc. Am. (95), 745–750 (2005)
3. Fan, T.-y., Yang, C.-c.: 3D adaptive seismic edge-preserving smoothing and its application. Oil Geophysical Prospecting (5), 558–563 (2009)
4. Luo, Y., Higgs, W.G., Kowalik, W.S.: Edge detection and stratigraphic analysis using 3-D seismic data. Expanded abstracts of 66th SEG Ann Internat Mtg., 324–327 (1996)

Trotting Gait Planning and Implementation for a Little Quadruped Robot

Bin Li[1,2], Yibin Li[1], Xuewen Rong[1], and Jian Meng[1]

[1] School of Control Science and Engineering, Shandong University, Jinan, 250061, China
ribbenlee@126.com, liyb@sdu.edu.cn
[2] School of Science, Shandong Polytechnical University, Jinan, 250353, China

Abstract. This paper presents a trotting pattern generation approach and online control strategy for a little quadruped robot. The trotting gait is scheduled based on the composite cycloid method in order to improve the stability of quadruped robot. The efficiency and performance of the proposed methods are verified by simulations and experiments by means of the little quadruped robot constructed by our laboratory. The experiment results show that the average speed of the little quadruped robot on even terrain is 0.1m/s and the proposed methods are suitable and simple in terms of controlling.

Keywords: Trot, Gait planning, Quadruped, Robot.

1 Introduction

Approximately half of the earth's land surface is inaccessible to either wheeled or tracked robots. Otherwise legged animals can move over most of earth's terrain, the quadruped robots are optimal for a multi-legged robots, because of the least complex mechanical design and a better stability of configuration [1].

The gait design and control methods for quadruped robots have been studied for several decades. The systematic planning based on mathematical analysis through forward and inverse kinematics equations and the incremental learning is two gait design methods commonly [2]. The gaits of quadruped robots are classified into static gaits and dynamical gaits. In dynamical gaits, Trot is the most energy efficient gait, which decouples pairs of legs and makes control simpler. And the research of quadruped trot will lead to understanding other dynamical gaits, such as pacing and galloping [3]. So, the focus of this paper is on trotting gait generation method and simple software control strategy of quadruped robots. As the most available locomotion pattern, the trotting pattern is generated so that the quadruped robot moves with a fast speed and stable body attitude.

The position control and force control are two main control methods for quadruped robots. The dynamical model of quadruped robots will be needed in the force control method, the dynamical model is difficult to obtain for the quadruped robot with many degrees of freedom. For decreasing the control complexity of quadruped robot, a simple control schedule is proposed through soft programming in Visual C++. The soft controlling platform controls the electronics rudder servos directly. Therefore, if the position control is precise for the quadruped robot based on the kinematics equations, the walking speed and attitude of quadruped robot may be very well with good foot trajectory method.

First of all, this paper proposes the forward kinematics (FK) and inverse kinematics (IK) equations of the quadruped robot, Secondly, based on the composite cycloid method of gait planning [4], a trotting gait generation method on even terrain is presented. Furthermore, for verifying the simple control strategy and good foot trajectory method, a little quadruped robot is constructed by our laboratory. The other intention to design the little quadruped robot is to study and verify the solutions of our forward and inverse kinematics and online control method for the hydraulic actuated quadruped robot in the future.

The paper is structured as follows: Section 2 describes the FK and IK equations of the quadruped robot with three degrees of freedom. Section 3 explains the reasons for choosing the trotting gait. Section 4 contains the trotting gait generation method in the swing period and support period. The simulations and experiments are proposed in Section 5. Finally, section 6 and 7 describe the conclusion and future work.

2 FK and IK Model of the Quadruped Robot

The quadruped robot used in this paper consists of a body frame and four legs as show in Fig. 1. Small size electronic rotary actuators are installed for the quadruped robot to carry heavy payload and move fast on even terrain. The size of this robot is 23cm x 18cm x 18cm (length, height and width, respectively). One leg module comprises one hip joint, one shoulder joint and one knee joint, so the little quadruped robot has 12 degrees of freedom, with three degrees of freedom per leg.

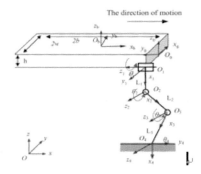

Fig. 1. The real model of quadruped robot   **Fig. 2.** The coordinates of quadruped robot

Fig. 2 shows the little quadruped robot's kinematic diagram of the right front leg [5]. Each leg consists of three degrees of freedom. The kinematic equation of the foot tips can be derived by making use of the coordinate frames established in Fig. 2.

The FK and IK equations are deduced based on the D-H method. The solution of forward kinematics for each leg is described as:

$$P_{RF} = \begin{Bmatrix} p_x \\ p_y \\ p_z \end{Bmatrix} = \begin{Bmatrix} b + L_2 s_2 + L_3(c_2 s_3 + s_2 c_3) \\ -w - (L_1 s_1 + L_2 s_1 c_2 + L_3(s_1 c_2 c_3 - s_1 s_2 s_3)) \\ -L_1 c_1 - L_2 c_1 c_2 + L_3(c_1 s_2 s_3 - c_1 c_2 c_3) - h \end{Bmatrix} \quad (1)$$

$$P_{RH} = \begin{Bmatrix} p_x \\ p_y \\ p_z \end{Bmatrix} = \begin{Bmatrix} -b + L_2 s_2 + L_3(c_2 s_3 + s_2 c_3) \\ -w - (L_1 s_1 + L_2 s_1 c_2 + L_3(s_1 c_2 c_3 - s_1 s_2 s_3)) \\ -L_1 c_1 - L_2 c_1 c_2 + L_3(c_1 s_2 s_3 - c_1 c_2 c_3) - h \end{Bmatrix} \quad (2)$$

$$P_{LH} = \begin{Bmatrix} p_x \\ p_y \\ p_z \end{Bmatrix} = \begin{Bmatrix} -b + L_2 s_2 + L_3(c_2 s_3 + s_2 c_3) \\ w + (L_1 s_1 + L_2 s_1 c_2 + L_3(s_1 c_2 c_3 - s_1 s_2 s_3)) \\ -L_1 c_1 - L_2 c_1 c_2 + L_3(c_1 s_2 s_3 - c_1 c_2 c_3) - h \end{Bmatrix} \quad (3)$$

$$P_{LF} = \begin{Bmatrix} p_x \\ p_y \\ p_z \end{Bmatrix} = \begin{Bmatrix} b + L_2 s_2 + L_3(c_2 s_3 + s_2 c_3) \\ w + (L_1 s_1 + L_2 s_1 c_2 + L_3(s_1 c_2 c_3 - s_1 s_2 s_3)) \\ -L_1 c_1 - L_2 c_1 c_2 + L_3(c_1 s_2 s_3 - c_1 c_2 c_3) - h \end{Bmatrix} \quad (4)$$

where LF, RF, RH and LH is left front, right front, right hind and left hind leg respectively, $\{p_x, p_y, p_z\}$ are the foot tips of the leg.

The inverse kinematics of the legs are also analyzed based the D-H method. For simplicity, the RF inverse kinematics is only as follows:

$$\theta_1 = \tan^{-1}\left(\frac{w + p_y}{p_z + h}\right) \quad (5)$$

$$\theta_2 = \tan^{-1}\frac{E - D\frac{\sqrt{D^2 + E^2 - K^2}}{K}}{D + E\frac{\sqrt{D^2 + E^2 - K^2}}{K}} \quad (6)$$

$$\theta_3 = \cos^{-1}\left(\frac{D^2 + E^2 - L_2^2 - L_3^2}{2 L_2 L_3}\right) \quad (7)$$

where $D = -s_1 p_y - c_1 p_z - w s_1 - h c_1 - L_1$, $E = p_x - b$, $L_2 + L_3 c_3 = K$. By inputting the foot-tips value, the respective angles, $\theta_1, \theta_2, \theta_3$ will be solved by the aid of IK equation. In reverse, by formulating the FK equations and knowing the foot tips position of the quadruped robot, we will be able find the angles of the quadruped robot. After the three angles have been found, it will be able to construct the pattern of gait planning for trotting.

3 Trotting Gait Planning

Gait means a pattern of discrete foot placements performed in a given sequence. The quadruped robots have two different kinds of gaits. Static gaits(e.g., crawl, wave) occur when the vertical projection of center of gravity remains always inside the polygon formed by the supporting legs. Dynamic gaits (e.g., trot, pace and gallop) mean that the vertical projection of center of gravity is not necessary to remain inside the polygon formed by the supporting legs with the dynamic balance to be maintained.

The trotting gait was chosen for the little quadruped robot in this paper because of its observed energy efficiency over a wide range of running speed and its wide use in nature [3], [6]. The trot, as shown in Fig. 3, is a symmetric gait during which the diagonal forelimb and hind limb move in unison, ideally contacting and leaving the ground at the same time. Fig. 3 shows the complete trotting stride with two transfer phases and two support phases.

Fig. 3. The complete trotting stride of horse [11]

In the description of gaits. the transfer phase of a leg is the period in which the foot is not on the ground. The support pahse of a leg is the period in which the foot is on the ground and the cycle time T is the time for a complete cycle of leg locomotion of a periodic gait.

Compared with other dynamic walking gaits, the trotting gait is a more practical way. The reasons are as follows:

1) The trotting gait has a high rate of energy efficiency and greater range of adaptation of speed.
2) Even if the two support diagonal legs overturned. with the help of touching the ground quickly, the other two tranfer diagonal legs also can prevent the robot overturning.
3) The trotting gait is adaptation to the static gait of crawling, and the gait transition from trot to crawl is esay and vice versa. The trotting gait has good adaptability to complex terrain.
4) The diagonal legs have the same motion phase. In theory, they strike the ground at the same time. They leave the ground at the same time, and they swing forward at the same time, So, the symmetry trotting gait can implement the symmetry motion of the robot, and can keep the self-stabilty and reduce the complexity of attitude control.

In trotting gait, the diagonal legs form pairs. The members of a pair strike the ground in unison and they leave the ground in unision. While one pair of legs provides support, the other pair of legs swings forward in preparation for the next step [7]. In order to simplify the complexicy of gait planning, the virtural legs introduced by Raibert are used, one virtural leg symbolizes two legs with simultaneous moton, as shown in Fig. 4. The dimension of each virtural leg is the same as those of the real legs. The use of virtual legs implies that the behavior of the two real legs in simultaneous motion is identical [8], [9], [10]. Thus, the method of gait planning is summaried as follows:

1) According to the working tasks (e.g., motion control, stability control) of quadruped robot is completed and the terrain information (uneven terrain, terrain slope, etc.) , the gait of virtual legs is planned.
2) The gait of the virtual legs is decomposed to the physical legs of the diagonal anterior and posterior, and the gait coordination and planning task of the physical legs is done.

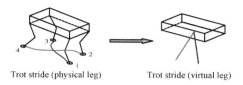

Trot stride (physical leg)   Trot stride (virtual leg)

Fig. 4. In the trot diagonal pairs of legs act in unison [7]

4 Gait Generation of Transfer and Support Period

The motion of the quadruped robot is determined by two factors, the trajectory generation of transfer legs and the angle computation of supporting legs. Based on the FK and IK equations of the quadruped robot, the trotting gait is constructed in terms of the foot trajectory of the transfer legs and the posture of the supporting legs. The foot trajectory is obtained by the stride and the height of leg. The stride of one step is determined by the desired body velocity of forward direction. And the height of foot trajectory is obtained by the height of obstacle on the ground or the terrain environment during walking.

4.1 Transfer Gait Generation

Quadruped robots need to have the ability to move on the uneven or slope terrains. In the dynamic trotting walking case, the relative speed between the foot and the ground causes loss of body balance. For reducing the impact between the foot and the ground and consideration is given to a trajectory where the foot velocity and acceleration can be continuous both at the time of departure and landing [4]. Therefore, this paper presents a pattern generation strategy based on the composite cycloid method for the trotting gait. The formula of the composite cycloid is as follows:

$$X(t) = a(\frac{t}{t_0} - \frac{1}{2\pi}\sin(2\pi\frac{t}{t_0})) \quad 0 \leq t \leq t_0$$
$$Z(t) = 2h(\frac{t}{t_0} - \frac{1}{4\pi}\sin(4\pi\frac{t}{t_0})) \quad 0 \leq t \leq \frac{t_0}{2}$$
(8)

where a is the stride length, h is the maximum foot height of the quadruped robot, t is the sampling time of the gait trajectory. t_0 is the transfer time. The x axis is taken on a linear line connecting the departing point and landing point and the z axis is on the vertical line. So $X(t)$ is the horizontal trajectory and $Z(t)$ is the vertical trajectory of transfer leg of the quadruped robot.

4.2 Support Gait Generation

Suppose the locomotion speed of the quadruped robot v is 0 in the z direction for keeping the stability of the body, as is shown in Fig. 5. The control objective of support legs for the quadruped robot is to maintain the constant body height and the desired body velocity. The desired body motion is obtained from the motion of the support legs. Therefore, when the initiate and end states of the support period are determined, the trajectories of each joint of support legs can be obtained in terms of the IK equation.

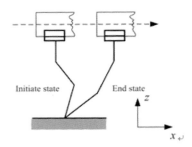

Fig. 5. Gait planning of the support legs

5 Simulation and Experiments

To test the trotting gait generation method and simple software controlling strategy proposed in this paper experimentally, a relatively simple forward trotting gait on even terrain was chosen. The cycle period is 1 s and the stride of one leg is set to 0.1 m. Software controlling platform of the quadruped robot is shown in Fig. 6.

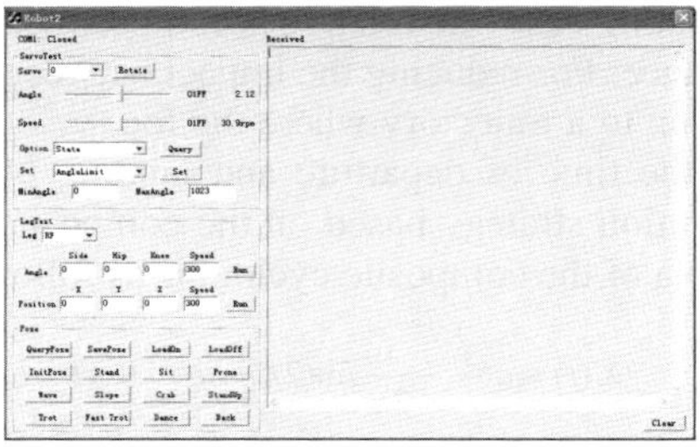

Fig. 6. The contolling platform of the little quadruped robot

The gait of the quadruped robot is controlled by the steps as follows:

1) The switching sequence and time of legs are scheduled based the trotting gait method;
2) The stride of the quadruped gait is obtained in terms of the landing position of legs;

3) The period of stride of the quadruped robot is computed by the desired motion velocity;
4) The motion trajectories and angle values of legs are planned based on the FK and IK equations;
5) The data frames are sent to the electrical actuated rudder servos with a delay time, the delay time is computed by the walking period divided by the key frames.

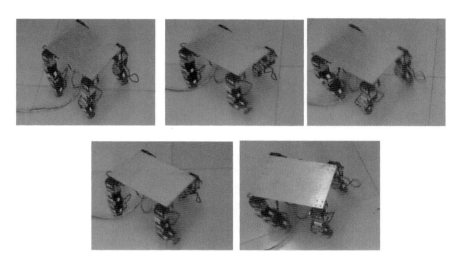

Fig. 7. The snapshots of a trotting gait period for the quadruped robot

Some successful experiments videos are obtained and Fig. 7 is the snapshots of a trotting gait from one video. Fig. 7 shows the trotting gait of the little quadruped robot in two walking periods. The little quadruped can walk with trotting pattern in a constant velocity 0.1 m/s.

6 Conclusion

A little quadruped robot model is developed based on the forward and inverse kinematics and electronic rudder servo controller in this paper. The trotting motion is scheduled based on the composite cycloid method. Trotting gait in the even terrain is simulated by our real little quadruped robot and a simple online algorithm with position control method successfully compels the robot to walk continuously over a flat terrain.

7 Future Work

The future advancement can be carried out in the project by going for embedded processor that can process and transmit the control signal faster to the actuators.

Remote control through wireless Ethernet mode can also be considered. Only forward experimental trotting walking on even terrain is presented in this paper, the different gaits and walking on different types of terrain will be constructed in our future work.

Acknowledgments

This paper is supported by the Independent Innovation Foundation of Shandong University (IIFSDU) under grant 2009JC010.

References

1. Song, S.M., Waldron, K.J.: Machine that walk: The Adaptive Suspension Vehicle. The MIT Press, Cambridge (1988)
2. Yi, S.: Reliable Gait Planning and Control for Miniaturized Quadruped Robot Pet. Mechatronics 20(4), 485–495 (2010)
3. Palmer III, L.R.: Intelligent Control and Force Redistribution for a High Speed Quadruped Trot. Ph. D. dissertation, The Ohio State University (2007)
4. Sakakibara, Y., Kan, K., Hosoda, Y., Hattori, M., Fujie, M.: Foot Trajectory for a Quadruepd Walking Machine. IEEE Internatioanal on Intelligent Robots and Systems 1, 315–322 (1990)
5. Song, H.J., Rong, X.W., Li, Y.B., Ruan, J.H.: The OpenGL and Ginac based Approach to Quadruped Robot gGait Smulation System. In: IEEE International Conference on Intelligent Computing and Intelligent Systems, vol. 3, pp. 203–207 (2010)
6. Palmer III, L.R., David, O.E.: Force Redistribution in a Quadruped Running Trot. In: 2007 IEEE International Conference on Robotics and Automation, Roma, Italy, pp. 4343–4348 (2007)
7. Raibert, M.H.: Trotting, Pacing and Bounding by a Quadruped Robot. Journal of Biomechanics 23, 79–81 (1990)
8. Raibert, M.H.: Legged Robots that Balance. The MIT Press, Cambridge (1986)
9. Raibert, M.H., Brown, H.B., Chepponis, J.M., et al.: Dynamically Stable Legged Locomotion. MIT Artificial Intelligence Laboratory, Technical Report (1989)
10. Kim, H.K., Won, D., Kwon, O., Kim, T.J., Kim, S.S., Park, S.: Foot Trajectory Generation of Hydraulic Quadruped Robots on Uneven Terrain. In: Proceedings on the 17th World Congress,The International Federation of Automatic Control, Seoul, Kerea, pp. 3021–3026 (2008)
11. Gambarian, P.P.: How Mammals Run Anatomical Adaptations. John Wiley & Sons, New York (1974)

Optimal Reducing the Solutions of Support Vector Machines Based on Particle Swam Optimization

Jih Pin Yeh[1] and Chiang Ming Chiang[2]

[1] Department of Multimedia and Entertainment Science, Asian-Pacific Institute of Creativity,
Toufen Township, Miaoli County, Taiwan 35153, ROC
yehbin@ms.apic.edu.tw
[2] Department of Computer Science and Information Engineering, Tamkang University,
Tamsui, Taipei County, Taiwan 25137, ROC
cmchiangtw@gmail.com

Abstract. SVM has been winning increasing interest in areas ranging from its original application in pattern recognition to other applications such as regression analysis due to its noticeable generalization performance. One of the SVM problems is that SVM is considerably slower in test phase caused by large number of the support vectors, which greatly influences it into the practical use. To address this problem, we proposed an adaptive particle swarm optimization (PSO) algorithm which based on a reasonable fitness to optimally reduce the solutions for an SVM by selecting vectors from the trained support vector solutions, such that the selected vectors best approximate the original discriminant function. Experimental results show that adaptive particle swarm optimization algorithm is an optimal algorithm to simplify the solution for support vector machines and can reduce the solutions for an SVM by selecting vectors from the trained support vector solutions.

Keywords: Support vector machine, Particle swarm optimization, Optimization, Discriminant function.

1 Introduction

Support vector machines (SVMs) were developed by Vapnik [1], and are winning popularity resulting from their many attractive features and their promising empirical performance. Since the discriminant function of a SVM is parameterized by a large number of support vectors and corresponding weights, the machine is much slower in the test phase than other learning machines such as neural network and decision trees [1,2,14,16,18]. A lot of approaches have been proposed to address this disadvantage by approaching the discriminant function using a reduced set of vectors. Burges [2] and Thies et al. [23] each provided an explicit way to build a reduced set of vectors, which are usually not support vectors, and their approaches can only be applied effectively to SVMs using quadratic kernels. Downs et al. [7] reduced the solution of support vectors base on their linear dependency without any loss of generalization, but their reduction rate is not high enough. Guo et al. [8] evaluated the contribution of each support vector to the shape of the discriminant function of a SVM by iteratively retraining the SVM using the training sample excluding that support vector. The result was a new set of support vectors, which is, as they claimed without any theoretic proof, of smaller size. Li et al. [15] built a reduced set of so-called feature vectors, which were chose from the support vector solutions based on the use of the vector correlation principle and a greedy algorithm. However, the solutions produced by their greedy strategy were not guaranteed to be optimal. In our work, we proposed an adaptive approach for optimally reducing the support vector solutions using a particle swarm optimization (PSO) algorithm based on a more reasonable fitness. It was theoretically and empirically proven that the reduced set of support vectors produced by the proposed approximation algorithm much better approximates the original discriminant function.

The remainder of this paper is organized as follows. We briefly describe the SVM Section 2. Section 3 we describe the theory of Simplifying the Solution of Support Vector Machines, and the mathematic model of the selection of support vectors problem. Section 4 we describe our proposed method that using particle swarm optimization algorithm to select the optimal vectors details. Moreover, the experiments are presented in Section 5. Finally, we give our concluding remarks in Section 6.

2 Support Vector Machines (SVMs)

For the sake of simple discussion, we shall concentrate on the two-class pattern recognition problem. SVM uses Support vectors kernel to map the data in input space to a high-dimension feature space in which we can solve the problem in linear form [4, 10, 17, 20, 23, 26].

Given the training sample $\{(x_i, y_i) : x_i \in R^d, y_i \in \{-1,+1\}\}_{i=1}^{N}$, then the two-class pattern recognition problem can be cast as the primal problem of finding a hyperplane:

$$w^T x + b = 1 \qquad (1)$$

where w is a d-dimensional normal vector, such that these two classes can be separated by two margins both parallel to the hyperplane; that is, for each x_i, i = 1, 2, ..., N:

$$w^T x_i + b \geq +1 + \zeta_i, \text{ for } y_i = +1 \qquad (2)$$

$$w^T x_i + b \leq -1 - \zeta_i, \text{ for } y_i = -1 \qquad (3)$$

where $\zeta_i \geq 0, i = 1, 2, ..., N$, are slack variables and b is the bias. This primal problem can easily be cast as the following quadratic optimization problem ([4, 6, 20]):

$$\begin{aligned}&\text{Minimizing } \psi(w, \zeta) = \frac{1}{2}\|w\|^2 + C\left(\sum_{i=1}^{N}\zeta_i\right)\\&\text{s.t. } y_i[(w \cdot x_i) - b] \geq 1 - \zeta_i,\\&\text{where } \zeta = (\zeta_1, \zeta_2, ..., \zeta_N).\end{aligned} \qquad (4)$$

The objective of a support vector machine is to determine the optimal w and optimal bias b such that the corresponding hyperplane separates the positive and negative training data with maximum margin and it produces the best generation performance. This hyperplane is called an optimal separating hyperplane. Using Lagrange multiplier techniques leads to the following dual optimization problem [4, 6, 20, 21].

2.1 Dual Optimization Problem

Given the training sample $\{(x_i, y_i) : x_i \in R^d, y_i \in \{-1, +1\}\}_{i=1}^{N}$, find the Lagrange multipliers $\{\alpha_i\}_{i=1}^{N}$ that minimize the objective function

$$Q(\alpha) = -\sum_{i=1}^{N} \alpha_i + \frac{1}{2} \sum_{i=1}^{N} \sum_{j=1}^{N} \alpha_i \alpha_j y_i y_j K(x_i, x_j) \quad (5)$$

subject to the constraints:

$$\sum_{i=1}^{N} \alpha_i y_i = 0 \quad 0 \le \alpha_i \le C \quad \text{for } i = 1, 2, 3, \dots, N \quad (6)$$

where C is a user-specified positive constant and K(•, •) is a positive semidefinite kernel function. For any positive semidefinite kernel K on a sample space X, there is a Hilbert space H with inner product <•, •> and a feature mapping $\phi : X \to H$ such that

$$K(x, x') = <\phi(x), \phi(x')>, \quad (x, x' \in X) \quad (7)$$

Let $\{\alpha_i\}_{i=1}^{N}$ be an optimal solution of the above problem, then the discriminant function takes the form:

$$f(x) = \sum_{i=1}^{N} \alpha_i y_i K(x_i, x) + b \quad (8)$$

It is apparent from the constraint (6) that $0 \le \alpha_i \le C$ holds for i = 1, 2,..., N. All training samples x_i such that $\alpha_i > 0$ are called support vectors. To distinguish between support vectors with $0 < \alpha_i < C$ and those with $\alpha_i = C$, the former are called unbounded support vectors while the latter are called bounded ones [4, 10, 11]. In the following, we assume without loss of generality that $0 < \alpha_i \le C$ for i = 1, 2,..., n, and $\alpha_i = 0$ for i = n + 1, n + 2, . . . , N. Thus, the discriminant function in (8) can be written as follows:

$$f(x) = \sum_{i=1}^{n} \alpha_i y_i K(x_i, x) + b. \quad (9)$$

From Eq. (9), we can see that if there are more support vectors then the decision function will be more complicated and the classification speed of the SVM will be slower. Therefore it is important to develop a way to simplify the decision function (9) effectively without reducing the generalization capability of the SVM.

For simplicity, in what follows, we let $\lambda_i = \alpha_i y_i$ so that $\lambda_i \in [-C, C] \setminus \{0\}$ and that the discriminant function given in (6) can be written in a further simpler form:

$$f(x) = b + <\phi(x), w> \qquad (10)$$

Where $w = \sum_{i=1}^{n} \lambda_i \phi(x_i)$ is the normal vector.

3 Simplifying the Solution of Support Vector Machines

3.1 Approximation of Support Vectors

For improving the testing speed of an SVM while remaining the accuracy acceptable, we propose to find a subset of support vectors that best approximates the discriminant function formed by the original support vectors.

Let $S = \{x_1, x_2, ..., x_n\}$ be the support vectors for a sample obtained by an SVM and let $X_F = \{x_{F_1}, x_{F_2}, ..., x_{F_m}\} \subseteq S$ ($m \leq n$) be a subset of vectors selected from S. If the mapping selected support vectors $\{\phi(x_{F_1}), \phi(x_{F_2}), ..., \phi(x_{F_m})\}$ span the vector space spanned by the mapping support vectors $\{\phi(x_1), \phi(x_2), ..., \phi(x_n)\}$, then any mapping support vector $\phi(x_i)$ can be exactly approximated by the mapping selected vectors or exactly expressed as a linear combination of them: $\phi(x_i) = \sum_{j=1}^{m} \beta_{ij} \phi(x_{F_j})$.

3.2 Error of the Reduced Set of Support Vectors

For a reduced set of selected support vectors $X_F = \{x_{F_1}, x_{F_2}, ..., x_{F_m}\}$, we are looking for coefficients $\beta_{ij} \in R, 1 \leq i \leq n, 1 \leq j \leq m$, such that each mapping support vector $\phi_i = \phi(x_i)$ can be best approximated by a linear combination of the mapping selected vectors $\phi(X_F): \phi_i \cong \tilde{\phi}_i = \sum_{j=1}^{m} \beta_{ij} \phi(x_{F_j})$. The coefficients β_{ij} are the least squares solution of the linear system $\phi_i = \sum_{j=1}^{m} \phi(x_{F_j}) x_{ij}$, $i = 1, 2, ..., n$. The discriminant function $f_{X_F}(x)$ determined by the selected vectors X_F can be expressed as:

$$f_{X_F}(x) = b + <\phi(x), w_{X_F}> \qquad (11)$$

Where $w_{X_F} = \sum_{i=1}^{n} \lambda_i \tilde{\phi}_i = \sum_{i=1}^{n} \lambda_i \sum_{j=1}^{m} \beta_{ij} \phi(x_{F_j}) = \sum_{j=1}^{m} \gamma_j \phi(x_{F_j})$ and $\gamma_j = \sum_{i=1}^{n} \lambda_i \beta_{ij}$. For searching the optimal reduced set of vectors we define the approximation error δ_{X_F} as follows:

$$\delta_{X_F} = \sum_{x_i \in S} \alpha_i^2 \delta_{F_i} \qquad (12)$$

Where

$$\delta_{F_i} = \|\alpha_i \phi_i - \alpha_i \tilde{\phi}_i\|^2$$
$$= \alpha_i^2 \phi_i^T \phi_i - 2\alpha_i^2 \phi_i^T \Phi_F \vec{\beta}_i + \alpha_i^2 \vec{\beta}_i^T \Phi_F^T \Phi_F \qquad (13)$$

And the coefficient vector $\vec{\beta}_i$ such that $\tilde{\phi}_i = \Phi_F \vec{\beta}_i$ best approximates ϕ_i is just a least squares solution to the linear system $\phi_i = \Phi_F \vec{\beta}$, which has a unique solution as given in (14) if $(\Phi_F^T \Phi_F)^{-1}$ exists.

$$\vec{\beta}_i = (\Phi_F^T \Phi_F)^{-1} \Phi_F^T \phi_i \qquad (14)$$

3.3 The Selection of Support Vectors Problem

As we have shown in above, our goal is to find out an optimal reduced subset of support vectors such that the discriminant function has the minimal error, therefore we propose to select optimal vectors to approximate the original support vectors with error δ_{X_F}.

In other words, we can think the question as that we want to select m support vectors from n support vectors. This problem is a combinatorial optimization problem. Therefore, we can apply the PSO algorithm to the combinatorial optimization problem, so the selection of support vectors is formulated as the following optimal problem.

$$\text{Minimize } \delta_{X_F}$$
$$\text{subject to } X_F \subset S \qquad (15)$$

3.4 Evaluation of the Approximation Error

The error δ_{X_F} can be used for performance evaluation and its expression given in (12) can be expanded as follows:

$$\delta_{X_F} = \sum_{x_i \in S} \alpha_i^2 (\phi_i^T \phi_i - \phi_i^T \Phi_F (\Phi_F^T \Phi_F)^{-1} \Phi_F^T \phi_i)$$
$$= \sum_{x_i \in S} (\alpha_i^2 \phi_i^T \phi_i) - \sum_{x_i \in S} \alpha_i^2 (\phi_i^T \Phi_F (\Phi_F^T \Phi_F)^{-1} \Phi_F^T \phi_i) \qquad (16)$$

Note that minimizing δ_{X_F} is equivalent to maximizing the summation $\sum_{x_i \in S} \alpha_i^2 (\phi_i^T \Phi_F (\Phi_F^T \Phi_F)^{-1} \Phi_F^T \phi_i)$ since $\sum_{x_i \in S} (\alpha_i^2 \phi_i^T \phi_i)$ is a constant. Thus, we may modify the sum of the approximation errors δ_{X_F} to:

$$\sum_{x_i \in S} \alpha_i^2 (\phi_i^T \Phi_F (\Phi_F^T \Phi_F)^{-1} \Phi_F^T \phi_i). \tag{17}$$

And we may replace Equation (12) in a further simpler form:

$$\delta_{X_F} = \sum_{x_i \in S} \alpha_i^2 (\phi_i^T \Phi_F (\Phi_F^T \Phi_F)^{-1} \Phi_F^T \phi_i). \tag{18}$$

In this work, we use PSO for searching an optimal solution. The PSO searching algorithms adopt the objective function $fit(X_F)$ given in (18).

4 The PSO Process Mechanism

In above describing, we can always encode m support vectors to one particle from support vectors set **X** for initial generation and the population size for each generation can set up as we need. Thus, one particle can express one feasible solution of the above optimization problem. Therefore, if we can use PSO to find the best particle, then the optimal solution then will be obtained.

Particle Swarm Optimization (PSO) is a randomized searching technique and originated from the simulation of social behavior of birds in a flock, which was developed by [17]. PSO updates a population of candidate solutions, called swarm. Each candidate solution in the swarm is called a particle. In PSO, each particle flies in the search space with a velocity adjusted by its own flying memory and its companion's flying experience. Its general steps are specified in the following.

A PSO algorithm first randomly initializes a swarm of particles. Each particle is represented as a string, say $C_i^k = c_{i1} c_{i2} \dots c_{im}$, $i = 1, 2, \dots, d$, where d is the number of particles or the swarm size. Thus, each particle is randomly placed in the m-dimensional space as a candidate solution. Each particle adjusts its trajectory toward its own previous best position and the global best position, namely $pbest_i$ and $gbest$. In each iteration, the swarm is updated by the following equations:

$$v_{ij}^k = \hat{\omega} \cdot v_{ij}^{k-1} + \theta_1 \eta_1 (pbest_{ij}^k - c_{ij}^k) + \theta_2 \eta_2 (gbest_j^k - c_{ij}^k), \tag{19}$$

$$c_{ij}^{k+1} = c_{ij}^k + v_{ij}^k, \tag{20}$$

where k is the current iteration number, v_{ij} is the updated velocity on the j-th dimension of the i-th particle, $\hat{\omega}$ is the inertia weight, θ_1 and θ_2 are acceleration constants, η_1 and η_2 are real numbers drawn from two uniform random sequences of $U(0, 1)$.

In the discrete binary version [13], a particle moves in a state space restricted to zero and one on each dimension, where each v_{ij} represents the probability of bit c_{ij} taking the value 1. Thus, the step for updating v_{ij} as shown in (19) remains unchanged, except that $pbest_{ij}$ and $gbest_j$ are integers in {0, 1} in binary case. The resulted changes in position are defined as follows:

$$s(v_{i,j}^k) = 1/(1 + \exp(-v_{i,j}^k)), \tag{21}$$

$$\text{if } (\tau < s(v_{i,j}^k)) \text{ then } c_{i,j}^k = 1 \text{ else } c_{i,j}^k = 0 \tag{22}$$

where τ is a random number drawn from uniform sequence of $U(0,1)$.

A discrete particle swarm works by adjusting trajectories through manipulation of each coordinate of a particle[13]. In a binary space, the velocity of a particle may be described by the change of some bits. Following this idea, we propose a modified version of discrete PSO to perform the search of optimal subset of support vectors, as described in the following. First, we take the velocity updating formula given in (19) and set $\hat{\omega}=0$, $\theta_1 = \theta_2 = \eta_1 = \eta_2 = 1$, to form a tentative formula:

$$U_i^k = (u_{i1}^k, u_{i2}^k, ..., u_{im}^k) = pbest_i^k + gbest^k - 2C_i^k \tag{23}$$

Let n_1 and n_2 denote the number of negative components and the number of positive components of U_i^k, respectively, and let $n_0 = \lceil \min(n_1, n_2)/2 \rceil$. For each component $u_{i,j}^k$ of U_i^k, we store its index j into the array Γ_1 if $u_{i,j}^k > 0$; otherwise, into the array Γ_2 if $u_{i,j}^k < 0$. From each of the sets Γ_1 and Γ_2, we randomly select n_0 elements to remain so that $|\Gamma_1| = |\Gamma_2| = n_0$. We give the velocity updating formula in (24), with which we simply take (19) as the position updating formula. We use T to denote the operation transform U_i^k into $V_i^k = (v_{i1}^k, v_{i2}^k, ..., v_{im}^k)$ and denote $V_i^k = T(U_i^k)$.

$$v_{i,j}^k = \begin{cases} 1 & \text{if } j \in \Gamma_1 \\ -1 & \text{if } j \in \Gamma_2 \\ 0 & \text{otherwise} \end{cases} \tag{24}$$

Each particle in a swarm, representing a candidate solution, is expressed as a binary vector as follows:

$$C_i^k = (c_{i1}^k, ..., c_{in}^k), \text{ subject to } \sum_{j=1}^{n} c_{i,j}^k = m, \tag{25}$$

where $c_{i,j}^k$ is 1, if the j-th support vector is selected and is 0, otherwise. We define the fitness for the particle C corresponding to the reduced set X_F as $fit_{PSO}(C) = fit(X_F)$, where the fitness function fit was given in (18). The procedure of the proposed binary PSO algorithm for searching the optimal subset of support vectors is described as follows:

Algorithm PSO Reduction of Solutions for SVM

Step 1. $k := 0$;
//Initialization
  for $i := 1$ to d
    randomly generate initial position C_i^0; $pbest_i := C_i^0$;
  $r := \arg\max_{1 \leq j \leq d} fit_{PSO}(C_j^0)$; $gbest := C_r^0$;
Step 2. for $i := 1$ to d
//updating
  $U_i^k := pbest_i^k + gbest^k - 2C_i^k$; $V_i^k := T(U_i^k)$; $C_i^{k+1} := C_i^k + V_i^k$;
  if $fit_{PSO}(C_i^{k+1}) > fit_{PSO}(pbest_i)$ then $pbest_i := C_i^{k+1}$;
  if $fit_{PSO}(C_i^{k+1}) > fit_{PSO}(gbest)$ then $gbest := C_i^{k+1}$;
  $k := k + 1$;
Step 3. if the stopping criteria are met **then stop else goto** Step 2
//repeat or not

By above describing, we point out how to apply the PSO mechanism this probabilistic search technique to as a postprocessing method to select optimal feature vectors to approximate original support vectors and use these optimal feature vectors as the final solutions so that we can improve the sparsity of the solutions and reduce the scale of the test computation complexity.

5 Experimental Results and Comparisons

Our experiments were implemented on a PC with 2.8 GHz Pentium Dual-Core processor and 4 GB RAM using Borland C++ Builder 6.0 complier. We carried out experiments on spirals [19] and cubic polynomials. Each test includes randomly generated 3000 samples, 300 of them were randomly chosen as training data, and the remaining as test data. For this test we set σ = 8 and C = 10 for spirals and σ = 40 and C = 1 for cubic polynomials. The population size was set to 50, no. of iterations for PSO is set to 200 and *RBF* kernel $K(x,y) = \exp(-\|x-y\|^2/2\sigma^2)$ is used. The recognition rates presented in Tables 1-2 refer to the mean of 30 executions of the algorithm. Tables 1 and 2 we compare the proposed methods with Li's method [15]. The experimental results show that the Our PSO methods can effectively reduce the solutions for an SVM. The parametric equations for each of the testing examples are given in the following.

$$\begin{cases} \text{Spiral-1:} \ (x,y) = ((4\theta+10)\cos(\theta),(4\theta+10)\sin(\theta)), \\ \text{Spiral-2:} \ (x,y) = ((4\theta+1)\cos(\theta),(4\theta+1)\sin(\theta)) \end{cases} \quad (26)$$

$$\begin{cases} \text{Cubicpolynomial-1:} \ (x,y) = (t,\ 200+2t-0.031t^2+0.001t^3), \\ \text{Cubicpolynomial-2:} \ (x,y) = (t,\ 150+1.8t-0.0312t^2+0.001t^3) \end{cases} \quad (27)$$

Table 1. Recognition rates for two spirals.

m	Reduction rate (%)	Recognition rate		
		Li's method	Previous work(GA)	PSO
5	94.8	0.5000	0.5030	0.7485
10	89.8	0.5000	0.7375	0.7895
15	84.6	0.5315	0.7440	0.8005
25	74.4	0.5470	0.9300	0.9350
n=98	0.0	Recognition rate of the original support vectors = 0.9995		

Table 2. Recognition rates for two waveform graphs.

m	Reduction rate (%)	Recognition rate		
		Li's method	Previous work(GA)	PSO
5	95.1	0.4880	0.4965	0.6215
20	80.6	0.5185	0.6535	0.6535
30	70.9	0.5805	0.7045	0.7145
40	61.2	0.5600	0.7085	0.8290
n=103	0.0	Recognition rate of the original support vectors = 0.9700		

6 Conclusions

We have proposed algorithms for reducing solutions for SVMs. Experimental results have shown that PSO outperforms our previously work [16] and Li's method [15]. And the PSO method is fast converge performance to find the best solution and take advantage of the easily implementing performance. The simulation results also demonstrated that the proposed PSO based algorithm can be easily realized with high efficiency and can obtain higher quality solution than our previously work [16] and Li's method [15]. Although PSO finds good solutions much faster than other evolutionary algorithms, it usually can not improve the quality of the solutions as the number of iterations is increased; that is, PSO usually suffers from premature convergence. Future work will concentrate on solving these problems. For the problem, we shall consider some a more reasonable fitness to optimally reduce the solutions for an SVM or other swarm intelligence (SI) technique further improves the effectiveness of PSO.

References

1. Boser, B.E., Guyon, I.M., Vapnik, V.N.: A training algorithm for optimal margin classifiers. In: Proceedings of the 5th Annual ACM Workshop on Computational Learning Theory, pp. 144–152 (1992)
2. Burges, C.J.C.: Simplified support vector decision rules. In: Proceedings 13th International Conference on Machine Learning, Bari, Italy, pp. 71–77 (1996)

3. Brank, J., Grobelnik, M., Milic-Frayling, N., Mladenic, D.: Feature selection using linear support vector machines, Microsoft Research Technical Report (MSR-TR-2002-63), 12 (June 2002)
4. Burges, C.J.C.: Geometry and invariance in kernel based method. In: Advance in Kernel Method-Support Vector Learning, pp. 86–116. MIT Press, Cambridge (1999)
5. Chakraborty, B., Chaudhuri, P.: On the use of genetic algorithm with elitism in robust and nonparametric multivariate analysis. Austrian Journal of Statistics 32(1&2), 13–27 (2003)
6. Cortes, C., Vapnik, V.: Support Vector Networks. Machine Learning 20, 273–297 (1995)
7. Downs, T., Gates, K.E., Masters, A., Cristianini, N., Shaw-taylor, J., Williamson, R.C., Masters, A.: Exact simplification of support vector solutions. Journal of Machine Learning Research 2, 293–297 (2001)
8. Guo, J., Takahashi, N., Nishi, T.: A Learning Algorithm for Improving the Classification Speed of Support Vector Machines. In: Proceedings of 2005 European Conference on Circuit Theory and Design, ECCTD 2005 (August-September 2005)
9. Goldberg, D.E.: Genetic Algorithms in Search, Optimization and Machine Learning. Addison Wesley, New York (1989)
10. Graepel, T., Herbrich, R., Shawe-Taylor, J.: Generalisation error bounds for sparse linear classiers. In: Proceedings of the Thirteenth Annual Conference on Computational Learning Theory, pp. 298–303 (2000)
11. Joachims, T.: Estimating the generalization performance of an SVM efficiently. In: Proceedings ICML 2000,17th International Conference on Machine Learning, pp. 431–438 (1999)
12. Kennedy, J., Eberhart, R.: Particle Swarm Optimization. In: Proceedings of IEEE International Conference on Neural Networks, Perth, Australia, vol. 4, pp. 1942–1948 (1995)
13. Kennedy, J., Eberhart, R.C.: A discrete binary version of the particle swarm algorithm. In: Proceedings of the 1997 Conference on Systems, Man, and Cybernetics, pp. 4104–4109. IEEE, Piscataway (1997)
14. Lecun, Y., Jackel, L.D., Eduard, H.A., Bottou, N., Cartes, C., Denker, J.S., Drucker, H., Sackinger, E., Simard, P., Vapnik, V.: Learning algorithms for classification: A comparison on handwritten digit recognition. Neural Network, 261–276 (1995)
15. Li, Q., Jiao, L., Hao, Y.: Adaptive simplification of solution for support vector machine. Pattern Recognition 40(3), 972–980 (2007)
16. Lin, H.J., Yeh, J.P.: Optimal Reduction of Solutions for Support Vector Machines. Applied Mathematics and Computation 214(2), 329–335 (2009)
17. Lin, H.J., Yen, S.H., Yeh, J.P., Lin, M.J.: Face Detection Based on Skin Color Segmentation and SVM Classification. In: The Second IEEE International Conference on Secure System Integration and Reliability Improvement (SSIRI 2008), Yokohama, Japan, July 14-17 (2008)
18. Liu, C., Nakashima, K., Sako, H., Fujisawa, H.: Handwritten digit recognition: Benchmarking of state-of-the-art techniques. Pattern Recognition 36, 2271–2285 (2003)
19. Lang, K.J., Witbrock, M.J.: Learning to tell two spirals apart. In: Proceedings of 1989 Connectionist Models Summer School, pp. 52–61 (1989)
20. Platt, J.: Sequential minimal optimization: A fast algorithm for training support vector machines. Technical report, Microsoft Research, Redmond (1998)
21. Ruszczynski, A.P.: Nonlinear Optimization, p. 160. Princeton University Press, Princeton (2006)
22. Scholkopf, B., Smola, A.: Learning with Kernels. MIT Press, Cambridge (1999)

23. Thies, T., Weber, F.: Optimal reduced-set vectors for support vector machines with a quadratic kernel. Neural Computation 16, 1769–1777 (2004)
24. Vapnik, V.N.: The Nature of Statistical Learning Theory. Springer, New York (1995)
25. Vapnik, V.N.: Statistical Learning Theory. Wiley, New York (1998)
26. Yeh, J.P., Pai, I.C., Wang, C.W., Yang, F.W., Lin, H.J.: Face Detection using SVM-Based Classification. The Special Issue of Far East Journal of Experimental and Theoretical Artificial Intelligence (FEJETAI) 3(2), 113–123 (2009)

Improved Relaxation Algorithm for Passive Sensor Data Association

Cheng Ouyang, Hong-Bing Ji, and Jin-Long Yang

School of Electronic Engineering, Xidian University, Xi'an, 710071, China

Abstract. The Lagrangian relaxation algorithm is widely used for passive sensor data association. However, there are two major problems about it. Firstly, the cost function of the algorithm is computed by using least square estimation of the target position without taking the estimation errors into account. To solve this problem, a modified cost function is derived which can reflect the correlation between measurements more reasonably owing to the integration of estimation errors. Secondly, due to the fact that building the candidate assignment tree would take a lot of CPU time, we propose a statistic test based on indicator function with a great improvement of the computational efficiency. Simulation results show that both the correlation accuracy and the computational efficiency of the improved relaxation algorithm are higher than that of the traditional one.

Keywords: multidimensional assignment, Lagrangian relaxation, passive sensor, data association.

1 Introduction

In the passive sensor data association problem, it is a key issue to determine from which target, if any, a particular measurement originates. Such a problem can be solved by a wide range of algorithms including greedy heuristics, tabu search, simulated annealing, genetic algorithms, neural networks, and Lagrangian relaxation 1. Among these algorithms, Lagrangian relaxation-based methods have been found to perform well in tracking applications 2. The advantage of relaxation approach is that the resulting dual optimal cost is a lower bound of the feasible cost and, hence, provides a measure of how close the feasible solution is to the optimal solution. For the passive sensor data association problem, the feasible solution costs are typically within 1% of their corresponding dual optimal costs 3.

The Lagrangian relaxation algorithm introduced in 3 solved a 3-D assignment problem as a series of 2-D subproblems, by relaxing the third constraint and appending it to the cost function using Lagrange multipliers. However, when $S \geq 4$, multiple sets of constraints have to be relaxed. In the method used by Poore 4, the constraints are relaxed one set a time, and then the S -1 dimensional subproblem can be solved iteratively. However, the $S-1$ dimensional subproblem is not optimally solvable in polynomial time if $S \geq 4$, so Poore's approach will take unacceptable computational resources for dense graphys. Pattipati and Deb solved this problem by succesive relaxation and constraint enforcement, so all the $S-2$ dimensional constraints are relaxed simultaneously, and then the subproblem is a 2-D assignment problem that can be optimally solved in pseudo-polynomial time 5.

These algorithms above are built upon the same cost function which represents a generalized likelihood so that track association becomes a generalized likelihood ratio test (GLRT). Unfortunately, the GLRT does not necessarily maximize the probability of correct track associations. Recently, Kaplan derived a cost function in 6 as the limit of the likelihood when the target state is random and uniformly distributed over a space that approaches infinite support. However, such a cost function is derived for active sensors, but not for passive sensors. This paper proposed a modified cost function derived for passive sensors and the performance of which is better than that of the generalized likelihood function owing to the integration of estimation errors.

Another key issue to the Lagrangian relaxation algorithm is that building the candidate assignment tree would take a lot of CPU time. Chummun presented a fast data association algorithm which combined clustering techniques with a multidimensional assignment method in a systematic manner 7. Because the assignment problem is partitioned into small subproblems, the CPU time required for building the candidate assignment tree is reduced. However, the designed parameters used to estimate the variable cluster size are selected empirically, and the dihedral angles used in clustering can not sufficiently reflect the possibility that two lines of sight (LOS) come from the same target. A statistic test based on indicator function is proposed in this paper, which can achieve the same result as that of Lagrangian relaxation algorithm in some special cases, thus a part of correct pairs can be selected directly without redundant relaxation and enforcement processes, and then the possible set can be simplified according to the constraint conditions so that more correct pairs can be picked out by repeating such a process.

2 Problem Formulation

Suppose that the actual position of target j is $\mathbf{X}_j = (x_j, y_j, z_j)^T$, and the position of sensor s is $\mathbf{X}_s = (x_s, y_s, z_s)^T$, then the measurements from sensor s are $\mathbf{z}_{si_s}$.

$$\mathbf{z}_{si_s} = \begin{bmatrix} \beta_{si_s} \\ \varepsilon_{si_s} \end{bmatrix} = \begin{bmatrix} \arccos\left(\dfrac{x_j - x_s}{\sqrt{(x_j - x_s)^2 + (y_j - y_s)^2}}\right) \\ \arctan\left(\dfrac{z_j - z_s}{\sqrt{(x_j - x_s)^2 + (y_j - y_s)^2}}\right) \end{bmatrix} + \begin{bmatrix} v_{\beta_{si_s}} \\ v_{\varepsilon_{si_s}} \end{bmatrix} = h(\mathbf{X}_j, \mathbf{X}_s) + \mathbf{v}_{si_s} \quad (1)$$

where $\mathbf{v}_{si_s}$ is the measurement noise of sensor s. Suppose that the measurement error is zero mean Gaussian white noise, then $E(\mathbf{v}_{si_s}) = 0$ and $E(\mathbf{v}_{si_s} \mathbf{v}_{si_s}^T) = \mathbf{R}_s$. Taking a measurement from each sensor, we can build a S-tuple set of $\{\mathbf{z}_{1i_1}, \mathbf{z}_{2i_2}, ..., \mathbf{z}_{Si_S}\}$, which has a joint Gaussian probability density function as follow,

$$f(\mathbf{z}_{1i_1}, \mathbf{z}_{2i_2}, \cdots, \mathbf{z}_{Si_S} | \mathbf{X}_j)$$

$$= \prod_{s=1}^{S} \left(\frac{P_{D_s}}{|2\pi \mathbf{R}_s|^{1/2}} \exp\left\{ -\frac{1}{2} \left[\mathbf{z}_{si_s} - h(\mathbf{X}_j, \mathbf{X}_s) \right]^T \mathbf{R}_s^{-1} \left[\mathbf{z}_{si_s} - h(\mathbf{X}_j, \mathbf{X}_s) \right] \right\} \right)^{u(i_s)} \left[1 - P_{D_s} \right]^{1-u(i_s)} \quad (2)$$

where $u(i_s)$ is an indicator function, defined as

$$u(i_s) = \begin{cases} 0, & \text{if } i_s = 0 \\ 1, & \text{otherwise} \end{cases} \quad (3)$$

Suppose that clutters are uniformly distributed in the field of view, the joint probability density function of measurements from clutters is

$$f_c = \prod_{s=1}^{S} \left(\frac{1}{V_s} \right)^{u(i_s)} \quad (4)$$

where V_s is the volume of the field of view of sensor s.

The cost function of S-tuple $\{\mathbf{z}_{1i_1}, \mathbf{z}_{2i_2}, \ldots, \mathbf{z}_{Si_S}\}$ associated with target j is given by the negative likelihood ratio as follow

$$c_{i_1 i_2 \cdots i_S} = -\ln\left(f(\mathbf{z}_{1i_1}, \mathbf{z}_{2i_2}, \cdots, \mathbf{z}_{Si_S} | \hat{\mathbf{X}}_j) / f_c \right)$$

$$= \sum_{s=1}^{S} [u(i_s) - 1] \ln(1 - P_{D_s}) - u(i_s) \ln\left(\frac{P_{D_s} V_s}{|2\pi(\mathbf{R}_s)|^{1/2}} \right) \quad (5)$$

$$+ u(i_s) \left\{ \frac{1}{2\mathbf{R}_s} \left[\mathbf{z}_{si_s} - h(\mathbf{X}_j, \mathbf{X}_s) \right]^T \left[\mathbf{z}_{si_s} - h(\mathbf{X}_j, \mathbf{X}_s) \right] \right\}$$

So the problem of passive sensor data association can be reformulated as the following S-D assignment problem,

$$\min \sum_{i_1=0}^{n_1} \sum_{i_2=0}^{n_2} \cdots \sum_{i_S=0}^{n_S} \rho_{i_1 i_2 \cdots i_S} \times c_{i_1 i_2 \cdots i_S} \quad (6)$$

subject to

$$\left. \begin{array}{l} \sum_{i_2=0}^{n_2} \sum_{i_3=0}^{n_3} \cdots \sum_{i_S=0}^{n_S} \rho_{i_1 i_2 \cdots i_S} = 1; \quad \forall i_1 = 1, 2, \cdots, n_1 \\ \sum_{i_1=0}^{n_1} \sum_{i_3=0}^{n_3} \cdots \sum_{i_S=0}^{n_S} \rho_{i_1 i_2 \cdots i_S} = 1; \quad \forall i_2 = 1, 2, \cdots, n_2 \\ \vdots \qquad \qquad \vdots \\ \sum_{i_1=0}^{n_1} \sum_{i_2=0}^{n_2} \cdots \sum_{i_{S-1}=0}^{n_{S-1}} \rho_{i_1 i_2 \cdots i_S} = 1; \quad \forall i_S = 1, 2, \cdots, n_S \end{array} \right\} \quad (7)$$

where $n_s, s = 1, 2, ..., S$, represent the measurement number of sensor s, and $\rho_{i_1 i_2 \cdots i_S}$ are binary association variables such that $\rho_{i_1 i_2 \cdots i_S} = 1$ if the S-tuple $\{z_{1i_1}, z_{2i_2}, \cdots, z_{Si_S}\}$ is associated with a candidate target, otherwise, it is set to zero.

The computational cost of the S-D problem grows exponentially with the increase of the number of sensors and measurements. It is impossible to find the optimal solution in polynomial time even under the assumption of unity detection probability and no spurious measurements, hence is commonly referred to as NP-hard problem. The generalized S-D assignment algorithm proposed by Pattipati and Deb solved this problem by successive relaxation and constraint enforcement 5, so all the ($S-2$)D constraints are relaxed simultaneously, and then the subproblem is a 2-D assignment problem that can be optimally solved by the auction algorithm as follows 8:

Consider N persons wishing to divide among themselves N objects. For each person i there is a nonempty subset $A(i)$ of objects that can be assigned to i. An assignment S is a set of person-object pairs (i, j) such that $j \in A(i)$ for all $(i, j) \in S$, for each person i there is at most one pair $(i, j) \in S$, and for each object j there is at most one pair $(i, j) \in S$. Let a_{ij} represent a given integer value that a person i associates with an object $j \in A(i)$, and p_j represent the price of object j, then the auction algorithm proceeds iterative and terminates when a complete assignment is obtained. There are two phases in each iteration, i.e., the bidding phase and the assignment phase.

Bidding Phase: For each person i that is unassigned under the assignment S:
Compute the "current value" of each object $j \in A(i)$ given by

$$v_{ij} = a_{ij} - p_j \tag{8}$$

Find a "best object" j^* having maximum value

$$v_{ij^*} = \max_{j \in A(i)} v_{ij}, \tag{9}$$

and find the best value offered by objects other than j^*

$$w_{ij^*} = \max_{j \in A(i), j \neq j^*} \{a_{ij} - p_j\}. \tag{10}$$

(If j^* is the only object in $A(i)$ we define w_{ij^*} to be $-\infty$, or, for computational purposes, a number that is much smaller than v_{ij^*}.)

Compute the "bid" of person i for object j^* given by

$$b_{ij^*} = p_{j^*} + v_{ij^*} - w_{ij^*} + \varepsilon = a_{ij^*} - w_{ij^*} + \varepsilon. \tag{11}$$

Assignment Phase: For each object j:
Let $P(j)$ be the set of persons from which j receives a bid in the bidding phase of the iteration.

If $P(j)$ is nonempty, increase p_j to the highest bid

$$p_j = \max_{i \in P(j)} b_{ij}, \qquad (12)$$

remove any pair (i, j) (if one exists) from the assignment S, and add the pair (i^*, j) to S where i^* is some person in $P(j)$ attaining the maximum above.

3 Improved Relaxation Algorithm

3.1 Modified Cost Function

The cost function is a key factor to the Lagrangian relaxation algorithm especially for passive sensor data association, where due to the nonlinearity in the measurement equation, the actual position of targets can not be observed directly, and a least square estimation would bring about some estimation errors.

At a given instance of time, the likelihood function is given by

$$f\left(\mathbf{z}_{1k_1}, ..., \mathbf{z}_{Sk_S} \big| \mathbf{X}_j\right) = f\left(\mathbf{z}_{1i_1} \big| \mathbf{X}_j\right) f\left(\mathbf{z}_{2i_2} \big| \mathbf{Z}_{1:1}, \mathbf{X}_j\right) \cdots f\left(\mathbf{z}_{Si_S} \big| \mathbf{Z}_{1:S-1}, \mathbf{X}_j\right) \qquad (13)$$

If $\mathbf{X}_j$ is the actual position of target j, then the likelihood function is shown in Equation (2). However, the actual position of target j is unknown, so it is replaced by the least squares estimation $\hat{\mathbf{X}}_j$, then the estimation errors should be taken into account. If letting $\mathbf{C}_s$ represent the covariance matrix, then the modified likelihood function can be expressed by

$$f'(\mathbf{z}_{1i_1}, \mathbf{z}_{2i_2}, \cdots, \mathbf{z}_{Si_S} \big| \hat{\mathbf{X}}_j)$$
$$= \prod_{s=1}^{S} \left(\frac{P_{D_s}}{|2\pi\mathbf{C}_s|^{1/2}} \exp\left\{ -\frac{1}{2} \left[\mathbf{z}_{si_s} - h(\hat{\mathbf{X}}_j, \mathbf{X}_s)\right]^T \mathbf{C}_s^{-1} \left[\mathbf{z}_{si_s} - h(\hat{\mathbf{X}}_j, \mathbf{X}_s)\right] \right\} \right)^{u(i_s)} \left[1 - P_{D_s}\right]^{1-u(i_s)}$$
(14)

where

$$\mathbf{C}_s = E\left[\left(\mathbf{z}_{si_s} - h(\hat{\mathbf{X}}_j, \mathbf{X}_s)\right)\left(\mathbf{z}_{si_s} - h(\hat{\mathbf{X}}_j, \mathbf{X}_s)\right)^T\right] \doteq \mathbf{R}_s + \mathbf{H}_\mathbf{X} \mathbf{R}_f \mathbf{H}_\mathbf{X}^T \qquad (15)$$

where $\mathbf{R}_f$ is the fusion covariance matrix of estimation $\hat{\mathbf{X}}_j$, and $\mathbf{H}_\mathbf{X}$ is the Jacobian matrix represented by

$$\mathbf{H}_\mathbf{X} = \frac{\partial h(\hat{\mathbf{X}}_j, \mathbf{X}_s)}{\partial \mathbf{X}} \bigg|_{\mathbf{X} = \hat{\mathbf{x}}_j} \qquad (16)$$

Similar to the generalized likelihood function, the modified cost function can also be represented by the negative likelihood ratio as follow,

$$c'_{i_1 i_2 \cdots i_S} = -\ln\left(f'(\mathbf{z}_{1i_1}, \mathbf{z}_{2i_2}, \cdots, \mathbf{z}_{Si_S} \mid \hat{\mathbf{X}}_j)/f_c\right)$$

$$= \sum_{s=1}^{S}[u(i_s)-1]\ln(1-P_{D_s}) - u(i_s)\ln\left(\frac{P_{D_s}V_s}{|2\pi(\mathbf{C}_s)|^{1/2}}\right) \quad (17)$$

$$+ u(i_s)\left\{\frac{1}{2(\mathbf{C}_s)}\left[\mathbf{z}_{si_s} - \mathbf{h}(\mathbf{X}_j, \mathbf{X}_s)\right]^T\left[\mathbf{z}_{si_s} - \mathbf{h}(\mathbf{X}_j, \mathbf{X}_s)\right]\right\}$$

As seen from the comparison of Equations (5) and (17), the modified cost function is different from the original one only in the covariance matrix. It is more reasonable because the least squares estimation of the actual position would bring about some estimation errors which should be considered in the covariance matrix.

3.2 Statistic Test Based on Indicator Function

As seen from Equation (11), if j^* is the only object in $A(i)$, w_{ij^*} is defined to be $-\infty$, then the "bid" of person i for object j^* is given by

$$b_{ij^*} = p_{j^*} + v_{ij^*} - w_{ij^*} + \varepsilon = \infty, \quad (18)$$

which means that no matter how many iterations it is, the "bid" of person i for object j^* is the highest, so the object j^* can be only assigned to person i in this case.

Consider the 2-tuple $\{\gamma_k, i_r\}$ as a candidate person-object association, if i_r^* is the only object in $A(\gamma_k)$, then the object i_r^* can be only assigned to person γ_k.

Recall that each S-D assignment problem consisted of S lists with n_s measurements in list $s = 1,...,S$. Each S-tuple $(i_1,...,i_S)$ represents a candidate pair, and all the possible correct pairs are put into a possible set P,

$$P = \bigcup_{k=1}^{M}\{(i_{1_k},...,i_{S_k})\} \quad (19)$$

where M is the total number of candidate pairs.

We define the indicator function $\xi(i_s)$ as

$$\xi(i_s) = NUM(P, i_s), \quad s = 1,...,S \quad (20)$$

where the function $NUM(\cdot)$ means counting the number of element i_s in set P.

If $\xi(i_s^*)$ equals one, it indicates that i_s^* is the only object in $A(\gamma_k)$, where

$$\gamma_k = \{i_{1_k},...,i_{s_k},...,i_{S_k}\} \setminus i_s^*, \quad i_{s_k} = i_s^* \quad (21)$$

then we can assign i_s^* to γ_k directly without relaxation and enforcement, and the possible set can be simplified according to the constraint Equation (7). After that, the indicator function will be calculated again, and there may be new $i_s^{**}, i_s^{**} \neq i_s^*$, making $\xi(i_s^{**}) = 1$, resulting in a chain reaction.

According to this principle, a new method of statistic test based on indicator function is proposed. Firstly, a two-sensor statistic test using the distance between two lines of sight from two different sensors has been made 9, and a part of correct pairs is selected according to the indicator function. If there is no i_s^* making $\xi(i_s^*)=1$, then a three-sensor statistic test based on the relationship between the angle measurements from three different sensors will be made 9, and another part of correct pairs can be selected until there is no i_s^* making $\xi(i_s^*)=1$ again. After these two statistic tests, the size of the possible set will become very small, in some cases all the correct pairs can even be picked out directly, so the redundant relaxation and enforcement processes are avoided, a great improvement of the computational efficiency. A brief approach of the algorithm is presented as follows:

Step 1. Initialize, construct the possible set P. Let the correct set $Z = NULL$ and set $flag = 0$, where $flag$ is used to mark whether the three-sensor statistic test has been made or not.

Step 2. Delete the impossible candidate pairs from the possible set P through the two-sensor statistic test.

Step 3. If the possible set P is not empty, go to Step 4, otherwise, the algorithm finishes and the associated pairs in the correct set Z are the final results.

Step 4. Compute the indicator function $\xi(i_s)$ of P according to Equation (20). If there is i_s^* making $\xi(i_s^*)=1$, put the candidate pair including i_s^* into the correct set Z and simplify the possible set P according to Equation (7), then go to Step 3. If there is no i_s^* making $\xi(i_s^*)=1$ and $flag = 0$, go to Step 5, otherwise, go to Step 6.

Step 5. Delete the impossible candidate pairs from the possible set P through the three-sensor statistic test, then set $flag = 1$ and go to Step 3.

Step 6. Compute the modified cost functions of candidate pairs in P according to Equation (17).

Step 7. Pick out the remaining correct pairs using the Lagrangian relaxation algorithm according to the modified cost functions, then the algorithm finishes and the associated pairs in the correct set Z are the final results.

4 Simulation Results

A. The comparison of different cost functions.

In this simulation, a simple illuminative case with the probability of detection $P_d = 1$ and the false alarm density $P_f = 0$ is considered. The number of sensors is three and the number of targets is two, so the total number of possible assignment hypotheses is only four. Then we can compare the performances of different cost functions without considering the assignment algorithm in this case. The positions of the three sensors are $s_1(0,10,0), s_2(10,0,0), s_3(0,0,0)$, respectively, and the measurement standard deviation is $\sigma_s = 0.005$. For a given target separation, 10000 independent Monte Carlo simulations were run. In each simulation, two targets with fixed distance d are generated randomly in the filed of view. d changes from 0.005km to 1km with a step of 0.005km. The resulting percent of correct assignment $p(\%)$ as the function of target separations $d(km)$ are plotted in Fig. 1.

As seen from Fig. 1, when the target separation is too big or too small, in which cases it is either too easy or too difficult to distinguish different targets. Thus, the two different cost functions can get almost the same result. However, when the target separation is appropriate, the modified cost function results in a significantly better performance than the original one. This is due to the fact that the modified function integrates the least squares estimation errors, which can reflect the correlation between measurements more reasonably.

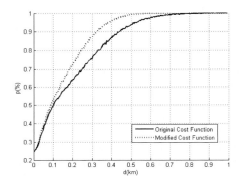

Fig. 1. Percent of correct assignment versus target separation with $\sigma_s = 0.005$.

B. The performance of the algorithms with different target numbers.

In this simulation, N targets in a queue are observed by three sensors, where $N \in \{5, 10, 15, 20\}$, and the positions of the targets are $\mathbf{t}_j (5 + j \cdot d, 5, 1)$, $j = -\left[\frac{N-1}{2}\right], -\left[\frac{N-1}{2}\right] + 1, \ldots, \left[\frac{N-1}{2}\right]$, where $[\cdot]$ means taking the integer part of the numbers and the target separation is $d = 1km$. The positions of the three sensors are $\mathbf{s}_1 (0, 10, 0), \mathbf{s}_2 (10, 0, 0), \mathbf{s}_3 (0, 0, 0)$, respectively, and the measurement standard deviation is $\sigma_s = 0.01$. A comparison of the two algorithms with different target numbers is made in the perfect detection case. The average results of 1000 Monte Carlo simulations are shown in Fig. 6, Fig. 7 and Fig. 8, respectively.

The percent of correct assignment versus target number is shown in Fig. 2. An obvious trend in the results is that the large target number results in unreliable data association, which is due to the fact that a large number of targets may create a large number of ghosts, making deghosting difficult. However, the percent of correct assignment of the improved algorithm is always higher than that of the original one, the reason behind which lies in our modified cost function discussed before.

The simulation time versus target number is shown in Fig. 3. It is obvious that when the target number is small, the running time of the two algorithms is close to each other. However, when the target number is large, the running time of the improved algorithm is much less than that of the original one, and the larger the target number is, the greater the gap is. This is due to the fact that when the target number is small, the candidate assignment tree is small, so the traditional Lagrangian relaxation algorithm requires only a few iterations to get the correct results and the advantage of the indicator function is not that obvious. However, when the target number is large, the candidate assignment tree is large, so the traditional Lagrangian relaxation algorithm requires a large number of iterations to get the correct results, while the improved relaxation algorithm can select a part of correct pairs by using the indicator function without iterations, and then the possible set can be simplified, a significant improvement of the computational efficiency.

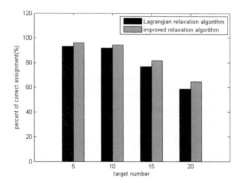

Fig. 2. Percent of correct assignment.

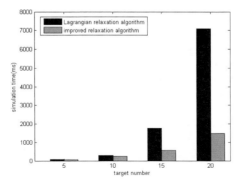

Fig. 3 Simulation time.

5 Conclusion

In this paper we present an improved relaxation algorithm for passive sensor data association, which improves the traditional one in two aspects.

First, the cost function is a key factor to the Lagrangian relaxation algorithm especially for passive target tracking problems, where due to target state-to-measurement nonlinear, the actual position of targets can not be observed directly, and the least square estimation would bring about some estimation errors. However, the generalized likelihood function is computed by using least square estimation of the target position without taking the estimation errors into account. So a modified cost function is derived which can reflect the correlation between measurements more reasonably and the performance of which is better than that of the generalized likelihood function owing to the integration of estimation errors.

Second, due to the fact that building the candidate assignment tree would take a lot of CPU time. A statistic test based on indicator function is proposed which can improve the computational efficiency. It is proved that the method based on indicator function can get the same result as that of Lagrangian relaxation algorithm in some special cases, so a part of correct pairs can be selected directly without redundant relaxation and enforcement processes, then the possible set can be simplified according to the constraint conditions so that more correct pairs can be picked out by repeating such a process.

The simulation results show that the correlation accuracy and computational efficiency of the improved relaxation algorithm are both higher than that of the traditional one, implying good application prospect.

Acknowledgements

This work was supported by the National Natural Science Foundation of China No. 60871074.

References

1. Popp, R., Pattipati, K.R., Bar-Shalom, Y.: An M-best multidimensional data association algorithm for multisensor multitarget tracking. IEEE Transactions on Aerospace and Electronic Systems 37(1), 22–39 (2001)
2. Poore, A., Robertson III, A.J.: A new class of Lagrangian relaxation based algorithms for a class of multidimensional assignment problems. Computational Optimization and Applications 8, 129–150 (1997)
3. Pattipati, K., Deb, S., Bar-Shalom, Y., Washburn, R.: A new relaxation algorithm and passive sensor data association. IEEE Transactions on Automatic Control 37(2), 197–213 (1992)
4. Poore, A., Rijavec, N.: A Lagrangian relaxation algorithm for multidimensional assignment problems arising from multitarget tracking. SIAM Journal of Optimization 3(3), 544–563 (1993)
5. Deb, S., Yeddanapudi, M., Pattipati, K., Bar-Shalom, Y.: A generalized S-D assignment algorithm for multisensor-multitarget state estimation. IEEE Transactions on Aerospace and Electronic Systems 33(2), 523–538 (1997)

6. Kaplan, L., Bar-Shalom, Y., Blair, W.D.: Assignment costs for multiple sensor track-to-track association. IEEE Transactions on Aerospace and Electronic Systems (2008)
7. Chummun, M.R., Kirubarajan, T., Pattipati, K.R., Bar-Shalom, Y.: Fast Data Association Using Multidimensional Assignment with Clustering. IEEE Transactions on Aerospace and Electronic Systems 37(3), 898–913 (2001)
8. Bertsekas, D.P.: The auction algorithm: A distributed relaxation method for the assignment problem. Annals of Operat. Res. 14, 105–123 (1988)
9. Liu, H., Dou, L.-h., Pan, F., Dong, L.-x.: Research on data association in three passive sensors network. In: IEEE International Conference on control and Automation, Guangzhou, China, May 30-June 1, pp. 3235–3238 (2007)

A Novel Integrated Cmos Uhf Rfid Reader Transceiver Front-End Design

Chao Yuan, Chunhua Wang, and Jingru Sun

School of Computer and Communication
Hunan University
Changsha, Hunan Province, P.R. China
yuanchao_hn@163.com, wch1227164@sina.com, sjrbird@163.com

Abstract. A novel integrated UHF RFID reader transceiver front-end operating from 860 MHz to 960 MHz supporting the EPC global™ Class-1 Generation-2 and ISO-18000-6A/B/C standards is presented. The transceiverer front-end consists of up-conversion mixer, linear power amplifier (PA), LNA and down-conversion mixer. The up-conversion mixer provides 5.5 dB power gain and IIP3 of 12 dB. The power amplifier exhibits an output power of 21 dBm and a power-added efficiency of 32%. Noise figure of the LNA is 4.07 dB, S11 is -15.7dB.The down-mixer also provides 11.13 dB power gain and NF of 11.18 dB. The transceiver front-end consumes a total power of 262mW when the output power is 20.5dBm.The chip simulated in 0.18-μm CMOS process has area of 3.2mm X 1.2mm.

Keywords: RFID, Reader, UHF, CMOS, Transceiver.

1 Introduction

Radio frequency identification (RFID) application is increasing rapidly with the advancement in IC technologies and decrease of size and cost of the RFID tags. The reading/writing range of a RFID system mainly depends on the choice of frequency, radiated power from the reader, sensitivity of tag, data rate and so on[1]. Compared with near field RFID systems, such as 125kHz or 13.56MHz, which are limited by the short communication distance and low data rate, UHF band RFID system has the advantages of a longer communication distance and a higher data rate [2],[3]. More and more researches are reported about mobile UHF RFID Reader transceiver. Some proposed transceivers circuit uses off-chip inductor which allows the circuit has high efficiency, but is not easy to integrated[4],[5]. In [6]and[7], two fully integrated transceivers are proposed. However, its output power is low and it may not appear smooth communication in a complex environment.

In this paper, a novel fully-integrated CMOS transceiver front-end is proposed. It consists of up-conversion mixer, linear power amplifier (PA), LNA and down-conversion mixer. In the up-mixer, a complementary transconductance current injection technique is employed to improve the circuit linearity and reduce noise. Implementing a highly efficient PA in standard CMOS is a challenging task due to the low breakdown voltage. To solve this problem, self-biased cascode techniques are used in PA circuit. It is easy to integration for all inductors in this circuit are on-chip. A 2nd-order intermodulation current injection structure is adopted in proposed LNA to obtain a good linearity, and it achieves center operating frequency adjustable by adding switch capacitor at the output terminal. A novel mixer is proposed in this paper, a dynamic current injection structure is employed at the output terminal, which so as to obtain a good linearity and noise characteristics. Compared to the previous paper[4]-[7], the proposed transceiver front-end has characteristic of the high output power, linearity and appropriate supply voltage.

2 Transceiver Architecture and Circuit Implementation

A complete UHF RFID system comprises of tags, reader and antenna. When the reader communicates with the tags, the tag's energy is supplied by the unmodulated carrier. At present the UHF reader chip are realized mainly in the 0.18-um CMOS process.

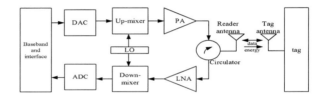

Fig. 1. Diagram of the RFID reader

Fig.1 shows a block diagram of the typical transceiver for RFID reader. This transceiver proposed in the paper adopts the zero-IF structure.

2.1 Up-Mixer

The characteristic of the mixer is that complementary transconductance current injection technique is used. Shown in Fig.2, M3 and M4 are complementary transconductance.

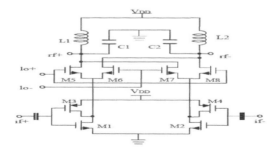

Fig. 2. Schematic of up-mixer

Fig.2 is the schematic of the proposed CMOS up-mixer. The transistors M1,M3 and M2, M4 serve as the complementary transconductance stages. Gain of the mixer circuit can be simply expressed as follows[8]:

$$CG = \frac{2}{\pi}(g_1 + g_3)wL_1 \qquad (1)$$

As can be seen from above equation, the transconductance gain will be increased compared with the transconductance stage of a single MOS transistor.

2.2 Power Amplifier

The proposed power amplifier is shown in Fig.3 Transistors M1-M4 and the resistors R1-R4 form the pre-amplification stage circuit. As the inductor will occupy a large area in the chip, so resistors R1-R4 are used instead of the inductors. Direct inter-capacitive coupling is used.

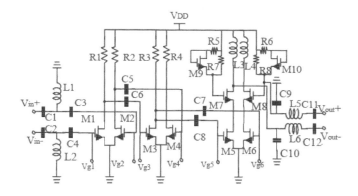

Fig. 3. Schematic of PA

A cascode structure is used in the output stage. When the output power reaches the maximum, gate-drain voltage of M7 and M8 will exceed breakdown voltage under normal condition. So the self-biased techniques is used in common-gate structure in order to overcome low breakdown voltage. The principle of self-biased circuit is that one can adjust the gate voltage of the common-gate transistor in a certain range of changes to make sure that gate-drain voltage is not more than the breakdown voltage while the drain voltage is kept constant[8].

2.3 LNA

Fig.4 shows the proposed LNA circuit of this paper.

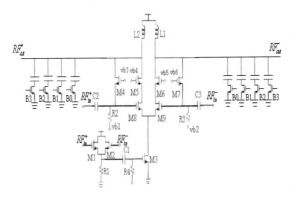

Fig. 4. Schematic of LNA

In order to improve the linearity of LNA, so we adopted 2nd-order intermodulation current injection to improve the linearity. As shown in Fig.5.

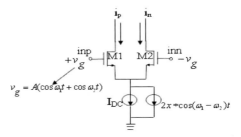

Fig. 5. 2nd-order intermodulation structure

The small-signal output current of transistors can be expressed as a Taylor series:

$$i_d = g_1(v_g - v_s) + g_2(v_g - v_s)^2 + g_3(v_g - v_s)^3 + \ldots \ldots \tag{2}$$

Where g_i is ith-order transconductance of the transistor, v_g is the gate voltage of the transistor, and v_s is the source voltage. When we input two different frequencies but same amplitude signals ω_1, ω_2, and at the same time we inject a low frequency 2nd-order-intermodulation current $2x * \cos(\omega_1 - \omega_2)t$, now we can get:

$$i_p + i_n = 2x * \cos(\omega_1 - \omega_2)t \tag{3}$$

Together with equation (2) and equation (3), we can get the small-signal source voltage v_s at frequency ($\omega_1 - \omega_2$) is:

$$v_s |_{\omega_1 - \omega_2} = \frac{g_2 A^2 - x}{g_1} \cos(\omega_1 - \omega_2)t \tag{4}$$

At the other three 2nd-order frequency point as $\omega_1+\omega_2$, $2\omega_1$ and $2\omega_2$. The amplitude of v_s is $(1,1/2,1/2)*\frac{g_2 A^2}{g_1}$. Now together with equation (4) and equation (2), when the output frequency is ($2\omega_1-\omega_2$) or ($2\omega_2-\omega_1$), the 3th-order-intermodulation current is:

$$-g_2 A \frac{3g_2 A^2 - 2x}{2g_1} + \frac{3g_3 A^3}{4} \tag{5}$$

So, we can eliminate the 3th-order-intermodulation current through letting equation (5) be zero, and we can get:

$$x = \left(-\frac{3g_1 g_3}{4g_2} + \frac{3}{2}g_2\right) A^2 \tag{6}$$

So the injection current is:

$$i_{inj} = 2\left(\frac{-3g_1 g_3}{4g_1} + \frac{3}{2}g_2\right) A^2 \cos(\omega_1 - \omega_2)t \tag{7}$$

It can eliminate the 3th-order-intermodulation current (IM3) effectively by using 2nd-order-intermodulation injection structure, and improve the linearity of LNA. In the output terminal, it achieves center operating frequency adjustable by controlling the switch capacitance B1, B2, B3 and B4.

2.4 Down-Mixer

Fig.6 shows the scheme of the mixer, which belongs to a Gilbert-cell mixer. M7 and M8 is the 2nd-order-intermodulation injection structure which being added into the input, MP1、MP2 and MP3 is the dynamic current injection structure.

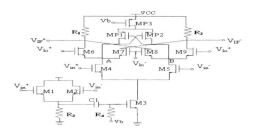

Fig. 6. Schematic of down- mixer

The dynamic current injection structure is adopted at the output terminal to reduce the noise.

3 Simulated Results

The transceiver front-end was simulated in standard 0.18- um CMOS process. The layout is shown in Fig. 7 which occupies an area of 3.2 mm× 1.2 mm.

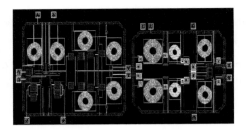

Fig. 7. Layout of the transceiver

Fig. 8(a) and (b) show the up-mixer input IP3 and NF with a 10 MHz base-band signal and 850MHz -950 MHz LO signal.

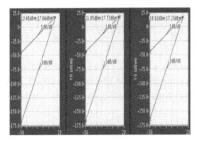 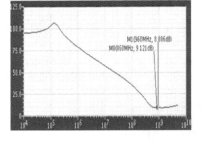

Fig. 8(a). Input IP3 of the up-mixer   **Fig. 8(b).** NF of the mixer

From 860MHz to 960MHz of RFID frequency band, the curves of Fig.9 show the output power and PAE of PA.

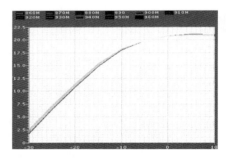 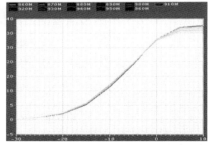

Fig. 9(a). Pout of the PA   **Fig. 9(b).** PAE of the PA

Fig.10(a) and (b) show simulation results of LNA.

A Novel Integrated Cmos Uhf Rfid Reader Transceiver Front-End Design   233

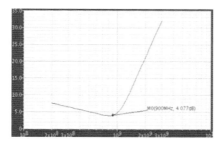
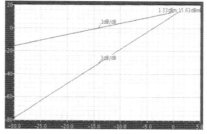

Fig. 10(a). NF of the LNA   Fig. 10(b). IIP3 of the LNA

Fig.11 shows the IIP3 simulation results of the down-mixer:

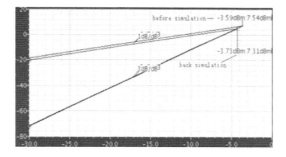

Fig. 11. IIP3 of the down-mixer

The output waveform of the transmitter is shown in Fig.12. It is clear from the output waveform that the output voltage exceeds 4V, the output power is about 20.5dBm when the input signal is 0dBm. The simulated results summary is given in Table 1.

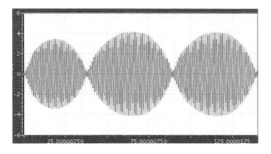

Fig. 12. Output voltage of the transceiver front-end

Table 1. The simulated results

Performance \ Modules	IIP3	Gain	NF
Up-mixer	11.5 dBm	6.15 dB	9.05 dB
Power amplifier	Pout	PAE	P1dB
	21dBm	30%	19dBm
LNA	S11	NF	IIP3
	-15.7dB	4.07 dB	1.33 dBm
Down-mixer	IIP3	Gain	NF
	2.78dBm	13.71 dB	8.63 dB
Transceiver front-end	Supply Voltage	Pout	Frequency
	3V	20.5dBm	860-960MHz

4 Conclusion

A novel integrated UHF RFID reader Transceiver Front-end operating from the 860 MHz to 960 MHz is presented. The transceiver Front-end is integrated with a CMOS on-chip PA output voltage exceeds 4V, which is suitable for mobile applications.

Acknowledgments

This work was supported by the Science and Technology Project of Hunan Province (No. 2010GK3052).

References

1. Finkenzeller, K.: RFID Handbook. Wiley, New York (2003)
2. EPC UHF radio frequency identity protocols: Class 1 generation 2 UHF RFID, ver.I.2.0, EPC global (2007)
3. ISO-IEC_CD 18000-6C, version 2.1 c2 (July 2005)
4. Jang, J.-E.: A 900-MHz Direct-Conversion Transceiver for Mobile RFID Systems. In: IEEE Radio Frequency Integrated Circuits Symposium J., pp. 277–280 (2007)
5. Shim, S., Han, J., Hong, S.: A CMOS RF polar transmitter of a UHF mobile RFID reader for high power efficiency. IEEE Microwave and Wireless Letters 18(9), 635–637 (2008)
6. Khannur, P.B., Chen, X., Yan, D.L., et al.: An 860 to 960MHz RFID Reader IC in CMOS. In: IEEE Radio Frequency Integrated Circuits Symposium, pp. 269–272 (2007)
7. Ye, L., Liao, H., Song, F., et al.: A 900MHz UHF RFID Reader Transceiver in 0.18μmCMOS Technology. In: 9th International Conference on Solid-State and Integrated-Circuit Technology, pp. 1569–1572 (2008)
8. Sowlati, T., Leenaerts, D.M.W.: A 2.4-GHz 0.18-um CMOS Self-Biased Cascode Power Amplifier. IEEE Journal of Solid-State Circuits 38(8), 1318–1324 (2003)

Hammerstein-Wiener Model Predictive Control of Continuous Stirred Tank Reactor

Man Hong[1,2,*] and Shao Cheng[1]

[1] Institute of Advanced Control Technology,
Dalian University of Technology, Dalian 116024, China
[2] Information Engineering Institute, Dalian Jiaotong University,
Dalian 116052, China
manhong74@gmail.com

Abstract. A nonlinear Hammerstein-Wiener model predictive controller based on LSSVM is built to describe the dynamic characteristic of a continuous stirred tank reactor (CSTR), which is made up by a linear optimal component and radial basis function neural networks in series, using BP neural network to train the input sequences of the predictive control, solving the nonlinear predictive control laws by the quasi-Newton algorithm, and a neural network predictive control algorithm is achieved based on LSSVM Hammerstein-Wiener model. The simulation results of CSTR show that this approach is effective tracking and controlling product concentration.

Keywords: Hammerstein-Wiener model, Least squares support vector machines, BP neural network, Nonlinear predictive control.

1 Introduction

In the area of chemical process control, because of the high nonlinearity and time-varying characteristics [1], for instance, continuous stirred-tank reactor (CSTR) and the control of pH processes, the performance of a linear model predictive control (LMPC) can be poor [2]. Many nonlinear predictive control algorithm was proposed based on some nonlinear models, such as Hammerstein model [3-4], Wiener model [5], Volterra model [6], Neural network model [7] and Least squares support vector machines (LSSVM)model [8]. In [3], the identification methods of the model parameters are given based on polynomial function to describe the nonlinear part of Hammerstein model. The nonlinear model predictive controller has been designed on the basis of autoregressive model with exogenous input (ARX) in the linear part. The simulation results of pH neutralization processes show that this algorithm is effective. The nonlinear predictive controller is designed based on the improved Hammerstein model, the control effect of which is better than PID controllers' in [4]. A nonlinear model predictive controller (NMPC) for pH neutralization and CSTR processes using Wiener model method is used in [9].However, these methods have certain limitations because which are presented based on the approximate nonlinear models.

---

* This work is supported by the National Key Basic Research Program of China (973 Program)(2007CB714006) and the National Natural Science Foundation of China (61074020).

Hammerstein-Wiener model that is closer than either of two models in the nonlinear systems contains Hammerstein and Wiener nonlinear modules. In [8], the two-stage method is adopted to set up the predictive controller, which uses the equivalent subspace algorithm to identificate Hammerstein's parameters at the first step, and then to identificate the output based on least square process at the next step. This method used in the ionosphere dynamic system model is effective.

In the paper, the two-stage method is adopted to set up the predictive controller based on Hammerstein-Wiener model to describe the dynamic characteristic of CSTR, using BP neural network to train the input sequences of the predictive control in the two stages respectively, solving the nonlinear predictive control laws by the quasi-Newton algorithm, and then a BP-NMPC algorithm based on LSSVM is realized.

2 Hammerstein-Wiener Model Structure and Identification

2.1 Hammerstein-Wiener Model Structure

Hammerstein-wiener model structure of CSTR based on LSSVM is developed in the section. Hammerstein-Wiener model consists of a dynamic linear block (G) and two static nonlinearity blocks (F and H), as shown in Fig.1.

Fig. 1. Hammerstein-Wiener model structure

Mathematical representation of the Hammerstein-Wiener model can be stated as follows [10]:

$$y(k) = \sum_{i=1}^{n} a_i g(y(k-i)) + \sum_{j=1}^{m} b_j f(y(k-j)) + e(k) \qquad (1)$$

To use LSSVM function, $f(u) = w^T \varphi(u) + d$ is used to replace the nonlinearity $f(u)$ and $g(u)$ in the Hammerstein model [11]. Assume $\omega_i^T = a_i \omega^T \square \omega_j^T = b_j \omega^T \square$ and $\sum_{i=1}^{n} a_i d_1 = c_1 \square \sum_{j=1}^{n} b_j d_0 = c_2$. We can derived the equation (2) established.

$$y(k) = \sum_{i=1}^{n} \omega_i^T \varphi(y_{k-i}) + \sum_{j=1}^{m} \omega_j^T \varphi(x_{k-j}) + c_1 + c_2 + e(k) \qquad (2)$$

2.2 Identification Method

Hammerstein-wiener model structure of CSTR based on LSSVM is represented by (1). The nonlinearity blocks (F and H) and linear parameters a_i, b_j can be identified.

To estimate a_i, b_j, define the following optimization problem:

$$\min_{w_i, w_j, c_1, c_2, e} J(w_i, w_j, e) = \frac{1}{2} \sum_{i=1}^{n} w_i^T w_i + \frac{1}{2} \sum_{j=1}^{m} w_j^T w_j + \frac{1}{2} \gamma \sum_{k=r}^{N} e_k^2 \qquad (3)$$

$$\text{s.t.} \begin{cases} \sum_{i=1}^{n} \omega_i^T \phi(y_{k-i}) + c_1 + \sum_{j=1}^{m} \omega_j^T \varphi(x_{k-j}) + c_2 + e(k) - y(k) = 0 \\ \sum_{k=r}^{N} \omega_i^T \phi(y_k) = 0 \\ \sum_{k=r}^{N} \omega_j^T \varphi(x_k) = 0 \end{cases} \quad (4)$$

With $k = r, \cdots, N$ and $r = \max(m, n) + 1$.

The Lagrangian of the constrained optimization problem is constructed as:

$$L(w_i, c_1, w_j, c_2, e, \alpha_k, \beta_j, \gamma_i, \gamma) = J(w_i, w_j, e) - \sum_{j=1}^{m} \beta_j \left[\sum_{k=r}^{N} \omega_j^T \varphi(x_k) \right] - \sum_{i=1}^{n} \gamma_i \left[\sum_{k=r}^{N} \omega_i^T \phi(y_k) \right] \quad (5)$$

$$- \sum_{k=r}^{N} \alpha_k \left[\sum_{i=1}^{n} \omega_i^T \phi(y_{k-i}) + c_1 + \sum_{j=1}^{m} \omega_j^T \varphi(x_{k-j}) + c_2 + e(k) - y(k) \right]$$

where $\alpha_k, \beta_j, \gamma_i$ Lagrange multipliers.

Solving the optimum conditions is as follows:

$$\begin{bmatrix} 0 & 0 & 1_{N-r+1}^T & 0 & 0 \\ 0 & 0 & 1_{N-r+1}^T & 0 & 0 \\ 1_{N-r+1}^T & 1_{N-r+1}^T & K + \Re + I/C & \Delta & \Omega \\ 0 & 0 & \Delta^T & 1_N^T \Gamma_x 1_N \cdot I_m & 0 \\ 0 & 0 & \Omega^T & 0 & 1_N^T \Gamma_y 1_N \cdot I_n \end{bmatrix} \begin{bmatrix} c_1 \\ c_2 \\ \alpha_k \\ \beta_j \\ \gamma_i \end{bmatrix} = \begin{bmatrix} 0 \\ 0 \\ Y_f \\ 0 \\ 0 \end{bmatrix} \quad (6)$$

As far as c_1 and c_2 are solved by (6), a_i and b_j are determined by singular value decomposition method (SVDM). So (7) can be built as:

$$\begin{bmatrix} b_1 \\ b_2 \\ \vdots \\ b_m \end{bmatrix} \begin{bmatrix} \hat{f}(x_1) \\ \hat{f}(x_2) \\ \vdots \\ \hat{f}(x_N) \end{bmatrix}^T = \begin{bmatrix} \alpha_N & \cdots & \alpha_r & & 0 \\ & \alpha_N & \cdots & \alpha_r & \\ & & \ddots & & \ddots \\ 0 & & & \alpha_N & \cdots & \alpha_r \end{bmatrix} \times \begin{bmatrix} K_{N,1} & K_{N,2} & \cdots & K_{N,N} \\ K_{N-1,1} & K_{N-1,2} & \cdots & K_{N-1,N} \\ \vdots & \vdots & & \vdots \\ K_{r-m+1,1} & K_{r-m+1,2} & \cdots & K_{r-m+1,N} \end{bmatrix} \begin{bmatrix} \beta_1 \\ \vdots \\ \beta_m \end{bmatrix} + \sum_{k=1}^{N} \begin{bmatrix} \Gamma_{xk,1} \\ \vdots \\ \Gamma_{xk,N} \end{bmatrix}^T \quad (7)$$

Because the right-hand side of (7) is known, b_j can be solved by SVDM. Assume $\hat{f}(x_k)$ is the estimation of $\underline{f}(x_k) = f(x_k) - c_2$, then $f(x) = \hat{f}(x) + c_2 \big/ \sum_{j=1}^{m} b_j$. Similarly a_i can be solved like the above method, too.

RBF kernel that described by (8) is selected as SVM kernel function in the study. We have the output of SVM as (9).

$$K(u_k, u_l) = \exp\left(-\|u_k - u_l\|^2 \big/ (2\sigma^2) \right) \quad (8)$$

$$f(u) = \sum_{i=1}^{N} \alpha_i K(u, u_i) + d \quad (9)$$

So the expression of Hammerstein-wiener model can be described as:

$$y_k = \sum_{i=1}^{n} a_i \left(\sum_{sv=1}^{N} \alpha_{sv} \exp\left(-\|v_{k-j}-v_{sv}\|^2 / (2\sigma^2)\right) + c_1 \right) + \sum_{j=1}^{m} b_j \left(\sum_{sv=1}^{N} \alpha_{sv} \exp\left(-\|x_{k-j}-x_{sv}\|^2 / (2\sigma^2)\right) + c_2 \right) + e_k \qquad (10)$$

Now all the dynamic parameters and the static nonlinearity of the Hammerstein-wiener model have been identified, so the model can be used to represent the characteristics of CSTR.

3 Nonlinear Predictive Controller

The block diagram of NMPC based on Hammerstein-Wiener model is given in Fig.2. Dashed part in Fig.2 is viewed as Hammerstein-Wiener model that be decomposed into one dynamic linear model and two static nonlinear models.

where G_c is the controller in CSTR,

NL^{-1}, Gm is the nonlinear part and linear part of Hammerstein-Wiener model respectively,

y_{sr} is the given input,

$y_r(t+k)$ is the reference input of the future k steps,

$y(t)$ is the current output.

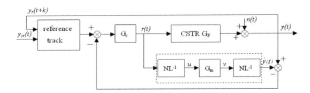

Fig. 2. Block diagram of NMPC based on Hammerstein-Wiener model

In the study, BP neural network to train the input sequences of the predictive control is adopted for improving the control accuracy. Nonlinear BP predictive control algorithm based on LSSVM is given in Fig.3.

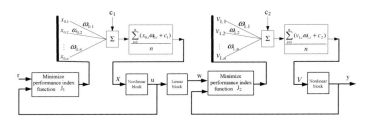

Fig. 3. BP-NMPC algorithm of LSSVM Hammerstein-Wiener model

Because the Hammerstein-Wiener model is that can be decomposed into two same nonlinear structures, we obtain the performance index functions minimized as (11) by two-stage method in every control cycle.

$$J_1 = \sum_{i=N_1}^{N_2}[r(t+i)-\hat{u}(t+i)]^2 + \rho\sum_{i=1}^{N_u}[\Delta x(t+i-1)]^2$$
$$J_2 = \sum_{i=N_1}^{N_2}[w(t+i)-\hat{y}(t+i)]^2 + \rho\sum_{i=1}^{N_u}[\Delta v(t+i-1)]^2 \quad (11)$$

To solve the control input $X(t) = [x(t)\cdots x(t+N_u-1)]^T$ and $V(t) = [v(t)\cdots v(t+N_u-1)]^T$ respectively, they must satisfy the following conditions after N_u control cycle:

$$\begin{cases}\Delta x(t+i)=0, N_u \leq i \leq N_2-d\\ \Delta v(t+i)=0, N_u \leq i \leq N_2-d\end{cases}$$

where N_1 and N_2 are the minimum and the maximum predictive domain respectively. N_u is the control domain. ρ is control weight coefficient. d is the delay time.

3.1 Predictive Model

At t times, k step ahead predictive outputs of two structures can be described as:

$$\hat{u}(t+k|t) = \hat{f}\begin{bmatrix}\hat{u}(t+k-1),\cdots,\hat{u}(t+k-\min[k,n]), u(t-1),\cdots,\\ u(t-\max[n-k,0]), x(t+k-d),\cdots, x(t+k-d-m)\end{bmatrix}$$

$$\hat{y}(t+k|t) = \hat{g}\begin{bmatrix}\hat{y}(t+k-1),\cdots,\hat{y}(t+k-\min[k,n]), y(t-1),\cdots,\\ y(t-\max[n-k,0]), v(t+k-d),\cdots, v(t+k-d-m)\end{bmatrix} \quad (12)$$

We adopt the nonlinear Hammerstein-Wiener model of (10) to realize $F(\cdot)$ and $H(\cdot)$ functions that is as follows:

$$\hat{u}(t+k) = \hat{u}(t+k|t) = \sum_{i=1}^{n}a_i u_{t+k-i} + \sum_{j=0}^{m}b_j f(x_{t-j}) + e_t = \sum_{i=1}^{\min(k,n)}a_i\hat{u}(t+k-i) + \sum_{i=1}^{\max[n-k,0]}a_i u(t-i) +$$
$$\sum_{j=0}^{m}b_j\left(\sum_{sv=1}^{N}\alpha_{sv}\exp\left(\frac{-\|x(t+k-d-j)-x_{sv}\|^2}{2\sigma^2}\right)+c_1\right)+e_t \quad (13)$$

$$\hat{y}(t+k) = \hat{y}(t+k|t) = \sum_{i=1}^{n}a_i y_{t+k-i} + \sum_{j=0}^{m}b_j g(v_{t-j}) + e_t = \sum_{i=1}^{\min(k,n)}a_i\hat{y}(t+k-i) + \sum_{i=1}^{\max[n-k,0]}a_i y(t-i) +$$
$$\sum_{j=0}^{m}b_j\left(\sum_{sv=1}^{N}\alpha_{sv}\exp\left(\frac{-\|v(t+k-d-j)-v_{sv}\|^2}{2\sigma^2}\right)+c_2\right)+e_t \quad (14)$$

3.2 Performance Index Functions

The NMPC matrix of the performance index functions can be described as follows:

$$J(X(t)) = E^T(t)E(t) + \rho\Delta X^T(t)\Delta X(t) = (R(t)-\hat{U}(t))^T(R(t)-\hat{U}(t)) + \rho\Delta X^T(t)\Delta X(t)$$
$$J(V(t)) = E^T(t)E(t) + \rho\Delta V^T(t)\Delta V(t) = (W(t)-\hat{Y}(t))^T(W(t)-\hat{Y}(t)) + \rho\Delta V^T(t)\Delta V(t) \quad (15)$$

3.3 Computing the Predictive Control Law

The processes to compute the predictive control law are that the nonlinear optimization processes of (15) are solved circularly in each control cycle based on the predictive model of (13) and (14). It can be described as follows:

$$\min_{X(t)} J(X(t)) \quad \text{and} \quad \min_{X(t)} J(V(t))$$

$$s.t. \quad \hat{u}(t+k) = \sum_{i=1}^{n} a_i u_{t+k-i} + \sum_{j=0}^{m} b_j f(x_{t-j}) + e_t \\ \hat{y}(t+k) = \sum_{i=1}^{n} a_i y_{t+k-i} + \sum_{j=0}^{m} b_j g(v_{t-j}) + e_t \quad (16)$$

In the whole process, we use neural network to train the input sequences and can obtain the output sequences, and then the obtained outputs are averaged and viewed as the predictive control input. The same optimization will be proceeded sequentially in the next control cycle. The nonlinear optimization problem is solved by Quasi-Newton algorithms. $X^{(i)}$ and $V^{(i)}$ are as the approximate optimal solution when they satisfy the key conditions of $|X^{(i)} - X^{(i-1)}| < \delta$ and $|V^{(i)} - V^{(i-1)}| < \delta$. Where $\delta > 0$ is tolerant error.

4 Simulation Results

4.1 Description of CSTR

The simulated CSTR is a typical nonlinear dynamic process used in chemical and biochemical industry. The mathematical model of CSTR can be shown by two nonlinear differential equations, which parameters are shown in [5, 12]. The following equations is available to describe the CSTR process:

$$\dot{C}_A = \frac{q}{V}(C_{Af} - C_A) - k_0 e^{-E/(RT)} C_A \quad (17)$$

$$\dot{T} = \frac{q}{V}(T_f - T) + \frac{\rho_c C_{pc}}{V \rho C_p} q_c [1 - \exp(-\frac{U_A S}{\rho_c C_{pc} q_c})] \cdot (T_c - T) - \frac{\Delta H}{\rho C_p} k_0 e^{-E/(RT)} C_A \quad (18)$$

4.2 Simulation Results and Analysis

Assume to use a group of random sequence from 80 to 115 as the inputs of CSTR, we can collect 800 sample data of the input/output data in 0.1 second intervals. The front 400 data is used as model identification and we may get out 200 data from the later 400 data respectively as training sets and checking sets. The tracking between the actual values and the predictive values is shown as follows Fig.4. It can be observed that the model identification performance is good.

In order to verify the effectiveness of the nonlinear predictive control algorithm for CSTR in the study, the comparison between NMPC and PI controller for product concentration C_A of CSTR is shown as Fig.5. The parameters of PI controller[12]: K_p=52, T_i= 0.46. The output comparison of them about coolant flow rate q_c of CSTR is described in Fig.6. The comparison of the errors' absolute values about some performance indexes is shown in Table 1.

Prediction results indicate that the nonlinear predictive controller built in the paper has higher control accuracy than PI controller, and it can approximate the dynamic behavior of CSTR.

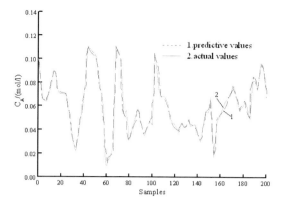

Fig. 4. Tracking of the predictive values and the actual values

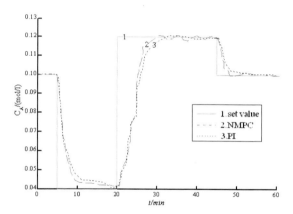

Fig. 5. Product concentration C_A set-point tracking

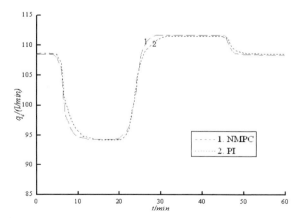

Fig. 6. Output tracking of coolant flow rate q_c

Table 1. Tracking performance indexes comparison of NMPC and PI.

	NMPC	PI
MAE	0.0045	0.0054
SAE	2.5634	3.2334
MSE $\times 10^4$	1.2375	1.3125
SSE	0.0743	0.0789

5 Conclusions

A BP-NMPC algorithm based on LSSVM Hammerstein-Wiener model has been presented. It is shown that the Hammerstein-Wiener model of CSTR in the study is more attractive and competitive in that BP neural network is used to train input sequences for improving the control accuracy of the system. The simulation results show that the algorithm in the paper is better effective than PI algorithm tracking and controlling product concentration of CSTR.

References

1. Vielhaben, T., Schulz-Ekloff, G., Neimeier, R., et al.: Design of Fuzzy Control for Maintaining a CSTR in the Steep Section of Its Nonlinear Characteristic. Chemical Engineering and Technology 24(2), 205–212 (2001)
2. Rouhani, R., Mehra, R.K.: Model algorithm control (MAC)basic theoretical properties. Automatica 18, 401–414 (1982)
3. Fruzzetti, K.P., Palazoglu, A., McDonald, K.A.: Nonlinear model predictive control using Hammerstein models. Journal of Process Control 7, 31–41 (1997)
4. Zhu, X.F., Seborg, D.E.: Nonlinear predictive control based on Hammerstein models. Control Theory and Applications 11, 564–575 (1994)
5. Cervantes, A.L., Agamennoni, O.E., Figueroa, J.L.: A nonlinear model predictive control system based on Wiener piece wise linear models. Journal of Process Control 13, 655–666 (2003)
6. Doyle, F.J., Ogunnaike, B.A., Pearson, R.K.: Nonlinear model based control using second order Volterra models. Automatica 31, 697–714 (1995)
7. Venkates warlu, C., Rao, K.V.: Dynamic recurrent radial basis function network model predictive control of unstable nonlinear process. Chemical Engineering Science 60, 6718–6732 (2005)
8. Zhang, R.D., Wang, S.Q.: Support vector machine based predictive functional control design for output temperature of coking furnace. Journal of Process Control 18, 439–448 (2008)
9. Gomez, J.C., Jutan, A., Baeyens, E.: Wiener model identification and predictive control of a pH neutralization process. IEE Proceeding of Control Theory Appl. 151, 3–4 (2004)
10. Song, H., Gui, W., Yang, C.: Identification of Hammerstein-Wiener Model with Least Squares Support Vector Machine. In: Proceedings of the 26th Chinese Control, pp. 260–263 (2007)
11. Yan, W., Shao, H.: The research and comparison of the support vector machine and least squares support vector machine. Control and Decision 18, 358–360 (2003)
12. Morningred, J.D., Paden, B.E., Seborg, D.E., et al.: An adaptive nonlinear predictive controller. Chemical Engineering Science 47, 755–762 (1992)

The Relativity Analysis between Summer Loads and Climatic Condition of Ningbo

Yiou Shao, Xin Yu[*], and Huanda Lu

Ningbo Institute of Technology, Zhejiang University, Ningbo 315100, China
shaoyiou881103@gmail.com,
yuxin@nit.net.cn,
hundalu@nit.net.cn

Abstract. The climatic factor is one of the main factors that affect the summer load. Through the study of the relativity between the summer load and the climatic factor, based on the data of daily load and temperature of Jiangdong District, Ningbo, in July and August, we get the correlation coefficients between the loads on 288 time points and the average temperature, the maximum temperature, the minimum temperature and the T8 (temperature at 8 o'clock a.m.), respectively, which provide a theoretic support for forecasting of the summer load and the Demand Side Management (DSM).

Keywords: daily load, peak load, relativity analysis, correlation coefficient.

1 Introduction

Electrical loads may be broadly classified as residential, commercial and industrial. Among three broad classes of loads, residential loads have the most evident fluctuations with season variations. Commercial loads are also characterized by seasonal changes. These fluctuations are mainly due to extensive use of air conditioning and space heating appliances. Industrial loads are considered as the base loads that contain very little dependence on weather variation [1].

Ningbo is the subtropics climate and Jiangdong District is the center of Ningbo city. The residential loads and the commercial loads account for very great proportion of total power consumptions. On the other hand, the change of temperature during summer and winter is large in Ningbo [2]. Especially, due to many uses of the air conditioners, the electrical load demand is significantly increased during summer.

The power system expansion planning begins with a forecast of anticipated load requirements [3-5]. Estimation of load demand is crucial to effective system planning. Thus it is important to study the relativity analysis of the summer loads and the climatic conditions since it is valuable for promoting the accuracy of forecasting the electrical load. Moreover, relativity analysis [6-7] is helpful for the Demand Side Management (DSM), which ensures the power system to work safely and economically.

---

[*] Corresponding author.

In this paper, we use the load data and the four climatic factors, including the average temperature, the maximum temperature, the minimum temperature and the T8, of Jiangdong District, Ningbo, in July and August, to analysis the relativity between summer load and climatic factor. We get the correlation coefficients between the loads on 288 time points and the average temperature, the maximum temperature, the minimum temperature and the T8, respectively.

2 Relationship between Load and Temperature

It is obviously that the behaviors of the load consumptions in weekdays and weekends are very different. In this paper, we only study the relativity of weekdays between the summer load and the climatic conditions. Thus after deleting the data of weekends, we can only take use of the data of 45 days. Another fact that should be paid attention to is that the load data for every day have 288 data because of one sample point per 5 minute. In the following, we take the 45 weekday's data to analyze the relativity between the loads on 288 time points and the climatic conditions.

We will first observe the relationship between the daily load and the average temperature, the maximum temperature, the minimum temperature and the T8 (refer to the fig.1).

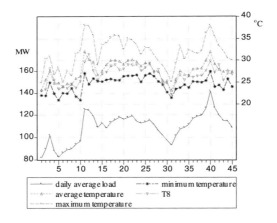

Fig. 1. The curve of the daily load and the climatic factors.

From the fig.1, we can see easily that: the daily average load are closely relative to the average temperature, maximum temperature, minimum temperature and the T8.

More concisely, according to the data of the daily average load, the average temperature, maximum temperature, minimum temperature and the T8, we can get the fitting curve below, respectively:

The Relativity Analysis between Summer Loads and Climatic Condition of Ningbo 245

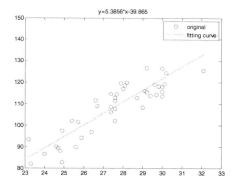

Fig. 2. Fitting curve of the daily load and average temperature

Fig.2 is the fitting curve of the daily average load and the average temperature. By the method of OLS (Ordinary Least Square), we can get the regression equation between the daily average load and the average temperature as follows:

$$y = 5.3856x - 39.865 \tag{1}$$

By above equation, we can see that for each 1^oC in average temperature change will result in about $5.3856MW$ change in daily average load.

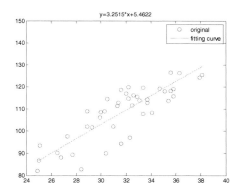

Fig. 3. Fitting curve of maximum temperature and daily load.

Fig.3 is the fitting curve of the daily average load and the maximum temperature. The regression equation between the daily average load and the maximum temperature is as follows:

$$y = 3.2515x + 5.4622 \tag{2}$$

By above equation, we can see that for each 1^oC in maximum temperature change will result in about $3.2515MW$ change in daily average load.

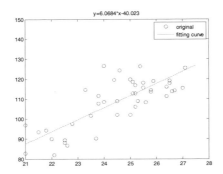

Fig. 4. Fitting curve of the daily load and the minimum temperature

Fig.4 is the fitting curve of the daily average load and the minimum temperature. The regression equation between the daily average load and the minimum temperature is as follows

$$y = 6.0684x + 40.023 \tag{3}$$

By above equation, we can see that for each 1^oC in minimum temperature change will result in about $6.0684MW$ change in daily average load.

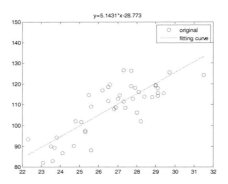

Fig. 5. Fitting curve of daily load and T8

Fig.5 is the fitting curve of the daily average load and the T8. The regression equation between the daily average load and the T8 temperature is as follows

$$y = 5.1431x - 28.773 \tag{4}$$

By above equation, we can see that for each 1^oC in T8 change will result in about $5.1431MW$ change in daily average load.

From the above figures and the regression equations, we can get that the summer load is heavily dependent on the average temperature, the maximum temperature, the minimum temperature and the T8. In the following, we will analyze the relativity between the summer load and the climatic factors, and get their correlation coefficients, respectively.

3 Relativity Methods

3.1 The Calculation Method of Correlation Coefficient

We take the load on every time point and climatic factors as the random variables, and take the historical data of the summer load and the temperature as the samples. By the calculation method of correlation coefficient, we can calculate the correlation coefficients between the summer load and the average temperature, maximum temperature, minimum temperature and the T8, respectively. For two random variables ξ and η, the correlation coefficient can be defined as

$$\rho_{\xi\eta} = \frac{E(\xi - E\xi)(\eta - E\eta)}{\sqrt{Var\xi Var\eta}}. \tag{5}$$

Moreover, we take the average values of samples

$$\bar{\xi} = 1/n \sum_{i=1}^{n} \xi_i, \quad \bar{\eta} = 1/n \sum_{i=1}^{n} \eta_i \tag{7}$$

as the expected values $E\xi$ and $E\eta$, respectively, and the values of

$$\frac{1}{n}\sum_{i=1}^{n}\left(\xi_i - \bar{\xi}\right)^2, \quad \frac{1}{n}\sum_{i=1}^{n}\left(\eta_i - \bar{\eta}\right)^2 \tag{8}$$

as the variances $Var\xi$ and $Var\eta$, respectively, where n is the sample size.

3.2 Relativity between Load and Temperature Factors

By taking the weekday data of Jiangdong District in Ningbo, including the load data and the temperature, and using the function 'corrcoef' of Matlab tool, we get the correlation coefficient between the load at the 288 time point and the four temperature factors:

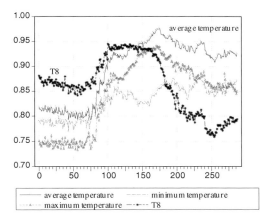

Fig. 6. Correlation coefficient between the daily load and climatic factors.

Now, we use the denotations $r_{Average}$, r_{High}, r_{Low}, r_{T8} to stand for the correlation coefficient of the load and the four temperature factors. From fig. 6, we can get the following results:

(1) From 0:00 to 11:25(the x-coordinate from 0-138 in above figure), the correlation coefficient r_{T8} is the biggest, that is to say, the load from 0:00 to 11:25 is the most relative to T8.

(2) From 11:30 to 23:55 (the x-coordinate from 139-287 in above figure), correlation coefficient $r_{Average}$ is the biggest, that is to say, the load from 11:30 to 23:55 is the most relative to the average temperature.

(3) The average correlation coefficient between load and average temperature is 0.8958, the maximum temperature's is 0.8456, the minimum temperature's is 0.8314, and the T8's is 0.8626. As a whole, the average temperature most heavily influences the load, and the second is the T8.

3.3 Relativity between Peak Load and Temperature Factors

The peak load is the biggest load in some period in one day. The value of the peak load heavily affects the safety of Ningbo power system and is the main indicator for Demand Side Management (DSM) [3][8].

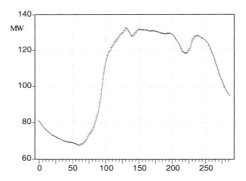

Fig. 7. The typical curve of load.

Table 1. Correlation coefficients between the peak load and climatic factors

Climatic factors	Morning	Midday	Night
Average	0.9356	0.9530	0.9473
Maximum	0.8964	0.9412	0.8792
Minimum	0.8464	0.8188	0.8734
T8	0.9402	0.9261	0.8047

The fig.7 is the typical curve of load of Ningbo. From fig.7, we can see that the typical curve of load has three clear peak times. The first appears at 10:55(the 131st point); the second appears at 13:05(the 157th point); the third appears at 19:30(the 234th point). In the following, we call the load at the peak time morning peak load, midday peak load, night peak loads, respectively.

From the table 1, we conclude that:

(1) The morning peak load appears at 10:55. Table 1 also shows that the correlation coefficient between the morning peak load and the T8 is the biggest, and the second is the average temperature.

(2) The midday peak load appears at 13:05. Table 1 shows that the correlation coefficient between the midday peak load and the average temperature is the biggest; the influence of maximum temperature is nearly equal to it.

(3) The night peak load appears at 19:30. Table 1 also shows that the correlation coefficient between the night peak load and the average temperature is the biggest; the influences of maximum temperature and the minimum temperature are at the same level.

4 Conclusion

Through the in-depth analysis and the data mining of the vast data of load and temperature of Ningbo, we study the relativity of the summer load and the climatic condition. We get the correlation coefficients between the loads on 288 time points and the average temperature, the maximum temperature, the minimum temperature and the T8, respectively. The obtained results are valuable for promoting the accuracy of forecasting the electrical load. Moreover, it is helpful for the power system to work safely and economically. Owing to the development of the economic and the change of the industrial structure in the past few years, the behavior of the power consumption take places great changes. The proportions of load consumption of service trade, commercial trade, port trade, tertiary industry, resident and etc increase fast, which lead to the relationships between load and climatic factors being closely and inseparably related. Then climatic factors exert more and mare important impact on load consumption.

Acknowledgment

Authors want to thank the support of National Natural Science Foundation of China (No. 10871154 and 10501039) and the National Science Foundation for Post-doctoral Scientists of China (No. 20070420167).

References

1. Wang, R., Zhang, Z., ShuchunWang, J.: Discussion on Relationship between Load and Climatic Condition. Jilin Electric Power (5), 37–39 (2004)
2. Tu, X., Chen, Z.: Analysis on Distribution and Variation Characteristic of Temperatures in Ningbo City. Atmospheric Science Research and Application (2), 76–83 (2007)

3. McSharry, P.E., Bouwman, S., Bloemhof, G.: Probabilistic Forecasts of the Magnitude and Timing of Peak Electricity Demand. IEEE Trans on Power Systems 20(2), 1166–1172 (2005)
4. Hyde, O., Hodnett, P.F.: An A dap table Automated Procedure for Short-term Electricity Load Forecasting. IEEE Trans. on PWRS 12(1), 84–94 (1997)
5. Gupta, P.C., Keigo, Y.: Adaptive Short-term Forecasting of Hourly Loads Using Weather Information. IEEE Trans. on PAS 92(5), 2085–2094 (1972)
6. Yan, Z.: Research on Program Analysis on Sensitivity of Climate-Power Load in Shanghai Area. Huadong Electric Power (9), 4–8 (2000)
7. Heinemann, G.T., Nordman, D.A., Plant, E.C.: The Relationship between Summer Weather and Summer Loads—A Regression Analysis. IEEE Trans on PAS 85(11), 1144–1154 (1966)
8. Li, Y., Wang, Z., Yi, L.: Characteristic Analysis of Summer Air Temperature—Daily Peak Load in Nanjing. Power System Technology 25(7), 63–66 (2001)

Application of the Modified Fuzzy Integral Method in Evaluation of Physical Education Informationization in High School

Tongtong Liu and Ting Li

School of P.E, Shandong University of Technology, China, 255049
yuxiuyan123456@163.com

Abstract. The indexes of physical education informationization in high school are related. The traditional evaluation methods will affect the objectivity of the result. The paper improved the fuzzy integral evaluation method by means of the triangle fuzzy data, integration of subjective and objective weight, the choosing of parameter. The improved method can choose the different evaluation parameters to deal with the different evaluation goals for offering the decision information. At last, the paper tested the improved method.

Keywords: Physical education informationization; Fuzzy integral; Evaluation.

1 Introduction

Physical education informationization is an important part of physical education. In order to manage and improve it, we need to evaluate its developing level [1, 2]. On the basis of the literatures, the paper built the evaluation indexes and improved the traditional fuzzy integral method to enrich the existing researches.

2 The Fuzzy Integral Evaluation Method

2.1 The Traditional Fuzzy Integral Evaluation Method

The fuzzy integral evaluation method is a non-linear method based on the fuzzy measure, which is a new integration technique of the evaluation indexes. The evaluation principle of the pervasive CHOQUET integration is the following:

Suppose: f is the performance of one of index; g is the weight of its index; $f(x_1) \geq f(x_2) \cdots \geq f(x_i) \cdots \geq f(x_n)$, Then fuzzy measure of g of f on X is:

$$E = \{fdg = f(x_n)g(X_n) + [f(x_{n-1}) - f(x_n)]g(X_{n-1}) + \cdots + [f(x_1) - f(x_2)]g(x_1) \quad (1)$$

E is the evaluation result; $g(X_i)$ is the weight of considering the feature of $x_1, x_2, \cdots, x_i$, that is,

$$g(X_1) = g(\{x_1\}), g(X_2) = g(\{x_1, x_2\}), \ldots, g(X_n) = g(\{x_1, x_2, \cdots, x_n\})$$

The paper applies λ-fuzzy measure, g_λ is decided with the following formula

$$g_\lambda(\{x_1, x_2, \cdots, x_n\}) = \frac{1}{\lambda}\left|\prod_{i=1}^{n}(1 + \lambda g(x_i)) - 1\right| \qquad (2)$$

2.2 Improved Fuzzy Integral Evaluation Method

(1) The triangle fuzzy of expert information

The qualitative indexes need to be quantified by the experts. The experts always use the natural word "about", "or so" and "up and down" to express the qualitative indexes and their weights. The paper deals the expert's information with the triangle fuzzy data. The corresponding relationship between the natural language of the indexes and the triangle fuzzy data is in Table 1. The corresponding relationship between the natural language of the weights of the indexes and the triangle fuzzy data is in Table 2.

Table 1. The corresponding relationship between the natural language of indexes and the triangle fuzzy data

Natural language	the triangle fuzzy data
Very low	(0,0.2,0.4)
Low	(0.2,0.4,0.6)
average	(0.4,0.6,0.8)
Good	(0.6,0.8,1)
Very good	(0.8,1,1)

Table 2. The corresponding relationship between the natural language of the weights and the triangle fuzzy data

Weight	the triangle fuzzy data
Very high	(8,9,9)
Between very high and high	(6,7,8)
high	(4,5,6)
Between high and average	(2,3,4)
average	(1/2,1,2)
same	(1,1,1)
Between average and low	(1/4,1/3,1/2)
low	(1/6,1/5,1/4)
Between low and very low	(1/8,1/7,1/6)
Very low	(1/9,1/9,1/8)

(2) Integration of subjective and objective weight

The weight of indexes includes subjective and objective weight. The subjective weight is affected by the expert information and the objective weight neglects the expert information. The paper puts forward a new method of integration of subjective and objective weight.

Suppose: every evaluation unit has m evaluation indexes, z_{ij} represents the index j, w_{qj} represents the weight of the index j.

The evaluation function $y_i = \sum_{j=1}^{m} w_{qj} z_{ij}$. In the function, $q = z$ represents the subject weight and $q = k$ represents the object weight, $q = l$ represents the theoretic weight.

The subjective weight is w_{zj}, $j = 1, 2, \cdots, m$.

Among it, $\sum_{j=1}^{m} w_{zj} = 1$, $w_{zj} \succ 0$;

The objective weight is w_{kj}, among it: $\sum_{j=1}^{m} w_{kj} = 1$, $w_{kj} \succ 0$;

The integrated weight is w_{lj}, Among it: $\sum_{j=1}^{m} w_{lj} = 1$, $w_{lj} \succ 0$,

A linear goal planning model:

$$\min Z = \alpha \sum_{i=1}^{n} \sum_{j=1}^{m} [(w_{zj} - w_{lj})z_{ij}]^2 + \beta \sum_{i=1}^{n} \sum_{j=1}^{m} [(w_{kj} - w_{lj})z_{ij}]^2 \quad (3)$$

$$s.t. \quad \sum_{j=1}^{m} w_{lj} = 1 \qquad w_{kj} \succ 0 \quad j = 1, 2, \cdots, m$$

Among it, α、β represents the relative significance of subject and object weight and $\alpha + \beta = 1, \alpha$、$\beta \succ 0$. in order to solve the model(3), a Lagrange function is built:

$$L = \alpha \sum_{i=1}^{n} \sum_{j=1}^{m} [(w_{zj} - w_{lj})z_{ij}]^2 + \beta \sum_{i=1}^{n} \sum_{j=1}^{m} [(w_{kj} - w_{lj})z_{ij}]^2 + 2\lambda (\sum_{j=1}^{m} w_{lj} - 1)$$

Among it, λ is Lagrange multiplier. we make $\partial L / \partial w_{lj} = 0$,

$$(\alpha w_{zj} + \beta w_{kj}) \sum_{i=1}^{n} z_{ij}^2 = w_{lj} \sum_{i=1}^{n} z_{ij}^2 + \lambda \quad (4)$$

Because of $\sum_{j=1}^{m} w_{zj} = 1$, $\sum_{j=1}^{m} w_{kj} = 1$, $1 + \sum_{j=1}^{m} \dfrac{\lambda}{\sum_{i=1}^{n} r_{ij}^2} = 1$ and $\lambda = 0$ is tenable

$$\text{Then } w_{lj} = \alpha w_{zj} + \beta w_{kj} \quad (5)$$

$g(X_n)$, $g(X_{n-1})$, $\cdots g(X_1)$ can be computed using w_{lj}, the evaluation value based on w_{lj} can be computed.

(3) The value of λ

In order to satisfy the different evaluation goal, the different values of λ can be chosen. The goals and the values are list in table 3.

Table 3. The principle and value of λ

Evaluation request and destination	λ
special single or double indexes performance better	close to −1
some single or double indexes performance better	close to −1
Every index performance smooth	Large value
Considering the better performance of the special single or double indexes and the smooth of every index	Between −1 and 0
Considering the better performance of every single or double indexes and the smooth of Every index	Between −1 and 0

3　The Evaluation Indexes of PE Informationization in High School

The core of PE informationization in high school is to develop and offer information resource to serve for physical science, physical teaching and physical management. The base of physical education informationization in high school is infrastructure and human resources [3, 4]. The evaluation indexes are list in Table 4.

Table 4. Evaluation indexes of physical education informationization

first indexes	second indexes	Meaning of indexes
Information infrastructure	Net infrastructure	Net facilities, net structure
	Services	Net speed, multimedia system
	Basic infrastructure	Computers, sum of multimedia classroom
Information resource	Electronic information resource	Electronic books, electronic database, sum of database
	Net resource	Sum of website, sum of http and ftp
	multimedia Software	Multimedia software, teaching software
IT application	in the class	Frequency and level of IT in the sport class
	in the management	Automation level
	Optional class system	Automation of optional class
IT persons	Qualities of manager	Usage ability
	Qualities of teachers	Usage ability
	Qualities of students	Operation ability

4　Case Study

In order to test the evaluation method, ten experts were chosen to decide the subjective weight and entropy method to decide the objective weight. It evaluated the developing level of physical education informationization in five high schools. The last results were in Table 5- Table 8.

From Table 8, the sort of evaluation value was different with the different λ. The No 1 High school descended from 1 to 5.The No 3 high school had a ascending tendency and the No 2 and 4 high school changed little. And we could also analyze the critical factors of PE informationization from Table 5-7.

Table 5. Evaluation value $\lambda = -0.99$

Index	$\lambda = -0.99$				
	1	2	3	4	5
Information infrastructure	0.5993	0.5942	0.5224	0.8260	0.6401
Information resource	0.7608	0.7004	0.5403	0.8660	0.5461
IT application	0.6124	0.6239	0.5751	0.8094	0.5576
IT persons	0.6352	0.5856	0.5142	0.7526	0.6425

Table 6. Evaluation value $\lambda = 0$

Index	$\lambda = 0$				
	1	2	3	4	5
Information infrastructure	0.5531	0.5854	0.5020	0.8026	0.5935
Information resource	0.7055	0.6793	0.5087	0.8514	0.5219
IT application	0.5771	0.6029	0.5590	0.7856	0.5387
IT persons	0.5143	0.6241	0.5753	0.7241	0.5642

Table 7. Evaluation value $\lambda = 1$

Index	$\lambda = 1$				
	1	2	3	4	5
Information infrastructure	0.5123	0.5794	0.5794	0.7861	0.5613
Information resource	0.6600	0.6579	0.4902	0.8379	0.5071
IT application	0.5519	0.5883	0.5467	0.7689	0.5256
IT persons	0.5462	0.6153	0.5642	0.6852	0.4521

Table 8. The sort of evaluation value according to the different λ value

	Evaluation value				
	1	2	3	4	5
$\lambda = -0.99$	0.6276	0.6178	0.6147	0.8215	0.5942
sort	2	3	4	1	5
$\lambda = 0$	0.5625	0.5692	0.5642	0.7855	0.5480
sort	4	2	3	1	5
$\lambda = 1$	0.5058	0.5317	0.5624	0.7577	0.5169
sort	5	3	2	1	4

5 Conclusions

The paper constructed the evaluation indexes of physical education informationization in high school from 4 dimensions. According to the feature of indexes, the paper put forward the improved fuzzy integral method to evaluate the developing level of physical education informationization.

References

1. Feng, W., Bin, F.: integration of physical course and IT. Physical Technology of Anhui 26(6), 66–69 (2005)
2. Ying, Z.: Construction of Physical materials room informationalization. Journal of Adult Physical Education 20(5), 42–43 (2004)
3. Yong, H.: Discussion about values of physical teachers. Physical Technology of Fujian 22(6), 52–54 (2003)
4. Junhua, P., Kai, N.: Studys of informationalization indexes of physical colleges. Physical Journal of Shanxi Normal University 21(4), 68–70 (2006)

Research on the IT Governance Framework of Sports Products Industry

Baozeng Qu and Jinfeng Li

School of P.E, Shandong University of Technology, China, 255049
Shandong Zibo No 17 middle shool, China, 255036
yuxiuyan123456@163.com

Abstract. In the paper, we analyzed the status quo of informationization of Sports products industry and set up an IT governance framework. The framework includes three components: IT decision mechanism, IT risks management process and IT performance evaluation. The detailed contents of the components were analyzed and their relation was described.

Keywords: Sport products industry; Informationization; IT governance.

1 Introduction

Information technology (IT) has changed the ways of business operation and management. Sports products industry also has been improved by information technology. And lots of resources were input in IT infrastructure and application systems. But because of the complication of IT project and scarcity of IT persons, some problems exist in the process of informationization of Sports products industry.

(1) Not having integrated IT strategy with enterprise strategy.

IT strategy not only contributes to IT strategic value but also shuns the risks of informationization. Its importance has been realized by lots of enterprises. But usually, IT strategy is driven by enterprise strategy and the enabling effect of IT strategy on enterprise strategy in the sports products industry is neglected [1,2]. The no-integration of IT strategy with enterprise strategy affects the realization of IT strategic value.

(2) Not coming into being the norm of evaluation.

The process of informationization is complicated and it needs an evaluation norm to control the scope and goal of IT project. But most of IT value is indirect and difficult to be quantified, which leads to the scarcity of the norm of evaluation. In turn, the scarcity leads to the exceeded time, the lost data and the deferred response time of IT project and IT application systems.

(3) Information islands exist

Recently, because some kinds of the systematic structure of system, operation system platform and database platform exist, information islands of sports products industry come into being. For example, whether China sports industry net or China sports products net, they apply the platform of suppliers and exhibition. They do not integrate with the exchange and payment platform [3]. Information islands affect the efficiency of information resources.

2 Definition and Goal of IT Governance

According to ISACA, IT governance is a structure of relationship and processes to direct and control the enterprise in order to achieve the enterprise's goals by adding value while balancing risk versus return over IT and its processes [4]. IT governance can realize enterprise strategy, management innovation and control the risks and construct the sustainable developing mechanism to realize IT value. The existing questions of sports products industry need to be solved by IT governance to make management with rules and regulations and execution with norm.

3 IT Governance of Sports Products Industry

On the basis of COBIT, ITIL, Prince, we set up an IT governance framework of sports products industry (fig 1).

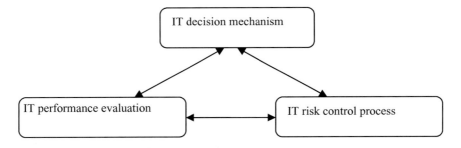

Fig. 1. IT governance framework of sports products industry

IT governance framework includes three components: IT decision mechanism, IT risk management process and IT performance evaluation. The scientific development depends on the scientific decision and it is very important for the sports products industry to decide what the decision mechanism is. The implementation of the decision mechanism depends on the normative, fine, initiative IT risk management process .IT risk management process should be evaluated and improved. The goal of IT governance is to realize IT performance. The three components form the cycle of PDCA and should be considered at the same time.

3.1 IT Decision Mechanism

IT decision mechanism includes two dimensions: what are the decisions and who should make the decisions. IT decision includes IT steering principle decision, IT system framework decision, IT infrastructure decision, business application decision, and IT investment decision.

IT steering principle decision means how to apply IT in the sports products industry. IT system framework decision deals with the norm and integration of data, information, IT application and IT infrastructure. IT infrastructure decision deals with the sharable and dependent service. Business application decision reflects the business requirement of sports products industry and decides the interaction and coordination of sports products industry. IT investment decision decides to how to realize IT investment.

The decision bodies include enterprise leaders, CIO (IT chief), business leaders and they make up of IT strategic committee, IT steering committee, IT framework committee and IT organization units.

(1) IT strategic committee

IT strategic committee is composed of board of directors and directs the integration of enterprise and IT. It helps the board of directors to finish the IT activities governance and monitoring to ensure IT governance to be included and evaluated in the agenda.

(2) IT steering committee.

IT steering committee answers for the conformity of IT project and enterprise strategy, cost-benefit analysis, priority of IT project decision, the critical success factors of IT projects.

(3) IT framework committee

IT framework committee builds the construction of the technological norm, framework and infrastructure. It tracks the development of IT and gives some advice for IT application to ensure the conformity of IT framework and law.

(4) IT organization units

IT organization units offer the service and support for the other business units.

The suitable decision bodies should be chosen according to the factual status quo of sports products industry.

3.2 IT Risks Management Process

The informationization of Sports products industry needs to card the relation and construct the management environment. And the IT risks management process is the most important topic. The risks management process includes basic process, supporting process and controlling process (fig 2).

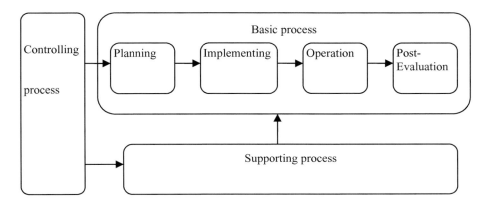

Fig. 2. IT risks management process

The basic process is defined by the lifecycle of IT construction. The supporting process supports the basic process and the controlling process .It controls the cost of the basic and supporting process to ensure the right direction of the basic and supporting process.

In order to realize the normative, scientific process, the risks management process includes

(1) the process diagram;(2)the description of process diagram details;(3)the scope and responsibility of every organization unit;(5)mature degree;(6)critical success factors;(7)key goal indexes; (8) key performance indexes; (9)the documentary management.

The process and the process description are the main thread. The mature degree is the description of the mature degree level for the process. The goal is the directional description of the process. Critical success factors are the key control points. Key goal indicators are the result measurement of the process and key performance indicators are the process measurement.

3.3 IT Performance Evaluation of Sports Products Industry

The ultimate goal is to get IT performance from IT investment. IT value is the contribution of IT to the organization which includes the financial performance, the process performance and the contribution to competitive advantage. Based on balance score card, IT performance evaluation can be evaluated from four dimensions (fig 3).

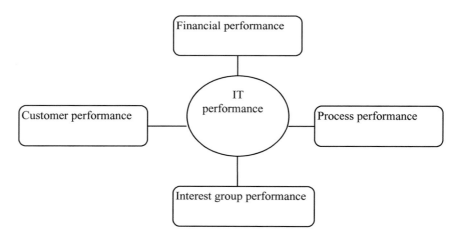

Fig. 3. IT performance dimension

(1) Financial performance

Financial performance measures the contribution of IT to the ultimate benefits which include payment, function and development indicators and so on.

(2) Customer performance

Customer performance measures the contribution of IT to the customers. The customers are critical factors and the organizations should define their customers and offer the suitable service for them. Customer performance includes market share, customer satisfaction indicators and so on.

(3) Process performance

Process performance measures the contribution of IT to the internal process. The internal process evaluation can find the questions and improve the informationization of the Sports products industry.

(4) The others interest group performance

The others interest groups include the suppliers, the government and the community, who are very important for the Sports products industry and the contribution of IT should be evaluated.

4 Conclusions

IT governance framework is a system made up of IT decision mechanism, IT risks management process and IT performance evaluation. The implementation of IT governance framework should be dealt with systematic method. And the IT governance framework should change with the internal and external environment.

References

1. Henderson, J.C., Venkatraman, N.: Strategic alignment: leveraging information technology for transforming organizations. IBM Systems Journal 32(1), 4–16 (1993)
2. Luftman, J.N., Lewis, P.R., Oldach, S.H.: Transforming the enterprise: the alignment of business and information technology strategies. IBM Systems Journal 32(1), 198–221 (1993)
3. Yin, B.: Research on the sports management informationization. Journal of Beijing Sports University 29(5), 611–613 (2006)
4. Chen, W., Yuan, R.b.: COBIT and its tips. Friends of Accouting 1, 18–20 (2006)

An Efficient Incremental Updating Algorithm for Knowledge Reduction in Information System

Changsheng Zhang[1] and Jing Ruan[2,*]

[1] College of Physics and Electronic Information Engineering, Wenzhou University, Wenzhou, 325035, P.R. China
[2] Wenzhou Vocational and Technical College, Wenzhou, 325035, P.R. China
{jsj_zcs,ruanjing1979}@126.com

Abstract. Knowledge reduction is one of challenging problems in rough set theory, many static knowledge reduction algorithms have been proposed, however, since the objects in the actual information system are often changed, these static algorithms are computationally time-consuming. In order to overcome this shortcoming, the concept of simplified information system is introduced, and it is proved that knowledge reduction based on the simplified discernibility matrix is equivalent to that based on skowron's discernibility matrix, On this condition, an efficient incremental updating algorithm for knowledge reduction is designed, which can be used to improve the efficiency of knowledge reduction, when a new object is added to information system. Finally, Theoretical analysis and example results show that illustrate the efficiency and feasibility of the algorithm.

Keywords: Rough set; Incremental updating; Knowledge reduction; Simplified discernibility matrix; Information system.

1 Introduction

Rough Set theory was proposed by Pawlak[1] in 1982, which is an efficient mathematic tool dealing with vagueness, imprecision and incomplete information. The main advantage of rough set theory in data analysis is that it does not need any preliminary or additional information about data, it can find the hiding and potential knowledge, that is decision rules. Knowledge reduction has been successfully applied to many areas such as pattern recognition, decision analysis, data mining and machine learning. In recent years, we encounter information system in which the number of objects becomes larger. So designing efficient algorithm for knowledge reduction is a significant work in information system. At present, classical models often are divided into the following three categories: knowledge reduction based on positive region[2], knowledge reduction based on discernibility matrix[3], knowledge reduction based on information entropy[4].

In the past years, many static algorithms of knowledge reduction have been developed in rough set theory[5]-[7], however, but we often find out that the objects in the

---

* Corresponding author.

actual information system are changed, because the information systems are changed when new objects insert or delete to systems, static knowledge reduction algorithms are computationally time-consuming. At present, some researchers also have proposed some algorithms of dynamic knowledge reduction, whose algorithms are not ideal. To deal with the above issue, In order to overcome this shortcoming, The concept of simplified information system is introduced, and it is proved that knowledge reduction based on the simplified discernibility matrix is equivalent to that based on skowron's discernibility matrix, On this condition, an efficient incremental updating algorithm for knowledge reduction is designed, theoretical analysis and example results show that illustrate the efficiency and feasibility of the algorithm.

2 Preliminaries

In this section, we will review several basic concepts in rough set theory.

Definition 1. For an information system (decision table) $S = (U, C, D, V, f)$, where $U = \{x_1, x_2, ..., x_n\}$ is the universe of objects, called domain, $C = \{c_1, c_2, ..., c_r\}$ is a nonempty set of condition attributes, D is a nonempty set of decision attribute, and $C \cap D = \emptyset$; $V = \bigcup_{a \in C \cup D} V_a$, that is, where V_a is the value range of attribute a, $f: U \times C \cup D \to V$ is an information function, that is $f(x,a) \in V_a$ holds. With nonempty each attribute subset $B \subseteq (C \cup D)$, determines an equivalence relation $IND(B) = \{(x,y) \in U \times U \mid \forall a \in B, f(x,a) = f(y,a)\}$, denoted by U/B in short.

Definition 2. For an information system $S = (U, C, D, V, f)$, for $P \subseteq C \cup D$, if $X \subseteq U$, denote $U/P = \{P_1, P_2, ..., P_m\}$, then $P_-(X) = \{P_i \mid P_i \in U/P, P_i \subseteq X\}$ is called lower approximation of X with respect to P.

Definition 3. For an information system $S = (U, C, D, V, f)$, let $U/D = \{D_1, D_2, \cdots, D_k\}$ be the partition of D with respect to U, for $\forall P \subseteq C$, let $U/P = \{P_1, P_2, \cdots, P_m\}$ be the partition of P with respect to U, $POS_P(D) = \bigcup_{D_i \in U/D} P_-(D_i)$ is called positive region P with respect to U.

Definition 4[3]. For an information system $S = (U, C, D, V, f)$, let $M = (m_{ij})_{n \times n}$ be skowron's discernibility matrix, whose elements are defined as follows:

$$m_{ij} = \begin{cases} \{c_k \mid c_k \in C, f(x_i, c_k) \neq f(x_j, c_k) \wedge f(x_i, D) \neq f(x_j, D)\} \\ \emptyset \quad \text{else} \end{cases}$$

Definition 5[3]. For an information system $S = (U, C, D, V, f)$, let $M = (m_{ij})_{n \times n}$ be skowron's discernibility matrix, for $R \subseteq C$, there exists $\exists m_{ij} \in M \wedge m_{ij} \neq \emptyset$ and

$R \cap m_{ij} \neq \emptyset$; for $\forall c_k \in R$, there exists $R - \{c_k\} \cap m_{ij} = \emptyset$, then R is called the attribute reduction of C for D based on skowron's discernibility matrix.

Definition 6[5]. For an information system $S = (U, C, D, V, f)$, let $U/C = \{[x_1']_C, [x_2']_C, ..., [x_m']_C\}$ and $U' = \{x_1', x_2', ..., x_m'\}$. By the definition of positive region, there is $POS_C(D) = [x_{i_1}']_C \cup [x_{i_2}']_C \cup ... [x_{i_t}']_C$, where $\{x_{i_1}', x_{i_2}', ..., x_{i_t}'\} \subseteq U'$ and $|[x_{i_s}']_C / D| = 1$ ($s = 1, 2, ..., t$), let $U'_{pos} = \{x_{i_1}', x_{i_2}', ..., x_{i_t}'\}$ and $U'_{neg} = U' - U'_{pos}$. It is said that $S' = (U', C, D, V, f)$ is a simplified information system.

Definition 7. For an information system $S = (U, C, D, V, f)$, let $S' = (U', C, D, V, f)$ is the simplified information system, $M' = (m'_{i'j'})_{n \times n}$ be the simplified discernibility matrix, whose elements are defined as follows:

$$m'_{i'j'} = \begin{cases} \{c_k \mid c_k \in C, f(x_i', c_k) \neq f(x_j', c_k) \wedge f(x_i', D) \neq f(x_j', D) \wedge x_i' \in U'_{pos} \wedge x_j' \in U'_{pos}; \\ c_k \in C, f(x_i', c_k) \neq f(x_j', c_k) \wedge ((x_i' \in U'_{neg} \wedge x_j' \in U'_{pos}) \vee (x_j' \in U'_{neg} \wedge x_i' \in U'_{pos}));\} \\ \emptyset \quad else \end{cases}$$

Definition 8. For a simplified information system $S' = (U', C, D, V, f)$, let $M' = (m'_{i'j'})_{n \times n}$ be skowron's discernibility matrix, for $R' \subseteq C$, there exists $\exists m'_{i'j'} \in M' \wedge m'_{i'j'} \neq \emptyset$ and $R' \cap m'_{i'j'} \neq \emptyset$; for $\forall c_k \in R'$, there exists $R' - \{c_k\} \cap m'_{i'j'} = \emptyset$, then R' is called the attribute reduction of C for D based on simplified discernibility matrix.

3 Equivalence Analysis

Theorem 1. For an information system $S = (U, C, D, V, f)$, let $M = (m_{ij})_{n \times n}$ be skowron's discernibility matrix, and For the simplified information system $S' = (U', C, D, V, f)$, let $M' = (m'_{i'j'})_{n \times n}$ be simplified discernibility matrix. Then they have the same non-empty elements.

Proof: For $x_i \in U$, $x_j \in U$, if $x_i' \in U'$, $x_j' \in U'$ and $m_{ij} \neq \emptyset$ holds, then there exists $x_i' \in [x_i]_C$, $x_j' \in [x_j]_C$ and $m'_{i'j'} \neq \emptyset$. If $|[x_i]_C / D| = 1$ and $|[x_j]_C / D| = 1$ holds, then there exists $x_i' \in U'_{pos}$, $x_j' \in U'_{pos}$, according to the definition 4, there at least exist elements $f(x_i', D) = f(x_i, D)$, $f(x_j, D) = f(x_j', D)$ and $f(x_i', D) \neq f(x_j', D)$, and according to the definition 7, there exists element $m'_{i'j'} = m_{ij}$; If there exists $|[x_i]_C / D| = 1$

and $|[x_j]_C/D|=1$ are not true at the same time, then one of x_i and x_j is in U'_{neg} at least, according to the definition 7, there is $m_{i'j'}=m_{ij}$, Since m_{ij} is selected arbitrarily, it is known that any non-empty element in Skowron's discernibility matrix is also in simplified discernibility matrix.

For $x'_i \in U'$, $x'_j \in U'$, if $x_i \in U$, $x_j \in U$ and $m_{i'j'} \ne \emptyset$ holds, then there exists $x'_i \in [x_i]_C$, $x'_j \in [x_j]_C$ and $m_{ij} \ne \emptyset$, where there $x'_i \in U'_{pos}$, $x'_j \in U'_{pos}$, according to the definition 7, there exists element $f(x'_i,D)=f(x_i,D)$, $f(x'_j,D)=f(x_j,D)$ and $f(x_i,D) \ne f(x_j,D)$, according to the definition 4, there exists $m_{ij}=m_{i'j'}$. Where one of x'_i and x'_j is in U'_{neg} at least. Without loss of generality, we suppose $x'_i \in U'_{neg}$. If there exists $f(x_i,D)=f(x'_i,D)$, $f(x_j,D)=f(x'_j,D)$ and $f(x_i,D) \ne f(x_j,D)$, according to definition 4, there exists element $m_{ij}=m_{i'j'}$; Where there exists $f(x_i,D=f(x_j,D)$, since $x'_i \in U'_{neg}$ and $x'_i \in [x_i]_C$, there exists $x_{i_1} \in [x_i]_C$, such that $f(x_{i_1},D) \ne f(x_i,D)$, so $f(x_{i_1},D) \ne f(x_j,D)$, according to the definition 4, there exists element $m_{ij}=m_{i'j'}$, Since $m_{i'j'}$ is selected arbitrarily, it is known that any non-empty element in simplified discernibility matrix is also in skowron's discernibility matrix.

As analyzed above, the proposition is true.

Theorem 2. For an information system $S=(U,C,D,V,f)$, let R be knowledge reduction based on skowron's discernibility matrix $M=(m_{ij})_{n \times n}$ and for a simplified information system $S'=(U',C,D,V,f)$, let R' be knowledge reduction based on simplified discernibility matrix $M'=(m_{i'j'})_{n \times n}$, Then R is equivalent to R'.

Proof: For M be skowron's discernibility matrix, if there exists element $c_k \in R$, according to the definition 5, then there exists $m_{ij} \in M \wedge c_k \in m_{ij}$, such that $R \cap m_{ij} \ne \emptyset \wedge R-\{c_k\} \cap m_{ij}=\emptyset$, according to the Theorem 1, for M' be simplified discernibility matrix, there must exists corresponding $m_{i'j'} \in M'$, such that $m_{i'j'}=m_{ij}$, so there exists element $c_k \in m_{i'j'}$, such that $R' \cap m_{i'j'} \ne \emptyset \wedge R'-\{c_k\} \cap m_{i'j'}=\emptyset$, according to the definition 8, then there exists $c_k \in R'$, hence $R \subseteq R'$. For M' be simplified discernibility matrix, if there exists element $c_k \in R'$, according to the definition 8, then there exists $m_{i'j'} \in M' \wedge c_k \in m_{i'j'}$, such that $R' \cap m_{i'j'} \ne \emptyset \wedge R'-\{c_k\} \cap m_{i'j'}=\emptyset$, according to the Theorem 1, for M be skowron's discernibility matrix, there must exists corresponding $m_{ij} \in M$, such that $m_{ij} = m_{i'j'}$, so there exists element $c_k \in m_{ij}$, such that $R \cap m_{ij} \neq \emptyset \wedge R - \{c_k\} \cap m_{ij} = \emptyset$, according to the definition 5, then there exists $c_k \in R$, hence $R' \subseteq R$. As analyzed above, the proposition is true.

4 Incremental Updating Algorithm for Knowledge Reduction

According to the Theorem 1 and Theorem 2 show that knowledge reduction based on skowron's discernibility matrix is equivalent to that based on simplified discernibility matrix. Since many repetitive elements are deleted, which also save time-consuming and storage space, on this condition, incremental updating algorithm for knowledge reduction based on simplified discernibility matrix have been proposed, which can be used to accelerate efficiency of knowledge reduction algorithm.

Algorithm 1: Incremental Updating Algorithm for Knowledge Reduction
Input: simplified information system $S' = (U', C, D, V, f)$, simplified discernibility matrix M', original Red, new object x_0;
Output: new Red;
Step1: $if (\exists x_j \in U'_{neg} \wedge \forall c_k \in C \wedge f(x_0, c_k) = f(x_j, c_k))$
    Red remain the same;
Step2: $if (\exists x_j \in U'_{pos} \wedge \forall c_k \in C \wedge f(x_0, c_k) = f(x_j, c_k) \wedge f(x_0, D) = f(x_j, D))$
    Red remain the same;
Step3: $if (\exists x_j \in U'_{pos} \wedge \forall c_k \in C \wedge f(x_0, c_k) = f(x_j, c_k) \wedge f(x_0, D) \neq f(x_j, D))$
    Step3.1: $if (\forall x_i \in U'_{pos} \wedge f(x_i, D) = f(x_j, D))$
        { $m_{i'j'}$ be added between x_i and x_j in M';
        $if (|m_{i'j'}| = 1)$ Red remain the same;
        else
                { Let $P = m_{i'j'} \cap Red$;
                $if (P \neq \emptyset)$ Red remain the same;
                else //count is the frequency of Attribute
                    { $if (\exists c_r \in m_{i'j'} \wedge c_r.count \geq \forall c_w \in m_{i'j'} - \{c_r\})$
                    $Red = Red \cup c_r$;
                    According to the definition 8, redundant attributes may be removed;
                    }
                }
        }
        x_j be deleted in U'_{pos};
    }

Step3.2: $if\ (\forall x_t \in U'_{neg})$

  { $m_{tj'}$ be deleted between x_t and x_j in M';

  $if\ (\exists c_r \in Red \wedge c_r.count = |M'|)$

  $Red = \{c_r\}$;

  else Red remain the same;

  x_j be added in U'_{neg};

  }

Step4: $if\ (\forall x_j \in U'_{pos} \wedge \exists c_k \in C \wedge f(x_0, c_k) \neq f(x_j, c_k))$

  { $if\ (\forall x_i \in U')$

  $m_{0i'}$ be added between x_0 and x_i in M';

  $if\ (|m_{0i'}| == 1)$ Red remain the same;

  else

  { Let $P = m_{0i'} \cap Red$;

  $if\ (P \neq \emptyset)$ Red remain the same;

  else

  { $if\ (\exists c_r \in m_{0i'} \wedge c_r.count \geq \forall c_q \in m_{0i'} - \{c_r\})$

  $Red = Red \cup c_r$;

  According to the definition 8, redundant attributes may be removed;

  }

  }

  }

  x_0 be added in U'_{pos};

  }

The Performance Analysis and Comparison of Algorithm 1:

According to the algorithm 1, the worst time complexity of Step 1 is $O(|C \parallel U'_{neg}|)$, the worst time complexity of Step 2 is $O(|C \parallel U'_{pos}|)$, the worst time complexity of Step 3.1 is $O(|C \parallel U'_{pos}| + |C \parallel U'_{neg}|)$, the worst time complexity of Step 3.2 is $O(|C \parallel U'_{pos}|) + O(|Red|^2 |U'_{pos} \parallel U/C|)$, the worst time complexity of Step 4 is $O(|C \parallel U'|) + O(|Red|^2 |U'_{pos} \parallel U/C|)$, so the worst time complexity of algorithm is $\max\{O(|Red|^2|U'_{pos}\parallel U/C|), O(|C\parallel U/C|)\}$. Clearly, the worst space complexity of the algorithm is $O(|C \parallel U'_{pos} \parallel U/C|)$. However, the time and space complexity are $O(|R| \times |C \parallel U_1|(|U_1| + |U'_2|))$ and $O(|C \parallel U|^2)$ in [8] respectively, the time complexity and space complexity are $O(|C|^2|U|^2)$ and $O(|C||U|^2)$ in [9] respectively. Therefore the time complexity and space complexity of the algorithm are better than [8] [9].

5 Example Analysis

For information system as shown in Table 1 to illustrate efficiency of the new algorithm, we first use the algorithm in [5] to simplify the information system, hence we can get $U'_{pos} = \{x2, x5, x8, x9\}$, $U'_{neg} = \{x1, x6\}$.

Table 1. The Information System

U	a	b	c	d	D
x1	3	0	2	2	1
x2	2	1	2	1	2
x3	3	0	2	2	3
x4	2	1	2	1	2
x5	2	0	1	2	2
x6	2	1	3	1	3
x7	2	1	1	2	2
x8	1	2	3	2	1
x9	1	2	2	1	3
x10	1	2	3	2	1

According to the definition 7, the simplified discernibility matrix can be obtained as shown in Table 2.

Table 2. Simplified discernibility matrix M' of table 1

	x5	x8	x9	x1	x6
x2	∅	{abcd}	{ab}	{abd}	{c}
x5	∅	{abc}	{abcd}	{abc}	{bcd}
x8		∅	{bcd}	{abc}	{abd}
x9			∅	{abd}	{abc}

According to the definition 7, definition 8 and the Theorem 2 we can get that original knowledge reduction $Red = \{a, c\}$.

Now we illustrate different cases of the change of knowledge reduction by adding new objects to simplified information system.

1). When The new object x_0 =[2,1,3,1,1] is added in the simplified information system, according to step 3 of the algorithm, $Red = \{a,c\}$ remain the same.

2). When The new object x_0 =[2,0,1,2,2] is added in the simplified information system, according to step 4 of the algorithm, $Red = \{a,c\}$ remain the same.

3). When The new object x_0 =[2,1,2,1,3] is added in the simplified information system, according to step 5 of the algorithm know that the element (x2,x5)={bcd} should be added and (x2,x1)={abd}, (x2,x6)={c} should be deleted in M' as shown, so we can obtain $Red(C) = \{a,b\}$.

4). When The new object x_0 =[1,3,3,1,3] is added in the simplified information system, according to Step 6 of the algorithm know that the element (x0,x2)={abc} (x0,x5)={$abcd$}, (x0,x8)={bd}, (x0,x1)={$abcd$}, (x0,x6)={ab} should be added in M', so we can obtain $Red(C) = \{a,c,d\}$ or $Red(C) = \{a,b,c\}$.

6 Conclusion

To overcome the limitations of the existing static knowledge reduction algorithms, in this study, the concept of simplified information system is introduced, and it is proved that knowledge reduction based on the simplified discernibility matrix is equivalent to that based on skowron's discernibility matrix, On this condition, an efficient incremental updating algorithm for knowledge reduction has been presented, which can significantly reduce computing time of knowledge reduction. Theoretical analysis and example results show the incremental updating algorithm for knowledge reduction based on the simplified discernibility matrix is an effective algorithm and can efficiently obtain a knowledge reduction.

Acknowledgments. This research work is supported by the Education Department Foundation of ZheJiang Province, P. R. China (Y200907161).

References

1. Pawlak, Z., Skowron, A.: Rudiments of rough sets. Information Science 117(1), 3–37 (2007)
2. Skowron, A., Rauszer, C.: The discernibility matrix and function in information systems. In: Slowinski, R. (ed.) Intelligent Decision Support Handbook of Applications and Advances of the Rough Sets Theory, pp. 331–362. Kluwer Press, Dordrecht (1992)
3. Wang, G.Y., Yu, H., Yang, D.C.: Decision table reduction based on information entropy. Chinese Journal of Computers 25(7), 759–766 (2002)
4. Xu, Z.Y., Yang, B.R.: A quick attribution reduction algorithm with complexity of max {O(|U||C|), O(|C|²|U/C|}. Chinese Journal of Computer 29(3), 391–399 (2006)

5. Wang, J., Wang, J.: Reduction algorithms based on discernibility matrix: the ordered attributes method. Journal of Computer Science and Technology 16(6), 489–504 (2001)
6. Li, F., Liu, Q., Ye, M., Yang, G.W.: Approaches to Knowledge Reduction s in Decision Tables. Control and Decision 21(8), 857–862 (2006)
7. Miao, D.Q., Cheng, Y.M.: Knowledge Reduction Algorithm under Graph View. Acta Electronic Sinca 38(8), 1952–1957 (2010)
8. Yang, M.: An incremental updating algorithm for attribute reduction based on improved discernibility matrix. Chinese Journal of Computers 30(5), 815–822 (2007)
9. Hu, F., Dai, J., Wang, G.Y.: Incremental algorithms for attribute reduction in decision table. Control and Decision 22(3), 268–272 (2007)

Principle of Radial Suspension Force Generation and Control System Simulation for a Bearingless Brushless DC Motor

Leigang Chen, Xiaodong Sun, and Huangqiu Zhu

School of Electrical and Information Engineering,
Jiangsu University, Zhenjiang 212013, China
clgujs@163.com

Abstract. The dynamic decoupling control among electromagnetic torque and radial suspension forces is the basic condition of stable operation for a bearingless brushless DC motor. In the paper, the generation principle of radial suspension forces is expounded, mathematical models of electromagnetic torque and radial suspension forces are deduced. A control system based on rotor field oriented control method is designed and then simulated with Matlab/Simulink. The simulation results show that the torque and radial suspension force subsystems can be controlled independently, the validity of the proposed control method is verified, and the whole control system has good static and dynamic performance.

Keywords: Bearingless brushless DC motor, Mathematical model, Radial suspension force, Decoupling control, Simulation.

1 Introduction

Depending on its outstanding characteristics of good speed regulation, high efficiency and high torque-inertia ratio, the brushless DC motor has been widely used in aerospace, medical treatment, industrial automation and so on [1]. However, with the development of modern power electronics and control theories, brushless DC motors with conventional mechanical bearings can't meet the requirement for higher speed or super power operation any longer. To solve problems in traditional bearings, magnetic bearings have been widely studied and already achieved many major breakthroughs. The magnetic bearings can suspend rotor stably with electromagnetic force without any mechanical contact. Nevertheless, microminiaturization and critical velocity of high speed motors are limited owing to occupation of certain axial space for magnetic bearings [2]. The bearingless drive technique was proposed by virtue of structural similarities between magnetic bearing and motor. Suspension force windings and torque windings are wound together in the stator core, so that the air-gap flux density distribution is unbalanced by the superposition of fluxes produced by suspension force windings and torque windings. The principle of electromagnetic torque generation of bearingless motors is similar to that of traditional motors [3]. The appearance of the technique greatly reduces dimensions of motors, making it possible for motors to attain super high speed and to realize microminiaturization. Due to its non-mechanical wear, high utilization factor of shaft length and excellent function of rotor suspension, the bearingless technique becomes one of the important research directions in special motors. The bearingless brushless DC motor is a new-type motor, which applies magnetic bearings to the conventional brushless DC motor. Therefore, it not only has the fundamental features of high speed, maintenance-free and long life of magnetic bearings, but also possesses the advantage of high efficiency and accuracy of brushless DC motors [4].

In the paper, mathematical models of radial suspension forces and motor rig are built. The dynamic decoupling control is implemented based on the rotor field oriented control method. Aiming to test the motor's performance indexes of speed, electromagnetic torque and rotor's suspension, a simulation system is built with Matlab/Simulink toolbox. From the simulation results, the proposed control strategy is verified and the rotor can be stably suspended by suspension forces along the central axis.

2 Principle of Radial Suspension Forces Generation

Fig. 1 shows the principle of suspension force generation of a bearingless brushless DC motor with 12 slots and 8 poles [5], [6].

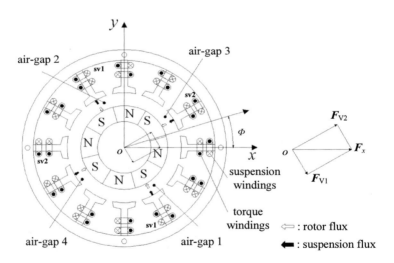

Fig. 1. The principle of suspension force generation.

When the rotor angular position Φ is within the range of 0 to 15 mechanical degrees, the suspension force windings sv1 and sv2 of V phase are excited, which results in unbalance of flux density in air-gap 1 and air-gap 2. The flux density increases in air-gap 1 and decreases in air-gap 2 oppositely. Eventually, a suspension force F_{V1} is generated in the negative direction of V_1. Accordingly, F_{V2} is generated in the negative direction of V_2. By adjusting the amplitude and direction of the suspension force windings sv1 and sv2 can generates the suspension force F_x along the x axis, which is the resultant force of F_{V1} and F_{V2}. Similarly, when the suspension force windings su1, su2 or sw1, sw2 of U or W phase are excited, the corresponding suspension forces are generated. Which of the suspension force windings are excited depends on the rotor angular position.

3 Mathematical Models

To build mathematical models of the bearingless brushless DC motor, electromagnetic torque and suspension forces can be analyzed separately.

3.1 Mathematical Models of Electromagnetic Torque

Mathematical models of electromagnetic torque can be analyzed on the basis of a three-phase brushless DC motor. Assuming that magnetic circuit is unsaturated, eddy current and hysteresis losses are ignored, the three-phase windings are symmetrical completely and Y connection is adopted, then voltage balance equation is

$$\begin{bmatrix} U_u \\ U_v \\ U_w \end{bmatrix} = \begin{bmatrix} R & 0 & 0 \\ 0 & R & 0 \\ 0 & 0 & R \end{bmatrix} \begin{bmatrix} i_u \\ i_v \\ i_w \end{bmatrix} + \begin{bmatrix} L-M & 0 & 0 \\ 0 & L-M & 0 \\ 0 & 0 & L-M \end{bmatrix} p \begin{bmatrix} i_u \\ i_v \\ i_w \end{bmatrix} + \begin{bmatrix} e_u \\ e_v \\ e_w \end{bmatrix}. \qquad (1)$$

where U_u (i_u, e_u), U_v (i_v, e_v), U_w (i_w, e_w) denote phase voltages (currents, back EMFs) of torque windings, respectively. L is self inductance, R is internal resistance and M is mutual inductance of each windings, p is the differential operator.

The equation of electromagnetic torque is the same as that of general brushless DC motors, which is the result of interaction of the currents between the stator and the PM rotor. Note that electromagnetic torque T_e is proportional to the fluxes and currents, there is

$$T_e = \frac{1}{\omega}(e_u i_u + e_v i_v + e_w i_w) . \qquad (2)$$

So torque can be controlled by regulating the amplitude of square wave current output by the inverter correspondingly.

3.2 Mathematical Models of Suspension Force

The equivalent currents in the four-pole stator equivalent windings N_a and N_b are i_{ap} and i_{bp}, respectively. I_p is the component amplitude of the equivalent torque current and I_q is that of the PM. Assuming that the motor is in no-load operation, current components of stator windings can be expressed as [7]

$$i_{ap} \approx I_p \cos 2\omega t . \qquad (3)$$

$$i_{bp} \approx I_p \sin 2\omega t . \qquad (4)$$

Where ω denotes rotor angular speed.

The flux linkages of N_a, N_b and the two-pole radial force windings N_x, N_y are denoted by ψ_{ap}, ψ_{bp}, ψ_x, ψ_y, respectively. The relation between fluxes and currents is

$$\begin{pmatrix} \psi_{ap} \\ \psi_{bp} \\ \psi_x \\ \psi_y \end{pmatrix} = \begin{pmatrix} L_4 & 0 & M'x & -M'y \\ 0 & L_4 & M'y & M'x \\ M'x & M'y & L_2 & 0 \\ -M'y & M'x & 0 & L_2 \end{pmatrix} \begin{pmatrix} i_{ap} \\ i_{bp} \\ i_x \\ i_y \end{pmatrix}. \tag{5}$$

Where L_4 and L_2 are self induction of the four-phase stator windings and two phase suspension windings, respectively; M' is derivative of mutual induction relative to radial displacement between the two windings. If irrespective of saturation,

$$M' = \frac{\mu_0 \pi n_2 n_4 l}{8} \cdot \frac{r - (l_m + l_g)}{(l_m + l_g)^2}. \tag{6}$$

Where n_2 and n_4 are the equivalent turns of torque and suspension windings, respectively; l is length of the rotating shaft; r is the semi-diameter of the stator's inner circle; l_m is the thickness of the PM; l_g is air-gap length; $l_m + l_g$ is the distance between the stator's inner circle and the rotor's external surface.

$$W_m = \frac{1}{2} \begin{pmatrix} i_{ap} & i_{bp} & i_x & i_y \end{pmatrix} \begin{pmatrix} L_4 & 0 & M'x & -M'y \\ 0 & L_4 & M'y & M'x \\ M'x & M'y & L_2 & 0 \\ -M'y & M'x & 0 & L_2 \end{pmatrix} \begin{pmatrix} i_{ap} \\ i_{bp} \\ i_x \\ i_y \end{pmatrix}. \tag{7}$$

Neglecting saturation, radial force components F_x, F_y can be

$$F_x = \partial W_m / \partial x, \quad F_y = \partial W_m / \partial y. \tag{8}$$

The current is a square-wave, so the current i_{ap} can be replaced by I_p. Given $i_{bp} = 0$, applying (3) and (4), there is

$$\begin{pmatrix} F_x \\ F_y \end{pmatrix} = M' I_p \begin{pmatrix} i_x \\ -i_y \end{pmatrix}. \tag{9}$$

3.3 Motor Motion Equations

Rotor mass, pole-pairs, rotational inertia and load torque are denoted by m, p_1, J, and T_L, respectively. F_{zx}, F_{zy} are the components of disturbing force from outside. Thus, motor motion equations can be expressed as [8]

$$F_{zx} + F_{sx} - F_x = m \frac{d^2 x}{dt^2}. \tag{10}$$

$$F_{zy} + F_{sy} - F_y = m\frac{d^2 y}{dt^2} .\qquad(11)$$

$$T_e - T_L = \frac{J}{p_1} \cdot \frac{d\omega}{dt} .\qquad(12)$$

4 Design of the Integrated Control System

The first sinusoidal harmonic of the back EMF plays a leading role in all harmonics. If in spite of the difference of motor structure, the proposed motor can be approximate to a PM synchronous motor. Therefore, suspension forces can still be computed by d-q coordinate transformation. From (2) and (9), the control aim can be achieved simply by using magnetic orientation according to control methods of general brushless DC motors. This paper presents the block diagram in Fig. 2.

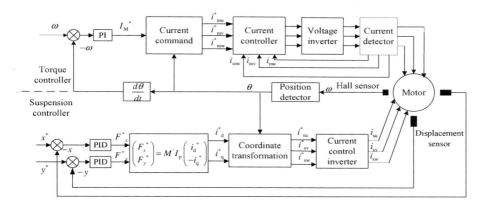

Fig. 2. Block diagram of the integrated control system.

Fig.2 shows a new method of modeling for the bearingless brushless DC motor, which makes the control system modularization, such as motor rig module, hysteresis current control module, speed control module, reference current module, torque calculation module and voltage inverter module, etc [9]. This approach not only can shorten the design period of control schemes and detect the control algorithm rapidly, but also can make full use of the advantage of computer simulation [10]. Electromagnetic torque is calculated by (2) directly. The rotor radial position along two the perpendicular axes is detected by three eddy current sensors. The displacement commands x^*, y^* in the x- and y- directions are zero. When there is an error occurring between the command and the feedback, the radial force commands F_x^* and F_y^* are generated by PID regulators, then the current commands i_{dr} and i_{qr} along the x- and y-axes are determined by transformation in two phases. Finally, the current commands

i_a^*, i_b^* and i_c^* are obtained. Therefore, the three actual currents i_a, i_b, i_c can be controlled through a current regulator.

5   Simulation Design and Experimental Results

After organic integration of the functional modules proposed, simulation models of motor rig and the whole control system are created according to the block diagram with Matlab/Simulink software.

By amending the parameters of these versatile modules or exerting disturbance on the system artificially, the dynamic and static performance in different experimental conditions can be examined [11]. An S function is written with piecewise-linear method, which can reduce problems resulted from the unsatisfactory back EMFs, such as higher torque ripple and the undesirable phase currents. The ode23t is chosen as variable step in the system, which is very appropriate to settle the problem of stiffness. The starting and the ending time are 0s and 2s, respectively. The simulation parameters are as follows:

(a) Torque windings: resistance $R_s=1\Omega$, mutual inductance $L_m=-0.0067H$, self-inductance $L=0.02H$, pole-pairs $p_1=2$, rotary inertia $J=0.005kg·m^2$, back EMF factor $k_e=0.3821$.

(b) Suspension windings: resistance $R_s=1\Omega$, self-inductance $L_s=0.015H$, pole pairs $p_2=1$. Rotor mass $m=1kg$.

The suspension force control system subsystem is shown in Fig. 3.

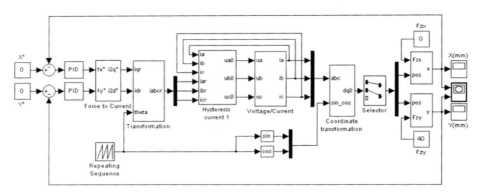

Fig. 3. The simulation model of the suspension force subsystem with MATLAB/Simulink.

The control system is simulated after established. The given speed is 5000r/min. Starting torque is 0N·m, added to 1N·m at 0.07s. Both the initial values in the x- and y- axes are 0.05mm. The simulation results are shown in Fig. 4. Fig. 4(a) is the output characteristic of speed. The overshoot is below 0.5% and steady-state error is below 0.1r/min, so the system has a good performance of speed regulation. Fig 4(b) is the displacement along the x-axis. It shows that the larger of axial offset, the greater of radial force. Finally, the suspension force changes in a small range. Fig 4(c) is the displacement curve when the rotor endures the interference force of 100N in the y direction. The shaft is rotating stably after transitorily vibrating. Fig 4(d) shows that the rotor shaft moves towards the center in a spiral trajectory and the system can reach steady state quickly.

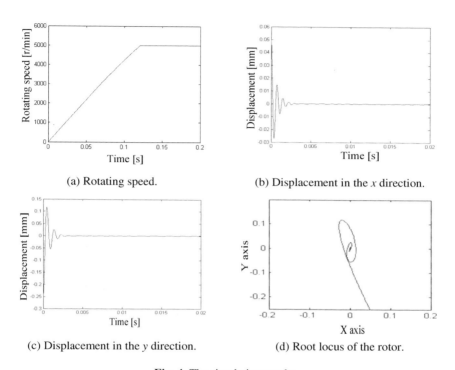

Fig. 4. The simulation results.

6 Conclusion

In this paper, the principle of suspension force generation of a bearingless brushless DC motor is analyzed. Mathematical models of the radial suspension forces and electromagnetic torque are established, and the decoupling control by using the control strategy of the rotor field oriented method is realized. Finally, the control system is designed and simulated with Matlab/Simulink toolbox. The simulation results verify the effectiveness of the proposed control algorithm and show that the control system has excellent dynamic and static performance.

Acknowledgment

This work is supported by Natural Science Foundation of Jiangsu Province under Grant BK2009204.

References

1. Chen, Z.: Principles and Application of Brushless DC Motors. China Machine Press, Beijing (2004)
2. Bosch, R.: Development of a Bearingless Electric Motor. In: Proc. of Int. Conf. Electric Machines, Pisa, pp. 373–375 (1988)
3. Salazar, A., Chiba, A., Fukao, T.: A Review of Developments in Bearingless Motors. In: 7th International Symposium on Magnetic Bearings, Zurich, pp. 335–401 (2000)
4. Fengxiang, W., Jiqiang, W., Zhiguo, K.: Study on Brushless DC Motor with Passive Magnetic Bearings. Proceedings of the CSEE 24, 94–99 (2004)
5. Ooshima, M.: Analyses of Rotational Torque and Suspension Force in a Permanent Magnet Synchronous Bearingless Motor with Short-pitch Winding. In: Power Engineering Society General Meeting, Florida, pp. 1–7 (2007)
6. Ooshima, M., Takeuchi, C.: Magnetic Suspension Performance of a Bearingless Brushless DC Motor for Small Liquid Pumps. IEEE Transactions on Industry Applications 47, 72–78 (2011)
7. Chiba, A., Fukao, T., Ichikawa, O., Oshima, M., Takemoto, M., Dorrell, D.: Magnetic Bearings and Bearingless Drives. Newnes Publishers, Burlington (2005)
8. Xiaoting, F., Jiangkang, L., Yang, G., Yikai, Z.: Magnetic Suspension Bearingless BLDCM and the Simulation. Micromotors Servo Technique 39, 66–68 (2006)
9. Jiankang, L., Huiyun, F., Xiaoting, F.: The Decoupling of Magnetic Suspension Bearingless Motors. Micromotors Servo Technique 39, 15–68 (2006)
10. Zhicheng, J., Yanxia, S., Jianguo, J.: A Novel Method for Modeling and Simulation of BLDC System Based on Matlab. Acta Simulata Systematica Sinica 12, 1745–1749 (2003)
11. Salem, T., Haskew, T.A.: Simulation of the Brushless DC Machine. In: 27th Southeastern Symposium on System Theory, Mississippi, pp. 18–22 (1995)

Based on Regularity District Scheduling of Resources for Embedded Systems

Min Rao[1], Jun Xie[1,2], Di Cai[1], and Minhua Wu[1]

[1] Information Engineering College, Capital Normal University, Beijing, 100048, China
[2] Department of Computer Science and Technology, Tsinghua National Laboratory for Information Science and Technology, Tsinghua University, Beijing 10084, China

Abstract. Embedded computers have become pervasive and complex. This approach entails at least three serious consequences: firstly, profligate and rigid usage of resources necessitated by the binding of subsystems to hardware platforms; secondly, significantly more difficult system integration because individually developed and tested systems are not guaranteed to work in combination; thirdly, the lack of higher-level system control because conceptually indivisible functions are isolated on the hardware level. These problems have caused not only financial loss such as unusable systems due to the high cost of unjustified resource redundancy and integration failures, but also the loss of human lives as exemplified by a number of fatal accidents induced by the third problem. It is critical that system engineers have a solid basis for addressing these fundamental design problems in large-scale real-time embedded systems. Towards this end, in this dissertation we introduce the notion of a Real-Time Virtual Resource (RTVR) which operates at a fraction of the rate of the shared physical resource and whose rate of operation varies with time but is bounded. Tasks within the same task group are scheduled by a task level scheduler that is specialized to the real-time requirements of the tasks in the group. The scheduling problems on both task level and resource level are analyzed.

Keywords: Embedded System, Resource Locking, Scheduling.

Regular District

It is critical that system engineers have a solid basis for addressing these fundamental design problems in large-scale real-time embedded systems. Towards this end, in this dissertation we introduce the notion of a Real-Time Virtual Resource (RTVR) which operates at a fraction of the rate of the shared physical resource and whose rate of operation varies with time but is bounded. Tasks within the same task group are scheduled by a task level scheduler that is specialized to the real-time requirements of the tasks in the group. The scheduling problems on both task level and resource level are analyzed.

Because of has the unique nature based on the regularity district as a result of its entire number field, therefore separates to their scheduling of resources's discussion with other district.

Theorem 1

A regular district except the counter-balance, determined only by its use factor.

Theorem 2

Assigns a set $\{n_i / p_i, 1 \leq i \leq m\}$ to represent the regular district the use factor, whether to have a dispatcher to be able to contain all districts the determination question is a NP question.

Although this question is the NP question, so long as but the new use factor is higher than the old use factor, always may transform the use factor to be possible to dispatch the set.

Theorem 3

If the regular district's use factor is some digital involution, and the total use factor is smaller than is equal to 1.0, then this regular district may dispatch.

The use factor respectively is $1/2, 1/4, 1/8, 1/8$ the regular district $\prod_1, \prod_2, \prod_3, \prod_4$ may very easily in the cycle be 8, assignment time slot for (1,2,1,3,1,2,1,4) the private resource on dispatches.

Theorem 4

Assigns a set $\{n_i / p_i, 1 \leq i \leq m\}$ to represent the regular district the use factor, if $\sum_{i=1}^{m} \frac{n_i}{p_i} \leq 0.5$, then these districts may transfer.

Irregular District

Theorem 5

Comes from the same resources two regular district $\prod_1, \prod_2$ union to compose a new supplies regularity is 2 new districts together $\prod_3$.

Proved: Supposes $I_1(t), I_2(t), I_3(t)$ distinguishes representative's $\prod_1, \prod_2, \prod_3$ immediate regularity function. Regarding $\forall a, \forall b, a < b$, has:

$$|I_1(b)-I_1(a)|<1$$
$$|I_2(b)-I_2(a)|<1$$
$$|I_3(b)-I_3(a)|$$
$$=|I_1(b)+I_2(b)-I_1(a)-I_2(a)|$$
$$\leq|I_1(b)-I_1(a)|+|I_2(b)-I_2(a)|$$
$$<1+1$$
$$=2$$

Requests two regular districts is comes from the same resources the reason is for the time slot which guarantees them not to be impossible to have conflicts. If the district is comes from the different resources to look like in the distributed environment situation such, then that kind of conflict occurs on the possibility. In this paper does not do to this question further studies.

When in the example district $\prod_1$ and $\prod_3$ combines in together, a use factor is $5/8$, the supplies regularity will be 2 new districts will produce.

Theorem 6

When k regular district combines in then can compose a supplies regularity together is the k district.

Theorem 7

Assigns a set $\{a_k, 1 \leq k \leq n\}$ to take the 2-supply irregular district the use factor, if $\sum_{k=1}^{n} a_k \leq 0.75$, the district may dispatch.

Proved: Each ak recomposition is:
$$a_k = \frac{1}{2^i} + \frac{x}{2^j}$$

Among

$$i = \left\lceil \log_{\frac{1}{2}}(a_k - \frac{1}{2^i}) \right\rceil \quad \text{and} \quad x = \frac{a_k - \frac{1}{2^i}}{\frac{1}{2^j}} \quad \text{when} \quad (a_k - \frac{1}{2^i}) \neq 0;$$

them, $j = i+1$, and $x = 0$ when $a_k - \frac{1}{2^i} = 0$.

so, when $x \neq 0$, $1.0 \leq x \leq 2.0$ and $i < j$.

Constructs one for each ai $b_k = \dfrac{1}{2^i} + \dfrac{1}{2^{j-1}}$. Very easy to see, $b_k \geq a_k$. Therefore, if b_k May dispatch, that a_k May also dispatch.

$$\dfrac{a_k}{b_k} = \dfrac{\dfrac{1}{2^i} + \dfrac{x}{2^j}}{\dfrac{1}{2^i} + \dfrac{1}{2^{j-1}}}$$

$$= \dfrac{2^{j-i} + x}{2^{j-i} + 2} > \dfrac{2^{j-i} + 1}{2^{j-i} + 2}$$

$$\geq \dfrac{2+1}{2+2}$$

$$= 0.75$$

so, $\sum_{k=1}^{n} a_k \leq 0.75$, $\sum_{k=1}^{n} b_k \leq 1$.

Considered that three 2-supply irregular districts $\prod_1, \prod_2, \prod_3$, their use factor respectively is 0.36,0.29,0.08,0.36+0.29+0.08=0.73<0.75, therefore they may dispatch. Next, how regarding to let construct the dispatcher, because of 1/4+1/16<0.36<1/4+1/8, presently respectively is 1/4 and 1/8 two districts assigns the use factor gives $\prod_1$. Likewise, 1/4 and 1/16 assigns gives $\prod_2$, 1/16 and 1/32 assigns gives $\prod_3$. Very easy to see the total use factor not to surpass 1. Therefore an effective dispatch constructed.

Theorem 8

Assigns a set $\{a_i, 1 \leq i \leq n\}$ to take the k+1-supply irregular district the use factor, if $\sum_{i=1}^{n} a_i \leq 1 - \dfrac{1}{2^{k+1}}$, the district may dispatch.

Proved: Each a_i The recomposition is:

$$a_i = \dfrac{1}{2^{i_1}} + \dfrac{1}{2^{i_2}} + \ldots + \dfrac{1}{2^{i_k}} + \dfrac{x}{2^j}$$

In which

$$i_1 = \left\lceil \log_{\frac{1}{2}} a_i \right\rceil$$

$$i_2 = \left\lceil \log_{\frac{1}{2}} (a_i - \frac{1}{2^{i_1}}) \right\rceil \text{ when } (a_i - \frac{1}{2^{i_1}}) \neq 0$$

$$\cdots$$

$$i_m = \left\lceil \log_{\frac{1}{2}} (a_i - \sum_{z=1}^{m-1} \frac{1}{2^{i_z}}) \right\rceil \text{ when } (a_i - \sum_{z=1}^{m-1} \frac{1}{2^{i_z}}) \neq 0$$

$$\cdots$$

$$i_k = \left\lceil \log_{\frac{1}{2}} (a_i - \sum_{z=1}^{k-1} \frac{1}{2^{i_z}}) \right\rceil \text{ when } (a_i - \sum_{z=1}^{k-1} \frac{1}{2^{i_z}}) \neq 0$$

$$j = \left\lceil \log_{\frac{1}{2}} (a_i - \sum_{z=1}^{k} \frac{1}{2^{i_z}}) \right\rceil \text{ when } (a_i - \sum_{z=1}^{k} \frac{1}{2^{i_z}}) \neq 0$$

$$x = \frac{(a_i - \sum_{z=1}^{k} \frac{1}{2^{i_z}})}{\frac{1}{2^j}} \text{ when } (a_i - \sum_{z=1}^{k} \frac{1}{2^{i_z}}) \neq 0$$

Because $0 \leq x \leq 2.0$ and $i_1 < i_2 < \ldots < i_k < j$, to each a_i Structure

Then $b_k = \dfrac{1}{2^{i_1}} + \dfrac{1}{2^{i_2}} + \ldots + \dfrac{1}{2^{i_k}} + \dfrac{2}{2^j}$

$$\frac{a_i}{b_i} = \frac{\dfrac{1}{2^{i_1}} + \dfrac{1}{2^{i_2}} + \ldots + \dfrac{1}{2^{i_k}} + \dfrac{x}{2^j}}{\dfrac{1}{2^{i_1}} + \dfrac{1}{2^{i_2}} + \ldots + \dfrac{1}{2^{i_k}} + \dfrac{2}{2^j}}$$

$$= \frac{\sum_{m=1}^{k} 2^{j-i_m} + x}{\sum_{m=1}^{k} 2^{j-i_m} + 2}$$

$$= 1 - \frac{2-x}{\sum_{m=1}^{k} 2^{j-i_m} + 2}$$

$$\geq 1 - \frac{2-x}{\sum_{m=1}^{k} 2^m + 2}$$

$$> 1 - \frac{1}{2^{k+1}}$$

Therefore, $\sum_{i=1}^{n} a_i \leq 1 - \frac{1}{2^{k+1}}$, so $\sum_{i=1}^{n} b_i < 1$.

In fact, so long as assigns for each district use factor sum total surpasses 1, these districts may dispatch. If from 0.08 enhances its use factor to 0.31, this will cause the total use factor to climb to 0.96, but the district still might dispatch.

Mix District

After the discussion has the same regularity district scheduling problem, below gives about has different regularity district scheduling problem some conclusions.

Theorem 9

Assigns a set $\{(a_i, k_i), 1 \leq i \leq n\}$ to take the district the use factor and the supplies regularity, if $\sum_{i=1}^{n} (\frac{a_i \times 2^{k_i}}{2^{k_i} - 1})_i \leq 1$, the district may dispatch.

Theorem 10

Assigns a set $\{(a_i, k_i), 1 \leq i \leq n\}$ to take the district the use factor and the supplies regularity, if $\sum_{i=1}^{n} AFF(a_i, k_i) \leq 1$, then the district may dispatch.

Attention theorem 13 and theorem 14 differences. The theorem 3.13 is derives from the worst situation, therefore is pessimistic, but the theorem 3.14 is leads from the ordinary circumstances, therefore obtained more ideal results.

4 districts $\prod_1, \prod_2, \prod_3, \prod_4$, respective use factor and the supplies regularity respectively is (0.43,3),(0.12,1),(0.31,2),(0.11,2). It has according to the theorem:

$$\sum_{i=1}^{4}(\frac{a_i \times 2^{k_i}}{2^{k_i}-1})$$
$$= 0.34 \times 8/7 + 0.12 \times 2/1$$
$$+ 0.31 \times 4/3 + 0.11 \times 4/3$$
$$\approx 1.29 > 1$$

As if these districts may not dispatch, however It also has according to the theorem:

$$\sum_{i=1}^{n} AFF(a_i, k_i)$$
$$= AAF(0.43, 3) + AAF(0.12, 1)$$
$$+ AAF(0.31, 2) + (0.11, 2)$$
$$= (1/4 + 1/8) + (1/16 + 1/8)$$
$$+ (1/4 + 1/16) + (1/16 + 1/16)$$
$$= 1$$

Therefore, in fact they may dispatch.

Conclusion

In this paper, we proposed an interesting question is if the operating system is not higher than the present operation by one is with the clock rate support interrupt can be what kind. But, this question will not bring a true question. Because in the reality, the computer hardware's real-time clock by is higher than 100Hz obviously the speed movement. This kind of huge frequency disparity may because in the dispatch expenses. The technology uses this attribute to raise the use factor does a higher level.

Acknowledgement

This research was supported by China National Key Technology R&D Program (2009BADA9B02), the Beijing Nature Science Foundation (4092011) and by Science Foundation of Beijing Municipal Commission of Education (KM200910028018). We would like to acknowledge our sponsors for their financial support, and as well as thank the reviewers for their valuable feedback.

References

1. Bettati, R.: End-to-End Scheduling to Meet Deadlines in Distributed Systems. PhD thesis, the University of Illinios at Urbana-Champaign (1994)
2. Burchard, A., Liebeherr, J., Oh, Y., Son, S.H.: Assigning real-time tasks to homogeneous multiprocessor systems. IEEE Transactions on Computers 44(12), 1429–1442 (1995)

3. Burns, A., Tindell, K., Wellings, A.: E®ective analysis for engineering real-time fixed priority schedulers. IEEE Transactions on Software Engineering 21(5), 475–480 (1995)
4. Caccamo, M., Lipari, G., Buttazzo, G.: Sharing resources among periodic and aperiodic tasks with dynamic deadlines. In: IEEE Real-Time Systems Symposium (December 1999)
5. Caccamo, M., Sha, L.: Aperiodic servers with resource constraints. In: IEEE Real-Time Systems Symposium, pp. 161–170 (2001)
6. Chan, M., Chin, F.: General schedulers for the pinwheel problem based on double-integer reduction. IEEE Transactions on Computers 41(6), 755–768 (1992)
7. Chan, M.Y., Chin, F.: Schedulers for larger classes of pinwheel instances. Algorithmica 9, 425–462 (1993)
8. Chen, D.: Real-Time Data Management in the Distributed Environment. PhD thesis, The University of Texas at Austin (1999)
9. Frisbie, M.: A Unified Scheduling Model for Precise Computation Control, Master's Thesis, University of Kansas (June 2004)
10. Frisbie, M., Niehaus, D., Subramonian, V., Gill, C.: Group Scheduling in Systems Software. In: The 12th International Workshop on Parallel and Distributed Real-Time Systems (WPDRTS 2004) at IPDPS 2004, Santa Fe, New Mexico, April 26-27 (2004)
11. Gill, C.D.: Flexible Scheduling in Middleware for Distributed Rate-Based Real-Time Applications, Ph. D. Dissertation, Department of Computer Science, Washington University (May 2002)
12. Gupta, A., Ferrari, D.: Resource partitioning for multi-party real-time communication. Technical Report TR-94-061, Berkeley, CA (1994)
13. Sen, P., Namata, G.M., Bilgic, M., Getoor, L., Gallagher, B., Eliassi-Rad, T.: Collective classification in network data. AI Magazine 29(3), 93–106 (2008)

ICAIS: Improved Contention Aware Input Selection Technique to increase routing efficiency for Network-On-Chip

Ebrahim Behrouzian Nejad[1,*], Ahmad Khademzadeh[2], Kambiz Badie[2], Amir Masoud Rahmani[3], Mohammad Behrouzian Nejad[4], and Ahmad Zadeali[1]

[1] Computer Engineering Department, Islamic Azad University, Shushtar Branch, Shushtar, Iran
[2] Iran Telecommunication Research Center, Tehran, Iran
[3] Computer Engineering Department, Islamic Azad University, Science & Research Branch, Tehran, Iran
[4] Computer Engineering Department, Islamic Azad University, Dezful Branch, Dezful, Iran
Member of Young Researcher Club (YCR)
{Behrouzian.e,Rahmani}@srbiau.ac.ir,
{Zadeh,K_badie}@itrc.ac.ir

Abstract. Network-on-Chip(NoC) has been proposed as a solution to provide better modularity, scalability, reliability and higher bandwidth compared to bus-based communication infrastructures. The performance of Network-on-Chip largely depends on the underlying routing techniques. A routing technique has two constituencies: output selection and input selection. this paper focuses on the improvement of input selection. Two traditional input selections have been used in NoC, First-Come-First-Served (FCFS) input selection and Round-Robin input selection. also, recently a contention-aware input selection (CAIS) technique has been presented for NOC, But there is some problem and defection in this technique. In this paper we improve the problems and defections of contention-aware input selection (CAIS) technique to develop a simple yet effective input selection technique named ICAIS. The simulation results with different traffic patterns show that ICAIS can achieves better performance than the FCFS and CAIS input selections, when combined with either deterministic or adaptive output selection.

Keywords: Network on chip, Routing Algorithm, Input selection.

1 Introduction

In the past decades, System on Board (SOB) has been the dominant methodology for designing complex digital systems. As the complexity of applications and their required algorithms have grown so rapidly, SOB has been replaced by System on Chip (SOC) methodology[1]. SOC consist of a number of pre-designed Intellectual property (IP) assembled together using electrical bus to form large chips with very complex functionality. But future generations of systems-on-chip (SoC) will consist of hun-

---

[*] Corresponding author.

dreds of pre-designed IPs assembled together to form large chips with very complex functionality. As technology scales and chip integrity grows, on-chip communication is playing an increasingly dominant role in System-on-Chip (SoC) design. To deal with the increasingly difficult problem of on-Chip communication, it has been recently proposed to connect the IPs using a Network-on-Chip (NoC)architecture. In NOC each core is connected to a switch by a network interface. Cores communicate with each other by sending packets via a path consisting of a series of switches and inter-switch links [2,3,4].

Fig. 1 shows an abstract view of a NOC in this architecture. As shown in fig.1, A typical NoC consists of four major components: Cores (C), Network Interface (NI) Units, Switches (S) and Physical Links. Each core can be a processing Element (PE), embedded memory, DSP or etc. Other components constitute the communication fabric. The router is connected to the four neighboring tiles and its local resource via channels. Each channel consists of two directional point-to-point links between two routers or a router and a local resource.[5,6,7,9]

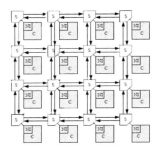

Fig. 1. Abstract view of a 4*4 2-D mesh-based NOC.

The performance of NoC largely depends on the underlying routing technique, which chooses a path for a packet and decides the routing behavior of the switches. Routing algorithms can be generally classified into two types: deterministic and adaptive. In deterministic routing, the path is completely determined by the source and the destination address. On the other hand, a routing technique is called adaptive if, given a source and a destination address, the path taken by a particular packet depends on dynamic network conditions (e.g. congested links due to traffic variability)[6,7,8,9,10]. A routing technique has two constituencies: output selection and input selection. A packet coming from an input channel may have a choice of multiple output channels. The output selection chooses one of the multiple output channels to deliver the packet. Similarly, multiple input channels may request simultaneously the access of the same output channel; the input selection chooses one of the multiple input channels to get the access [15]. Almost all of the researches on routing techniques for NoC have focused on the improvement of output selection. Two input selection methods have been used in NoC, First-Come-First-Served (FCFS) input selection and Round-Robin input selection. In [15], this paper investigates the impact of input selection, and presents a contention-aware input selection (CAIS) technique for NoC that improves the routing efficiency. ut there is some problem and defection in this technique. The motivation of this paper is to improve this input selection technique to develop a simple yet effective input selection technique.

The reminder of this paper is organized as follows. In Section 2, we review the related work. In Section 3, we present the proposed improved contention-aware input selection technique (ICAIS). The simulation results are presented in Section 4. Finally, Section 5 contains the summary and the conclusion of the paper.

2 Related Works

Routing strategies have a key role on communication and performance in on-chip interconnection networks. several efforts have been done attempting to improve the performance of them in on-chip interconnection networks. In [7], a partially adaptive routing algorithm, called turn model which is based on prohibiting certain turns during routing packets to prevent deadlock is presented. In [8] a routing algorithm called odd-even was proposed based on turn model. It restricts some locations where turn can be taken so that deadlock can be avoided. In comparison with previous methods, the degree of routing adaptiveness provided by the model is more even for different source destination pairs. A routing scheme called DyAD was proposed in [9]. This algorithm is the combination of a deterministic routing algorithm and an adaptive routing algorithm. The router can switch between these two routing modes based on the network's congestion. Another adaptive routing named DyXY along with an analytical model based on queuing theory for a 2D mesh has been proposed [10]. The authors claim that DyXY ensured deadlock-free and livelock-free routing and it can achieve better performance compared with static XY routing and odd-even routing. In [11], [12] some fully fault tolerant routing algorithms are explained, one of them is named directed flooding algorithm. In this algorithm a message is sent to each outgoing link with probability p which is not fixed but varies based on the destination of the packet. In [13] a source routing algorithm called Predominant Routing was proposed which exploits the advantages of both deterministic and adaptive routing algorithms. Also in [14] a routing algorithm for avoiding congested areas using a fuzzy-based routing decision is proposed.

All of these routing techniques focused on the output selection. Two input selections have been used in NoC, First-Come-First-Served (FCFS) input selection and Round-Robin input selection. In FCFS, the priority of accessing the output channel is granted to the input channel which requested the earliest. Round-robin assigns priority to each input channel in equal portions on a rotating basis. FCFS and Round-robin are fair to all channels but do not consider the actual traffic condition. In [15], this paper investigates the impact of input selection, and presents a novel contention-aware input selection (CAIS) technique for NoC that improves the routing efficiency. But there is a starvation possibility in this technique. The motivation of this paper is to improve input selection to develop a simple yet effective input selection technique.

3 Improved Contention Aware Input Selection Technique(ICAIS)

Each routing algorithm has two constituencies: output selection and input Selection. In this section we will present a novel input selection technique for NOC. The proposed input selection technique can be combined with an output selection, either deterministic or adaptive, to complete the routing function. In this paper, the XY routing [8] is used as a representative of deterministic output selection for its simplicity and popularity in NoC. To avoid deadlock, the minimal odd-even (OE) routing [8] is used as a representative of adaptive output selection.

Multiple input channels may request simultaneously the access of the same output channel, e.g., packets p0 of input_0 and p1 of input_1 can request output_0 at the same time. The input selection chooses one of the multiple input channels to get the access. Two input selections have been used in NoC, First-Come-First-Served (FCFS) input selection and Round-Robin input selection. In FCFS, the priority of accessing the output channel is granted to the input channel which requested the earliest. Round-Robin assigns priority to each input channel in equal portions on a rotating basis. FCFS and Round-Robin are fair to all channels but do not consider the actual traffic condition. This section presents an input selection that performs more intelligent, by considering the actual traffic condition, leading to higher routing efficiency. In this paper we consider NoCs with 2D mesh topology . Wormhole switching is employed because of its low latency and low buffer requirement. Similar to [15], the basic idea is to give the input channels different priorities of accessing the output channels. The priorities are decided dynamically at run-time, based on the actual traffic conditions of the upstream switches. More precisely, each output channel within a switch observes the contention level (CL)(the number of requests from the input channels) and sends this contention level to the input channel of the downstream switch, where the contention level is then used in the input selection. When multiple input channels request the same output channel, the access is granted to the input channel which has the highest contention level acquired from the upstream switch. Fig 2 shows the concept of Contention Level(CL) in a switch. This input selection removes possible network congestion by keeping the traffic flowing even in the paths with heavy traffic load, which in turn improves routing performance. Such an input selection helps reduce the number of waiting packets in congested areas. This removes possible network congestions and leads to better NoC performance.

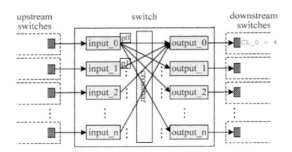

Fig. 2. Concept of Contention Level(CL) in a switch

Based on this observation CAIS input selection is developed. With a little attention to above input selection technique (CAIS), we notice an important problem. If an input channel which has lower CL continuously competing with channels which have higher CL, obviously will be defeated any time. The packets in this channel won't be able to get their required output channel and face with starvation and this will cause the problem of decreasing network efficiency. Thus, there is a starvation possibility in this input selection technique, because it performs input selection only based on the highest contention level (CL) and the channels with low CL have a little chance for winning. So, now we try to consider priority parameter in a way that input channels with low CL, have the opportunity to win. Therefore, in addition to CL, another parameter with the name of AGE for every input channel is taken into consideration and measure of priority will be a compound of CL+AGE (Fig 3).

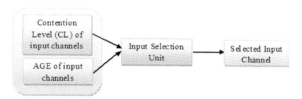

Fig. 3. Priority parameters in ICAIS

The initial value of AGE for every channel is zero. When some input channels compete each other to achieve a specific output channel, finally only one channel will succeed. After this, the AGE of the winner channel will reset to zero and AGE of the other channels entering in the competition will increase for one unit. With this new criteria(CL+AGE) each time that an input channel compete with other input channels to achieve specific output channel, in case of failure, it's AGE increase one unit and this increase its priority for the following competitions. This itself increase the opportunity of success and finally this channel be able to gain its desired output channel. In competition the following conditions may occurs:

a) if the priority(CL+AGE) of an input channel be higher than of other input channels, then the desired output channel will be granted to it and then its AGE will reset to zero. Then the AGE of all other input channel will increase for one unit.

b) If multiple output channels have the equal priority, the output channel will be granted to that input channel which has higher AGE. Then its AGE will be reset to zero and the AGE of other input channels will increase for one unit.

Fig 4 shows Pseudo code of ICAIS input selection technique.

Fig. 5 illustrates the detailed architecture of a switch with new input selection technique(ICAIS). As can be seen, the structure of this switch is similar to CAIS switch, with slight difference. Here, a very small unit to compute the AGE parameter is added to each port.

```
Input selection :
IS function :   for ( ; ; ) {
CL= # of access request to j th output channel;
send CL to downstream switch;
for each input channel i do{    AGE i = # of unsuccessful try ;
M i = observed (CLi )+ AGE i;}
In each competition do {
    if M i > M j,Mk,... then { output channel granted to channel i;
    AGE i = 0; increament AGE of M j,Mk,....}
    if M i= =M j then
        if AGE i > AGE j then { grant the access to channel i;
                AGE i = 0;    increament AGE of M j,Mk,...; }
        else if AGE i < AGE j then{
            get the access to channel j;    AGE j=0;
            increament AGE of    M i.Mk....; }
            else   { get acces to a chanel randomly;
                    winner AGE = 0;
                    increament AGE of other channel.}   } }
```

Fig. 4. Pseudo code of proposed input selection technique (ICAIS)

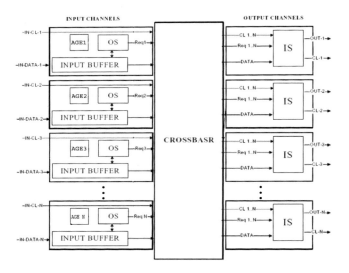

Fig. 5. Switch architecture with ICAIS input selection technique

4 Experimental Results

To evaluate the performance gains that can be achieved with our new input selection technique(ICAIS), we developed a C++ based simulator. Experiments are conducted to evaluate the performance of the ICAIS input selection technique and give a comparison between ICAIS ,CAIS and traditional input selections. Due to the advancement of FCFS over round-robin, FCFS is selected to compare with CAIS. All input selection techniques are combined with a deterministic output selection (XY routing) and an adaptive output selection (OE routing). The network size during simulation is fixed to be 6*6 tiles. It is assumed that the packets have a fixed length of 5 flits and the buffer size of input channels is 5 flits. The efficiency of each type of routing is evaluated through latency-throughput curves. Similar to other work in the literature, we assume that the packet latency spans the instant when the first flit of the packet is created, to the time when last it is ejected to the destination node, including the queuing time at the source. For each simulation, the packet latencies are averaged over 50,000 packets. Latencies are not collected for the first 5,000 cycles to allow the network to stabilize. Since the network performance is greatly influenced by the traffic pattern, in this set of experiments we consider three synthetic traffic patterns: uniform, transpose, and hot spot. In the uniform traffic pattern, a core sends a packet to any other cores with equal probability. In the transpose traffic pattern, a core at (i, j) only send packets to the core at (5-j, 5-i). In the hot spot traffic pattern, the core at (3, 3) is designated as the hot spot, which receives 10% more traffic in addition to the regular uniform traffic. Fig. 6 shows the performance of the six routing schemes under uniform traffic. As can be seen from the figure, the four schemes have almost the same performance at low traffic load (<0.040 packets/cycle). As the traffic load increases, the packet latency rises dramatically due to the network congestion. Comparing the curves of OE+FCFS, OE+CAIS and OE+ICAIS it can be seen that, using the OE output selection, ICAIS performs better than CAIS and FCFS. Similarly, the curves of XY+FCFS, XY+CAIS and XY+ICAIS show that ICAIS also outperforms FCFS and CAIS when using XY output selection. Fig. 7 shows the performance of the six routing schemes under transpose traffic. It can be seen that FCFS and CAIS and have the same performance when using the XY output selection; FCFS works slightly better than CAIS and ICAIS when using the OE output selection. This is because with transpose traffic, it is rarely the case that more than one input channels compete for the same output channel. Therefore, the input selection policy has little impact on the routing performance. Fig. 8 shows the routing performance under hot-spot traffic. Once again, it can be seen that ICAIS significantly outperforms FCFS and CAIS, either using XY or OE output selection.

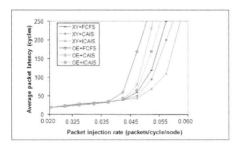 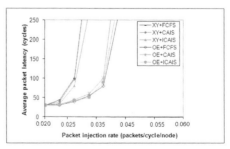

Fig. 6. Performance of routing schemes under uniform traffic

Fig. 7. Performance of routing schemes under transpose traffic

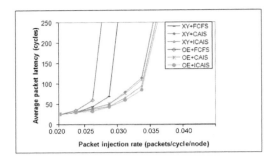

Fig. 8. Performance of routing schemes under hot spot traffic

5 Conclusion

The performance of Network-on-Chip largely depends on the underlying routing techniques. A routing technique has two constituencies: output selection and input selection. This paper has shown the importance of input selection in routing efficiency. In this paper an new efficient input selection technique, ICAIS, is presented which performs more intelligent ,by considering the actual traffic condition of the network, leading to higher routing efficiency. In this paper we improve the problems and defections of contention-aware input selection (CAIS) technique to develop a simple yet effective input selection technique, ICAIS. The simulation results show the effectiveness of ICAIS by comparing it with CAIS and traditional input selections.

References

1. Mehran, A., Saeidi, S., Khademzadeh, A.: Spiral: A heuristic mapping algorithm for network on chip. IEICE Electronic Express 4(15), 478 (2007)
2. Driankov, D., Hellendoom, H., Reinfrank, M.: An introduction to fuzzy control. Springer, Berlin (1993)
3. Benini, L., De Micheli, G.: Networks on-chip: a new soc paradigm. Journal of IEEE Computer 35, 70–78 (2002)
4. Kumar, S., Jantsch, A., Soininen, J.P., Forsell, M., Millberg, M., Oberg, J.A.: Network on chip architecture and design methodology. In: ISVLSI, USA, pp. 117–124 (2002)
5. Ivanov, A., Micheli, G.D.: The Network-on-Chip Paradigm in Practice and Research. IEEE Design and Test of Computers 22(5), 399–403 (2005)
6. Behrouzian-Nejad, E., Khademzadeh, A.: BIOS: A New Efficient Routing Algorithm for Network on Chip. Journal of Contemporary Engineering Sciences 2(1), 37–46 (2009)
7. Glass, C.J., Ni, L.M.: The Turn Model for Adaptive Routing. In: ISCA 1992: Proceeding of the 19th Annual International Symposium on Computer Architecture (1992)
8. Chiu, G.M.: The Odd-Even Turn Model for Adaptive Routing. IEEE Trans. on Parall. and Dist. Sys. 1(7) (2002)
9. Hu, J., Marculescu, R.: DyAD-Smart Routing for Networks-on-Chip. In: Proceeding of DAC 2004, San Diego, California, USA, pp. 260–263 (2004)

10. Li, M., Zeng, Q.A., Jone, W.B.: DyXY- A Proximity Congestion-Aware Deadlock-Free Dynamic Routing Method for Networks-on-Chip. In: Proceedings of ACM|IEEE Design Automation Conf., pp. 849–852 (2006)
11. Pirretti, M., Link, G.M., Brooks, R.R., Vijaykrishnan, N., Kandemir, N.M., Irwin, M.J.: Fault tolerant algorithms for network-on-chip interconnect. In: Proceeding of IEEE Computer society Annual Symposium on VLSI, February 19-20, pp. 46–51 (2004)
12. Dumitras, T., Kerner, S., Marculescu, R.: Towards on-chip fault tolerant communication. In: Proceedings of Asia and South Pacific Design Automation Conference (2003)
13. Asad, A., Seyrafi, M., Ehsani Zonouz, A., Seyrafi, M., Soryani, M., Fathy, M.: A Predominant Routing for On-Chip Networks. In: Proceeding of IDT, Riyadh, pp. 1–6 (2009)
14. Salehi, N., Dana, A.: A fuzzy-based power-aware routing algorithm for network on-chip. In: Proceeding of ICACT 2010, Phoenix Park, pp. 1159–1163 (2010)
15. Wu, D., Al-Hashimi, B.M., Schmitz, M.T.: Improving Routing Efficiency for Network-on-Chip through Contention-Aware Input Selection. In: Proceeding of 11th Conference, ASP-DAC 2006 (2006)

A New Method of Discriminating ECG Signals Based on Chaotic Dynamic Parameters

Canyan Zhu, Aiming Ji, Lijun Zhang, and Lingfeng Mao

School of Urban Rail Transportation, Soochow University,
Suzhou 215006, China
cyzhu@sohu.com

Abstract. Using chaotic dynamic parameters to analyze the ECG signal is the hotspot of research in recent year. However the result of previous studies using features such as the maximal Lyapunov exponent and power spectrum can't get a satisfactory result. In this paper, a new method of discriminating ECG signal has been put forward. This method uses scatter diagram and Kolmogorov entropy to distinguish different kinds of cases. Experiment using the MIT-BIH Arrhythmia database shows the result of improvement.

Keywords: ECG, chaos, Lyapunov exponent spectrum, phase space reconstruction.

1 Introduction

Although there are a number of studies in discriminating ECG signals using chaotic dynamic parameters like power spectrum, correlation dimension and complexity, the results are not satisfactory[1]. There are some other researchers use the largest lyapunov exponent to analyze the ECG signal, but it can't reflect the whole characteristic of human heart[2]. The ECG signals inevitably contain noise, which affect the calculation precision. In this paper we first use WFDB (waveform database) to convert the database to R-R interval time series and reconstruct the scatter diagram of the interval as well. From the diagram we can find the difference between normal and abnormal beats. Secondly, we calculate the embedding dimension m and delay timeτ by CC (Cross Correlation) method. Thirdly we reconstruct the phase space. At last we calculate the Lyapunov exponential spectrum and Kolmogorov entropy of each case by matrix algorithm.

2 The Source of Test Sample and Choice of Case

The source of the ECG included in the MIT-BIH Arrhythmia Database is a set of over 4000 long-term Holter recordings that were obtained by the Beth Israel Hospital Arrhythmia Laboratory between 1975 and 1979. Approximately 60% of these recordings were obtained from inpatients [3]. The database contains 48 records. Each of the 48 records is slightly over 30 minutes long. After the analysis of the database we know that the base contain 4 kinds of typical cases: left or right bundle branch block beat, paced rhythm, Atrial premature beat and nearly normal beat (arrhythmic beat account about 0.2%), we select 12 patients from the 4 kinds of records. The classification of cases is shown as the table 1 below.

Table 1. Classification of cases.

Case Type	Recording Number	Normal Beat Number	Abnormal Beat Number
Class A: Bundle Branch Block	214	1003	1000
	118	1096	1166
	212	923	1825
	124	1256	1003
Class B: Atrial Premature Beat	209	2621	383
	106	1507	527
	119	`543	444
	207	1457	105
Class C: Paced Rhythm	102	99	2028
	217	244	1542
Class D: Normal	101	1860	3
	103	2002	2

There are annotating files for every record in the MIT-BIH database, they annotated the starting and ending time of every beat, besides it contains the classification of each beat too. A software package WFDB(waveform database), based on UNIX, can be downloaded in the base[3], they can convert the annotating file to R-R interval time series.

3 The Scatter Diagram of R-R Interval

The R-R interval scatter diagram is to mark every data of R-R interval time series in the rectangular coordinate system, which can reveal the overall characteristic and instantaneous variety of the ECG signal. It can reveal the nonlinear characteristics of the signal. The mean to draw the scatter diagram is to take the adjacent R-R interval as the abscissa and ordinate respectively till reach the end of data. We choose 4 typical time series from the 4 classes and draw the scatter diagram as shown in Figure 1 below.

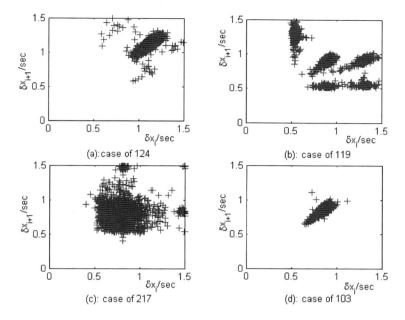

Fig. 1. Scatter diagram of R-R interval.

The distribution of interval time series in the diagram represents the trend of heart rate. The clustering degree on 45°line illustrates the instantaneous heart rate change. The length of the distribution along the 45°line represents change of heart rate of long-term.

From figure 1 we can find that class D's distribution centralizes on 45°line, and is shortest of all. The long term variation of class D is not very obvious, but a few of instantaneous R-R interval change. The class B and C are easy to distinguish because its long term and instantaneous variation.

4 CC Method and Phase Space Reconstruct

The reconstruction of phase space is the base of computing any other parameters. The choosing of τ is of extraordinary important as it directly influences the quality of reconstruction. To estimate the delay time τ, there are several algorithms like correlation function algorithm, cross correlation function method[6][8], CC method and so on[4]. The correlation algorithm has a low algorithm complexity, suitable for small data set, but is error prone and isn't fit nonlinear problem. The cross correlation method makes up for the disadvantages of the first method, but is complex. In this paper we use CC method, it is obtained by statistics and hasn't solid theoretical foundation, but its calculation result uniforms the result of cross correlation algorithm.

In CC method, it is mainly used for the statistic of correlation integral defined as:

$$S(m,N,r,t) = C(m,N,r,t) - C^m(1,N,r,t) \tag{1}$$

To describe the correlation of nonlinear time series, and calculate delay time τ and embedding dimension m by statistic $S(m,N,r,t)$, so call this method as CC method. The correlation integral function $C(m,N,r,t)$ is defined as:

$$C(m,N,r,t) = \frac{2}{M(M-1)} \times \sum_{1 \le i \le j \le M} H(r - \|x_i - x_j\|) \tag{2}$$

In Equ.(2), $H(r)$ is a Heaviside step function, $\|\cdot\|$ is Euclidean distance, and $M = N - (m-1)\tau$ is the embedded point of m dimensional phase space.

Then define $S(m,N,r,t)$ as:

$$S(m,N,r,t) = \frac{1}{t}\sum_{s=1}^{t}\left[C_s\left(m,\frac{N}{t},r,t\right) - C_s^m\left(1,\frac{N}{t},r,t\right)\right] \tag{3}$$

While $N \to \infty$ we can get:

$$S(m,r,t) = \frac{1}{t}\sum_{s=1}^{t}\left[C_s(m,r,t) - C_s^m(1,r,t)\right] \tag{4}$$

Program to calculate the variables $\overline{S}(t)$, $\Delta\overline{S}(m,t)$ and $S_{cor}(t)$ according to equations shown as below:

$$\overline{S}(t) = \frac{1}{m \cdot j}\sum_m \sum_j S(m,r_j,t)$$

$$\Delta\overline{S}(m,t) = \max\{S(m,r,t)\} - \min\{S(m,r,t)\} \tag{5}$$

$$S_{cor}(t) = \Delta\overline{S}(t) + |\overline{S}(t)|$$

Then plot the figure of $\overline{S}(t)$, $\Delta\overline{S}(m,t)$ and $S_{cor}(t)$ respectively, the delay time τ is where makes $\Delta\overline{S}(t)$ get its minimum point. Make T_{scor} be the minimum point of $S_{cor}(t)$. Then calculate m, through the formula:

$$m = round(T_{scor}/\tau) \tag{6}$$

Make the R-R interval time series as the input data, and after calculation we can get m=6 and τ=3.

Then reconstruct the phase space based on parameters m and τ[5]:

$$X_i = \{x_i, x_{i+\tau}, x_{i+2\tau} \cdots x_{i+(m-1)\tau}\}$$

According to our calculation, the dimension of phase space reconstruction is 6. We map it to three dimensional space, and plot it.

We choose four typical cases from the four classes, and plot its 3 dimensional phase space shown as figure 2. We can find that their chaotic attractor have been expanded well. So we can say that the selection of m andτ is appropriate.

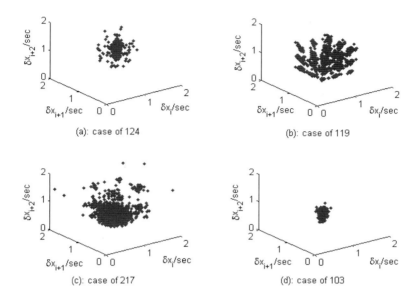

Fig. 2. Reconstructed phase space

From the phase space we can find the difference between different cases. The attractor of normal case 103 is the smallest, which means the normal heart's chaotic characteristics don't vary obviously. The difference between normal case 103 and bundle branch block case 124 isn't in its beat rate but its waveform, so the phase space can't distinguish it from normal beat. The artial premature beat and paced rhythm can be discriminated easily.

5 The Computing of Lyapunov Exponent Spectrum

Reconstruct a series {x} to a phase space [5]:

$$X_i = \{x_{i+\tau}, x_{i+2\tau}, x_{i+3\tau} \cdots x_{i+(m-1)\tau}\}$$

Assume X_j and X_i are neighborhood[6], and meet that $\|X_i - X_j\| \leq \varepsilon$, where ε is a very small positive number. Hypothesize Y_j as:

$$\{Y_j\} = \{x_i - x_j / \|x_i - x_j\| \leq \varepsilon\} \qquad (7)$$

After m times iteration, we could get $X_i \to X_{i+m}$ and $Y_i \to Y_{i+m}$. Hypothesize Z_j as:

$$\{Z_j\} = \{X_{j+m} - X_{i+m} / \|X_j - X_i\| \le \varepsilon\} \tag{8}$$

If ε is small enough, we could assume Y and Z are vectors of tangent space. And $Y_j \to Z_j$ could be approximately expressed as:

$$Z_j = A^j Y_i \tag{9}$$

Where A is a matrix of $d \times d$. To estimate the elements in A, we could use the least area method. So the elements should meet that:

$$\min S = \min \frac{1}{N} \sum_{j=1}^{N} \|Z_j - A^i Y_j\|^2 \tag{20}$$

So the Lyapunov exponent could be obtained from the formula below:

$$\lambda_i = \lim_{n \to \infty} \frac{1}{n} \sum_{j=1}^{n} \ln \|A^j e_j^i\| \tag{31}$$

In Equ.(11) e_i^j is the base vector of tangent space. The Lyapunov exponent describes the divergent rate of two adjoining trajectories in every dimension. A positive exponent means the trajectory is divergent in this dimension, and a negative exponent means convergence.

This paper computed the Lyapunov exponent spectrum of every RR interval time series of every case, using the algorithm above. Besides, this paper calculated the average exponent of each class, shown as table2:

Table 2. Lyapunov Exponent Spectrum

	Mean of Class A	Mean of Class B	Mean of Class C	Mean of Class D
λ1	0.154	0.171	0.112	0.063
λ2	0.074	0.081	0.048	0.026
λ3	0.019	0.024	0.006	0.004
λ4	-0.016	-0.026	-0.032	-0.045
λ5	-0.936	-0.088	-0.076	-0.091
λ6	-0.219	-0.231	-0.228	-0.228
Entropy	0.247	0.276	0.166	0.085

From table 2 we could find that: the maxim exponent is positive and the summation of all exponent, λ_1 to λ_6, is negative. This result is coincidence with the definition of deterministic chaos. That proves the RR interval series is a chaotic sequence.

Then we introduce the conception of Kolmogorov entropy, K-entropy for short. The K-entropy is a chaotic parameter that measures the degree of confusion, which equal the summation of all positive Lyapunov exponent in numerical. We calculate the K-entropy of every case and show as table 3.

Table 3. Kolmogorov Entropy

Class A	K-entropy	Class A	K-entropy
214	0.273	209	0.241
118	0.425	106	0.243
212	0.092	119	0.399
124	0.203	207	0.213
Class C	K-entropy	Class D	K-entropy
102	0.173	101	0.102
217	0.158	103	0.067

From the table 3 we discovery that class A can't be distinguished from other classes. Main reason is that the difference between normal ECG and bundle branch block case isn't in RR interval time but in the wave form. The K-entropy of class B centralizes at 0.2 to 0.4, it is much bigger than class D, so we can say the premature beat is more chaotic than normal beat.

Class C's K-entropy is between B and D. The K-entropy of class D is below 0.1, so the normal heart is the least chaotic.

6 Conclusion

This paper analyzed the RR interval time series of 12 cases, the analysis include reconstruction of phase space , Lyapunov exponent spectrum and Kolmogorov entropy. Through the analysis we found that the K-entropy could discriminate, to a certain extent, cases include normal, artial premature and paced beat. The result proves that the chaotic dynamic parameters works in the analysis of ECG signal.

The common used method in ECG analyzing is to analyze every beat's wave form. But the wave form changed after the occurrence of lesion in heart. Research shows that the ECG signal will change in macroscopic before heart attack. Our research is to analyze the ECG signal at a whole. So we hope the combination of method introduced in this paper and existing methods could create a new approach to early diagnosis.

References

1. Owis, M.I., Abou-Zied, A.H.: Study of features based on nonlinear dynamical modeling in ECG arrhythmia detection and classification. IEEE Transaction on Biomedical Engineering, 733–736 (2002)

2. Acharya, R., Kumar, A.: Classification of cardiac abnormalities using heart rate signals. Medical & Biological Engineering & Computing 42, 288–293 (2004)
3. http://www.physionet.org/physiobank/database/html/mitdbdir/records.htm
4. Lv, J., Lu, J., Chen, S.: Chaotic time series analysis and its application, pp. 57–66. Publishing house of Wuhan university, Wuhan (2002) (in Chinese)
5. Jovic, A., Bogunovic, N.: Feature Extraction for ECG Time-series Mining Based on Chaos Theory. In: Proceeding of the ITI 2007, 29th Int. Conf. on Information Technology Interfaces, June 25-28, pp. 63–68 (2007)
6. Wang, H., Lu, S.: Nonlinear time series analysis and its application, pp. 27–28. Pulishing House of Science, Beijing (2006)

Research on Iterative Multiuser Detection Algorithm Based on PPIC

Canyan Zhu, Lijun Zhang, Aiming Ji, and Yiming Wang

School of Urban Rail Transportation, Soochow University,
Suzhou 215006, China
cyzhu@sohu.com

Abstract. In this paper, an iterative multiuser detection (IMUD) algorithm based on partial parallel soft interference cancellation (PPSIC) structure is proposed. In reference [8], Divsalar et al proposed a partial parallel interference cancellation multiuser detection algorithm, it was introduced a partial cancellation weight in each stage to alleviate the cost of improper interference estimations in parallel interference cancellation (PIC) structure. Based on this PPIC multiuser detection algorithm, we develop a soft input soft output (SISO) multiuser detection algorithm to fit the iterative multiuser detection process, and develop a new iterative structure of the multiuser detection algorithm in this paper. Due to that the veracity of the output bit information in fore-stage affects the correctness of detection in next stage in PPIC, we let the output bit information in each stage in PPIC process iteratively with the maximum a posteriori (MAP) channel decoder, so to improve the veracity of the output bit information in each stage in PPIC. The simulating results show that our proposed structure really does improve the performance of iterative multiuser detection based on PPIC. Also we can conclude that our iterative structure should be applied in other iterative multiuser detection algorithms based on PIC.

Keywords: Iterative Multiuser detection; Partial parallel soft interference cancellation; Soft input soft output.

1 Introduction

The iterative multiuser detection (IMUD) combining with channel decoding is a new kind of multiuser detection technology, it can improve the system performance greatly through the iterative exchange of the soft information between multiuser detector and channel decoder[1]-[3]. However, the computation complexity of optimal IMUD algorithm based on the MAP criterion is rising by exponentially with the number of users $o(K^2)$[1] (K is the number of users), so the research of low computation complexity algorithm will become inevitable. Recently, the low complexity IMUD algorithm based on nonlinear interference cancellation has become a new hotspot, such as in [2]-[7]. In [3], firstly completely soft eliminate the output vectors of matched filter, and then eliminate multi-access interference through MMSE matched filter. In [7], soft interference cancellation is realized by PIC, it realizes the iterative multiuser detection between PIC and channel decoding. In fact the idea of IMUD algorithm based on interference cancellation is the same as the parallel interference cancellation (PIC) of multiuser detector used before, the former's information used for interference cancellation in soft input and soft output (SISO) multiuser detector is computed by the prior information which is sent by channel decoder, the latter's information used for interference cancellation in the first stage can be acquired through amplitude estimation and pseudo-noise (*PN*) restructure of bits information outputted by matched filter, while in the latter stages, the information that used for interference cancellation is acquired through amplitude estimation and *PN* restructure of bits information outputted by the former stages. In this paper we propose an iterative multiuser detection algorithm based on PPIC, soft interference cancellation is implemented by PPIC after matched filter. The complexity is low and linear with user numbers. We also develop the iterative structure based on PPIC and get a better performance of bit error ratio. The structure we propose can be also used in other iterative multiuser detection algorithms based on PIC.

2 System Model

This paper considers the ideal synchronous CDMA channel model, shown in figure 1.

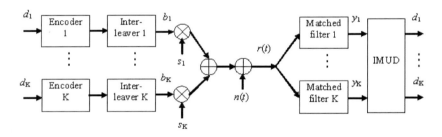

Fig. 1. System model

We assume no inter-symbol interference (ISI) exists among coded bits, and the channel only exists additive Gaussian white noise, then the received signal can be expressed as:

$$r(t) = \sum_{i=1}^{L} \sum_{k=1}^{K} A_k b_k(i) s_k(t-iT) + n(t) \tag{1}$$

In equation (1), L is the number of coded bits in one frame of each user; K is the number of users; A_k is the signal amplitude of the kth user; $b_k(i) \in \{+1,-1\}$ is the coded bits sent by the kth user; $s_k(t)$ is the spreading frequency wave of the kth user; T is the coded bits duration, $n(t)$ is the Gaussian white noise with zero mean and variance σ^2. Firstly let received signal $r(t)$ through K matched filters, then sample the output of the matched filters, so the output vector of K matched filters in ith bit duration can be expressed as:

$$y(i) = RAb(i) + n(i) \quad (2)$$

In equ.(2), R is the normalized spreading frequency coefficient cross-correlation matrix, $A=\text{diag}\{A_1,\ldots,A_K\}$, $b(i)=[b_1(i),b_2(i),\ldots b_K(i)]^T$, $n(i)$ is the Gaussian white noise vector with zero mean and variance matrix $R\sigma^2$.

3 IMUD Algorithm Based on Partial Parallel Interference Cancellation Structure

In reference [8], *Divsalar et al* have proved that completely eliminating the multi-access interference (MAI) on each user is not the only best way. And as the number of user increasing, the performance of system is not improved obviously. So *Divsalar et al* proposed a partial parallel interference cancellation (PPIC) multiuser detection algorithm for further eliminating the MAI. The measuring criterion of this algorithm for kth user in mth PIC is expressed as follows:

$$\tilde{b}_k^{(m)} = p^{(m)}(Y_k - \hat{I}_k^{(m)}) + (1 - p^{(m)})\tilde{b}_k^{(m-1)} \quad (3)$$

$$\hat{b}_k^{(m)} = \text{sgn}\{\tilde{b}_k^{(m)}\} \quad (4)$$

$$\hat{I}_k^{(m)} = \sum_{j=1, j\neq k}^{K} A_j \rho_{jk} \hat{b}_j^{(m-1)} \quad (5)$$

In equation (5), $\hat{I}_k^{(m)}$ denotes the estimated MAI value on kth user from other users' hard decision in $(m-1)$th stage; $p^{(m)}$ denotes partial interference cancellation coefficient in mth stage; Y_k denotes the output of the kth matched filters; $\tilde{b}_k^{(m)}$ denotes the kth user soft information bits value in mth stage; ρ denotes the cross-correlation coefficient of each user's characteristic wave.

Based on the inputs of information Y_k, $\hat{I}_k^{(m)}$, $\tilde{b}_k^{(m-1)}$ in mth stage, the output extrinsic information of kth user in mth stage can be expressed:

$$\lambda_{1,ex}^{(m)}[b_k] = \log\frac{p(Y_k, \hat{I}_k^{(m)}, \tilde{b}_k^{(m-1)} | b_k = +1)}{p(Y_k, \hat{I}_k^{(m)}, \tilde{b}_k^{(m-1)} | b_k = -1)} = \log\frac{p(Y_k, \tilde{b}_k^{(m-1)} | b_k = +1, \hat{I}_k^{(m)}) p(\hat{I}_k^{(m)} | b_k = +1)}{p(Y_k, \tilde{b}_k^{(m-1)} | b_k = -1, \hat{I}_k^{(m)}) p(\hat{I}_k^{(m)} | b_k = -1)} \quad (6)$$

Under ideal condition, the interference $\hat{I}_k^{(m)}$ on kth user coming from other users is independent on b_k. so $p(\hat{I}_k^{(m)} | b_k) \approx p(\hat{I}_k^{(m)})$, thus:

$$\log\frac{p(Y_k, \hat{I}_k^{(m)}, \tilde{b}_k^{(m-1)} | b_k = +1)}{p(Y_k, \hat{I}_k^{(m)}, \tilde{b}_k^{(m-1)} | b_k = -1)} = \log\frac{p(Y_k, \tilde{b}_k^{(m-1)} | b_k = +1, \hat{I}_k^{(m)})}{p(Y_k, \tilde{b}_k^{(m-1)} | b_k = -1, \hat{I}_k^{(m)})} \quad (7)$$

By considering the residual noise and $\tilde{b}_k^{(m-1)}$ being approximates to Gauss with zero mean as assumed in [8], equation (7) can be expressed as:

$$\log \frac{p(Y_k, \tilde{b}_k^{(m-1)} | b_k = +1, \hat{I}_k^{(m)})}{p(Y_k, \tilde{b}_k^{(m-1)} | b_k = -1, \hat{I}_k^{(m)})} \approx (Y_k - \hat{I}_k^{(m)}) \cdot 2A_k / \sigma_{k,m}^2 + \tilde{b}_k^{(m-1)} \cdot 2A_k / \tilde{\sigma}_{k,m}^2 \approx \tilde{b}_k^{(m)} \cdot 2A_k / p^{(m)} \sigma_{k,m}^2 \qquad (8)$$

In equation (8), $\tilde{b}_k^{(m)} = p^{(m)}(Y_k - \hat{I}_k^{(m)}) + (1 - p^{(m)})\tilde{b}_k^{(m-1)}$; $\tilde{\sigma}_{k,m-1}^2$ denotes the variance of $\tilde{b}_k^{(m-1)}$; $p^{(m)} = \tilde{\sigma}_{k,m-1}^2 / (\sigma_{k,m}^2 + \tilde{\sigma}_{k,m-1}^2)$, it denotes the partial interference cancellation coefficient of the mth stage of PPIC; $\sigma_{k,m}^2$ denotes the variance of residual noise which interfering kth user in mth stage, the its value can be estimated through various methods. To be simple, the value of $\sigma_{k,m}^2$ is approximated to be white noise variance σ^2. So the extrinsic information output by PPIC in mth stage is expressed as:

$$\lambda_{1,ex}^{(m)}[b_k] \approx \tilde{b}_k^{(m)} \cdot 2A_k / (p^{(m)} \cdot \sigma^2); m = 1, \cdots, M \qquad (9)$$

Change the MAI estimated value which is caused by the other users' hard decision in the $(m-1)$th stage to the kth user for soft decision value, that is:

$$\tilde{I}_k^{(m)} = \sum_{j=1, j \neq k}^{K} A_j \rho_{jk} \tilde{b}_j^{(m-1)} \qquad (10)$$

$$\tilde{b}_k^{(m)} = p^{(m)}(Y_k - \tilde{I}_k^{(m)}) + (1 - p^{(m)})\tilde{b}_k^{(m-1)} \qquad (11)$$

Being instead of $\hat{I}_k^{(m)}$ by $\tilde{I}_k^{(m)}$, it can't be affected the expression of extrinsic information in equation (8).

According to the IMUD principle, MAP channel decoder's input is the extrinsic information output of the PPIC, and it is updated by the log-likelihood ratio (LLR) of coded bits. The output LLR of the coded bits is again feedback as the extrinsic information λ_2 to PPIC. According to the extrinsic information λ_2 provided by the MAP channel decoder, we can get soft estimated value of all users' coded bits:

$$\tilde{b}_k(i) = E\{b_k(i)\} = \sum_{b \in \{-1, +1\}} b \cdot p[b_k(i) = b] = \tanh[\lambda_2[b_k(i)]/2] \qquad (12)$$

Specific implementing steps are as follows:

Step 1: in the first iteration, we make coded bits value which is outputted by matched filter be the input of PPIC, that is $\tilde{b}_k^{(0)} = Y_k$;

Step 2: for a M-stage PPIC, completes the PPIC calculation with M times according to equations (10) and (11), and get the extrinsic information according to equation (9);

Step 3: update LLR of coded bits by MAP channel decoder[8], i.e., calculating the extrinsic information λ_2 based on the extrinsic information λ_1;

Step 4: we can get soft estimated value of the coded bits according to equation (12). Then it is input to PPIC for second iteration.

And so on. When certain iterative times are finished, the LLR value of information bits are output from the MAP channel decoder, then carry on the hard decision.

4 System Simulation and Analysis

We will evaluate the performance of the above algorithm by computer simulation. The result is shown in figure 2. In figure 2, there are 4 users, which have same power, the correlative coefficient of the characteristic wave of different users is equal. We choose $\rho=0.5$. All the users choose the convolutional codes which is generated by polynomial decimal (7,5), and with same ratio 1/2 and restriction length 3. Each user uses a random interleaver with length of 256. PPIC structure is of two stages. According to reference [8], partial interference cancellation coefficient value should increase with information bits for more reliable, so we set $[p^1\ p^2]$ as [0.5 0.55] for the first iteration, [0.6 0.65] for the second, the third [0.7 0.75], the forth [0.8 0.85].

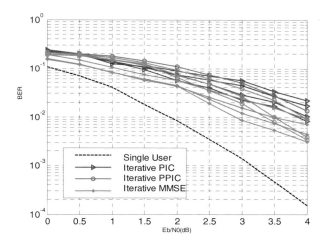

Fig. 2. Performance comparison of different IMUD algorithm in different iterative times.

In figure 2, it shows the application results of three iterative algorithms, PIC, PPIC, MMSE respectively corresponding to triangle, circle and star. Each algorithm has shown four curves, which denote the iterative times 1,2,3,4 respectively.

From figure 2, we can see that, the performance of algorithm presented in this paper is better than that in [7]. Because PPIC algorithm introduces a partial interference cancellation coefficient in each stage of PIC, it alleviates the performance loss which is caused by the inappropriate interference estimation in PIC structure. So the iterative multiuser detection algorithm based on the PPIC can acquire better performance than that based on the PIC. Compare to [3], the performance is relatively worse. But the computational complexity in [3] is $O(K^2)$, the computational complexity in [8] and this paper is $O(MK)$ (M is the number of stages of interference cancellation). Complexity is only linear with user numbers K. From figure 2 we can also find that, with the increase of the iterative times, although the bits error ratio performance of the algorithm in [8] and this paper can be enhanced, but the extent of enhancement is very limited. The reason is that, PIC or PPIC multiuser detector is multi-stages in series, the accuracy of information output from the former stage will affect the detection in latter stage. Errors transmission will deteriorate the performance of the PIC or PPIC. So in order to further enhance the performance of PPIC IMUD detector, it is necessary to improve the output information accuracy in each stage of PPIC. Next, we will improve the iterative structure of the IMUD detector based on PPIC.

5 The Improvement of Iterative Structure Based on PPIC Iterative Multiuser Detector

In order to improve the accuracy of coded bits output from each stage in PPIC, we process the output bit information in each stage of PPIC iteratively combined with MAP channel decoder separately. When certain iterative times are finished, the coded bit information input to the next stage of PPIC; in the next stage, it is also iteratively processed combining with MAP channel decoder. So the entire structure is M iterative multiuser detectors in series. The improved structure is shown in figure 3.

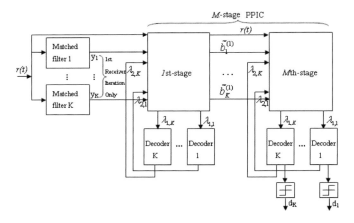

Fig. 3. Performance comparison of PIC IMUD algorithm and PIC IMUD algorithm with improved structure

We apply the algorithms in [7] and presented in third part of this paper to this improved iterative structure. The simulation results are shown in figure 4.

In figure 4, each has two groups of curves respectively represent performances of PIC or PPIC IMUD algorithms without and with improved structure of Fig.3. Each group has four curves. From top down to bottom, they respectively denote performance of four situations, they are one stage of PPIC(PIC) with one iterative time; two stages of PPIC(PIC), each stage takes one iterative time; two stages of PPIC(PIC), the former stage take two iterative times, and the latter takes one time; two stages of PPIC(PIC), each stage both take two iterative times. Considering the same PPIC(PIC) stages and the same decoding times, we can see this improved iterative structure can achieve better performance. Because PIC itself is worse than PPIC in each stage detection, the performance enhancement of the PIC algorithm with improved structure is

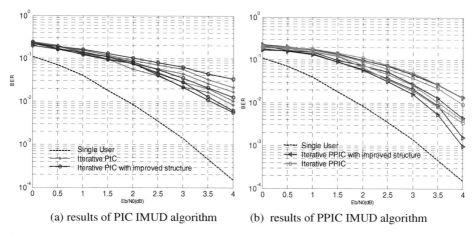

(a) results of PIC IMUD algorithm (b) results of PPIC IMUD algorithm

Fig. 4. Performance comparison of PIC and PPIC IMUD algorithms between with and without improved structure.

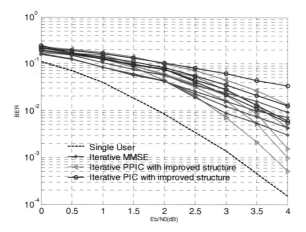

Fig. 5. Performance comparison of PIC IMUD algorithm, PPIC IMUD algorithm with improved structure and MMSE algorithm.

very limited. We also compare the MMSE algorithm in [3] with the PPIC and PIC algorithm with improved structure, as shown in figure 5.

Although the PIC IMUD algorithm performance with improved structure is yet worse than the MMSE algorithm, the PPIC IMUD algorithm performance with improved structure is significantly better than MMSE algorithm. In figure 5, every group of four curves show the performances correspoding to the four situations listed as in figure 4. But except the red group of curves, its last curve shows the performance in PPIC IMUD algorithm with improved structure correspoding to two stages of PPIC, the former and the latter take three iterative times respectively. From figure 5, we see that the improved structure has great performance enhancement for PPIC IMUD algorithm.

6 Conclusion

In this paper, an iterative multiuser detection algorithm based on PPIC structure is proposed. Its computational complexity of each stage in PPIC is linear with the user numbers (the computational complexity of M-stage is $O(MK)$). And the iterative structure of multiuser detection based on PPIC is also developed. The simulation results show that the iterative multiuser detection based on PPIC with improved structure can acquire better performance. And this iterative structure can be used in other iterative multiuser detection algorithms based on PIC.

References

1. Moher, M.: An iterative multiuser decoder for near-capacity communications. IEEE Trans. on Communications 46(7), 870–880 (1998)
2. Alexander, P.D., Grant, A.J., Reed, M.C.: Iterative detection in code-division multiple-access with error control coding. Eur. Trans. Telecommunications 9(5), 419–426 (1998)
3. Wang, X., Poor, H.V.: Iterative (Turbo) soft interference cancellation and decoding for coded CDMA. IEEE Trans. on Communications 47(7), 1046–1061 (1999)
4. Hsu, J.M., Wang, C.L.: A Low-complexity iterative multiuser receiver for turbo-coded DS-CDMA systems. IEEE Journal on Selected Areas in Communications 19(9), 1775–1783 (2001)
5. Morosi, S., Fantacci, R., Bernacchioni, A.: Improved iterative parallel interference cancellation receiver for future wireless DSCDMA systems. Eurasip Journal on Applied Signal Processing (5), 626–634 (2005)
6. Tan, S., Xu, L., Chen, S., Hanzo, L.: Iterative soft interference cancellation aided minimum bit error rate uplink receiver beamforming. In: 2006 IEEE 63rd Vehicular Technology Conference, pp. 17–21 (2006)
7. Pan, J.Y., Soh, C.B., Gunawan, E.: Iterative Soft Parallel Interference Cancellation for Convolutional-coded DS-CDMA System. Wireless Personal Communications 25(3), 177–186 (2003)
8. Divsalar, D., Simon, M.K., Raphaeli, D.: Improved parallel interference cancellation for CDMA. IEEE Trans. on Commun. 46(2), 258–268 (1998)

Research of Aero-Engine Robust Fault-Tolerant Control Based on Linear Matrix Inequality Approach

Xue-Min Yang[1], Lin-Feng Gou[1], and Qiang Shen[2]

[1] School of Power and Energy, Northwestern Polytechnical University, Xi'an 710072, China
[2] School of Aeronautics and Astronautics, Zhejiang University, Hangzhou 310027, China
yangxuemin534@163.com

Abstract. Associated with the problem of fault tolerant control against actuator fault happened to the aero-engine, a D-stabilized robust tolerant control approach was proposed based on linear matrix inequality (LMI). To satisfy the regional poles constraints and performance in actuator fault condition and structured perturbations, a state feedback fault tolerant controller was designed in terms of feasible solutions to the LMIs. Simulation results illustrate that, the proposed approach can maintain the performance of robust stability in specified pole region as well as integrity of the control system in actuator fault condition, with allowable perturbations and fault bounds.

Keywords: aero-engine; actuator fault; linear matrix inequality; robust fault tolerant control.

1 Introduction

Stimulated by the growing demand for high reliability and high security of aero-engine control systems, robust fault-tolerant control has become a critical issue since its goal is to make the system stable and retain acceptable performance under the system faults. In comparison with the costly implementation of hardware redundancy, or on-line component cut which may lead to serious oscillation, the application of robust fault-tolerant control is a good deal by modifying the commands to the actuators and reconfiguring the control law. Nevertheless, the uncertainties between routine slight drift linearization model and turbine engine dynamics is also considered through the whole process of controller design, which highlights practical significance to engineering.

A broad class of sensor and actuator faults tolerant methods makes explicit use of a mathematical model of the plant. Due to analytical redundancy, which makes the reconfigured control law possible, the fault component can be compensated by the rest. Survey papers by [1] [2] present excellent overviews of advances in the research of fault-tolerant control. In [3]-[5] the authors studied sensor/actuator failure and derived tolerant methods via a iterative solution of Riccati matrix equations. However, some dynamic performance must be sacrificed to guarantee stability due to the short of optimizing undetermined parameters. Here the linear matrix inequality (LMI) process based on interior point method helps establish conditions of solvability of the problem. Then a set of feasible solutions are obtained as well as in the condition of multi-objective control (e.g., see the discussion in [6]-[8]).

In this paper, Section 2 discusses general form of turbine engine linear dynamic models, and transforms required performance into the solvability of LMI. Section 3 elaborates on designing a state feedback robust D-stabilized fault-tolerant controller with closed loop pole constraint, based on Lyapunov stability theorem and LMI approach. On occurrence of actuator failures, all closed loop system pole still allocated to a specified region, with allowable perturbations and fault bounds. Simulation of turbine engine control system is conducted and robust stability as well as dynamic performance are all illustrated in Section 4. Some corresponding conclusions are given in Section 5.

2 Problem Statement

To develop a robust fault-tolerant engine controller an accurate representation of the engine dynamics is desired. Although turbine engine dynamics are inherently nonlinear, aerodynamic nonlinearities and inertial coupling effects generally are smooth enough in the operating regions so that linear design techniques are applicable [9].

In this section a linear normalized state-space model of a certain type of turbofan engine around an operating point can be described as

$$\dot{x}(t) = (A + \Delta A)x(t) + (B + \Delta B)u(t) \tag{1}$$

where $x(t)=[n_L, n_H]^T, u(t)=[m_f, A_8]^T, A, B \in R^{2\times 2}$, and n_L, n_H, m_f, A_8 respectively represent the fan rotor speed, the core rotor speed, the lord combustor fuel flow, and the exhaust nozzle area. Perturbation matrixes ΔA, ΔB reflect the mismatch between original model structure and the real one when some condition changes in the system. Given that norm bounded perturbations takes the form

$$[\Delta A \quad \Delta B] = E\Sigma(t)[F_1 \quad F_2] \tag{2}$$

where E, F_1 and F_2 are matrixes of proper dimensions and usually assumed to be known as uncertain structural information. $\Sigma(t)$ is unknown but with content $\Sigma^T(t)\Sigma(t) \leq I$.

Suppose (A, B) is stabilizable, ensuring system asymptotic stability. With the state feedback control law

$$u(t) = Kx(t) \tag{3}$$

the state space model becomes

$$\dot{x}(t) = A_c x(t) \tag{4}$$

where $A_c = A + BK + \Delta A + \Delta B K$.

The eigenvalue of A_c is defined as $\lambda = \zeta\omega_n + j\omega_d$, and allocated to a circular region $D(r, q)$ of which centre is $(-q, 0)$ and radius r in the phase plane. It is equivalent to ensure minimum disturbance attenuation α and damping ratio $\zeta = \cos\theta$, as well as maximum damped oscillation frequency $\omega_d = r'\sin\theta$ of the system. By this means we can satisfy the required dynamics of overshoot, rise time and accommodation time. Such the circular region $D(r, q)$ is shown as Fig.1.

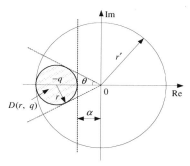

Fig. 1. Region $D(r,q)$

Lemma 1[6]. If all eigenvalues of matrix A lies in a circular region $D(r, q)$ of which centre is $(-q, 0)$ and radius r, $\sigma(A) \subset D$ in mathematical expression, matrix A is so-called D-stable, if and only if a symmetric positive definite matrix P exists, so that

$$\begin{bmatrix} -rP & qP+AP \\ qP+PA^T & -rP \end{bmatrix} < 0 \tag{5}$$

Suppose possible actuator fault model as

$$u_f(t) = Mu(t) \tag{6}$$

where the fault matrix is $M = diag\{m_1, m_2, \cdots, m_r\}$, $0 \le m_{il} \le m_i \le m_{iu} \le 1$ ($i = 1, 2, \ldots, r$), with each element corresponding to a specific fault. Here $m_i = 1$ indicates actuator i is in normal operation, $m_i = 0$ in total failure state, $0 < m_{il} \le m_i \le m_{iu} < 1$ partial deteriorated. Cite the following symbols:

$$M_0 = diag(m_{01}, m_{02}, \cdots, m_{0r})$$
$$J = diag(j_1, j_2, \cdots, j_r)$$
$$L = diag(l_1, l_2, \cdots, l_r)$$

$m_{0i} = \dfrac{m_{il}+m_{iu}}{2}$, $j_i = \dfrac{m_{iu}-m_{il}}{m_{iu}+m_{il}}$, $l_i = \dfrac{m_i - m_{0i}}{m_{0i}}$. Easy to see that $M = M_0(I+L)$, $|L| \le J \le I$.

Hence forms the state equation of closed-loop system with actuator faults

$$\dot{x}(t) = A_{cf} x(t) \tag{7}$$

where $A_{cf} = A + BMK + E\Sigma(F_1 + F_2 MK)$.

In summary, the design of robust fault-tolerant controller under region pole constraint can be described as follows: In terms of actuator fault model (6), system (7) with the parameter perturbations can obtain a controller (3), which allocates all poles of fault closed-loop system to a certain circular region $D(r, q)$, then the controller (3) is called as the robust D-stabilized fault-tolerant controller. Thus the solvability of controller can be transferred into that of a set of LMIs.

3 Design of Robust D-Stabilized Fault-Tolerant Controller

Before processing norm bounded perturbations and actuator faults, we need the following lemma:

Lemma 2[7]. For any matrixes X, Y with appropriate dimensions and scalar $\varepsilon > 0$, it follows that $X^T Y + Y^T X \leq \varepsilon X^T X + \varepsilon^{-1} Y^T Y$.

Lemma 3[7]. Foe any matrixes Y、D and E with appropriate dimensions, where Y is symmetrical, then for all matrix F that meet $F^T F \leq I$, $Y + DFE + E^T F^T D^T < 0$ holds, if and only if there exists scalar $\varepsilon > 0$ such that $Y + \varepsilon^{-1} DD^T + \varepsilon E^T E < 0$.

Theorem 1. Given a circular region $D(r, q)$, a D-stabilized routine control law for system (1) is achievable, if and only if there exist scalar $\varepsilon_1 > 0$, matrix $P > 0$ and matrix V such that LMI

$$\begin{bmatrix} -rP + \varepsilon_1 EE^T & (A+qI)P + BV & 0 \\ [(A+qI)P + BV]^T & -rP & (F_1 P + F_2 V)^T \\ 0 & F_1 P + F_2 V & -\varepsilon_1 I \end{bmatrix} < 0 \tag{8}$$

then (P, V, ε_1) is a feasible solution of control law as follows:

$$u^*(t) = VP^{-1} x(t) \tag{9}$$

Proof. Given a circular region $D(r, q)$, it's easy to verify that inequality (5) is equivalent to

$$\begin{bmatrix} -rP & A_c \\ A_c^T & -rP \end{bmatrix} < 0$$

Substitute A_c in above

$$\begin{bmatrix} -rP & (A+qI)P + BKP \\ [(A+qI)P + BKP]^T & -rP \end{bmatrix} + \begin{bmatrix} E \\ 0 \end{bmatrix} \Sigma \begin{bmatrix} 0 & (F_1 + F_2 K)P \end{bmatrix} + \begin{bmatrix} 0 & (F_1 + F_2 K)P \end{bmatrix}^T \Sigma^T \begin{bmatrix} E \\ 0 \end{bmatrix}^T < 0$$

From lemma 3, for all uncertain matrix Σ that meet $\Sigma^T \Sigma \leq I$, the above inequality holds, if and only if there exists scalar $\varepsilon_1 > 0$ such that

$$\begin{bmatrix} -rP & (A+qI)P + BKP \\ [(A+qI)P + BKP]^T & -rP \end{bmatrix} + \varepsilon_1 \begin{bmatrix} E \\ 0 \end{bmatrix} \begin{bmatrix} E^T & 0 \end{bmatrix} + \varepsilon_1^{-1} \begin{bmatrix} 0 \\ [(F_1 + F_2 K)P]^T \end{bmatrix} \begin{bmatrix} 0 & (F_1 + F_2 K)P \end{bmatrix} < 0$$

that is

$$\begin{bmatrix} -rP + \varepsilon_1 EE^T & (A+qI)P + BKP \\ [(A+qI)P + BKP]^T & -rP \end{bmatrix} - \begin{bmatrix} 0 \\ [(F_1 + F_2 K)P]^T \end{bmatrix} (-\varepsilon_1^{-1} I) \begin{bmatrix} 0 & (F_1 + F_2 K)P \end{bmatrix} < 0$$

By application of Schur complement theory, an alternative way of writing this is

$$\begin{bmatrix} -rP + \varepsilon_1 EE^T & (A+qI)P+BKP & 0 \\ [(A+qI)P+BKP]^T & -rP & (F_1P+F_2KP)^T \\ 0 & F_1P+F_2KP & -\varepsilon_1 I \end{bmatrix} < 0$$

Set $V = KP$, then completes the proof.

Theorem 2. Given a circular region $D(r, q)$ and the occurrence of actuator failures, a robust D-stabilized fault-tolerant control law for system (1) is achievable, if and only if there exist scalar $\varepsilon_1 > 0$, $\varepsilon_2 > 0$ and positive definite matrix P, V such that LMI with symmetrical structure

$$\begin{bmatrix} H_{11} & H_{12} & H_{13} & 0 \\ * & -rP & H_{23} & V^T J^{1/2} \\ * & * & H_{33} & 0 \\ * & * & * & -\varepsilon_2 I \end{bmatrix} < 0 \qquad (10)$$

where

$$H_{11} = -rP + \varepsilon_1 EE^T + \varepsilon_2 BM_0 JM_0 B^T$$
$$H_{12} = (A+qI)P + BM_0 V$$
$$H_{13} = \varepsilon_2 BM_0 JM_0 F_2^T$$
$$H_{23} = (F_1 P + F_2 M_0 V)^T$$
$$H_{33} = \varepsilon_2 F_2 M_0 JM_0 F_2^T - \varepsilon_1 I$$

Proof. Consider fault matrix $M = M_0(I + L)$, the inequality (8) can be written as

$$\begin{bmatrix} -rP + \varepsilon_1 EE^T & (A+qI)P+BMV & 0 \\ [(A+qI)P+BMV]^T & -rP & (F_1P+F_2MV)^T \\ 0 & F_1P+F_2MV & -\varepsilon_1 I \end{bmatrix} < 0$$

that is

$$\begin{bmatrix} -rP + \varepsilon_1 EE^T & (A+qI)P+BM_0V & 0 \\ * & -rP & (F_1P+F_2M_0V)^T \\ * & * & -\varepsilon_1 I \end{bmatrix} + \begin{bmatrix} BM_0 J^{1/2} \\ 0 \\ F_2 M_0 J^{1/2} \end{bmatrix} \begin{bmatrix} 0 & J^{1/2}V & 0 \end{bmatrix} + \left(\begin{bmatrix} BM_0 J^{1/2} \\ 0 \\ F_2 M_0 J^{1/2} \end{bmatrix} \begin{bmatrix} 0 & J^{1/2}V & 0 \end{bmatrix} \right)^T < 0$$

According to Schur complement property and lemma 2, the above formula can be transferred into inequality (10) with further consolidation, then completes the proof.

Furthermore, given system (1) and region D, with respect to the following optimal problem

$$\min_{P,V} Trace(S)$$

such that: (i) linear matrix inequality (10) holds; (ii) $\begin{bmatrix} S & I \\ I & P \end{bmatrix} > 0$

where the constraint (ii) is equal to $S > P^{-1} > 0$. The minimum of $Trace(S)$ will ensure the minimum of $Trace(P^{-1})$, that is, the minimum allowable bound of system performance. With solution (P, V), the formula (9) is called as the optimal robust fault-tolerant control law of the system.

Remark. According to (10), the fault-tolerant controller design problem of the system is equivalent to determine the optimum solutions $(P, V, \varepsilon_1, \varepsilon_2)$ of LMI, by utilizing LMI Control Toolbox which implements interior-point algorithm. This algorithm is significantly faster than classical convex optimization algorithms [10][11].

4 Simulation Example

Consider the operating condition of X type of turbofan engine in a state of $H = 0km$, $Ma = 0$, the double variant normalized linear model is built with parameter matrixes

$$A = \begin{bmatrix} -2.3642 & -0.3014 \\ 3.0958 & -2.8747 \end{bmatrix}, \quad B = \begin{bmatrix} 0.6978 & 0.9386 \\ 0.6834 & 1.7602 \end{bmatrix}$$

the uncertain decomposition matrix as

$$E = \begin{bmatrix} 0.236 & 0.381 \\ 0 & 0.252 \end{bmatrix}, \quad F_1 = \begin{bmatrix} 0.6 & 0 \\ 0 & 0.8 \end{bmatrix}, \quad F_2 = \begin{bmatrix} 0.2 & 0 \\ 0.3 & 0.6 \end{bmatrix}$$

and with the nominal system poles $-2.6194 \pm j0.9316$.

Given the dynamic performance index: attenuation $\alpha \geq 3$, overshoot $\sigma\% \leq 5\%$, accommodation time $t_s \leq 4.6s$, the pole of closed-loop system can be assigned into the circular region $D(2, 5)$.

According to theorem 1, the robust D-stabilized routine control law $u^*(t) = K_1 x(t)$ can be obtained by the solution of LMI (8) where the variable is (P, V). One state feedback gain matrix is

$$K_1 = VP^{-1} = \begin{bmatrix} -2.8806 & 3.9930 \\ -0.6422 & -2.7353 \end{bmatrix}$$

In the case of fuel controlling, such throttle switch and fuel distribution device closely related to oil-supplied characteristics are vulnerable to oil leakage caused by assembly quality or sealing failure. On the other hand, hydraulic fluid valve contaminated by work pollutants will easily lead to jet actuator stuck tightly. Here lord fuel metering valve and jet actuator fault are simulated respectively as fashions $m_1 \in [0.8,1]$ and $m_2 \in [0.4,1]$. Work out

$$M_0 = diag\{0.9, \ 0.7\}, \quad J = diag\{0.1111, \ 0.4286\}$$

According to theorem 2, calculate the robust D-stabilized fault-tolerant control law $u^*(t) = K_2 x(t)$, where the state feedback gain matrix is

$$K_2 = VP^{-1} = \begin{bmatrix} -1.6814 & -0.3413 \\ -1.4738 & -0.5560 \end{bmatrix}$$

4.1 Robust Stability Analysis

Take into account 10 groups of model with structural parameter norm bounded perturbations $\|\Delta A\| \le 0.4$, $\|\Delta B\| \le 0.3$. Given that

(1) $M = I$, i.e. actuators in normal condition;
(2) $M = diag\{1, \ 0.4\}$, i.e. lord fuel metering valve works normally, but jet actuator totally stuck.

The poles of closed-loop system corresponding to two control laws distribution are shown as Fig. 2 and Fig.3. Note that closed-loop poles utilized by both control laws are all in the designated circular region under normal state. While on occurrence of actuator faults, partial poles by routine control law deviate from the designated circular region to positive complex plane, which may lead to system unstable. Yet by adopting robust fault-tolerant control law, all poles are still remain in the designated circular region, ensuring the system stability.

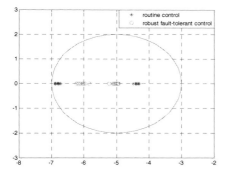
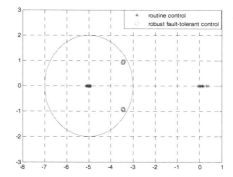

Fig. 2. Distribution of closed loop poles under condition (1)

Fig. 3. Distribution of closed loop poles under condition (2)

4.2 Robust Fault-Tolerant Performance Analysis

Consider the actuator faults at $t = 5s$,
 (1) $M = diag\{1, \ 0.6\}$, i.e. jet actuator totally stuck faults;
 (2) $M = diag\{0.8, \ 1\}$, i.e. lord fuel metering valve leak faults.

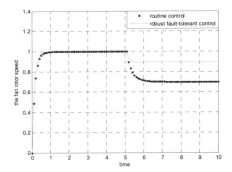 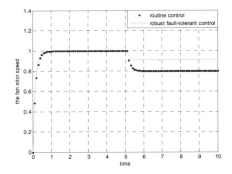

Fig. 4. Step response of the fan rotor speed under fault (1)

Fig. 5. Step response of the fan rotor speed under fault (2)

By Application of two control laws, the step response cure of the fan rotor speed n_L under the assumption of above faults are shown as Fig.4 and Fig.5. Note that the robust fault-tolerant control law proposed is obviously superior to routine control law. With strong fault-tolerant capability for both actuator faults, the dynamic performance remains approximate adjust to normal state.

5 Conclusion

A general form of turbine line dynamic model with norm bounded perturbations is in discussion. By adoption of robust fault-tolerant control approach based on LMI, a state feedback fault tolerant controller is designed to satisfy the regional poles constraints and performance in actuator fault condition and structured perturbations. Simulation results illustrate that, the proposed approach possesses strong fault-tolerant capability and maintains the performance of robust stability in designated circular region. Furthermore, this approach can be expanded to sensor faults situation.

References

1. Blallke, M., Staroswieeki, M., Wu, N.E.: Concept and methods in fault-tolerant control. In: Proceedings of the American Control Conference, Arlington, VA, pp. 2606–2620 (2001)
2. Diao, Y.: Stable fault-tolerant adaptive fuzzy/neural control for a turbine engine. IEEE Contr. Syst. Dyn. 9(3), 494–509 (2001)

3. Zhang, H., Tan, M.: Deisgn of fault-tolerant controller to state feedback control systems. Control and Decision 15(6), 724–726 (2000)
4. Xu, Q., Li, h.: Robust fault tolerant controll for aeroengine based on solution of riccati equation. Journal of Aerospace Power 18(3), 440–443 (2003)
5. Nazli Gundes, A.: Stability of feedback systems with sensor or actuator failures analysis. International Journal of Control 56(4), 735–753 (1992)
6. Yu, L.: Robust control-an linear matrix inequality approach. Tsinghua University Press, Beijing (2002)
7. Wang, F., Yao, B., Zhang, S.: Reliable control of regional stabilizability for linear systems. Control Theory and Applications 21(5), 835–839 (2004)
8. Yu, L., Chen, G., Yang, M.: Robust control of uncertain linear system with disk pole constraints. Acta Automatica Sinical 26(1), 116–120 (2000)
9. Fan, S., Li, H., Fan, D.: Aero engine control. Northwestern Polytechnical University Press, Xi'an (2008)
10. Boyd, B., Ghaoui, L.E., Feron, E., Balakrishnan, V.: Linear matrix inequalities in systems and control theory. SIAM, Philadelphia (1994)
11. Gahinet, P., Nemirovskii, A., Laub, A.J., Chilali, M.: LMI control toolbox. Decision and Control (1994)

Research on m Value in Fuzzy c-Means Algorithm Used for Underwater Optical Image Segmentation

Shilong Wang*, Yuru Xu, and Lei Wan

National Key Laboratory of Science and Technology on Autonomous Underwater Vehicle, Harbin Engineering University, Haerbin, China
wangshilong@hrbeu.edu.cn

Abstract. In the underwater environment, gathered images would have low SNR and the detail is fuzzy due to scattering and absorption of a variety of suspended matter in water and the water itself. The good segmentation results are obtained by the introduction of fuzzy C means clustering algorithm to the processing of underwater images. Of fuzzy C means clustering algorithm, the fuzzy index- m, affecting the clustering result, is further studied. With underwater image segmentation effects, taken the fuzzy partition coefficient, fuzzy partition entropy and XB validity as the basis for segmentation quality evaluation, systematic demonstration and analysis on m are carried out. Comparative experimental results show that when m is taken 1.5, it can obtain better segmentation quality and the higher timeliness, which provide an important prerequisite for autonomous underwater vehicle (AUV) to complete a special mission.

Keywords: underwater image; fuzzy C means; image segmentation; fuzzy index; autonomous underwater vehicle (AUV); timeliness.

1 Introduction

The vast ocean of water space bears the rich resources and has an important strategic position in modern warfare[1]. Autonomous Underwater Vehicle (AUV) is an important carrier for the exploitation of marine resources. The vision system of AUV is the key to the perception on the environment, fast locating and tracking of the targets[2]. It is fundamental for AUV to complete underwater tasks.

The underwater image segmentation, as the key and premise technology for image analysis and object recognition, is one of the classic research topics in computer vision research field, and is one of the difficulties in underwater image processing.

Access to underwater target information mostly depends on the optical and sound vision [2], however, research and development of underwater optical vision is much later than land vision, basically still at the primary stage. Underwater images are much sensitive to various noises and other interference and what's more, poor lighting condi-

---

* Supported by the National Natural Science Foundation of China under Grant No.50909025/E091002 and the Open Research Foundation of SKLab AUV, HEU under Grant No.2008003.

tions under the water will get underwater images to be false details, such as self-shadow, false contour, etc. Additionally, for some other reason, such as the existence of a variety of suspended solids, impact of water flowing and shake of the camera lens, underwater images are with strongly fuzzy.

Thousands of current image segmentation algorithms are mostly for specific issues, and although every year, some new image segmentation algorithms were proposed, but no general algorithm can be applied to all images segmentation[3]. Fuzzy C means (FCM)[4-5] method has been widely used in various areas of image segmentation[6-7] because it can keep more original image information than traditional hard segmentation algorithms.

When using fuzzy C means algorithm, how to select the fuzzy index m has been an open question. Bezdek[9] gave the experience of a range of m is from 1.1 to 5, and derived from the physical interpretation that when m is equal to 2, the most meaningful segmentation can be obtained. Through the study on Chinese character recognition application, Chan et al.[10] drew that the optimal value of m should be between 1.25-1.75. Bezdek and others[11] obtained from the perspective of convergence that value of m is related to the number n of samples and proposed that m should be greater than $n/(n-2)$. Pal et al.[12] concluded from the cluster validity of the experiment that the best m interval is [1.5,2.5], and further proposed that in the case of no special requirements, m can be taken the median value 2 of the interval. GAO Xin-bo[13-14] et al. and LIU Yi-ping[15] et al. proposed the preferred method of m based on fuzzy decision. YU Jian[16] proved that how to select the fuzzy index theoretically depends on the data itself, and gives the rule for the theoretical selection of m.

Above all, so far, many scholars and experts have studied on the range of m, and get a lot of interesting results. But due to experimental constraints and theoretical analysis of the data defects, a uniform standard for the selection of fuzzy index m has not yet been given. In this paper, taken the typical representative artificial targets shot by professional underwater video CCD camera in tank as the object of study, with underwater image segmentation effects, taken several validity functions as the basis for segmentation quality evaluation, systematic research on m in FCM algorithm for underwater image segmentation is carried out.

Although this algorithm is based on underwater images, but the evaluation principle for dealing with other types of image segmentation is also a good reference and the selection method on fuzzy clustering index m will be more abundant.

2 The Traditional FCM Algorithm for Underwater Image Segmentation

The central idea of FCM clustering segmentation algorithm is to measure the weighted similarity between image pixel and cluster center, iteratively optimizing the objective function to determine the best clustering. The implementations are: for the $L \times H$ image (L and H represent the width and height of the image respectively), according to the weighted membership degree ($n = L \times H$) pixels in image belong to each of c clustering center, iteratively optimize the objective function to obtain the fuzzy partition matrix U and the class center matrix V when the objective function is minimum. The objective function is:

$$J_m(U,V) = \sum_{i=1}^{c} \sum_{j=1}^{n} (u_{ij})^m d(x_j, v_i)^2 \quad (1)$$

In Eq.(1), $U = \{u_{ij}\}$ are sets of membership degree value that j-sample belongs to i-class and $\sum_{i=1}^{c} u_{ij} = 1$; $V = \{v_1, v_2, ..., v_c\}$ is the clustering center set and $2 \leq c \leq n$; m is the fuzzy index, which controls the division fuzzy degree[17]; $d(x_j, v_i)$ is the distance of j-pixel to i-clustering center, there are a lot of traditional measurement methods for distance[18], in this paper we use the Euclidean distance. That is to say, let $d(x_j, v_i)$ be $\|x_j - v_i\|$.

By Lagrange multiplier method, a necessary condition for the establishment of Eq. (1) can be derived:

$$u_{ij} = \left[\sum_{k=1}^{c} \left| \frac{(x_j - v_i)^2}{(x_j - v_k)^2} \right|^{\frac{1}{m-1}} \right]^{-1} \quad (2)$$

$$v_i = \left| \sum_{j=1}^{n} (u_{ij})^m x_j \right| / \sum_{j=1}^{n} (u_{ij})^m \quad (3)$$

This typical FCM image segmentation is to obtain the minimum objective function by iteratively searching for optimal clustering centers and membership degree.

3 Several Evaluation Function

3.1 PC and PE Validity Function

To evaluate the clustering validity, many papers proposed solutions. Among those, partition coefficient PC and partition entropy PE are the typical evaluation methods for fuzzy partition efficiency.

PC[19] is defined as:

$$PC = \frac{1}{n} \sum_{i=1}^{c} \sum_{j=1}^{n} u_{ij}^2 \quad (4)$$

PE[20] is defined as:

$$PE = -\frac{1}{n}\{\sum_{i=1}^{c}\sum_{j=1}^{n}u_{ij}\log_a(u_{ij})\} \; a \in [1,+\infty] \quad (5)$$

When the clustering is to achieve the best results, PC has the highest value, while the PE has a minimum.

3.2 XB ValidityFunction

XB validity function is given by Xie and Beni[21] in 1991. It considers the segmentation result based on the distance between classes and the distance within each class. When the clustering result is closer to the real structure, the distance between classes is larger, while the distance within the class is smaller.

XB is defined as follows:

$$V_{XB} = \frac{J_m(u,v)/n}{Sep(v)} = \frac{\sum_{i=1}^{c}\sum_{j=1}^{n}u_{ij}^m \|x_j - v_i\|^2}{n \min_{i \neq j}\|v_i - v_j\|^2} \quad (6)$$

Where, the molecule is compact metric, that is, the distance within the class. The denominator is a measure of separation, that is, the distance between classes. Normally, the best structure of data sets takes place when XB function value is smallest.

4 Experimental Results and Analysis

In order to make FCM underwater segmentation algorithm more generality and practicality, in a computer with XP operating system, the main frequency being 2.60 GHz, the memory being 2G, for representative man-made objects such as three-prism, sphere and four-prism shot in the pool (image size is 768×576), the accuracy and real-time effectiveness are compared with different m value taken from 1.1 to 3 of the increment of 0.1. Figure 1-3 show the comparison results. (in the experiments, the number of clustering centers is taken 2 and the iterative threshold is taken 1.0×10^{-9}. And then, further demonstration of m value is given

4.1 The Underwater Image Segmentation Results with Different m

When using FCM algorithm for the underwater sphere, three-prism and four-prism image segmentation, A-E in Figure 1-3 is the segmentation results respectively when m is taken as 1.1, 1.5, 2.0, 2.5, 3.0.

It can be seen from Fig.1-3, the underwater sphere and four-prism segmentation results used by FCM clustering algorithm with different m are all with better clustering features to different degree, well to eliminate the noise impact and extract enough image details, and thus the complete extraction of the target area is get.

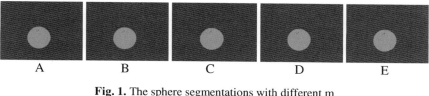

Fig. 1. The sphere segmentations with different m

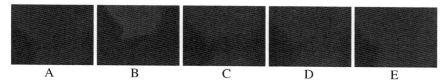

Fig. 2. The three-prism segmentations with different m

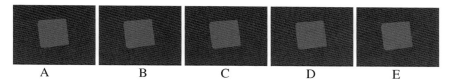

Fig. 3. The four-prism segmentations with different m

It can be seen from Figure 2, to the three-prism in which the light collection process is uneven and there is part of edge blur, when m is taken different value, the segmentation results are obvious different. And the clustering effect is ideal when m is between 1.5 and 2.5, what's more, the image segmentation quality is best with m value being 1.5.

To get more accurate comparison of the clustering effect and the quality of segmentation, the following will give professional analysis combined with the three evaluation function mentioned in the previous section.

4.2 Analysis on Effectiveness of Clustering

The following will compare and illustrate the segmentation result for different values of m combining with figures. Since when m=2.3, the sphere can not be separated out and when m=2.7, the three-prism can not be separated out, so to further analysis the segmentation results does not make sense and then to discard the two situation. The segmentation evaluation functions are as shown Fig. 4-6.

From Fig. 4-6, it can be seen that using FCM clustering algorithm for the three types of underwater targets segmentation, the basic trend is that the segmentation results is better when m is smaller, and with the increase of the value of m, the segmentation results firstly become better and better and then gradually decreased. For the sphere and four-prism images, there is sufficiently high partition coefficient, partition entropy and XB effectiveness value are also ideal low. It can be seen from the Fig. 5 and Fig. 7 that the change scope of the effective function values are small, and

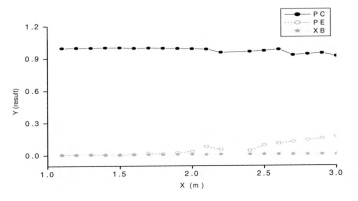

Fig. 4. The segmentation evaluation function of the sphere with different values of m

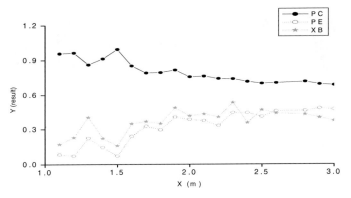

Fig. 5. The segmentation evaluation function of the three-prism with different values of m

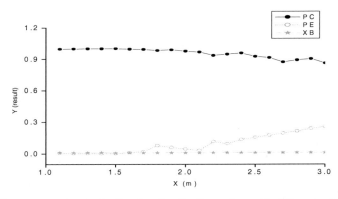

Fig. 6. The segmentation evaluation function the four-prism with different values of m

the overall trend that the results is better when m is less than 2.0 than when m is greater than 2.0. But form Fig.5, it can be seen that for the three-prism segmentation, the partition coefficient is higher, while XB values and the partition entropy are all lower when the value of *m* is near to 1.5. The higher partition coefficient demonstrates that the algorithm can be able to clearly divide the target from background, and lower XB value and partition entropy illustrate that the obtained segmentation results are with smaller loss of information.

4.3 Analysis on Timeliness of the Clustering Algorithm

The following will compare and illustrate the segmentation timeliness for different values of m combining with figures. Similarly, since when m=2.3, the sphere can not be separated out and when m=2.7, the three-prism can not be separated out, so to further analysis the segmentation results does not make sense and then to discard the two situation. The segmentation timeliness is as shown Fig. 7

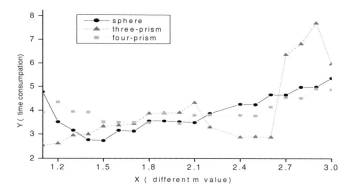

Fig. 7. The time consumption with different values of *m* to three types of underwater objects

It can be seen form Fig.8 that for the sphere segmentation, when *m* is 1.4 or 1.5, the time consumption is less, and for the three-prism segmentation, when *m* is 1.1, 1.2, 2.4 or 2.6, the timeliness is better, and for four-prism segmentation, when m is from 1.5 to 1.8, the time consumption is less.

Comprehensive analysis based on Fig. 1-7, it can be seen that for the three types of underwater images, it can meet that when *m* value is 1.5, PC is larger, while PE and XB are smaller, and the time consumption is less for the sphere and three-prism segmentation, and the time consumption is slightly higher for the four-prism segmentation, but from the view of the combination of the quality of image segmentation, the best value of m should be 1.5.

Above all, it had been fully descript that using FCM algorithm for underwater image segmentation, when the value of m is 1.5, the fuzzy classification result is more reasonable and more effective.

5 Conclusions

For the three types of underwater targets, through the introduction of FCM clustering segmentation algorithm, good segmentation results are achieved. Comprehensively considering different values of fuzzy index m affecting the clustering results, based on the fuzzy partition coefficient, fuzzy partition entropy and XB validity as segmentation quality evaluation, by comparative experiments, the optimum value of m is discussed, and thus further expanded the scope of application of FCM clustering algorithm and the best argument of fuzzy index. For the special requirements and mission of AUV, the usefulness of the operation is improved, which provide a strong guarantee for the further feature extraction and target tracking.

References

1. Xu, Y.-r., Xiao, K.: Technology Development of Autonomous Ocean Vehicle. Acta Automatica Sinica 33(5), 518–521 (2007)
2. Yuan, X.H., Qiu, C.C., et al.: Vision System Research for Autonomous Underwater Vehicle. In: Proceedings of the IEEE International Conference on Intelligent Processing System, vol. 2, pp. 1465–1469 (1997)
3. Zhang, M.-j.: Image Segmention. Science Press, Peking (2001)
4. Dunn, J.C.: A fuzzy relative of the ISODATA process and its use in detecting compact, well-separated clusters. J. Cybern. 3, 32–57 (1974)
5. Bezdek, J.C.: Pattern recognition with fuzzy objective function algorithms. Plenum Press, New York (1981)
6. Udupa, J.K., Saha, P.K.: Fuzzy connectedness and image segmentation. Proceedings of the IEEE on Emerging Medical Imaging Technology 91(10), 1649–1669 (2003)
7. Martino, G., Petrosino, A.: Fuzzy connectivity and its application to image segmentation. In: Apolloni, B., Marinaro, M., Nicosia, G., Tagliaferri, R. (eds.) WIRN 2005 and NAIS 2005. LNCS, vol. 3931, pp. 197–206. Springer, Heidelberg (2006)
8. Naz, S., Majeed, H., Irshad, H.: Image segmentation using fuzzy clustering: A survey. In: 6th International Conference on Emerging Technologies (ICET), 181-186 (2010)
9. Bezdek, J.C.: A physical interpretation of fuzzy ISODATA. IEEE Transactions on Systems, Man and Cybernetics 6(5), 387–389 (1976)
10. Chan, K.P., Cheng, Y.S.: Modified fuzzy ISODATA for the classification of handwritten Chinese characters. In: Proc. Int. Conf. Chinese Comput., Singapore, pp. 361–364 (1986)
11. Bezdek, J.C., Hathaway, R., et al.: Convergence theory for fuzzy c-means: counter examples and repairs. IEEE Trans., PAMI 17(5), 873–877 (1987)
12. Pal, N.R., Bezdek, J.C.: On clustering validity for the fuzzy c-means model. IEEE Fuzzy Systems 3(3), 370–379 (1995)
13. Gao, X.b., Pei, J.h., Xie, W.x.: A Study of Weighting Exponent m in a Fuzzy c Means Algorithm. Acta Electronica Sinica 28(4), 80–83 (2000)
14. Gao, X.B., Li, J., Xie, W.: Parameter optimization in FCM clustering algorithms. In: International Conference on Signal Processing, vol. 3, pp. 1457–1461 (2000)
15. Liu, Y., Shen, Y., Liu, Z.: Improvement and Optimization of a Fuzzy C-Means Clustering Algorithm. Systems Engineering and Electronics 22(4), 1–3 (2000)
16. Yu, J.: On the Fuzziness Index of the FCM Algorithms. Chinese Journal of Computers 26(8), 968–973 (2003)

17. Bezdek, J.C., Dubois, D., Prade, H.: Fuzzy sets in approximate reasoning and information systems. Fuzzy Sets and Systems 123(3), 405–406 (2001)
18. Yang, S.: Image pattern recognition with VC++ technology, pp. 161–162. Tsinghua University Press, Jiaotong University Press, Beijing (2005)
19. Bezdek, J.C.: Numerical Taxonomy with Fuzzy Sets. J. Math. Biol. 1(1), 57–71 (1974)
20. Bezdek, J.C.: Cluster Validity with Fuzzy Sets. Journal of Cybernetics 3(3), 58–73 (1974)
21. Xie, X.L., Beni, G.: A validity measure for fuzzy clustering. IEEE Trans. Pattern Anal. Mach. Intell. 13(8), 841–847 (1991)

A Novel Method on PLL Control

Youhui Xie and Yanming Zhou

Lushan College of Guangxi University of Technology, Liuzhou, China
youyou302@foxmail.com, z_lzdq@163.com

Abstract. Phase-locked technique is widely used in the field of automation. A new design method of a phase-locked loop (PLL) is presented here. It's based on artificial neural network. It takes RBF network into that close-loop control of Phase Locked Loop. Then set electric network voltage as the expected output and current as training sample. After neural network self-learning it can gradually reduce the error of output between the sample and the expected target, and achieve the synchronization and tracking of the expected output. This paper introduces the principle and the realization of phase-locked loop, and its performance is also be analysised and simulated with MATLAB Simulink Power System. This simulation results suggest a good performance of tracking and adaptive capacity.

Keywords: BP network, Phase tracking, Phase Locked Loop (PLL), self-leaning.

1 Introduction

In the power system, the phase locked loop is used for achieving the purpose of the phase tracking control. Phase-locked system is essentially a closed-loop phase control system. In another word, it refers to a circuit that makes output signal can be synchronized with the input signal in the frequency and phase. As the good character of phase-locked loop, it is widely used in many fields such as the produce and clock distribution of high-performance processors, synthesis and conversion of system frequency, automatic tracking of frequency tuning, bit synchronous extraction in digital communication, phase-locked frequency stabilization and so on [1]. With the development of phase-locking technology, as well as a wide range of applications and in-depth study, at present phase-locked theory has been paid a more and more attention.

Some research shows the information and application of three-phase Phase Locked Loop (PLL) [2-3]. But the single-phase locked loop is used in some circumstance such as AC/DC/AC electric locomotive traction system. So the research and application of single-phase PLL has its practical significance. A phase detection method of single-phase based on least-square estimate [4], which is related to many matrix operations of division, will have to pay higher price in the achieving such a phase-detection. And also its anti-interference capacity for non-white noise is limited. There is a single-phase PLL control method based on the filter [5]. When with this method in the general low-pass filter, the system response time will be affected [6-7]. But if in the high-order one it will inevitably increase the PLL price [8-9].

This paper presents a control method of phase tracking based on artificial neural network. BP network has added into Phase Locked Loop. It takes electric network voltage as the expected output and current as training sample. Simulation shows that it can keep in phase of the expected output.

2 Phase Locked Loop

PLL is a closed loop phase control system. And by using PLL it can get the output signal not only in-phase but also with a frequency of integer or fraction times. Basic structure of PLL is showed in Fig.1, which consists of phase detector (PD), loop filter (LF), voltage-controlled oscillator (VCO).

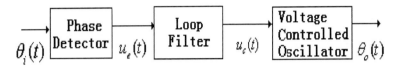

Fig. 1. The Basic Structure of PLL

Phase detector is a phase comparison device, which can convert the phase-change into the amplitude change. In another word, it is used to detect the phase difference between the two input signals; in some digital phase detectors it detects the difference of the rising edge of the two signals. The error signal, which is as the output from PD, is a function of the phase difference. Phase detector plays a role of adjusting and giving the needed variable value. In another word, it can be instead of a controller which can get the tracking signal well.

Loop filter is low-pass filter, which is to average the DC signal of the phase detector output with the ripple, and then transform it into DC signal with little AC signal.

Voltage controlled oscillator is a voltage-frequency convert device. As a controlled oscillator, its frequency changes linearly along with its input voltage. The convert mode is $\omega_v(t) = \omega_o + K_o u_o(t)$. Where, $\omega_v(t)$ is the instantaneous angle frequency of voltage controlled oscillator; K_o is control sensitivity or called gain and its unit is $[rad/s \cdot V]$. As the mentioned above, VCO is a frequency modulation oscillator with linear control character.

3 Neural Network Theory

The basic structure of a BP network is showed in Fig.2. A neural network composes of input layer, hidden layer and output layer.

The network exact algorithm of the learning process consists of two directions of forward and backward.

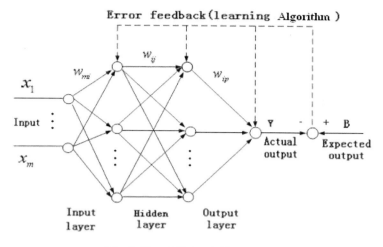

Fig. 2. Neural Network Structure

The neural network shown here is a BP network with two hidden layers of multi-input and single-output. $X = [x_1 \cdots x_m]$ is the training sample and the network input. Y is the actual output and B is the expected output. w_{mi} is the weight between input layer and hidden one. w_{ij} is the weight between the first and the second hidden layer. w_{jp} is the weight between hidden layer and output one. If this output velum number is more then one, the No. p neuron signal error of in output layer is $e_p(n) = b_p(n) - y_p(n)$, where n is the iteration. The weights and the actual output are the function of n. If there are p samples, the square error between the expected output B and the actual output Y in output layer is

$$E(n) = \frac{1}{2}\sum_{p=1}^{P} e_p^2(n) = \frac{1}{2}\sum_{p=1}^{P}\left[b_p(n) - y_p(n)\right]^2 \tag{1}$$

This BP network learning algorithm is the steepest gradient descent learning algorithm. Its theory is like this: W is a weight vector, E(w) is a multi-function. With the network learning, it can get the minimum W value. At first choose randomly the initial value of the point W. Then the step and the movement direction of the point $W(k+1)$ are got according to the gradient of the point $W(k)$. It is in order to make the error function $E(W(k+1))$ more rapidly declined to the minimum in this direction than $E(W(k))$. Then the point $W(k+1)$ is got. Then set $W(k+1)$ as the starting point and repeat that again to fix $W(k+2)$ with the method of steepest gradient decline. At last, it can find out the point $W(k+n)$ which makes the error function get the minimum. It can be the solution of network weights. The iterative formula of the steepest gradient descent learning algorithm is:

$$W(k+1) = W(k) - \alpha(k)\nabla E(W(k)) \qquad (2)$$

Where $\alpha(k)$ is a positive and is called as step. Its velum in each iteration step is solved with a one-dimensional linear function.

4 PLL Control with BP Network

BP network is added into the phase detector control of PLL, for achieving adaptive adjusting. The structure of PLL should be changed as shown in Fig.3.

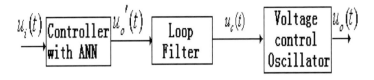

Fig. 3. Structure of PLL with RBF Network

The input of Phase-Locked Loop $u_i(t)$ comes from the electric sources of network or the inverter of transformer. The tracking output $u_o'(t)$ will be gotten after the network is trained and adjusted. For the BP network, $u_i(t)$ which is available to be the training sample of network, is also the input of the input layer. A single-output should be chosen for output layer. The expected output comes from the electric network.

The neural network can adaptively adjust the value of thresholds and weights according to the change of input. And at last it fixes all the values. Please, follow our instructions faithfully; otherwise you have to resubmit your full paper. This will enable us to maintain uniformity in the journal. Thank you for your cooperation and contribution.

5 Simulation

Set up a BP network with two layers [10]: The training samples of input layer are a linear proportion function. The expected output of output layer is the voltage wave, which is a sine wave. The transmission function of two layers is respectively hyperbolic S-type (Tan-Sigmoid) and linear (Purelin). The training optimization algorithm based on the gradient descent is used as the network learning algorithm. Some other parameters are set as Table 1.

When the expectation is given, for keeping on approaching the expected voltage wave, the network began to self-train till the error is less than the given square error maximum. And the simulation result is showed in fig.4. Most values of error are less than 0.005, which is about 5‰.

Table 1. Parameter Setting

Parameter	Value
Expected Wave's Amplitude	1
Number of neurons in hidden layer	5
Initial value of learning rate	0.001
Maximum value of square error between actual output and the expected	0.01

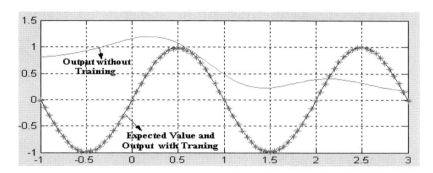

Fig. 4. The Expected Output and the Ouput with Training

The training process of BP network is shown in Fig.5. Where, wave1, wave2 and wave3 are the training wave results of time-last. It can be seen that the output changes step by step in order to approach the expected wave.

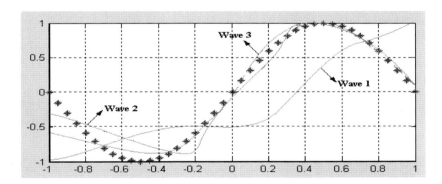

Fig. 5. Training Process of BP Network

In accordance with the structural plans given in Fig. 3, the system simulation model can be designed as the structure in fig.6. Loop filter and Voltage Control Oscillator are important components, but the role of the control of the ANN is only considered here. So the model of LF is represented as a differential operator, and VCO is represented as a model with proportion and integral [10]. The output waveform of Graph in Fig.8, which contains the expected output sinusoid and the system output, are shown in Fig.7.

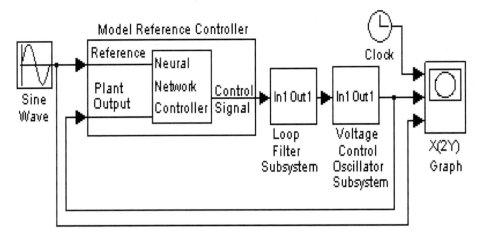

Fig. 6. PLL simulation structure

From the Fig.7, it can be seen that the whole process of adjusting.

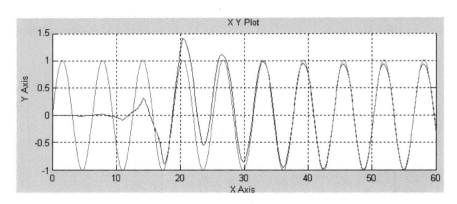

Fig. 7. The System Output of Expected and Actual

After this system finishes adjusting and begins output regular wave, choose a period output data for Fourier Transformed. This Amplitude-Frequency graph is shown in Fig.8. Seem from the graph, this output wave contains fundamental wave mainly and has a few high power harmonies. Then those harmonies are two-power, three-power and four-power ones. Where, amplitude of two-power harmony is 0.032, then 0.02 for three-power harmony and 0.008 for four-power one.

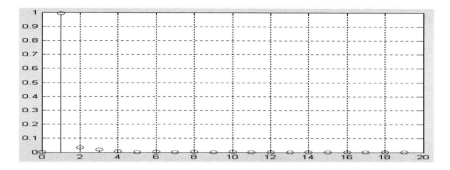

Fig. 8. Amplitude-Frequency Graph of Output afeter Adjusting

The Phase-Frequency graph of system output wave (in red line) is shown in Fig.9, which is from 0~20 of cross axis. Where, that character of expected wave (in blue line) is also shown in the same figure. There are quite differences in those harmonies of two-power, three-power and four-power. Because of their small amplitudes known in Amplitudes-Frequency character before, they will have nothing impacted. It can keep in-phase with the expected in higher-power harmonies, nearly no error.

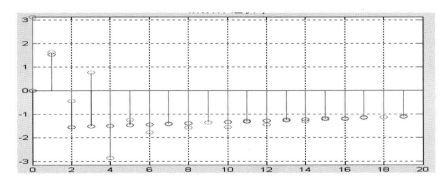

Fig. 9. Phase-Frequency Graph of Output afeter Adjusting

6 Conclusion

In this paper, for single-phase PLL, the BP network is put into the phase detector as control algorithm. And the simulation has been carried out by software. The results suggest that this method can track the target wave well with adaptive capacity and little error.

References

1. Liang, Y.-w., Hu, Z.-j., Chen, Y.-p.: A Survey Of Distributed Generation and its Application in Power System, pp. 72–77. Power System Technology (2003)
2. Lee, S.-J., Kang, J.-K., Su, S.-K.: A New Phase Detecting Method for Power Conversion Systems Considering Distorted Conditions In Power System. In: Industry Applications Conf. Record of the 1999, pp. 2167–2172 (1999), IEEE Transl. J.
3. Shu, Z.-l., Guo, Y.-h.: Implementation of FPGA based Three Phase Phase-Locked Loop System. Power Electronics 39(6), 126–128 (2005)
4. Seok Song, H., Nam, K., Mutschler, P.: Very Fast Phase Angle Estimation Algorithm for a Single-Phase System Having Sudden Phase Angle Jumps. In: 2002 37th IAS Annual Meeting, pp. 925–937 (2002), IEEE Transl.
5. Djuric, P.M., Begovic, M.M., DoroslovaEki, M.: Instantaneous Phase Tracking in Power-Networks by Demodulation. In: IEEE Trans. on Instrumentation and Measurement, pp. 963–967 (1992)
6. Endo, T., Chuu, L.O.: Synchronization of Chaos in Phase-Locked Loops. IEEE Transactions on Circuits and Systems, 1580 (1991)
7. Best, R.E.: Phase-locked loops: design, simulation and appli- cations, 4th edn., pp. 7–100. McGraw-Hill, New York (1999)
8. Chow, H.C., Yeh, N.L.: A new phase-locked loop with highspeed phase frequency detector. In: Proc of MWSCAS, Cincinnati, Ohio, USA, vol. 2, pp. 1342–1345 (2005)
9. CML Microcircuits.: Communication Semiconductors. CMX589A GMSK Modem Application Notes. D/589A/4 (April 2002)

Analysis and Design of Mixer with DC-Offset Cancellation

Jianjun Song[*], Shuai Lei, Heming Zhang, and Yong Jiang

Key Lab of Ministry of Education for Wide Band-Gap Semiconductor Materials and Devices,
School of Microelectronics, Xidian University, Xi'an, China, 710071
jianjun_79_81@xidian.edu.cn

Abstract. This paper designed a novel DC offset cancellation down conversion mixer with SiGe BiCMOS technology, based on the offset problem of direct-conversion receiver. The extra differential mode feedback circuitry (DMFB) and common mode feedback circuitry (CMFB) were used to cancel the differential DC offset and common DC offset, respectively. The mixer was designed and simulated in a 0.35 μm BiCMOS technology with the supply voltage of 3V for 5GHz applications by ADS, and it has the conversion gain of 11dB, NF of 7.7dB and -8.9dBm of IIP3. The addition of DC offset correction circuitry reduces the differential DC offset voltage to less than 3% of the uncorrected offset. Moreover, the proposed mixer consumes 17.07 mA of current.

Keywords: Mixer, DC offset, Common mode feedback, Differential mode feedback, Zero-IF.

1 Introduction

With the development of wireless communication systems, direct-conversion architecture have attracted a great deal of attention over the passed few years, but several problems still exist in the direct-conversion architecture[1] such as DC offset, second order intermodulations (IM2), LO leakage, self-mixing, flicker noise and amplitude and phase mismatch. One of the most critical problems is DC offset. Because of the characteristics of high linearity and low noise, SiGe BiCMOS technology was used to design a DC offset and common DC offset respectively, with which high performance and low cost are obtained compare to the conventional digital method. The mixer was simulated by ADS, and it has a conversion gain of 11 dB, NF of 7.7dB and -8.9dBm IIP3. The DC offset with correction circuitry is less than 3% of uncorrected DC offset.

2 Design of DC Offset Cancellation Direct Down-Conversion Mixer [2,3]

DC offset include common mode DC offset and differential mode DC offset. The mismatch of mixer components and the self-mixing effect are two main source of DC

---

[*] Project supported by National Ministries and Commissions (Nos. 51308040203, 6139801), the Fundamental Research Funds for the Central Universities (Nos. 72105499, 72104089) and Natural Science Basic Research Plan in Shaanxi Province of China (No.2010JQ8008).

offset. The generation of self-mixing is shown in Fig.1. In direct-conversion receiver, the down-converted spectrum is centered at 0Hz, which is called DC point. The existence of DC offset will degrade the signal quality. In fact the DC offset may be larger than signal, which will degrade the following stages. AC coupling circuit is widely used to cancel DC offset in circuit-level design [4]. Mixer presented in this research consists of core circuit of Gilbert cell, common mode feedback (CMFB) circuitry and differential mode feedback (DMFB) circuitry [5].

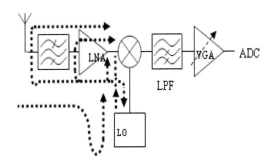

Fig. 1. Mechanism of self-mixing

2.1 Mixer Core Circuit Design

Gilbert double-balanced mixer structure was used in this paper to achieve minimum output DC offset, because it can suppress the leakage of LO signal [6]. Large load resistance (R5, R6 in Fig.2) can be connected to the Gilbert cell to obtain higher conversion gain, circuit is shown in Fig.2. RF signals from LNA enter transistors Q2, Q3 from Rfin+ and Rfin-, these two transistors determine the conversion gain and linearity of the mixer. Loin+ is applied to the base of transistors Q4, Q7, while Loin- is applied to the base of transistors Q5, Q6. These four transistors constitute the core circuit of the mixer. Baseband signal is outputted from BBout+ and BBout- after mixing, at the same time, R2 and R3 constitute a common mode voltage sampling circuit to sample common mode DC voltage and then cancel the common mode DC offset through feedback loop circuit. Similarly, the differential mode voltage is sampled by R1 and R4 and then outputted to another feedback loop circuit to cancel differential mode DC offset.

2.2 Common Mode Feedback Loop Circuit

Common mode feedback circuitry designed in this paper is shown in Fig.3. Common mode signal VCMsense sampled by the common mode voltage sampling circuit constituted by R2 and R3 perform operation with standard voltage Vref, which leads to the drain current of M4 adjusted by VCMsense. M2, M3 and M4 constitute a mirror current source. R5, R6 are the load resistances of M2, M3. The current flowing through the load resistance can regulate collector voltage, when the loop gain is large enough, Vref and VCMsense approximately equal. The relation will exist between

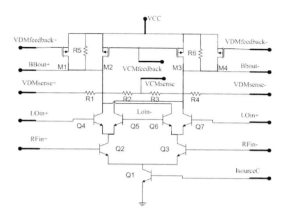

Fig. 2. Mixer with Gilbert cell and current steering

Vref and mixer core collector voltage. When common mode DC offset appearance, the feedback loop circuit will make common mode signal approximately equal to Vref. Vref is used to optimize the linearity and the maximum output voltage amplitude.

2.3 Differential Mode Feedback Loop Circuit

Model of DC offset cancellation is shown in Fig.4, where Vos1 and Vos2 are offset voltage of Gm1 and Gm2. Gm1 is the differential pair transconductance of mixer, Gm2 is the differential pair transconductance of negative feedback loop, R is a derivative of transconductance amplifier. Low frequency output DC offset voltage of mixer is detected and stored in C1 and C2, which is converted to correction current by Gm2. The correction current have an opposite phase with mixer core DC offset, Gm2 injects the correction current to mixer to correct the DC offset current. According to the literature[7], input offset voltage can be expressed as

$$V_{os} = \frac{V_{os1}}{G_{m2}R} + \frac{V_{os2}}{G_{m2}R} \qquad (1)$$

Where, Gm2R>>1, if Gm1R and Gm2R are very large, Vos will be very small. The differential offset correction circuit is shown in Fig.5, the correction circuit include a differential negative feedback loop, which have a low-pass filter (R7,C1 or R8,C2 in Fig.5). R1 and R4 sample the differential mode offset and input them to VDMsense+ and VDMsense-, low frequency DC offset is obtained by LPF consisted of R7,C1 or R8,C2,and then enter M6 and M7. M8,M1 and M9,M4 constitute mirror current source, which feed back the correction current to mixer core to correct the differential mode DC offset. DC offset often appear in the narrow-band applications with the bandwidth from a few hundred KHz to several MHz. According to the cutoff frequency, the value of C1,C2 should be 60pf, R7,R8 are 50K Ohm.

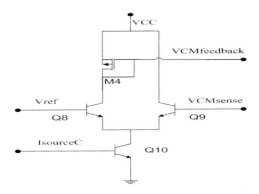

Fig. 3. Common Mode Feedback Circuit

3   Simulation Results and Analysis

Circuit in Fig.2, Fig.3 and Fig.5 were simulated in a 0.35 μm BiCMOS technology by ADS. In fact, the mismatch of mixer components and the self-mixing effect are two main sources of DC offset, but from simulation point of view, mismatch of collector resistance was employed to simulate the DC offset with the difference of 10% between R5 and R6. Output DC offset was 31mV with open-loop simulation and 873 μV with closed-loop simulation. Noise feature of mixer is shown in Fig.6, double sideband noise was 7.7dB, simulation results are shown in table 1.

From table1 we can notice that DC offset is improved obviously and the mixer performance degraded not too much. But extra current and noise are introduced by feedback loop. AC coupling is not suitable for integrated circuits, digital solution will complicate the design and increase the power dissipation. Method in this paper has a good performance without much additional cost.

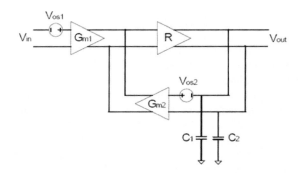

Fig. 4. Principle of offset cancellation technique

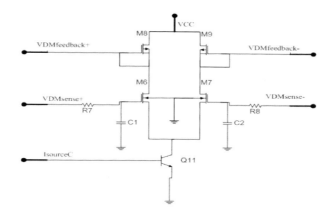

Fig. 5. Differential mode feedback loop

Table 1. Simulated parameters of mixer with ADS

Parameters	Simulation Results	Remarks
Frequency	5.25GHz	
Conversion gain G	11dB	With input signal 5.25GHz@-50dBm
NF	7.7dB	DSB NF@150kHz offset
IIP3	-8.9dBm	Dual audio test frequency interval of 5MHz
DC offset correction rate	2.8%	Relative to the uncorrected
Total current I	17.07mA	

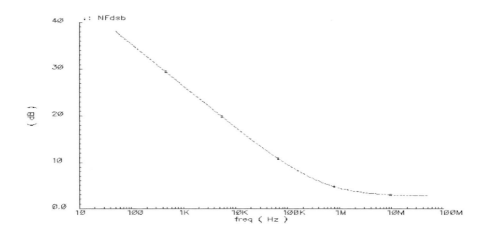

Fig. 6. DSB NF of the mixer

4 Conclusions

DC offset is a critical problem which constrains the applications of direct-conversion receivers, conventional correction method have high cost or bad effect. DC offset cancellation down-conversion mixer method presented in this paper is a compromise between cost and effect. The DC offset with correction circuitry is less than 3% of uncorrected DC offset which improve the performance of direct-conversion receiver.

References

1. Razavi, B.: Design considerations for direct conversion receivers. IEEE Trans. Circuits Syst. 44, 428–435 (1997)
2. Feng, H., Wu, Q., Guan, X., Zhan, R., Wang, A.: A 5GHz Sub-Harmonic Direct Down-Conversion Mixer for Dual-Band System in 0.35 μm SiGe BiCMOS. In: IEEE Radio Frequency Integrated Circuits Symposium, pp. 4807–4810 (2005)
3. Johansen, T.K., Vidkjær, J., Krozer, V.: Analysis and Design of Wide-Band SiGe HBT Active Mixers. IEEE Transactions on Microwave Theory and Techniques 53, 2389–2397 (2005)
4. Namgoong, W.: Performance of a direct-conversion receiver with AC coupling. IEEE Trans Circ. and Syst-II: Analog and Digital Signal Processing 47, 1556–1559 (2000)
5. Enz, C., Temes, G.: Circuit Techniques for Reducing the Effects of Op-amp Imperfection: Autozeroing, Correlate Double Sampling and Chopper Stablization. Proc. IEEE 84(11), 1584–1614 (1996)
6. Gilbert, B.: A precise four-quadrant multiplier with subnanosecond response. IEEE J. Solid-State Circuits SC-3(11), 365–373 (1968)
7. Razavi, B.: Design of analog Integrated circuit in Nonlinearity and Mismatch. McGraw-Hill, New York (2001)

UWB Low-Voltage High-Linearity CMOS Mixer in Zero-IF Receiver

Huiyong Hu[*], Shuai Lei, Heming Zhang, Rongxi Xuan, Bin Shu,
Jianjun Song, and Qiankun Liu

Key Lab of Ministry of Education for Wide Band-Gap Semiconductor Materials and Devices,
School of Microelectronics, Xidian University, Xi'an, China, 710071
huhy@xidian.edu.cn

Abstract. This paper presents and designs a UWB, low voltage, high linearity, folded CMOS direct down-conversion mixer in zero-IF receiver. The proposed mixer uses inverter, self-biasing and constant G_m biasing circuit. Also, inductor in LC resonating network is used to get rid of the negative effect of parasitic capacitor in this paper. Consequently, all performances of the mixer are effectively improved. This design is based on 0.18 μm CMOS technology of SMIC (Semiconductor Manufacturing International Corporation), the simulation results show the conversion gain of 7dB, the NF of 13.8dB (the IF frequency is 20MHz), the IIP3 of 13.4dBm. The circuit operates at the supply voltage of 1.8V and dissipates 11mW. The dynamic range of total receiver can be effectively improved because of high linearity of the mixer.

Keywords: UWB, Folded topology, Inverter, Self-biasing, Constant Gm.

1 Introduction

UWB technology is used to realize high-speed wireless communication, since it provides low power consumption, wide band and low complexity. At the same time, with the characteristics of low cost, low power consumption, excellent device performance and being able to integrate with digital system[1], CMOS technology is increasingly used to achieve RF circuit.

A double-balanced Gilbert-type Mixer is the core module of RF circuit used for wireless communications, compare to the single tube and single-balanced mixer, it have a higher linearity and a well anti-jamming capability to the background noise[2]. However, this topology consists of four stacked stages, it difficult to ensure all transistors operate at saturation region with lower supply voltage, which will decrease the performance of the mixer. So conventional circuit topology of mixer can not meet the low supply voltage applications [3].

The design of folded topology mixer is presented in this paper with the spectrum from 3.268GHz-4.752GHz. This design is based on 0.18 μm CMOS technology of

---

[*] Project supported by National Ministries and Commissions (Nos. 51308040203, 6139801), the Fundamental Research Funds for the Central Universities (Nos. 72105499, 72104089) and Natural Science Basic Research Plan in Shaanxi Province of China (No.2010JQ8008).

SMIC(Semiconductor Manufacturing International Corporation), the simulation results show that the mixer have a good performance and the dynamic range of total receiver can be effectively improved because of high linearity of the mixer.

2   Optimized Mixer Design

In this paper we presents an optimized topology for designing mixer, as shown in Fig.1, optimized topology of mixer consists of folded topology, inverter, self-biasing and constant Gm biasing circuit.

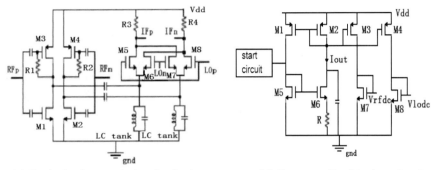

(a) Optimized central circuit of mixer   (b) Constant Gm biasing circuit

Fig. 1. Topology of mixer presented in this paper

A comparison between stacked and folded topology[4] is shown in Fig.2. (a) and (b) are DC and RF equivalent circuits of stacked topology, which is often employed in conventional RF IC. This topology use DC current sharing to realize necessary biasing and functionality. If Von is the minimum voltage to turn on an element in the topology, the minimum supply voltage Vdd should be 2Von, so this topology has a good performance only when the supply voltage is high. With the supply voltage become smaller, especially more stacked active devices were employed; stacked topology can not guarantee all the active devices biasing in the saturation region. Folded topology can be used to solve this problem, as shown in Fig.2, DC current sharing does not used in the topology. This scheme uses two "RF traps" and a coupling capacitor, in the ideal conditions, RF trap becomes short circuit at DC and open circuit at RF, which is opposite to the coupling capacitor. DC equivalent circuit is shown in Fig.2(c), the minimum supply voltage required is only Von, and so this topology can operate at a lower supply voltage. Fig.2 (d) is the RF equivalent circuit, which is the same with conventional stacked topology. With addition of coupling capacitor, DC bias of transconductance and switch stage can be adjusted respectively, which ensure a high transconductance for transconductance stage and a right operating point for switch stage. This topology improves the linearity of the mixer. And also it can decrease the bias current of switch and increases the load resistance to improve the conversion gain of the mixer.

For the supply voltage is not too low, to improve the conversion gain of the mixer, PMOS transistors M3 and M4 are used to replace the RF trap 1, which is shown in Fig.1(a). A transconductor is replaced by an inverter, so total transconductance of the mixer become the sum of two parts, which improves the gain of the mixer. Conversion gain of the proposed scheme can be expressed as:

$$G = 20\log(\frac{2}{\pi}(g_{mn} + g_{mp})R) \quad (1)$$

Where, g_{mn} and g_{mp} are transconductance of NMOS and PMOS transistor, respectively, R is load. Conversion gain of conventional mixer can be expressed as:

$$G = 20\log(\frac{2}{\pi}g_{mn}R) \quad (2)$$

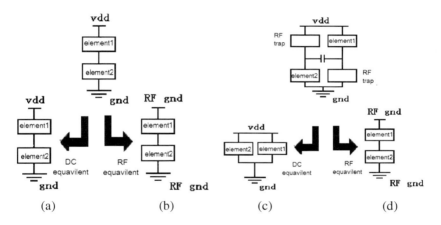

Fig. 2. The equivalent circuits of stacked and folded topology

In the direct down-conversion mixer, flicker noise is a critical problem which is mainly determined by switch transistor [5]. Switch transistor have two mechanisms to generate noise at the output end, direct and indirect mechanism. In an indirect mechanism, flicker noise is determined by frequency and the parasitic capacitor at common source of switch transistor. LC resonant circuit is used to replace RF trap2, which is shown in Fig.1 (a). Inductor in LC resonant circuit is used to cancel the effect of parasitic capacitor, which will reduce flicker noise. The linearity and conversion gain of the mixer are improved by the LC resonant circuit as well. So LC resonant circuit presented in this paper not only works as RF traps but also improves the performance of the mixer.

Self-biasing structure (R1,R2 shown in Fig.1(a)) is used to make the circuit more stable and reduce bias circuit, save chip area. At the same time, the circuit can be adjusted easily.

Constant Gm biasing circuit is employed in this paper to make the circuit more stable which is shown in Fig.1(b). The biggest advantage of this circuit is that output current I_{out} is independent of supply voltage[6], which is verified by the simulation as well. Start circuit is introduced to the biasing circuit structure to avoid the circuit operating at zero current state. M1 and M2 have same width to length ratio, so DC bias V_{rfdc} of RF transistor and V_{lodc} of LO transistor can be expressed respectively as:

$$V_{rfdc} = V_{TH} + \sqrt{\frac{2I_{out} \cdot (W/L)_{M3}}{\mu_n C_{ox} (W/L)_{M2} \cdot (W/L)_{M7}}} \qquad (3)$$

$$V_{lodc} = V_{TH} + \sqrt{\frac{2I_{out} \cdot (W/L)_{M4}}{\mu_n C_{ox} (W/L)_{M2} \cdot (W/L)_{M8}}} \qquad (4)$$

3 Simulation Results and Analysis

The circuit designed in 0.18 μm CMOS technology of SMIC(Semiconductor Manufacturing International Corporation) in this paper is simulated with Cadence Spectre RF. LO power is -1dBm, according to the requirement of UWB system, LO frequency is 3.432GHz, 3.960GHz and 4.488GHz, respectively. The input frequency of RF is around the center frequency changes in a range of ±264MHz, so IF output frequency is ±264MHz.

Simulation is Simulation is performed with a load of 50Ω and 10KΩ respectively, simulated conversion gain curves are shown in Fig.3. The conversion gain is only -4dB with a load of 50Ω (shown in Fig3.(a)) for that there is no source follower used as buffer in the output port. The conversion gain is 7dB with a load of 10KΩ(shown in Fig.3(b)). From the figure, one can notice that conversion gain of the mixer changes no more than 1dB, which means the gain change little with the variation of RF frequency, which can be used in the UWB system.

As shown in Fig.4(a), IIP3 of this mixer is 13.4dBm, which indicate a very high linearity. This topology increases the dynamic range of the mixer. NF of the mixer are shown in Fig.4(b), where curve1 is NF with broadband matching, NF is 13.8dB (the IF frequency is 20MHz). Curve 2 is NF without broadband matching, NF is 10.6dB (the IF frequency is 20MHz).

As the constant Gm structure is introduced, when the supply voltage down to 1V, the mixer can still work normally. The conversion gain is 5dB, NF of 15.8dB (the IF frequency is 20MHz) with broadband matching, IIP3 of 11.5dB, power consumption of 4.5mW.

According to theory of RF design, if IIP3 is larger 10dBm than LO power, the designed mixer have a good linearity, E can be used to measure the intermodulation performance, which can be expressed as:

$$E = [IP3(dBm) - LO\_Power(dBm)]/10 \qquad (5)$$

If E large than 1, the mixer have a good intermodulation characteristic[7]. The design presented in this paper show a LO power of -1dBm, IP3 of 13.4dB, E of 1.44, so the mixer have a high linearity.

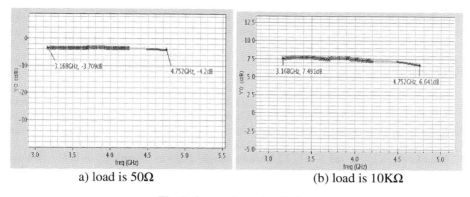

a) load is 50Ω (b) load is 10KΩ

Fig. 3. Conversion gain of mixer

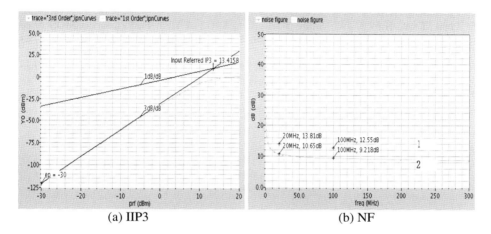

(a) IIP3 (b) NF

Fig. 4. IIP3 and NF of mixer

4 Conclusions

With the 0.18μm CMOS technology of SMIC(Semiconductor Manufacturing International Corporation),this paper presents and designs a UWB, low voltage, high linearity, folded CMOS direct down-conversion mixer in zero-IF receiver. Optimized methods include inverter; self-biasing and constant G_m biasing circuits are used to improve the performance of the mixer. Also, inductor in LC resonating network is used to get rid of the negative effect of parasitic capacitor in this paper. Consequently, all performances of the mixer are effectively improved. In addition, for the use of constant Gm circuit, when the supply voltage reduces the circuit can operate normally without adjusting any parameters. The simulation results show the conversion gain of 7dB, the NF of 13.8dB (the IF frequency is 20MHz), the IIP3 of 13.4dBm. The circuit operates at the supply voltage of 1.8V and dissipates 11mW. E is 1.44, which indicate that the mixer have a high linearity. The dynamic range of total receiver can be effectively improved because of high linearity of the mixer. From above, we can see that this topology can meet the requirement of UWB IF receiver.

References

1. Chen, J., Shi, B.: CMOS Radio Frequency Integrated Circuits: Achievements and Prospect. Microelectronics 10, 323–328 (2001) (in Chinese)
2. Silver, J.P.: Gilbert Cell Mixer Design Tutorial: RF, RFIC & Microwave, Theory, Design, 1–20 (2003), http://www.rfic.co.uk
3. Vidojkovic, V., van der Tang, J., Leeuwenburgh, A., van Roermund, A.H.M.: A Low-Voltage Folded-Switching Mixer in 0.18-μm CMOS. IEEE Journal of Solid-State Circuits 40(6), 1259–1264 (2005)
4. Manku, T., Beck, G., Shin, E.J.: A Low-Voltage Design Technique for RF Integrated Circuits. IEEE Transactions on Circuits and systems—II: Analog and digital signal Processing 45(10), 1408–1413 (1998)
5. Phan, T.-A., Kim, C.-W., Shim, Y.-A., Lee, S.-G.: A High Performance CMOS direct Down Conversion Mixer for UWB System. IEICE Trans. Electron. E88-C(12), 2316–2320 (2005)
6. Razavi, B.: Design of Analog CMOS and Integrated Circuits. Translated by Guican Chen, pp. 310–311. xi'an jiaotong university press (2002)
7. Shevchuk, E., Choi, K.: Folded Cascode CMOS Mixer Design and Optimization in 70 nm Technology, pp. 943–946. IEEE, Los Alamitos (2005)

Design and Simulation of Reserved Frame-Slotted Aloha Anti-collision Algorithm in Internet of Things

Jun-Chao Zhang and Jun-Jie Chen

College of Computer Science and Technology
Taiyuan University of Technology
030024 Taiyuan, China
zjc6691018@163.com,
chenjj@tyut.edu.cn

Abstract. Tag collision is an obstacle in the population process of Internet of Things (IoT). Based on Dynamic Frame-Slotted Aloha (DFSA) algorithm, an improved algorithm was presented and designed, namely Reserved Frame-Slotted Aloha (RFSA) algorithm. The improved reserve mechanism enables reader access to data reading status on each slot in advance, so that subsequent communication can directly performed on normal slots. Then, key parameters of RFSA were determined. Process model, node model and network model of RFSA were also presented with OPNET 14.5 tool. Simulation experiment results based on OPNET show that the improved algorithm is superior to DFSA in overall performance.

Keywords: Internet of Things; anti-collision; Aloha; reserved mechanism.

1 Introduction

In the population process of IoT, one of the biggest obstacles is low recognition rare caused by tag collision. The tag is also called as responder, which is usually attached to the object to be identified to record related information of the object. Main modules of tag are integrated in a chip. External antenna is connected to the chip to help tag communicate with reader with manner of radio. The reader of IoT is also known as interrogator, which has built-in antenna to communicate with tags. Reader connects to computer system with wire or wireless way to send received tag information to data processing subsystem. A major advantage of IoT is that it can identify multiple targets simultaneously. Such kind of ability to correctly identify multiple targets at the same time is called anti-collision algorithm. In many applications, there may be multiple tags or multiple readers in the action scale of identification system simultaneously. As reader shares wireless channel with tags, the signal sent by reader or tag may overlap to disable reader identify tags correctly, namely collision occurs.

In many non-deterministic anti-collision algorithms based on Aloha, such as pure algorithm [1], slot Aloha algorithm [2], frame-slotted Aloha algorithm [3] and DFSA algorithm [4, 5], frame-slotted Aloha algorithm that can dynamically adjust time frame has best performance. In DFSA, the probability of tag to send data obeys binomial distribution with parameters n and $1/N$. From Poisson's Theorem we know that when tag number approach infinity and product of two parameters in binomial distribution is a constant bigger than zero, the binomial distribution will approximate the Poisson distribution. In order to optimal system performance, time frame is set to number of tag that has not been identified. At this moment, the product of two parameters in binomial distribution is 1. If there are many tags that have not been identified, the performance of DFSA is similar to that of slotted Aloha algorithm. While the advantages and disadvantages of anti-collision algorithm are mainly reflected in case of large tag number. We can know that the performance of DFSA is still unsatisfactory. Therefore, the paper presents an improved algorithm on DFSA, namely RFSA. Simulation analysis on the improved algorithm with OPNET is also provided. The paper is organized as follows. Section 2 gives working flow of RFSA algorithm. Detail design on RFSA is presented in section 3. Section 4 conducts simulation analysis on RFSA and section 5 concludes our work.

2 Working Flow of RFSA

RFSA achieve some improvements on DFSA, the biggest of which is that it introduced a reservation mechanism based on original algorithm. With this reserved mechanism, the tag can enable access to data read status on each slot in advance, namely idle, normal communication and collision. In this way, subsequent transmission of tag can skip idle and collision slot to directly sent on normal communication slot. Improvement on DFSA can avoid time waste on idle slot and collision slot, so as to improve system performance.

The specific communication process of RFSA is as follows.

Step 1: Reader detects tags that have not been identified in read scope by sending query command. If there is, these tags will send response signal to reader on receiving query command.

Step 2: If reader detected response signal collision, it indicates that multiple tags in the reading field. Reader will send a time frame size to all tags. After tag received the size, it will randomly select number of a micro-slot and send a random number based ob selected order in the header of micro-slot to reserve slot for the outgoing transmission.

Step 3: Reader communicate with tags on normal slot skipping detected idle and collision slot in the header of time frame. If the received data was calibrated correct, reader sends a silent command to this tag. Then the tag will be in sleep mode, namely it will not answer any command sent by reader in the subsequent period until the tag out reading field.

Step 4: After a reading period, the reader will estimate number of tags that have not been identified in reading field and re-set time frame size in next reading period according number of tags that have not been identified.

Step 5: The identification enters into next period and repeat *Step* 2 to *Step* 5 according to time frame until all tags in reading field are identified.

3 Related Design of RFSA

3.1 Time Frame Header Design

Time frame header is mainly used by tag to reserve wanted transmission slot to reader before formal transmission of serial. This method is similar to the basic bit-map method of the reservation protocol [6]. For difference of tags located in reading field, the power of reflect wave to reader is not same. So it is difficult for reader to determine whether collision occurs by setting threshold to filed strength in the 1 bit data sent by tag. The RFSA algorithm reserves slot for serial to be sent by sending a short random number in the micro-slot of time frame header. Using special encoding, such as Manchester encoding, the reader can determine data reading status from the received random number to reserve corresponding slot.

However, there is a flaw to reserve slot with random number, namely reading status in micro-slot can not represent reading status of whole slot completely. There is a situation in actual communication that two tags send random number at same micro-slot and same random number to reserve slot. At this time, it is difficult for reader to determine collision occurrence from signal received. It will mistake to think data reading status in corresponding state as normal communication to transmit serial at this slot. For the detection failure, collision occurs at the slot will cause useless waste of time. By selecting length of random number, the probability of such condition can be controlled so that effect of it on system performance can be reduced to minimum.

3.2 Selection of Time Frame Size

Although RFSA algorithm can skip all idle slots and most collision slot with reserve mechanism, communication process of time frame header still should go through three statuses of idle, collision and normal communication as that in DFSA. In order to optimize system performance, it is an indispensable step to select an appropriate time frame size in RFSA.

From introduction of RFSA we can know that system delay is sum of time consumed in time frame header and micro-slot reservation as well as that consumed in successful transmission. As probability of collision in slot only relate to length of reserve random number, it means consumed time of collision in slot only related to length of random number. As tag number in reading field is fixed, time for successful transmission tag serial is also fixed. Therefore, time frame size only impact on time consumed in micro-slot reservation. Communication process between reader and tag in time frame header is performed in accordance with DFSA. The only difference is data to be transmitted and micro-slot can not send silent signal after normal communication of reader. This does not affect selection of time frame size. Thus, to optimize system performance of RFSA algorithm, time frame size in the algorithm should also be set as number of tags that have not been identified in the reading field.

As probability of tag to send data at some micro-slot or slot in RFSA also obey binomial distribution, which is same as DFSA algorithm. Although same tag estimation method was adapted, slot number statistics method to read data status also changes in RFSA. Assume C_0 is slot number that has not tag to send, namely idle slot number; C_1 is slot number of successful transmission and C_k is slot number of collision. Then, it is relatively easy to obtain C_0. The calculation method of C_1 is micro-slot number for normally reading data by subtracting slot number of collision. C_k is micro-slot number of collision added by slot number of actual collision.

4 Performance Simulation and Result Analysis

4.1 Establishment of Simulation Model

OPNET network simulation platform was used to simulate anti-collision algorithm. The simulation models include process model, node model and network model.

(1) Process model
The process models of tag and reader were shown in figure 1 and figure 2.

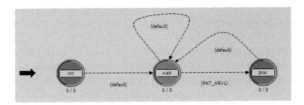

Fig. 1. Process model of tag

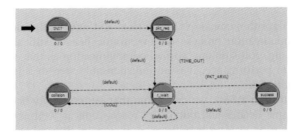

Fig. 2. Process model of reader.

Tag process model initialized tags in the beginning of simulation, namely set initial value to parameters in tags. After initialization, tag enters into wait state to wait for commands from reader in that communication manner of RFSA is reader speak firstly. After tag received signal from reader, simulation kernel will generate a flow interrupt to tag process model, when interrupt condition PKT_ARVL was satisfied. The flow interrupt drive tag from wait state into proc state. In the proc state, tag determines command from reader according to improved algorithm, and then conducts corresponding processing, such as changing parameters or sending data.

As initialization, reader process model will automatically enter into pkt_req state. In the state, reader sends request command including slot number and time frame size. In the simulation of this paper, initial slot number was set 0 and time frame size set minimum 4. After first command is sent, reader enters into r_wait state to wait for answer from tags. When tag receives request command whose slot number is 0, it will select a random number in scope from 1 to time frame as slot number to send data. Aster waiting out, reader will enter into pkt_req state again and plus 1 on slot number, then send request command. Then reader re-enter into r_wait state.

After tag received request command whose slot number is not zero, it compares slot number in command with the selected one. If there are different, it will not answer. Otherwise, it should send serial of this tag. At the moment, if only one tag sending serial, simulation kernel generates a flow interrupt to transfer reader into success state. If more than one tag sending serial, simulation kernel generates state interrupt of collision to transfer reader into collision state. If no tag answer, reader has to keep waiting until time out to return pkt_req state.

(2) Node model
The tag node model and reader node model were shown in figure 3 and figure 4.

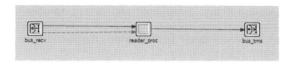

Fig. 3. Node model of reader.

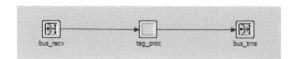

Fig. 4. Node model of tag.

Node models of reader and tag consist of three modules, namely bus receiver, bus transmitter and processor. Bus receiver and transmitter are used for duplex communication between reader and tags. The difference between two node models is that each calls different process model and node model of reader has an additional state line between receiver and processor. It is used for transmitting channel status to processor. When there is data collision in channel, simulation kernel sends state interrupt to processor.

(3) Network model
Simulation network model consists of a reader node and multiple tag nodes, which are interconnected according to bus topology. To verify system performance, it needs to count simulation result of different tag number. We built multiple scenarios in a simulation project. Each scenario established a network model with different tag number. Figure 5 showed a network model with 10 tags whose number is node_1 to node_10.

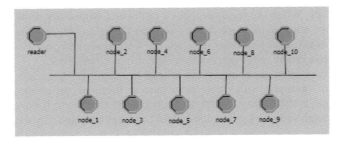

Fig. 5. Network model with 10 tags.

4.2 Simulation Result Analysis

(1) Simulation result of improved algorithm

To measure performance of different tag anti-collision algorithms, three indexes were given as follows: system delay, throughput and repeat number. System delay here means consumed time in whole recognition process. Throughput refers to average times of successful communication in unit time. Repeat number is times of same operation for successful reading all tags in the reading field.

Figure 6 and 7 are change curve of system delay and throughput as tag number increases, which was plotted by OPNET tool according to collected scale file in simulation process of RFSA. The former is system delay curve, where unit of system delay is unit time. We can know that system delay rise with tag number increases. Nevertheless, the curve is not smooth. From the latter figure, we can see that throughput jitter back and forth in a certain range with tag number increases, which is not consistent with result of theoretical analysis.

4.3 Algorithm Performance Comparison

To eliminate randomness of single simulation, we made 10 group simulations on improved RFSA algorithm and existing DFSA algorithm. Average value of 10 group simulations were concluded to access to statistical characteristics of these algorithm to some extent. Figure 7 and figure 8 show comparison of system delay and throughput between these algorithms. The former shows difference of system delay between two algorithms. It is obvious that the curve is not very smooth for randomness in simulation process can not be eliminated completely. We can see that consumed time of improved algorithm is less than the original one. The latter figure shows throughput comparison between two algorithms.

In summary, we can arrive at following result:

(1) As randomness of single simulation instance caused fluctuation in result curve, times of simulation on algorithms can eliminate its randomness to obtain more stable statistical characteristics.

(2) Taking system delay as index, we can know from simulation result that system delay of improved algorithm decreases down to 42% of DFSA. So the performance of improved algorithm is superior.

(3) Taking throughput as evaluation index, we can conclude from result data that the throughput of improved algorithm is 1.4 times of DFSA in steady decline phase. The superiority of improved algorithm is proved once again.

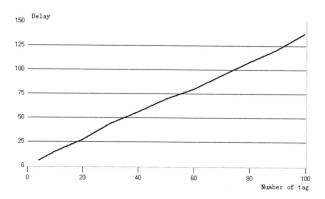

Fig. 6. Relationship between system delay and tag number in a simulation instance.

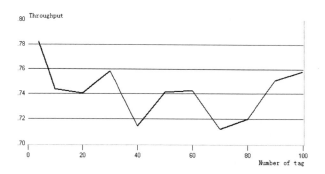

Fig. 7. Relationship between throughput and tag number in a simulation instance.

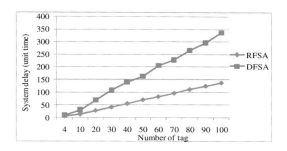

Fig. 8. System delay comparison between two algorithms.

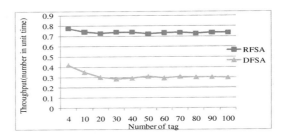

Fig. 9. Throughput comparison between two algorithms.

5 Conclusion

An improved algorithm RFSA was presented for IoT in the paper, which enables reader access to data reading status on each slot in advance, so that subsequent communication can directly conducted on normal slots. Key parameters of RFSA were determined. Process model, node model and network model of RFSA were also presented with OPNET. Simulation experiment results show that the improved algorithm is superior to DFSA in overall performance. Although we can arrive at the conclusion, the system also should pay some price compared with current anti-collision algorithms, which is rightly out research focus in the near future. Corresponding author of this paper is Chen Jun-jie.

References

1. Khan, J.: ALOHA Protocol. ELEC350, ALOHA note
2. Zhao, L.: ALOHA Random Multiple Access Communication Technique–from Pure ALOHA to Spread ALOHA. Mobile Communications 23, 1–5 (1999)
3. Chen, X., Xue, X.-p., Zhang, S.-d.: Studies on Tag Anti-collision Algorithms. Modern Electronics Technique 5, 13–15 (2006)
4. Lee, S.-R., Joo, S.-D., Lee, C.-W.: An Enhanced Dynamic Framed Slotted ALOHA Algorithm for RFID Tag Identification. In: Proceedings of the Second Annual International Conference on Mobile and Ubiquitous Systems, Networking and Services, pp. 166–172 (2005)
5. Wu, J., Xiong, Z., Wang, Y.: Multiple object anti-collision RFID technology with dynamic slot allocation. Journal of Beijing University of Aeronautics and Astronautics 31, 618–622 (2005)
6. Tanenbaum, A.S.: Computer networks, 4th edn. Tsinghai Univ. Pr. (2003)

A Hop to Hop Controlled Hierarchical Multicast Congestion Control Mechanism

Jun-Chao Zhang[1], Rong-Xiang Zhao[2], and Jun-Jie Chen[1]

[1] College of Computer Science and Technology
Taiyuan University of Technology
030024 Taiyuan, China
[2] Shanxi Tiangang Construction Project Management Co., Ltd.
030024 Taiyuan, China
zjc6691018@163.com, zyx6014882@163.com, chenjj@tyut.edu.cn

Abstract. Hierarchical multicast can effectively achieve multicast communication of heterogeneous network. RED algorithm and HTH algorithm were analyzed and discussed. RED can keep low delay while pursuit larger throughput, but it has shortcoming of sensitive to parameters. HTH algorithm can rapidly respond to congestion and effectively improve throughput as well as utilization of link, but it occupies too much router resources in case of many streams. On the basis, a kind of hop-to-hop controlled hierarchical multicast congestion control mechanism combining RED and HTH was presented. With this mechanism, each router determines its congestion status based on current network state. If congestion occurs, it discards with certain probability and report congestion status to upstream router. After router received congestion state from other routers, it will take appropriate measures to increase or decrease transmission rate to some downstream router. At the same time, packet discard number of each link was considered in congestion control to ensure fairness. Simulation result based on NS-2 shows that the proposed algorithm is superior to RED in aspects of throughput and packet loss rate.

Keywords: hierarchical multicast; congestion control; hop-to-hop control.

1 Introduction

Hierarchical multicast is effective means to achieve multicast communication under heterogeneous network [1]. Now the data hierarchy of hierarchical multicast is accumulated. It requires receiver firstly receive data of basic layer and then data of enhancement layer from low to high. Then original data can be decoded, so it can not achieve rate adjustment in detail and can not adapt to heterogeneous network environment, which waste available network bandwidth. Based on research of accumulated hierarchical multicast, adaptive layered strategy was added into data stream of hierarchical multicast. It means receiver can randomly join these multicast layers and does not need to receive according to specific order, which enhance flexibility of receiver joining multicast layer and adequately utilize network bandwidth. Based on analysis on RED and HTH algorithms, a hop-to-hop controlled hierarchical multicast congestion control mechanism combining RED and HTH was presented. With this mechanism, each router determines its congestion status based on current network state. If congestion occurs, it discards with certain probability and report congestion status to upstream router. After router received congestion state from other routers, it will take appropriate measures to increase or decrease transmission rate to some downstream router. At the same time, packet discard number of each link was considered in congestion control to ensure fairness. The paper is organized as follows. Section 2 analyzes advantages and disadvantages of RED and HTH algorithms. On the basis of RED and HTH, a hop-to-hop controlled hierarchical multicast congestion control mechanism combining RED and HTH is presented in section 3. Section 4 performs simulation and performance analysis on the presented algorithm with NS-2. Section 5 concludes our work.

2   Analysis on RED and HTH

The basic idea of RED is that all router nodes with multicast function can compute congestion state in advance. When congestion threshold was triggered, it sends traffic control information to adjacent upstream router with multicast function. After upstream node received traffic control information, it will determine whether to implement flow control to avoid or reduce congestion. In short, multicast router node of each network layer needs to maintain buffer of its downstream route ports. The disadvantage of RED is that it is too sensitive to parameter settings. It should be further researched that how to balance relationship between throughput and delay so as to find optimal parameters. Furthermore, RED algorithm can not effectively estimate severity of congestion. In addition, RTT difference of different TCP flow is one of reasons to damage fairness. As network is widely deployed and has complex structure, there is large difference of time from some router to different terminal system [2].

Hop-to-hop congestion control manner is implemented on all routers along the path [3]. Therefore, HTH control mechanism has obvious advantages, especially in the delay-bandwidth product of a wide range network. First, it can quickly get feedback information, First, it can quickly get feedback messages, because the distance between each hop is much shorter compared with the distance between multi-hop. After propagation delay of feedback message was decreased, control mechanism can quickly play role to decrease possibility of packet loss in congestion. Meanwhile, under the control of the HTH, packet not only is stored in the bottleneck node, but also stored in nodes that go through before bottleneck node. So if the capacity of the bottleneck node speed increases, HTH control method may be faster to use these additional capacities. Secondly, with the increase of network bandwidth, delay bandwidth product of network also growing. The method to control sudden flow is to aggregate them into a group and allow that group to accept control based on message feedback. As the aggregation is carried out at the gateway, so HTH control is the best choice. Finally, as HTH control flow load at the edge of network, it can protect flow has restrict behaviors and limit flow that hostile to seize bandwidth [4, 5]. However, HTH also has disadvantages that it needs to keep state of flow. In case of too much flow, it will take up too much resource.

3 Design of HTHRED

3.1 Basic Idea

From above analysis we can know that RED and HTH have their own advantages and disadvantages. On this basic, we presented a hop-to-hop random early detection algorithm. The basic idea is hop-to-hop congestion control method. Each router determines its congestion status based on current network state. If congestion occurs, it discards with certain probability and report congestion status to upstream router. If congestion occurs, it discards with certain probability and report congestion status to upstream router. After router received congestion state from other routers, it will take appropriate measures to increase or decrease transmission rate to some downstream router. At the same time, packet discard number of each link was considered in congestion control to ensure fairness.

3.2 Main Implementation Principle

HTHRED algorithm use improved RED algorithm to measure length of queue and find congestion to be taken place, and then upstream router take flow control measures to alleviate and eliminate congestion. The HTHRED was made up of three parts, namely improved RED congestion control, flow control and hop-to-hop congestion control message protocol among routers. Congestion detection was obtained by improvement on classical RED algorithm, which is called per-hop random early detection. It is used to detect current status of node and to determine whether there is possibility of congestion, and then activate other corresponding measures. Flow control is designed to adjust current packet transmission according to requirements of congestion notification, which is called per-hop traffic flow control. Per-hop congestion control message protocol is responsible for computing and generating key parameters of congestion notification to be set after received congestion occurrence signal form HRED, and then send notification.

3.3 Per-Hop Random Early Detection

Router detected early congestion with low pass filter, the structure and related parameters of which are same as RED algorithm. The difference is that HTHRED transmit congestion control information among any two routers. Compared with RED, the time to find congestion and take response is relatively shorter. Therefore, setting of $\max th$ and $\min th$ has changed. The optimal value of $\max th$ and $\min th$ is determined by avg. If flow is often sudden, the setting of $\min th$ should be a bit more so that link utilization at a higher level. Setting of $\max th$ is determined by maximum average delay permitted by gateway to some extent.

3.4 Per-Hop Flow Control

It computes discard probability of new arrival packets with same manner as RED to determine accept or discard new arrival packet. It also notifies terminal take flow control measures alleviate network congestion in this way. If terminal system does not take measures, the algorithm can still play role. With the message transmission, the sender will eventually reach the nearest router, because they do not occupy a larger bandwidth, the sender in the closest router with HTHRED algorithms keep discarding packets of the flow. The congestion notification will then generated so as to limit influence of non-standard flow within a small range to some extent.

After some node received congestion notification sent by downstream adjacent node, router will control flow with leaky bucket algorithm. The leaky bucket is equivalent to a finite internal queue. If there is no leaky when packet arrives, the packet can only wait. That is, when the leaky in bucket has been used up, if there is new packet arrival, these packets can only waiting for generation of new leaky. The leaky can not cause packet loss, but it can limit packet transmission. If HTHCCMP packet of reducing traffic was received, router will reduce transmission traffic by half leaky generation rate or the method of a fixed value. The specific value was determined by some field in the congestion message. Every T seconds, the leaky generation rate will add 1 average packet size. But the maximum value will not exceed initial leaky generation rate, namely the maximum capacity of link. In this way, it can both improve link utilization and find the flow caused congestion, so as to control transmission of these flow to indicate certain fairness. When the resume flow HTHCCMP packet was received, HTHRED will immediately initial leaky generation rate so as to send data packets as soon as possible and improve network throughput and link utilization.

3.5 Per-Hop Congestion Control Message Protocol

HTHCCMP is mainly used to send congestion control messages between adjacent routers or router and upstream hosts. When edge router sends congestion control notification to host, if the host can not recognize HTHCCMP, it can directly discard. In the TCP/IP, HTHCCMP is in the middle of network layer IP protocol and transport layer protocol. In the network transmission, it should refer to IP protocol for encapsulation.

In each router, when data packet enters into the queue, congestion detection part will detect whether current node will occur congestion. If so, HTHRED algorithm generate congestion control message and put it into HTHCCMP queue and then send to corresponding upstream node. After node received congestion message from downstream router, it will either reduce transmission to relieve congestion of downstream node or resume transmission of current link to improve network resource utilization according to different message content.

4 Performance Simulation

4.1 Simulation Settings

The paper compared performance of HTHRED and original RED with NS-2. As RED is typical representative of AQM algorithm, so RED was used as standard for comparison of HTHRED. With network simulation tool NS-2, part of tcl simulation script code is as follows:

set p($i) [new Application/Traffic/Pareto]
$p($i) set bust_time_ 400ms
$p($i) set idle_time_ 100ms
$p($i) set shape_ 1.3
$p($i) set rate_ 1200k

The sender includes S1-S8 and receiver D1-D3. Connection rate of all terminals with router are 10Mbps. There are four routers of R1-R4, which is shown in Fig. 1.

4.2 Simulation Results

Congestion control in the network is to improve network throughput and link utilization. Therefore, congestion control algorithm should not only try to prevent network congestion, but also increase network throughput as possible. In the simulation network, congestion control problem will concentrated at R3-R4. The throughput and packet loss of HTHRED and RED were shown in Fig. 2 and Fig. 3.

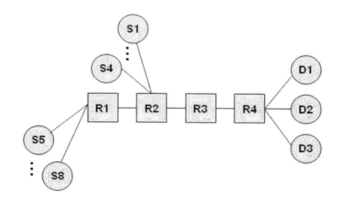

S1-S8: Data transimitter   R1-R4: Router   D1-D4: Data receiver

Fig. 1. Simulation network topology.

From Fig. 2 we can see that in most cases, the throughput of HTHRED is significantly higher than that of RED. In the first 15 seconds of beginning, network produced a longer duration of sudden flow, which mainly caused by rapid growth of TCP transmission window. In the slow start-up phase of TCP, transmission window grows in exponential. In addition, UDP data source randomly send packets, so network sudden traffic dramatically increased, which eventually leads to HTHRED and RED consider network will fundamentally congested. So packets were discarded to suppress sudden traffic and prevent severe network congestion. At the beginning of congestion detected, the number of discarded packet of HTHRED is more than RED. Subsequently, RED will continuously discard packets, while HTHRED will not discard packets in a relative long time. In the event of congestion, HTHRED algorithm sent messages to adjacent routers and upstream adjacent router will respond to limit

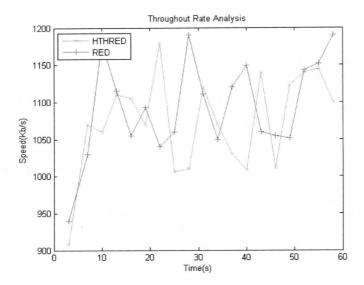

Fig. 2. Throughput comparison of HTHRED and RED.

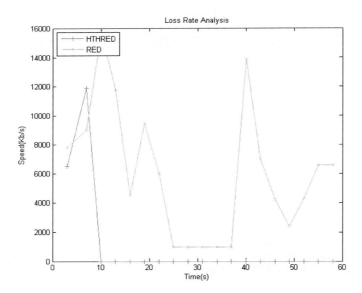

Fig. 3. Packet loss comparison of HTHRED and RED.

packet number sent to router to be congested, so as to buffer part of packets to be sent to downstream routers. Meanwhile, after upstream router limited transmission rate, local HTHRED can detect the will-be congestion and discard packets. This is the reason that HTHRED discard more packets than RED in the beginning. When RED detect congestion, it sent notification of limiting packet transmission in the manner of packet discarding or packet mark. The subsequent terminals will adjust packet transmission after congestion notification, it will influence respond time of congestion nodes and lead to absolute number of discarded packet of RED were times of HTHRED.

5 Conclusion

Hierarchical multicast is an effective means of multicast communication in heterogeneous network environment. Based on analysis on RED and HTH algorithms, a hop-to-hop controlled hierarchical multicast congestion control mechanism combining RED and HTH was presented. With this mechanism, each router determines its congestion status based on current network state. If congestion occurs, it discards with certain probability and report congestion status to upstream router. After router received congestion state from other routers, it will take appropriate measures to increase or decrease transmission rate to some downstream router. At the same time, packet discard number of each link was considered in congestion control to ensure fairness. Simulation result based on NS-2 shows that the proposed algorithm is superior to RED in aspects of throughput and packet loss rate. Corresponding author of this paper is Jun-jie Chen.

References

1. Byers, J.W., Horn, G., Luby, M.: FLID-DL. Congestion Control for Layered Multicast. IEEE Journal of Selected Areas in Communications 20, 1558–1570 (2002)
2. Wang, H.A., Schwartz, M.: Achieving Bounded Fairness for Multicast and TCP Traffic in the Internet. In: Proceedings of ACM SIGCOMM, pp. 81–92 (2000)
3. Liu, J.-c., Li, B., Zhang, Y.-q.: A hybrid adaptation protocol for tcp-friendly layered multicast and its optimal rate allocation. In: Proceedings of IEEE INFOCOM, pp. 1520–1528 (2001)
4. Deng, Y.-q., Liu, W.-y.: Research and Application of a Congestion Control Algorithm for Multimedia Data Streaming. Computer Engineering and Applications 38, 143–146 (2002)
5. Liu, K.-j., Cheng, Z.-q., Zhao, Y.-p.: Multicast congestion control based on hop to hop. Computer Engineering 33, 99–101 (2007)

A Moving Mirror Driving System of FT-IR Spectrometer for Atmospheric Analysis

Sheng Li, Yujun Zhang, Minguang Gao, Liang Xu, XiuLi Wei,
JingJing Tong, Ling Jin, and Siyang Cheng

Anhui Institute of Optics and Fine Mechanics,
Chinese Academy of Sciences, Hefei 230031, China
shengli@aiofm.ac.cn

Abstract. A moving mirror driving system of FT-IR spectrometer for atmospheric analysis was developed. It uses a tilt-compensated interferometer, a He-Ne laser reference interference system, and a driving system based on microcontroller with the control mechanism of adaptive adjustment Fuzzy-PI. The structure is simple and easy alignment. The controller is robust and high precision. It can be used in many applications of atmospheric analysis, such as open path or solar occultation flux FT-IR methods.

Keywords: FT-IR; moving mirror; driving; control, atmospheric; analysis, Fuzzy-PI.

1 Introduction

Optics remote sensing technology has developed very fast those years, of which FT-IR spectrometry is the dominant technique with substantial advantages in SNR, resolution, speed and detection limits over other ones. Methods include measurements over open paths in situ, sampling and measurement in closed cells, remote sensing using the sun, sky or natural hot objects as an IR radiation source [1]. Usually, the field conditions are more complicated and instability over laboratory, especially in the situations of vehicle, ball-borne, airborne, space born applications [2], [3]. They need the spectrometer be more stable, reliable and anti-interference which traditional ones can't stratify.

Mirror driving is the most important part in a FT-IR spectrometer. To get the correct spectrum, the moving mirror must be driven with high precision. But it is easily affected by the factors such as vibration and temperature. For FT-IR spectrometer used in the atmospheric, the moving mirror driving system should be carefully designed in order to fulfill the requirements of measurements.

With the need of FT-IR spectrum for the application of atmospheric analysis, we developed a moving mirror driving system. The system contains a tilt-compensated interferometer, a He-Ne laser reference interference system, and a driving system based on microcontroller with the control mechanism of adaptive adjustment Fuzzy-PI. Its structure is simple and easy alignment and the controller is robust and high precision. It has been used in some applications based on the open-path and solar occultation methods.

2 Principles of FT-IR Spectrometry

The structure of a typical FT-IR spectrometer is show in figure 1. It is consist of Michelson interferometer, infrared source, He-Ne laser interference system, detector, data acquirement and process systems.

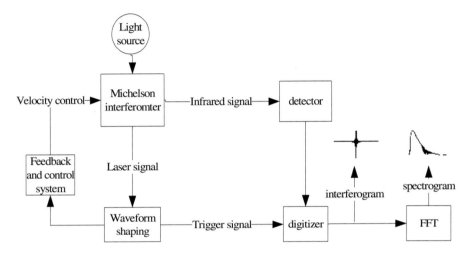

Fig. 1. Fourier transform infrared spectrometer system

The principle is based on the interference characteristic of light and this process is completed by Michelson interferometer [4]. Let suppose a beam with wavenumber $\tilde{v}_0$, and the intensity is $I(\tilde{v}_0)$, the intensity at the detector is expressed below.

$$I'(\delta) = 0.5I(\tilde{v}_0)(1+\cos 2\pi\tilde{v}_0\delta) \tag{1}$$

δ is the optical path difference (OPD). After detection and amplification, a wavenumber-dependent correction factor is added to Eq. (1).

$$I(\delta) = 0.5H(\tilde{v}_0)I(v)\cos 2\pi\tilde{v}_0\delta = B(\tilde{v}_0)\cos 2\pi\tilde{v}_0\delta \tag{2}$$

When the source is a continuum, the interferogram can be represented by the integral.

$$I(\delta) = \int_{-\infty}^{+\infty} B(\tilde{v})\cos 2\pi\tilde{v}\delta d\tilde{v} \tag{3}$$

And the spectrum can be gotten by the cosine fourier transform.

$$B(\tilde{v}) = \int_{-\infty}^{+\infty} I(\delta)\cos 2\pi\tilde{v}\delta d\delta \tag{4}$$

Eq. (3) and (4) are the basic integrals of FT-IR spectrometry. From which we can see that to get the correct spectrum, the interferogram must be measured at the distance of equally OPD which is controlled by the moving mirror. In the whole scan process, the moving mirror must be well aligned.

3 Interferometer Structure

There are many kinds of interferometer structure. The simplest type consists of a fixed and a moving plane mirror with a beamsplitter held at an angle bisecting the planes of these two mirrors [5]. The plane of the moving mirror must not tilt by an amount that will cause the OPD for the ray that is reflected from one edge of the moving mirror to differ from the OPD for the ray reflected from the diametrically opposite edge by more than about $\lambda_{min}/10$. This level of precision is very hard to achieve.

To eliminate the effect of tilt, cube-corner retroreflector can be used to replace the flat mirror. It is the three-dimensional equivalent of the roof and compensated for tilt in any direction. But the use of it has the disadvantage of requiring quite delicate initial alignment and introducing some polarization effects.

The FT-IR spectrometer for atmospheric analysis need a simple and easy alignment design. To meet this requirement, we use a tilt-compensated interferometers design with four mirrors as shown in figure 2.

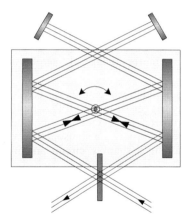

Fig. 2. Tilt-compensated interferometer with four mirrors

In the interferometer, two mirrors are mounted on a common base plate which is rotated to give rise to the path difference in two arms of the interferometer. Any misalignment of the tilt table will cause each beam to be affected in the same way, so the effect of the tilt of the mirrors on the tilt table is compensated. With these characteristics, it is very stable and anti-interference.

4 Driving and Control System

The moving mirror driving and control system diagram is shown in figure 3. It consists of He-Ne laser detection and process circuit, digital signal process circuit, microcontroller system and motor driving circuit.

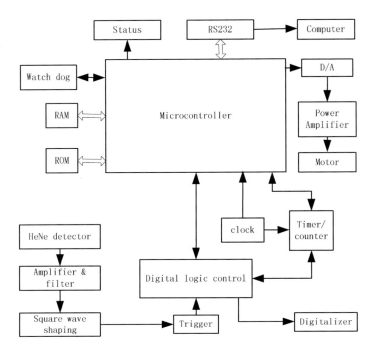

Fig. 3. Moving mirror driving and control system diagram

In the system, He-Ne laser which wavenumber is 632.8nm is used to achieve detection of scan velocity and as trigger of equal OPD sampling. The laser interferogram after the Michelson interferometer is detected by a photodiode detector. The signal is then be amplified and filtered to produce a sinusoidal wave. It is then reshaped to become square wave after passing a shaping circuit and inputs to the microcontroller system.

In the digital signal process circuit, the He-Ne square signal is used to produce zero cross pulse signal which has twice the frequency over the original signal. This signal has three usages. First, the digital logical control unit put it into the digitizer's trigger source. Second, the pulse signal inputs to a timer/counter (C/T) which uses the system clock as reference source. The C/T counts the clock number during one signal cycle time. The microcontroller reads this number and used gets current velocity after calculated. Third, the microcontroller counts the pulse signal and used it to calculate the scan distance.

The control algorithm runs on the microcontroller that uses the calculated velocity as reference input. It produces a control value. After a D/A converter, this value becomes analogous signal which then is amplified and used to drive the motor. This will complete a close loop moving control.

The microcontroller has a 64KB ROM and RAM with system extension. Its status is monitored by an auxiliary circuit whose has the function of watchdog. It also accepts the control parameters from the computer through a RS232 interface. So the computer can control the maximum OPD and scan velocity of moving mirror.

5 Control Method

Mirror driving with high precision is significant for the system, and the control method must be carefully designed. In the control theory, classical PID controller as a linear controller can form the control value by the difference between the reference value and output value with composing the proportion (P), integration (I), and differentiation (D) in a linear way. But in the moving mirror driving system, there exist some non-linear characteristics because of the precision of components and some uncertain factors such as change of load and outside interference [6]. Thus, the simple PID will not suitable for this application.

Fuzzy control as an intelligent control method is suitable for the control model that has non-linear and uncertain characteristics. By combining the fuzzy and PID control method, we get the Fuzzy-PI controller which can improve dynamic response and control precision. But because the control parameters are fixed, they don't have the ability of adaptive adjustment. To do this, an adaptive adjustment Fuzzy-PI controller is realized base on the Fuzzy-PI controller. The control diagram is shown in figure 4.

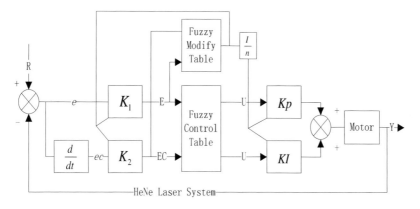

Fig. 4. Adaptive adjustment Fuzzy-PI control diagram

In the controller, Kp, KI, K_1, K_2 can be modified based on the change of E and EC, so they has the adaptive ability and can change with the different e and ec. Let the language variable N of n has the fuzzy subset N= {AB, AM, AS, OK, CS, CM, CB}. The universe of N is N= (1/8, 1/4, 1/2, 1, 2, 4, 8). Design the assigned table of degree of membership of N. The assigned table of degree of membership and fuzzy subsets of E and EC are the same as the fuzzy controller. The method of parameter adjustment is expressed by the modify rule. The fuzzy modify table of parameters adaptive adjustment can be calculated by the compound algorithm as simple fuzzy controller and manual amendment.

6 Experiment and Application

The response curve of control system is shown in figure 5. It shows that the adaptive adjustment Fuzzy-PI controller has a faster response speed, and stable response curve

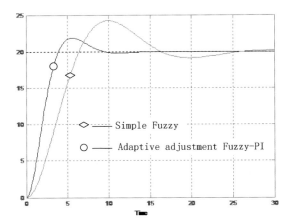

Fig. 5. Response of two kinds of control methods

compare to simple fuzzy controller. The performance of control precision and anti-interference is also butter.

A FT-IR spectrometer system is developed based on the mirror control system whose spectrum resolution is 1cm-1. The frequency of He-Ne laser interferogram is 6 KHz. By the method of double side scan, 16384 data point of infrared interferogram is sampled and shown in figure 6.

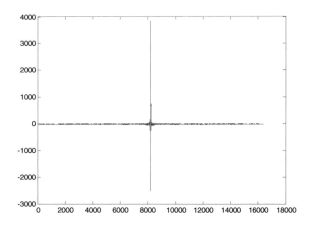

Fig. 6. Interferogram of developed FT-IR spectrometer

Atmospheric monitoring systems include open-path and solar occultation flux methods were also developed with the FT-IR spectrometer above. These systems were used in the World Exposition in ShangHai and Asian Games in Guangzhou in 2010. The results show that the systems are reliable and stable and are suitable for these kinds of applications.

7 Conclusion

With the need of FT-IR spectrum for the application of atmospheric analysis, a moving mirror driving system was developed. It contains a tilt-compensated interferometer, a He-Ne laser reference interference system, and a driving system based on microcontroller. The paper describes the designs of interferometer structure, driving and control system and control method. In the end, some experiments and applications are given. The results show that this system is simple and easy alignment. The controller is robust and high precision. It is suitable for the applications in the field environment.

Acknowledgments. The system is supported by National Natural Science Foundation of China(No.40905011).

References

1. Gao, M., Liu, W., Zhang, T., et al.: Passive Remote Sensing of VOC in Atmosphere by FTIR Spectrometry. Spectroscopy and Spectral Analysis 25(7), 1042–1044 (2005)
2. Gao, M., Liu, W., Zhang, T., et al.: Remote Sensing of Atmospheric Trace Gas by Airborne Passive FTIR. Spectroscopy and Spectral Analysis 26(12), 2203–2205 (2006)
3. Xu, L., Liu, J., Gao, M., et al.: Application of Long Open Path FTIR System in Ambient Air Monitoring. Spectroscopy and Spectral Analysis 27(03), 448–451 (2007)
4. Griffiths, P.R., de Haseth, J.A.: Fourier Transform Infrared Spectrometry, pp. 21–26. John Wiley & Sons, New York (1986)
5. Rippel, H., Jaacks, R.: Performance data of the double pendulum interferometer. In: Proceedings of the 6th International Conference on Fourier Transform Spectroscopy, vol. II (1-6), pp. 303–306 (1988)
6. Liu, R., Yin, D.: Driving Control Technology for Moving Mirror of Fourier Transform Spectrometer. Infrared 30(9), 20–25 (2009)

Research on Intelligent Cold-Bend Forming Theory and Fitting Method of Actual Deformation Data[*]

Guochang Li

School of Economics and Management Hebei University of Science & Technology
Shijiazhuang, Hebei Province, China
lgckdjg@163.com

Abstract. Cold-formed steel caliber design of complex profile is a difficult problem of plastic processing technology, existing cold-bend forming theory couldn't very well to guide the algorithm of caliber design, Thus it lead to the problem that existing caliber design algorithm cannot precisely design caliber. How to use the method of artificial intelligence, let forming theory rationally guide practical groove design, eventually get more precise caliber design method, this is a relatively new strategy to solve the current caliber design which is not high accuracy, make caliber design of complex profile more reasonable and precision ,reduce cycle of roll manufacturing and the modification of roll.

Keywords: Complex profile, Precise caliber design, Forming theory, Artificial intelligence, Cold-formed steel.

1 Preface

Cold-formed steel is common use in the national construction, it widely used in automobile manufacturing, aerospace equipment manufacturing, decoration, and agriculture, etc. As people living standard rise, they have more and more demand to the function of cold-formed steel, this makes the cross-section geometry of cold-formed steel becoming more and more complicated, rolling the caliber design of Cold-formed steel increasingly difficult. How to accurately design complex profile of caliber design became the worldwide problem cold-bend forming industrial. Now the research of caliber design basically divided into two trends: One is the mathematical modeling and theoretical analysis that use of the existing physical and mathematical methods for cold-bend forming, hope to use this method to find the rule of accurate caliber design. Another method is deduced actual caliber design method according to the actual deformation data. the last method can only derived the caliber design method of simple sections. But have no specific general regularity for complex profile of caliber design. How to find the precise caliber design method using the existing deformation theory and actual deformation data, especially the caliber design method of complex profile cold-formed steel, is the common long-expected desire of plastic deformation

---

[*] This research is supported by National Natural Science Foundation (No: 50375135); International Science and Technology Cooperation (No: 2006DFA72470); Natural Science Foundation of Hebei Province (No: E2011208014), (F2007000643).

380   G. Li

theorists and production enterprise. Based on the study of spline finite strip deformation theory, Using artificial intelligence method to fit the actual deformation parameters, seek the accurate method of actual caliber design.

2 Deformation Theory Analysis and Research

There are a variety of cold-bend forming theories, spline finite strip theory is a relatively practical method. Spline finite strip method is a kind of stress-strain analysis method of cold-bend forming based on the finite element method, considering the characteristics of cold-formed steel which have no longitudinal deformation, that cancel the analysis of longitudinal dimensions. In order to describe the limbic tensile deformation and simplified equation of steel strip in forming process, which could make the change from original series description to spline function description, that constitute the basic displacement function is:

$$\begin{Bmatrix} \Delta u \\ \Delta v \\ \Delta w \end{Bmatrix}_s = [N]_s [\varphi]_s \{\Delta \delta\}_s \tag{2-1}$$

In the formula:

$$[N]_s = \begin{bmatrix} N_{s1} & 0 & 0 & 0 & N_{s2} & 0 & 0 & 0 \\ 0 & N_{s1} & 0 & 0 & 0 & N_{s2} & 0 & 0 \\ 0 & 0 & N_{s3} & N_{s4} & 0 & 0 & N_{s5} & N_{s6} \end{bmatrix} \tag{2-2}$$

$$\{\Delta \delta\}_s^T = \{\Delta u_i \quad \Delta v_i \quad \Delta w_i \quad \Delta \theta_i \quad \Delta u_j \quad \Delta v_j \quad \Delta w_j \quad \Delta \theta_j\} \tag{2-3}$$

Thereinto, $N_{s1} = 1 - \xi_s$, $N_{s2} = \xi_s$, $N_{s3} = 1 - 3\xi_s^2 + 2\xi_s^3$, $N_{s4} = b_s \xi_s \left(1 - 2\xi_s + \xi_s^2\right)$, $N_{s5} = 3\xi_s^2 - 2\xi_s^3$, $N_{s6} = b_s \xi_s \left(\xi_s^2 - \xi_s\right)$, $\xi_s = \dfrac{y}{b_s}$, that is said b_s show the s th finite strip width

Using the above formula can obtain the concrete expression of commonly used cubic spline function is:

$$\varphi_3(x) = \frac{1}{6} \begin{cases} (x+2)^3, & x \in (-2,-1) \\ (x+2)^3 - 4(x+1)^3, & x \in (-1,0) \\ (2-x)^3 - 4(1-x)^3, & x \in (0,1) \\ (2-x)^3, & x \in (1,2) \\ 0, & |x| > 2 \end{cases} \tag{2-4}$$

3 The Analysis of Actual Caliber Design Method

The actual caliber design is based on forming method, according to the geometry dimension of cold-formed steel special-section, make sure reasonable deformation lane and the deformation task should be completed of every lane, final based on simple calculation, finish caliber design and mechanical parameters of ralls every lane. Now we explain the caliber design method with the centre sill caliber design of cold-bend as an example.

Van with cold-formed steel is atmospheric corrosion resisting steel plate. Its related performance shall be in conformity with the B/T1979-2003 and transporting the loading van <2003> 387 file regulation. The specifications of product is 600mm×312mm×12mm, that uses continuous rolls of cold-bend forming, the sectional drawing of product see Fig. 1.

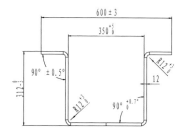

Fig. 1. Sectional drawing of cold-formed steel

3.1 The Springback Problem of Van with Cold-Formed Steel Molding

We can found that the springback amount of the van with cold-formed steel material should to exceed 2 °than ordinary in 90 °range through the compares in shape springback amount of two materials above. So in design of the rolls caliber we must consider the influence of springback amount.

3.2 Van with Cold-Formed Steel Caliber Forming Method and Computer Aided Caliber Design

3.2.1 Determine the Forming Lane

The process of van with cold-formed steel molding, every lane or every pair have limited bending deformation only for deformation materials. It can increase costs with too much molding lane, but the cold-formed steel would easy reach generation size outoftolerance and distortion which have too little molding lane. According to that the performance of van with cold-formed steel material and complexity of the steel shapes and to the experience for determining molding lane of van with cold-formed steel are: ten horizontal rolls, two vertical rolls, two universal mill feed rolls, two straightening rolls.

3.2.2 Choices of Deformation Amount

The correct choice of deformation amount is the key of molding which van with cold-formed steel, in order to prevent rebound, and avoid the strip local anomalies deformation and the scratch on material surface due to unevenly distribute of bend angle, ensure accuracy of product size, the example adopts equally bend Angle method in early forming and medium-term forming. And it adopts lesser deformation amount method in molding late to prevent the rebound, the distribution of every bend angle as table 1.

Table 1. Distribution of bend angle

sorties	1	2	3	4	5	6	7	8	9	10	11	12
β°	0	15	30	30	30	30	45	60	75	85	90	92
α°	0	15	30	50	70	92	90	90	90	90	90	90

3.2.3 Create Mathematical Model

The calculation of Van with cold-formed steel width and middle deformation lane all by the computer caliber design software, below it's the introduction to calculation method.

3.2.4 Calculation of Van with Cold-Formed Steel Width

The calculation chart of van with cold-formed steel width as Fig.2, its appearance dimensions are shown in Fig.1. when calculation would on the basis of neutral layer, select A, B, C, or D for position changeless points on steels strip, including B and C for arc midpoint, R_{mid} size is changeless, α, β for deformation angle.

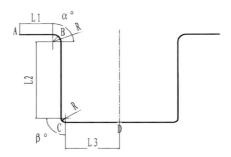

Fig. 2. Calculating chart of steel strip width

(1) Calculating the arc size of steel strip R_{mid}, R_W:

$$R_{mid} = R_n + K \times T = 13 + 0.41 \times 12 = 17.92 \quad (3\text{-}1)$$

$$R_W = R_n + T = 13 + 12 = 25 \quad (3\text{-}2)$$

(2) Calculating each line size of steel strip L1,L2,L3

L1=(600-350)/2-25=100   (3-3)
L2=312-25×2=262   (3-4)
L3=(350-13×2)/2=162   (3-5)

(3) Calculating each size of steel finished product AB, BC, CD:

AB=L1+ R_{mid} ×α/2=100+17.92×π/4=114.07   (3-6)

BC=L2+ R_{mid} ×α/2+ R_{mid} ×β/2 =262+17.92×π/4+17.92×π/4=290.15   (3-7)

CD=L3+ CD=L3+ CD=L3+ R_{mid} ×β/2=162+17.92×π/4=176.07   (3-8)

(4) Calculating steel width W
W=2(AB+BC+CD)=2(114.07+290.15+176.07)=1160.58   (3-9)

it is: take W=1160 R_{mid} — radius of arc neutral layer, mm; R_n — radius within arc, take R_n=13, mm;

R_W — radius outer arc, mm; L1,L2,L3 — each line size of Van with cold-formed steel, mm;

T — steel strip thickness, mm; K — coefficient of neutral layer, In this example take K=0.41.[1];
W — Strip width, mm.

3.2.5 Calculation of among the Deformation Lane (see Fig.3)

Computation formula is as follows:

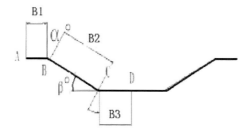

Fig. 3. Calculation of among the deformation lane

$$B1=AB- R_{mid} ×α/2 \quad (3\text{-}10)$$

$$B2=BC- R_{mid} ×α/2-R×β/2 \quad (3\text{-}11)$$

$$B3=CD- R_{mid} ×β/2 \quad (3\text{-}12)$$

Using computer caliber design software design caliber and optimization, the calculation results data intermediate lane see Table 2.

Table 2. Data sheet

deformation lane	2	3	4	5	6	7	8	9	10	11	12
B1	111.72	109.38	106.25	103.12	99.68	100	100	100	100	100	100
B2	285.46	280.77	277.64	274.51	271.1	269	266.7	264.3	262.8	262	262
B3	173.72	171.38	171.38	171.38	171.38	169	166.7	164.3	162.8	162	162

The product specs is 600 mm×312 mm×12 mm, materials is Q450, elastic modulus 210000, uses continuous roller for roll bending forming, forming units are nine horizontal rolls, two vertical rolls, three universal mill feed rolls, two straightening rolls. Flat roll is active roller, vertical roll is passive roller, and will not participate in deformation. we only consider nine horizontal rolls and three universal mill feed roll in calculation. Through rolling of two component type parameters, the first group is proved the parameters that can effective produce qualified products. The difference between two groups of parameters is the different of bend Angle at β corner in the second frame, the first group parameters second frame bend Angle is 15 degrees, the second group parameters second frame bend Angle is 30 degrees.

4 Intelligent Theory and Practical Critical Path Method

Function modules of microscopic more Agent of the conceptual model (as Fig.4 shows), the conceptual model of more Agent technology with the spline finite strip method need each function module has the following functions:

(1)Reasoning mechanism realize the independent agency functions of Agent;
(2) Knowledge repository completing independent agency functions together with reasoning mechanism;
(3) Displacement function mathematical analysis capabilities of displacement polynomials and spline function, etc;
(4) Strip element line parameters social bond of dynamic deformation between strip element;
(5) Self-interest mechanism operation for realizing the independent and agent of monomer Agent;
(6) Function set operation function sets.

Agent$_0$		...		Agent$_k$			Agent$_n$
reasoning mechanism		...		reasoning mechanism			reasoning mechanism
knowledge repository		...		knowledge repository			knowledge repository
displacement function		...		displacement function			displacement function
strip element line parameters		...		strip element line parameters			strip element line parameters
function set		...		function set			function set

⬅——— steel strip width direction ———➡

finite spline 0		...		finite spline k		...		finite spline n
		...				...		
		...				...		

Fig. 4. The conceptual model of many agent and spline finite strip
The "Y, T, Q" in Fig.4 expresses 'spline area'.

Due to the deformation of cold-bend centre sill can only limit to A,B , C and symmetric point, so only set the Agent in this part, we would send deformation lane and deformation Angle in Fig.4 into the Agent, let corresponding Agent reasoning and determine caliber design data.

5 Conclusion

Caliber design is the difficulty of cold-formed steel equipment manufacture, due to the factors influence cold-bend forming are many, in the design, so we must balance the weight when design caliber, especially for the cold-formed steel with complex cross-section, that along with the increase of cross-section complexity, the difficulty of design also increased. Some caliber be to finished design must by trying many times. If we can use cold-bend forming theory to guide caliber design algorithm, the caliber designed will be optimization. But now the forming theory and practical caliber design algorithm have larger difference, cannot be used in the actual caliber design, if we can find the method that use cold-bend forming theory guide practical caliber design, which be able to solve the caliber design problem complex profile.

References

1. Li, G., Yang, X.: Research on pass design expert system for hypertext knowledge base. Computer Engineering and Application (21), 246–250 (2002)
2. Yang, X., Zhang, W., Li, Y.: The computation and computer-aided design of cold-bend centre sill roll caliber. Mechanical Design and Manufacturing (August 2009)
3. Wooldridge, M.: An Introduction to Multi-Agent System, pp. 1–13. John Wiley & Sons, Chichester (2003)
4. Wang, X.: Production and application of cold-formed steel, vol. 26(3). Metallurgical Industry Press, Beijing (1994)

Research on Cold-Belt Deformable Characteristic Eigenfinite Strip Filter and Characteristic Agent Model

Guochang Li

School of Economics and Management Hebei University of Science & Technology
Shijiazhuang, Hebei Province, China
`lgckdjg@163.com`

Abstract. cold-bent section is one kind economic section molding of low energy consumption and high benefit, it is widely applied in the construction, the astronautics, the light industry and so on many fields. Deformation mechanism of the cold- bent section is extremely complex, relations of deformation strength and energy and steel belt deformation can not described by the simple function. Thus leads to complexity of the roll pass design. Although eigenfinite strip theory has removed the factor of longitudinal deformable dimension, it has abbreviated the massive redundant data proceeding than the finite element method. But it still had the massive redundant data in the deformation process analysis. Existing styles of these redundant data are difference, styles of manifestation are also different, moreover changing along with change of time and logic. These redundant data have directly affected speed and precision of information processing. Seeking the primary playing factor in the steel belt deforming process is solving redundant data'good method, because thus may place study emphasis on the deformable primary factor, few consider other factors. Studying shape characteristic of thing information may remove the redundant data. Establishing multi-Agent system corresponding with the steel belt characteristic eigenfinite strip sets, Introducing characteristic multi- Agent model strategy can intelligentize the condensing data, thus can realize the roll pass design intelligentization.

Keywords: Characteristic Agent, Characteristic finite strip, Filtration, Pass design, Roller band shaping.

1 Introduction

Cold curved steel is an application extensive economic material, for adapting the usage request of various customer, cold curved model steel have different geometrical section shape. One of the important methods that completes the figuration of cold curved steel is roller band shaping, namely make the roller band extrusion figuration by several roller systems .The shape of cold curved steel decide by pass composed of several roller systems. Therefore, the quality of cold curved steel is decided by pass. The pass design is a complicated and difficult things, this is because there is having no very mature theories system assurance the quality optimization of pass design so far. The relation between sheet materials subjected to and sheet materials deformation is relatively complicated, the region of elasticity and plasticity transition lack definition. The finite numeral expression can't reflect the mechanism of sheet materials really etc., various factor restrict the development of pass design theorization. Still, the act of people for looking for the pass design theories hasn't been stopping all along. Adopting big elasticity and plasticity deformation finite strip method is an effective method. Finite strip method takes the rule to displacement function of the strip materials choice as follows: the selective displacement function, must make the contingency which is needed in energy equation can keep limited on the interface of strip, namely to Take the steel band to be divided into some finite strip, the displacement function of strip is selected by the contingency which is needed in energy equation can keep limited on the interface of strip, so to called finite strip method. Make use of finite strip method research the application of steel band, need to take in to discuss to all contingency of finite strip, this makes the processing data is big so that microsoft computer in common use can not competent for this work. In numerous finite strip, some have a characteristic. These characteristics have important influence function to the deformation of finite strip and steel band. Study the characteristic finite strip and its contingency control to non-characteristic ones', solved the whole deforming mechanism of steel band basically. Adopt what principle to divide the characteristic finite strip, is the most important problem in post the deforming mechanism of steel band by use of limited of characteristic finite strip. Furthermore, characteristic finite strip how to constitute the deforming mechanism of steel band is also an important problem that wants to discuss.

2 Filtration Principle of Characteristic Finite Strip

2.1 Finite Strip Contact to Roller Is Defined as Characteristic Finite Strip

In the process of rolling cold curved steel, to the whole gather, some finite strip contact to the extrusion surface of roller, and some not. The first ones accept roller's extrusion power directly, they not only producing contingency themselves, but also still controlling not contact finite strip carry on contingency under the complicated stress. So, that contact with roller have characteristic. Studying them would bring the simplification calculate way for study whole deformation of finite strip.

2.2 Finite Strip with Displacement Change Exquisite Is Defined as Characteristic Finite Strip

In the shaping process of cold curved steel, some finite strip's displacement change exquisite, but some ones' almost don't change. This has great relation with the shaping section of cold curved steel. The shaping of cold curved steel mainly concentrate in finite strip changing exquisite. Defined the finite strip with exquisite change as characteristic finite strip, the study of distortion to characteristic finite strip is the study of the whole shaping of cold curved steel. So makes the research contract to a pimping scope inside, simplify the research method.

2.3 Finite Strip with Big Distribution Density can be Defined as Characteristic Finite Strip

In the deforming process of cold curved steel, the distribution of finite strip maybe asymmetric, in the part of big displacement change and energy centralized. For the sake of the accurate reflection contingency circumstance, the finite strip's distribution density increased. So, finite strip of big distribution density have the characteristic of whole distortion process of cold curved steel.

Of course, we can define finite strip with other characteristics as characteristic finite strip. This decided by frondose application situation.

3 Mathematics Model Analysis of Characteristic Finite Strip

3.1 Characteristic Finite Strip Contact with Roller

Characteristic finite strip contacts with roller accept the roller's extrusion power directly. They not only have distortion of elasticity and plasticity, but also pass the roller's extrusion power to those finite strips not contact with roller, makes these finite strip distort after accepting indirect power, and distortion accord to the request. So, this characteristic finite strip has deforming control function. Such as Fig.1 shows, the stress of non- characteristic finite strip is not from the roller, but from the characteristic finite strip.

Suppose that the extrusion power to steel band by left roller is $\vec{F}_1$, the right is $\vec{F}_2$, the displacement equation of characteristic finite strip is:

$$\begin{Bmatrix} \Delta u \\ \Delta v \\ \Delta w \end{Bmatrix} = \sum_{m=1}^{r} [N]_m [\varphi]_m \{\Delta \delta\}_m \qquad (1)$$

Suppose the power of characteristic finite strip creating displacement by left and right roller is $\vec{F}_{11}$ and $\vec{F}_{21}$, the extrusive force pressed by non-characteristic finite strip is $\vec{F}_{12}$ and $\vec{F}_{22}$, then have:

$$F_1 = F_{11} + F_{12} \qquad F_2 = F_{21} + F_{22} \qquad (2)$$

The finite strip pressed by different power in the region of characteristic finite strip. The above part subject to F_{11} bigger, the below is F_{21} bigger, in the center part is subjected to the function of the common power. So, on the above of area of non-characteristic finite strip, the change grads of F_{12} is $\dfrac{\partial F_{12}}{\partial x}$, the change grads is $\dfrac{\partial F_{22}}{\partial x}$ on below.

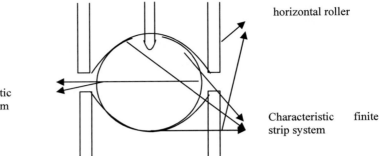

Fig. 1. Stress appearance figure of characteristic finite strip and non-characteristic finite strip

3.2 Characteristic Finite Strip with Displacement Change Exquisite

Displacement with exquisite change, show that the coefficient of every coordinate orientation is big. Suppose characteristic finite strip's X, Y, Z orientation displacement is U, V, W, then have:

$$\frac{\partial U}{\partial x} \geq C_1, \quad \frac{\partial V}{\partial y} \geq C_2, \quad \frac{\partial W}{\partial z} \geq C_3 \qquad (3)$$

In the formula, C_1, C_2, C_3 is bigger numerical value.

Then the deforming stress is basically acting to characteristic finite strip, the non-characteristic finite strip stress little because the displacement change is small. Suppose the power added to steel band is F_1 and F_2 (F_1 is contingency power added by the above roller, F_2 is the nether) ,then have:

$$F_1 = F_2 \qquad (4)$$

Such as Fig.2 shows, only point A, B, C change exquisite, the rest parts is basically not change.

3.3 Finite Strip with Big Distribution Density

The reason that characteristic finite strip has big distribution density at somewhere is because use less finite strip at this point can't indicate the deforming mechanism. Define the finite strip here as characteristic finite strip is advantage of the change analysis of cold curved model. The numeral description of big distribution density finite strip as follow:

$$N/l_i > C \qquad (5)$$

In the formula, N is the number of finite strip, l_i is the i segment of bandwidth orientation, C is the big finite numerical value.

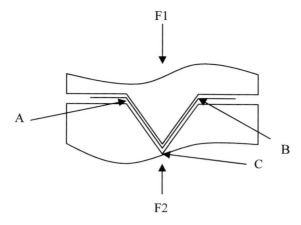

Fig. 2. The parallelism situation between displacement change exquisite and not exquisite

In figure 2, because point A, B, C change exquisite, increase the density of finite strip in this place can reflect deforming mechanism exactly, other place can decline the density of finite strip.

4 Structure of Characteristic Agent

On research the deformation of finite strip making use of Agent technology, each finite strip replies a Agent. The research for finite strip changes in pairs the research of Agent. Each characteristic finite strip is no exception, it replies a Agent, this Agent called characteristic Agent. Based on the theory of characteristic finite strip analysed above, the characteristic Agent its three kinds characteristic finite strip replies as follows:

4.1 Characteristic Agent Contacted with Roller

Adding a control deforming function on the original Agent function, the model is as follows:

```
class Agent
{
    public: function fair
            // operating function function fair
            Agent( int max1,int max2,int max3,float ε)
            //characteristic finite strip judge initial value;
    strip element knot link function
    //social relation of dynamic deformation between //finite strip;
    spline interpolate function
    //spline deforming function replies this Agent;
    consequence mechanism
    //implementing the intelligence of characteristic //finite strip theory combine with knowledge //container;
```

the judgement of contacting model characteristic finite strip and transaction function
　　　//making the characteristic finite strip intelligence;
　private: knowledge containerfunction
　　　//storing the knowledge of characteristic finite //strip;
　　　displacement multinomial
　　　//implementing the analysis function of //displacement multinomial and increment //function;
　　　independence mechanism
　　　//implementing this Agent's operation of //autocephaly and deputize;
　　　int max1,max2,max3;
　　　float ε;
}

4.2 Characteristic Finite Strip with Exquisite Change

Adding a exquisite deforming function on the original Agent, the model is as follows:

class Agent
{
　public: function fair
　　　// operating function function fair
　　　Agent (int max1,int max2,int max3,float ε)
　　　//characteristic finite strip judge initial value;
　　　strip element knot link function
　　　//social relation of dynamic deformation //between finite strip strip;
　　　spline interpolate function
　　　//spline deforming function replies this //Agent;
　　　consequence mechanism
　　　//implementing the intelligence of //characteristic finite strip theory combine //with knowledge container;
　　　the judgement of deforming exquisite model characteristic finite strip and transaction function
　　　//making the characteristic finite strip //intelligence;
　private: knowledge container function
　　　//storing the knowledge of characteristic //finite strip;
　　　displacement multinomial
　　　//implementing the analysis function of //displacement multinomial and increment //function;
　　　independence mechanism
　　　//implementing this Agent's operation of //autocephaly and deputize;
　　int max1,max2,max3;
　　float ε;
}

4.3 Finite Strip with Big Distribution Density

Adding a big distribution density transaction function on the original Agent, the model is as follows:

class Agent
{
    public: function fair
        //operating function function fair
        Agent (int max1,int max2,int max3,float ε)
        //characteristic finite strip judge initial value;
        strip element knot link function
        //social relation of dynamic deformation //between finite strip;
        spline interpolate function
        //spline deforming function replies this //Agent;
        consequence mechanism
        //implementing the intelligence of //characteristic finite strip theory combine //with knowledge container;
        the judgement of big distribution density model characteristic finite strip and transaction function
        //making the characteristic finite strip //intelligence;
    private: knowledge container function
        //storing the knowledge of characteristic //finite strip;
        displacement multinomial
        //implementing the analysis function of //displacement multinomial and increment //function;
        independence mechanism
        //implementing this Agent's operation of autocephaly and deputize;
        int max1,max2,max3;
        float ε;
}

5 Conclusion

Though the memory of the modern computer is very big, operating is speed very high, if there are a lot redundancy data of the operation, will influence the calculation speed and quality of computer. The idea of characteristic finite strip is to exchange the research on each finite strip's displacement deformation on whole steel band to the research on the deformation of representational finite strip only. Under the situation that guarantee the research quality, simplified the emphases research and reduced the calculation quantity. But the definition of characteristic finite strip isn't unchangeable, but settle according to circumstance. The same theory, characteristic Agent is the intelligentize process unit replies to characteristic finite strip, it deputizing the character of characteristic finite strip and adding to intellective process mechanism, made the strip material deforming theory has intelligence in guiding pass design. Characteristic Agent changes with the definition of characteristic finite strip. When structuring characteristic

Agent, the mathematics model and process method of characteristic finite strip is very important.

References

1. Li, G.-c., Liu, C., Jia, G.-m.: The program model reseach of multiagent system's intelligentization of spline finite strip method. Journal of Hebei University of Science & Technology 3, 43–46 (2005)
2. Wooldridge, M.: An Introduction to Multi-Agent System, pp. 1–13. John Wiley & Sons, Chichester (2003)
3. Li, G.-C., Yang, X.: Research on pass design expert system for hypertext knowledge base. Computer Engineering and Application (21), 246–250 (2002)

TPMS Design and Application Based on MLX91801

Song Depeng[1,2], Wan Xiaofeng[1], Yang Yang[1,2,3], Gan Xueren[2],
Lin Xiaoe[2], and Shen Qiang[2]

[1] Institute of Information Engineering, Nanchang University, Nanchang 330031
[2] Jiangxi Kysonic Inc, Nanchang 330096, China
[3] Institute of Microelectronics, Tsinghua University, Beijing 100084, China
sdp-qdcy@163.com

Abstract. The MLX91801 is a System in a Package (SIP) pressure sensor launched by MELEXIS company. It combines the silicon pressure sensor, the temperature sensor and voltage sensor with a low consumption 16-bit microprocessor and the system can be made compliant with existing Remote Keyless Entry (RKE) systems. TH series high-speed single wireless transceiver chips provide a good solution to the short-distance wireless communication. They are mainly used in the application of TPMS system. This paper mainly introduces the hardware of the tire pressure sensor module based on the MELEXIS 91801 chip and wireless transceiver chip TH72015 and discusses their specific application in TPMS system.

Keywords: MLX91801, TH72015, sensor, wireless communication, TPMS.

1 Introduction

Automobile electronic is closely related to people's everyday life, especially in Europe, where cars are served as necessities. Cars are almost the only alternative for most people's daily travelling. With the rising emerging markets in China and India, car market is in rapid growth as well. In recent years, the related components in automobile electronic system and its surrounding updated production are in a rapid growth along with the increasing demand of safety, energy saving, environmental protection, comfortable and entertainment. Many design challenges in the automobile electronic field need us to take components into consideration. Safety is the most important factor to be taken into consideration and it will become nonsense without stable engine management system. Sensors, which are widely used device, is also applied in automobile security system.

Meanwhile, the application of wireless communication has been involved in almost all fields, including small wireless network, wireless metering, entrance guard system, district pager, industrial data acquisition system, wireless remote control system, wireless tags identification, non-contact RF CARDS, tires pressure monitoring system, etc, brought great convenience to people's daily life. People's demand for wireless communication application has further promoted the rapid development of wireless communication technology. At present, many big chip manufacturing companies have launched a series of powerful, excellent performance of wireless transceiver products.

This paper mainly introduces Melexis company's latest sensor chip ---- Melexis 91801 and the high-performance single chip wireless transceiver chip TH72015, discusses the hardware design of the tire pressure sensor module which is mainly consisted of these two components. The specific applications with the practical examples in TPMS system are discussed as well.

2 Tire Pressure Monitoring System

Tire pressure monitoring system (TPMS) is composed of pressure monitoring (temperature, battery voltage monitoring) and the signal processing chip of intelligent sensor, MCU, RF launch chips, wireless receiver, display composition. The system integrated the pressure sensor with signal generator through advanced pressure sensor technology .By measuring tire pressure directly using the pressure sensor installed in tire device .when a tire is flat, pressure sensor will send signal to the roof of the cab through wireless transmission on the dash and display the flat tire. This system can provide real-time monitoring for each tire's pressure through automobile driving process, drivers can intuitively judge whether each tire pressure is normal. When a tire is flat, it can be automatic alarm in a form of sound or light, effectively preventing accidents and achieve safety.

The principle of the system is showed as follows :

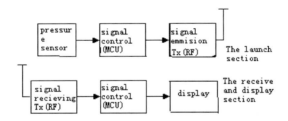

Fig. 1. Theory structure of tire pressure monitoring system

3 Hardware Design of Transmitters

The requirements for transmitters TPMS system is extremely high, especially for internal installation in the tire. Because the internal tire working environment is very terrible, the highest temperature can be reached more than 100 °C and the heart acceleration can reach to 2000g while the vibration can reach to 5g. This puts forward higher request for device performance. In order to meet the practical application requirements, we choose auto/military level chips whose working temperature is range from - 40 °C to + 125 °C.

In hardware circuit design, we should still consider reducing power consumption, facilitate to MCU control and management system. This thesis mainly discusses the most important parts of the TPMS transmitters - sensor chip and RF launch chips.

3.1 Sensor

Sensor module bears the functions of collecting tire pressure data. Due to the follow-up data transmission and processing are based on the acquisition of the data sensor module, sensor module will directly influence the performance of the whole system's stability, accuracy and availability.

We choose MLX91801 from MELEXIS company.MLX91801 is an integration of a pressure sensor,a temperature sensor and a battery voltage monitor, its accuracy can reach to 1%. It can set different pressure range (topped 1400kPa), the default value is 100KPa - 800KPa. At 25 ° C, the sleep current is less than 0.5 uA and the external PZT testing vibration can reach to 4000G. TPMS won't be work when vehicle is standstill , which can be helpful to extend the life of batteries.

The working temperature of MLX91801ranges from- 40 ℃ to + 125 ℃, satisfying the requirements of TPMS. It also has a standby current 26uA (+ 125 ℃), which is suitable for low power applications.

MLX16 is the core of MLX91801 devices, with seven programmable IO pins and can fully meet the demand of the tire pressure tests. It includes the following peripherals: software control sensor interface, the communication interface and power management module. The pressure sensor unit is a pressure resistance bridge type based on MEMS. The following diagram displays the internal functions of MLX91801 chip.

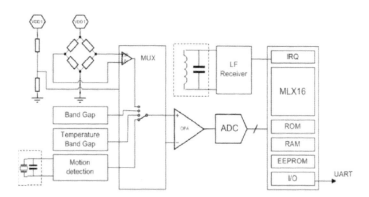

Fig. 2. The internal functions of 91801 chip MLX diagram

3.2 RF Transceiver

3.2.1 Function Description of TH72015

TH72015 is the monolithic programmable wireless transceiver chip launched by Melexis company. It can be applied in low power consumption multi-channel or single channel half-duplex data transmission system. It can work in ISM band (380-450MHz). In programmable user's mode, through the use of an external transfiguration diode VCO, its lowest working frequency can reach to 27MHz. The working temperature ranges from - 40 ℃ to + 125 ℃, satisfying the requirements of TPMS. It also has a standby current 200nA (only for + 85℃), which is very suitable for low power applications.

3.2.2 Internal Structure of TH72015

The internal structure of TH72015 includes crystal oscillators (XOSC), crystal oscillator XBUF), buffer (Low Voltage Detector (Low by Detector), Mode Control (Mode Control), and phase lock loop (PLL) frequency synthesizer, transmit power amplifier (PA), wireless Matching Network (Antenna Matching a) circuits, etc. PLL is composed by a devide-by-32 divider (% 32), PFD, VCO, CP. Its main module is a programmable PLL synthesizer, which can be used to produce the carrier frequency by FSK/ASK modulation mode. It can produce the local oscillating signal at the receiving mode, adopting the specialized superheterodyne receiving mode. Its internal structure is shown in Fig.3(the left one).

3.2.3 Output Power Selection of TH72015

TH72015 provides single FSK modulation, launching 433.92 MHz carrier frequency. At FSKDTA=0, the FSK switch is shut-off, frequency is calculated by the following forum: $f_{min} = f_c - \Delta f$; at FSKDTA=1, FSK switch is on, frequency is calculated by the following forum: $f_{max} = f_c + \Delta f$. Doppler frequency shift can be adjusted by CX1, CX2, the scope of the frequency wide capacitance can be range from + 2.5 KHz to + 40KHz. The implementation of the output power control logic is shown in fig.4(the right one).

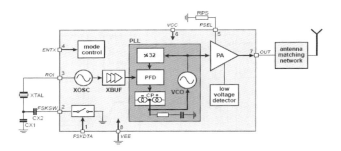

Fig. 3. Block diagram with external components of TH72015 chip

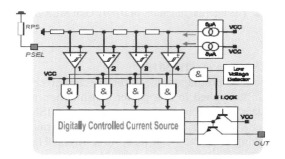

Fig. 4. Block diagram of output power control circuitry of TH72015

As Fig.4 shows, there are four kinds of optional output power, ranging from -12dBm 10dBm, therefore, the output working current is also be divided into four segments, ranging from 1.5 mA to 13.3% mA. There are two ways to select the desired output power step.Firstly, by applying a DC voltage at the pin PSEL, then the voltage directly select the desired output power. This kind of power selection can be used when the transmission power must be changed during operation; Secondly, a fixed power application, a resistor can be used which is connected from the PSEL pin to ground. The voltage drop across the resistor selects the desired output power level. The output power is largest with the resistance not connected.

3.2.4 Timing Diagram

TH72015 provides wide voltage power supply range from 1.46 V to 5.5 V and provide low voltage detection function, if the power supply voltage is below 1.85 V, power amplifier (PA) will be closed. This can stop power radiation when the power supply voltage is too low. The chip provides two working mode: sending modes and standby mode. The two modes can be switched through the ENTX pin. When ENTX is 0, the chip works at the standby mode, right now PSEL is under a high resistance state and the whole chip circuit will not work; When EXTX is enabled, PA begin to work within the time duration T_{ON}, After the successful PLL LOCK ,the lock signal turns on the power amplifier, and then the RF carrier can be FSK or ASK modulated. Timing diagram is shown in Fig 5.

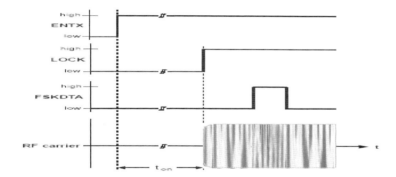

Fig. 5. Timing diagrams for FSK and ASK modulation

4 The Specific Application of MLX91801 and TH72015 in Tire Pressure Monitoring System

Based on the above discussion of performance and characteristics, the application of MLX91801 and TH72015 is wide, including: two-way half-duplex digital/simulation of communication, low power remote sensing detection, security and warning systems, RKE system, tire pressure monitoring system (TPMS), intelligent remote control system, etc. This paper will discuss wireless RF application in TPMS. The following diagram is the principle of Melexis TPMS application.

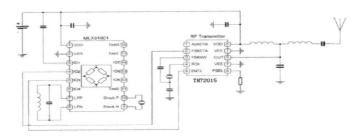

Fig. 6. Composition structure of Melexis TPMS

TPMS system is a set of high-tech products based on the tire pressure and temperature of the car in the dynamic real-time monitoring, which provide warnings whenever dangerous conditions occurs, thus effectively preventing a blowout to guarantee safe driving. TPMS includes two parts-- emission module and receiving module. Emission module is consisted of sensors, MCU and RF IC. Receiving module is consisted of RF IC, MCU, displaying control and interface control parts. The basic principle of TPMS is that, the tire pressure, temperature, acceleration information in the launching module measured by sensors are sent to MCU through the interface circuit by MCU control, then emitted wireless after sending to TH72015. In receiving module, TH72015 receives signals information from launch module and sends to MCU, after the corresponding processing of MCU, it will drive the display to receive information. The TPMS emission module and receiving modules is composed of TH72015 ,which is shown in Fig 7.

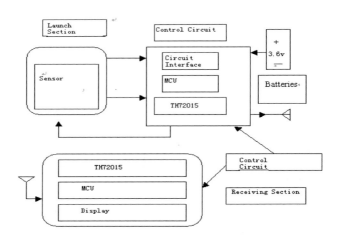

Fig. 7. TPMS launch module and the receiving modules system

5 Closing

As the popularity of the car and the continuous consciousness of the automotive safety, TPMS will get fast development in our country. MLX91801 and TH72015 have has a wide range of applications in many European countries, it provides convenient power management and very low power consumption. As Melexis entering the Chinese market and the promotion of tire pressure monitoring system, the excellent performance sensor MLX91801 and monolithic wireless transceiver chip TH72015 series products could be also widely used in domestic.

References

1. Chen, F., Chen, w.: Wireless interface circuit design and its application in TPMS. In: The 7th Single-chip Microcomputer and Embedded System, p. 24 (2005)
2. Fan, c. (ed.): Communication principle, 5th edn., pp. 130–137,142–150. Defense industry press (2002)
3. Ke, Y.: Development of Automobile tire pressure monitoring system. Thesis of Master's degree in University of electronic science and technology (September 23, 2005)
4. Li, W., Zhou, B., Tong, Z.: Tire pressure monitoring system design and implementation. In: 2004 the Phase Automobile Technology, vol. 2, pp. 23–27 (2004)

Research on Control Law of Ramjet Control System Based on PID Algorithm

Duan Xiaolong[1], Mao Genwang[1], Xu Zhongjie[2,*], and Wu Baoyuan[2]

[1] College of Astronautic, Northwestern Polytechnical University, Xi'an 710307, China
[2] Xi'an Aerospace Propulsion Institute, Xi'an 710100, China
`juanwang168@yahoo.com.cn`

Abstract. Control system is used to satisfy the requirements of the ballistic trajectory of ramjet and missile with different laws. In this paper, based on the analysis of the control design requirements of the ramjet, the control laws for the starting, acceleration and cruising stages are brought forward by introducing different control input parameters., Specially a PID algorithm of the ramjet is designed for controls in different phases according to the laws. And to validate our digital simulation, a hardware-in-the-loop simulation system is built up. The experiment results show that the digital simulation coincides with the semi-physical simulation very well with the requirements for controlling satisfied.

Keywords: control law, control algorithm, simulation.

1 Introduction

As one of the most critical systems of engine, ramjet control system is progressing with the continuous development of propulsion systems. This progressing can be roughly summarized as follows: it has already been transferred from the classical single constraint control theory to the modern multi-constrained control theory; it has also been transferred from the hydraulic control system to digital electronic control system [1-5].

Projectile characteristics and engine performance vary with flight conditions, so the flight trajectory relies on aircraft engine control system to ensure strict oil supply In the multi-constrained digital engine control system, we also need to study the robustness and good performance of the engine control law to adapt to the changes. Only in this way, the engine can provide more thrust and larger range effectively. Currently, most engine parameters are based on external flow control law [1-2], which generally can only be verified in a free jet test or flight test. However the control law based on inlet output parameters can be verified in dynamic engine test without any influence of projectile parameters, reducing the difficulty and risk of the development effectively. As far as we know, there is few research about the control law based on inlet output parameters.

---
[*] Corresponding author.

At present a variety of advanced control algorithms such as adaptive control, fuzzy control, multivariable decoupling control and multivariable optimal control etc have been widely used in many applications in industry [6], however such kinds of algorithms are relatively complex and rarely applied in military system control. As a special case, PID algorithm is widely used in the field of aero-engine systems [7-8] because of its maturity. Since its application in the ramjet is not so much, we introduce the PID algorithm into ramjet control community.

In this paper we present a control law based on inlet output parameters to meet the demand of engine thrust and the reliable and easy-to-implement PID algorithm is applied to realize the controlling.

2 Control Law and Control Algorithm Design

2.1 The Requests of Control Law Design

The requirement for control is different for different working stages of the ramjet. During the later stage of the boost phase, the control system is required to work in advance to establish an appropriate hydraulic flow and oil pressure to ensure fast and reliable ignition. After falling off of the booster, missile speed will climb, requiring the ramjet to provide the greatest possible net thrust, so correspondingly the ramjet control system has to supply enough fuel gas to provide the greatest amount of thrust by adjusting the equal residual gas coefficient (α = constant) or the given residual gas coefficien. During cruise process, ramjet thrust is required to overcome the resistance; In addition, due to the continuous fuel consumption, the mass and flight attitude of the missile changes constantly, resulting in thrust changing, so the equal Mach number (Ma = constant) is used to adjust fuel supply.

2.2 Start the Ignition Control Law

In general, ramjet needs to be ignited within a short period to meet the relay condition. When ramjet control system receives the ignition instruction, it will accelerate solenoid valve to action, so that in the original location it can provide maximum flow output at first and then move to the minimum with the configured residual gas coefficient or oil to complete engine start ignition. The ignition control law could be written as in Equation 1.

$$\{q_{mf} = q_{mf0}\} or \{\alpha = \alpha_0\} \tag{1}$$

where qmf is the fuel flow, α is the coefficient of residual gas, qmf0 and α0 are the fuel flow and the residual gas factor respectively.

After startup, the signal disappears, and the fuel amount could be calculated with the normal working control.

At this point the control and regulation system should response quickly to provide reliable ignition oil-gas ratio to ensure timely and reliable ignition, after the success of fast ignition fuel delivery scheduled arrival.

2.3 Design of Speed Control Law Section

The calculations based on the inlet output parameters are as follows:

$$\alpha = \frac{q_{ma}}{L q_{mf}} \quad (2)$$

$$q_{ma} = m_2 \frac{P_{t2} q(Ma_2) A_2}{\sqrt{T_{t2}}} \quad (3)$$

$$Ma_2 = f(P_{t2}/P_2) = \sqrt{5((P_{t2}/P_2)^{\frac{1}{3.5}} - 1)} \quad (4)$$

Equation (2) is the definition of coefficient of residual gas coefficient α, qma as the air flow, qmf as fuel flow, L as the equivalent mixing ratio; Equation (3) is used to calculate the inlet flux with the inlet exit section. Here the total pressure Pt2, total temperature Tt2, Mach number Ma2, size A2 and constant m2, m2=0.041 which is calculated by the gas thermodynamics. Equation (4) is used to calculate the inlet exit Mach number Ma2. From Equation (2)-(4), we can derive:

$$q_{mf} = \frac{1}{\alpha L} A_2 q(Ma_2) \frac{P_{t2}}{\sqrt{T_{t2}}} \quad (5)$$

Based on ground test and empirical data, Equation (5) can be simplified into Equation (6):

$$q_{mf} = f(\alpha, T_{t2}, P_2) \quad (6)$$

As shown in Equation (6), given a residual gas coefficient, the total fuel flow can be determined by measuring the inlet temperature Tt2 and static pressure P2 around the exit section.

2.4 Design of Cruise Control Law

In cruise phase, PID is used to control the Mach number deviation to calculate the fuel flow (or residual gas factor) as shown in Equation (7), which will be discussed in detail in Section 1.5.

$$q_{mf} = f_{PID}(\Delta Ma) \quad (7)$$

2.5 Control Algorithm

In order to find an effective control technique for ramjet, PID control algorithm is used since PID control has the advantages of strong robustness, good adaptability and easy to implement in engineering etc [6]. Figure 1 gives PID control system block diagram.

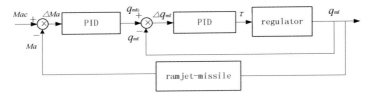

Fig. 1. PID control system block diagram

Design of PID algorithm is as follows:

$$\begin{cases} u(k) = u(k-1) + \Delta u(k) \\ \Delta u(k) = K_p \{e(k) - e(k-1) + \dfrac{T}{T_i} e(k) + \dfrac{T_d}{T}[e(k) - 2e(k-1) + e(k-2)]\} \\ \quad\quad = K_p \Delta e(k) + K_i e(k) + K_d [\Delta e(k) - \Delta e(k-1)] \\ e(k) = r(k) - y(k) \end{cases} \quad (8)$$

where e(k) is the control deviation (corresponding to the block diagram of the fuel flow deviation △qmf and the Mach number deviation △Ma), r(k) as input parameters (corresponding to the input Mach number Mac and the input fuel flow qmfc), the actual feedback value (corresponding to the measured Mach number Ma and the measured fuel flow qmf), Kp for the scaling factor, T for the sampling period, Ti as the integration time, Td as the differential time, Ki = KpT/ Ti for integral coefficient, Kd = KpTd/ T as the differential coefficient, u(k) to control the volume (corresponding to qmfc and duty cycle τ).

In the PID algorithm design, in the regulation of large deviations, in order to reduce the overshoot algorithm using integral separation; when the deviation is less than a certain value, in order to reduce the impact of frequent adjustment, PID algorithm

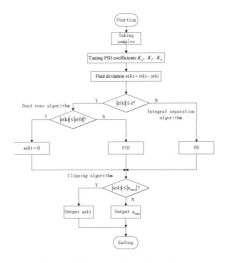

Fig. 2. PID algorithm flow chart

with dead zone is applied; by limiting algorithm limits saturated output; the same time, consider using variable gain PID algorithm to reduce the overshoot and steady increase in accuracy, PID algorithm flow shown in Figure 2.Luminescent properties of as-synthesized.

3 Simulation Model and Simulation Programs

3.1 Simulation Model

Based on modern control theory and Laplace transform principle, a fuel regulator and the ramjet - missile mathematical model are established [1]: $\frac{q_{mf}(s)}{\tau(s)} = \frac{k_a}{Ts + k_b} e^{-T_1 s}$ and $\frac{Ma(s)}{q_{mf}(s)} = \frac{k_{qmf}}{T_w s + 1}$, where qmf(s) is the fuel flow to Laplace transform output, τ(s) for input duty cycle change to Laplace transform; ka is the proportional coefficient; kb is the bias constant, T is the regulator time constant, T1 is the delay time, Tw is the time constant for the aircraft, kqmf is Mach number for the oil to the transmission coefficient, Ma(s) is Mach number to Laplace transform.

On this basis, the control law and control algorithm are studied in simulation testing.

3.2 Simulation Scheme

Ramjet fuel-supply control loop-simulation test system is shown in Figure 3, mainly composed of the supply system, fuel regulators, electronic controllers and simulation machines and so on. The simulation is done in two steps. The first step is to establish the mathematical models of missile-ramjet, fuel regulator, control laws and control algorithms etc other parts, and then complete the simulation of various parts in Matlab / Simulink environment digitally Secondly, input and output signals of the simulation model are replaced with the actual A/D, D/A, and I/O board, and then describe the objectives of the hardware to generate real-time code, the code will be downloaded to the local real-time simulation platform, fuel regulator, the system piping etc are brought into the control loop, the loop simulation is conducted and the simulation model can be modified online.

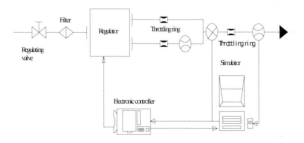

Fig. 3. Loop simulation system principle scheme

4 Simulation Results

4.1 Simulation Test of Start the Ignition Control Law

In the simulation, simulating engine start process, after simulation computer starts receiving instructions, the regulator accelerating valve moves, fuel flow from the largest to the smallest transition, the solenoid valve closed in the set point 1kg/s, the electromagnetic fast PWM regulator valve action, achieve closed loop fuel control. Experimental curve shown in Figure 4, accelerating fuel solenoid valve closes the set, then the role of the PWM valve, the oil from 0.88kg / s up to the Relay Point 3kg/s.

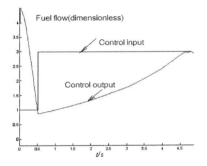

Fig. 4. changes in the fuel oil as accelerating solenoid valve action

4.2 The Simulation for Speed Control Law

The results shown in Figure 5, the figure "dotted line" is the result of digital simulation, "thin solid line" is a section of semi-physical simulation measured the result of angular displacement sensor output by the control law based on inlet output parameters in the accelerating period, it can be seen the two results have basic agreement, achieve the desired control effect.

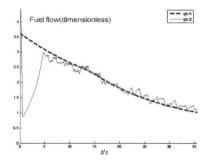

Fig. 5. Oil supply curve of the control law based on inlet output parameters

4.3 Cruise Control Law and Control Algorithm Testing

Such Mach number cruise segment in the large closed-loop feedback control, PID algorithm is used to control the Mach number and to calculate fuel flow. The results shown in Figure 6 and 7, the figure "dotted line" is a digital simulation results, the oil level is due to appear simulation results as the control law switched; "thin lines" is a semi-physical simulation test results, it can be seen that the digital simulation results of the fuel flow and Mach number are consistent with the semi-physical simulation test results, to achieve the desired results.

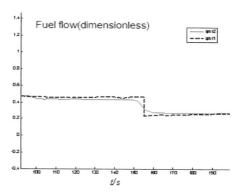

Fig. 6. Cruise fuel flow section

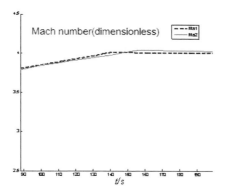

Fig. 7. Mach number cruise segment

5 Conclusion

The research on control law of ramjet control system based on PID algorithm is established, and the results are summarized as follows:

(1) Based on the analysis of control requirements, the control law of each phase of the trajectory is determined.

(2) On the basis of the establishment of missile and ramjet mathematical models, different stages of the ramjet control law for the digital simulation and semi-physical simulation are studied, both in good agreement, realized the calculation and control of fuel flow on the different stages.

(3) The PID algorithm is designed to ensure the regulator to supply the required fuel flow at different control stages.

References

1. Smith, T.F., Waterman, M.S.: Identification of Common Molecular Subsequences. J. Mol. Biol. 147, 195–197 (1981)
2. May, P., Ehrlich, H.-C., Steinke, T.: ZIB structure prediction pipeline: Composing a complex biological workflow through web services. In: Nagel, W.E., Walter, W.V., Lehner, W. (eds.) Euro-Par 2006. LNCS, vol. 4128, pp. 1148–1158. Springer, Heidelberg (2006)
3. Foster, I., Kesselman, C.: The Grid: Blueprint for a New Computing Infrastructure. Morgan Kaufmann, San Francisco (1999)
4. Czajkowski, K., Fitzgerald, S., Foster, I., Kesselman, C.: Grid Information Services for Distributed Resource Sharing. In: 10th IEEE International Symposium on High Performance Distributed Computing, pp. 181–184. IEEE Press, New York (2001)
5. Foster, I., Kesselman, C., Nick, J., Tuecke, S.: The Physiology of the Grid: an Open Grid Services Architecture for Distributed Systems Integration. Technical report, Global Grid Forum (2002)
6. National Center for Biotechnology Information, http://www.ncbi.nlm.nih.gov

Layout Design and Optimization of the High-Energy Pulse Flash Lamp Thermal Excitation Source

Zhang Wei, Wang Guo-Wei, Yang Zheng-Wei, Song Yuan-Jia,
Jin Guo-Feng, and Zhu Lu

203 room, The Second Artillery Engineering College, Xi'an China, 710025

Abstract. In the non-destructive detection of infrared thermal wave, the uniformity of thermal excitation not only affects the quality of thermal-image, but also influenced the measurement accuracy of surface temperature-difference, which affect the last quantitative identification of defects. Therefore, the spatial and illuminance distribution characteristics of the heat stimulating were simulated using TracePro, the array of four lamps and the model of shadow mask were designed and simulated. The results show that the radiation effect of the array of four lamps is better than the single, the evenness degree of heat and energy utilization ratio are greatly improved and the even radiation is gained on the irradiated area, which lay the foundation for the subsequent hardware design and actualization.

Keywords: Thermal Excitation, Infrared Thermal Wave, Illumination distribution, Uniformity.

1 Introduction

The basic concept of the infrared thermal wave nondestructive testing technology is to choose different thermal excitation for different specimens, use of modern infrared imaging technology, detective timing thermal wave signal and collect data under the controlling of the computer. According to the thermal wave theory and modern images processing technology and special computer software developed for real-time image signal processing and analysis. The means of the thermal excitation currently are a plasma jet, direct flame, high-energy pulse heating, infrared light, heat ultrasound, etc, but the most commonly used is high-energy pulse flash heating method. This method is use of light (visible) radiation emitting from high-energy pulsed xenon lamp or other high-energy gas discharge lamp when it discharging, used the instantaneous release heat energy for transient excitation to the measured object [1]. In the actual testing process, the different detected objects beacusing of its different density, thermal conductivity and specific heat, the internal heat transfer process is different.

The linchpin to the success is the effective temperature difference coming into being on the workpiece surface beacusing of different thermal diffusion process in the internal heterogeneity, under the external thermal excitation. The effective temperature difference includes two aspects: one is the surface temperature difference reaches or exceeds the heat sensitivity of the Infrared; the other is the duration of the surface temperature difference should be longer than the response output time of the infrared camera[2]. Generally, the effective temperature difference can be achieved by increasing thermal excitation energy, but in the high-energy pulse infrared thermal wave nondestructive testing, in addition to the effective temperature difference of thermal excitation, the uniformity is one factor that should be consideration important.

The thermal excitation uniformity refers to the uniformity of the energy inflicted to the tested surface or blazing uniformity energy on the tested surface by external thermal excitation, so that the heat spread along the thickness direction only as much as possible. Thermal excitation uneven will directly cause large ups and downs of the infrared thermal image background, and bring a lot of negative impacts to the follow-up data and image processing, and then affect the last quantitative identification of defects, so that the thermal stimulating-uniformity becomes a important factor of active pulse thermal wave nondestructive testing.

Therefore, this paper based on linear pulse flash structure, applying lighting design software TracePro, from radiation uniformity and energy utilization ratio, firstly researched radiation illumination distribution characteristics of the single lamp, and then optimized the layout of multi-lamp array, designed corresponding components to achieve the purpose that it got the uniform thermal excitation on the object surface.

2 Basic Radiation Theory

2.1 Surface Radiation Principle [3--4]

If the surface area source size and the distance between the lighted plane and surface source is relatively small, it can be called little area source, shown in the figure 1, supposing the area of surface area source is A_m, radiation luminance is L, the illuminated surface area is A_n, the distance is l, the angle between A_m, normal of A_n and l is respectively θ_m and θ.

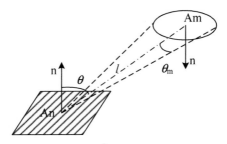

Fig. 1. The radiation illumination produced by point source

Radiation intensity of the point source: $I = \dfrac{dP}{d\Omega} = L \cos\theta_m A_m$ (1)

Radiation illumination produced by point source:

$$E = \frac{I \cos\theta}{l^2} = L \cdot A_s \cdot \frac{\cos\theta m \cdot \cos\theta}{l^2} \quad (2)$$

Tubular xenon lamp belongs to line radiation source, but in practice in order to get uniform radiating, generally we used the method that plus the reflector, so that it no longer treated as line radiation, but expanded into a surface emitter, as shown in the figure 2.

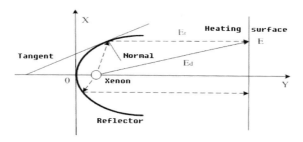

Fig. 2. The cutaway illustration of the parabolic reflector

Clearly, the radiation to heat radiation on the plane from two parts, the first part is the direct irradiation from the pulse xenon lamp; and the other part is the reflection by the reflector. Therefore, the radiation intensity E of the xenon lamp illumination equals the summation of the direct illumination E_d and the reflection illumination E_r.

$$E = E_d + E_r \tag{3}$$

The surface of the pulsed xenon lamp was split into many small elements, so it can be approximately treated as a small surface radiation, we calculated the irradiance from dA_s to the point p, and then we got its integral and obtained the irradiance of the point p.

The directly irradiance at the point P of heating plane is,

$$E_d = \int_D \frac{M}{L} \cdot \frac{\cos\theta_s \cdot \cos\theta}{l^2} \cdot dA_s \tag{4}$$

In the formula, D is surface area of the pulsed xenon lamp that can irradiate directly the point p.

The reflected irradiance is:

$$E_r = \int_D \frac{M}{\pi} \cdot \frac{\cos\theta_s}{\sqrt{\left|\frac{D(x,y)}{D(\alpha,\beta)}\right|^2 + \left|\frac{D(y,z)}{D(\alpha,\beta)}\right|^2 + \left|\frac{D(z,x)}{D(\alpha,\beta)}\right|^2}} \cdot dA_s \tag{5}$$

In the formula, D' is the surface area of the pulsed xenon lamp that irradiates the point p by the reflector.

2.2 Characterization of the Irradiance Uniformity

In the actual testing process, due to the high sensitivity of the thermal infrared, it is easy to find subtle changes of the temperature, so nonuniform heat load may generated non-normal temperature difference on the heating plane, even leaded to the distortion of the image signal, and produced false estimation. Therefore, the thermal excitation uniformity is an important parameter. To characterize the homogeneity, we used the following formula to calculate the radiation uniformity on the heating plane.

$$N = 1 - \frac{E_{max} - E_{min}}{E_{max} + E_{min}} \times 100\% \qquad (6)$$

N ——radiation uniformity of the source on the heating plane;

E_{max} ——The maximum irradiance on the heating plane;

E_{min} ——The minimum irradiance on the heating plane[5]

In considering the uniformity of the radiation, we must ensure that the total radiation flux and optical efficiency in the permitted range.

3 Simulation and Design

3.1 Radiation Illumination Emulation and Analysis of the Single-lamp[6]

We built an accurate pulse xenon lamp single lamp irradiance distribution model, and first we accomplished the emulation and analysis of the Single-lamp. The figure 3 showed a model of the pulse xenon lamp.

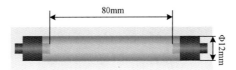

Fig. 3. The simulation model of the pulse xenon lamp

The radiation spectrum of the pulsed xenon lamp was continuous full-spectrum, since mainly analyzed the radiation uniformity and the capacity usage ratio only, we simplified the xenon lamp model: the temperature of the system was 25℃, the calculated light number was 2,000,000, the average power was 1000W, and the distance between the 40cm × 40cm heating plane and the xenon was 45cm. The figure 4 showed the irradiance distribution of the single lamp, the left showed the irradiance distribution on the heating plane; the right showed the radiation intensity distribution curve in the horizontal / vertical direction.

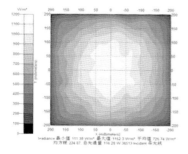 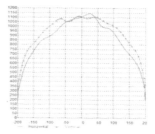

Fig. 4. The illumination distribution and distribution curve of the single-pulse flash lamp

From the illumination distribution curve, it indicated that the maximum illumination appeared in the central region, and declined rapidly to the around, so it showed that the spatial gradient of a single lamp was large. This would induced the uneven heating, increased the speed of the horizontal heat flow and weaken the ability of the vertical heat transfer, the surface temperature information could not accurately reflect the information of the internal defects, and it also made the distortion of the infrared image. So it was generally that we did not use a single lamp, and we used the multi-lamp array instead.

3.2 The Layout Design and Analysis of the Four-Light Array

As a single light had a drawback for spatial gradients large, as well as it could not satisfy the actual testing needs in the heating energy and the uniformity, so we should considered to increase the number of the xenon lamp, and in the actual detection we mostly adopted the multi-light plane array form to provide enough energy and large radiant surface with high uniformity. According to the literature [7], the multi-lamp array usually used the symmetrical layout which could improve the large spatial gradient of the illumination and help to form a uniform heating surface. There was four pulse xenon lamps in the hardware system, we should consider the distance between the lamps. If we made a poor choice, it would cause the heat radiation intensity too concentrated or scattered on the heating plane, too concentrated energy might burn surface; too scattered energy would make more radiation energy leaking to the surrounding environment, and it resulted in lacking of energy on the heating plane, and sequentially it reduced the energy efficiency and affected the heating uniformity.

As the pulse xenon lamp was a linear radiation source with a larger aspect ratio, so we mainly considered the vertical distance M in the layout, it was shown in the figure 5, and we selected a thermal imager with a criteria lens, whose diameter was 30mm.

The reference plane was 30cm × 30cm, and the heating distance was 45cm, the distance between the xenon lamp and the reflector was 39mm, , we respectively selected the space M in 100mm ~ 200mm, the spacing was 5mm, and then we did multiple calculation and analysis, the detailed data was shown in the table 1.

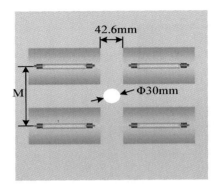

Fig. 5. The layout of four-lamp array

Table 1. Simulation and analysis of data spacing

Space M	Illumination uniformity N		Energy efficiency	Space M	Illumination uniformity N		Energy efficiency
	Horizontal	Vertical			Horizontal	Vertical	
100mm	93.78%	56.32%	65.11%	155mm	90.75%	83.77%	37.79%
105mm	94.45%	58.37%	64.14%	160mm	89.14%	85.55%	35.51%
110mm	94.98%	60.51%	62.87%	165mm	88.51%	86.78%	32.66%
115mm	95.01%	62.12%	59.44%	170mm	87.92%	90.47%	29.14%
120mm	94.25%	64.56%	56.78%	175mm	86.35%	91.32%	26.29%
125mm	93.77%	66.98%	53.69%	180mm	87.34%	88.12%	23.88%
130mm	93.61%	70.01%	50.14%	185mm	85.34%	87.41%	20.94%
135mm	93.24%	72.16%	47.98%	190mm	83.63%	85.69%	17.41%
140mm	92.55%	74.28%	44.02%	195mm	81.77%	83.71%	15.36%
145mm	91.65%	77.11%	43.23%	200mm	79.76%	82.79%	14.22%
150mm	91.13%	80.98%	40.83%	-----	-----	-----	-----

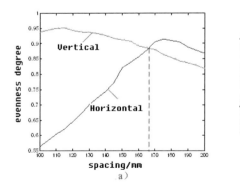 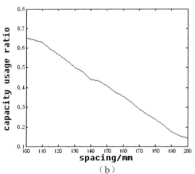

Fig. 6. The relationship between the energy efficiency, illumination uniformity and the space

According to the mentioned data above, we made a curve for the irradiance uniformity, the energy efficiency and the center illumination with the space changing, as it was shown in the figure 6, the figure (a) was the energy efficiency curve, and the figure (b) was the irradiance curve.

From the table 1 and the figure 6, we could find that the uniformity on the vertical direction firstly increased, and then decreased as the space increased, it was because that the irradiance of the center heating surface area became lower in the vertical direction when the space between xenon lamps exceeded a certain value, while the irradiance of the central region around increased; In the horizontal direction, the evenness degree decreased, but it changed with a lesser range, it was because that the radiation uniformity in the horizontal mainly related with the lateral space, but smaller related with the vertical spacing; the energy efficiency decreased with the space, it was because that the greater of the space the more flux leaked to the surrounding environment, and the less radiation flux entrance into the heating plane. So we selected the space which was vicinity with the intersection point (168.7mm) as the actual space, and then we got a better result.

3.3 The Design of the Mask

In the detection process, in order to reduce the interference to the heat load and the thermal imager from the external environment, at the same time collect more energy, reduce the spread loss and play a role in concentrating, a mask was appended around the xenon lamp, in the case of a constant heat source. That not only engendered a uniform light environment inside the mask, but also prevented the bright light from irradiating into the eye directly, so it played a protective role on the human eye. We considered to use the most simple and standard structure- rectangular mask, as shown in the figure 7, the figure (a) was the mask model, the figure (b) was the xenon lamp radiation model which added a mask. The size of the mask structure was 40cm × 40cm × 52cm, the area out the tube was 40cm × 40cm, the material was aluminum, the heating distance was 45cm, the average power was 4000w, the total number of light was 200 million.

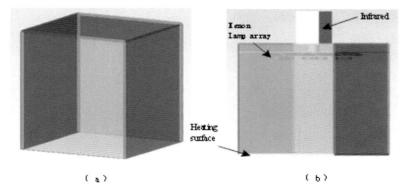

Fig. 7. The model of the rectangular mask and the flash lamp

In order to explain the advantages of the spotlight effect and the energy collected better, the paper verified it by contrast. As shown in the Figure 8, it was the illumination distribution of the four-lamp array without mask, the vertical spacing between the lights was 172mm, and the horizontal distance was 42.6mm.

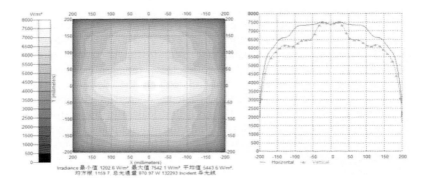

Fig. 8. The illumination distribution of the xenon lamp array without the mask

Through the simulation analysis, the results was shown in the figure 9, the figure (a) was the three-dimensional display for irradiance distribution, the figure (b) was the radiation intensity distribution curve. Which could be seen directly from the figure (a) was that the uniformity of the irradiance in the central region was high, and the high uniformity radiation area was relatively large; the figure (b) was the illumination distribution curve and the peak field was relatively wide. The energy utilization was 80.41%. In the reference plane, the illumination curves on the vertical and horizontal were basic coincidence. The illumination uniformity was 98.13%, the spatial gradient of the heat source was small enough to satisfy the requirements for the defect detection, so in the actual detection, we should mainly collected the temperature change image of the center region where the radiation uniform was high.

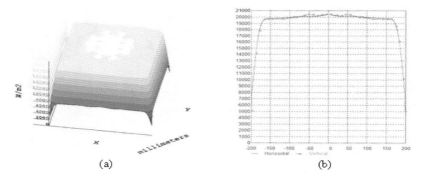

Fig. 9. The illumination distribution of the xenon lamp array with the mask

4 Summary

Based on the fundamental concept radiation theory, this paper had done much detailed simulation analysis with TracePro for the single linear pulsed xenon lamp, four-lamp array and the mask, etc. It mainly focused on improving the thermal excitation uniformity and the energy utilization of the xenon lamp on the heating plane. The results showed that the radiation effects of four-lamp array was far superior to a single lamp, and we could significantly improve the heating uniformity and the energy efficiency by using the mask, and formed a relatively uniform radiation on the irradiated area.

References

1. Lv, Z.-c., Tian, Y.-p., Zhou, K.-y.: Study on the Thermal Excitation in the Infrared Thermal Wave Nondestructive Testing. Nondestructive Testing 29(6) (2005)
2. Yang, X.-l., Lv, B.-p., Xian, M.-l.: Heat Stimulating Methods in the Infrared Thermal Image Testing of Airplane composite. Nondestructive testing
3. Chen, H.: Infrared Physics. National Defence Industry Press, Beijing (1985)
4. Cheng, X.-J., Xu, X.-h.: Discussion for Improving Infrared Iryer of Irradiation Uniformity. Beijing Printing Institute Transaction 8(4), 29–34 (2000)
5. Li, W.-t.: Study on Illumination Uniformity of the New Type White Light LED Lamp. Qinghai Science and Technology (2007)
6. Wang, Y.: Design and Implement in Hardware for the System of the Infrared Nondestructive Testing on Missile-engine. Second Artillery Engineering College thesis (2010)
7. Ma, Y.-L.: Application of the Infrared Lamp-array. Environmental Technology 4(53), 25–30 (1997)

Effective Cross-Kerr Effect in the N-Type Four-Level Atom

Liqiang Wang[1] and Xiang'an Yan[2]

[1] School of Science, Xi'an University of Post and Telecommunications,
Xi'an 710061, China
[2] School of Science, Xi'an Polytechnic University, Xi'an 710048, China

Abstract. A theoretical investigation was carried out into the program for obtaining effective cross-Kerr effect for the four-level atom interacting with three electromagnetic fields. Using density-matrix equations and perturbative iterative method, the variation dependence of the third-order susceptibility on the detunings of the applied electromagnetic fields was derived. The results show that a large cross-Kerr effect can be generated when the detunings of the two weak probe fields are Raman resonant, and the corresponding linear and self-Kerr interactions vanish. The realization of the effective Kerr effect with vanishing absorption is possible in a Raman resonant four-level system. Moreover, the large cross-Kerr interaction with vanishing absorption can be used to realize single-photon nonlinear devices and optical Kerr shutters.

Keywords: nonlinear optics; the third-order effect; cross-Kerr nonlinearity; Raman resonant.

1 Introduction

The exploration of nonlinear optics developed rapidly after the demonstration of the first laser by Maiman in 1960. However, the cross-Kerr nonlinearity was discovered much earlier in 1875 by the Scottish physicist Rev. John Kerr. Originally, the refractive index experienced by a light ray propagating in the Kerr medium was controlled by applying a strong dc electric field. The advent of high-intensity laser light enabled the substitution of the dc field with an optical electromagnetic field. The cross-modulation of the refractive index experienced by one electromagnetic caused by the intensity of the other is called the optical cross-Kerr effect. The large optical Kerr nonlinearities have numerous applications in the fields of low-light-level and quantum optics [1-5]: for instance, the low-light-level switching [6], quantum nondemolition measurements of photon number [7, 8], entangling optical wavepackets [3, 9]. It has also been proposed as a deterministic quantum logic gate [10, 11].

In most materials, however, the cross-Kerr interaction is always accompanied by competing nonlinear effects, such as self-Kerr nonlinearity and sum- and difference-frequency generation. In this letter, we show that a pure cross-Kerr nonlinearity can be generated in the four-level atom in the N configuration (Fig. 1) by tuning the detunings of the probe and the pump field are Raman resonant, where the absorptive component of the linear and self-Kerr susceptibilities will also vanish.

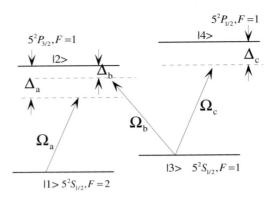

Fig. 1. Four-level Rb atomic system interacting with three electromagnetic field. We use the hyperfine components of 87Rb D lines to form the four-level system.

2 Atomic Model and Motion Equations

We consider a four-level N-type atomic system, as shown in Fig. 1. One strong pump field Ω_b and two weak probe field (Ω_a and Ω_c), respectively, are applied to |2⟩↔|3⟩, |1⟩↔|2⟩, and |3⟩↔|4⟩ transitions. ω_a, ω_b, and ω_c are carrier frequencies of the corresponding fields; $\Delta_a = \omega_{21} - \omega_a$, $\Delta_c = \omega_{43} - \omega_c$ and $\Delta_b = \omega_{23} - \omega_b$ are the frequency detunings of the two weak probe field and the pump field.

Under rotating-wave approximation, systematic density matrix in the interaction picture can be written as

$$\begin{aligned}
\dot{\rho}_{11} &= -i(\rho_{12} - \rho_{21})\Omega_a + 2\gamma_1\rho_{22} \\
\dot{\rho}_{12} &= i\Delta_a\rho_{12} + i(\rho_{22} - \rho_{11})\Omega_a - i\rho_{13}\Omega_b - \gamma_1\rho_{12} \\
\dot{\rho}_{13} &= i(\Delta_a - \Delta_b)\rho_{13} + i\rho_{23}\Omega_a - i\rho_{12}\Omega_b - i\rho_{14}\Omega_c \\
\dot{\rho}_{14} &= i(\Delta_a - \Delta_b + \Delta_c)\rho_{14} + i\rho_{24}\Omega_a - i\rho_{13}\Omega_c \\
\dot{\rho}_{22} &= i(\rho_{12} - \rho_{21})\Omega_a - i(\rho_{23} - \rho_{32})\Omega_b - 2\gamma_1\rho_{22} \\
\dot{\rho}_{23} &= -i\Delta_b\rho_{23} + i\rho_{13}\Delta_a - i(\rho_{22} - \rho_{33})\Omega_b - i\rho_{24}\Omega_c - \gamma_1\rho_{23} \\
\dot{\rho}_{24} &= -i(\Delta_b - \Delta_c)\rho_{24} + i\rho_{14}\Omega_a + i\rho_{34}\Omega_b - i\rho_{23}\Omega_c - (\gamma_1 + \gamma_3)\rho_{24} \\
\dot{\rho}_{33} &= i(\rho_{23} - \rho_{32})\Omega_b - i(\rho_{34} - \rho_{43})\Omega_c + 2\gamma_3\rho_{44} \\
\dot{\rho}_{34} &= -i\Delta_c\rho_{34} + i\rho_{24}\Omega_b - i(\rho_{33} - \rho_{44})\Omega_c - \gamma_3\rho_{34} \\
\rho_{11} &+ \rho_{22} + \rho_{33} + \rho_{44} = 1
\end{aligned} \quad (1)$$

In above equations, $2\gamma_1$, $2\gamma_2$ denote the spontaneous decay rates from level |2⟩→|3⟩ and |4⟩→|3⟩, respectively. In the present letter, we aim to investigate the feasibility of enhancing cross-Kerr nonlinearity while simultaneously inhibiting the self-Kerr nonlinearity and linear absorptions. Therefore, we need to derive the expressions for the linear, self-Kerr, and cross-Kerr susceptibilities. By Taylor expanding the off-diagonal density-matrix elements in terms of the fields Ω_a and Ω_c, for the $|1\rangle \leftrightarrow |2\rangle$ transition, the linear and self-Kerr susceptibilities are

$$\chi^{(1)} = \frac{(\Delta_a - \Delta_b)|\mu_{12}|^2}{\hbar \varepsilon_0 [(\Delta_a - \Delta_b)(\Delta_a - i\gamma_1) - \Omega_b^2]} \tag{2}$$

$$\chi_s^{(3)} = \frac{-2|\mu_{12}|^4 (\Delta_a - \Delta_b)^3 \Omega_b^2}{3\hbar^3 \varepsilon_0 [(\Delta_a - \Delta_b)(\Delta_a - i\gamma_1) - \Omega_b^2]^3} \tag{3}$$

and the cross-Kerr susceptibility is found to be

$$\chi_c^{(3)} = \frac{|\mu_{12}|^2 |\mu_{34}|^2 \Omega_b^2}{6\hbar^3 \varepsilon_0 (\Delta_a - \Delta_b + \Delta_c - i\gamma_3)[(\Delta_a - \Delta_b)(\Delta_a - i\gamma_1) - \Omega_b^2]} \tag{4}$$

The linear, self-Kerr and cross-Kerr responds of the atomic medium to the probe field can be described by $\chi^{(1)}$, $\chi_s^{(3)}$ and $\chi_c^{(3)}$, respectively.

3 Discussion and Results

In what follows, we will focus on how the linear, self-Kerr, and cross-Kerr susceptibilities are affected by the detunings of the two weak probe fields. Based on (2) and (3), Figures 2 and 3 show the linear and self-Kerr susceptibilities as functions of the probe detuning Δ_a. Given a resonant control field $\Delta_b=0$, we set the parameters $\gamma_1=7$MHz, $\gamma_3=7$MHz, $\Omega_b=4$MHz. These graphs illustrate the electromagnetically induced transparency (EIT) response of the Λ subsystem. On the Raman resonance $\Delta_a=\Delta_b=0$, the system forms an atomic dark state and the real and imaginary parts of both susceptibilities vanish. Under the same conditions, the imaginary part of the self-Kerr susceptibility show three EIT points and the transparency window for the self-Kerr susceptibility is slightly narrower than four the linear response. At the same time, we also find that the strength of the self-Kerr susceptibility is several orders of magnitude weaker.

Figure 4 shows the dependence of the cross-Kerr susceptibility on the detuning of the probe field Ω_c for the Raman resonance $\Delta_a = \Delta_b = 0$. It is observed that off resonance the real part of the cross-Kerr susceptibility decays much more slowly than the imaginary part. Therefore, when the field Ω_c is relatively far off resonance we can achieve an appreciable refractive cross-Kerr interaction with negligible absorption. From the graph it can be seen that a cross-Kerr susceptibility of about $10^{-8} m^2 V^{-2}$ is possible at the position of $\Delta_c=100$ MHz. Importantly, this is achieved on Raman resonance, where the linear and self-Kerr interaction vanish.

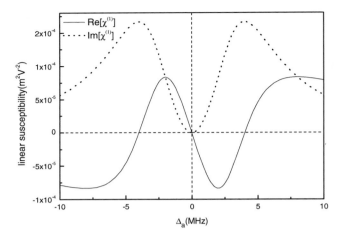

Fig. 2. Variation of Re[$\chi^{(1)}$] (solid line) and Im[$\chi^{(1)}$] (dotted line) of the linear susceptibility(m^2V^{-2}) as a function of the robe detuning Δ_a(MHz)

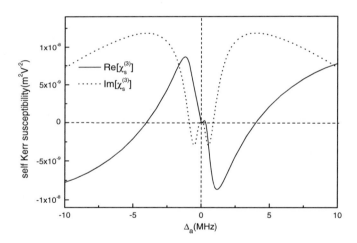

Fig. 3. Variation of Re[$\chi_s^{(3)}$] (solid line) and Im[$\chi_s^{(3)}$] (dotted line) of the self Kerr susceptibility(m^2V^{-2}) as a function of the robe detuning Δ_a(MHz)

Given that cross-Kerr susceptibility makes the only appreciable contribution on resonance, it is found that the phase shift [12] experience by the field Ω_a is given by

$$\Delta\phi = \frac{3}{2}\frac{\pi d}{\lambda_0}|E_c|^2 \text{Re}[\chi_c^{(3)}] \tag{5}$$

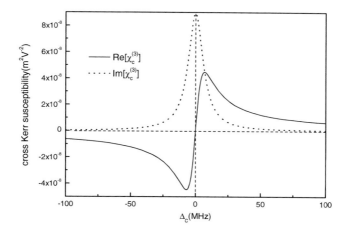

Fig. 4. Variation of Re[$\chi_c^{(3)}$](solid line) and Im[$\chi_c^{(3)}$](dotted line) of the cross Kerr susceptibility(m²V⁻²)as a function of the robe detuning Δa(MHz). The fields Ω_a and Ω_b are assumed to be Raman resonant ($\Delta_a = \Delta_b = 0$)

Here, l is the interaction length and λ_0 is the wavelength of the field Ω_a in free face. Using the experimental parameters given above, we find that an optical phase shift of $\Delta\phi = 0.1$rad is possible.

4 Conclusions

We have shown that the four-level atom in the N-type atomic system can give rise to a pure cross-Kerr nonlinearity. The analytical expression shows that the strength of the cross-Kerr interaction depends on the detuings of all three fields. In general condition, the cross-Kerr interaction will be accompanied by a linear an self-Kerr response of atom. However, in our considered system, the absorptive component of the linear and self-Kerr susceptibilities will vanish when tuning the detunings of the fileds Ω_a and Ω_b are Raman resonant. Although the absorption related to the cross-Kerr interaction will remain, one can achieve a giant cross-Kerr effect by increasing the detuning of the weak probe filed Ω_c. This system may potentially be applied to all-optical switch, polarization phase gates, and other information processes.

Acknowledgment

This work was supported by the National Natural Science Foundation of China (No.61078020), the Specialized Research Fund for Doctoral Program (No.BS1008), and the Education Office of Shaanxi Province (No. 2010JK556).

References

1. Hariss, S.E., Yamamoto, Y.: Photon switching by quantum interference. Phys. Rev. Lett. 81, 3611–3614 (1998), doi:10.1103/PhysRevLett.81.3611
2. Fleischhauer, M., Lukin, M.D.: Dark-state polaritons in Electromagnetically induced transparency. Phys. Rev. Lett. 84, 5094–5097 (2000), doi:10.1103/PhysRevLett.84.5094
3. Petrosyan, D., Kurizki, G.: Symmetric photon-photon coupling by atoms with Zeeman-split sulevels. Phys. Rev. A. 65, 033833(4 pages) (2002), doi:10.1103/PhysRevA.65.033833
4. Fleischhauer, M., Imamoglu, A., Marangos, J.P.: Electromagnetically induced transparency. Rev. Mod. Phys. 77, 633–673 (2005), doi:10.1103/RevModPhys.77.633
5. Ottaviani, C., Reic, S., Vitali, D., Tombesi, P.: Quantum phase-gate operation based on nonlinear optics: Full quantum analysis. Phys. Rev. A 73, 010301(R) (2006), doi:10.1103/PhysRevA.73.010301
6. Kang, H., Zhu, Y.: Observation of large Kerr nonlinearity at low light intensities. Phys. Rev. Lett. 91, 093601(4 pages) (2003), doi:10.1103/PhysRevLett.91.093601
7. Beausoleil, R.G., Munro, W.J., Spiller, T.P.: Applications of coherent population transfer to quantum information processing. J. Mod. Opt. 51, 1559–1601 (2005), doi:10.1080/09500340408232474
8. Munro, W.J., Nemoto, K., Beausoleil, R.G.: High-efficiency quantum-nondemolition single-photon-mumber-resolving detector. Phy. Rev. A 71, 033819(4 pages) (2005), doi:10.1103/PhysRevA.71.033819
9. Lukin, M.D., Imamoglu, A.: Nonlinear optics and quantum entanglement of ultraslow single photons. Phys. Rev. Lett. 84, 1419–1422 (2000), doi:10.1103/PhysRevLett.84.1419
10. Meter, R.V., Nemoto, K., Munro, W.: Communication links for distributed quantum computation. New J. Phys. 56, 137–140 (2007), doi:10.1088/1367-2630/56/12/137
11. Shapiro, J.H., Rahzavi, M.: Continuous-time cross-phase modulation and quantum computation. New J. Phys. 9, 16–33 (2007), doi:10.1088/1367-2630/9/1/016
12. Schmidt, H., Imamoglu, A.: High-speed properties of a phase-modulation scheme based on electromagnetically induced transparency. Opt. Lett. 23, 1007–1009 (1996), doi:10.1364/OL.23.001007

The Design and Implementation of Groundwater Monitoring and Information Management System

Yujie Zhang and Yuanyuan Zhang

College of Electrical and Information Engineering,
Shaanxi University of Science and Technology
Xi'an, China
zyy19870215@163.com

Abstract. Groundwater monitoring is the only effective method to get the dynamic data of groundwater quality and water quantity directly. This paper designed and implemented a low-power, GSM-based and GIS-based remote sensing of groundwater conditions and information management system. The system used GIS to represent the hydrological trends. Use the data transmission of GSM, GPRS and Internet and other ways to realize the centralized and hierarchical management about the information for regional hydrological. The lower-powered machine used the way of the selective separation, which greatly reduced the energy consumption of the system. The practical application showed that the system has a high visibility, stability and practicality and low energy consumption.

Keywords: GIS; GSM; groundwater monitoring; hierarchical management; separation of power.

1 Introduction

With the increasingly serious destruction of the ecological environment, the importance of monitoring of groundwater resources has becoming increasingly prominent. For reasonable development and utilization of groundwater resources and in order to contain further deterioration of groundwater and prevent new environmental problems of groundwater, we must monitor for the changes of groundwater. In order to achieve the purpose of sustainable development and utilization of groundwater resources, we need to understand the distribution and dynamic change in time and space and develop appropriate strategies in a timely. The traditional methods of artificial groundwater monitoring measurements not only can not guarantee the accuracy, and on-site operator's labor intensity is great, which cannot be managed automatically and real-time data is poor, and even it cannot be measured in some geographically remote or dispersed monitoring sites. In this paper, we combined the computer technology, database technology, GIS technology and water resource management business to develop a GIS and GSM-based monitoring of groundwater conditions and information management system with which we get the dynamic data acquisition on the spread of groundwater, unify management and make comparative analysis.

2 The Work Principle of System

GIS and GSM-based monitoring of groundwater conditions and information management system takes full advantage of powerful spatial analysis and mapping functions of GIS technology to represent the hydrological conditions intuitively. It uses GSM short message service to realize the remote transmission of hydrological information of the data from a single point, which solves the problem that measure the water temperature and water level under the harsh environment and carry out the wireless data communication remotely and real-timely. It is low investment and easy to be universalized on a large scale. It uses the GPRS and the Internet and other means of transmission to realize the centralized hierarchical management of regional hydrological information.

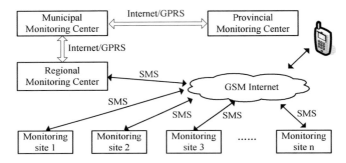

Fig. 1. The Schematic Diagram of Groundwater Monitoring and Information Management System

Shown in Figure 1, the system is mainly consisted of a more decentralized system of monitoring stations, regional monitoring centers, municipal and provincial monitoring center and there are three levels of municipal and provincial monitoring and management center.

Monitoring stations achieved the automatically collecting, computing and storage of water level and water temperature data, and sent them to the district-level monitoring center through GSM-SMS service and it also received information from the monitoring center. The three levels of monitoring and management center used GPRS or Internet to interconnect. District monitoring center dealt with and stored the data from the scattered monitoring sites and then uploaded to the municipal monitoring center. After processing, it sent the comprehensive data analysis to the provincial monitoring center for further treatment, and then understood the distribution and dynamic change in time and space and developed appropriate strategies in a timely to achieve the purpose of sustainable development and utilization of groundwater resources.

3 Analysis and Design of System

The design of groundwater monitoring and information management system mainly included the design of the monitoring terminal hardware and software of the management center of the system.

3.1 The design of Monitoring Terminal Hardware

The power of traditional groundwater monitoring terminal consumed largely, which is generally ranging from 200mW ~ 600mW. The volume of lower computer is too big, and it needs extra batteries and solar panels, but installation of which is complex and easy to be theft and vandalism. It does not apply where the conditions are more demanding, because the cost of setting up is higher in the absence of power.

In this system, the hardware uses selective separation of power of each functional module. Procedures adopt the way of regular collection and uploading regularly to reduce power consumption, and then to determine the feasibility of battery supplying the power long. To make it easier to be replaced and maintain the terminal, we used common 5 alkaline batteries. Integrated GSM communication module reduces the size and it is easy to install and post maintain.

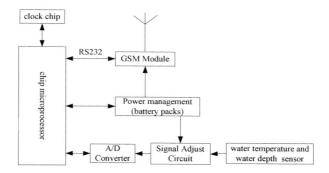

Fig. 2. The Block Diagram of the Terminal Monitoring

Terminal monitoring collects several data in different time points one day and uploads once data. Supply power to the module which will be used, and at the same time, other modules are off. After opening the terminal and initializing the function module, read a time per minute from the clock chip. Determine whether the data collection is over time. If yes, give the acquisition module power to make it work and after which make the module power off. Determine whether the data uploaded is over time. If yes, open the GSM module, and after the access network, send data through SMS. If the data uploaded failed to upload the next time, then the module power can be off. System can also receive commands to modify acquisition time and upload time.

3.2 The Design of Function of Monitoring Software

Groundwater monitoring and information management system composes the user management module, GIS module, data acquisition module, query management module, transaction management module and other modules such as help modules, the core module of which is GIS module, data acquisition module and the transaction management module. We can mark location information correspond to the site and hydrological information on the map in the GIS module according to collected intelligence information and data from the data acquisition module in all monitoring sites of groundwater. And generate a variety of chart information dynamically in the transaction management module after statistical analysis of the data, which can present to users visually and vividly. System has the function of user interaction, GIS and spatial analysis capabilities. The structure of the whole function module of the system is shown in Figure 3.

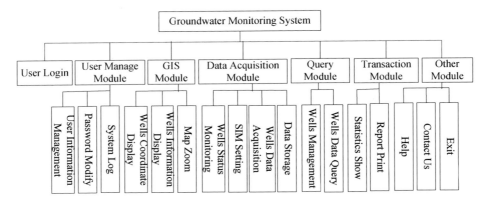

Fig. 3. The Design of the Overall Function of Groundwater Monitoring and Information Management System

GIS Module

The module is based on the secondary development of ArcGIS, and realizes the management of the monitoring site more intuitive and easier, through spatial data processing.

GIS module is mainly responsible for labeling and showing the collecting intelligence data back from the groundwater monitoring sites on the loading spatial map, as the mark of the new monitoring sites and so on. By clicking on the monitoring site marked on the map, we can easily see all kinds of information about the site, and draw diagrams of the site's hydrological information, which can show the trend of hydrological information more intuitively. GIS module also includes features of the usual zoom and dragging and has a basic view of Eagle Eye.

Data Acquisition Module

The module is the core module, which controls terminal state of work or gets information about the groundwater condition information depending PC software communicating with the monitor terminal of the next crew via the serial port and GSM modules. The key functions of the module include: 1) condition monitoring of monitoring site: requiring to monitoring points to report the state at any time; 2) SIM card settings: set SIM card number to receive data, which includes temporary and permanent sets; 3) the data collection of site: including historical data collection and data acquisition, the data collection process of the site is shown in Figure 3; 4) data storage: the collected data into a database of water resources.

Management Module

The module is mainly responsible for statistical analysis of hydrological information and the data report printing of system. Statistical analysis is to analyze the water temperature and water level data trends of a monitoring site but at different time or make the comparative analysis of the hydrologic data of different monitoring sites but at the same time. And draw the corresponding graph information to indicate the results more intuitively. Report printing is to generate reports about the data of a monitoring site in a time period according to the format required by the user in order to facilitate a deeper analysis of them.

4 Key Technology

4.1 Mark the New Monitoring Site on Map

When the user needs to add a new monitoring site, add and set the relevant information of the new monitoring site involving Device ID, Well Number, Area, Longitude, Latitude, altitude, SIM card number and so on in the wells management module. Save it, and the system will call the site marking function in the GIS module automatically to mark the new monitoring site on the space map by its Longitude and Latitude information.

The key code of the site marking as follows:

Fig. 4. The Block Diagram of the Terminal Monitoring

wellRow.BeginEdit();

ESRI. ArcGIS. Mobile. Geometries. Point ptGeometry = wellAction. Geometry as ESRI. ArcGIS. Mobile. Geometries. Point;

wellRow["Longitude"] = Math. Round(ptGeometry. Coordinate.X, 4, MidpointRounding.ToEven);

wellRow["Latitude"] = Math. Round(ptGeometry. Coordinate.Y, 4, MidpointRounding.ToEven);

wellRow.Geometry = ptGeometry;

wellRow.EndEdit();

4.2 Hydrologic Data Charting

When the user selected a site, a start date and a end date, the system can draw out the curve of the water level and the water temperature during the specified time period. The curve can make the user find the hydrological information trends intuitively. Shown in Figure 5, system drew out the chart of the hydrological information by the collected data.

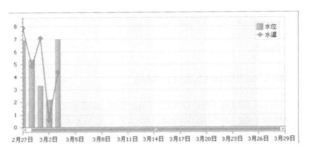

Fig. 5. The Hydrological Information Trends Curve of the Specific Monitoring Site

The key code of drawing out the chart as follows:

chartControl1.DataSource = dataTable;

XYDiagram xyDiagram = chartControl1.Diagram as XYDiagram;

Series[] series = chartControl1.Series.ToArray();

series[0].DataSource = dataTable;

series[0].DataFilters[0].Value = _Well.ID;

series[0].DataFilters[1].Value = minDate;

series[0].DataFilters[2].Value = maxDate;

series[1].DataSource = dataTable;

series[1].DataFilters[0].Value = _Well.ID;

series[1].DataFilters[1].Value = minDate;

series[1].DataFilters[2].Value = maxDate;

5 Conclusion

The Groundwater monitoring and information management system makes full use of the powerful spatial analysis and mapping functions of GIS to express the hydrological conditions intuitively. And to implement the remote hydrological information data transmission from the distributed monitoring site with less amount of data by the SMS business of GSM. This system not only has the advantages of stable and reliable performance of the single chip but also the advantages of convenient data communication, easy installation, low running cost of the GSM SMS business. By selective separation of power for each functional module, the energy consumption has been reduced greatly. And it also has the powerful spatial analysis and mapping functions of GIS module. In a word, the system will have a good market prospects.

The development of the system provides a reference for strengthening the management of groundwater resources, protecting groundwater resources and water environment, optimizing the groundwater monitoring networks, and the new method using in groundwater monitoring work. It creates good economic and social benefits.

Acknowledgments

This work was supported by a grant from the Programs for Science and Technology Development of Shaanxi Province (2006K05-G18), the Education Department Special Fund of Shaanxi Province (05JK159) and the Graduate Innovation Fund of Shaanxi University of Science and Technology. My tutor, Yujie Zhang, provided valuable comments that helped to improve this paper.

References

1. Peng, B., Zhang, Y.C.: Design and implementation of dynamic signal acquisition system for groundwater level. Network and Communication 16, 50–53 (2009)
2. Liu, J.: The design and implementation of groundwater information management system. Geotechnical Investigation and Surveying 12, 70–73 (2009)
3. Zhang, W.-h., Zhao, Y.-s., Di, Z.-q., Guo, X.-d.: The Design and Realization of Groundwater Resources and Geological Environment Information System Based on ArcGIS Engine. Journal of Jilin University 36(4), 574–577 (2006)
4. Liu, J., Shan, X.-f., Duan, Z.-d.: The design and realization of a shallow groundwater system based on GIS. Engineering of Surveying and Mapping 18(4), 62–65 (2009)
5. Huo, D.-y., Guo, R.-h., Chen, Y.-g.: The Design and Implementation of the Dynamic Display System of Groundwater Resources in Heze City. EWRHI 29, 180–181 (2008)
6. Dai, C.-l., Chi, B.-m., Lin, L., Shi, F.-z.: The Analysis and Design of the Groundwater Monitoring and Management Information System based on GIS. Remote Sensing Technology and Application 20(6), 625–629 (2005)
7. Huang, Y.-x., Liu, A.-g., Yang, J., Yang, W.-g., Hu, Y.-l., Wang, H.: Design and Application of Drought-Monitoring System Based on WebGIS for Hubei Province. Chinese Journal of Agricultural Meteorology 31, 112–116 (2010)

Design of Hydrological Telemetry System Based on Embedded Web Server

Yujie Zhang, Yuanyuan Zhang, Sale Xi, and Jitao Jia

College of Electrical and Information Engineering,
Shaanxi University of Science and Technology
Xi'an, China
zyy19870215@163.com

Abstract. Regarding the problem of poor real-time status and high maintenance costs of the data transmission in hydrological telemetry system, the paper proposes a design of hydrological telemetry system based on embedded Web server, describes the overall structure principle of the system in great detail and introduces the design of hardware and software of the embedded Web server. The system can not only improve the real-time status of the transmission of hydrological data and the accuracy of hydrological data, but it can also greatly reduce system cost and maintenance expenses.

Keywords: embedded Web server; hydrological telemetry; LPC2378; μClinux.

1 Introduction

Hydrological telemetry system is mainly for real-time monitoring the monitoring area's surface water, groundwater and rainfall, and on-site monitoring is widely distributed in the region. Traditionally, hydrological monitoring equipment and monitoring management center communicate through radio station and other wireless means of communication, so that communication distance is limited and the maintenance costs are high. In recent years, with the embedded technology's development, Internet's popularization as well as the GPRS wireless network's maturity, the remote monitoring system based on Internet has become the tendency of surveillance.

This paper presents a hydrological telemetry system based on embedded Web server. By using μclinux operating system, this system completes the work of real-time monitoring, management controlling and real-time hydrological data processing and transmission of the monitoring area's surface water, groundwater and rainfall. It uses the existing Internet network and the GPRS network, economizing costs on wiring and line maintenance. It also realizes real-time monitoring of any site anywhere at any time and greatly reduces the installation costs at the same time.

2 Working Principle and Function Design of the System

Hydrological telemetry system based on embedded Web server is divided into three modules: the hydrological monitoring module, the embedded Web server module and the remote client module, which are shown in Figure 1.

Hydrological monitoring module is responsible for collecting the on-site hydrological data and sending the collected data to the embedded Web server. Embedded system platform processes the collected data, and then it sends the collected data to the remote client through the Internet network to users for real-time monitoring and inquiry.

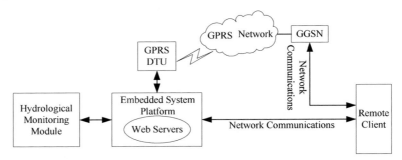

Fig. 1. the Schematic Diagram of Groundwater Monitoring and Information Management System

There are two different ways of communication between the embedded system platform and remote client. In areas covered by Internet network, users can directly log Internet through the embedded Web server to connect with the remote client. In areas Internet network can not reach, users can log mobile gateway GGSN through GPRS DTU to connect with the remote client.

This system does the work of real-time online monitoring of the monitoring area's surface water, groundwater and rainfall and its main functions are: monitoring real-time hydrology, inquiring historical data, setting system parameters, and inquiring system log.

Hydrological telemetry system based on embedded Web server realizes the remote access of hydrological telemetry equipment, parameters allocation and management through constructing Web server in the embedded system. It is simple and without installing a dedicated PC software in the client. Using the existing Internet network and GPRS networks, the system reduces the network costs and network maintenance expenses.

3 Hardware Design of the System

The hardware platforms of this system are made up of 32-bit embedded processor chips LPC2378, signal conditioning circuits, SD card, RS232 interfaces, I2C, SPI, keyboard and display circuit, Ethernet interfaces and Flash memory circuit. The hardware platform structure diagram of this system is shown in Figure 2.

Choose NXP Company's LPC2378 processor, which is responsible for the overall operation of the system. LPC2378 internal resources are abundant, which integrated UART, CAN-bus, and I2C, SPI / SSP, SD / MMC interface. One important reason of using this processor is that the inner part of the processor integrates a fully functional

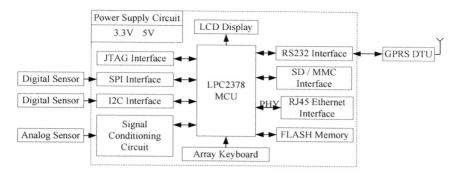

Fig. 2. Hardware Diagram of the System

10/100M Ethernet MAC controller, which can provide optimum performance by using the accelerating DMA hardware. We can conveniently connect it with the off chip using RMIII and simplify the Ethernet interface design through MIIM (independent media interface management) serial bus. Ethernet PHY interface chip chooses the integrated and triggered interrupt MIII management interface DM9161A, which can be easily connected with LPC2378 interface.

Signal conditioning circuit is used for connecting output signals such as, 4-20mA, 0-5V, 1-5V analog sensors of the standard signal. I2C, SPI interfaces are used for connecting the output signal matches with I2C, SPI standard digital sensor. RS232 interface is used for connecting GPRS Modem to achieve connection of the monitoring terminal with a wireless Internet. SD card interface is used to store hydrologic data, for hydrologic information management center to inquire. External Flash is used to store page data of Web server and working parameters of the system.

4 Software Design of the System

When the user needs to add a new monitoring site, add and set the relevant information of the new monitoring site involving Device ID, Well Number, Area, Longitude, Latitude, altitude, SIM card number and so on in the wells management module. Save it, and the system will call the site marking function in the GIS module automatically to mark the new monitoring site on the space map by its Longitude and Latitude information.

Software design of the system uses the model of layered software structure, dealing with different levels of modular design. The module is divided according to different events and duties: each functional module has a single task and different tasks use the message mechanism to communicate.

Software design of the system has five levels, including the bottom drive layer, the operating systems layer, the component layer, the interface layer and the application layer from bottom to top. The software structure diagram of this system is shown in Figure 3.

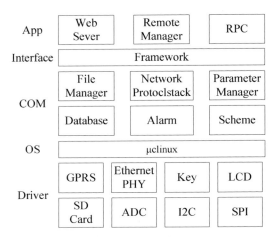

Fig. 3. New monitoring sites marking

Driver layering mainly realizes LPC2378 processor peripheral hardware device driver, so that applications and hardware devices can communicate with each other. All sorts of underlying hardware device drivers are SD card, GPRS, RS232, LCD, KEY specific, Ethernet PHY, etc.

The system uses μclinux operating system, and kernel version is 2.6. This operation system has good real-time capability, stability, and excellent network support. It is free for the source code and relatively easy to develop, which can effectively shorten the development cycle and reduce development costs.

Component module is the core of the telemetry system, and the realization of all functions of monitoring system is completed in this module. Component module achieves document management, embedded network protocols, system parameters management, alarm management, hydrologic data storage, and database management, etc.

The framework layer, the interactive interface between each functional module and the application procedure, realizes message scheduling and message types' distribution and management between the application procedure and system. The framework layer is mainly responsible for two direction news's distribution in the application procedure and system, in order to coordinate the relationship between each function modules' news. The advantage of using this design method is, for the top-level applications, that designers can ignore the format of all the news' interface in function module layer, using the unified call mode.

The application layer is the concrete realization of user application function, which falls into three parts: Web Server, Remote Manager and RPC. RPC is a remote procedure call protocol. Its function is to enable a remote application to call a local process function through RPC. From the view of execute effect, RPC is the same with the local call. RPC is divided into two parts: the server and the client. The server provides process function for remote calls, and the client sends remote call request to the server. In this system, the server is located in µclinux operating system, while the client is located in Windows system. They communicate through the Internet network. The Remote Manager mainly functions to monitor the remote client's data requests, controlling requests and response remote request, providing the corresponding server-side interface to remote Web management software, responsible for hydrologic data and control command transmitting and management. Web Server module constructs an embedded Web Server in monitoring device terminal, so that remote client can directly access and control monitoring equipment through the browser.

This kind of hierarchical design method makes a clear development division of each functional module. The development work of different modules can be done at the same time. This method not only reduces the dependence between layers, making each functional module have good scalability and reusability, but also shortens the development cycle and improves work efficiency.

5 Software Design of Web Client

5.1 Realization of GPRS Accessing to Internet

There are many methods for the realization of GPRS accessing to Internet, such as DDN dedicated line access, APN dedicated line access, DNS dedicated line access, and DDN, APN access methods require that the IP address of the client is fixed. While the remote client of the hydrological telemetry system can be any computer of Internet, the system uses DNS access.

It needs applying for a tertiary or secondary domain name and then assigns the dynamic DNS service in the domain management centre. The process of establishing the Internet is as the following. Remote client computer will get a dynamic IP when it opens, then through the work of the DDNS server and the Client software to refresh the domain name and compare with IP. Monitoring terminal gets dynamic Internet IP address and establishes a connection with the Internet through GPRS DTU successfully logging GGSN and then through the method of DNS interprets the positive and negative domain name and establishes a connection with the remote client.

5.2 The Software Design of Embedded Web Server

The implementation of the basic function between embedded web server and the common web server is the same. They both must be able to receive the client's request, analysis and response to requests and finally to return the requested. Embedded Web services are based on the HTTP protocol. Web server stores the services which are set in HTML in the internal FLASH memory in the form of a web page.

After remote client and the embedded Web server establish the connection through the handshake, the client sends the request to the Web server by way of port 80 which is defaulted by the browsing web page, and the system will start the Web server. Then according to the request of the Get data packet the system sends a corresponding page to Client. The flow chart of Embedded Web server receives and sends data packets is shown in Figure 4.

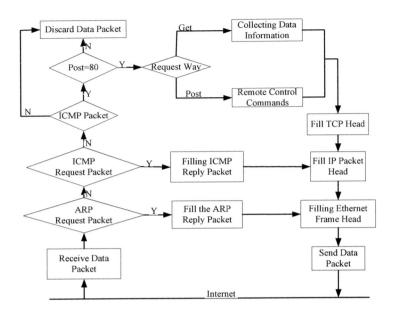

Fig. 4. Embedded Web Server Packet Processing Flow

Embedded Web server handles two kinds of requests: Get request is that the client requires the server to upload Water Data; Post request is to send control commands to the server. Clients and Web servers exchange information through the CGI scripting, and the client can complete remote monitoring hydrological data according to the options provided on the receiving page.

Java Applet technology is adopted on data calculation and graphical display: translate and edit the graphic displayed code into class files and store them in ROM, which can be downloaded from the server when used. For the client browser, hydrological data can be displayed graphically in real time.

6 Conclusion

Hydrological telemetry system based on embedded web server uses the internal resource-rich 32-bit ARM processor LPC2378 as the control center for design. It can realize function of all parts, and is remote networked and can real-time monitor the area's reservoirs, groundwater, rivers and lakes water level, temperature and rainfall. In addition, it has advantages. of strong real-time performance, wide range of monitoring, high reliability, scalability and so on.

References

1. Wang, B.: GPRS technologies in the application of automatic hydrologic information telemetry system. Water and Water Engineering Journal 20(03), 132–134 (2009)
2. Li, J.: Based on GPRS river levels real-time monitoring system. Journal of HeiLongjiang Technical Information (32), 51–51 (2008)
3. Gao, X.: MCU Embedded Internet Technology Web application implementation. Microcomputer & its Applications (11), 55–57 (2010)
4. Huang, B., Li, D., He, C.: Dynamic Web technology's realization in real-time monitoring system. Microcomputer information (14), 37–39 (2007)
5. Xie, S., Xu, B.: Embedded Web server design and CGI implementation. Computer Engineering and Design 28(7), 1598–1600 (2007)
6. Liu, Z.: Based on embedded Web of remote real-time monitoring technology research. Computer Engineering and Design 28(15), 3734–3736 (2007)
7. Fu, B., Wang, Z., Ban, J., et al.: Based on CGI embedded monitoring system dynamic data exchange realization 31(24), 196–197 (2005)

Passive Bistatic Radar Target Location Method and Observability Analysis

Cai-Sheng Zhang[1,2], Xiao-Ming Tang[1], You He[1], and Jia-Hui Ding[2]

[1] Research Insti. of Infor. Fusion, Naval Aeronautical and Astronautical Univ., Yantai, China 264001
[2] Nanjing Research Institute of Electronics Technology, Nanjing China 210039
caifbi2008@yahoo.com.cn

Abstract. Passive bistatic coherent radar system operates with distinct non-cooperative transmitter and receiver located at different sites. In such spatial separation, the critical issues in attempting to operate a bistatic system with a noncooperative illuminator are target location and its observability, which were explored and solved in this paper. The solution to the bistatic triangle for target range from the receiver requires calculating the azimuth of the direct path pulse, which allows accurate measurement of the bistatic range to the target and computation of target slant range. Finally, the observability of the system was demonstrated.

Keywords: Passive Bistatic Radar (PBR); Passive Coherent Location (PCL); Observability Analysis.

1 Introduction

Since the navigation radar system was explored as a non-cooperative illuminator in our passive bistatic radar application, the baseline between radiation source and the receiving site is unknown. In order to resolve the bistatic triangle, we need to estimate the angles of illuminator and target, as well as beam scanning angle [1,2]. If the transmitter and receiver travel at a constant velocity, for the uniform cycling scanned antenna, a simple tracker can be used to record the instant that the main beam of the transmitting antenna meets the main lobe of the receiver to estimate the scanning angle.

Reference [2] solves the bistatic triangle by law of cosines when the baseline is prior to know. In this paper, we will first introduce the scheme of a non-cooperative bistatic receiving system as well as its bistatic geometry, and on this basis, the estimation method of beam scanning angle is given in section 3.1. Performance analysis of this method is presented in section 3.2. The basic principle of positioning is proposed in the following section and observability analysis is completed in section 4. All these analysis demonstrate the feasibility of non-cooperative bistatic detection system.

2 System Configuration

A scenario with non-cooperative radar illuminator considered in this paper is illustrated in Figure 1. Assume that the transmitter is mechanically scanned in azimuth, and the receiver is stationary while the target is moving. Both the transmitter and receiver are focusing on the moving target. The receiving system intercepts the direct-path waveform transmission through a reference antenna when it tunes to the transmitting frequency, and target reflection echoes are intercepted by the target antenna. Time and phase synchronization need to be completed via direct-path signal. Then, optimum detection and parameters estimation are commonly involved in passive coherent processing [3,4].

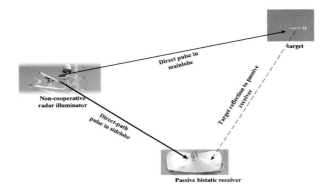

Fig. 1. Signal Path in Passive Bistatic Radar System

3 Solution to Bistatic Triangle

Measurement of certain illuminator's parameters is required to solve for the range of the target from the receiver in the bistatic triangle. The parameters that must be measured depend upon the geometry, which parameters are known or can be measured, and the equation chosen for solving the bistatic triangle. A generalized coordinate system, along with all kinds of parameters, is depicted in Figure 2. we have

$$\tau = N\Delta t . \qquad (1)$$

then

$$R_t + R_r - L = c\tau . \qquad (2)$$

where c is the speed of light, N is the number of range bins between the direct path pulse and the target detection range bin, and Δt is the width in seconds of each range bin. Obviously, the equation that we can chose to solve target range is largely depends on the type of illuminator used in the system and the relative location of the targets with respect to illuminator and receiver. Under normal circumstances, for this system, the azimuth θ_r and elevation ϕ_r, path delay $\Delta \tau$ can be obtained more accurately.

Then, the necessity to obtain the target range is that one parameter of θ_t and L must be measurable. Since the navigation radar system explored in the passive bistatic radar is non-cooperative illuminator, the baseline L is unknown. So, we need to measure the value of θ_t.

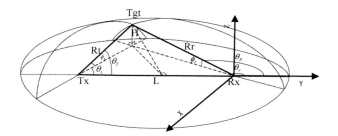

Fig. 2. Schematic of typical geometry relationship of passive bistatic radar system

3.1 Estimation of Transmit Azimuth θ_t

The use of a non-cooperative illuminator requires estimating the lobe direction which is a function of time [5]. As illustrated in Figure 3, a peak is seen at the receiver Rx at time τ_0 when it is illuminated by the circular scanning illuminator. Subsequently, a target Tx is illuminated and the scattered path reflection is detected at the receiver at time τ_1. Relative to scan periods, transmission delays are regarded as negligible at this point, which will illustrated in section 3.2. Then

$$\theta_t = \frac{\tau_1 - \tau_0}{T_S} \times 360° . \tag{3}$$

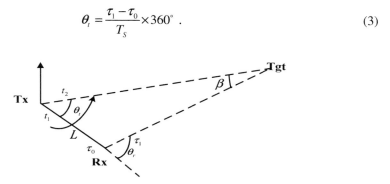

Fig. 3. Typical geometry of passive bistatic radar system in plane

3.2 Feasibility Analysis

Let the scan period of transmitter antenna be T_s, and sweep in counterclockwise as shown in Figure 3. Assume that the pulse radiate at the moment t_1, and the main beam is in the direction of the target T, then the receiving station Rx will intercept this pulse at time $t_1 + (R_t + R_r)/c$. With the scanning of the Tx antenna, the main beam will illuminate the Rx station at the moment t_2, then the pulse will arrive at Rx at time $t_2 + L/c$. Therefore, the estimation of θ_t will be

$$\theta_t = \frac{t_2 - t_1}{T_s} \times 360° . \quad (4)$$

However, due to the system is non-cooperative, the pulse emission time t_1 and t_2 are unknown to Rx. So azimuth estimation will be conducted by signals intercepted by Rx. As it well known, signal processing method will estimate the echo signal relative to the direct wave of the delay effectively. Then time delay will be

$$\tau = \left(t_2 + \frac{L}{c}\right) - \left(t_1 + \frac{R_t + R_r}{c}\right) = (t_2 - t_1) - \left(\frac{R_t + R_r}{c} - \frac{L}{c}\right) . \quad (5)$$

Here, we assume that

$$\Delta \tau = \frac{R_t + R_r}{c} - \frac{L}{c} . \quad (6)$$

Let $R_t + R_r - L = 90$ km, it is easy to get $\Delta \tau = 300$ μs. And the general scanning rate is approximate to 21 r/min for navigation radar, and then its corresponding scan period will be 3s. Therefore, the estimation error introduced by τ will be

$$\Delta \theta_t = \frac{360°}{T_s} \Delta \tau \approx 3.6° \times 10^{-2} . \quad (7)$$

Judging from Eq.(11), the current angle measurement accuracy achieved is acceptable for our application of surveillance. Furthermore, we must note that the results calculated in the following are relative to the time of $t_1 + R_t/c$. If the target flies at twice the speed of sound, the target distance its movement is of about 7m, which is also acceptable.

3.3 Equations for Calculating Target Range

We can estimate θ_t by Eq.(7), while θ_r can be measured by other high-precision sensor in Rx. By the geometric relationship of the passive bistatic system as shown in Fig.2, according to the law of Sine, we have

$$\frac{L}{\sin(\theta_r - \theta_t)} = \frac{R_{t\perp}}{\sin \theta_r} = \frac{R_{r\perp}}{\sin \theta_t} . \quad (8)$$

From Fig.2, it is easy to get

$$R_r = \frac{R_{r\perp}}{\cos \phi_r} \Rightarrow R_{r\perp} = R_r \cos \phi_r . \quad (9)$$

$$H = R_{r\perp} \tan \phi_r = R_r \sin \phi_r . \quad (10)$$

$$R_{t\perp}^2 + H^2 = R_t^2 . \tag{11}$$

$$\cos\theta_R = \cos\phi_r \cos\theta_r . \tag{12}$$

$$L = \frac{R_r \cos\phi_r}{\sin\theta_t} \sin(\theta_r - \theta_t) . \tag{13}$$

Then, we have

$$R_{t\perp} = \frac{R_r \cos\phi_r \sin\theta_r}{\sin\theta_t} . \tag{14}$$

$$R_t = c\tau + \frac{R_r \cos\phi_r}{\sin\theta_t} \sin(\theta_r - \theta_t) - R_r . \tag{15}$$

$$H = R_r \sin\phi_r . \tag{16}$$

Substitute Eq.(18), (19), (20) into (15), we can get

$$\left[\frac{R_r \cos\phi_r \sin\theta_r}{\sin\theta_t}\right]^2 + \left[R_r \sin\phi_r\right]^2 = \left[c\tau + \frac{R_r \cos\phi_r}{\sin\theta_t} \sin(\theta_r - \theta_t) - R_r\right]^2 . \tag{17}$$

The Solution to Eq.(21) will be

$$R_r = c\tau \frac{\left\{\left[1 - \frac{\cos\phi_r}{\sin\theta_t}\sin(\theta_r - \theta_t)\right] + \sqrt{\sin^2\phi_r + \left(\frac{\cos\phi_r \sin\theta_r}{\sin\theta_t}\right)^2}\right\}}{\left[\frac{\cos\phi_r}{\sin\theta_t}\sin(\theta_r - \theta_t) - 1\right]^2 - \sin^2\phi_r - \left(\frac{\cos\phi_r \sin\theta_r}{\sin\theta_t}\right)^2} . \tag{18}$$

Then, the height of the target will be

$$H = R_r \sin\phi_r . \tag{19}$$

Thus, the passive bistatic radar system has the potential capacity of altimetry.

4 Observability Analysis

According to references [6-9], there are two kinds of method to demonstrate the observability of a maneuvering target trajectory. One is to prove that the system has the tracking solution by showing the uniqueness criterion for the solution of functional equations directly. The other is to analysis the observability matrix from a set of non-linear measurement equations or the matrix of FIM. Since the equations for TDOA or Doppler shift are nonlinear, one cannot transform them to linear and cannot solve them directly. We adopt the second approach in this paper.

In fact, for the moving target in the three-dimensional, it can be completely restricted by the movement track and observation stations onto the plane described by the observation station and moving objects [9]. Therefore, we discuss the problem in two-dimensional plane. The geometry relationship of the passive bistatic system is shown in Figure 4. Let the receiver locate at the origin of the Descartes coordinate, while the location of the illuminator and target be $[x_T, 0]$ and $[x_{tgt}, y_{tgt}]$, respectively. Without loss of generality, assume that the azimuth in counter-clockwise is positive.

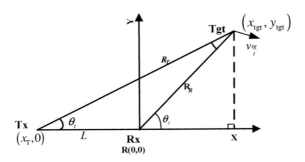

Fig. 4. Cartesian Coordinates of passive bistatic radar system

Suppose the state vector for the moving target be $\mathbf{X} = [x_0 \ y_0 \ v_x \ v_y]^T$, then the corresponding equation of the target is linear, i.e.

$$\mathbf{X}_k = \mathbf{\Phi}_k \mathbf{X}_0 . \tag{20}$$

where

$$\mathbf{\Phi}_k = \begin{bmatrix} 1 & 0 & kdt & 0 \\ 0 & 1 & 0 & kdt \\ 0 & 0 & 1 & 0 \\ 0 & 0 & 0 & 1 \end{bmatrix}.$$

that is to say

$$x_{tgt}(k) = x_0 + kdt v_x \qquad y_{tgt}(k) = y_0 + kdt v_y . \tag{21}$$

In order to facilitate the process of derivation, assume that the platform is stationary. According to [10], the bistatic Doppler frequency shift will be

$$f_d(k) = \frac{1}{\lambda} \left[\frac{dR_T(k)}{dt} + \frac{dR_R(k)}{dt} \right] = \frac{1}{\lambda} \left[\frac{x_{tgt} - x_T}{R_T} \frac{dx_{tgt}}{dt} + \frac{y_{tgt}}{R_T} \frac{dy_{tgt}}{dt} + \frac{x_{tgt}}{R_R} \frac{dx_{tgt}}{dt} + \frac{y_{tgt}}{R_R} \frac{dy_{tgt}}{dt} \right]$$

$$= \frac{k}{\lambda} \left[v_x \cos \theta_t(k) + v_y \sin \theta_t(k) + v_x \cos \theta_r(k) + v_y \sin \theta_r(k) \right]. \tag{22}$$

Where $R_T(k) = \sqrt{(x_{tgt} - x_T)^2 + (y_{tgt} - y_T)^2}$, $R_R(k) = \sqrt{x_{tgt}^2 + y_{tgt}^2}$, and c is the speed of light. Let the Doppler observation sequence be $\mathbf{Z} = [f_d(1) \; f_d(2) \; f_d(3) \; f_d(4)]$, and then the corresponding observation matrix will be $\mathbf{H}(1,4) = \left[\dfrac{\partial f_d(k)}{\partial \mathbf{X}}\right]_{\mathbf{X}=\mathbf{X}_k}$. The eigenvalues and eigenvectors of Gramer matrix $\mathbf{H}^T\mathbf{H}$ will present a complete observability description of the system.

They will reflect the strength of observability in quantity. However, it is not easy to obtain the analytical solution to the Gramer matrix. Since the value of $|\det(\mathbf{H})|^2$ is the reciprocal of uncertainty ellipsoid volume measurement, it is maximum likelihood estimation of the accuracy measurement under the conditions of the Gaussian. Then $\rho = \det(\mathbf{H})$ can be defined as a measurement parameter of observability. Therefore, we can judge the observability by the determinant of $\mathbf{H}$ [9]. With respect to the state vector, the partial derivative of $f_d(k)$ will be

$$\frac{\partial f_d(k)}{\partial v_x} = \frac{1}{\lambda}(\cos\theta_t(k) + \cos\theta_r(k)) = \frac{1}{\lambda}\cos\alpha(k)\cos\beta(k) = c_k;$$
$$\frac{\partial f_d(k)}{\partial v_y} = \frac{1}{\lambda}(\sin\theta_t(k) + \sin\theta_r(k)) = \frac{1}{\lambda}\sin\alpha(k)\cos\beta(k) = d_k.$$
(23)

$$\frac{\partial f_d(k)}{\partial x_0} = \frac{\partial f_d(k)}{\partial v_x}\frac{\partial v_x}{\partial x_0} = -\frac{c_k}{k\mathrm{dt}}; \quad \frac{\partial f_d(k)}{\partial y_0} = \frac{\partial f_d(k)}{\partial v_y}\frac{\partial v_y}{\partial y_0} = -\frac{d_k}{k\mathrm{dt}} \quad (24)$$

Then, we get the corresponding determinant of $\mathbf{H}$ as follow

$$\det(\mathbf{H}) = \begin{vmatrix} -\dfrac{c_1}{\mathrm{dt}} & -\dfrac{d_1}{\mathrm{dt}} & c_1 & d_1 \\ -\dfrac{c_2}{2\mathrm{dt}} & -\dfrac{d_2}{2\mathrm{dt}} & c_2 & d_2 \\ -\dfrac{c_3}{3\mathrm{dt}} & -\dfrac{d_3}{3\mathrm{dt}} & c_3 & d_3 \\ -\dfrac{c_4}{4\mathrm{dt}} & -\dfrac{d_4}{4\mathrm{dt}} & c_4 & d_4 \end{vmatrix} \quad (25)$$

then

$$\det(\mathbf{H}) = -\frac{\Lambda}{24\lambda^4(\mathrm{dt})^2}. \quad (26)$$

where

$\Lambda = \cos\beta(1)\cos\beta(2)\cos\beta(3)\cos\beta(4) \times$
$[\sin(\alpha(3)-\alpha(2))\sin(\alpha(4)-\alpha(1)) + 3\sin(\alpha(3)-\alpha(4))\sin(\alpha(2)-\alpha(1))]$

From Eq.(31), it can conclude that

1) if $\beta(k) = -\frac{\pi}{2}$, then $\theta_t(k) = \theta_r(k) + \pi$, $\det(\mathbf{H}) = 0$. At this point, the target will be in the baseline, then the system is unobservable;

2) if $\beta(k) \neq -\frac{\pi}{2}$, while $\alpha(i) = \alpha(j) = 0, i, j = 1,2,3,4, i \neq j$, then $\det(\mathbf{H}) = 0$, the target will appear in the extension of baseline at this case, the system is unobservable, either;

3) if $\beta(k) \neq -\frac{\pi}{2}$, while $\alpha(i) = \alpha(j) \neq 0, i, j = 1,2,3,4, i \neq j$, then $\det(\mathbf{H}) = 0$. In fact, exclude the case when the illuminator and receiver still are static, if the bistatic geometry satisfy this condition, the target must travel back and forth between the hyperbolic with radiation source and receiving station as foci, then the system will be unobservable. In practice, we do not need to consider such kind target.

Therefore, the system observable necessity will be

$$\begin{cases} \theta_t(k) \neq \theta_r(k) + \pi; \\ \theta_t(k) + \theta_r(k) \neq \theta_t(k+n) + \theta_r(k+n), n = 1,2,3. \end{cases} \quad (27)$$

Furthermore, if the time difference of arrival TDOA is used for observational analysis, the same conclusions will be achieved. If the signal frequency and the observation time interval are determined, the strength of observability depends on the sign of Λ, i.e. the related transmit and receiving angle, as shown in Eq.(27).

5 Conclusions

The geometric location method for passive bistatic radar system is proposed in this paper. We first established the schematic of a common non-cooperative bistatic receiving system, as well as its bistatic geometry, and on this basis, the estimation method of transmit azimuth is proposed. Performance analysis of this method is also presented. Observability analysis is completed in detail. The system is unobservable when the target travels along the baseline or its extension. All these analysis demonstrated the feasibility of these kinds of non-cooperative bistatic radar systems.

References

1. Thompson, E.C.: Bistatic Radar Noncooperative Illumination Synchronization Techniques. In: IEEE Radar Conference, pp. 29–34 (1989)
2. Hawkins, J.M.: An Opportunity Bistatic Radar. In: IEE Radar Conference, pp. 318–322 (1997)
3. Racal Outlines Non cooperative Bistatic Radar. Microwave J. (7), 45 (June 2000)
4. Triton: Non Co-operative Bistatic Radar System. Thales Sensors (2001)
5. Terje, J., Karl, E.O.: Bi- and Multistatic Radar. Norwegian Defense Research Establishment (FFI). ADA470685 (2006)

6. Li, W.C., Wei, P., Xiao, X.C.: TDOA and T^2/R Radar Based Target Location Method and Performance Analysis. In: IEE Proc. Radar, Sonar and Navig., vol. 152(3), pp. 219–223 (2005)
7. Sun, Z.K., Zhou, Y.Y., He, L.X.: Single/Multiple Station for Active and Passive Location Technology, pp. 199–225. National Defense Industry Press, Beijing (1996) (in Chinese)
8. Sun, Z.K., Guo, F.C., Feng, D.W., et al.: Single Observer Passive Location and Tracking Technology, pp. 206–208. National Defense Industry Press, Beijing (2008) (in Chinese)
9. Zhou, Y.Y., Sun, Z.K.: Positioning and Observability for Passive Detection Radar. Electronics 22(3), 51–57 (1994) (in Chinese)
10. Yang, Z.Q., Luo, Y.J.: Bistatic/Multistatic Radar System, pp. 3–12, 75–76. National Defense Industry Press, Beijing (1999)

Analysis of Coherent Integration Loss due to FSE in PBR*

Cai-Sheng Zhang[1,2], Xiao-Ming Tang[1], You He[1], and Jia-Hui Ding[2]

[1] Research Institute of Information Fusion, Naval Aeronautical and Astronautical Univ., Yantai, China 264001
[2] Nanjing Research Institute of Electronics Technology, Nanjing China 210039
caifbi2008@yahoo.com.cn

Abstract. In passive bistatic radar system, the receiver needs to independently solve the problem of frequency synchronization via the fluctuating direct-path signal. Under the circumstance, it is difficult to achieve highly accurate frequency synchronization since the instantaneous frequency measurement accuracy depends on the signal to noise ratio of the direct-path signal. In this paper, we focused on the impact of FSE on passive bistatic radar system. Since the imbalance between the in-phase(I) and quadrature(Q) component is a function of frequency, the imperfection arises from FSE in digital I and Q components has been introduced. The integration loss was calculated with respect to amplitude and phase imbalance factors. An analysis that explains how the extraneous components generated is presented. Then NIP was defined to evaluate the impact of FSE in quantity. Simulation results demonstrated that the interference frequencies caused by FSE will deteriorate the performance of weak target detection and parameter estimation.

Keywords: Passive Bistatic Radar(PBR);Coherent Integration Loss; Frequency Synchronization Error (FSE); Normalized Interference Power (NIP).

1 Introduction

Passive bistatic receiving system based on non-cooperative transmitters needs to independently solve time, frequency and space synchronization[1]-[2]. Frequency synchronization between the transmitting platform and receiving platform can be only achieved by the frequency estimation of direct-path signal. A fixed value of intermediate frequency is added to the estimated frequency, which is the frequency of local oscillator[3]-[4]. If the illuminator frequency changes every coherent dwell, real time frequency measurement is required[5]. In practice, the accuracy of frequency estimation is limited by the signal to noise ratio of direct-path signal as well as its multipath effect. Therefore, it is difficult to obtain accurate frequency estimation of the transmitted signals. Then, there exists FSE between these two platforms. In practice, since the degree of amplitude and phase imbalance is a function of frequency. The actual

---
* This work is support by National Natural Science Foundation of China (60672139, 60972160 & 61002045).

amplitude and phase imbalance is deteriorated by the FSE, and then the mirror-frequency component will become large.

In theory, the static clutter echo should only appear in the vicinity of zero Doppler frequency cells after passing through the Doppler filter bank. However, the frequency that the receiver tuned into may not be the actual carrier frequency of the non-cooperative illuminator. The static clutter echo will not appear in the zero-Doppler clutter cell, and the nominal frequency of the target in the spectra is not the actual Doppler frequency of the target. Cross-correlation in wide sense is commonly used in non-cooperative bistatic radar for target detection and parameter estimation[4]-[7]. However, due to the effect of image frequency components, the output of cross-correlation processing give rise to a number of new interference frequency components, which will affect the target detection in certain frequency cells.

2 Effect of FSE on the Quadrature Processing in PBR

The main task of PBR is to intercept the signal and then down convert it to baseband. A typical block diagram of digital IF quadrature processing is shown in Fig.1[8]. The input IF signal digitalized by AD is mixed with the quadrature local oscillator, then passing through the FIR low band passing filter (LBP), and then amplified to the signal processor.

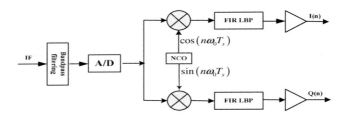

Fig. 1. IF quadrature processing block diagram

Let the digital input IF signal be $\tilde{s}(n)\cos\big((\omega+\omega_0)n+\phi(n)\big)$, where $\tilde{s}(n)$ is the signal amplitude, and ω_0 is the carrier frequency, ω is FSE caused by inaccurate frequency estimation, $\omega < \omega_0$ and $\phi(n)$ is the signal phase. Let the output of orthogonal components I and Q be

$$I(n) = \tilde{s}(n)\cos(\omega n+\phi(n))$$
$$Q(n) = \tilde{s}(n)(1-\varepsilon)\sin(\omega n+\phi(n)+\delta). \qquad (1)$$

where ε is amplitude imbalance, and δ is the degree of phase imbalance. The ideal output of the orthogonal sequences can be expressed as

$$I(n) = \tilde{s}(n)\cos\phi(n)$$
$$Q(n) = \tilde{s}(n)\sin\phi(n). \qquad (2)$$

Let $A(n) = \tilde{s}(n)$, $B(n) = \tilde{s}(n)(1-\varepsilon)$, then

$$I(n) = A(n)\cos(\omega n + \phi(n))$$
$$Q(n) = B(n)\sin(\omega n + \phi(n))\cos\delta + B(n)\cos(\omega n + \phi(n))\sin\delta. \quad (3)$$

denote the orthogonal output sequences as the real part and imaginary part, i.e.

$$Z(n) = I(n) + jQ(n)$$
$$= \frac{A(n) + jB(n)\sin\delta + B(n)\cos\delta}{2} e^{j(\omega n + \phi(n))} \quad (4)$$
$$+ \frac{A(n) + jB(n)\sin\delta - B(n)\cos\delta}{2} e^{-j(\omega n + \phi(n))}.$$

Regardless of the FSE and the amplitude-phase imbalance, then the IF quadrature output sequences $Z(n)$ can be expressed as

$$Z(n) = \tilde{s}(n)e^{j\phi(n)}. \quad (5)$$

Due to the effect of image frequency components, the actual power of signal component will be lower than the ideal output. Thus, the power loss of the desired signal is defined as

$$L = \frac{\left|\dfrac{A(n) + jB(n)\sin\delta + B(n)\cos\delta}{2}\right|^2}{|\tilde{s}(n)|^2} = \frac{1}{2}(1-\varepsilon)(1+\cos\delta) + \frac{\varepsilon^2}{4}. \quad (6)$$

Then the power ratio between the image frequency component and the actual signal component is

$$r = \frac{2(1-\varepsilon)(1-\cos\delta) + \varepsilon^2}{2(1-\varepsilon)(1+\cos\delta) + \varepsilon^2}. \quad (7)$$

Judging from Eq.(4) and Eq.(5), $-\omega$ is the image frequency caused by the amplitude-phase imbalance in IF quadrature processing. The spectra of the quadrature signal will not appear in the vicinity of zero frequency cells. Even if the ground clutter is static, the FSE will make the clutter appear at nonzero Doppler cell. According to Eq.(6), it seems that the power loss is only related to the degree of amplitude and phase imbalance, while have nothing to do with the FSE. However, the degree of amplitude and phase imbalance is a function of frequency [9]. The FSE will deteriorate the degree of amplitude and phase imbalance, and then the power of mirror frequency rate will become large.

3 Effect of FSE on Coherent Detection

In practice, only a few pulses intercepted in the space synchronous case are available for coherent integration, and the gain obtained in the cross-correlation process will not that large. The following qualitative analysis will illustrate how the nuisance frequency components come out. Although the cross-correlation detection is implemented after digital sampling, the following analysis will be based on their analog form for facilitating the derivation process. Without loss of generality, let the direct-path reference signal $\tilde{X}_T(t)$ be

$$\tilde{X}_T(t) = Ge^{j\omega t} + G_\varepsilon e^{-j\omega t}. \tag{8}$$

where the amplitude of the actual frequency ω is G, while the amplitude of image frequency is G_ε. Similarly, the target echo contained only one moving target can be expressed as

$$\tilde{X}_R(t) = Je^{j(\omega+\omega_d)(t-t_d)} + J_\varepsilon e^{-j(\omega+\omega_d)(t-t_d)} + He^{j\omega(t-t_d)} + H_\varepsilon e^{-j\omega(t-t_d)}. \tag{9}$$

where $\omega+\omega_d$ is the actual frequency of the target with amplitude J, while the frequency of clutter component is ω with amplitude J_ε, and t_d is time delay, the clutter amplitude in target channel is H, and the magnitude of the clutter's image frequency is H_ε. According to [4]-[7], the commonly used detection algorithm is to calculate the Range-Doppler two-dimensional cross-correlation function between the target signal and the reference signal in passive bistatic radar, that is

$$y(t_d, \omega_d) = \int_0^{t_i} \tilde{X}_T^*(t-t_d)\tilde{X}_R(t)e^{-j\omega_d t} dt. \tag{10}$$

where t_i is the coherent integration time, the superscript * is complex conjugate operator. Let

$$\tilde{y}(t_d) = \tilde{X}_T^*(t-t_d)\tilde{X}_R(t). \tag{11}$$

Substituting Eq.(8) and (9) into Eq.(11), we have

$$\tilde{y}(t_d) = \left(G^* e^{-j\omega(t-t_d)} + G_\varepsilon^* e^{j\omega(t-t_d)}\right)\left(Je^{j(\omega+\omega_d)(t-t_d)} + J_\varepsilon e^{-j(\omega+\omega_d)(t-t_d)} + He^{j\omega(t-t_d)} + H_\varepsilon e^{-j\omega(t-t_d)}\right). \tag{12}$$

Expanding Eq.(12), we have

$$\tilde{y}(t_d) = G^*\left(Je^{j\omega_d(t-t_d)} + J_\varepsilon e^{-j(2\omega+\omega_d)(t-t_d)} + H + H_\varepsilon e^{-j2\omega(t-t_d)}\right) \\ + G_\varepsilon^*\left(Je^{j(2\omega+\omega_d)(t-t_d)} + J_\varepsilon e^{-j\omega_d(t-t_d)} + He^{j2\omega(t-t_d)} + H_\varepsilon\right). \tag{13}$$

Obviously, there are seven output frequency components after Fourier transform of Eq.(13) during the coherent integration time. Besides to the target's Doppler frequency, there are other nuisance frequency components as shown in Fig.2.

If FSE is zero, the effect of amplitude and phase imbalance degree of processing can be ignored. However, as can be seen from Fig.2, if the powers of these nuisance frequencies caused by FSE are strong enough, weak targets echoes will be easily swamped in most Doppler cells, even cause false alarms.

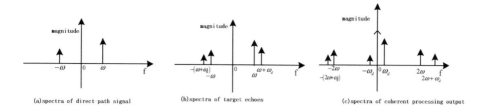

(a) spectra of direct path signal (b) spectra of target echoes (c) spectra of coherent processing output

Fig. 2. Spectra of cross-correlation output with FSE effect

4 Quantity Analysis

Under the circumstance of accurate frequency synchronization, if the phase imbalance is 1°, as long as the amplitude imbalance is less than -12 dB, the power of the image frequency is 30 dB smaller than the desired signal's, as illustrated in Fig.3. The interference of image frequency can be neglected. In general, the effect of amplitude and phase imbalance can be ignored in the active coherent radar. However, in the passive bistatic radar, the amplitude and phase imbalance will be deteriorated due to the effect of FSE. Thus, the impact of FSE needs to discuss further.

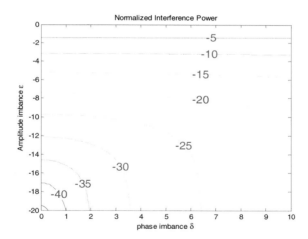

Fig. 3. NIP contours in the synchronous frequency case

According to [10-11], the NIP is defined to assess the impact of FSE, i.e.

$$\text{NIP} = \frac{1}{P(\omega_d)}\left(P(-\omega_d) + P(-2\omega) + P(-2\omega - \omega_d) + P(2\omega + \omega_d) + P(2\omega) + P(0)\right) \quad (14)$$
$$= \left(G^*J\right)^{-2}\left[\left(G_\varepsilon^*J_\varepsilon\right)^2 + \left(G^*H_\varepsilon\right)^2 + \left(G^*J_\varepsilon\right)^2 + \left(G_\varepsilon^*H\right)^2 + \left(G_\varepsilon^*J\right)^2 + \left(G^*H\right)^2 + \left(G_\varepsilon^*H_\varepsilon\right)^2\right].$$

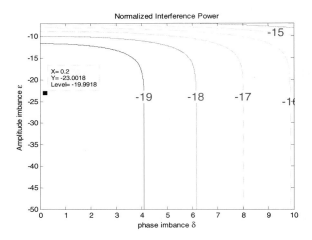

Fig. 4. NIP contours in dB for $SCR = 20$ dB

where * denotes complex conjugation. Let $r_G = (G_\varepsilon/G)^2$, $r_J = (J_\varepsilon/J)^2$, $r_H = (H_\varepsilon/H)^2$ then in the target channel $r_J = r_H$, and Eq.(14) can be simplified as

$$\text{NIP} = r_G r_J + r_J + r_G + (r_G+1)(1+r_J)\frac{1}{\text{SCR}}. \tag{15}$$

where $\text{SCR} = (J/H)^2$ is signal to clutter ratio in the target channel. In order to simplify the process of evaluation, we assume that the variation principle of amplitude and phase imbalance with respect to frequency in the double-channel system are the same, that is $r_J = r_G = r$, then, we have

$$\text{NIP} = r^2 + 2r + (r+1)^2 \frac{1}{\text{SCR}}. \tag{16}$$

According to Eq.(16), the NIP contours for different SCR condition are illustrated as Fig.4 to Fig.6. In fact, the energy loss of the desired signal component makes the noise level rise. The NIP can be manifest as SCR loss due to the effect of FSE. Therefore, it is rational to use NIP as a specification to evaluate the performance of coherent integration in passive bistatic radar. According to Fig.4 to Fig.6, it can be found that the small deterioration in amplitude imbalance will make the NIP increasing. With the decreasing of SCR in the target channel, NIP becomes larger and larger. SCV is reduced gradually at the same time.

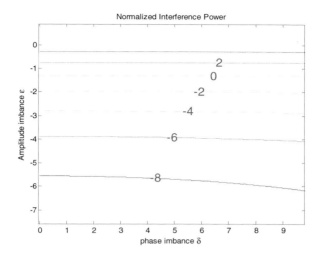

Fig. 5. NIP contours in dB for $SCR = -10$ dB

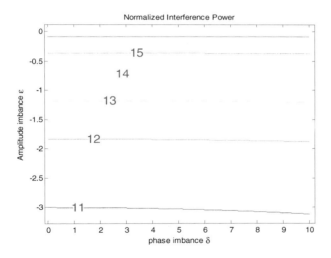

Fig. 6. NIP contours in dB for $SCR = 10$ dB

5 Conclusions

In this paper, we focused on the impact of FSE on the target detection in the ambiguity surface which generated by cross-correlation in wide sense. Since the imbalance in I and Q channels is a function of frequency, the imperfection arises from FSE in digital I and Q components has been introduced. The loss is calculated with respect to amplitude and phase imbalances. An analysis that explains how the nuisance frequency components generated is presented. Then NIP was defined to evaluate the impact of FSE in quantity. If the powers of these nuisance frequencies caused by FSE are strong enough, many frequency components will easily cover the weak targets in most Doppler cells; even cause false alarms. Simulation results confirmed that multiple frequency components caused by FSE would seriously deteriorate the detection and Doppler frequency estimation. Thus, highly accurate frequency synchronization is one premise of coherent detection in passive bistatic radar.

References

1. Wang, X.M., Kuang, Y.S., Chen, Z.X.: Surveillance Radar Technology, pp. 366–369. Publishing House of Electronics Industry, Beijing (2008) (in Chinese)
2. Yang, Z.Q., Zhang, Y.S., Luo, Y.J.: Bistatic/Multistatic Radar System, pp. 209–214. Publishing House of Electronics Industry, Beijing (1998) (in Chinese)
3. Wang, W.Q., Ding, C.B., Liang, X.D.: Time and Phase Synchronization via Direct-path Signal for Bistatic Synthetic Aperture Radar Systems. IET Radar Sonar Navigation 2(1), 1–11 (2008)
4. Daniel, D.T.: Synchronization of Non-cooperative Bistatic Radar Receivers. Syracuse University, N.Y (1999)
5. Howland, P.E., Maksimiuk, D., Reitsma, G.: FM Radio Based Bistatic Radar. IEE Radar, Sonar and Navigation 152(3), 107–115 (2005)
6. Kulpa, K.S., Czekała, Z.: Masking Effect and its Removal in PCL Radar. IEE Radar, Sonar and Navigation 152(3), 174–178 (2005)
7. Griffiths, H.D., Baker, C.J.: Passive Coherent Location Radar Systems. Part 1 Performance Prediction. IEE Radar, Sonar and Navigation 152(3), 153–159 (2005)
8. Yi, W.: Radar Receiver Technology, pp. 106–107. Publishing House of National Defense Industry, Beijing (2005)
9. James, T.: Digital Techniques for Wideband Receiver, pp. 254–257. Artech House Inc., M.A (2001)
10. Choi, Y.S., Voltz, P.J., Casara, F.A.: On Channel Estimation and Detection for Multicarrier Signals in Fast and Selective Rayleigh Fading Channels. IEEE Trans. Communication 49(8), 1375–1387 (2001)
11. Su, F., Cheng, S.X., Li, C.Y., et al.: Analysis of TSE and FSE in OFDM Wireless Communication System. Science in China Series E: information Science 35(2), 135–149 (2005)

Research on Double Orthogonal Multi-wavelet Transform Based Blind Equalizer

Han Yingge[1], Li Baokun[1], and Guo Yecai[1,2]

[1] Anhui University of Science and Technology, Huainan 232001, China
[2] Nanjing university of information Science and Technology, Nanjing 210044, China
han_ying_ge@126.com

Abstract. Aiming at the slow convergence of Constant Modulus Algorithm based Decision Feedback blind Equalizer(CMA-DFE), a new Double orthogonal Multi-Wavelet Transform based Constant Modulus Algorithm Decision Feedback blind Equalizer(DMWT- CMA-DFE) is proposed on the basis of analyzing on orthogonal Multi-Wavelet Transform based Constant Modulus Algorithm Decision Feedback blind Equalizer(MWT-CMA-DFE). In the proposed equalizer, the orthogonal multi-wavelet transform is placed not only in the front of the forward filter but also in the front of the feedback filter to improve the convergence rate. Simulation tests with underwater acoustic channel have indicated that the proposed equalizer has the faster convergence rate and the smaller residual mean square error compared with the CMA-DFE and MWT-CMA-DFE.

Keywords: constant modulus algorithm based decision feedback blind equalizer; orthogonal multi-wavelet transform; mean square error; convergence rate.

1 Introduction

Equalizing a channel with requiring a training mode is known as blind equalization(BE)[1].The most popular adaptive BE algorithms are the so-called Constant Modulus Algorithms(CMA). However, the equalization performance of the CMA is no good for the channel with the severe ISI and nonlinear distortion. On the basis of this, the CMA based decision feedback blind equalizer(CMA-DFE) is proposed. The CMA-DFE have a feedback filter component which introduce the nonlinear characteristics, so it can better adapt to the channel of different types. However, the CMA-DFE suffers from slow convergence rates and the big Mean Square Error (MSE).

The convergence rate of the equalizer can be advanced by normalizing orthogonal wavelet transform for the equalizer input signal[2-7]. For example, the momentum term and orthogonal wavelet transform based blind equalization algorithm is proposed in[3], in which the momentum term and orthogonal wavelet transform is introduced into the blind equalization algorithm and the convergence rate of the equalizer can be advanced by means of normalizing for the equalizer input signal and introducing the momentum; the orthogonal wavelet transform based LE+DD algorithm and orthogonal wavelet packlet transform based BE algorithm is proposed in [4] and [5]; the balanced orthogonal multi-wavelet transform based BE algorithm is proposed in [6]. Although these algorithms improve the equalizer convergence rate, the common ideas of this kind of equalizer design is carry out normalized orthogonal wavelet transform for the equalizer input signal (ie, feed-forward filter input signal) only.

In addition, the multi-wavelet consist of several scaling functions and its support is without overlapping at any scale compared with the single wavelet. So it can overcome the boundary effect effectively[6].

On the basis of analyzing on the above these literatures, a new orthogonal multi-wavelet transform based constant module algorithm decision feedback blind equalizer is proposed.

2 CMA-DFE

Fig.1 shows the structure of CMA-DFE, It consists of forward filter and feedback filter, where the feedback filter component is being used to counteract the interference from the previous symbol. Qu, g(·), C and $v(n)$ stand for the decision device, the nonlinear transformation device, the channel frequency response and the additive noise of the channel respectively.

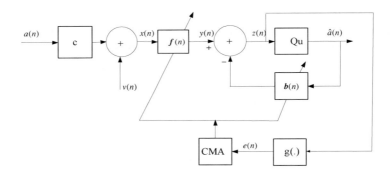

Fig. 1. The structure of CMA-DFE

In the Fig.1, it is supposed that the tapping coefficient vector of the forward filter and feedback filter is $f(n)$ and $b(n)$ respectively. Such $f(n)$ and $b(n)$ can be represented as:

$$f(n) = [f(0) , f(1) , \cdots , f(N_f - 1)]^T \quad (1)$$

$$b(n) = [b(0) , b(1) , \cdots , b(N_b - 1)]^T \quad (2)$$

Where N_f and N_b stand for the weight length of the forward and feedback filter respectively. Assume that the input recursion vector of the forward and feedback filter is $X(n)$ and $A(n)$ respectively. Thus, $X(n)$ and $A(n)$ can be expressed as:

$$X(n) = [x(n) , x(n-1) , \cdots , x(n - N_f + 1)]^T \quad (3)$$

$$A(n) = [\hat{a}(n) , \hat{a}(n-1) , \cdots , \hat{a}(n - N_b + 1)]^T \quad (4)$$

Thus, according to the Fig.1, the decision device input can be represented as:

$$z(n) = \sum_{i=0}^{N_f-1} f(n)x(n-i) - \sum_{i=0}^{N_b-1} b(n)\hat{a}(n-i) \qquad (5)$$
$$= f^T(n)X(n) - b^T(n)A(n)$$

Where, $\hat{a}(n)$ stand for the decision device output.

Now, we use the CMA to adjust the weights of the forward and feedback filter in the minimum mean square criteria, the weight coefficients update formula can be shown as:

$$\begin{cases} f(n+1) = f(n) - \mu_1 e(n) y^*(n) z(n) \\ b(n+1) = b(n) + \mu_2 e(n) \hat{a}^*(n) z(n) \end{cases} \qquad (6)$$

Where μ_1 and μ_2 stand for the step size. $e(n)$ stands for the error function, and

$$e(n) = |z(n)|^2 - R_2 \qquad (7)$$

Where $R_2 = E[|a(n)|^4]/E[|a(n)|^2]$. We define the algorithm consisted of Eq.(1-7) as constant module based decision feedback BE algorithm. In the Eq.(6), μ_1 and μ_2 is constant, if the step size is small, the convergence is slow; if the step is big, the steady-state error is large.

3 DMWT-CMA-DFE

3.1 MWT-CMA-DFE

The [2-7] show that the convergence rate of equalizer can be improved by normalizing orthogonal wavelet transform for the equalizer input signal, in which the orthogonal wavelet transform is placed in the front of the forward filters in [2-7]. According to the [2-7] ideas, we gives orthogonal Multi-Wavelet Transform based Constant Modulus Algorithm Decision Feedback blind Equalizer (MWT-CMA-DFE), in which the adaptive algorithm adopt CMA and the equalizer adopt decision feedback structure. Fig.2 gives the structure of MWT-CMA-DFE.

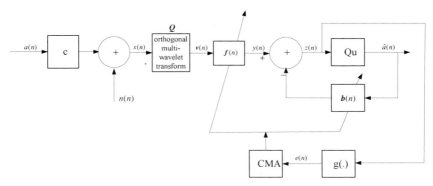

Fig. 2. The structure of MWT-CMA-DFE

In the Fig.2, the forward filter input signal vector is $X(n)$, it can be expressed by Eq.(3). The output signal vector by orthogonal multi-wavelet transform can be represented as

$$v(n) = [r_{1,0}(n), r_{1,1}(n), \cdots, r_{J,k_J}(n), s_{J,0}(n), \cdots s_{J,k_J}(n)]^T \qquad (8)$$

Where J represent the maximum scaling of wavelet decomposition; k_j is the maximum translation of wavelet function under the scaling j and k_j can be given by $k_j = N/2^j - 1$ ($j=1,2,\cdots, J$). Such,

$$v(n) = QX(n) \qquad (9)$$

Where, Q express the orthogonal multi-wavelet transform matrix, it is a $N_f \times N_f$ orthogonal matrix, Q can be given by $Q=[Q_1; Q_2P_1; Q_2P_2P_1;\cdots; Q_JP_{J-1}\cdots P_2P_1; P_J P_{J-1}\cdots P_2P_1]^{[7]}$. Therefore, under the minimum mean square criteria, the weight coefficients update formula of orthogonal multi-wavelet transform based constant modulus decision feedback blind equalizer can be shown as:

$$\begin{cases} f(n+1) = f(n) - \mu_1 R^{-1}(n)v^*(n)e(n)z(n) \\ b(n+1) = b(n) + \mu_2 e(n)\hat{a}^*(n)z(n) \end{cases} \qquad (10)$$

Where $R(n) = diag[\sigma_{j,0}^2(n), \sigma_{j,1}^2(n), \cdots, \sigma_{J,k_J}^2(n), \sigma_{J+1,0}^2(n) \cdots, \sigma_{J+1,k_J}^2(n)]$, $\sigma_{j,k}^2(n)$ and $\sigma_{J+1,k}^2(n)$ express the average power value of the $r_{j,k}(n)$ and $s_{J,k}(n)$ respectively, $\sigma_{j,k}^2(n)$ and $\sigma_{J+1,k}^2(n)$ can be gain by the following formula recursively,

$$\begin{cases} \sigma_{j,k}^2(n+1) = \beta\sigma_{j,k}^2(n) + (1-\beta)|r_{j,k}(n)|^2 \\ \sigma_{J+1,k}^2(n+1) = \beta\sigma_{J+1,k}^2(n) + (1-\beta)|s_{J,k}(n)|^2 \end{cases} \qquad (11)$$

According, we define the algorithm consisted of Eq.(3) and (8-11) as a common orthogonal multi-wavelet transform based constant modulus decision feedback BE algorithm. In the algorithm, the orthogonal multi-wavelet transform is introduced into CMA-DFE and the convergence rate of the CMA-DFE can be improved by normalizing orthogonal wavelet transform for the input signal of the forward filter.

3.2 DMWT-CMA-DFE

In order to further improve the convergence rate of the equalizer, a new Double orthogonal Multi-Wavelet Transform based variable structure Constant Modulus Algorithm Decision Feedback blind Equalizer(DMWT-CMA-DFE) is advanced on the basis of analyzing on the structure of the [2-7]. The structure of the advanced equalizer is shown in Fig.3.

Research on Double Orthogonal Multi-wavelet Transform Based Blind Equalizer 465

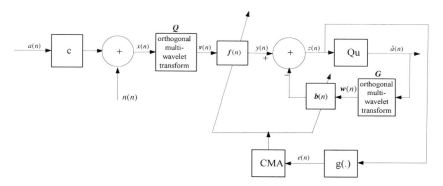

Fig. 3. The structure of DMWT-CMA-DFE

In Fig.3,

$$v(n) = QX(n) \tag{12}$$

Where, Q, $X(n)$ and $v(n)$ express as the mentioned above.

$$w(n) = GA(n) \tag{13}$$

Where, G is the orthogonal multi-wavelet transformation matrix, it is a $N_b \times N_b$ orthogonal matrix, G can be given by $G = [G_1;\ G_2P_1;G_2P_2P_1;\cdots;\ G_{JJ}P_{JJ-1}\cdots P_2P_1;\ G_2P_1;G_2P_2P_1;\cdots;\ P_{JJ}P_{JJ-1}\cdots P_2P_1]$, JJ represent the maximum scaling of wavelet decomposition. $A(n)$ is the output recursive vector of the decision device, it is given by Eq.(4); $w(n)$ is the output recursive vector by orthogonal multi-wavelet transformation.

As the same as, under the minimum criteria, the weight coefficients update formula can be shown as:

$$\begin{cases} f(n+1) = f(n) - \mu_1 R^{-1}(n) v^*(n) e(n) z(n) \\ b(n+1) = b(n) + \mu_2 RR^{-1}(n) w^*(n) e(n) z(n) \end{cases} \tag{14}$$

Where $R(n)$ represent as the above, $RR(n) = diag[\sigma_{j,0}^2(n), \sigma_{j,1}^2(n), \cdots, \sigma_{JJ,k_{JJ}}^2(n),\ \sigma_{JJ+1,0}^2(n)$, $\cdots,\ \sigma_{JJ+1,k_{JJ}}^2(n)]$, $\sigma_{j,k}^2(n)$ and $\sigma_{JJ+1,k_j}^2(n)$ express the average power value of the $r_{j,k}(n)$ and $s_{JJ,k}(n)$ respectively, $\sigma_{j,k}^2(n)$ and $\sigma_{JJ+1,k_j}^2(n)$ can be gain by the following formula recursively,

$$\begin{cases} \sigma_{j,k}^2(n+1) = \beta \sigma_{j,k}^2(n) + (1-\beta)|r_{j,k}(n)|^2 \\ \sigma_{JJ+1,k}^2(n+1) = \beta \sigma_{JJ+1,k}^2(n) + (1-\beta)|s_{JJ,k}(n)|^2 \end{cases} \tag{15}$$

we define the algorithm consisted of Eq.(3,4,12~15) as a new double orthogonal multi-wavelet transform based constant modulus decision feedback BE algorithm. In the algorithm, in order to advance the algorithm convergence rate, the orthogonal multi-wavelet transform is placed not only in the front of the forward filter but also in the front of the feedback filter.

4 Simulation and Analysis

To demonstrate the performance of DMWT-CMA-DFE, the computer simulation is carried out. In this simulation, The simulation channel adapt the underwater acoustic channel, the length of forward filter is 33; the length of feedback filter is 16; the source signal is16-QAM.

4.1 Simulation with the Complex Channel

The channel impulse response is given by
$$c=[1,\ 0,\ 0.3e^{-0.7i},\ 0,\ 0,\ 0.2e^{-0.8i}] \quad (16)$$

The signal to noise ratio is 25 dB, the value of the others simulation parameters are shown in Tab.1. The simulation results are shown in Fig.4.

Table 1. The value of simulation parameters

Equalizer	The size-step of the forward and feedback filter		Initial weight		β	The $R(n)$ initial value
			Forward filter	Feedback filter		
CMA-DFE	0.00002	0.00002	The initial value of the first tap is 1, the others is 0.	The initial value of the sixteen tap is 1, the others is 0.		
MWT-CMA-DFE	0.001	0.001			0.9	1
DMWT-CMA-DFE	0.00004	0.003			0.9	1

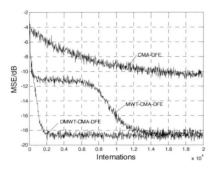
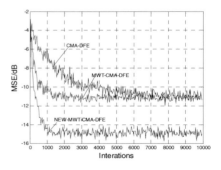

Fig. 4. Comparisons of MSE   Fig. 5. Comparison of the MSE

In Fig.4, 20000 different realizations are performed with independent input sequences, It is clear that the convergence rate of the DMWT-CMA-DFE is faster nearly 17,000 steps than CMA-DFE and its residual mean square error is smaller nearly 8dB than CMA-DFE, it is faster nearly 5,000 steps than MWT-CMA-DFE under the nearly same mean square error, so, it is obvious that the convergence rate of DMWT-CMA-DFE is superior to MWT-CMA-DFE and CMA-DFE.

4.2 Simulation with Real Channel

The channel impulse response is given by

$$c=[1,\ 0.51,\ 0.1] \qquad (17)$$

The signal to noise ratio is 20 dB, the value of the others simulation parameters is shown in Tab.2.The simulation results are shown in Fig.5.

Table 2. The value of the simulation parameters

equalizer	The size step of the forward filter and the feedback filter		Initial weight		The Value of β	The initial value of $R(n)$
			forward filter	feedback filter		
CMA-DFE	0.00008	0.00008	The initial value of the first tap is 1,the others is 0.	The initial value of the sixty tap is 1, the others is 0.		
MWT-CMA-DFE	0.0005	0.0005			0.9	1
DMWT-CMA-DFE	0.000008	0.005			0.9	1

In Fig.5, 10,000 different realizations are performed with independent input sequences, It is clear that the residual MSE of DMWT-CMA-DFE is smaller nearly 4dB than MWT-CMA-DFE and CMA-DFE, thus, it is obvious that the MSE of DMW T-CMA-DFE is superior to MWT-CMA-DFE and CMA-DFE. The convergence rate of DMWT-CMA-DFE is faster nearly 6,000 steps than the CMA-DFE. It is nearly the same as the MWT-CMA-DFE.

5 Conclusions

The slow convergence and mean square error is the major index to measure the communication system performance. On the basis of analyzing on Multi-Wavelet Transform based Constant Modulus Algorithm Decision Feedback blind Equalizer(MWT-CMA-DFE), we have designed a new Double MWT-CMA-DFE (DMWT-CMA-DFE), in DMWT-CMA-DFE, the orthogonal multi-wavelet transform is placed not only in the front of the forward filter but also in the front of the feedback filter in order to improve the convergence rate. Simulation tests with underwater acoustic channel have indicated that the DMWT-CMA- DFE have not only the faster convergence rate but also the smaller residual mean square error compared with the CMA-DFE and MWT-CMA-DFE.

References

1. Godard, D.: Self-recovering equalization and carrier tracking in two dimensional data communication systems. IEEE Transaction on Communication 28(11), 1867–1875 (1980)
2. Attallah, S., Najin, M.: On the convergence enhancement of the wavelet transform based LMS. In: Acoustics, Speech, and Signal Processing, pp. 973–976. IEEE, Detroit (1995)
3. Yingge, H., Yecai, G., Baokun, L., Qiaoxi, Z.: Momentum term and orthogonal wavelet-based blind equalization algorithm. Journal of System Simulation 20(6), 1559–1562 (2008)
4. Chao, Y., Yecai, G.: A OWT-LE+DD Algorithm Based on Orthogonal Wavelet Transform. Acta Armaentarii 31(2), 189–203 (2010)
5. Rui, H.L.: Adaptive equalization algorithm based on wavelet packlet transform. Acta Electronica Sinica 31(8), 1205–1208 (2003)
6. Guo, Y., Liu, Z.: Blind Equalization Algorithms Based on Balance Orthogonal Multi-wavelet Transform. Acta Armaentarii 31(3), 279–284 (2010)
7. Erol, N., Basbug, F.: Wavelet transform based adaptive filters: analysis and new results. IEEE Trans on Signal Processing 9(44), 2163–2171 (1996)

A Novel Approach for Host General Risk Assessment Based on Logical Recursion

Xiao-Song Zhang[1], Jiong Zheng[1], and Hua Li[2]

[1] School of Computer Science & Engineering
University of Electronic Science and Technology of China
Chengdu 611731, China
[2] Unit 78155 of People's Liberation Army, Chengdu 610016, China
johnsonzxs@uestc.edu.cn,
k7zjmost@163.com

Abstract. Propagable attacks can cause more serious harm to the network. These attacks are often conducted by vulnerabilities. Traditional vulnerability detection tools always care only about the threat of the assessed host itself ignoring the further attack. This paper presents a novel approach to assess the general risk of host based on the vulnerability and network information using logical recursion.

Keywords: network security, vulnerability, risk assessment.

1 Introduction

In recent years, the attack incidences led by vulnerabilities become prevalent day by day. For example, the Stuxnet worm is a "groundbreaking" piece of malware so devious in its use of unpatched vulnerabilities. It was first publicly identified in June, it exploited more than one unpatched, or "zero-day," vulnerability in Windows and spread through infected USB flash drives[1]. For another example, CONFICKER worm took advantage of the MS08067 infected 9 million machines till January, 2009. The threats spread from one to another for the unpatched vulnerability[2][3]. The attacker controlled a huge botnet using this kind of worm. The vulnerability is the key to launch attack. Especially, exploitable vulnerabilities could be used to invade systems and do a lot of damage to users' computer without authorization. And it could be worse if the vulnerability could cause further attack that the invader can reach the system connected by the victim host.

It is so necessary to avoid the attack before it takes place. So, in the paper, we present a novel approach to assess the general risk of host using logical recursion. In virtue of the network and vulnerability information, attack relationship between hosts could be acquired. Then a logical recursion will be conducted and the attack paths will be obtained[4]. With the attack paths, take the individual risk and the predecessor risk into consideration. We can calculate the general risk of the host finally. In this paper, we discuss modules of the framework in detail.

2 Architecture and Details

2.1 Framework

We present a framework consisted of three main modules which are shown in Fig. 1.

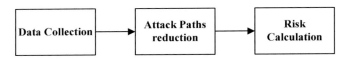

Fig. 1. Framework.

It can be seen from the figure above that data collection module is the first function module which is responsible for the input of the whole system. It is deployed on the target hosts in the network. And it takes charge of collection of vulnerability information, network information and configuration of these hosts. The collected data will be sent to the server which analyzes them. And then the second module which in the server will derive the attack paths of them based on the collected data. Finally, the last module will calculate the risk of each host using attack paths and collected information. Each module will be discussed in detail in the next sections.

2.2 Data Collection

The deployment structure of data collection module is shown in Figure 2. Each collect agent is setup on the target host. It follows the instructions from server to conduct data collection.

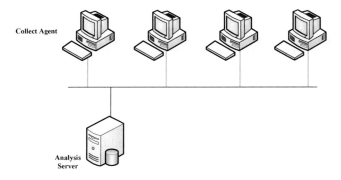

Fig. 2. Data Collection Structure

Collect agent takes in charge of collecting three kinds of data as following:

Vulnerability information collection: We develop a vulnerability information collecting agent which is based on OVAL. OVAL is an international information security community standard to promote open and publicly available security content[5]. The language standardizes the three main steps of the assessment process: representing configuration information of systems for testing; analyzing the system for the presence of the specified machine state; and reporting the results of the assessment. The vulnerability database of OVAL is opening to public, which means we can always get the newest vulnerability data. And the data files are organized in a certain XML format, so we can easily add new vulnerability definitions into our database. For example, CVE-2006-4847 means multiple buffer overflows in Ipswitch WS_FTP Server 5. 05. The original database of OVAL does not contain any detection rules about this vulnerability. So to extend the ability to detect vulnerabilities caused by third party application, we add corresponding rules to the database. For example, we add a rule which check the registry item SOFTWARE\Ipswitch\iFtpSvc\Version. In such a rule, if the value is equal to 5.05, the vulnerability is true.

And we also improve the vulnerability scanning. The scanning process of OVAL takes too long for the large amount of vulnerabilities definitions. Therefore, we split the definitions into parts, and set multiple threads for scanning which greatly improves the speed. We define $t(n)$ which denotes the time cost by assessing n definitions. So by a given count of all the definitions, we can conclude the formula as follow:

$$T(m)_{min} = m \times \min(\frac{t(n)}{n}) \tag{1}$$

$T(m)_{min}$ represents the least time cost of scanning with m definitions. From the formula (1) we can see that a property $\frac{t(n)}{n}$ will make the time cost minimal.

Network information collection: We have to set some limitations of network information collection. The connections between hosts are determined by the situation of the ports. Because our system discuss mainly about the threat transmission, the port is associated with some certain vulnerability. Ports data will be configured in a special file when collecting. There are several methods to get network information, such as scanner, firewall, and etc. We prefer choosing network scanner to collect the network information. We design network scanner as an individual executable function module. We can set different parameters to control it while scanning with different require.

Hosts configuration collection: the module collect information as following:

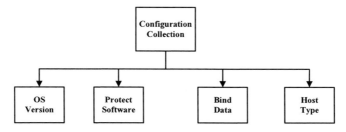

Fig. 3. Configuration Collection

OS version: OS versions can be acquired by calling system application interface such as GetVersionEx, GetNativeSystemInfo and etc. These APIS return enumerable value which helping determine the version of operation system. The OS version is so important for the higher version provides a more powerful ability to protect itself from attack by means of vulnerability. This is useful when assessment the risk of the host.

Protect software: the protect software name, version can be got through examination of the registry and the file version. The characteristics information of protect software is defined in a profile in advance. The collected information can be matched with the profile and the protect software information is determined finally. And we also define scores of these protect software which is helpful for our assessment. For different protect software provides different ability of defense against attack caused by vulnerability such as remote buffer flow, malicious shellcode and etc.

Bind data: different server application stores opening data in different ways. For example, ws_ftp which is a ftp software. Its opening data storing information is saved at profile security\$index.acc in its installation directory. This file describes the ftp date directories and the corresponding user. For another example, winftp save these information in data\user.xml in the installation directory. Bind data is used to sensitive information detection which will help the administrator being aware of the key data to protect.

Host type: host type can be determined by matching the port. For example, POP3 server opens 110 port to receive the requests from clients. Using the PMIB_TCPTABLE returned by calling GetTcpTable, the existed ports can be got. The host type is very important. For the threat spreads through a server is much wider.

2.3 Attack Paths Reduction

The vulnerability and network information obtained from last module can be used as the input for attack paths reduction. Firstly, we need to construct a direct attack relationship between two hosts. For example, host a can access host b at port 25. Port 25 is often used for SMTP service, so if there is a corresponding SMTP software vulnerability, then it means that a can attacks b. The attack relationship is the key factor of attack paths reduction.

In the paper, we take DATAOG as the tool to construct the attack graphs. DATALOG is a query and rule language, and it is a kind of logic programming language which is a sequence of facts and rules[6][7]. A rule is of form as follow:

$$facts(\arg s) := facts(\arg s_1), ..., facts(\arg s_2) \qquad (2)$$

All the collected information need to be transformed into the form which can be identified by DATALOG[8][9]. The vulnerability information is transformed into available format to fit the facts. See the fact below:

$$HaveVul(hostname, vul\_id) \qquad (3)$$

This rule represents that there is a vulnerability in one host. The vul_id is a CVE number which means a real vulnerability.

Attack between hosts is determined by a remote vulnerability, and therefore, we care only about the remote vulnerability when constructing these facts. In addition to the previous fact, we also need a fact to describe the vulnerability and the port. So a fact is as follow:

$$VulInfo(vul\_id, port) \qquad (4)$$

Besides, the connections between hosts are necessary. The following fact helps us to describe the connections.

$$Access(srchost, dsthost, port) \qquad (5)$$

According to the facts above, we can get rule to derivate the attack relationship between two hosts.

$$\begin{aligned}&Attack(SRCHOST, DSTHOST, []):- \\ &HaveVul(DSTHOST, VUL\_ID), \\ &VulInfo(VUL\_ID, PORT), \\ &Access(SRCHOST, DSTHOST, PORT).\end{aligned} \qquad (6)$$

There may be several middle hosts between the source host and destination host. Middle attack paths may appear in such a situation. So we optimize the rules by marking the visited hosts. We use the assert function of DATALOG to mark a host that is visited. The rule is shown as follow:

$$\begin{aligned}&AttackPath(SRCHOST, DESHOST, [], VLIST):- \\ &not(visited(SRCHOST, VLIST)), \\ &attack(SRCHOST, DESHOST, []).\end{aligned} \qquad (7)$$

We need to resolve the recursion when construct the attack paths. So we present a MIDHOST as a middle host when constructing the attack path between SRCHOST, DSTHOST. The middle hosts will be appended in list PLIST. And all the visited hosts will be marked and appended in list VLIST to avoid resolving repeatedly. RECLIST and NEWVLIST are temporary list for the recursion that is visited. The rule is shown as follow:

$$\begin{aligned}&AttackPath(SRCHOST, DSTHOST, PLIST, VLIST):- \\ &not(visited(SRCHOST, VLIST)), \\ &attack(SRCHOST, MIDHOST, []), \\ &append([SRCHOST], VLIST, NEWVLIST), \\ &AttackPath(MIDHOST, DSTHOST, RECPLIST, NEWVLIST), \\ &append([MIDHOST], RECPLIST, PLIST).\end{aligned} \qquad (8)$$

Finally, the $AttackPath(X, Y, L, [])$ includes all the attack paths.

2.4 Risk Calculation

Before we discuss the risk calculation, the item individual risk should be understand. Individual Risk (IR): The risk of the assessed host itself. It is calculated by the property of host and the vulnerabilities on the host. It is an important factor which does damage to the entire network system. IR(x) represents the individual risk in host x. IR is based on the following factors.

Host Configuration (HC): the configuration of the host, such as host type, OS version and etc.

Access Factor (AF): access amount, access frequency and user amount.

Vulnerability Risk (VR): the Threat of vulnerabilities that exist on the host.

These three factors determine the individual risk of the host. Below we discuss the score definition of these factors.

For host configuration, different host has different ability to defense the threat, so we define different score for the operation system, host type and etc. For example, the score of XP operation system is 0.5 but it is 0.7 of VISTA. And the score of a server is obviously larger that the none-server host. In our system, we will define these scores in a profile. And the profile can be updated and modified.

For the access factor: we can detect the amount of AA, AF, UA when we scanning the host. These amount is useful for the individual risk calculation.

For vulnerability risk: vulnerability is the key important standard for assessing the individual risk, so an international criteria is necessary. We decide to use the Common Vulnerability Scoring System (CVSS). This system is developed and maintained by NIAC and FIRST. It is an open and free standard for the product manufactures. This standard is outstanding to score the vulnerability risk.

CVSS considers three areas for the assessment:

Base metrics: base metrics is concerned about the features and possible impact of inherent of vulnerability. Such as the assess vector, a remote vulnerability or local vulnerability, the complexity to exploit the vulnerability and the impact on the network environment.

Temporal metric: temporal metric refers to the life cycle of the vulnerability which from exploited to reported and fixed. For example, worm CONFICKER invades thousands of hosts using MS08067. It lasted for a long time in the network. So the score of such serious vulnerability is much larger.

Environmental Metrics: environmental metrics refers to the computer environment and potential prevalence of it.

So based on the factors above, a formula is represented as follow:

$$IR(x) = HC * AF * VR \tag{9}$$

Now we will introduce another item general risk (GR), as its name, it represented the whole risk of a host. In this paper GR is determined by the individual risk and predecessor.

$$GR(x) = PR(x) + IR(x) \tag{10}$$

Predecessor attack risk (PR) is risk from the host which is as an attacker. It is determined by the general risk of predecessor hosts and transfer probability.

$$PR(x) = \sum_{i=1} GR(i) * P_i \qquad (11)$$

The item Pi is determined by three factors. As we have discussed in the previous section. Different OS version provides different ability to protect itself from attack by vulnerability. Remote vulnerability, a subscore system of CVSS provides a method to calculate the threat of remote vulnerability. Protect software: different protect software offers different ability to defense against attack caused by vulnerability such as remote buffer flow, malicious shellcode and etc. In out system, we also defined corresponding profile to the factors.

3 Conclusion

This paper presents an approach a novel approach to assess the general risk of host using logical recursion. This method takes advantage of vulnerability and network information as the inputs of the assessment. This method can help assessing the potential risk of the network.

But there is much more to explore to make our method better. In the future, we will do research on optimizing this method this paper proposed.

References

1. Stuxnet, from http://en.wikipedia.org/wiki/Stuxnet
2. Fitzgibbon, N., Wood, M.: Conficker.C: A technical Analysis, April 1. Sophoslabs, Sophon Inc. (2009)
3. Charkraborty, R.: Detailed Study of W32. Downnadup or Win32/Conficker, March, 28 (2009), http://www.malwareinfo.org
4. Chen, X.-z., Li, J.-h.: A Novel Vulnerability Assessment System Based on OVAL. Journal of Chinese Computer Systems 28(9), 1554–1557 (2007)
5. Wojcik, M., Bergeron, T., Roberge, R.: Introduction to OVAL: A new language to determine the presence of software vulnerabilities, MITEE Corporation (November 2003)
6. Datalog, http://en.wikipedia.org/wiki/Datalog
7. Baral, C., Gelfond, M.: Logic programming and knowledge representation. Journal of Logic Programming 19, 73–148 (1994)
8. Ou, X., Govindavajhala, S., Appel, A.W.: MulVAL: A logic-based network security analyzer. In: 14th USENIX Security Symposium, Baltimore, Maryland, U.S.A (August 2005)
9. Ou, X.: A logic-programming approach to network security analysis. PhD dissertation, Princeton University (2005)

A Time Difference Method Pipeline Ultrasonic Flowmeter

Shoucheng Ding[1,2], Limei Xiao[1,2], and Guici Yuan[1]

[1] College of Electric and Information Engineering, Lanzhou University of Technology, Lanzhou, Gansu, China
dingsc@lut.cn
[2] Key Laboratory of Gansu Advanced Control for Industrial Processes, Lanzhou, China

Abstract. When the liquid inside the pipes was in the downstream or upstream, the ultrasound would have a different spread. Because of different downstream or upstream speed, resulting in ultrasonic propagation time was not equal. From the transmitter to the receiver of the time difference, it can obtain the pipe liquid flow. It was a hardware and software combination system. AT89C51 microcontroller as the system core, and the ultrasonic transmitter circuit, receiver circuit, signal conditioning circuitry, LCD display circuit together had constituted the system's hardware. Application of C language developed system software, to achieve signal acquisition and processing, the measurement results of the conversion and display. The system is low cost, high speed, high precision, reliability, adaptability, easy to operate.

Keywords: Ultrasonic flowmeter, Time difference method, Microcontroller.

1 Introduction

With the automation of production, pipeline development, flow meter instrument in the proportion of the total is growing. At present, China on the flow device compared with the international is still a big gap there, especially in ultrasonic flowmeter technology development, small flow and micro flow measurement needs are growing, increasing the accuracy of measurement. This paper studies and implements an ultrasonic pipe flowmeter based on AT89C51.

2 An Ultrasonic Flowmeter

In unit time through a pipe or equipment somewhere, the amount of fluid is called flow. The volume is called volume flow Q; by mass calculated as mass flow G.

Fluid through a pipe or equipment located somewhere in a small cross-section area is dF, and through the small area of the average velocity is taken as v_d, through the small area dF, the volume flow dQ is:

$$dQ = v_d \cdot dF \tag{1}$$

In fact, the points on the interface speed is not equal, the average velocity is:

$$v_d = \frac{Q}{F} \tag{2}$$

Then, mass flow G:

$$G = Q \cdot \rho \tag{3}$$

Within a certain period of time, the total amount of fluid flow is called the cumulative flow, cumulative flow and corresponding flow is called the instantaneous flow.

2.1 Fluid Velocity Distribution

Cross-sectional area in the pipe axial component of fluid velocity on the distribution pattern known as the velocity distribution. This is because the actual viscosity of the fluid has caused. In general, the more close to the wall, the impact of the viscous flow rate will be smaller, the wall on the flow rate is zero; more close to the pipeline center, this effect will be smaller the viscosity, the greater the flow rate, pipe center Maximum flow rate value. Different flow conditions inside the tubes, the velocity distribution presented is different. Such as fluid flow distribution model:

Laminar flow:

$$v_x = v_{max}\left[1 - \left(\frac{r_x}{R}\right)^2\right] \tag{4}$$

Turbulent flow:

$$v_x = v_{max}\left(1 - \frac{r_x}{R}\right)^{\frac{1}{n}} \tag{5}$$

In the formula (4), (5) and, r_x is the radial distance of the center pipe, v_x is the r_x from the pipe at the center of the velocity, v_{max} is the maximum velocity of the center pipe, R is the pipe radius, n is as reynolds number varies in the index.

2.2 Fluid Average Velocity

The average velocity is the average velocity on the pipe section. The implication is that the fluid when the tube to a uniform velocity v, through the pipes of a pipe section fluid flow is equal to one when the velocity distribution of the flow through the pipeline section, then v is the velocity distribution of the cross section. Its mathematical expression is:

$$\bar{v} = \frac{q_v}{A} = \frac{\int v_x dA}{A} \tag{6}$$

For a circular pipe, the v_x into the formula (6), the flow average velocity is:

$$\overline{v} = \frac{\int_0^R v_x 2\pi r dr}{\pi R^2} = \frac{2\pi v_{max} \int \left[1-\left(\frac{r}{R}\right)^2\right] r dr}{\pi R^2} = \frac{1}{2} v_{max} \quad (7)$$

In the turbulent state, the average velocity is:

$$\overline{v} = \frac{\int_0^R v_x 2\pi r dr}{\pi R^2} = \frac{2\pi v_{max} \int (1-\frac{r}{R})^{\frac{1}{n}} r dr}{\pi R^2} = \frac{2n^2}{(2n+1)(n+1)} v_{max} \quad (8)$$

Because ultrasound is not a cross section through the pipe, and often is the result of the central axis of pipe, so it is a pipe flow velocity measured central axis direction of the average velocity, and its mathematical expression is:

$$\overline{v}_d = \frac{2\int_0^R v_x dr}{2R} \quad (9)$$

For a circular pipe, the v_x into (9), then in the laminar flow state, the center of the pipe axis, the average velocity is:

$$\overline{v}_d = \frac{2\int_0^R v_x dr}{2R} = \frac{v_{max} \int \left[1-\left(\frac{r}{R}\right)^2\right] dr}{R} = \frac{2}{3} v_{max} \quad (10)$$

$$\overline{v} = \frac{3}{4} \overline{v}_d \quad (11)$$

In the turbulent state, the central axis of the tube, the average velocity is:

$$\overline{v}_d = \frac{2\int_0^R v_x dr}{2R} = \frac{v_{max} \int (1-\frac{r}{R})^{\frac{1}{n}} dr}{R} = \frac{n}{n+1} v_{max} \quad (12)$$

$$\overline{v} = \frac{2n}{2n+1} \overline{v}_d \quad (13)$$

2.3 Time Difference Method Ultrasonic Flowmeter

It is the use of ultrasound in the downstream or upstream transmission fluid the same distance, there is time difference. Between the propagation time difference and the fluid flow rate there is a certain relationship. Therefore, according to the time difference, the fluid flow rate can be obtained, so it can calculate the fluid flow. The basic principle is shown in Fig. 1. Ultrasonic transducer A, B is a pair can take turns transmitting or receiving ultrasonic pulse transducer, the mounting clip is mounted outside the pipe.

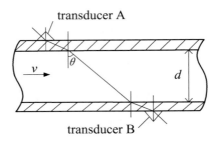

Fig. 1. Working principle of time difference method ultrasonic flowmeter

Assuming the fluid in the measured ultrasonic speed of sound is c. In the downstream, from A to B in time T_1, at the counter when the time from B to A is T_2. Arrangement of the transducer in the tube, the ultrasonic transducer and the wall in the spread takes time, and also a delay circuit, the total travel time as the delay time τ_0, then:

$$T_1 = \frac{d/\cos\theta}{c + v\sin\theta} + \tau_0, \quad T_2 = \frac{d/\cos\theta}{c - v\sin\theta} + \tau_0 \tag{14}$$

$$\Delta T = T_2 - T_1 = \frac{2dv\tan\theta}{c^2 - v^2\sin^2\theta} \tag{15}$$

τ_0 is much smaller than the ultrasonic wave propagation time in the fluid, the time difference is:

$$\Delta T = T_2 - T_1 = \frac{2dv\tan\theta}{c^2} \tag{16}$$

Therefore, the basic equations of the transit-time ultrasonic flowmeter can be expressed as:

$$v = \frac{c^2}{2d\tan\theta}\Delta T \tag{17}$$

$$Q = \frac{\pi d^2}{4}v \tag{18}$$

3 Flowmeter Hardware Design

Pipeline flowmeter ultrasonic hardware system consists of AT89C51 microcontroller, TCT40 series of ultrasonic probes, operational amplifiers and 12864 LCD display module. AT89C51 mainly completes the occurrence of square wave sequence and system control, and ultrasonic transducer achieves a sound electrical signal conversion, the main operational amplifier to the transducer output voltage for processing, 12864LCD real-time data display. The hardware block diagram of the system is shown in Fig. 2.

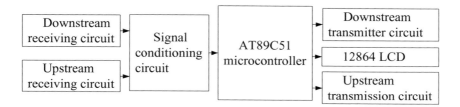

Fig. 2. Block diagram of system hardware

Working process of the hardware: When 89C51 SCM operates in mode 2, can produce a square wave sequence, and then amplified through the transistor, the transformer step-up, and reach a certain power, you can drive the transducer transmitting ultrasonic pulses. In this way, through the ultrasonic emission sensor, the electrical signal into acoustic signals, and then received by ultrasonic probe, in turn acoustic signals into electrical signals. Since then the signal is only a few millivolts, it is to go through the amplification link, and then by the rectifier filter and voltage comparators, ultrasonic signals can be received accurately by the time, according to downstream, upstream of the time can be considered as the time difference. From formula 17 can be considered as the flow through a cross-section through 12864LCD real-time display. The system flow rate data can also be output through the serial port.

Microcontroller is used in AT89C51 of MC-51 series. AT89C51 hardware resources: 4KB Flash memory, 128 bytes of RAM, 32I / O lines, two 16-bit timer / counter, interrupt structure, two five-source, full-duplex serial port, on-chip oscillator And clock circuitry and so on. In addition, AT89C51 from static logic design, selection of crystal frequency 12MHz. Obviously, this single chip on the development of equipment requirements is very low, shorten development time.

3.1 Ultrasonic Transducer

Transducer has an electric energy storage device and a mechanical vibration system. When the transducer is fired state, the power output stage output from the excitation signal caused by oscillation of the transducer in the energy storage component of the electric field or magnetic field changes, this change in electric or magnetic field through some effect on the transducer oscillation system to produce a driving force to make it into the vibration-like concept, so as to promote and mechanical vibration transducer system vibrate in contact with the media, the medium wave radiation.

The process of receiving sound exactly the opposite. This oblique probe used piezoelectric ultrasonic transducers TCT40-16, and it mainly consists of piezoelectric chips, wedges, etc. then it is resonant with the piezoelectric crystal to work. The basic parameters are: frequency, electromechanical coupling coefficient, the transducer's mechanical quality factor, transducer impedance characteristics, frequency characteristics, the direction of the transducer characteristics.

3.2 Ultrasonic Generator

Ultrasonic generator is also known as ultrasonic power supply. It is used to generate ultrasonic energy to the ultrasonic transducer device to provide energy. Specific generator circuit is shown in Fig. 3.

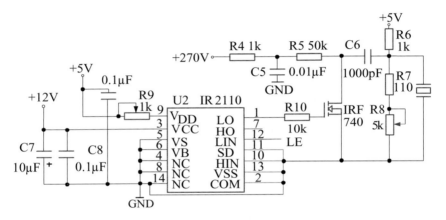

Fig. 3. Ultrasonic generator

When the control terminal inputs high voltage, fast turn-on RF740 and connected to ground, and negative pulse generated by capacitive coupling. The negative pulse will to stimulate the ultrasonic probe that will launch ultrasound. IR2110 chip hardware circuit can make the RF740 turn faster. IR2110 driver chip output voltage is 12V. +270 V power supply through R4, R5 charge on the capacitor C6, IRF740 electronic switch, when it turns on, C6 will quickly transfer the charge, thus forming a pulse voltage signal. The ultrasonic probe pulse can stimulate ultrasonic transmitter. LE is Micro SCM trigger signal. When LE is 1, IR2110 outputs +12 V voltage. So FET turns on, immediately start the process of ultrasonic transmitter. Resistors R7, R8 attenuator in the circuit from the role, the role of the pulse trailing reduced, and R8 is set to adjustable way.

4 Ultrasonic Flowmeter Software

Ultrasonic Flowmeter software system Mainly completes the set parameters, ultrasonic emission, propagation time measurement and control, velocity and flow calculation, measurement results of the display and storage.

System main program achieved Timer 0 and Timer 1 initial value setting. By Order TMOD = 0X12, making the work in the way of a timer T1, T0 timer in Mode 2. P1.0 pressed the start button, because the SCM is powered-on, so the system prompts on the display screen: "After placing the sensor measurement, press the Start button to start! " until the start button is pressed P1.0, System begins to clear the screen, and start the timer and counter, or else the interface has been prompted, until the start key is pressed. The microcontroller to calculate the flow, and pipe diameter d is an important parameter, for different pipe diameters need to manually input. Due to the general range of pipe diameter to be measured is 0 ~ 99cm, so the diameter bits, ten were modified to follow the additive zero. After the success of change to detect whether the first external interrupt, if any, off timers and counters, can be obtained downstream time T1. Otherwise, continue to wait, and then testing whether a second interruption, reflux can be obtained if the time T2, this time difference can be obtained, and then obtain the velocity and flow.

5 System Error Analysis

As an industrial measuring instruments, the measurement accuracy is a more important indicator. Factors of the system error are: channel length, the installation of ultrasonic transducer angle measuring accuracy of pipe radius, signal distortion and loss, the measured fluid temperature, etc.

Channel Length: According to the speed of sound formula (17), channel length is inversely proportional to the speed of sound. Thus d is an important factor of the speed of sound. In actual measurement, as long as the use of a micrometer, millimeter ruler and other measuring instruments to measure the length of the higher channel length of each road, this error is always kept less than 0.5%.

Pipe radius of the measurement accuracy: the fluid in the pipe, the pipe diameter is a factor in calculating the fluid and thus the measurement error caused it will directly affect the fluid flow. Assuming a circular pipe, the pipe diameter measurement error generated by 1% of the fluid flow will produce a 2% error.

Signal distortion and loss: measured in the fluid medium, generally can not be absolutely pure, which will contain large quantities of solid particles and air bubbles and other debris. Ultrasonic signal in the transmission process will inevitably have to receive the fluid in the pipe boundary impurities and reflection, resulting in multi-path effect, the signal distortion. The signal detection circuit, it is difficult to ensure that the distortion of the signal is detected the first transition is the emission signal is the first positive pulse, if the signal detection errors will directly affect the determination of acoustic wave propagation time, but also will test There was a flow rate of error. To ensure correct signal detection circuit detects the signal, the meter test the effectiveness of adding the signal part of the circuit, that is, the comparison voltage 2.5V, the maximum possible reduce the signal distortion caused by system errors.

The actual use of the instrument, the precision of measuring the low number, in addition to set the diameter, speed of ultrasound propagation error caused by, there are estimates of the distribution of fluid flow and other factors are not accurate.

6 Conclusion

Ultrasonic flow meter is ultrasonic transducer converts electrical energy to ultrasonic energy and the emission to the measured fluid, is received after receiving transducer converts the representative electrical signal flow and easy to detect, so that traffic can be achieved Detection and display. Ultrasonic flowmeter designed in this paper the main function is to require non-contact measurement of liquid pipe flow and flow velocity measurements and can live interactive manner using the appropriate parameter settings, data processing and data display. Because of its non-contact flow measurement, high precision, wide measurement range, easy installation, simple operation and its advantages are widely used in electric power, petroleum, chemical industry in particular water supply system.

References

1. Ren, X., Zhao, H.: Portable Ultrasonic Meter Based on Nios, Microcontrollers & Embedded Systems (2010)
2. Zhang, X.-h., Zhang, H., Wang, X.-q., Feng, J.-q., Wang, S.-b.: Design of the A/D Conversion Circuit for Ultrasonic Flowmeter Based on ARM. Automation & Instrumentation (2010)
3. Li, G.-h., Sheng, L., Liu, L.-n.: Design and implementation of ultrasonic flowmeter based on MSP430F447, Technical Acoustics (2010)
4. Wu, Y.-l., Yao, J., Li, B.: Application of High Precision Interval Measuring Chip TDC-GP2 in Transit-time Ultrasonic Measurement, Instrumentation Technology (2009)
5. Shi, W.-z., Li, A.-h., Wang, X.-b.: Design of Portable Ultrasonic Flowmeter Based on DSP and FPGA, Instrumentation Technology (2010)
6. Tong, J.-p., Sui, C.-h., Wei, G.-y., Xu, L.-d.: Development of equipment for measuring sound velocity with time difference method. Journal of Transducer Technology (2004)
7. Liu, X.-y., Yang, J.: Application of Time Scale Amplifying in Ultrasonic Flowmeter. Chinese Journal of Sensors and Actuators (2001)
8. Wang, Y.: The Application of CPLD Technique in Time Difference Method Ultrasonic Flowmeter. Control & Automation (2005)
9. Li, G., Liu, F., Gao, Y.: Accurate Measurement Technology of the Ultrasonic Flowmeter. Chinese Journal of Scientific Instrument (2001)
10. Cao, X.-h., Yang, Y.-b., Cao, X.-p.: Design and Realization of Ultrasonic Gas Flowmeter, Instrumentation Technology (2005)

A Study of Improvements on the Performances of Load Flow Calculation in Newton Method

Xiangjing Su[1], Xianlin Liu[1], Ziqi Wang[2], and Yuanshan Guo[2]

[1] School of Electrical Engineering, Zhengzhou University, Zhengzhou 450001, China
[2] Henan Electric Power Dispatching and Communication Center, Zhengzhou 450052, China
sxjyx865432@126.com

Abstract. Newton method is the most widely used algorithm of load flow calculation. With the rapid developments of power systems, performance requirements of the algorithms are also increasing. From the aspects of model and solution of the revised equation group, this paper studies deeply most of improved algorithms of load flow calculation in Newton method, compares and analyzes the effects of improvements respectively by some examples. Then the advantages and disadvantages of each algorithm are summarized. Finally, considering prospects and future requirements of load flow calculation, we give some suggestions on how to improve the load flow calculation in Newton method further.

Keywords: power system, load flow calculation, Newton method, improvement.

1 Introduction

Based on the given structure, parameters and operating conditions, load flow calculation (LFC) can figure out the steady states of power system, including the voltages, the power and loss etc. Generally, LFC is not only essential to the operation and planning, but also fundamental to the analyses of transient and static stabilities. It generally has the following basic requirements: 1) reliable convergence, 2) high calculation speed, 3) less memory demand, 4) flexibility and convenience.

In essence, LFC is solving the nth-order nonlinear algebraic equation group $\mathbf{f(x)} =\mathbf{0}$, mainly achieved in iterative methods, among which Newton method is the most popular. For the tth iteration, Newton revised equation group is:

$$\mathbf{\Delta f}^{(t)} = \mathbf{J}^{(t)}\mathbf{\Delta x}^{(t)} \qquad (1)$$

Where, $\mathbf{\Delta f}^{(t)}$ 、 $\mathbf{J}^{(t)}$ and $\mathbf{\Delta x}^{(t)}$ are the column vectors of unbalance and incremental amount of unknown **x** and Jacobian matrix(**J**) respectively. With initial values $\mathbf{x}^{(0)}$ and the given precision ε, when $t = k$, if $\mathbf{\Delta f}^{(t)} \leq \varepsilon$ or $\mathbf{\Delta x}^{(t)} \leq \varepsilon$, the iteration converges and the solution of equation group is:

$$\mathbf{x} \approx \mathbf{x}^{(k)} + \mathbf{\Delta x}^{(k)} \qquad (2)$$

Newton method has a moderate memory demand, square convergence, high speed and precision. But there are also some defects, such as sensitivity to initial values, excessive calculation for single iteration and divergence of ill-conditioned networks. With the developments of power grids, the performance requirements of LFC in Newton method will be improved. Therefore researchers have proposed lots of improved Newton load flow algorithms. Here, the analyses of improvements of these algorithms are made from different points of view.

2 Improvements on the Revised Equation Group

Newton method changes solutions of non-linear equation group into iterative solutions of linear revised equation group and the revised equation directly affects the calculation. So people proposed several improved revised equation groups.

2.1 Fast Decoupled (FD) Method

J of Newton equation usually has a large dimension and must be reformed each iteration, which significantly influences the computing efficiency. So the FD method is proposed on the basis of weak coupling of active and reactive power. It has two assumptions: 1) the reactance of component is much larger than the resistance. Accordingly, the active and reactive power are mainly affected by the voltage phase angle and amplitude respectively; 2) the phase angle difference across the line is small. The simplified revised equation group is:

$$\Delta P / V = -B' V \Delta \theta \qquad \Delta Q / V = -B'' \Delta V \qquad (3)$$

Where, B', B'' are usually the imaginary parts of the node admittance matrix. We can also have other different forms of B', B''. In this case, FD method has XB, BX and other modes, and their performances are also different. Researches indicate the XB mode with the following modifications has the best performances: 1)when forming B', overlook the factors regarding reactive power and voltage amplitude, including shunt branches, transformer tapping .etc; 2) when forming B'', omit resistances.

By the IEEE30 system, a comparison of Newton method, the basic and XB modes of FD method is done. The results show that although FD method increases the number of iteration but the time of single iteration significantly decreases and the total time is reduced. Due to further improvements on B', B'', the XB mode have better performances. The results are consistent with theoretical expectations.

Table 1. Comparison of basic Newton method and FD method (iteration num/ cal time/s)

Newton	basic FD	XB FD
4/0.109380	28/0.046875	5/0.015000

FD method is simple, fast and also saves memory, which makes it widely used. But the FD method also has some defects: 1) with a constant coefficient matrix, it only has linear convergence. For the same precision, FD method usually has a larger number of iteration than Newton method with square convergence; 2) when the system has high R/X ratio or other ill-conditioned branches, FD method usually can not converge well; 3) FD method is still sensitive to initial values.

2.2 Newton Method Retaining-Nonlinearity (NMR)

For ill-conditioned systems, Newton method becomes hard to converge. To overcome this defect, NMR is proposed and its revised equation group is show as (4).

$$\begin{bmatrix} P \\ Q \\ V^2 \end{bmatrix} = \begin{bmatrix} P^{(0)} \\ Q^{(0)} \\ V^{(0)^2} \end{bmatrix} + \underbrace{\begin{bmatrix} H & N \\ M & L \\ R & S \end{bmatrix}}_{J} \begin{bmatrix} \Delta f \\ \Delta e \end{bmatrix} + \begin{bmatrix} SP \\ SQ \\ SV \end{bmatrix} \quad (4)$$

Where, on the right are the Taylor expansion's constant term, 1st-order term and 2nd-order term respectively. As the load flow equation is square, truncation error can be eliminated by retaining to the 2nd-order term. Thus, in theory, NMR is more accurate and convergent and can handle ill-conditioned systems well. Because of constant J obtained from the initial values, NMR also should be faster.

Besides the basic form, NMR has some improved forms. Based on the basic form, [1] uses the FD method instead of Newton method to get initial values of the NMR, which further fastens the calculation. But it has a larger memory demand; [2] introduces a fast algorithm retaining the 2nd-order term in rectangular coordinate, it is simplified as follows: 1) change the diagonal elements of admittance matrix; 2) all the nodes take the slack node voltage as their initial values. So the memory demands are significantly reduced for the symmetry of J and the convergence is also reliable; [3] improves performances by combining a fast flow algorithm retaining-nonlinearity with FD method. By increasing the load of node 7 of the IEEE30 system, we compare NMR with the basic Newton method. The results are shown in Table 2:

Table 2. comparison of NMR and basic Newton method (iteration num / cal time /s)

load	basic Newton	NMR	load	basic Newton	NMR
L_7	4/0.10938	5/0.09375	21 L_7	6/0.16410	17/0.3180
5 L_7	4/0.10938	6/0.11250	24 L_7	7/0.19141	28/0.5150
11L_7	5/0.13672	8/0.15000	26 L_7	8/0.21870	divergent

For the constant J, computation time reduces and thus NMR is faster than the basic Newton method. But for ill-conditioned systems, the former is not as stable as the latter. In some cases, the convergence of NMR may be not better than the basic Newton method. Besides, it is still sensitive to the initial values. To improve the convergence, reformation of J after several iterations has been proved to be effective.

3 Improvements on Solving the Revised Equation Group

The Newton revised equation group is usually solved in direct methods with sparse techniques. But for large systems, computation and memory significantly increase for large matrix and the introduction of more non-zero elements. Further more, because of the forward and backward characteristic, direct methods can not realize parallel solution, making it difficult to solve large systems. So the solution of revised equation group is modified to improve performances of Newton method and meet the requirements of large systems.

3.1 Improvements on the Direct Methods

When solving the revised equation group in direct methods, the basic Newton method often takes lots of time to reform the **J** each iteration. So how to simplify the **J** has become important. Such as, [4] proposes a new method to reduce the computing time. It forms an approximate **J** for the initial iterations before striking certain precision and then turns to the normal form.

3.2 Iterative Methods for Solving the Revised Equation Group

Iterative methods, which solve the revised equation group by iteration, are introduced to deal with large systems in recent years. Generally, they can be divided into classical iterative methods and Krylov iterative methods. Classical iterative methods are rarely used now; According to the coefficient matrix, Krylov methods are also divided into two kinds. One is for symmetric problems, including Conjugate Gradient method [5] and Minimal Residual method etc; the other is for non-symmetric problems, including Generalized Minimal Residual method [6] and Bi-Conjugate Gradient method etc.

The convergence of iterative methods depends on the conditioner of coefficient matrix. So it is necessary to carry out preconditions of **J** [7] to reduce its conditioner and improve the convergence. Currently, preconditioned methods include: ILU decomposition method, block diagonal matrix method and PQ decomposition method, etc. In addition, [6] proposes two new sparse approximate inverse preconditioners. Combining them with Newton-GMRES, the convergence of LFC is significantly improved; [8] introduces an orthogonal preconditioning method. Compared with the most effective PQ method this method is easier and has better local convergence in theory.

Combined with preconditioners, iterative methods need less storage and computation and can realize parallel computing. It is very suitable for solving large and sparse revised equations of large systems. But they also have some defects: 1) limited application, such as the CG method only applies to positive and symmetrical **J**; 2) imperfect preconditioners, such as the choice of filling amount of ILU method is difficult; 3) how to choose the preconditioner reasonably is still in doubt. It is still in the theoretical stage.

4 Other Improvements

4.1 Improvements on the Initial Values

Newton methods are sensitive to the initial values. Generally, flat start [9] is acceptable in most cases. But it may get unreliable to complex systems. So many improvements on initial values have been proposed.

Considering Gauss-Seidel(GS) method is insensitive to initial values and fast at start, [10] uses the load flow results of the first iteration obtained by GS method as the initial values of Newton method and the computing time is shorten; [11] applies the Direct Current (DC) method to determine the phase angle initial values. It is proven effective in reducing the computing time and the number of iteration; [12] proposes an integrated method to get the initial values. Specifically, get phase angle initial values by DC method at first and then amplitude initial values are obtained by approximation of the imaginary part of GS method.

To compare these methods, we calculate the IEEE30 system and the 5 nodes system in [13] respectively. The results in Table3 show that the DC method can provide initial values reliably while the integrated method is not ideal because of too much approximation. As for the first iteration of GS method, the calculation does not necessarily point to the real solutions, so it is not reliable to take the results of the first iteration as initial values. As the iteration progresses, the results will be increasingly reliable, but not necessarily better than DC method and flat start.

Table 3. Number of iteration of different methods

	GS 1 iterations	GS 3 iterations	DC	Integrated	flat start
5nodes	9	8	9	10	10
IEEE30	31	29	28	32	28

4.2 Improvements on the Small Impedance Branches

For systems with branches of small impedance, the convergence of Newton method will deteriorate. So many improvements are proposed: [14] eliminates the impacts of small impedance branches by forming voltage initial values to-Zero Power(ZP) method to ensure the power of small impedance branch is zero. The convergence is improved with the reduction of unbalanced power injections; [15] proposes a Varied-Jacobian Newton (VJN) method to deal with small impedance branches. The **J** elements are formed by the known power injections instead of calculating. Small impedance branches influence both the precision of calculation and the requirements of initial values, so [16] uses double type variations and GS method twice to choose initial values for Newton method.

Here, we use a 6 node system in [15] with small impedance branches to verify the methods above. The results in Table 4 show that all the methods can deal with small impedance branches well.

Table 4. Comparison of methods solving small impedance branches(iteration num/ cal time/s)

ZP	VJN	GS &double type	Newton	FD
4	6	4	divergent	divergent
0.015	0.016	0.015		

5 Conclusion

The performances of load flow calculation in Newton method have been constantly improved since its presentation. They not only meet the requirements of speed, convergence and memory to some degree but also broaden applications of Newton method. However it still has some defects: 1) all the existing improvements mainly tackle local problems and there are few methods with improvements in all aspects; 2) as the calculation diverges, it is difficult to determine the exact reason; 3) the performances in dealing with ill-conditioned or large systems have to be further improved.

According to existing problems and future requirements, some suggestions are given: 1) new Newton methods, which have better performances in all aspects, should be proposed by combining existing methods with modern optimization techniques; 2) as for ill-conditioned systems, there are already some good methods, such as the optimal multiplier method. So, we can combine these methods with existing improved Newton methods to realize the overall optimization of performances in both ill and normal conditions; 3) we must continuously develop and improve Newton method to meet the requirements of effective and real-time calculation of large-scale systems.

References

1. Kou, Q., Li, B., Lu, B.: Improvement of Retaining-nonlinearity Load Flow Algorithm. Journal of Liaoning University of Technology 28(1), 10–12 (2008)
2. Nagendra, R.P.S., Prakasa, R.K.S., Nanda, J.: An Exact Fast Load Flow Method Including Second Order terms in Rectangular Coordinates. IEEE Trans. PAS 101(9), 3261–3268 (1982)
3. Hou, B., Zhang, R.: The Method of Fast PQ Decoupled Load Flow Calculation Retaining Non-linearity. Journal of Electrical Engineering 4(1), 20–25 (1984)
4. Yang, D., Zhou, B., Du, X.: A Simplification Method of Power Flow Calculation Based on Newton-Raphson. Journal of Shenyang Institute of Engineering (Natural Science) 4(1), 37–40 (2008)
5. Liu, Y., Zhou, J., Xie, K., et al.: The Preconditioned CG Method for Large Scale Power Flow Solution. Proceedings of the CSEE 26(7), 89–94 (2006)
6. Wang, F., He, Y., Ye, J.: Load Flow Calculation of Newton-GMRES Method With Sparse Approximate Inverse Preconditioners. Power System Technology 32(14), 50–53 (2008)
7. Li, X., Li, J., Zhang, L., et al.: Comparison and Study of Preconditioning Methods of Jacobian Matrix of Power Flow Calculation. Power System Protection and Control 33(15), 33–37 (2005)
8. Zhou, F., Gao, S.: Study on Orthogonal Preconditioning Method for Jacobian Matrix in Newton-Raphson Optimization. Power System Protection and Control 38(3), 20–24 (2010)
9. Wang, X.: Power System Calculation. China Water Power Press, Beijing (1995)
10. Li, M.: A Colligate Algorithm for Low Flow Calculation in Power System. Journal of Qinghai University 22(5), 78–81 (2004)
11. Yang, J.: A Method of Choosing Initial Values of Power Flow Calculation in Fast Decoupled Method. Central China Power 9(6), 4–6 (1996)
12. Stott, B.: Effective starting process for Newton-Raphson load flows. Proceedings of the Institution of Electrical Engineers 118(8), 983–987 (1971)

13. Chen, H.: Power System Stability and Analysis, 2nd edn. China Power Press, Beijing (1995)
14. Yao, Y., Lu, B., Chen, X.: A Method to Deal with the Effect of Small Impedance Branches to Prevented Divergence in Newton-Raphson Load Flow. Power System Technology 23(9), 27–31 (1999)
15. Yao, Y., Liu, D., Chen, X.: New Load-Flow Method to Deal with Small Impedance Branches. Journal of Harbin Institute of Technology 33(4), 525–529 (2001)
16. Han, P., Liu, W., Wu, Q.: Influence and Processing of Small Impedance Branches on the Convergence of Load Flow Calculation. Power System Protection and Control 37(18), 17–20 (2009)

A Design Approach of Model-Based Optimal Fault Diagnosis

Qiang Shen, Lin-Feng Gou, and Xue-Min Yang

School of Power and Energy, Northwestern Polytechnical University,
Xi'an 710072, China
shenqwin@gmail.com

Abstract. The main purpose of fault diagnosis is to detect faults rapidly and accurately, decide the types, sizes and trends of faults, furthermore, separate the fault and make a proper decision to avoid fault. A model-based optimal fault diagnosis method is proposed in this paper. Modeling for a class of typical faults whose dynamic characteristics were known and the initial conditions were unknown with methods in linear system theory, and a reduced-order state observer was designed for fault system. In order to realize fault state of on-line optimal estimation, an optimal fault diagnosis method was proposed by optimal control theory and duality principle, meanwhile, observation error and control energy was optimum. In addition, threshold value is used to decide whether the fault occurred. Simulation results demonstrate the optimal fault diagnosis system can detect the typical faults on-line.

Keywords: fault diagnosis, optimal observer, quadratic performance index, threshold.

1 Introduction

Various environmental changes, unknown disturbances, and changing operating condition are inevitable in many practical dynamical systems, thus sensors, actuators or components failure and faults are very common [1]. A fault [2],[3] in a dynamical system is a deviation of the system structure or the system parameters from the nominal situation. Fault diagnosis should detect system fault accurately, find in which component a fault has occurred, then estimate its magnitude and trend. At present, there are three main methods for fault diagnosis: model-based approach [4], signal processing approach [5],[6] and knowledge-based approach [7],[8].

In Ref. [9], the proportional-integral observer for unknown input descriptor systems is applied to fault estimation; Ref. [10], observer-based fault detection and estimation(FDI) problem using structured residual sets that allow fault isolation; The existence conditions and design algorithm of sliding mode observer for linear descriptor systems with faults are given in Ref. [11]. Ref. [12],[13] proposed nonlinear unknown input observer(UIO)-based FDI approaches, which extended UIO-based FDI from linear system to a respective class of nonlinear system. In this paper, the optimal fault diagnosis problems are studied by using the model-based approach, to a class of faults whose dynamic characteristics are known and the initial conditions are unknown, who realizes the on-line fault diagnosis and makes the observation error and control energy of the designed diagnostic system optimal.

The paper is organized as follows: in Section 2 the problem is formulated, in Section 3 a reduced dimensional observer for linear system is designed, in Section 4 the approach of model-based optimal fault diagnosis is proposed and a threshold value is used to decide whether the fault occurred or not, in Section 5 an example supporting effectiveness of the proposed approach is reported. Finally, some conclusions are given.

2 Problem Formulation

Considering the following linear time-invariant system [14],[15] :

$$\begin{cases} \dot{x}(t) = Ax(t) + Bu(t) + D_a f(t) + Gw(t) \\ y(t) = Cx(t) + D_s f(t) + v(t) \end{cases} \quad (1)$$

Where $x(t) \in R^n$ is the state vector, $u(t) \in R^m$ is the measurable input vector, $y(t) \in R^p$ is the output vector, $w(t)$ and $v(t)$ represent the system noise and measurement noise separately. Both of the noises are white Gaussian noise whose statistical property can be described as follows:

$$\begin{cases} E\{w(t)\} = 0 \quad E\{v(t)\} = 0 \\ E\{w(t)w^T(t+\tau)\} = Q_0 \delta(t-\tau) \\ E\{v(t)v^T(t+\tau)\} = R_0 \delta(t-\tau) \\ E\{w(t)v^T(t+\tau)\} = 0 \end{cases} \quad (\forall t, \tau) \quad (2)$$

$f(t) \in R^q$ is the fault vector, which is made up by actuator fault $f_a(t) \in R^{q_a}$ which happens at the time of t_a and sensor fault $f_s(t) \in R^{q_s}$ which happens at the time of t_s, where $q = q_a + q_s$.

$$f(t) = \begin{bmatrix} f_a(t) \\ f_s(t) \end{bmatrix} \quad (3)$$

If $\zeta_a \in R^{r_a}$ is the actuator fault state vector and $\zeta_s \in R^{r_s}$ is the sensor fault state vector, the dynamic behavior of fault is known as

$$\begin{cases} \zeta(t) = 0, \quad t \in [0, t_0) \\ \dot{\zeta}(t) = M\zeta(t), \quad t > t_0, \quad t_0 = \min\{t_a, t_s\} \\ f(t) = F\zeta(t) \end{cases} \quad (4)$$

Where $\zeta(t) = \begin{bmatrix} \zeta_a(t) \\ \zeta_s(t) \end{bmatrix}, M = \begin{bmatrix} M_a & 0 \\ 0 & M_s \end{bmatrix}, F = \begin{bmatrix} F_a & 0 \\ 0 & F_s \end{bmatrix}.$

Considering the fault system described above, how to design an optimal observer to diagnosis fault is the problem need to be solved.

3 Design of Reduced Dimensional Observer

According to the model of fault system and letting

$$z(t) = \begin{bmatrix} x(t) \\ \zeta(t) \end{bmatrix} \tag{5}$$

Combining (1) and (4), we can get the following state space expression in an augmented form

$$\begin{cases} \dot{z}(t) = A_1 z(t) + B_1 u(t) + G_1 w(t) \\ y(t) = C_1 z(t) + v(t) \end{cases} \tag{6}$$

Where

$$A_1 = \begin{bmatrix} A & D_a F \\ 0 & M \end{bmatrix}, B_1 = \begin{bmatrix} B \\ 0 \end{bmatrix}, C_1 = \begin{bmatrix} C & D_s F \end{bmatrix}, G_1 = \begin{bmatrix} G \\ 0 \end{bmatrix}$$

Lemma 1 [15]: The sufficient and necessary condition of completely observable for (C_1, A_1) is (C, A), $\left((C(\lambda I - A)^{-1} D_a F + D_s F), M \right)$, (DF, M) are completely observable.

If the system is completely observable, then we can design a reduced dimensional Luenberger observer to detect fault.

Forming a nonsingular matrix $T = \begin{bmatrix} T_1 \\ C_1 \end{bmatrix} = \begin{bmatrix} T_1^T & C_1^T \end{bmatrix}^T$, letting $T^{-1} = \begin{bmatrix} P_1 & P_2 \end{bmatrix}$, then we can get

$$\begin{cases} \overline{w}_c(t) = Tz(t) \\ w_c(t) = T_1 z(t) \end{cases} \Rightarrow \overline{w}_c(t) = Tz(t) = \begin{bmatrix} T_1 \\ C_1 \end{bmatrix} z(t) = \begin{bmatrix} w_c(t) \\ y(t) - v(t) \end{bmatrix}$$

$$z(t) = T^{-1} \begin{bmatrix} w_c(t) \\ y(t) - v(t) \end{bmatrix} = P_1 w_c(t) + P_2 \left[y(t) - v(t) \right] \tag{7}$$

$$\Rightarrow \begin{cases} \dot{w}_c(t) = T_1 A_1 P_1 w_c(t) + T_1 A_1 P_2 y(t) + T_1 B_1 u(t) + T_1 \left(G_1 w(t) - A_1 P_2 v(t) \right) \\ \dot{y}(t) = C_1 A_1 P_1 w_c(t) + C_1 A_1 P_2 y(t) + C_1 B_1 u(t) + C_1 \left(G_1 w(t) - A_1 P_2 v(t) \right) + \dot{v}(t) \end{cases} \tag{8}$$

Introducing the equivalent input $\overline{u}(t)$ and output $\overline{y}(t)$

$$\begin{aligned} \overline{u}(t) &= T_1 A_1 P_2 y(t) + T_1 B_1 u(t) + T_1 \left(G_1 w(t) - A_1 P_2 v(t) \right) \\ \overline{y}(t) &= \dot{y}(t) - C_1 A_1 P_2 y(t) - C_1 B_1 u(t) - C_1 \left(G_1 w(t) - A_1 P_2 v(t) \right) - \dot{v}(t) = C_1 A_1 P_1 w_c(t) \end{aligned} \tag{9}$$

Then we can get

$$\begin{cases} \dot{w}_c(t) = T_1 A_1 P_1 w_c(t) + \bar{u}(t) \\ \bar{y}(t) = C_1 A_1 P_1 w_c(t) \end{cases} \quad (10)$$

If $(C_1 A_1 P_1, T_1 A_1 H_1)$ is completely observable, reduced dimensional Luenberger observer can be formed as follow:

$$\dot{\hat{w}}_c(t) = (T_1 - LC_1) A_1 P_1 \hat{w}_c(t) + \bar{u}(t) + L\bar{y}(t) \quad (11)$$

Where L is observer gain, combining (9) and (11), we can get

$$\dot{\hat{w}}_c(t) = (T_1 - LC_1)\left[A_1 P_1 \hat{w}_c(t) + B_1 u(t) + A_1 P_2 y(t) + (G_1 w(t) - A_1 P_2 v(t)) \right] + L\left[\dot{y}(t) - \dot{v}(t) \right] \quad (12)$$

Using a special variable transformation

$$x_c(t) = \hat{w}(t) - L[y(t) - v(t)]$$

Then

$$\begin{aligned} \dot{x}_c(t) &= \dot{\hat{w}}(t) - L[\dot{y}(t) - \dot{v}(t)] \\ &= (T_1 - LC_1)\left[A_1 P_1 \hat{w}_c(t) + B_1 u(t) + A_1 P_2 y(t) + (G_1 w(t) - A_1 P_2 v(t)) \right] \end{aligned} \quad (13)$$

From (7) we can get

$$\hat{z}(t) = P_1 \hat{w}_c(t) + P_2[y(t) - v(t)] = P_1 x_c(t) + (P_1 L + P_2)[y(t) - v(t)] \quad (14)$$

Letting

$$P_1 = \begin{bmatrix} P_{11} \\ P_{21} \end{bmatrix}, P_2 = \begin{bmatrix} P_{12} \\ P_{22} \end{bmatrix}$$

Finally, the reduced dimensional Luenberger observer is

$$\begin{cases} \dot{x}_c(t) = (T_1 - LC_1)\left[A_1 P_1 x_c(t) + B_1 u(t) + G_1 w(t) + A_1 (P_1 L + P_2)(y(t) - v(t)) \right] \\ \hat{f}(t) = F\left[P_{21} x_c(t) + (P_{22} + P_{21} L)(y(t) - v(t)) \right] \end{cases} \quad (15)$$

4 Design of Optimal Fault Diagnosis

In order to realize optimal fault diagnosis, using the reduced dimensional observer designed above, observer gain L must satisfy a quadratic performance index to fulfill the observation error and control energy of the designed diagnostic system optimal. A new design approach is proposed by optimal control theory and duality principle.

Define observation error as

$$\tilde{w}_c = w_c - \hat{w}_c \quad (16)$$

Using (8) compare to (12), observation error equation is written to be

$$\dot{\tilde{w}}_c = \dot{w}_c - \dot{\hat{w}}_c = T_1 A_1 P_1 \tilde{w}_c - LC_1 A_1 P_1 \tilde{w}_c \quad (17)$$

So, we can get the dual system of observation error equation

$$\dot{\psi}(t) = (T_1 A_1 P_1)^T \psi(t) - (C_1 A_1 P_1)^T L^T \psi(t) \tag{18}$$

Obviously, the dual system (17) and error equation (18) have the same eigenvalues

$$\det\left[\lambda I - (T_1 A_1 P_1)^T + (C_1 A_1 P_1)^T L^T\right] = \det\left[\lambda I - T_1 A_1 P_1 + LC_1 A_1 P_1\right] \tag{19}$$

Above all, we can design an optimal state feedback gain matrix L of the dual system to achieve the observer optimal.

Letting

$$\varphi(t) = -L^T \psi(t) \tag{20}$$

Where, L^T is defined as the equivalent state feedback of dual system. So, the dual system can be written as the following open loop system

$$\dot{\psi}(t) = (T_1 A_1 P_1)^T \psi(t) + (C_1 A_1 P_1)^T \varphi(t) \tag{21}$$

In (21), $\varphi(t)$ is known as the equivalent input vector.

Taken together, the optimal estimate of fault diagnosis system (15) has transformed into optimal states feedback of the dual system (21).

For the dual system, considering the quadratic performance index as follow

$$J = \lim_{t \to \infty} \frac{1}{T} \int_0^T \left(\psi^T(t) Q_{obs} \psi(t) + \varphi^T(t) R_{obs} \varphi(t)\right) dt \tag{22}$$

Where, Q_{obs} is positive semidefinite matrix and R_{obs} is positive definite matrix. In (22), $\psi^T(t) Q_{obs} \psi(t)$ is observation error and $\varphi^T(t) R_{obs} \varphi(t)$ is the control energy. In a word, the quadratic performance index (22) makes the system in the whole process of dynamic observation error and consumption of control energy satisfies a tradeoff optimal.

Based on optimal control theory, we can get optimal control rule of the dual system,

$$\varphi^*(t) = -R_{obs}^{-1}(C_1 A_1 P_1) P_o(t) \psi(t) \tag{23}$$

Where, $P_o(t)$ is the solution of the following Riccati equation,

$$\dot{P}_o(t) = -P_o(t)(T_1 A_1 P_1)^T - T_1 A_1 P_1 P_o(t) + P_o(t)(C_1 A_1 P_1)^T R_{obs}^{-1} C_1 A_1 P_1 P_o - Q_{obs} \tag{24}$$

Consequently, the state feedback matrix of dual system in the sense of optimum is known as

$$L^T = R_{obs}^{-1}(C_1 A_1 P_1) P_o(t) \tag{25}$$

Then, the state feedback gain matrix of optimal fault diagnosis system is the inversion of (25)

$$L = P_o(t)(C_1 A_1 P_1)^T R_{obs}^{-1} \tag{26}$$

For the detected fault, threshold value can be confirmed based on the principle of maximum inconsistent [16]. Considering each unit which may have some fault, the basic idea is to determine allowed deviation under the worst situations, and according to the deviation to set threshold value of fault diagnosis system.

As for a system with two redundant channels, each unit which may have some fault with an allowable error E_i ($i=1,2,\cdots,n$), and the gain of each unit is R_i ($i=1,2,\cdots,n$), therefore, the maximum deviation can be described as

$$E_{\max} \approx \sum_{i=1}^{n}(|E_i|+|E_i|) = 2\sum_{i=1}^{n}|E_i| \tag{27}$$

Finally, we can define the threshold value *THR* as

$$THR \geq E_{\max} \tag{28}$$

For a fault system, if the final diagnosed results beyond the threshold periodically, we can decide the fault has happened.

5 Application Example

Considering the nominal model of aero-engine in the condition of $H = 0km$, $Ma = 0$, the input vector is $u(t) = \begin{bmatrix} q_{m,f}(t) & A_8(t) \end{bmatrix}^T$, state vector is $x(t) = \begin{bmatrix} N_H & N_L \end{bmatrix}^T$, output vector is $y(t) = \begin{bmatrix} N_H & \pi_P \end{bmatrix}^T$, where $q_{m,f}(t)$ is fuel flux, $q_{m,f}(t)$ is nozzle area, N_H and N_L are respectively engine high-pressure rotation speed and low-pressure rotation speed, π_P is the pressure ratio.

Establishing expand object model, considering the fuel metering device and tail spout actuator cylinder as inertial element, the characteristic time is 0.05s and 0.1s, their transfer function can be written as G_1 and G_2,

$$G_1 = \frac{1}{0.05s+1} = \frac{20}{s+20}, G_2 = \frac{1}{0.1s+1} = \frac{10}{s+10}$$

$\dot{q}_{m,f}$ and $\dot{A}_8$ can also be seen as state vector, then the extended object model is described as following [17]:

$$\dot{x}_{ex}(t) = \begin{bmatrix} -6.715 & 2.256 & 0.361 & 0.442 \\ 7.380 & -9.089 & -0.304 & 2.032 \\ 0 & 0 & -20 & 0 \\ 0 & 0 & 0 & -10 \end{bmatrix} x_{ex}(t) + \begin{bmatrix} 0 & 0 \\ 0 & 0 \\ 20 & 0 \\ 0 & 10 \end{bmatrix} u(t) + \begin{bmatrix} 1 & 0 \\ 1 & 0 \\ 0 & 0 \\ 0 & 0 \end{bmatrix} f(t) + \begin{bmatrix} 1 \\ 0 \\ 0 \\ 0 \end{bmatrix} w(t)$$

$$y(t) = \begin{bmatrix} 1 & 0 & 0 & 0 \\ 0.473 & 2.320 & 0.371 & -0.717 \end{bmatrix} x_{ex}(t) + \begin{bmatrix} 0 & 1 \\ 0 & 1 \end{bmatrix} f(t) + v(t)$$

Where $\dot{x}_{ex} = \begin{bmatrix} n_H & n_L & \dot{q}_{m,f} & \dot{A}_8 \end{bmatrix}^T$, $y(t) = \begin{bmatrix} N_H & \pi_P \end{bmatrix}^T$, $u(t) = \begin{bmatrix} q_{m,f}(t) & A_8(t) \end{bmatrix}^T$, the fault model is

$$M = \begin{bmatrix} 0 & 1 & 0 & 0 \\ -1 & 0 & 0 & 0 \\ 0 & 0 & 0 & 1 \\ 0 & 0 & -4 & 0 \end{bmatrix}, F = \begin{bmatrix} 1 & 0 & 0 & 0 \\ 0 & 0 & 0 & 1 \end{bmatrix}$$

The sensor fault is a sine wave fault, which happens at the time of $t_s = 20s$, and the frequency is $w_s = 2\,rad/s$; the actuator fault is also a sine fault, which happens at the time of $t_a = 30s$, and the frequency is $w_a = 1\,rad/s$, the amplitude of white Gaussian noise $w(t), v(t)$ are 0.2.

In optimal fault diagnosis system $T_1 = [0 \quad I_6]$, in quadratic performance index (22), $R_{obs} = 1$ and $Q_{obs} = I_6$, the allowable error of sensor is $E_s = 10\%$, and the allowable error of actuator is $E_a = 10\%$, based on the principle of maximum inconsistent we can get the maximum deviation is $E_{\max} = 10\% + 10\% = 20\%$. Hence, the threshold is $THR > 20\%$.

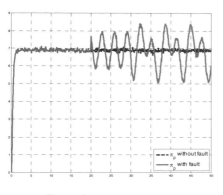

Fig. 1. System output π_p

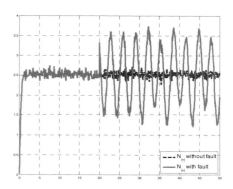

Fig. 2. System output N_H

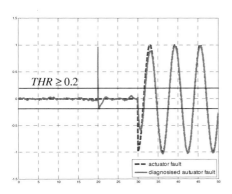

Fig. 3. True value and estimated value of actuator fault

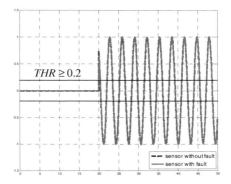

Fig. 4. True value and estimated value of sensor fault

As for the input is step signal $u(t) = \begin{bmatrix} 10 & 10 \end{bmatrix}^T$, we can get the optimal observer gain is

$$L = \begin{bmatrix} -0.0090 & 0.0525 & 0.0437 & 0.2748 & -0.9590 & 0.8423 \\ -0.0769 & 0.1554 & 1.2829 & 0.5260 & 0.0769 & -0.7671 \end{bmatrix}^T$$

Figure 1 and figure 2 are the system output π_P and N_H in the situation of happening fault and without fault. From figure 1 and figure 2, we can see dynamic performance has changed a lot after fault happens. In simulation, we set the threshold value $THR \geq 20\%$, through comparing the curves of true value and estimated value about sensor fault and actuator fault, we can see the diagnosed curves beyond the threshold periodically, actually, so we can judge the fault has happened. In figure 3, we can get the conclusion the actuator fault happens almost in the 30 seconds, similarly, in figure 4, the sensor fault happens almost in the 20 seconds.

6 Conclusion

A model-based optimal fault diagnosis method of aero-engine is proposed in this paper. Modeling for a class of typical faults whose dynamic characteristics are know and the initial conditions are unknown, with methods in linear system theory, a reduced-order state observer is designed for fault system. In order to realize fault state of on-line optimal estimation, an optimal fault diagnosis is proposed by optimal control theory and duality principle, at the same time, observation error and control energy are optimum. In addition, a threshold value is used to decide whether the fault occurred. Simulation results demonstrate the optimal fault diagnosis system can follow the fault on-line and ensure the following error smallest.

References

1. Liu, N., Zhou, K.: Optimal Robust Fault detection for Linear Discrete Time Systems. Journal of Control Science and Engineering (2008)
2. Chen, J., Patton, R.J.: Robust Model-based Fault diagnosis for Dynamic System. Kluwer Academic Publishers, Dordrecht (2009)
3. Mogens, B., Michel, K., Jan, L., Marcel, S.: Diagnosis and Fault-Tolerant Control. Springer, Heidelberg (2006)
4. He, H., Wang, G.Z., Ding, S.X.: A new parity space approach for fault detection based on stationary wavelet transform. IEEE Transactions on Automatic Control 49(2), 281–287 (2004)
5. Sun, R.X., Tsung, F., Qu, L.S.: Evolving kernel principal component analysis for fault diagnosis. Computers and Industrial Engineering 53(2), 361–371 (2007)
6. He, H., Ding, S.X., Wang, G.Z.: Integrated design of fault detection systems in time-frequency domain. IEEE Transactions on Automatic Control 47(2), 384–390 (2002)
7. Rajakarunakaran, S., Venkumar, P., Devaraj, D., et al.: Artificial neural network approach for fault detection in rotary system. Applied Soft Computing 8(1), 740–748 (2008)

8. Papadopoulos, Y.: Model-based system monitoring and diagnosis of failures using statecharts and fault trees. Reliability Engineering and System Safety 81(3), 325–341 (2003)
9. Damien, K.: Unknown Iinput Proportional Multiple-Integral Observer Design for Linear Descriptor Systems: Application to State and Fault Estimation. IEEE Transaction on Automatic Control 5(2), 212–217 (2005)
10. Commault, C., Dion, J.-M., Sename, O., Motyeian, R.: Observer-Based Fault Detection and Isolation for Structured Systems. IEEE Transactions on Automatic Control 47(12), 2074–2079 (2002)
11. Yu, J.-Y., Liu, Z.-Y.: Fault Reconstruction Based on Sliding Mode Observer for Linear Descriptor Systems. In: Proceedings of the 7th Asian Control Conference, Hong Kong, China, pp. 1132–1137 (August 2009)
12. Seliger, R., Frank, P.M.: Fault Diagnosis by Disturbance Decoupled Nonlinear Observers. In: Proc. IEEE CDC, Brighton, UK, pp. 2248–2253 (1991)
13. Seliger, R., Frank, P.M.: Robust component fault detection and isolation in nonlinear dynamic systems using nonlinear unknown input observers. In: Proc. IFAC/IMACS Symp. SAFEPROCESS 1991, Baden-Baden, Germany, pp. 313–318 (1991)
14. Li, J., Tang, G.-Y., Gao, H.-W.: Fault Detection and Self-Restore Control for Linear Systems. In: The Proceedings of Sixth International Conference on Intelligent Systems Design and Applications, pp. 873–878. IEEE Computer Society, Jinan (2006)
15. Ye, R.-H.: Study on Observer-Based Fault Diagnosis and Optimal Fault-Tolerant Control Approaches, Qingdao Agriculture University, China (2008)
16. Wen, X., Zhang, H.-Y., Zhou, L.: Fault Diagnosis and Fault-Tolerant Control for Control system. China Machine Press, Beijing (1998)
17. Fan, S.-Q., Li, H.-C., Fan, D.: Aero-engine control. Northwestern Polytechnic University Press, Xi'an (2008)

Dynamic Adaptive Terminal Sliding Mode Control for DC-DC Converter

Liping Fan[*] and Yazhou Yu

College of Information Engineering, Shenyang University of Chemical Technology,
Shenyang, 110142, P.R. China
flpsd@163.com

Abstract. Inherent time-varying and heavy nonlinearity make DC-DC converters have many difficulties in control. A kind of dynamic terminal sliding mode control method is designed in this paper. To make it suitable for DC-DC converter suffering unknown border disturbances, an adaptive term is introduced in the designing of sliding mode control law. This scheme can make the system reach steady state quickly without chattering, and have good robustness to external disturbances and inner parameter variations. Simulation results show the effectiveness of this approach.

Keyword: adaptive, sliding-mode control, dynamic, DC-DC converter.

1 Introduction

DC-DC converters have been widely used in most of the industrial applications such as DC motor drives, computer systems and communication equipments. Design of high performance control is a challenge because of nonlinear and time variant nature of DC-DC converters [1]. Generally, linear conventional control solutions applied to DC-DC converter failed to accomplish robustness under nonlinearity, parameter variation, load disturbance and input voltage variation [2-4].

The sliding-mode controller (SMC) is one of the effective nonlinear robust control approaches [5]. Its major advantages are the guaranteed stability and the robustness against parameter, line, and load uncertainties. But in practical systems, the discontinuous switch control of SMC may cause some "Chattering" problems. Chattering can arouse the unmodelled properties of the system and so affect the control performance of the system [6].

Dynamic SMC transfers discontinuous term in control system to first or higher derivative of control input, and this make the control law continuous essentially and thus reduce chattering. But if disturbances occur, the boundary values of disturbances must be known when designing such control method. To make the dynamic SMC can suit for control DC-DC converter suffering unknown border disturbances, a dynamic adaptive terminal SMC is designed in this paper.

---

[*] This work is supported by Key Lab Project of Education Department of Liao Ning Province, China (No. LS2010127).

2 Dynamic Sliding Mode Control

A basic DC-DC converter known as Buck converter can be modeled as

$$\begin{cases} \dot{x}_1 = x_2 \\ \dot{x}_2 = -\dfrac{x_1}{LC} - \dfrac{x_2}{RC} + \dfrac{E}{LC}d + F \end{cases} \quad (1)$$

where the state variables x_1 and x_2 are selected as the output voltage u_c and its derivative, d is the duty factor of a switching period, F denotes the whole disturbances the system suffered. Parameters such as L, C, R, E denote the given definite part [7].

Suppose the expected tracking voltage is r, then the tracking error and its derivative are

$$e = x_1 - r, \quad \dot{e} = x_2 - \dot{r} \quad (2)$$

then the switching function can be written as

$$s(t) = ce + \dot{e} = c(x_1 - r) + x_2 - \dot{r} \quad (3)$$

Construct a new dynamic switching function as

$$\sigma = \dot{s} + \lambda s \quad (4)$$

then

$$\dot{\sigma} = \lambda c(x_2 - \dot{r}) + (c+\lambda)f - (c+\lambda)\ddot{r} + (c+\lambda)\dfrac{E}{LC}d \\ + (c+\lambda)F + \dot{f} + \dfrac{E}{LC}\dot{d} + \dot{F} - \dddot{r} \quad (5)$$

where $f = -\dfrac{x_1}{LC} - \dfrac{x_2}{RC}$.

It is assumed that the border of the uncertain term is G(x), that is $|F| \le G(x)$, $\forall x \in R^2$. The border of the derivative of the uncertain term is $\overline{G}(x)$, that is $|\dot{F}| \le \overline{G}(x)$, $\forall x \in R^2$. The dynamic siding mode control law is designed as

$$\dot{d} = \dfrac{LC}{E}[-(c+\lambda)\dfrac{E}{LC}d - (c+\lambda)f - \dot{f} + (c+\lambda)\ddot{r} + \dddot{r} - \lambda c(x_2 - \dot{r}) - \varepsilon \operatorname{sgn}(\sigma) - h\sigma] \quad (6)$$

where ε is positive real number and $\varepsilon > (c+\lambda)G(x) + \overline{G}(x)$.

Lyapunov function is defined as

$$V = \dfrac{1}{2}\sigma^2 \quad (7)$$

Merge (6) to (5), we have

$$\dot{\sigma} = (c+\lambda)F + \dot{F} - \varepsilon \operatorname{sgn}(\sigma) - h\sigma \quad (8)$$

then the following equation can be derived

$$\dot{V} = \sigma\dot{\sigma} < \sigma[(c+\lambda)F + \dot{F}] - [(c+\lambda)G(x) + \overline{G}(x)]|\sigma| - h\sigma^2 \le 0 \quad (9)$$

3 Dynamic Adaptive Terminal Sliding Mode Control

To achieve better control effects, a dynamic terminal sliding mode control method is designed. The dynamic switching function is constructed as

$$\sigma = \dot{s} - \dot{p} + \lambda(s - p) \quad (10)$$

where $p \in C^n[0,\infty), \dot{p}, \cdots p^{(n)} \in L^\infty$, $C^{(n)}[0,\infty)$ denotes all the n-order differentiable continuous functions defined in $[0,\infty]$. Then

$$\dot{\sigma} = c\ddot{e} + \dddot{e} - \ddot{p} + \lambda(c\dot{e} + \ddot{e} - \dot{p})$$

$$= \lambda c(x_2 - \dot{r}) + (c+\lambda)f - (c+\lambda)\ddot{r} + (c+\lambda)\frac{E}{LC}d \quad (11)$$

$$+ (c+\lambda)F + \dot{f} + \frac{E}{LC}\dot{d} + \dot{F} - \dddot{r} - \ddot{p} - \lambda\dot{p}$$

Define Lyapunov function as

$$V = \frac{1}{2}\sigma^2 + \frac{1}{2}\alpha\tilde{v}^2 \quad (12)$$

where $\tilde{v} = \hat{v} - \overline{v}$. $\hat{v}$ denotes the estimation of disturbance border $\overline{v}$, $\tilde{v}$ denotes estimating error. α is gain of the adaptive term, and $\alpha > 0$.

It is assumed that the boundary condition $|F| \le G(x)$ and $|\dot{F}| \le \overline{G}(x)$ is satisfied, and there exist a term of $\overline{v}$ which satisfies the following condition

$$\overline{v} > (c+\lambda)G(x) + \overline{G}(x) \quad (13)$$

then

$$\dot{V} = \sigma\dot{\sigma} + \alpha\tilde{v}\dot{\tilde{v}} = \sigma[\lambda c(x_2 - \dot{r}) + (c+\lambda)f - (c+\lambda)\ddot{r} + (c+\lambda)\frac{E}{LC}d$$

$$+ (c+\lambda)F + \dot{f} + \frac{E}{LC}\dot{d} + \dot{F} - \dddot{r} - \ddot{p} - \lambda\dot{p}] + \alpha\tilde{v}\dot{\hat{v}} \quad (14)$$

The control law is designed as

$$\dot{d} = \frac{LC}{E}[-(c+\lambda)\frac{E}{LC}d - (c+\lambda)f - \dot{f} + (c+\lambda)\ddot{r} + \dddot{r} \quad (15)$$

$$- \lambda c(x_2 - \dot{r}) - \hat{v}\text{sign}(\sigma) - h\sigma + \ddot{p} + \lambda\dot{p}]$$

The adaptive law is designed as

$$\dot{\hat{v}} = \frac{1}{\alpha}|\sigma| \tag{16}$$

Then the following equation can be derived

$$\begin{aligned}\dot{V} &= \sigma[(c+\lambda)F + \dot{F}] - \hat{v}|\sigma| - h\sigma^2 + \alpha(\hat{v}-\bar{v})\frac{1}{\alpha}|\sigma| \\ &\leq \sigma[(c+\lambda)G(x) + \overline{G}(x)] - \bar{v}|\sigma| - h\sigma^2 \leq 0\end{aligned} \tag{17}$$

So the system is steady.

4 Simulation and Results

The proposed scheme was used in a DC-DC converter and simulation operation was carried out. The main parameters used in simulation are $L=80\mu H$, $E=24V$, $C=2000\mu F$. The expected tracking voltage is $r=20V$, and the initial state id the system is $\mathbf{x}=[x_1\ x_2]=[0\ 0]$. When $t<5$ or $t\geq 7$, the load $R=10\Omega$, and when $5\leq t<7$, $R=5\Omega$. Simulation results are shown in Fig. 1. The new scheme overcame the overshoot in the transient state process, and enable the system reach the steady state quickly, and also make the system have strong robustness to disturbances.

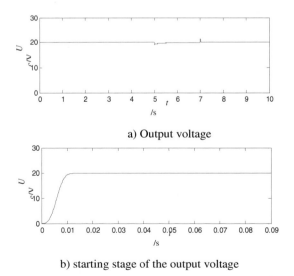

Fig. 1. Responding of dynamic adaptive terminal SMC

5 Conclusions

The dynamic terminal SMC with adaptive law can control DC-DC converter effectively, and it can eliminate chattering of the system.

References

1. Kadwane, S.G., Gupta, S., Karan, B.M., et al.: Practical iImplementation of GA Tuned DC-DC Converter. ACSE Journal 6(1), 89–96 (2006)
2. Sahbani, A., Saad, K.B., Benrejeb, M.: Chattering Phenomenon Supression of Buck Boost DC-DC Converter with Fuzzy Sliding Modes Control. International Journal of Electrical and Electronics Engineering 1(4), 258–264 (2008)
3. Guo, L., Hung, J.Y., Nelms, R.M.: Design and Implementation of a Digital PID Controller for a Buck Converter. In: Proceedings of the 36th Intersociety Energy Conversion Engineering Conference, vol. 1, pp. 187–195 (2001)
4. Racirah, V.S.C., Sen, P.C.: Comparative Study of Proportional-Integral Sliding Mode and Fuzzy Logic Controllers for Power Converters. IEEE Transactions on Industry Applications 33(2), 518–524 (1997)
5. Wai, R.J.: Adaptive Sliding-mode Control for Induction Servomotor Drive. IEE Proc. Electr. Power Appl. 147(6), 553–562 (2000)
6. Tsai, J.F., Chen, Y.P.: Sliding Mode Control and Stability Analysis of Buck DC-DC converter. International Journal of Electronic 94(3), 209–222 (2007)
7. Fan, L.P., Yu, Y.Z., Boshnokov, K.: Adaptive Backstepping based Terminal Sliding Mode Control for DC-DC Converter. In: International Conference on Computer Application and System Modeling, vol. 9, pp. 323–329 (2010)

Adoption Behavior of Digital Services: An Empirical Study Based on Mobile Communication Networks

Bing Lan[1] and Xuecheng Yang[2]

[1] International School
Beijing University of Posts and Telecommunications
Beijing, China
[2] School of Economics and Management
Beijing University of Posts and Telecommunications
Beijing, China
yangxuecheng@bupt.edu.cn

Abstract. Mobile communication has brought us into a network connecting with lots of people. This social network draws companies' attention because of it marketing value. Thus, it is important for companies to know the characteristics of this network and how user position will influence the product adoption process. This paper discusses the properties and what kind of people is more likely to adopt new products and services. Further, we use the Fetion case, which is one of the services of China Mobile, to conduct empirical analysis to test our hypothesis.

Keywords: mobile communication, social network analysis, adoption behavior, digital services.

1 Introduction

As the development of technology, mobile communication, in terms of SMS, emails, instant messages and etc, has brought us a lot of convenience by realizing social interaction, whenever and wherever. Via mobile phones, we can easily interact with others, as well as sharing product experience and asking advice for consumption. The large scale interaction can be viewed as kind of social network.

Many product and service in mobile communication industry, such as SMS, Color Ring Bach Tone (CRBT) and WAP, have shown great network effect. Such effect makes the decisions of users dependent on each other and a positive feedback mechanism will be built up as increasing followers adopt the same product. Large scale user base becomes the advantage for new users and then increases the switching cost for adopters. For an industry with such high network externality, it means more users in the network, more value will be brought to users and more users will be attracted.

Therefore, it is important for mobile companies to provide value-added product or service which is easy to form scale effect and difficult to be imitated. Some scholars find that if related marketing strategy is applied in an inappropriate manner, it will cause a negative effect to the product. For this reason, it has been a hotpot for scholars to explore how to utilize this kind of social networks for marketing. Based on the previous work on this topic, our research uses the adoption process of Fetion to research the properties of mobile communication network and who is more likely to try a new adoption.

2 Conceptual Framework

2.1 Small World Model

Previous research has shown that most real networks reflect the small world effect and cluster (the higher clustering coefficient).Small world effect means that most nodes are not adjacent with one another, but most nodes can be reached by any other through a small number of hops or steps. It realizes the transformation from a regular network to a random network by adjusting a parameter.

Duncan Watts and Steven Strogatz (1998) first identified certain category of small world network as a class of random graphs, which is classified according to two independent structural features, namely the clustering coefficient and average shortest path length. They proposed a novel graph model named the WS model with the property of a small average shortest path length, and a large clustering coefficient.

In mobile communication network, we set the user as a network node and the communication behavior between users as the node edges. Based on the above analysis, we infer small-world effect of the user's mobile communication network will appear.

Hypothesis 1: The relationship network between mobile users shows with small-world network effects.

2.2 Scale-Free Network

Researchers have indicated that for most real networks, the degree distribution follows a power law $p(k)$: k^{-x}, which is defined by by Barabási and etc. Degree is a measure of individual centrality in the network structure, which is defined as the number of edges the node links. If a person is the friend of a lot of people, in this sense, he has a high degree of centrality.

To explain the mechanism of scale free network, Barabási proposes the famous BA model. They find previous network model have not taken two important properties into consideration, which are growth and preferential connectivity. Growth means that new nodes will join in the network constantly while preferential connectivity means that the new nodes will prefer to select to link the nodes with high degree in the network. If the network's degree distribution can be described by power law form: $P(k)$ $\alpha k \wedge (-\gamma)$, it is called that the network follows a power law distribution.

Based on the above analysis, we infer that in the user's network of mobile communication, the degree distribution will obey the power law, the network will take on the characteristics of scale-free networks.

Hypothesis 2: The network of mobile communication users shows with features of scale-free.

2.3 Node Property

With the emergence of SNS, the role of traditional customer has gradually changed. They are no longer the product buyer, but more of a product reviewer and recommender.

Enterprises are faced with more challenges of managing the great power of word of mouth. Trusov et.al have found that 33% of the influencer will take on nearly 66% of the whole influence, which indicates that some user will perform as 'key user' from the perspective of marketing.

A media scholar Lazarsfeld first raised the concept of opinion leader, who have a strong influence on the majority of members of the SNS group. Based on this model, many researchers who study word of mouth marketing and viral marketing are contributed to search related content about opinion leaders and find some characteristics and properties of opinion leaders (1) they are more willing to expose their satisfied or dissatisfied emotion about using experience of product and service than general customer (2) it is more possible for them to adopt innovative product than general customer; (3) they invest more in one kind of product than general customer. It can be seen from these outcomes that the previous research mainly focuses on the attribute level while ignores the relationship level.

In social network analysis, network structure is represented by network centrality position, Wasserman and Faust (1994) pointed out that the network centrality means individual actors' position in the network, which is the level actors occupied an important strategic position because of being involved in many important links in the network. Consumers in the network center may not be the actual buyer, but have a decisive role in consumer decision-making. If the word of mouth of "opinion leader" is in search, this information is expected to influence the purchaser's purchase decision.

Previous research tells us that some individuals have more social connections. Thus, a fundamental question is to explore these relationships hub. According to two-step flow model (two steps flow), it has been widely accepted that the flow of information first arrives at the opinion leaders through the mass media, and then the opinion leaders affect public.

Hypothesis 3: In the course of the spread of Fetion, the more links users have, the higher the probability of adoption is.

3 Methodology

This research chose a municipal operator of China Mobile Communications Corporation in Jiang Xi province as the object of data extraction. The scope of data extraction for this article is all users of the same city; the final choice of the time period is October 1, 2009 ~ Dec. 31, 2009. The time lasts for three months, a total of 92 days.

3.1 The Average Distance between Nodes in Network

In the network, distance between two nodes i and j, d_{ij} is defined as the edges connecting the two nodes on the shortest path. The maximum distance between any two nodes is called network diameter, denoted by D, i.e.

$$D = \max\nolimits_{i-j} d_{ij}. \tag{1}$$

The network's average path length L is defined as the average distance between any two nodes, that is,

$$L = \frac{1}{\frac{1}{2}N(N+1)} \sum_{i \gg j} d_{ij} . \qquad (2)$$

where N is the number of nodes.

It is generally believed that, if the network average distance between two nodes L increases with the number of network nodes n, that is $L \propto \ln n$, we claim that the network has the small world phenomenon.

3.2 Network Clustering Characteristic

In general, assume that there are k_i edges between node i and other nodes in the network, where k_i nodes are called node i's neighbors. It's obvious that there are at most $k_i (k_i-1) / 2$ edges in these k_i nodes. The ratio of actual edges E_i between the k_i nodes compared with the total number of possible edge $k_i(k_i - 1)/2$ is aggregate coefficients C_i of node i, i.e.

$$C_i = 2E_i/(k_i(k_i - 1)) . \qquad (3)$$

The clustering coefficient C of the whole network is the average of clustering coefficient.

3.3 Degree Distribution

To describe the characteristics of scale-free network with degree distribution, we use the method of drawing the cumulative degree distribution function:

$$P_k = \sum_{k'=k}^{\infty} p(k') . \qquad (4)$$

It indicates the probability distribution of the nodes whose degree is not less than k.

3.4 User Individual Characteristic

Analysis of individual characteristics uses correlation analysis, regression analysis and other statistical methods and modules, and implements a series of hypothesis testing. The main part of the data analysis is finished using SPSS software, and SAS software.

Correlation analysis consists of two kinds of phenomena in changing direction and size of the development and can be used to study the uncertain relationship between variables. But these two phenomena are not sure about which is the cause, which is the outcome. Correlation coefficient r is the quantitative indicators from -1 to +1 describing correlation degree between two variables. The closer the absolute value of r is to 1, the higher degree of correlation between variables; r from 0 to 0.1 indicates a weak relationship, 0.1 to 0.4 means a low correlation; 0.4 to 0.7 shows moderate correlation, while r above 0.7 indicates a high degree of correlation.

4 Results and Discussion

4.1 Small World Phenomenon Testing

The result of our research shows that the average distance of user relationship network is 6.03, which is corresponding to the random network. It demonstrates that average path length of tested network is similar to that of random network.

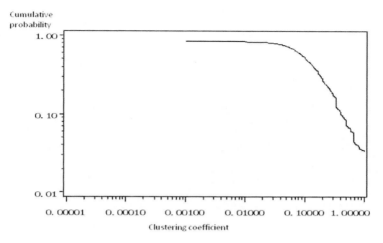

Fig. 1. Cumulative probability distribution of clustering coefficient

We gained the cumulative probability distribution of clustering coefficient relative to degree of nodes by calculating on clustering coefficient (presented in Fig. 2). We can concluded from the data that, above the level of 0.01, cumulative probability of clustering coefficient is 1, which means that all the clustering coefficient of nodes (no matter degree of these nodes) are larger than 0.01 level. The average clustering coefficient of the whole network is 0.21. This clustering coefficient is 26.25 times of the clustering coefficient of the corresponding random network.

Hence, we can prove that mobile communication user relationship network has small world phenomenon. Hypothesis 1 is supported by empirical data.

4.2 Scale-Free Network Testing

If we want to test scale-free characteristics of a network, we will transform power-law distribution $P(k) \sim k^{\alpha}$, and then test the linear relationship of logarithmic formation of it ($\log P(k) \sim \log k$) in the log-log figure. If $\log P(k) \sim \log k$ satisfies linear relationship, then $P(k) \sim k^{\alpha}$ is true.

In order to demonstrate the scale-free network characteristics of the whole network, we need to calculate degree distribution (number of links of each node). Under most circumstances, this distribution satisfies power-law distribution, and this network is scale-free network (Barabassi, 2003). The degree distribution is presented in Fig. 2. As the typical situation, distribution is highly tilt; some nodes are super nodes (have many links), but most of the nodes have few links.

We can conclude from Figure 2 that, the degree distribution of user relationship network satisfies the characteristics of power-law distribution. Apart from the extreme value of the tail, most of them are straight. Hence, we can prove that mobile communication network has the characteristics of scale-free network, and hypothesis 2 is supported by empirical data.

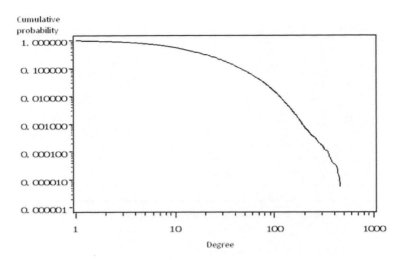

Fig. 2. Degree distribution of user relationship network.

4.3 Attributes of Nodes Analysis

Firstly, we draw the distribution figure of degree and cumulative adoption probability based on degree distribution of different users and mapping it with adoption probability data of Fetion (see Fig. 3). By observation, we can know that when the degree of user ranges from 0 to 150, the adoption probability of user increases very slow, which means that the degree has little influence on adoption probability under this situation; when the degree of user reaches 150, degree has a significant influence on adoption probability, which can be illustrate from this figure.

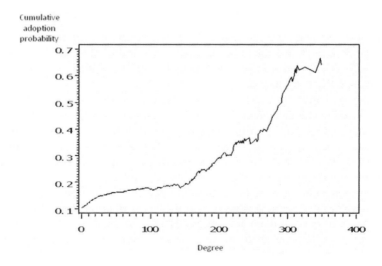

Fig. 3. Cumulative probability distribution of degree of user and Fetion adoption.

Next, we calculate the correlation coefficient of degree and adoption probability. We adoption Pearson correlation coefficient here, and conducts two-tailed test. The result is shown in Table 1.We can see from Table 1 that, the correlation coefficient of degree and adoption probability is 0.359, and it passes through the significant test of level 0.01. This demonstrates that the degree has positive correlation to adoption probability, i.e. the larger the degree is, the higher the probability of adoption.

Table 1. Correlation coefficient of degree and adoption probability.

		Degree	Adoption Probability
Degree	Pearson Correlation	1	.359(**)
	Sig.(2-tailed)	.	.000
	N	283	283
Adoption Probability	Pearson Correlation	.359(**)	1
	Sig.(2-tailed)	.000	.
	N	283	283

** Correlation is significant at the 0.01 level (2-tailed).

Further, we conduct the regression analysis of these two variables. To construct the regression formula, we use the degree of user as argument and adoption probability of Fetion as dependent variable. The result is presented in Table 2. The coefficient R^2 of regression formula here is 0.126, which shows that the power of degree over adoption probability is relatively strong. In addition, the result of regression analysis demonstrates that all the parameters pass through significance test.

Table 2. Result of regression analysis of degree and adoption probability

Model		Unstandardized Coefficients		Standardized Coefficients	t	Sig
		B	Std.Error	Beta		
1	Constant	.065	.026		2.490	013
	Degree	.001	.000	.359	6.444	000

From the analysis above, we can conclude that the higher degree of user, the larger probability of user adopts Fetion. Hypothesis 3 is supported by empirical data.

5 Conclusion

These empirical results have implications for understanding how the properties of mobile communication network, which has the small world and scale-free network characteristics. What we also obtained from the research is that the users who have more links with others are more likely to try new adoption. How can companies use such properties for marketing is not referred in this paper, which is for future research.

References

1. Watts, D.J., Strogatz, S.H.: Collective dynamics of 'small-world' networks. Nature 393, 440–442 (1998)
2. Wasserman, S., Faust, K.: Social Network Analysis: Methods and Applications. Cambridge University Press, Cambridge (1994)
3. Barabás, A.L.: Linked: how everything is connected to everything else and what it means for business, science, and everyday life. Plume, New York (2003)
4. Yang, S., Narayan, V., Assael, H.: Estimating the Interdependence of Television Program Viewership Between Spouses: A Bayesian Simultaneous Equation Model. Marketing Science 25(4), 336–349 (2006)
5. Trusov, M., Bodapati, A.V., Bucklin, R.E.: Determining Influential Users in Internet Social Networks. Journal of Marketing Research, 643–658 (2010)

Review of Modern Speech Synthesis

Cheng Xian-Yi and Pan Yan

College of Computer Science, Nantong University, Nantong Jiangsu 226019, China

Abstract. Speech synthesis and recognition technology has become a research focus in the field of intelligent computer. For nearly 50 years of study, speech synthesis has had huge development as an interdisciplinary, this paper reviewed the modern speech synthesis technology and research, analyzes the problems existing in the research in this field.

Keywords: Speech synthesis, Based on the semantic speech synthesis, Chinese speech synthesis.

1 Introduction

Speech synthesis is an important part of man-machine speech communication, which is called text to speech (TTS). It involves acoustical, linguistics, digital signal processing, computer science, etc. It is a frontier technology of Chinese information processing field [1]. Speech synthesis technologies give machine the function of "artificial mouth", working out how to let the machine talk like a man [2]. As early as 200 years ago, people began to study speech synthesis, with the development of the modern computer technology and digital signal processing technology speech synthesis technology has developed.

In 1939, Dudley Homer exhibited his speech synthesizer in New York world expo, called "Parallel Bandpass Vocoder "[3]. In 1960, the Swedish linguists and words engineers G.Fant introduced the speech production theory systematically in "Acoustic Theory of Speech Production", which promoted the development of the speech synthetic technology. Since 1970s, linear prediction technology started to use in speech coding and recognition [4].

In 1973, Holmes made parallel formant synthesizer. In 1980 Klatt designed strings/parallel hybrid formant **synthesizer** [5]. These two synthesizers can synthesize natural language by adjusting the parameters. **DECTALK of DEC of the United States was most representative in** 1987 [6]. This system uses Klatt's string/parallel formant **synthesizer**. It can provide all kinds of speech information services through standard interface, computer networking and separately receiving telephone network. Its pronunciation was vivid and it can produce 7 different tone of voice.

In recent years, a new speech synthesis method based on database is aroused people's attention. In this approach, the phonetic unit of synthetic statements are selected from **an advance recorded of huge speech database, as long as speech** database is enough big, including various possible speech units, it can stitch any statements.

Synthetic speech elements are from natural original pronunciation, the intelligibility and naturalness of speech statements will be very high [7].

2 Speech Synthesis Technology

2.1 Traditional Speech Synthesis Technology

Formant synthesis. Traditionally, the pole of track transmission frequency response is called formants, and the distribution characteristics of the resonant frequency (poles frequency) determine the timbre. It has the following three practical models [8].

①Cascade type formants model. In this model, the track is considered a group of series of second-order resonator. This model is mainly used in the synthesis of vowels.

②Parallel type formants model. Many researchers believe that the nasal vowels and other vowels and most consonants non-general, cascade model can not be described and simulated. Therefore, the parallel type formants model is produced.

③Mixed formants model., Formant filter first connected with tail in a cascade type formant model. Before amplitude adjustment the input signal through added to every resonant filter and superposed the output. Cascade type fulfills the acoustic theory speech production for the speech of synthetic source located at the end of track. Parallel model is more appropriate for the speech of synthetic source located at the middle of track, but its amplitude adjustment is very complex. So people combined both of them, mixed type formants model is proposed. It is an accurate simulation based on the track. It can be synthesized pronunciation of higher naturalness.

LPC rules synthesis LPC. Rules synthesis belongs to poles digital filter model of the whole linear source- channels speech production model. LPC rules synthesis technology is essentially encoding technology of time waveform. The synthesis process is essentially a simple decoding and stitching process. The advantages of LPC rules synthesis technology is simple, intuitive. Because speech of natural language and isolated streams has a great distinction, if simply put each isolated speech splicing together stiffly, the quality of the whole language will be very good. Therefore, the effect of whole continuous language of linear forecasting parameters synthesis is not good. It must combine with other technologies that can improve speech synthesis quality significantly.

PS0LA splicing synthesis. PSOLA is a kind of algorithm of modifying the rhythm of the synthesized speech, which was used for waveform edit synthesized speech technology. Main time-domain parameters of deciding speech waveforms rhythm included duration, intensity of a sound and pitch. According to the different methods of improving parameters, it mainly divided into LP-PSOLA, TD-PSOLA and FD-PSOLA. Its main characteristics are: Before stitching the requirements of speech waveforms PSOLA algorithm adjusted the prosodic feature of stitching unit according to the context requirements. The synthetic waveform maintains the main sound segment of the pronunciation and prosodic feature of stitching unit accord with context requirements; we can obtain high definition and naturalness. But PSOLA technology also has its flaws. Firstly, PSOLA technology is a kind of speech synthesis technology of pitch synchronous, it needs exact genes cycle and judgment of starting point. The error of pitch or the judgment of its starting point will influence the effect of PSOLA technology. Secondly, PSOLA technology is a kind of simple synthesis technology of waveform mapping splicing. We don't know that this Mosaic whether can keep smooth transition and it how to influence frequency domain parameters. Therefore, it may generate unsatisfactory result in the synthesis [9].

LMA channel model technology. The performance index of PSOLA synthesis systems is better than the other synthesis methods such as Formant synthesis and LPC rules. But PSOLA technology has the defects which are that the ability of prosodic parameter adjustment is weak and it is hard to handle collaborative pronunciation. Therefore, people proposed a kind of speech synthesis method based on the LMA channel model. Using this method, it can achieve various parameters adjustment needed by linguistic rules in high quality. LMA filter is a filter of exponent function form constructed a group of cascade according to certain formula using cestrum coefficients of input language signal. Its feature is that it can approach the logarithm spectrum amplitude of input signal under the least mean-square error rule. This kind of high precision simulation guaranteed the good timbre of speech synthesis output.

2.2 Modern Speech Synthesis Technology

Based on HMM the parameters speech synthesis. Based on HMM speech synthesis system is divided into two parts: Training phase and Synthesis stage(Fig.1).

During the training phase, it extracts spectrum parameters and fundamental frequency parameters of training corpus, establishes the hidden Markova models, and generates model predicts decision tree according to the context information. Due to the discontinuity of pitch, modeling is using multidimensionality probability distribution. In the stage of synthesis, it makes text analysis to the synthesis text, gets model state sequence according to the forecasting context information, generates speech parameters sequence using pronunciation parameters generation algorithm, and synthesizes target voice through parameter synthesizer sequence. **The method of parameter speech synthesis based on HMM using statistical modeling, characteristic forecast, and parameters** generated synthesis methods is different from the methods of traditional unit chosen and waveform splicing synthesis methods. This method can synthesize smooth fluent synthesized speech with several advantages of high automation degree of system constructing, small size of system, flexibility, etc. But this method also has some disadvantages [10]: **The sound quality of synthetic speech is not high;** Synthetic speech rhythms are insipid; Depending on data.

Based on AMR - WB the parameters speech synthesis technology. Linear forecast (LPC) parameters synthesis as a kind of simpler and practical method of speech synthesis receive special attention with its low data rate and low complexity, low cost. But parameters extracted from the original model is too simple, it is hard to get high quality synthetic speech. So in order to improve the effect of speech synthesis, we use these related technologies of AMR - WB in speech synthesis technology [11].AMR - WB adopts linear forecasting technology Generated by digital incentive. It expanded the traditional the bandwidth of 200 -3400 Hz into 50-7000Hz .It Calculated pitch delay using the method of combining closed loop and the open loop in the parameter extraction phase, reductive incentive using the way of combining adaptive yards and fixed yards. The effect of synthetic speech improved greatly for these. Compared with the traditional bandwidth signal, it strengthened the naturalness and the degree of distinction of fricatives speech, improving the comprehensible of speech. It is a kind of high quality digital broadband speech coding system. But the digital rate of speech synthesis of this method is larger and the vocabulary is limited.

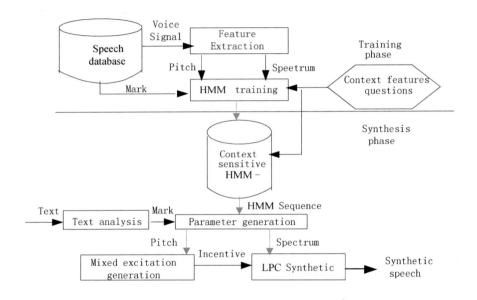

Fig. 1. Based on the HMM speech synthesis systems

Based on the semantic speech synthesis. In order to transmit semantic information completely and accurately, semantic description is a natural technology choice introduced in text analysis. Speech synthesis based on the semantic is the important research in the future. Speech synthesis system based on the semantic needs to implement three key function modules: semantic analysis, prosody prediction and high performance speech synthesizer [12].

2.3 Chinese Speech Synthesis

Domestic Chinese speech synthesis roughly also experienced three processes. They are formant synthesis, LPC synthesis and synthesis of applying PSOLA technology. Acoustic institute of Chinese Academy of Sciences first study the Chinese synthesis, then language institute of Scientific Socialism Academy, Tsinghua University, University of science and technology of China and Northern Transportation University began to research Chinese TTS.

In 1987, Li Linshan in national Taiwan University and ZhengQiuYu in historical linguistic institute developed a kind of system of" Chinese phonetic system," which looked LPC coefficient as parameters, establishing a speech parameter database with 411 syllables of level tone. In 1988,Acoustic institute of Chinese Academy of Sciences and Chinese computer service company constituted a parallel formant synthesis using 7720 type signal processing chip, made the speech synthesis board on the Great Wall computer 0520,and realized real-time synthesis from Chinese characters to syllables[13].

In 1988, University of science and technology of China successfully developed KD - 863 Chinese text to speech system supported by National 863 plan and national natural science foundation. KD - 863 used a new speech synthesis method based on the speech database, the basic idea of the technology is to quantize and merge protean Chinese syllables for listening, design the library of Chinese phonetic element. The library contains changing information of Chinese rhythm; synthesis of basic element can be realized rhythm control through the selection of the library samples. Meanwhile the sample of speech element library is directly intercepted from natural speech, to avoid the use of audio signal processing technology for the damage variable unit of timbre, So close to natural speech with synthesized speech timbre[14].

Chinese has a characteristic of single syllable, a Chinese characters corresponding to a syllable. Chinese syllables are composed by initials, finals and tones. **Syllable beginning consonants are initials, vowels, consonants part is behind all syllables of inflection is the tone.** Chinese mandarin has four tones. They are level tone, rising tone, falling-rising tone, falling tone, or called 1, 2, 3, 4 sound tones. Syllables constituted by the same initial and vowels have different meaning with different tones. Tone of mandarin undertakes the important role of composing words. Tone adjustment technology can be used to reduce the size of the speech database. The same sound vowels syllables simply store a tone of speech data, other tones pronunciation can be got by storage tones to transforming. Using tone adjustment technology can transform the storage of Chinese 1236 syllables shrunk into storing only 436 syllables, and greatly reduced the speech database capacity. Chinese speech synthesis including four processes: speech database establishing, text processing, dealing with rhythm, and speech synthesis [15].

3 Speech Synthesis Development Trend

Speech synthesis technology will develop in the following several directions [16]:

Improve the naturalness of speech synthesis. Improve the naturalness of speech synthesis is still the most urgent of high-performance text to speech. In terms of Chinese speech synthesis, currently in the words and phrases level the intelligibility and naturalness of synthesized speech has already been resolved basically. But to sentences and textual level the naturalness problem is greater. The summary of rhythm rules, especially the summary of rhythm rules of continuous speech has an great effect on naturalness of speech synthesis. And front-end text processing also has an effect on naturalness of speech synthesis.

Rich the expression of speech synthesis. Presently most speech synthesis research at home and abroad is aimed at text to speech system, these systems have been significantly improved in intelligibility, naturalness and other evaluation indicators, but it is still in the expressing speech level in essence. Along with the development of information society, it is required to live up to more and more demands in man-machine interactive, also begin to research man-machine dialogue system. In order to make the synthetic speech have ability of "communicates", it needs to make the conversion processing of text to speech take place in semantic level. Speech synthesis use semantic processing mechanism, implement text semantic analysis .On the one hand, it helps to solve the problem of recently system ubiquitous in phonetic and prosodic prediction. On the other hand, it solves the semantic words realization involved the modeling and predicting of semantic stress, function intonation, and pronunciation method. Based on this, it is to build the new generation of speech synthesis systems with accuracy and vivid performance ability.

Miniaturization corpus technology. At present Chinese TTS systems adopt the waveform splicing method based on large-scale speech corpus. Settlement method can be to compress speech data required capacity by speech compression coding, or to use smaller synthetic element, such as initials, finals or double phonemes, half a syllable, reducing synthesized speech required syllables element number and so on. Yet it can't increase the algorithm complexity, computation and system costs also will directly influence application of Chinese speech synthesis. Both must improve the quality of speech synthesis, and to reduce the complexity of the speech synthesis, which will always be a contradiction of two aspects.

Realize multilingual speech synthesis. Language is a tool for communication; people of different ethnic groups have their different languages. Communication of different language become very important in today's open information society and network times. Multilingual text to speech synthesis has a unique application value. The ideal multilingual synthesis system is the system which shares a synthetic algorithm or voice synthesizer in different languages. But the existing speech synthesis systems mostly developed aiming at a kind of language or a number of languages. Adopted the algorithm and the rules are closely related with some kind of language, so it is difficult to extended to other languages. Such as Chinese and western language there are great differences, domestic system are all doing Chinese language conversion, whose a set of rhythm control rules completely unsuitable for English, and most domestic systems are mandarin synthesis systems, which even promoting to the Cantonese and Shanghai dialect are rather difficult. It is clear that if you want to solve multilingual text to speech synthesis, you must have new ideas from text processing to speech synthesis.

Formulate industry standards, promote application of speech synthesis technology. Because speech technology is still far from the perfect state, you must optimize in the application level if you want get better application effect. Technology providers have their own set of specifications and standards, if such codes can be mastered by more users, which symbolizes that their key technologies and systems can get better effect in the application.

Therefore whether from phonetics key technologies are more likely to be integrated, or getting better effect in the application level, all these need a specification instruction. Who have greater discourse in industry standards, who will occupy greater advantages in industry promotion?

4 Conclusions

Speech is an important means of human-computer interaction, which has broad prospects. Speech synthesis and recognition technology has become a research focus in the field of intelligent computer. Now various commodities based on Chinese TTS has already appeared, such as intelligence game, SMS players, electronic dictionary, e-books, and telephone consultation response systems and so on. The synthetic technology of the pronunciation is a very important technology of human-machine interaction. Through to the speech synthesis technology research, we ought to develop Chinese speech synthesis systems with high intelligibility, high naturalness, and high visual as soon as possible, and translate this innovative and high technology to more powerful productivity. All this is to create greater social and economic benefits.

References

1. Li, Y.: The Research on Speech Rate in Speech Synthesis. Tianjin University (2007)
2. Li, Z.: Speech Signal Processing. Mechanical Industry Press (2009)
3. Bao, C., Ma, Z.: Speech Synthesis Technology and Its Research Progress. Inner Mongolia Science Technology &Economy (18), 31–33 (2010)
4. Liu, C., Yu, H.: Speech Synthesis Technology Research. Journal of Chinese Modern Education Technology (11), 64–66 (2008)
5. Klatt, D.H.: Software for a Cascade/Parallel Formant Synthesizer. JASA 87, 820–857 (1980)
6. Argente, J.A.: From speech to speaking styles. Speech Comm. 11(4-5), 325–335 (1992)
7. Sun, A., Wang, J.: Study & Application on Speech Synthesis System. Journal of Chinese Acoustic Technology 33(6), 48–51 (2009)
8. Liao, Z.: Analysis on the Voice Synthesis Technology. Journal of Chinese Science and Technology Information Development and Economy 16(18), 216–217 (2006)
9. Wu, Z.-y., Cai, L.-h.: The Principle of Voice synthesis technology. Products & Technology 3, 20 (2008)
10. Lei, M., Dai, L., Ling, Z.: Minimum Generation Error Training Based on Perceptually Weighted Line Spectral Pair Distance for Statistical Parametric Speech Synthesis. Journal of Chinese Pattern Recognition and Artificial Intelligence 23(4), 572–579 (2010)
11. Shu, C., Chen, M., Wang, M., Using, A.M.R.-W.B.: method to improve the results of speech synthesis. Journal of Chinese Information Technology (5), 239–244 (2009)

12. Zhu, W., Lv, S.: Semantic-Based Speech Synthesis—Survey and Perspective on the Speech Synthesis Technology. Transactions of Beijing Institute of Technology 27(5), 408–412 (2007)
13. Yang, S.: Speech Synthesis Technology Facing The Acoustic Phonetics Mandarin. Social Science Literature Press (1994)
14. Zhou, T.: Research on Text Normalization and Prosody Structure Prediction in Mandarin Text to Speech System. Beijing University of Posts and Telecommunications (2010)
15. Zhou, K.: Discussion on Process Analysis and Implementation of Chinese Speech Synthesis System. Journal of Chinese Modern Computer (4), 73–77 (2010)
16. Chi, M.: Studies on Techniques for Corpus-Based Text to Speech System Based on Cart. National University of Defense Technology of China (2008)

Design of the Stepper Motor Controller

Bo Qu and Hong Lin

School of Electronic &Information Engineering;
Soochow University,
Suzhou 215006, China
qubo@suda.edu.cn, lhqqm@suda.edu.cn

Abstract. The basic working principle of the stepper motor and its controlling method are introduced in this paper. The performance of the microprocessor STM32F103RBT6 and the driving principle of stepper motor driver chip L6208 are analyzed. A new type of stepper motor controller is designed. The input plus of the driver IC is provided by direct digital frequency synthesizer (DDS) chip. It describes how to control the speed, steering and rotation displacement of the stepper motor in detail. Then a real-time, accurate and reliable control of a four-phase stepper motor is achieved.

Keywords: ARM, stepper motor, controller, DDS.

1 Introduction

The stepper motor, transforms the electrical energy into mechanical energy and the electric pulse signals into open-loop control components of angular displacement or linear displacement by using the electromagnetic theory. In the case of non-overloaded, the motor speed and stop position depends on as much the pulse frequency and pulse number of pulse signal as the load changes, that is, when a pulse signal is added to the motor, it will turn a step-angle. Stepper motor can be grouped into two categories: variable reluctance stepper motors and permanent magnet stepper motor. This article focuses on permanent magnet stepper motor[1].

2 The Working Principle of Stepper Motor

The purpose of accurate positioning can be achieved by controlling angular displacement through controlling the pulses quantity; at the same time, the purpose of speed regulating can be achieved by controlling rotational speed of the motor through pulse frequency.

Figure1 shows a four-phase reaction stepper motor, using the unipolar DC power supply. As long as each phase winding of the stepper motor is energized in a suitable time series, the stepper motor can rotate. The working steps of four-phase reaction stepper motor are as below:

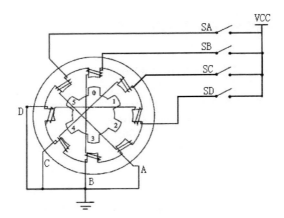

Fig. 1. Working principle of four-phase reaction stepper motor.

At the beginning, power on switch SB and SA, SC, SD disconnected, B-phase magnetic pole and No. 0,3 of rotor teeth aligned while staggered teeth is between No.1,4 teeth and C, D phasing winding magnetic pole and staggered teeth is between No.2,5 teeth and D, A phase winding magnetic pole.

When switch SC is on and SB, SA, SD is off, because of the interaction between the magnetic field lines of C phase windings and No. 1,4 teeth, the rotor turns and the magnetic pole between No.1,4 teeth and C phase winding is aligning. But staggered teeth is between No. 0, 3 teeth and A, B phase windings, and staggered teeth is between No. 2, 5 teeth and A, D phase windings. And so forth, A, B, C, D four-phase power winding take turns to supply power and the rotor turns along A, B, C, D direction [2].

The work modes of four-phase stepper motor can be divided into three modes: single four-beat, double four-beat, eight beats, in accordance with the power sequence. The step angles between single four-beat and double four-beat are the same, but the rotating torque of single four-beat is small. The step angle of eight-beat working mode is half of that of the single and double four-beat, therefore, eight-beat working mode can maintain higher torque and improve control accuracy [3].

The timing series and waveform of single four-beat, double four-beat and eight-beat is shown as a, b, c in figure 2 below.

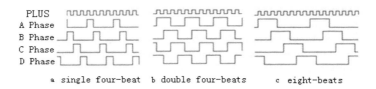

Fig. 2. Timing series and waveform of stepper motor.

3 The Overall Design of the Controller Module Structure

In this paper, core STM32F103RBT6 of ARM Cortex ™-M3 32-bit RISC produced by STMicroelectronics company is used as the control chip and L6208 is used as the stepper motor driver chips. The direct digital frequency synthesizer (DDS) AD9833 is used as the driver chip to input pulse. At the same time, using buttons and the infrared receiver TSOP34836 to control steps quantity and speed of the motor rotation and using LED to display the operation that is taken and the motor status.

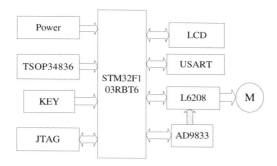

Fig. 3. The overall design of the controller module structure.

The direct digital frequency synthesizer (DDS) AD9833 is taken as the driver chip L6208 to provide clock in this system. AD9833's role is to change continuously as much as possible the generated waveform frequency and duty cycle, and the rotating speed of the stepper motor can be continuously changed. Although the PWM controller has been included within STM32, but the output waveform from PWM controller is still scalariform and the frequency and duty cycle cannot be changed continuously, so DDS is designed to substitute for PWM controller inside the processor.

4 Hardware Circuit Design

4.1 ARM Microprocessor and Peripheral Circuits

This system uses STM32F103RBT6 as the microprocessor, It include 128K bytes FLASH memory, 16K bytes SRAM, 7 channels of DMA controller, a 12-bit ADC, two SPI synchronous serial interfaces, two I2C interface, three USART interfaces, six-road PWM outputs, three 16-bit timers and each of them has up to 4 channels used for input capture/output compare/PWM or pulse counting channel and so on.

The user interface provided in this system includes buttons and infrared remote control. AD converter is used in buttons, thus the quantity of general GPIO ports is saved and the function of infrared remote control to replace the buttons can be used.

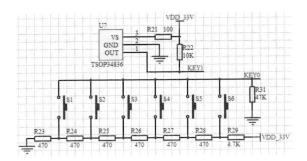

Fig. 4. Circuit diagram of buttons and infrared remote control

Circuit diagram of buttons and infrared remote control is as following.

KEY0 is connected to ADC interfaces PA0 of the microprocessor. There is no internal pull-up or pull-down resistor when PA0 is used as ADC interface, so a large 47K external pull-down resistor is added externally. KEY1 is connected to channel 3 (PB8) of Timer 4 to receive data sent by TSOP34836 by using the means of timer capture.

Meanwhile, the external circuit also includes power interface, JTAG port, serial port and LCD interface. The programmable voltage-stabilized source chip LT1940 is used by power source and it is converted to 3.3V through an external 12V power supply. At the same time, the 12V power supply to the stepper motor driver the power source and 3.3V is also the power supply for the processor and peripheral device power.

4.2 Driver Circuit Design

L6208 is used as the driver chips of stepper motor in the system. It is integrated by DMOS and has non-thermal dissipation over-current protection features. It also includes a dual DMOS full-bridge that can produce a stepping sequence constant off-time PWM current control circuit. L6208 integrates a constant off-time PWM current controller for each full bridge, and determines the full-bridge switch by detecting the source connected to two low-power MOS transistor and the pressure drop of external sense resistor between ground connections. When the voltage drop of sensitive resistor is greater than the voltage of reference input (VrefA or VrefB), the sensitive comparator will trigger the monostable circuit and turn off the whole bridge. Power MOS transistor remains off and the length of turn off time is set by the monostable circuit; at the same time, the electrical current defines re-cycle according to the selected decay mode. When the monostable time is ended, the full-bridge will break over again [4]. The circuit diagram of the driver hardware is shown in Figure 5.

Among this, AD9833 is DDS chip, FSYNC (control input, low level on) is connected to PB10 of the controller; SCLK (serial clock input) and SDATA (serial data input) is connected to bus SPI1 of the controller; X2 provides Clock to AD9833. Three programmable pins can use the software programming to achieve data transmission and to generate arbitrary waveforms.

Design of the Stepper Motor Controller   529

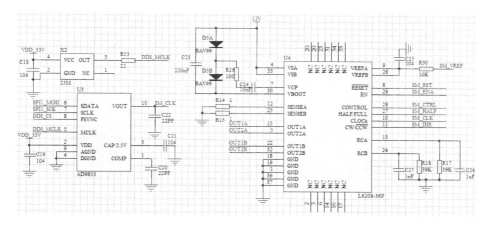

Fig. 5. Circuit diagram of the driver

SM_CLK is the pulse input provided by DDS; SM_DIR is Positive and negative selection controlled by PB15; SM_ENA is the enable pin controlled by PB12; SM_HALF is the whole step /half-step selection controlled by PB14; SM_RST is the restart chip signals, controlled by PB12; SM_CTRL is a fast / slow decay mode selection, controlled by PB13; OUT1A, OUT2A, OUT1B, OUT2B are the chip output end, accessing respectively into two-phase stepper motor windings. Another innovation: using PWM of the controller to supply the reference voltage to external sensitive resistors, thus, the reference voltage value can be set by using the software, which is used to control the output peak current.

5 Software Design

Man-machine interface control signals in this system, such as a keyboard, remote control signals are achieved through the interrupt function. The keyboard receives the keying information through ADC global interruption. The remote control signal is achieved through the capture / compare channel 3 of timer 4 [5]. The overall software process diagram of the system is shown in Figure6.

5.1 Control of Power on Commutation Order

The phase sequence generator of L6208 driver chip can generate three different stepping sequences: half-step mode, the two-phase full-step mode, single-phase full-step mode. By using these three different kind of stepping sequences, the phase control of stepper motor can be achieved. HALF / FULL signal is used to set stepping sequence mode [6].

(1) Set of half-step mode
HALF / FULL can be set to half-step mode when it is input with high level. This model uses two-phase eight-shot mode stimulated alternately by two-phase and single-phase to drive the stepper motor.

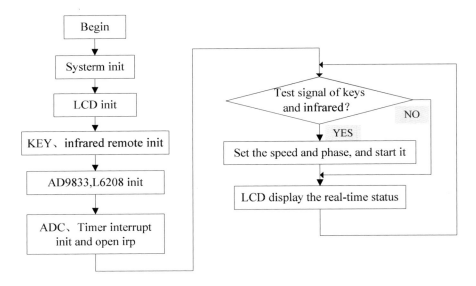

Fig. 6. The overall software flow chart of the system

(2) Set of two-phase mode
When phase sequence generator is in the odd state (state1, State3, state5, etc.), HALF / FULL can be set to two-phase full-step mode. The easiest way to set this mode is to make RESET and HALF / FULL at low level at the same time.

(3) Set of single-phase full-step mode
When the phase sequence generator is at the even number state (state2, state4, state6, etc.), HALF / FULL is the low level and can be set to single-phase full-step mode. The most commonly used method to set this model is to make RESET low level to reset the phase sequence generator so that it is in state 1, HALF/ FULL is high level keeping a pulse time. When the second pulse start, HALF/ FULL is low level [7].

5.2 Stepper Motor Steering Control

CW / CCW is used to control the turning direction of stepper motor, when input high level, the motor turns forward; when it is low level, the motor turns backward.

5.3 Stepper Motor Speed Control

CLOCK signal is as the clock of phase sequence generator, rising edge of CLOCK pulse trigger the phase sequence generator and make it jump to the next state. Thus, to control the phase sequence generator frequency by adjusting the frequency of CLOCK pulses to achieve the speed control of stepper motor. CLOCK of this system is generated directly by digital frequency synthesizer (DDS) AD9833. frequency and duty cycle of CLOCK is controlled by SPI1. There are five programmable internal register in AD9833, including a 16-bit control register, two 28-bit frequency register and two 12-bit phase registers. The functions the control registers have for the users to set are mainly mode selection, output waveform selection and specifying the frequency and phase registers, etc.; Frequency registers and phase registers are the frequency and phase to set the output waveform, and the output frequency is:

$$f = f_{MCLK} * \frac{FREQEG}{228} \tag{1}$$

Among above: FREQEG is the frequency word of selected frequency register; phase of output waveform is:

$$\phi = 2\pi * \frac{PHASEREC}{4096} \tag{2}$$

Among above: PHASEREC is the phase word of selected phase register.

6 Conclusion

In nowadays, ARM is used more widely in electronic products and the production costs has been getting lower and lower, so in this article, STM32F103RBT6 with higher cost performance is taken to substitute the 51 single chip machine to design the controller.

STM32F103RBT6 microprocessor can provide real-time control signals to L6208. AD9833 generates the stepping pulse signal required by stepper motor according to the control signals to control L6208 and to control the positive and negative turning and speed of the stepper motor. This control system can control the stepper motor in real-time, accuracy and reliability.

References

1. Liao, G.: Develop of High Performance control system for Stepping motor. Press of Xi'an University of Science and Technology, Xi'an (2004)
2. Zhang, G.: Design and Realization of AC Servo Motion Control System Based on Single-chip. Press of Inner Mongolia University of Technology, Inner Mongolia
3. Anonymous. SCM principles of stepper motor driver [EB /OL],
   http://www.elecfans.com/article/87/82/2009/2009091791653.html
4. Hai, L., Tinglei, H., Jie, L.: Study of Step Motor Driver Design Based on L6208. Instrumentation Technology (8), 49–50 (2007)
5. Chi, X., Zhao, L., Xu, G.: A Microcontroller-centered Steeping Motor Control and Drive Apparrtus. Journal of Southwest Forestry College (2), 68–69 (2005)
6. L6208 Datasheet[Z]. STMicroelectronics (2003)
7. Zhao, L., Liu, W., Sun, J.: Control system of stepper motor based on TMS320LF2407 and L6208. Electric Drive Automation 31(1), 32–34 (2009)

Design of Sensorless Permanent Magnet Synchronous Motor Control System

Bo Qu and Hong Lin

School of Electronic &Information Engineering;
Soochow University
Suzhou 215006, China
qubo@suda.edu.cn, lhqqm@suda.edu.cn

Abstract. The STM32F103RBT6 chip based on ARM Cortex-M3 is used as processor, and the FSBS3CH60 chip—one of Mini SPM series is used to drive motor in this paper. A kind of sliding mode observer (SMO) is designed to estimate motor rotor position and speed for sensorless permanent magnet synchronous motor (PMSM) control problems. The SMO adopts adjustable parameters of sigmoid function as the switch function, the oscillation of system is weakened effectively, and the improved SMO against the parameter variation and the uncertainties is of good robustness.

Keywords: sliding mode observer, permanent magnet synchronous motor, sensorless.

1 Introduction

In recent years, sensorless permanent magnet synchronous motor (PMSM) has been the research hot spot, because of its reasonable cost and favorable performance[1]. In the circumstances of not installing the electromagnetic or photoelectric sensor on the base and rotor of the motor, this technology uses the detected motor voltage, current and mathematical models to estimate the rotor position and velocity, which means that the characteristics of its mechanical motion is reflected by using electrical characteristics. In sensorless control strategy, the sliding mode observer (SMO) is of strong robustness to the parameter perturbation and external disturbance of PMSM. In traditional sliding mode observer, the position of the rotor and the discontinuous control function of the velocity information bring to the system the high-frequency "jitter" which can be weaken by the low pass filter (LPF) and subsequently a system time delay is produced[2]. Generally, the calculated rotor position plus corresponding offset will be used to compensate for time delay. In order to make the motor speed real-time and adjustable, cut-off frequency of LPF should also be adjusted according to the motor speed. The system will be complicated in this way. The SMO uses the sigmoid function with adjustable parameters as switching function to achieve continuous control. At the same time, "jitter" of the system is effectively weakened, LPF and cut-off frequency setting link are removed. The system structure is improved and the calculating amount is reduced on the premise of not losing robustness.

2 Sensorless PMSM Control System

The control structure of PMSM without speed sensor based on the SMO is shown in Figure1. Collect two-phase current i_b, i_c of PMSM and change them to i_α, i_β through the Clarke transformation. Make the current i_α, i_β and the calculated phase voltage u_α, u_β as the input of SMO. Its output is the estimated value of rotor position $\hat{\theta}$ and estimated value of rotor speed $\hat{w}$. $\hat{\theta}$ is transferred to the Park and inverse Park during the transformation process, the comparison take place between $\hat{w}$ and the given speed w^* directly, and the current reference value is generated after the tuning by PI controller.

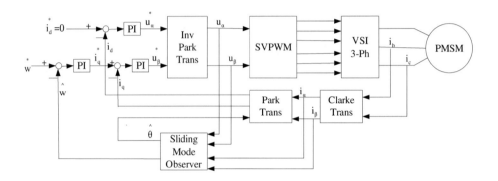

Fig. 1. Sensorless PMSM overall function module

3 Mathematical Model of PMSM

In order to obtain a simplified mathematical model of PMSM, we assume that: ignoring the motor magnetic saturation, eddy current and hysteretic losses; the motor current is symmetrical three-phase sine wave and ignoring the higher harmonics in the gap; the magnetic circuit is linear; the driver switch and freewheeling diode are ideal components; ignoring the effect of alveolar, reversing process and armature reaction. Then the mathematical model of PMSM in the stationary reference frame is as follows:

$$\begin{cases} \dot{i}_\alpha = -\frac{R}{L}i_\alpha - \frac{1}{L}e_\alpha + \frac{1}{L}u_\alpha \\ \dot{i}_\beta = -\frac{R}{L}i_\beta - \frac{1}{L}e_\beta + \frac{1}{L}u_\beta \end{cases} \tag{1}$$

$$\begin{cases} e_\alpha = -\lambda_0 w_e \sin\theta_e \\ e_\beta = \lambda_0 w_e \sin\theta_e \end{cases} \quad (2)$$

In these equations, R is stator electric resistance, L is the stator self inductance, i_α, i_β, u_α, u_β and e_α, e_β are respectively the phase current, phase voltage and back EMF in the stationary reference frame. w_e is the electrical angular velocity, λ_0 is the magnetic flux, and θ_e is the rotor position.

The response time of current changes in the stator is much faster than that of the rotate changes, so it can be assumed that:

$$\dot{w}_e \approx 0, \quad (3)$$

and the back EMF can be simply rewritten as:

$$\begin{cases} \dot{e}_\alpha = -w_e e_\beta \\ \dot{e}_\beta = w_e e_\alpha \end{cases} \quad (4)$$

Equation (4) show that the rotor position of PMSM is related only to the phase of the back EMF. The waveform of the back EMF is a sine wave, whose amplitude is proportional to the speed. Back EMF contains information of the rotor position and speed, and rotor position and velocity can be got by estimating the back EMF[3].

4 Design of SMO

The structure of SMO is shown in Figure2. It is formed by the current observer, continuous control function and calculation section of rotor position and speed[4] [5].

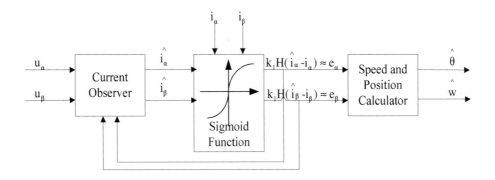

Fig. 2. Structure of sliding-mode observer

Design of current observer:

$$\begin{cases} \dot{\hat{i}}_\alpha = -\frac{R}{L}\hat{i}_\alpha + \frac{1}{L}u_\alpha - \frac{1}{L}k_1 H(\hat{i}_\alpha - i_\alpha) \\ \dot{\hat{i}}_\beta = -\frac{R}{L}\hat{i}_\beta + \frac{1}{L}u_\beta - \frac{1}{L}k_1 H(\hat{i}_\beta - i_\beta) \end{cases} \quad (5)$$

where superscript "^" represents the estimated quantities, k_1 is constant current observer gain, $H(\cdot)$ is the sigmoid function, the function form is as follows:

$$\begin{cases} H(\bar{i}_\alpha) = \left(\dfrac{2}{1+\exp(-a\bar{i}_\alpha)}\right) - 1 \\ H(\bar{i}_\beta) = \left(\dfrac{2}{1+\exp(-a\bar{i}_\beta)}\right) - 1 \end{cases} \quad (6)$$

Where a is a positive constant, and the slope of the sigmoid function can be changed by adjusting the size of a.

$$\begin{cases} \bar{i}_\alpha = \hat{i}_\alpha - i_\alpha \\ \bar{i}_\beta = \hat{i}_\beta - i_\beta \end{cases}, \quad (7)$$

Equation (7) is the current observation error. If the parameters are given, the current error equation can be obtained:

$$\begin{cases} \dot{\bar{i}}_\alpha = -\frac{R}{L}\bar{i}_\alpha + \frac{1}{L}u_\alpha - \frac{1}{L}k_1 H(\bar{i}_\alpha) \\ \dot{\bar{i}}_\beta = -\frac{R}{L}\bar{i}_\beta + \frac{1}{L}u_\beta - \frac{1}{L}k_1 H(\bar{i}_\beta) \end{cases} \quad (8)$$

Current error is generated by the harmonic of the electromotive force of PMSM. The sliding hyperplane is chose to build on the custom current error:

$$s_n = \begin{bmatrix} s_\alpha, s_\beta \end{bmatrix}^T, \quad (9)$$

Where

$$\begin{cases} s_\alpha = \bar{i}_\alpha \\ s_\beta = \bar{i}_\beta \end{cases} \quad (10)$$

Design the Lyapunov function:

$$V = \frac{1}{2} s_n^T s_n = \frac{1}{2}(s_\alpha^2 + s_\beta^2) \quad (11)$$

The existence conditions of SMO:

$$\dot{V} = s_n^T \dot{s}_n < 0 \quad (12)$$

i.e.,

$$s_n^T \dot{s}_n = -\frac{R}{L}(\bar{i}_\alpha^2 + \bar{i}_\beta^2) + \frac{1}{L}(e_\alpha \bar{i}_\alpha - k_1 \bar{i}_\alpha H(\bar{i}_\alpha)) + \frac{1}{L}(e_\beta \bar{i}_\beta - k_1 \bar{i}_\beta H(\bar{i}_\beta)) < 0 \quad (13)$$

$$k_1 > \max(|e_\alpha|, |e_\beta|), \quad (14)$$

When equation (14) is taken, we can ensure that the sliding mode observer is stable. Then the sliding surface becomes as

$$[s_\alpha, s_\beta]^T = [\dot{s}_\alpha, \dot{s}_\beta]^T = [0,0] \quad (15)$$

i.e.,

$$\begin{cases} (k_1 H(\bar{i}_\alpha))_{eq} = e_\alpha \\ (k_1 H(\bar{i}_\beta))_{eq} = e_\beta \end{cases} \quad (16)$$

Then the rotor position and speed can be calculated directly by combining the back EMF of PMSM:

$$\begin{cases} \hat{\theta}_e = -\tan^{-1}(\frac{e_\alpha}{e_\beta}) \\ \hat{w}_e = \frac{d\hat{\theta}_e}{dt} \end{cases} \quad (17)$$

5 Simulation

In order to verify the feasibility of the system, simulation model is established under MATLAB simulation environment and simulating is done. In the following figures, the solid line represents the actual value, while the dotted line indicates the estimated value.

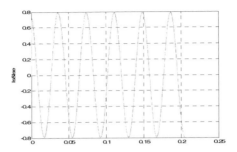

Fig. 3. Waveform of current actual and estimated value

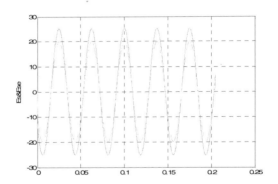

Fig. 4. Waveform of back EMF actual and estimated value

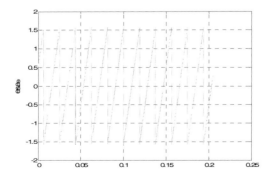

Fig. 5. Waveform of angle actual and estimated value

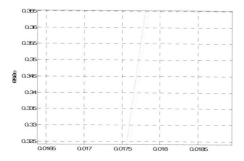

Fig. 6. Enlarged waveform of angle actual and estimated value

Seen from above charts, the estimated value can track the actual value accurately when the motor is running. Figure6 shows a little lag between the estimated angle and actual angle, and that is decided by inertia of the system.

6 Conclusion

In this paper, the STM32F103RBT6 of ARM Cortex-M3 is taken as the processing core whose cost is much lower than DSP chips and its property can meet the requirements. So it is completely feasible to use STM32F103RBT6 as the processing core. Combined with the designed SMO, the control system of sensorless PMSM is more advantageous, the cost is much lower and it is more beneficial for industrial production.

References

1. Wu, C., Chen, G., Sun, C.: Sliding mode observer for sensorless vector control of PMSM. Advanced Technology of Electrical Engineering and Energy 4, 1–3 (2006)
2. Liang, Y., Li, Y.: The State of art of Sensorless Vector Control of PM SM. Electric Drive 4, 4–9 (2003)
3. Paponpen, K., Konghirun, M.: Speed Sensorless Control of PMSM Using An Improved Sliding Mode Observer With Sigmoid Function. ECTI Transactions on Electrical ENG., Electronics, and Communications 5(1), 51–55 (2007)
4. Ding, H., Ji, Z.: A Control Method for PMSM Based on Sliding Mode Observer. Chinese Journal of Scientific Instrument 4, 363–366 (2008)
5. Guo, Q., Yang, G., Yan, P.: Application of SMO for sensorless driven and controlling system of PMSM. Electric Machines and Control 7, 354–358 (2007)

The Design of Synchronous Module in AM-OLED Display Controller

Junwei-Ma[1] and Feng Ran[1,2]

[1] School of Mechatronical Engineering and Automation Shanghai University, Shanghai, China
[2] Key Laboratory of advanced Display and System Application, Ministry of Education, Shanghai University, China

Abstract. On the basis of the analysis to the framework of AM-OLED display controllers, a key module — synchronous module is studied. In this paper, a FIFO circuit (first in first out) is designed using Gray convert technology and dual-port ram, which realizing scatheless transmit between different clock domain. The working principle of FIFO is analyzed, and the optimized FIFO circuit structure was put forward to improve the working stability of the FIFO. The synchronous module circuit was implemented by VerilogHDL language simulated through VCS (Verilog compiled simulator). The design had been applied successfully to a type of full colour TFT_OLED panel driver controller with gray levels 256, frame scanning frequency 60Hz ~ 100Hz. The testing results showed that this controller worked well with a stable and vivid display.

Keywords: AM-OLED, display controller, synchronous module, FIFO.

1 Introduction

Organic light-emitting diodes (OLEDs) have attracted increasing attention in recent years and are considered to hold the promise of the next generation of flat-panel displays due to their low-voltage operation, wide viewing angle, high contrast, and mechanical flexibility [1]. Currently, OLEDs have been successfully applied in mobile phones, MP3, digital cameras and other mobile devices, and the research of large size OLEDs has also been made great progress [2]. Moreover, excellent image quality for OLED is also closely bound up with the technological development of drive controller, which belongs to the input and output (IO) interface circuit of the display panels. Furthermore, the drive controllers on the one hand make drive and control independent of each other to increase the aperture ratio, on the other hand make displays have a more flexible application in intelligent system. However, the research of OLED external drive controllers is still in its early stages, and most of the OLED display panels have to use the improved LCD display drive controllers, which greatly restrict the development of high-performance OLED display technology. So, the study for high-performance OLED drive controllers, will become a hot topic in the field of integrated circuit design.

Designing high-performance AM - OLED drive controllers, two key problems need to be solved: one is how to design an efficient scanning module; another is how to deal with asynchronous clock domain. In this work, the system clock of controller and the pixel clock of DVI decoder chip are two completely different clock domain. Therefore, when the datas drived by pixel clock, are transmitted to the controller module drived by system clock, the question of asynchronous clock domain must be resolved.

On the basis of the analysis to the framework of AM-OLED display controllers, a key module —synchronous module is studied, In this module, a dual-port SRAM is used as asynchronous FIFO (first in first out) circuit to solve the problem of asynchronous clock domain. The working principle of FIFO is analyzed, and the optimized FIFO circuit structure is put forward to improve the working stability of the FIFO circuit. The synchronous module circuit was implemented by VerilogHDL language simulated through VCS (Verilog compiled simulator) software.

2 Overall Framework of the Controller Chip

The controller chip is divided into five main modules: input synchronization module (inputSyn), pre-processing module (preProc), frame buffer module (frameBuffer), scan-output module (scanOut) and I2C module (IIC). Chip block diagram is shown in Fig.1. InputSyn module is used to synchronize the digital video interface signals in the dCLK domain with the correlative signal in the gCLK domain; preProc module can generate read and write control signals for frameBuffer module, and has a function of image rotation; frameBuffer module includes a set of SRAM as the buffer of QVGA image, and can be expanded as the DRAM interface to support a larger resolution image; scanOut module is used to generate control signals required by AM-OLED display panel; IIC module is used to generate chip configuration signals to achieve rich functionality.

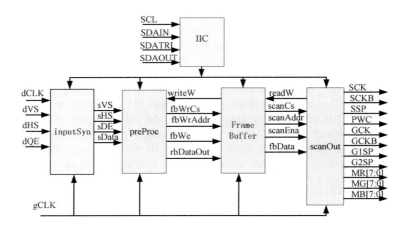

Fig. 1. Overall block diagram of the controller chip

3 Design of InputSyn Module

3.1 Synchronization Strategies

For single bit (or unit width) signals, when data signals generated under one clock are transmitted to another clock domain, a simple method for synchronization is using two triggers to latch signals; however, for multi-bit signal, there may be more than one bits changes at the same time, which may leads to sample incorrect datas, and cause circuit function errors if using the simple method above. Generally there are two ways to solve the problem that data signals transmit between different clock domains:

   a. Using handshaking signals. The more handshaking signals there are, the more complicated the system becomes. Accordingly, data transmission becomes time-consuming. So time-consuming and low transmission efficiency are the obvious drawbacks of the handshaking signals.

   b. Using asynchronous FIFO. An Asynchronous FIFO is a First-In-First-Out memory queue with control logic that performs management of the read and write pointers, and the generation of status flags. It can be used to transmit real-time datas quickly and easily between two different clock system. This is a relatively ideal method for designing synchronous circuit. A basic asynchronous FIFO block diagram is shown in Fig. 2.

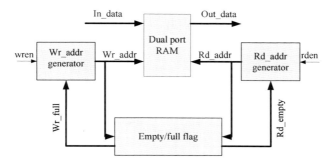

Fig. 2. The block diagram of asynchronous FIFO

3.2 Block Diagram of InputSyn Module

The block diagram of inputSyn module is shown in Fig.3. An asynchronous FIFO is designed using Gray conversion technology and dual-port SRAM, which realizing scatheless transmit between different clock domain. The dual-port SRAM is organized as 16 words by 26 bits, which used to cache 24 bits pixel datas (dQE), one bit horizontal (dHS)and vertical(dVS) synchronization signal respectively, and it has two separated port, each of which has standard independent SRAM control signals, working in their own clock domain synchronously. Moreover, wrData port writes datas from video source to SRAM drived by continuous pixels clock (dCLK), and rdData port reads datas from SRAM drived by continuous read clock (gCLK).

Control logic includes the address generator, the arbitration of read and write and the generation of full or empty flag. Write pointer (writePtr) and read pointer (readPtr) control module are constructed by the binary-gray conversion logic and binary code counter. Binary code is used to generate read (rdAddr) and write address (wrAddr) of SRAM, while gray code used to synchronize readPtr or writePtr signal into their opposite clock domain respectively. As can be seen from Fig. 3, ASYN_FIFO_FULL flag is generated by the Full Logic, the input of which includes synchronized readPtr and writePtr. Similarly, rdEmpty flag is generated by Empty Logic, the input of which includes synchronized writePtr and readPtr. The output of inputSyn module includes pixel data sData, sHS, sHV and data effective signal sDE. When the sDE is high-level, the datas from sData port is effective in this clock cycle.

In the inputSyn module, FIFO buffer can be seen as a circular array (realized by SRAM memory). The writePtr always points to the next word to be written; therefore, on reset, both pointers are set to zero, which also happens to be the next FIFO word location to be written. On a FIFO-write operation, the memory location that is pointed to by the write pointer is written, and then the write pointer is incremented to point to the next location to be written. Similarly, the readPtr always points to the current FIFO word to be read. Again on reset, both pointers are reset to zero, the FIFO is empty and the read pointer is pointing to invalid data (because the FIFO is empty and the empty flag is asserted). As soon as the first data word is written to the FIFO, the write pointer increments, the empty flag is cleared. Moreover, FIFO needs to compare the readPtr and writePtr in different clock domain, and detect the "full" or "empty" flag. FIFO full occurs when the write pointer catches up to the synchronized and sampled read pointer, and FIFO empty occurs when the read pointer catches up to the synchronized and sampled write pointer [3].

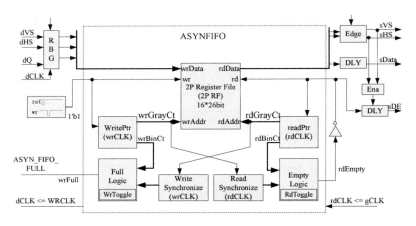

Fig. 3. The block diagram of inputSyn module

3.3 Control Logic of ReadPtr and WritePtr

Since the full flag is generated in the write-clock domain by running a comparison between the write and read pointers, one safe technique for doing FIFO design requires that the read pointer be synchronized into the write clock domain before doing pointer comparison. But trying to synchronize a binary count value from one clock domain to another is problematic because every bit of an n-bit counter can change simultaneously (example 7->8 in binary numbers is 0111->1000, all bits changed), which leads to metastability [4]. So in order to solve the question above, one can use gray code to substitute binary code. The first fact to remember about a Gray code is that the code distance between any two adjacent words is just 1 (only one bit can change from one Gray count to the next), which can greatly reduce the chance of metastability. The second fact to remember about a Gray code is that it is not a weighted code, and each bit of one code has no determinate size. Moreover, it can not carry out comparison and arithmetic operations directly [5]. Therefore, in the actual design of FIFO, one always uses a simple binary-to-Gray conversion logic to control the read and write pointers.

The control logic of writePtr is by and large similar to readPtr, the block diagram of readPtr logic is shown in Fig.4. Gray code counter assumes that the outputs of the register bits are the Gray code value itself (rdGrayCt). The Gray code outputs are then passed to a Gray-to-binary converter, which is passed to a conditional binary-value incrementer to generate the next-binary- count- value (bnext), which is passed to a binary-to-Gray converter that generates the next-Gray-count-value (gnext), which is passed to the register inputs. Then rdGrayCt or wrGrayCt is synchronized into the opposite clock domain before generating synchronous FIFO full or empty status signals.

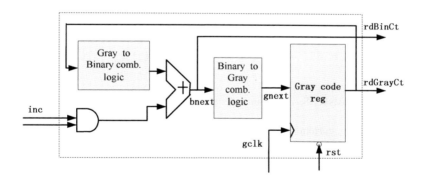

Fig. 4. The block diagram of readPtr

4 Synthesis and Simulation

The design was simulated under Linux environment by VCS (Verilog compiled simulator). VCS, with fast simulation and supporting multiple call way, developed by Synopsys Company, is specially used to simulate Verilog HDL language [6]. When using it, firstly do compile, then simulate and call the objects. The key signal waveforms of synchronous module circuit are shown in Fig.5.

As can be seen from Fig.5, gClk frequency generated by internal Controller is six times higher than external clock dCLK. During the data transmission, when the sDE is high-level, the datas from sData port is effective in this clock cycle, and under the control of sDE and system clock, the pixel datas can be transmitted to the next module correctly. From the simulation results, the timing is in accordance with the requirements of the design.

The controller has been applied successfully to a type of full colour TFT_OLED panel with gray level 256 levels, frame scanning frequency 60Hz ~ 100Hz. The testing results showed that this controller worked well with a stable and vivid display.

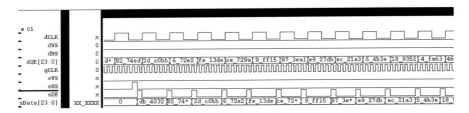

Fig. 5. Simulation waveforms of synchronous module circuit

Acknowledgements

The authors would like to acknowledge the financial supports of the 863 project (2008AA3A336), Shanghai Municipal Committee of Science and Technology under Grant No. 09530708600.

References

1. Tang, C.W., Vanslyke, S.A.: Organic electroluminescent diodes. Appl. Phys. Lett. 51(12), 913–915 (1987)
2. Li, X.-L.: The current situation and development of LED display technology. Optoelectronic Technology 16(4), 307–313 (1996)
3. Cummings, C.E.: Simulation and Synthesis Techniques for Asynchronous FIFO Design. In: SNUG 2002 (Synopsys Users Group Conference), San Jose, CA (2002), User Papers, Section TB2, 2nd paper (March 2002)
4. Fu, X.-C., Zou, X.-C., et al.: Application of FIFO circuit to liquid crystal display controllers. Journal of Huazhong University of Science and Technology (Nature Science Edition) 34(4), 8–10 (2006)
5. Gray, F.: Pulse Code Communication. United States Patent Number 2,632,058 (March 17, 1953)
6. Verification with VCS Workshop Lab Guide, 37489-000-S11, Sysnopsys (2002)

A Survey of Computer Systems for People with Physical Disabilities

Xinyu Duan and Guowei Gao

Multimedia Research Institute, Anyang Normal University, 455000, Henan, China
dxy@aynu.edu.cn, gaogw@yahoo.cn

Abstract. The development of the personal computer enabled everyone to hold a computer, except people with physical disabilities. The barriers to computer use of them is complex not only technology. In this paper, we focus on psychology, education and technology (software and hardware), and presents a high-level overview of computer systems for people with physical disabilities from our Educational and Psychological point of view. We discuss user and Products, and technology, highlighting challenges, open issues, and emerging applications for computer systems of disabled research.

Keywords: computer systems, disabled, education psychology.

1 Introduction

The computer have play more and more important role that become an integrated part of our lives. Educational, employment, social opportunities even recreation is dependent upon being able to make use of computers. For disabled people, computer use and skill can be particularly important. Computer system can help disabled people to reduce the isolation and social exclusion, the using of computer system helps them to increase their independence and confidence.

Some disabled people in home may be able to learn or work recreation or fun, make contact with service or friends with computer. The computer system provides them with an opportunity that to integrate into society. They also can be rehabilitating in hospital or another place based on hardware and software of computer system. While there are many barriers to overcome for the use of computer system, such as interest, awareness, access, cost, training and on-going support, the computer is becoming the strategic focus of global and is facing major challenges and opportunities for disabled people, but there are only one-quarter of people with disabilities own computers, and only one-tenth ever make use of the Internet. This is U.S. data [1], it will be much lower in China.

On the basis of the survey results, it is estimated that there are about 83 million persons with different types of disabilities in China. Improving the quality of life and the increasingly ageing population are leading to significant increases in the numbers of disabled people [2]. All the disable people mostly want to touch to computers include the older people, but they have different barrier for use. The key barrier is the HCI and software. All most the hardware and software is for normal person, and the designer consider the healthiness at first. And there are also some problems of disabled people, such as: Lack of interest, Lack of awareness, Difficulty of access, lack of content for disabled, high cost of ICT, Lack of training and Lack of on-going support.

Computer system includes hardware and software, special assistive hardware and special needs software. The special assistive hardware for disabled always means HCI of disabled peopled. The most HCI for disabled people draw support from professional tools. Based on the nature of different professional tools, they can be divided into three categories: Visual-Based, Audio-Based, and Sensor-Based. Disabled people do lots of things not only with hardware but also software. The software is designed for disabled people based on computer science, information technology, medical science, special education, psychology and other technology. And now it is usually based on multimedia technology, computer network technology and virtual reality technology. There are also known as the robot is a computer system in a sense. It is a set of hardware and software solutions to improve care cost and the quality of Life [3].

There are many technologies and applications of computer system for disabled people that posed by researcher from different country. In our paper, we focus on psychology, education and technology (Software and hardware), and present a high-level overview of computer systems for people with physical disabilities from our Educational and Psychological point of view.

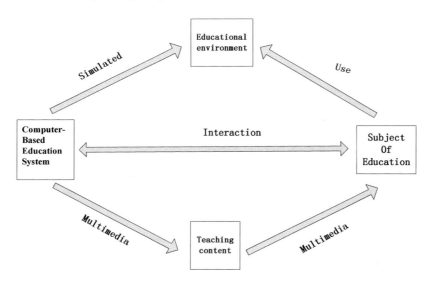

Fig. 1. Computer-based education concept

2 Computer-Based Education

Computer-based education based on the computer and network is a kind of education pattern appears in 1990's [4]. There is vibrant growth of computer-based education for disabled people into the 21st century at different country. It has such characteristic:

the teachers and students separate, rely on the media carrying on the teaching, resource sharing (Fig. 1). It takes the computer network technology, the communication technology as the foundation, it also take the multimedia, virtual reality and network technology as the main method. We may expect the education based on the computer will play the vital role in people's life and the study in the future information Society particular persons with disabilities.

Computer is a potentially powerful tool for more effective teaching and deeper learning. Computer-based education system for disabled people is working to create a barrier-free learning environment that includes special assistive hardware and special needs software. Learn with computers - particularly using the Internet - brings students valuable connections with teachers, other schools, other students, and a wide network of professionals around the globe and breaks limitations of the body. Those connections spice the school day with a sense of real-world pertinence, and broaden the educational resources. The important is this educational means to help the disabled breaking the barriers of time and space, if used wisely and well.

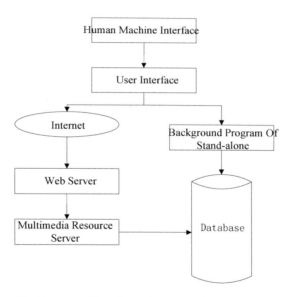

Fig. 2. Process flow chart of computer-based education

Traditional educators identify time and space as assets to classroom-based teaching. Every disable student wants to face-to-face, real-time interactions offer immediacy, personal contact, and community as normal man. But they have lots of barrier to go to school; the solution to extend time and to reconfigure space in the virtual classroom based on computer systems improves teaching and learning. Distance education and online education is more common, but a few are designed for disabled people. Most of them lack a component of the figure 2. Now the question is not the feasibility of the computer-based education, but to use the computer to more effective education of persons with disabilities. A good example of it is some scholars are studying the relationship between computer game and learning. Computer games have been attracting people of all kinds include the disabled people into using computers and give a better form of learning.

People with disabilities have been receiving a good public education after nearly decades in our China, but there are still more needs to be done. Since the launch of reform and opening-up 30 years ago, China has gradually built a special education system for its disabled people, which include fundamental education, vocational training, adult education, and provide them opportunities that receive higher education with normal people in every college. But there are only 20,000 disabled students studying in higher education institutions, accounting for less than one percent of the country's 82.9 million people with a disability. The government will continue to inject more funds and resources into education for disabled people. I think the distance education and online education should be the focus of future development.

3 Rehabilitation of the Disabled

The rehabilitation training is to train all categories of persons with disabilities to make fullest possible use of their capabilities and to provide them with employable skills by use of medical science, special education, psychology and other technology. Many research results of it have been put to use [5]. The prime objective of the rehabilitation training is to train all categories of persons with disabilities in employable skills to enable them live independent and dignified lives in society. And it is to contribute positively towards our country development by harnessing the human resource potentials of a segment of Society that otherwise would have been marginalized.

A computer system for rehabilitation training is in combination with the Information technology and the body weight supported treadmill training. When introduced into the rehabilitation training after spinal cord and another injury , hearing and speaking impaired people, the computer system performs with the characteristics of high accuracy, fast response and "tireless", and it is likely to avoid the drawbacks of the traditional body weight supported treadmill training. Meanwhile, the computer system will help to train the patients as well as record the training data, free the doctors from the heavy physical work so that they can have more time to focus on the superior work such as rehabilitation training evaluation and making recovery plans. And then, the quality and efficiency of the rehabilitation training will be highly enhanced. Thus far, the exoskeleton robot has been a focus on the development of neurons-rehabilitation technique in the worldwide.

Rehabilitation is a combination of medical and computer. In the 1980s it is the early stages of rehabilitation developing, the United States, Britain and Canada in terms of rehabilitation robot is a world leading position. Before 1990, the world's 56 research centers are distributed in five industrial areas: North America, the Commonwealth, Canada, continental Europe and Scandinavia and Japan. Since 1990 the rehabilitation of disabled people has been in the period of rapid growth. Now the focus of the research is manipulators in rehabilitation, hospital robot systems, intelligent wheelchairs, prosthetics and rehabilitation robotics and other aspects.

In Chinese since 1980s, thanks to importance attached by the government and all sectors of society, rehabilitation training for disabled people has developed into a new comprehensive science composed of medical science, special education, psychology and so on. Lots of hearing impaired people and upper or lower limb disabled have learned to speak and move through effective rehabilitation training. However, due to such factors as old education philosophy, inappropriate orientation, outdated teaching contents and methods, the effect of rehabilitation training is far from satisfaction. Among them, inappropriate orientation and outdated and unscientific teaching contents are two main reasons. The fundamental cause is the inefficient participation of special education and psychology.

4 Recreation or Fun for the Disabled

Recreation improves the life quality of the disabled people! Everyone benefits from recreation especially disable people. It makes us feel good and got a lot of vinegar. Everyone should get to choose something what they like to do, no matter risky or safe; noisy or quiet; together or alone; inside or outside…but the environment and limited of themselves of the possibilities can prevent disabled people to join with it. So the computer system of recreation for disabled people that are very necessary to research.

The computer has played an increasingly important role in leisure counseling and recreational activities for disabled people: to help people with disabilities through computer consultants to find the resources and opportunities for recreation, leisure interests of persons with disabilities assessed on this basis to develop for different disabilities recreation programs; persons with disabilities can be exempted from the restrictions of their disabilities to play golf on the computer simulation, to cave exploration, to experience the Great Wall hiking and other more exciting activities. It is say there are two applications of computer system in recreation for disabled people: virtual reality and online consulting.

Virtual reality of disable people for entertaining is PC games. Today the computer gaming industry has become larger than the other entertainment industry include world music and movie industries. Computer games have become part of everyday life for many people, and it make people more close to computer. all people of different ages can get lots fun from it. But there are few entertaining software provided by business and scholar for disabled people [6]. Research and development in the field of computer for disabled people has focused on education and rehabilitation rather than leisure. The entertaining software often tracks the position and orientation of the head in real time to control the computer games [7]. The setting up of guidelines for development of new entertaining software is not only for standard software but also designed especially for disabled users.

Some U.S. scholars have pointed out, when disabled people use computer simulation technology into a virtual world, they forgot they have a lot of inconvenience that caused by the disability, to communicate normally with others and communication. Perhaps only in the virtual world they fully felt to like a normal person.

5 Conclusion

This paper attempted to give an overview on computer systems for disabled people and present the current scenarios of computer system being used for the purposes of education rehabilitation and recreation. The application of computer systems for disabled people is not confined this several, but it is too few for demands of disabled people. In our country, the attention on it is deficient. Development of economic and technological did not bring enough immediate benefits to people with disabilities. Disabled people are vulnerable groups that have a low purchasing power. Pushed forward by business is not enough, but there are only a little funds be invested in it. The whole people should devote more attention to make all aspects of the world accessible for disabled people, not just computer.

References

1. Stephen Kaye, H.: Computer and Internet use among people with disabilities, Disability Statistics Report (2000)
2. McGregor, A.: A voice for the future. In: Proceedings of the European Conference on the Advancement of Rehabilitation Technology (ECART 1995), October 10-13, pp. 127–129. National Secretariat of Rehabilitation, Lisbon (1995)
3. Jia, S., Lin, w.: Network Distributed Multi-Functional Robotic System Supporting the Elderly and Disabled People. Journal of Intelligent and Robotic Systems (45), 53–76 (2006)
4. Boettcher, J.V.: Computer-Based Education: Classroom Application and Benefits for the Learning-Disabled Student. Annals of Dyslexia 33, 203–219 (1983)
5. Bugar, C.G., Lum, P.S., Shor, P.C., et al.: Development of robot for rehabilitation therapy: The PALO Alto VA/Stanford exPerienee. J. Rehabilitation Research and Development 37(6), 663–673 (2000)
6. Tollefsen, M.: Entertaining Software for Young Persons with Disabilities. In: Computers Helping People with Special Needs, pp. 626–626 (2004)
7. Meers, S., Ward, K., Piper, I.: Mechatronics and Machine Vision in Practice, 111–122 (2008)

A Monopole Scoop-Shape Antenna for 2.4GHz RFID Applications

Mu-Chun Wang[*] and Hsin-Chia Yang

Department of Electronic Engineering,
Minghsin University of Science and Technology, Hsinchu 30401, Taiwan
mucwang@must.edu.tw

Abstract. Through the assistance of Zeland IE3D software simulation, a high performance of a scoop-shape antenna integrating a monopole type and a PCB board using FR4 material as a substrate was fully and successfully represented. The relative dielectric constant of this FR4 material is 4.35 and its thickness in PCB board application is around 1mm. After the fabrication and the attentive measurement with Agilent E5071C network analyzer, we measured that the return loss, S_{11}, was about -17.1dB at 2.4GHz, -25.2dB at 2.45GHz and -15.5dB at 2.5GHz. The bandwidth was close to 290MHz, comparing them with the simulation results, S_{11}=-24dB at 2.45GHz and BW=300MHz. The performance between simulation and real measurement is almost identical. This range is adequate to many radio frequency identification (RFID) applications.

Keywords: Antenna, RF, Monopole, Scoop-shape, RFID.

1 Introduction

The precise design of antenna for communication and electrical consumer applications had become strictly demanded and desirable because the antenna was not only suitable to the efficiency of transmitting/receiving ability and the fashion shape in marketing consideration, but also satisfied several bandwidths in wireless communication technology. Furthermore, the antenna is the front-end component of the radio frequency (RF) system, responding to the feeble signal in the far/near field. There are several types of monopole antenna shape in the lectures [1,2]. In this work, we propose a monopole scoop-shape antenna to apply in RFID application. For the wireless communication market, the design of antenna technique is striding toward to the wide band, micro-miniaturization, and high transmission speed, etc. In order to develop a smaller size of antenna, an antenna element, monopole scoop-shape antenna (one half of a dipole antenna), was contrived and verified through a prototype through this research. This antenna is also able to be packaged as a hidden formation. With the assistance of Zeland IE3D (Imitative electromagnetic three dimensional, IE3D) software simulation and Agilent E5071C measurement, a scoop-shape antenna incorporating a monopole type and a PCB board using FR4 (expoxy resin fiber glass) material as a substrate is fully and successfully exhibited.

---

[*] Corresponding author.

2 Experimental Establishment

In Fig. 1, the block diagram of designed antenna, related to the position of the post-level circuit, was demonstrated. Here, SMA (SubMiniature version A) connector is a coaxial RF connector and has a 50Ω impedance. This component with the good impedance offers an excellent electrical response performance from DC to 18GHz.

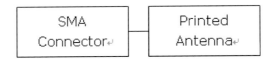

Fig. 1. Block diagram of designed antenna.

Continuously, the substantial printed antenna element is exhibited in Fig. 2 [3], which was fabricated with FR4 material as a substrate. The relative dielectric constant of this FR4 material is 4.35 and its thickness in PCB board application is around 1mm.

Fig. 2. A prototype of a scoop-shape antenna.

3 Measurement Results and Discussion

Through the assistance of Microwave Office 2003 software, how to design a suitable 50Ω transmission line simulation [4] on a PCB substrate is comprehensive with contour, as shown in Fig. 3. The left-hand-side information demonstrates the relative parameters of substrate as a waveguide. Here, the relative dielectric constant ε_r of this PCB material is about 4.3, and the ground connection in simulation is preset as "1" as well as the cover is set as "0". Moreover, for the loss tangent [5], it is approximate 0.02dB at 1GHz. This parameter is to absorb the noise at the high-frequency operation. Next, the right-hand-side information reveals the connecting points for two ports and the geometric parameters of an in-line transmission line.

In this simulation, the architecture of transmission line is coplanar wave guide (CPW). The PORT1 and PORT2 indicate the input terminal and the output terminal, respectively.

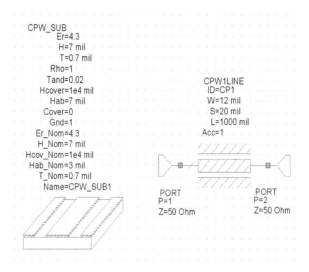

Fig. 3. Simulation of a 50Ω transmission line on a PCB substrate.

All of the impedances are designed as 50Ω. The line width is 12mil (1 mil = 10^{-3} inch) and the distance between the transmission line boundary and two side of GND are 20mil. The total line length is 1000mil. As a result, the return loss, S_{11}, is less than -25dB in real measurement. On the other word, as 1mW power is transmitted, there is just $10^{-2.5}$ power reflected.

While these precious parameters as a file are transferred to the Zeland IE3D software as a simulation [6,7], a possible and schematic antenna profile can be generated. Through some technical modification such as the empirical experience, a solid antenna in simulation is completed. Next, this antenna pattern can be transferred as a mask and printed on a PCB substrate. Besides, the measurement results of this real antenna can feedback the quality to the original designer to modify the error between the design and the measurement performance. This good relationship will shorten the next novel generation development in antenna. All of the simulation features with this IE3D software are exhibited in Fig. 4 and the return loss (S_{11}) is illustrated in Fig. 5. In this case, that the bandwidth is large enough about 300MHz and the central frequency is located at 2.45GHz.

As these parameters are converted into the smith chart to investigate the impedance matching, as shown in Fig. 6, the performance of the impedance matching is able to be adjusted to approach the design target. This action means that this chart reflects a good tool to tune the impedance matching for designed scoop-shape antenna in this study.

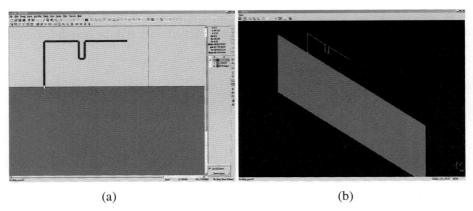

(a) (b)

Fig. 4. Schematic simulation profiles of a monopole antenna simulation with (a) designed contour and (b) 3-D simulation.

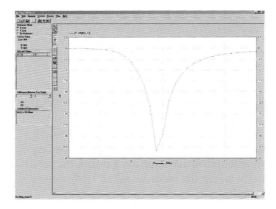

Fig. 5. Simulation performance of a return loss (S_{11}).

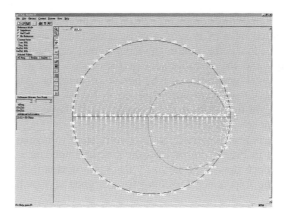

Fig. 6. Smith chart of impedance matching.

A Monopole Scoop-Shape Antenna for 2.4GHz RFID Applications

Moreover, when achieving the required return loss (S_{11}), the simulation and measurement results of an antenna radiation pattern simulation could be preceded [8]. In Fig. 7 and Fig. 8, the antenna radiation pattern in far field at Phi=0° and 90° are presented, respectively.

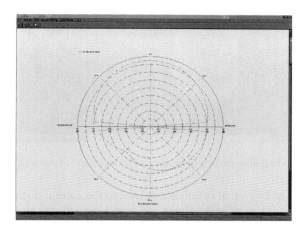

Fig. 7. Antenna radiation pattern in far field at Phi = 0°.

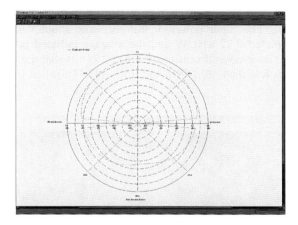

Fig. 8. Antenna radiation pattern in far field at Phi = 90°.

To probe a far-field performance in this antenna, a 3-D far-field radiation pattern simulation for a scoop-shape antenna was achieved and depicted in Fig. 9. The radiation pattern of an antenna design is a quite important consideration, correlating to the feeble signal of transmitting and receiving. There are several ways to influence the radiation pattern. For example, the wavelength frequency and phase, the geometric structure, and the relative position of GND are highly concerned.

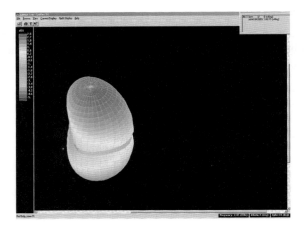

Fig. 9. 3-D far-field radiation pattern simulation of scoop-shape antenna (XY-Theta, YZ-Phi).

In addition, the end terminal of antenna related to the current distribution is an open circuit type that leads the highest voltage and a zero current, as shown in Fig. 10. In a short, we summarized some interesting simulation parameters in this prototype antenna. In the Table 1, the maximum radiation angle showed 5° and 110°, corresponding to Theta and Phi, respectively. The directivity was close to 2.52dB. The mismatch loss is lower than -7.76×10^{-3}dB. Its efficiency is approximate to 92.88% (-0.321dB). The input power at the port was about 9.98E-3W. The total radiated power depicted 9.27 $\times 10^{-3}$W and the average radiated power demonstrated 7.378×10^{-4} (W/s). The measured sensitivity for 802.11b is that the package error rate (PER) at 11Mbps is less than -90.1dBm and it for 802.11g is that the PER at 54Mbps is < -76.5 dBm.

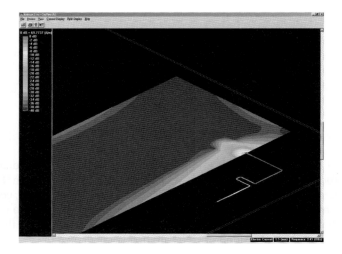

Fig. 10. Current distribution of a scoop-shape antenna.

Table 1. Simulation results of scoop-shape antenna radiation pattern

Scoop-shape antenna radiation Pattern	
Maximum radiation angle	(5, 110) degrees (Theta, Phi)
Directivity	2.51686 (dB)
Mismatch Loss	-7.761×10^{-3} (dB)
Efficiency	92.8769% (-0.320923 dB)
Input Power at Ports	9.982×10^{-3} (W)
Total Radiated Power	9.271×10^{-3} (W)
Average Radiated Power	7.378×10^{-4} (W)

4 Conclusion

A novel antenna prototype for 2.4GHz signal transmission was designed and fabricated. In Fig. 11, this prototype antenna was verified by a connection with Agilent E5071C network analyzer. The performance of the return loss in measurement depicted at Table 2. These results strongly demonstrate that this designed antenna is workable at 2.4GHz RFID applications. Comparing the measurement results with the simulation performance, they show the coincident quality. This evidence also supports this designed monopole scoop-shape antenna is a promising product in marketing applications.

Fig. 11. Return loss (S_{11}) measurement of a monopole antenna with Agilent E5071C instrument.

Table 2. S_{11} measurement results of Agilent E5071C.

Mode	2.4 GHz	2.45 GHz	2.5 GHz	Bandwidth
S_{11}	-17.1 dB	-25.2 dB	-15.5 dB	290 MHz

Acknowledgment

The authors thank the financial support coming from this university with an internal project number, MUST-100-Elecronics-05.

References

1. Chen, S.B., et al.: Modified T-shaped planar monopole antennas for multiband operation. IEEE Transactions on Microwave Theory and Techniques 54(8), 3267–3270 (2006)
2. Ruvio, G., Ammann, M.J.: From L-shaped planar monopoles to a novel folded antenna with wide bandwidth. In: IEE Proceedings of Microwaves, Antennas and Propagation, vol. 153(5), pp. 456–460 (2006)
3. Suma, M.N., et al.: A wideband printed monopole antenna for 2.4-GHz WLAN applications. Microwave and Optical Technology Letters 48(5), 871–873 (2006)
4. Raj, R.K., et al.: A New Compact Microstrip-Fed Dual-Band Coplanar Antenna for WLAN Applications. IEEE Transactions on Antennas and Propagation 54(12), 3755–3762 (2006)
5. : Novak: Reducing Simultaneous Switching Noise and EMI on Ground/Power Planes by Dissipative Edge Termination. IEEE Transactions on Advanced Package 22(3), 274–283 (1999)
6. Kretly, L.C., et al.: A hexagonal adaptive antenna array concept for wireless communication applications. In: 13th IEEE International Symposium Personal, Indoor and Mobile Radio Communications, vol. 1, pp. 247–249 (2002)
7. Wu, X.H., et al.: A Transmission Line Method to Compute the Far-Field Radiation of Arbitrarily Directed Hertzian Dipoles in a Multilayer Dielectric Structure: Theory and Applications. IEEE Transactions on Antennas and Propagation 54(10), 2731–2741 (2006)
8. Lin, Y.F., et al.: A Miniature Dielectric Loaded Monopole Antenna for 2.4/5 GHz WLAN Applications. IEEE Microwave and Wireless Components Letters 16(11), 591–593 (2006)

Parasitic Effect Degrading Cascode LNA Circuits with 0.18μm CMOS Process for 2.4GHz RFID Applications

Mu-Chun Wang[*], Hsin-Chia Yang, and Ren-Hau Yang

Department of Electronic Engineering,
Minghsin University of Science and Technology, Hsinchu 30401, Taiwan
mucwang@must.edu.tw

Abstract. Miniaturizing chip circuits in IC design is one of chief targets to promote chip production. However, accompanying the parasitic capacitance perhaps degrades the circuit performance. Here, a cascode low-noise amplifier (LNA) circuit incorporating parasitic capacitance of passive inductors was designed at 2.4GHz carrier frequency with low-cost and high-integration 0.18μm CMOS process. The Advanced Design System (ADS) software was used to probe the noise or power loss of LNA. The simulated results demonstrate the forward voltage gain is 14.289 dB, the input return loss is -11.027 dB, the output return loss is -4.240 dB, the reverse isolation is -14.121 dB, the minimum noise figure is 0.918 dB and the 3-dB gain bandwidth is 320 MHz. These interesting parameters verify that this contrived LNA with ultra-low noise performance is indeed suitable to be a candidate in the whole receiver integration of RFID applications even though the parasitic effect was considered.

Keywords: RFID, LNA, Cascode, Parasitic capacitance, Gain, Return loss, CMOS.

1 Introduction

Wireless communication systems [1] allow users communicating each other easily and, hence, this kind of systems or radio-frequency integrated circuits have developed rapidly. For radio-frequency identification (RFID), the past front-end RF circuits were always made with high cost efficiency and low integration manufacturing process due to the purpose of high operation frequency. Through the integrated circuit (IC) complementary-metal-oxide-semiconductor (CMOS) [2] process, this manufacturing is beneficial to mass-product RFID tag and reduce the previous issues.

With the demands of wireless communications increasing the amount of data, the data transfer rate for more complex applications is effective in increasing. In addition to the future of wireless communication services including voice, data, image transfer, storage management, and applications linking with the internet, these systems play the key role.

In this study, adding a parasitic capacitance between two neighboring passive inductors for a cascode LNA, which can provide the low noise figure and the good impedance (Z) match, probes the real signal performance including scattering (S)

---

[*] Corresponding author.

parameters and noise figures is the main objective, especially in noise consideration. The device models were supported by tsmc company in Taiwan. The general noise with random format is expressed with average power. There are two main sources: first of all, the noise source is provided by the outside environment mixing into the real signal to increase the complexity of signal processing; the other is generated by the internal circuits. Basically, many types of noise exist, such as thermal noise [3], flicker noise [4], diffusion noise, partition noise, shot noise and generated-recombined noise. Here, transistor noise including flicker noise and shot noise is more impressive.

2 Circuit Construction

In wireless communication systems, transceiver module is an important part. The operating principle is that the antenna receives the outside signal and the signal is amplified by a LNA circuit. Then, the amplified signal passes a band-pass filter (BPF) [5]. Through a mixer circuit, the RF carrier frequency is down-converted to the intermediate frequency (IF) range or called the baseband. Again, an IF BPF is employed to filter the extra-noise. Continuously, this signal passing a digital-analog converter (DAC) travels to the demodulator and the digital signal processor. The basic wireless communication system structure is shown in Fig. 1.

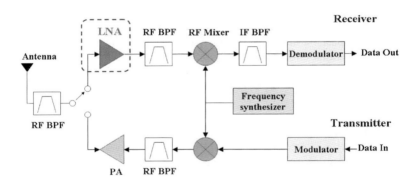

Fig. 1. Basic wireless front-end RF communication system structure.

An impressive LNA in Fig.1 should propose some good signal performance such as lower noise figure, good linearity, acceptable gain, impedance matching and integratable ability. In the Giga-range band frequency, the scattering parameters in circuit design are often adopted to analyze the signal transmission due to the obvious wave characteristics of the signal. If the Z-parameters or admittance (Y)-parameters are applied to the two-port network, as shown in Fig. 2, the definition for open or short circuit in high-frequency zone is not easy to be approached due to wave reflection. The other event is to avoid possible oscillation in circuit because of measurement. Therefore, adopting S-parameters in signal analysis is a helpful alternative and applying the power format to calculate the relationship between the incident power and the output power is more suitable in signal intensity consideration as well as the noise power interference.

Parasitic Effect Degrading Cascode LNA Circuits with 0.18μm CMOS Process   563

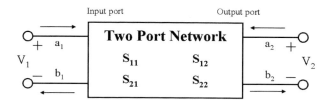

Fig. 2. Schematic of a two-port network.

Figure 2 shows the import voltage signal (a_1), the reflected signal (b_1), and the transmitted signal into the network through amplification or attenuation after the output signal (b_2) to the load. If the impedance of output port will not fully match the load, some output signal will be reflected as signal (a_2) generation. Generally, the low-noise amplifier can be presented as four kinds of matching structures depicted in Fig. 3. For instance, figure 3(a) is a resistive termination using the shunt resistance to reduce the circuit impedance, but this kind of structure proposes a higher noise value. Figure 3(b) is a $1/g_m$ termination with a source terminal of common gate to control input match, where g_m is device transconductance, but its minimum noise figure is still over 3 dB. Figure 3(c) is a shunt-series feedback contour with raising the power consumption, as well as the resistance making the noise figure increased. Figure 3(d) is an inductive degeneration, and applies an inductance to connect the gate terminal and the source terminal. The input impedance of this type is controlled well and the design for a lower noise figure is easier to be obtained [6].

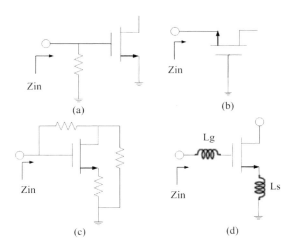

Fig. 3. Schematic circuits of four general kinds of matching structures for a LNA: (a) resistive termination, (b) $1/g_m$ termination, (c) shunt-series feedback, and (d) inductive degeneration.

In the IC design of low-noise amplifier circuit, this circuit can be roughly attributed to cascode and cascade structures. In this study, the cascode structure as a low-noise amplifier shown in Fig. 4(a) is adopted. A cascode structure not only provides the lower noise figure, but eliminates the Miller's effect to increase the output impedance and the gain, too. For an NMOS transistor as amplifier stage, the drain and gate terminals crossing a parasitic capacitance due to the neighboring passive inductors, as shown in Fig. 4(b), may cause a Miller effect increasing the load impedance generation and deteriorating the noise figure (F) defined as following

$$F = \frac{S_i / N_i}{S_o / N_o} = \frac{(SNR)_{in}}{(SNR)_{out}} \quad (1)$$

where S_i: input signal power, S_o: output signal power, N_i: input noise power, N_o: output noise power, $(SNR)_{in}$: input signal-to-noise ratio and $(SNR)_{out}$: output signal-to-noise ratio.

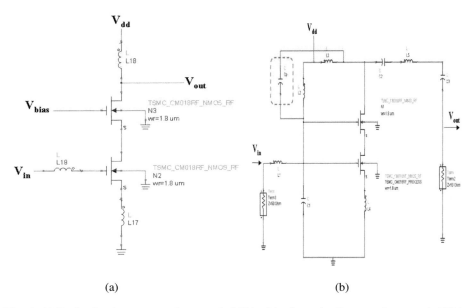

Fig. 4. (a) Basic circuit structure of a cascode LNA; (b) schematic diagram of a cascode LNA considering a parasitic capacitance, C_p.

For the relationship of noise and gain, this connection can be expressed in Fig. 5. Through this figure, the input signal intensity (S_i), the input noise intensity (N_i), the gain (G) in amplifier circuit, the output signal intensity (S_o) and the output noise intensity (N_o) can be defined well. Roughly, the S_o is equal to G · S_i, but N_o is greater than G□N_i because the output noise not only comes from the amplification, but includes the noise generated from this circuit (N_a). Therefore, the definition of a full noise figure (F) in this case is got by

$$F = \frac{S_i/N_i}{G \cdot S_i/N_o} = \frac{S_i/N_i}{\dfrac{G \cdot S_i}{G \cdot N_i + N_a}} \quad (2)$$

where the output noise is given as

$$N_o = G \times F \times N_i = G \times N_i + N_a \quad (3)$$

The internal noise in the circuit is demonstrated as

$$N_a = G \times F \times N_i - G \times N_i = (F-1) G \times N_i \quad (4)$$

The general noise figure (NF) can also be expressed as

$$NF = 10 \times \log(F): (dB) \quad (5)$$

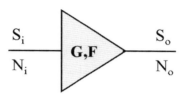

Fig. 5. Schematic profile of a single-stage amplifier.

3 Simulation Results and Discussion

In this work, the ADS simulation software was adopted and the device and process models in simulation were provided by tsmc company with 0.18 μm CMOS process. Through precise consideration and device choice in noise, gain and power consumption, the cascode LNA promisingly applying at 2.4 GHz RFID tags was completed. The final simulation results at 2.4 GHz adding parasitic capacitance shows the forward voltage gain is 14.289 dB, the input return loss is -11.027 dB, the output return loss is -4.240 dB, the reverse isolation is -14.121 dB, the minimum noise figure is 0.918 dB, and the 3-dB gain bandwidth is 320MHz. Figure 6 expresses two neighboring inductors inducing a parasitic capacitance, C_p causing some signal performance degradation, as shown in Fig. 7-9.

The added parasitic capacitance in circuit simulation can be shown as

$$C_P = \frac{\varepsilon_o k(\ell \cdot t)}{S} \quad (6)$$

where ε_o: vacuum permittivity, l: metal inductor length, t: metal inductor thickness, and S: space of two neighboring inductor metal lines.

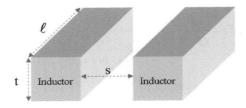

Fig. 6. Schematic of a cross-section contour for neighboring inductor metal lines.

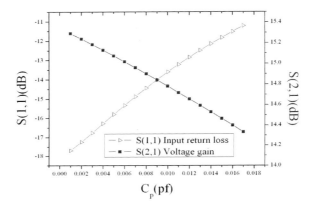

Fig. 7. Simulation results of input return loss and forward voltage gain, S_{11}, S_{21}.

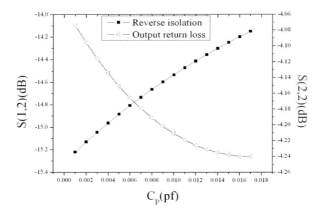

Fig. 8. Simulation results of reverse isolation and output return loss S_{12} and S_{22}, respectively.

Parasitic Effect Degrading Cascode LNA Circuits with 0.18μm CMOS Process 567

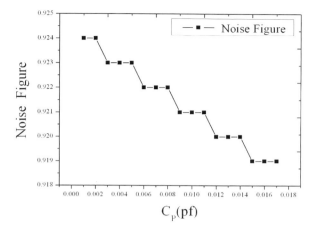

Fig. 9. Simulation result of noise figure, NF with the added C_p.

If the metal-line space is decreased, the C_p value is relatively increased. Thus, the input return loss, S_{11}, is increased and the forward gain, S_{21}, is decreased, shown in Fig. 7. In Fig. 8, we observed that the reverse isolation, S_{12}, is increased and the output return loss, S_{22}, is decreased, respectively. The noise figure is also reduced due to the decrease of forward gain. This simulation illustrates that if the real analysis in circuit will be come true, the parasitic effect must be included well, especially in the shrinkage of chip area. The full layout schematics of this LNA with C_p or without C_p effect are demonstrated in Fig. 10.

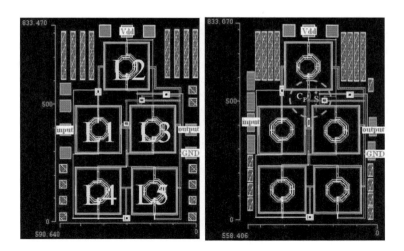

Fig. 10. Top-view layout of the cascode LNA: (a) without including parasitic capacitance, and (b) with parasitic capacitance.

Table 1. Performance comparison with other LNAs.

Parameters	This work w/C_p	This work w/o C_p	[7]	[8]	[9]
Supply voltage	1.8V	1.8V	1.8V	1.8V	1.8V
Frequency	2.4G	2.4G	2.4G	2.4G	2.4G
CMOS process technology	0.18μm	0.18μm	0.18μm	0.18μm	0.18μm
Voltage gain (S_{21}): dB	14.289	15.342	18.4	12.4	19
Reverse isolation (S_{12}): dB	-14.121	-15.314	NA	NA	NA
Input return loss (S_{11}): dB	-11.027	-18.119	-16.9	NA	NA
Output return loss (S_{22}): dB	-4.240	-4.051	-14.7	NA	NA
Noise Figure	0.918	0.924	2.1	4.4	2.1

4 Conclusion

The main functions of a LNA circuit in RFID application are to accurately detect the received input signal and support the sufficient power gain against the interference of noise. While designing the low-noise amplifier for wireless radio frequency applications, the quality integrity of low-noise amplifier is not easy to be handled well, especially with passive components. To obtain a high performance in low noise consideration, some of the circuit components are required to be traded off. In this work, the added parasitic capacitance can effectively reduce noise figure and area. When the parasitic capacitance in Eq. (6) is increased, other conditions remain unchanged, decreasing the space, S, can possibly reduce the chip area. Due to the shrinkage, the chip size can be reduced from 833.470μm x 590.640μm down to 833.070μm x 558.406μm. The chip area size is reduced around 4%. In the process constraint, the S minimum value is around 0.4~0.5μm. Here, we observed that even though the process constraint in metal line space can be shrunk to a minimum value, but the metal space in real case should be enlarged more to avoid the gain loss in signal operation. Table 1 shows the simulation result of cascode low-noise amplifiers comparing the circuit performance with other designs in the same process. In terms of this table, the result with C_p in noise figure exhibits the great performance. This cascode LNA, in deed, was verified as a good candidate integrating into the RFID tags.

Acknowledgments

The authors thank that National Chip Implementation Center (CIC) in Taiwan provides the simulation information and the financial support from this university with an internal project number, MUST-100-Elecronics-05.

References

1. Nair, V.: Heterogeneous wireless communication devices- present and future. Microwave Theory and Applications, 2 (2008)
2. Sundararajan, V., Parhi, K.K.: Low power synthesis of dual threshold voltage CMOS VLSI circuits. Low Power Electronics and Design, 139–144 (1999)
3. Ou, J.J., et al.: Submicron CMOS thermal noise modeling from an RF perspective. VLSI Technology, 151–152 (1999)
4. Zhaofeng, Z., Jack, L.: Experimental study on MOSFET's flicker noise under switching conditions and modeling in RF applications. IEEE/Custom Integrated Circuits, 393–396 (2001)
5. Zahabi, et al.: Design of a band-pass pseudo-2-path switched capacitor ladder filter. Quality of Electronic Design, 662–667 (2005)
6. Shaeffer, D.K., Lee, T.H.: A 1.5 V 1.5GHz CMOS Low Noise Amplifier. IEEE Journal of Solid-State Circuits 32, 745–759 (1997)
7. Chang, C.W., Lin, Z.M.: Design of a high gain LNA for wideband applications. Communications and Signal Processing, 1–3 (2009)
8. Chen, K.H., et al.: An Ultra-Wide-Band 0.4–10-GHz LNA in 0.18μm CMOS. IEEE Transactions, Circuits and Systems 54, 217–221 (2007)
9. Hwang, Y.S., et al.: A controllable variable gain LNA for 2 GHz band. In: Microwave Conference Proceedings, vol. 5, pp. 3–4 (2005)

Minimization of Cascade Low-Noise Amplifier with 0.18 μm CMOS Process for 2.4 GHz RFID Applications

Mu-Chun Wang, Hsin-Chia Yang, and Yi-Jhen Li

Department of Electronic Engineering,
Minghsin University of Science and Technology, Hsinchu 30401, Taiwan
mucwang@must.edu.tw

Abstract. The cascade LNA circuit adding parasitic capacitance (C_p) for neighboring passive inductors at 2.4GHz with low-cost and high-integration 0.18μm CMOS process was designed and studied. The simulation results adopting an ADS software as a simulator demonstrated the forward voltage gain is 11.908dB, the input return loss is -9.563dB, the output return loss is -21.153dB, the reverse isolation is -16.315dB, the minimum noise figure is 2.01dB and the 3-dB gain bandwidth is 240MHz. Comparing the performance with C_p and without C_p, we observed that they indeed had some change. The noise figure was slightly reduced and the voltage gain was increased a little. The benefit of chip size shrinkage was about 3%. Totally, these interesting parameters still verify that this contrived LNA with ultra-low noise performance plus parasitic capacitance effect is still suitable to be a candidate in the whole receiver integration for RFID applications.

Keywords: RFID, LNA, Cascade, Gain, Return loss, Parasitic effect, CMOS.

1 Introduction

In real commercial competition, minimizing chip size is an attracting idea. Through this effort, the chip cost is able to be greatly reduced. Combining the layout consideration with radio-frequency identification (RFID) tag, the tag can increase its influence in RFID marketing. Because the wireless communication system has been developed rapidly, incorporating the benefits of high integration and low power consumption of devices has really been implemented through the fabrications of metal oxide semiconductor field effect transistors (MOSFET). The whole hardware in the RFID system basically includes transmitter, antenna and receiver. When the signal emitted by a transmitter travels through the air, it is attenuated and then detected by antenna of pre-designed sizes in the other receiver. Therefore, to recover the signal without introducing extra unacceptable noise, the front-end low-noise amplifier (LNA) [1] thus predominantly takes the role providing a sufficient voltage gain. The specific low noise function is important because the subsequent signal processing is to be left alone from the unexpected interference of noises, even originated from other components in the receiver.

2 Circuit Design

In wireless communication systems, the receiver achieves the wireless signals from the antenna. Using a band-pass filter (BPF), the undesired noise is eliminated. Continuously, the weak signal will be amplified by a low-noise amplifier. If the noise intensity mixing in the desired signal is higher, the true signal is not easy to be recognized by the other back-end circuits in receiver. Thus, the receiving signal quality is deteriorated. The serious case is that the noise intensity is stronger than the real signal, inducing the receiving catastrophe. Therefore, providing an adequate LNA circuit in receiver is necessary to implement the signal amplification and reduce the noise magnification. The basic wireless receiver diagram is exhibited in Fig. 1.

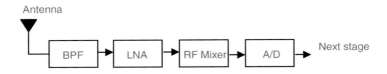

Fig. 1. Basic circuit blocks of a wireless communication receiver.

The LNA is usually treated as one of front-end RF circuit units. The concerned performance generally incorporates noise figure, linearity, gain, and impedance matching, etc. The relative considerations can be expressed as shown in Fig. 2.

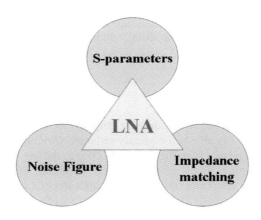

Fig. 2. The precise design keys to a low-noise amplifier.

In the microwave band, the signal transmission approaches to the wave characteristics. Because of the wave format, the forward gain is not cared well, but the reflection effect must be considered. Therefore, the impedance (Z) or admittance (Y) parameters are not fully suitable in the signal analysis. Adopting scattering parameters (S-parameters) is a good choice to design the high-frequency circuits [2]. If the Z or Y-parameters [3] are applied to the two-port network, as shown in Fig. 3, the definition for open or short circuit in high-frequency zone is not easy to be approached well due to wave reflection. The other event is to avoid possible oscillation in circuit because of measurement. Therefore, adopting S-parameters in signal analysis is a helpful alternative and applying the power format to calculate the relationship between the incident power and the output power is more suitable in signal intensity consideration as well as the noise power interference. Figure 3 illustrates the input voltage signal (a_1), the reflected signal (b_1), and the transmitted signal into the network through amplification or attenuation after the output signal (b_2) to the load. If the impedance of output port will not fully match the load, some output signal will be reflected as signal (a_2) generation. For the matching consideration, there are eight typical structures, as shown in Fig. 4.

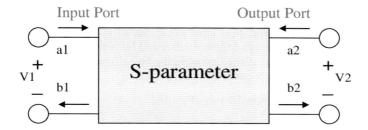

Fig. 3. Basic diagram of a two-port network.

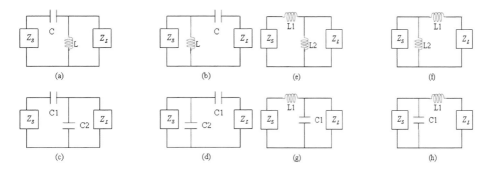

Fig. 4. Schematic blocks of alternative full-matching circuits.

The Z_S and Z_L represent the input or source impedance and load impedance, respectively. Choosing the feasible matching circuit, the reflection issue of signal transmission is reduced and the power consumption is also decreased due to the sufficient forward gain. Furthermore, the system cost is saved more and the system is more reliable.

In terms of an amplifier circuit, the gain value, as shown in Fig. 5, is the main consideration, but the noise coming from the input port is relatively amplified. Sometimes, the internal noise of the amplifier is also added to interfere in the desired signal and increase the complexity of signal processing.

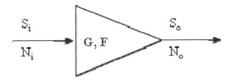

Fig. 5. Schematic diagram of a single-stage amplifier.

To figure out the noise interference degree, a noise figure (F) is defined as

$$F = \frac{S_i / N_i}{S_o / N_o} = \frac{N_o}{GN_i} = \frac{GN_i + N_d}{GN_i} \quad (1)$$

where S_i is input signal power, N_i is input noise intensity, S_o is output signal power, N_d is the internal noise intensity of the amplifier and N_o is output noise intensity [4].

If the gain in the first stage is high enough, the noise figure (NF) is preferred to use the logarithmic form to be represented as in Eq. (2).

$$NF = 10 \times \log(F): (dB) \quad (2)$$

If the parasitic capacitance (C_p) neighboring two inductors shown in Fig. 6 is included in simulation, the parasitic effect will affect the performance of this LNA when the chip size shrinkage is a critical concern in market competition.

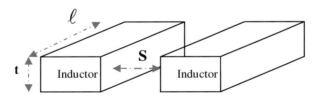

Fig. 6. Schematic contour of a cross-section of neighboring inductor metal lines.

Basically, the parasitic capacitance can be expressed as

$$C_P = \frac{\varepsilon_o k(\ell \cdot t)}{S} \quad (3)$$

where ε_o: vacuum permittivity, ℓ:: metal inductor length, t: metal inductor thickness, and S: space of two neighboring inductor metal lines.

Subsequently, figure 7 shows a prototype for a cascade LNA without C_p structure and figure 8 depicts a schematic circuit with the added parasitic capacitance effect between L3 and L4 inductor at 2.4 GHz operation.

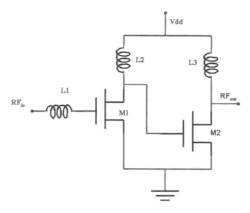

Fig. 7. Circuit diagram for a cascade LNA without parasitic effect.

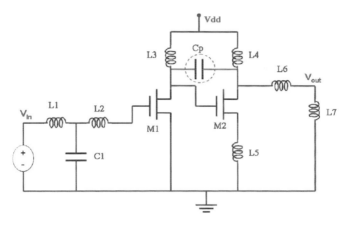

Fig. 8. Schematic of a cascade LNA with parasitic effect indicated with a dashed circle.

3   Results and Discussion

In this study, the simulator is the Advanced Design System (ADS) software supported by Agilent company. The device and process models with 0.18μm CMOS process were provided by tsmc company in Taiwan. The cascade circuit design [5] of the LNA promisingly applying at 2.4GHz shows the added parasitic capacitance effect proposing the forward voltage gain is 11.908 dB, the input return loss is -9.563 dB, the output return loss is -21.153 dB, the reverse isolation is -16.315 dB, the minimum noise figure is 2.01 dB, and the 3-dB gain bandwidth is 240 MHz. These interesting parameters still verify that this contrived LNA [6] with ultra-low noise performance is indeed suitable to be a candidate in the whole receiver integration of radio-frequency identification applications. Figures 9-10 exhibit the simulation results for the designed LNA circuit with C_p effect. According to Fig. 9(a), we observed that if the space distance, S, is shortened, the return loss is increased in the front-end part and decreased in the back-end part. However, the forward gain is increased and enters the saturation situation. Again, the output performance in Fig. 9(b) seems that the behaviors are similar to the input operation. While the C_p with shortening the space distance is increased, the noise figure is somewhat decreased. Finally, two full top-view layouts of this cascaded LNA without and with C_p are demonstrated in Fig. 11, respectively. The horizontal width (= 590.64 μm) is the same, but the vertical lengths are different (= 1164 μm and 1133 μm). The chip area in this case is indeed shrunk about 3%.

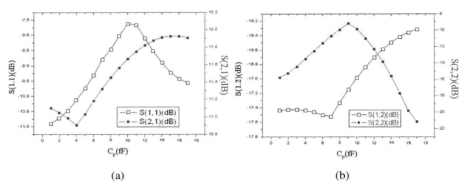

Fig. 9. Simulation results of (a) input return loss and forward voltage gain, S_{11}, S_{21}, respectively; (b) reverse isolation and output return loss, S_{12}, S_{22}, respectively.

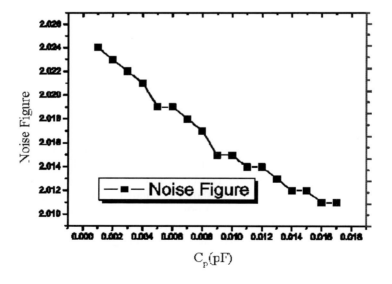

Fig. 10. Simulation result of noise figure, NF vs. C_p variation.

Minimization of Cascade Low-Noise Amplifier with 0.18 μm CMOS Process

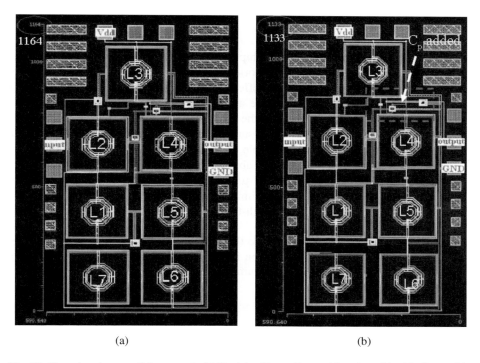

(a)　　　　　　　　　　　　　　　(b)

Fig. 11. Top-view layout of the cascade LNA: (a) without C_p consideration; (b) with C_p consideration shortening the vertical length from 1164 μm to 1133 μm.

Table 1. Performance comparison with other LNAs.

Parameters	This work w/C_p	This work w/o C_p	[7]	[8]	[9]
Supply voltage	1.8V	1,8V	0.9V	NA	NA
Frequency (GHz)	2.4	2.4	2.4	2.4	2.4
CMOS process technology	0.18μm	0.18μm	0.18μm	0.15μm	0.18μm
Voltage gain (S_{21}) (dB)	11.102	11.908	NA	12.1	15
Reverse isolation (S_{12}) (dB)	-17.439	-16.315	NA	NA	NA
Input return loss (S_{11}) (dB)	-10.892	-9.563	NA	-19	NA
Output return loss (S_{22}) (dB)	-15.773	-21.135	NA	-20.7	NA
Noise Figure	2.024	2.01	-111	NA	NA

4 Conclusion

In this work, we indicate that if the chip-size shrinkage of a cascade LNA is concerned well, the parasitic effect neighboring two passive inductors must be incorporated in simulation to reduce the side effect when the real electrical measurement is executed. In this simulation, the device and process models provided by tsmc with 0.18μm CMOS process. Prevailingly, the cascade LNA should have the larger gain due to the gain multiplication with two stages. However, the noise figure is relatively magnified a little. Integrating the feasible electrical components in IC design demonstrates good performance, shown in Table 1. The chip area is also reduced from 590.64 μm x 1164 μm down to 590.64 μm x 1133 μm and the benefit in this factor gains about 3%. Besides the process limitation in metal line space, this simulation provides the layout constraint due to high-frequency operation, too.

Acknowledgments. The authors thank that National Chip Implementation Center (CIC) in Taiwan provides the simulation information and the financial support from this university with an internal project number, MUST-100-Elecronics-05.

References

1. Cheng, W.C., et al.: Design of a Fully Integrated Switchable Transistor CMOS LNA for 2.1/2.4 GHz Application. In: European Microwave Integrated Circuits Conference, pp. 133–136 (2006)
2. Lavasani, S.H.M., Kiaei, S.: A New Method to Stabilize High Frequency High Gain CMOS LNA. In: IEEE Electronics, Circuits and System, vol. 3, pp. 982–985 (2003)
3. Larson, L.E.: Integrated circuit technology options for RFIC's present status and future directions. IEEE Journal of Solid-State Circuits 33(3), 387–399 (1998)
4. Shaeffer, D.K., Lee, T.H.: A 1.5 V 1.5GHz CMOS Low Noise Amplifier. IEEE Journal of Solid-State Circuits 32(5), 745–759 (1997)
5. Sundararajan, V., Parhi, K.K.: Low power synthesis of dual threshold voltage CMOS VLSI circuits. Low Power Electronics and Design, 139–144 (1999)
6. Lin, Y.T., et al.: 3–10-GHz Ultra-Wideband Low-Noise Amplifier Utilizing Miller Effect and Inductive Shunt–Shunt Feedback Technique. Microwave Theory and Techniques 55, 1832–1843 (2007)
7. Lee, H.I., et al.: An extremely low power 2 GHz CMOS LC VCO for wireless communication applications. In: IEEE European Conference on Wireless Technology, pp. 31–34 (2005)
8. Chandrasekhar, V., et al.: A Packaged 2.4GHz LNA in a 0.15μm CMOS Process with 2kV HBM ESD Protection. Solid-State Circuits Conference, 347–350 (2002)
9. Li, W.C., Wang, C.S., Wang, C.K.: A 2.4-GHz/3.5-GHz/5-GHz Multi-Band LNA with Complementary Switched Capacitor Multi-Tap Inductor in 0.18μm CMOS. VLSI Design, Automation and Test, 26–28 (2006)

Simulation and Analysis of Galileo System

Wei Zhang, Weibing Zhu, and Huijie Zhang

Beijing Institute of Petrochemical Technology
College of information Engineering
Beijing, China, 102617
zhangwei@bipt.edu.cn

Abstract. The paper presents the study on European global navigation satellite system (GNSS): Galileo system which is a European infrastructure of spatial information being implemented by European Community and European Space Agency mainly. GSSF is conceived as a simulations environment that reproduces the functional and performance behavior of the Galileo system to support the entire Galileo programmer lifecycle. In this paper, Simulation results with prime focus on UERE, DOP, NSP, Visibility Analyses and Integrity Analysis of the Galileo system.

Keywords: GALILEO; GSSF; UERE; NSP; DOP.

1 Introduction

Galileo is a global navigation satellite system (GNSS) currently being built by the European Union (EU) and European Space Agency (ESA). The €5.3 billion project is named after the famous Italian astronomer Galileo Galilei. One of the political aims with Galileo is to provide a high-accuracy positioning system upon which European nations can rely independent from the Russian GLONASS and US GPS systems, which can be disabled for commercial users in times of war or conflict. When in operation, it will use the two ground operations centre, one near Munich, Germany, and another in Fusion, Italy and will consist initially of 18 satellites by 2015. An additional €1.9 billion is planned to be spent bringing the system up to the full complement of 30 satellites (27 operational + 3 active spares)[1][2].he first experimental satellite, GIOVE-A, was launched in 2005 and was followed by a second test satellite, GIOVE-B, launched in 2008. The first four operational satellites for navigation will be launched in 2011 and once this In-Orbit Validation (IOV) phase has been completed, additional satellites will be launched. On 30 November 2007 the 27 EU transportation ministers involved reached an agreement that it should be operational by 2013[3], but later press releases suggest it was delayed to 2014 [4].

The navigation system is intended to provide measurements down to the meter range as a free service including the height (altitude) above sea level, and better positioning services at high latitudes compared to GPS and GLONASS (though with recent upgrades to GPS similar accuracy levels are reached[citation needed]). As a further feature, Galileo will provide a global Search and Rescue (SAR) function. To do so, each satellite will be equipped with a transponder, which is able to transfer the distress signals from the user's transmitter to the Rescue Co-ordination Centre, which will then initiate the rescue operation. At the same time, the system will provide a signal to the user, informing him that his situation has been detected and that help is on the way. This latter feature is new and is considered a major upgrade compared to the existing GPS and GLONASS navigation systems, which do not provide feedback to the user[5].The use of basic (low-accuracy) Galileo services will be free and open to everyone. The high-accuracy capabilities will be available for paying commercial users and for military use.

The GALILEO System Simulation Facility (GSSF) is a software simulator tool that reproduces the functional and performance behavior of the Galileo system in order to support the simulation needs during the Galileo programmer. GSSF version 2.1 supports only Service Volume Simulation capabilities. Raw Data Generation capabilities are available in the version 2.0 and will come again with the future version 2.2.The Service Volume Simulation (SVS) capability of GSSF allows analyzing the navigation and integrity performance over longer periods of time and over large geographical areas. GSSF allows the user to assess all relevant Figures of Merit on global or regional grids or for individual positions.

In this paper, we limit our discussion within on UERE, DOP, NSP, Visibility Analyses and Integrity Analysis of the Galileo system. The remainder of this paper is organized as follows. In Section II the GSSF simulation are introduced, in Section III is GSSF capabilities, Section IV is example simulation; Section IV concludes the paper.

2   Paper Preparation

GSSF provides a single simulator that uses alternative models depending upon the type of analysis the end-user wishes to perform.

The Service Volume Simulation (SVS) capability of GSSF allows the analysis of the navigation and integrity performance over long time periods and over large geographical areas. In particular, GSSF SVS allows the user to assess all relevant Figures of Merit on global or regional grids or for individual positions. Such Figures of Merit are Visibility, Coverage, Geometry, DOP, Navigation Precision, Integrity and Service (including Critical Satellites) as well as the associated availability and continuity figures. In addition GSSF provides GPS/Galileo global Interference analysis as well as Link Budget and Error Budget analyses. A comprehensive list of available analyses is provided in, while individual implementations are further detailed in.

The Raw Data Generation (RDG) capability of GSSF uses high fidelity models to generate GPS and Galileo observables acquired by Galileo Sensor Stations. This capability includes the definition of Feared Events and is suitable for the validation and tuning of OSPF and IPF algorithms.

GSSF offers flexibility to execute analysis both is standard and stand alone form. In standard form, the users can select analysis listed in the analysis catalogue and the recorded data is processed automatically. Some stand alone form, analysis is implemented in generic manner to manually process the recorded data, without running the simulation again.

2.1 The Following Is List of Standard Analyses Components

- Visibility
- Extended Visibility
- Coverage (Depth of Coverage)
- Inverse Coverage
- Geometry
- Dilution of Precision (DOP)
- Navigation System Precision (NSP)
- Independent Integrity Path
- Signal-In-Space Monitoring Accuracy (SISMA)
- Integrity

3 Example Simulation

3.1 Error Budget

The User Equivalent Range Error (UERE) Budget is calculated as the sum of errors resulting from tropospheric effects, ionospheric effects, receiver noise, ephemeris fit, satellite clocks, receiver clock and multipath. Each contribution is modeled using a user-defined distribution profile (e.g. obtained from a system simulation).Note, although the UERE budget calculation also represents errors due to environmental effects, it is from an architectural point of view part of the User Receiver Grid model.

$$UERE = \sqrt{\sum_i \sigma_i^2} \tag{1}$$

σ_i : Contribution due to tropospheric effects, ionospheric effects, receiver noise, ephemeris fit, satellite clocks, receiver clock and multipath [cm].

For applicability of UERE budgets and which prede-fined UERE budget tables (total values) are recommended and available together with the GSSF installation, please refer to Volume 3 - Reference Scenarios document. Please note that the user can also define his/her own UERE budget tables as necessary.

In GSSF, the standard deviation and mean for different values of elevation are calculated by linear interpolation.

Note, this UERE model also implements default total values that are specified in the table below. However, it is recommended that users load the predefined UERE tables mentioned above instead of applying the defaults below (Fig.1 shows Default UERE Budgets implemented in GSSF).

3.2 Dilution of Precision (DOP) Analyses

The DOP Analysis returns the mean, maximum, mini-mum, and standard deviation of DOP for each node of user receiver in a grid. The analysis computes and records the different DOP values namely GDOP, PDOP, TDOP, HDOP and VDOP as instantaneous recorded data, for each user receiver over entire simulation duration for post processing.

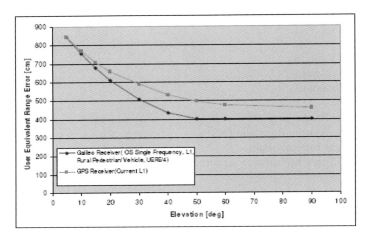

Fig. 1. Default UERE Budgets implemented in GSSF

These analyses can be run over an area or for some specific user locations. The following map illustrates the maximum HDOP expected for the Galileo constellation. The Global or Regional Availability of DOP Analysis returns the percentage of time that a DOP value is below a user specified threshold for each node in the grid. Availability of DOP is calculated for TDOP, PDOP, GDOP, HDOP and VDOP (Fig.2 shows Maximum Horizontal DOP).

3.3 Navigation System Precision Analysis

Computes the Navigation System Precision (NSP) for each user or ground segment element over the simulation time period. It returns the mean, maximum, and minimum

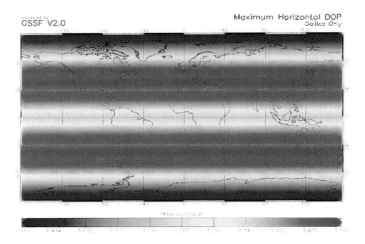

Fig. 2. Maximum Horizontal DOP

NSP for each user or ground segment element as well as the instantaneous data for each simulation time step. The analysis computes the NSP types Overall NSP (ONSP), Time NSP (TNSP), Horizontal NSP (HNSP), and Vertical NSP (VNSP).

As expected, Galileo yields an improved navigation system precision as compared to GPS, while a combined use of both systems allows for an even further improved performance (Fig.3 shows Mean NSP over Europe – Galileo & GPS).The NSP is computed from the UERE budget estimated by the Environment model.

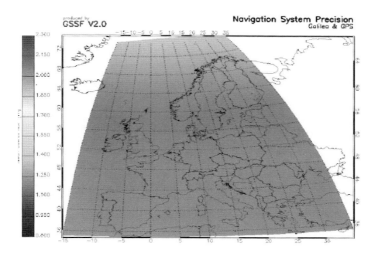

Fig. 3. Mean NSP over Europe – Galileo & GPS

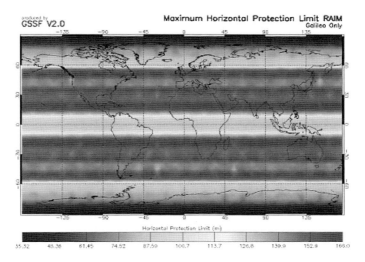

Fig. 4. Maximum Horizontal Protection Limit RAIM

3.4 Integrity Analysis

The Integrity Analysis returns statistics on the Vertical and Horizontal Protection Levels (PL) over an area. It provides maximum, minimum, average and 95-percentile PL for user receiver located in the area determined by the end-user. We can see Fig.4 and Fig.5.

Availability of Integrity can also be computed either including failures in the space segment or without failures.

A new release is under preparation to accommodate the GALILEO integrity concept.

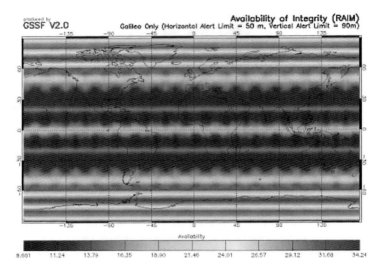

Fig. 5. Availability of Integrity (RAIM)

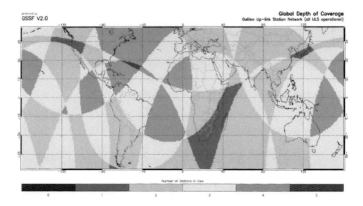

Fig. 6. Global Depth of Coverage for operational Uplink Station (ULS) network

3.5 Sign Coverage Analysis Subject to Ground Station Failures

GSSF provides a coverage analysis that returns the global number of ground stations in view at each satellite position over the simulation time period. The result of this analysis is typically visualized on a global map. For this analysis, the Galileo constellation was applied together with an reference Galileo constellation was applied together with an Uplink Station (ULS) network (Fig.6 shows Global Depth of Coverage for operational Uplink Station (ULS) network).

4 Conclusions

GSSF reproduces the functional and performance behavior of the Galileo system and offers the built-in flexibility to support Galileo system simulation needs during the entire program life cycle. GSSF provides a single simulator that uses alternative models depending upon the type of analysis the end-user wishes to perform. For this purpose, GSSF has to provide more flexibility and openness than traditional space system simulators. This is achieved by a component based design. The results presented in this paper demonstrate how GSSF SVS can be applied in particular for the justification of Galileo system design decisions. They show the capability of GSSF to analyze the complex satellite navigation system Galileo during its detailed definition phase.

References

1. The Galileo Project – GALILEO Design consolidation, European Commission (2003)
2. Hein, G.W., Godet, J., et al.: Status of Galileo Frequency and Signal Design. In: Proc. ION GPS 2002 (2002)
3. Issler, J.-L., Hein, G.W., et al.: Galileo Frequency and Signal Design. GPS World 14(6), 30–37 (2003)
4. Divis, D.A.: Military role for Galileo emerges. GPS World 13(5), 10 (2002)
5. North, R.: Galileo - The Military and Political Dimensions (2004)

High-Power Controllable Voltage Quality Disturbance Generator

Gu Ren and Xiao Xiangning

Key Laboratory of Power System Protection and Dynamic Security Monitoring and Control of Ministry of Education
North China Electric Power University, 102206, Beijing, China

Abstract. This paper proposes a high-power controllable voltage quality disturbance generator, which is composed of multiplex cascade H bridge modules, and simulates many types of voltage quality problems, such as voltage swell and sag, fluctuation and flicker and so on. One side of the device connects the power system with cascade H bridge to improve voltage level. The other side connects the system directly to aviod the problems caused by series transformer, such as additional phase shifting and voltage drop. Cascade inverter H bridge modules and PWM rectifier H bridge modules are all controlled by triangular carrier phase shifting to increase the equivalent switching frequency of inverter to simplify the control. Finally, a 10kV/2MVA controllable voltage quality disturbance generator has been created. The experiment result proves that the proposed main circuit topology, control strategy and EtherCat-based controller are feasible and efficient.

Keywords: controllable voltage quality disturbance generator, multiplex cascade, carrier phase shifting, EtherCat.

1 Introduction

The load structure in modern power system is changing greatly. The fast development of power electronics and the wide application of electricity comsumer based on power electronics cause serious distortion of voltage and current waveform in grid [1]-[2]. On the other hand, the popularity of equipments sensitive to power quality makes electricity comsumers put forward stricter requirements to power quality and power supply reliability [3]-[5]. To solve these power quality problems, DVR, UPQC, APF, DSTATCOM and other power quality improving devices have been used at home and abroad, and achieved great effects. Besides, new power quality theories and power quality analysers should be checked for research. Therefore, power quality disturbance generators, which can generate all types of power quality disturbances and have enough capacity, are needed to meet the requirements above.

This paper proposes a high-power controllable voltage quality disturbance generator, which can generate corresponding voltage disturbance according to demand. This device can output not only single voltage disturbance, but also compound one. The 10kV industrial prototype meets the indexes as following. Voltage sag range: 90%~50% of rating voltage, 10ms ~1min. Three phase unbalance factor: 20%. Harmonic voltage: maximan harmonic number not less than 25.

2   Main Circuit Topology

The present conventional disturbance generators can be devided into two types: the one with series transformer and the one without series transformer. There are two advantages for the one with series transformer: One is that adopting booster transformer can lower the voltage level of direct current part, so that swiching devices can be chosen more flexibly in high voltage level. The other is that it is easier to achive electrical isolation between inverter and grid. However, because of its nonlinearity, the series transformer inevitably bring in many disadvanges: (1)High harmonics of inverter make the design capacity of transformer larger. (2)Mutual influence between series transformer and filter inductance and capacitor cause additional phase shifting and voltage drop. (3)Short-circuit reactance of series transformer reduces the voltage precision of control system. (4)The use of transformer increases cost and loss of the device, as well as occupied area. If series transformer is removed, reliability and efficiency of the device will be improved, as well as the total investment. Therefore, the device proposed in this paper adopts the structure without series transformer. The main circuit topology is shown in Fig. 1.

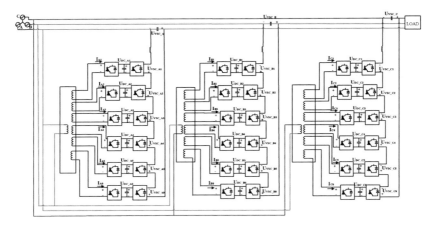

Fig. 1. The main circuit topology of the controllable voltage quality disturbance generator.

This topology adopts multiplex cascade structure, which means every phase of it is composed of several low voltage PWM power units. Every power unit gets power from a multi-windings isolation transformer, and is controlled by high-speed microprocessor, isolated and driven by optical fiber. Multiplex solves the harmonic problem caused by common six-pulse and twelve-pulse circuit [6]-[7]. This topology generates fundamental frequency disturbance by cascade H bridges, which are series connected with high frequency modules. Fundamental power module generates fundamental voltage disturbance, such as voltage swell and sag, fluctuation and flicker, voltage unbalance and so on. Harmonic power module generates harmonic voltage and other high frequency disturbance. High frequency disturbance and low frequency disturbance implement decoupling control in this way, and the requirement to swiching frequency of fundamental power module has been lowered.

3 Control Strategy

3.1 Carrier Phase Shifting Control

The multilevel converter with cascade H bridges in this device adopts carrier phase shifting control strategy, the advantage of which is that all the H bridge work under the same condition. The single power unit of cascade multilever inverter is shown in Fig. 2. Power unit PWM control strategy is shown in Fig. 3. U_{Tri} is triangular carrier wave. U_{ref} is modulation wave. δ is carrier shifting phase. d_{rv1} and d_{rv3} is are driving signals of insulated gate bipolar transistor(IGBT) Q_1 and Q_3, whose reference wave is perfect sine wave. The reference wave of Q_2 and Q_4 lead the one of Q_1 and Q_3 180°. The control signal of Q_1 in left bridge is obtained by the comparison of reference wave and triangular carrier wave. Q_1 turns on, Q_2 turns off, and outputs a high voltage when reference wave is higher than triangular carrier wave. In contrast, Q_1 turns off, Q_2 turns on, and outputs a low voltage. The control signal of Q_2 is obtained by reversing the one of Q_1. The control signal of Q_3 in left bridge is obtained by the comparison of reference wave and triangular carrier wave. Q_3 turns on, Q_4 turns off, and outputs a high voltage

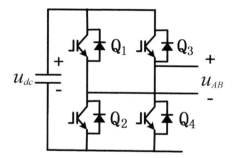

Fig. 2. Single power unit.

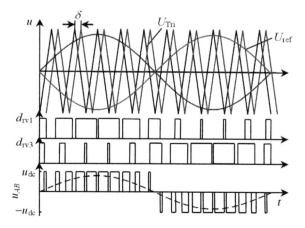

Fig. 3. PWM control strategy of power unit.

when reference wave is higher than triangular carrier wave. In contrast, Q_3 turns off, Q_4 turns on, and outputs a low voltage. The control signal of Q_4 is obtained by reversing the one of Q_3. The waveform obtaining from the difference between left and right bridges is the same as the one the all power unit outputs.

Cascade multilevel converter is composed of several cascade power units, every single phase of which contains N power units. The power units in the same cascade phase have the same reference wave. There are phase difference between two adjacent power units. Phase difference θ can be deduced by the expression blow:

$$\theta = 180° / N \tag{1}$$

3.2 Disturbance Generation Strategy

The disturbance of over-voltage, under-voltage, voltage swell and sag are controlled by reference voltage feedforward and inverter output voltage feedback. Voltage unbalance, fluctuation and flicker are controlled by open-loop in three-phase abc coordinates. These disturbances is generated by fundamental power module, and the control strategy is shown in Fig. 4. U_{sA}, U_{sB}, U_{sC} are the system voltage, U_{lA}, U_{lB}, U_{lC} are load voltage, U_l^* is the load expectation voltage.

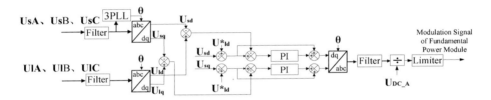

Fig. 4. Disturbance generation strategy of over-voltage, under-voltage, voltage swell and sag.

Harmonic voltage disturbance is generated by harmonic power module, controlled by open-loop in three-phase abc coordinates. Time-domain harmonic and unbalance voltage are given in three-phase abc coordinates as modulation waveforms. The control strategy is shown in Fig. 5.

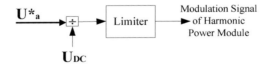

Fig. 5. Disturbance generation strategy of harmonic and unbalance voltage.

4 Controller Based on Series Real-Time EtherCat

EtherCat(Ethernet for Control Automation Technolog) is a real-time Ethernet technochgy. Initially developed by Beckhoff Automation GmbH in Germany, EtherCat builds new standard for real-time system capability and topology flexibility, and reduces fieldbus cost. In 2005, EtherCat formally become IEC standard-- IEC/PAS 62407. Unitl now, EtherCat has been widely used in the industrial fields such as wind power generation, automatic production lines, robots and so on. EtherCat transmits signals by series communication. EtherCat has features of good real-time communication, high tranfer rate, flexible network topology.

A layered and distributed controller based on EtherCat is adopted in the 10kV industrial prototype. The main controller deals with human-computer interaction, information collection and distrubance calculation. The unit controller is responsible for module protection and PWM wave generation. EtherCat transfers signals and synchronizes time clock between the two types of controllers.

5 Experiments of 10kV/2MVA Industrial Prototype

Based on the topology, control strategy and EtherCat controller, a 10kV/2MVA controllable voltage quality disturbance industrial prototype has been created. This device is composed of 18 power modules . In each phase, there are five fundamental power modules with the swiching frequency 5Hz, and one harmonic power module with the swiching frequency 2.5Hz. Expected effects have been achived. There are some of the results below.

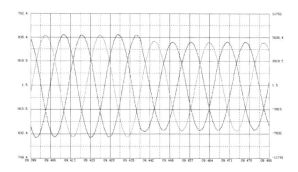

Fig. 6. Voltage sag, expected voltage 5000V

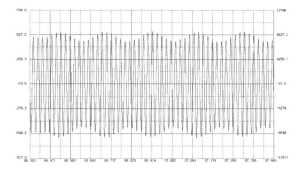

Fig. 7. Voltage fluctuation, 5Hz sine, 10%

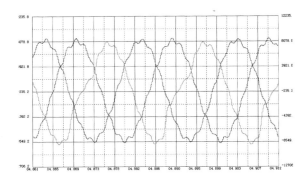

Fig. 8. Harmonic voltage, 11th, 20%.

6   Conclusion

This paper discusses the main circuit topology and control strategy of the high-power controllable voltage quality disturbance generator. It proposes a multiplex cascade structure without series transformer. Multilevel converter with cascade H bridge adopts carrier phase shifting control strategy. The control strategies of mangy voltage disturbances are given. EtherCat-based controller is proposed. Finally, several types of disturbance experiment waveforms are given to prove the effect. The emergence of this device will benefit the check of new power quality analysis theories and the test of power quality compensation devices, and will accelarate the development of power quality research field.

References

1. Xiao, X., Xu, Y.: Power Quality Analysis and its Development. Power System Technology, 66–69 (2001) (in Chinese)
2. Mishra, S.: Neural-network-based Adaptive UPFC for Improving Transient Stability Performance of Power System. IEEE Trans.on Neural Networks, 461–470 (2006)
3. Lauttamus, P., Tuusa, H.: Simulated Electric Arc Furnace Voltage Flicker Mitigation with 3-level Current-controlled STATCOM. In: Proc. IEEE Applied Power Electronics Conf. Expo., pp. 1697–1701 (2008)
4. Ladoux, P., Postiglione, G., Foch, H.: A Comparative Study of AC/DC Converters for High-power DC Arc Furnace. In: IEEE Trans.on Industrial Electronics, pp. 747–757 (2005)
5. Flores, P., Dixon, J., Ortuzar, M.: Static Var Compensator and Active Power Filter with Power Injection Capability, Using 27-level Iinvertersand Photovoltaic Cells. IEEE Trans.on Industrial Electronics, 130–138 (2009)
6. Jiang, Y., Cao, Y., Gong, Y.: Research on the Cascade Multilevel Inverter based on Different Carrier Phase-shifted Angle. Proceedings of the CSEE, 76–81 (2007)
7. Jiang, X., Xiao, X., Yin, Z., Ma, Y.: Harmonic Analysis for Cascade Multi-level Converter based on Carrier-Shifted SPWM. Power Electronics, 57–59 (2005) (in Chinese)

High-Power Controllable Current Quality Disturbance Generator

Gu Ren and Xiao Xiangning

Key Laboratory of Power System Protection and Dynamic Security Monitoring and Control of Ministry of Education
North China Electric Power University, 102206, Beijing, China

Abstract. This paper proposes a high-power controllable current quality disturbance generator, which is composed of multiplex cascade H bridge modules, and simulates current quality problems, such as positive and negative current distrubance, current fluctuation, harmonic current and so on. The device adopts a topology with decoupling fundamental and harmonic current. A power module parallel-connected on the fundamental connection reactance outputs harmonic current, while cascade fundamental power modules generate fundamental and low order harmonic current, which realizes decoupling control of fundamental and harmonic current. This paper proposes the control strategies of all types of current disturbances, and the EtherCat-based controller. Finally, a 10kV/1MVA controllable current quality disturbance generator has been created. The experiment result proves that the proposed main circuit topology, control strategy and EtherCat-based controller are feasible and efficient. The emergence of this device creates a fundamental platform for measurement, analysis, evaluation and control of current quality in grid.

Keywords: controllable current quality disturbance generator, multiplex cascade, decoupling control, EtherCat.

1 Introduction

The loads in modern power system develop with increasingly complicated categories and huge load capacity, especially the access of large wind turbines, high-speed locomotives and urban rail traffic to grid, causing new problems [1]-[2]. To analyze the influence of loads to modern power system in a more comprehensive and efficient way, it is of great necessity to research paralleling power quality disturbance generating technology [3]-[4].

With the improvement of power quality technology and the diversification of products, it seems necessary to analyze the device performance under power disturbance. For example, to check a new power quality analysis theory, to test a power quality analyzer, and to evaluate the endurance capability of a device or a production line to power quality problems, all need power quality disturbance generators, which can generate all types of power quality disturbances and have enough capacity.

2   Main Circuit Topology

Current quality disturbance generator parallelly connects power system. The main circuit topology is shown in Fig. 1, which is devided into fundamental current generating part and harmonic current generating one. The rectifier of the device is composed of rectifier transformer, rectifier inductance and H bridge rectifier, which stabilizes the DC voltage of modules by transferring power with system. The advantage of this main circuit topology is its decoupling control of fundamental and harmonic current. A power module parallel-connected on the fundamental connection reactance outputs harmonic current, while 12 cascade fundamental power modules generate fundamental and low order harmonic current, which realizes decoupling control of fundamental and harmonic current, and improves the efficiency and performance of device.

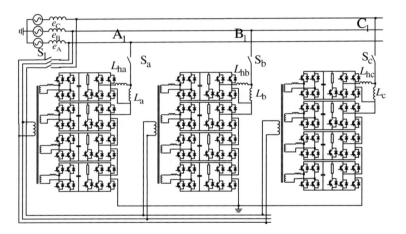

Fig. 1. The main circuit topology of the controllable current quality disturbance generator.

The A phase circuit is analyzed below. The main circuit adopts carrier phase shifting control with cascade H bridges, working in voltage source mode. The harmonic power module adopts hysteresis current control, working in current source mode. Therefore, the main circuit can be simplified into the equivalent circuit shown in Fig. 2. U_s is system voltage source. Z_s is system impedance. i_h is harmonic current source. U_f is fundamental voltage source. Z_f is fundamental connection impedance.

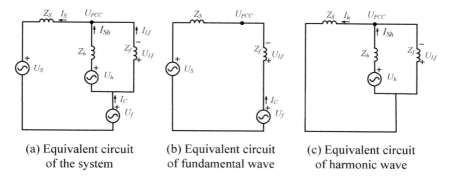

(a) Equivalent circuit of the system (b) Equivalent circuit of fundamental wave (c) Equivalent circuit of harmonic wave

Fig. 2. Equivalent circuit.

Fundamental current i_C and harmonic current i_h can be deduced respectively from the equivalent circuit of fundamental wave and harmonic wave. The total current i_s can be deduced by superposition theorem.

$$i_S = i_C + i_h = \frac{U_S - U_f}{Z_S + Z_f} + \frac{Z_f}{Z_f + Z_S} i_{Sh} \tag{1}$$

3 Control Strategy

3.1 Carrier Phase Shifting Control

The PWM modulation in this paper adopts cascade unipolar carrier phase shifting. Carrier phase shifting SPWM modulates multilevel converter by equally shifting carrier waves for a certain phase [5]. In this way, higher equivalent shifting frequency can be obtained in lower device shifting frequency. According to the difference among calculation methods of carrier phase shifting, there are two modulation strategies in practical application: phase shifting 180°/N and phase shifting 360°/N(N is cascade number). The paper [6] shows that phase shifting 180°/N has a smaller harmonic content of output voltage. Therefore, this paper adopts the former one.

3.2 Overall Control Strategy

According to the difference of module function, the main circuit of controllable current quality disturbance generator can be devided into three types of modules, including fundamental power modules, harmonic power modules and rectifier modules. Fundamental power module is to generate fundamental frequency current disturbances and those less than fundamental frequency. Harmonic power module is to generate current disturbances more than fundamental frequency. Rectifier module is to charge each DC capacitor of H bridge to stablize DC voltage, working in single phase rectifier mode.

Harmonic current injects system with inductance and H bridge modules series-connected. Harmonic module parallel-connects on the fundamental connection reactance, which realizes decoupling control of fundamental and harmonic current. As long as the voltage generated from harmonic module is bigger than the one of fundamental connection reactance, demand harmonic current can be generated by hysteresis current control or triangular comparison control.

3.3 Fundamental Power Module Control Strategy

The single phase electrical principle of inverter is shown in Fig. 2(a). Harmonic power module can stop working by locking its IGBT. Fundamental power module is in charge to output fundamental positive and negative sequence current and fluctuation current. The output current is controlled by dq decoupling method, which is fast and stable. When only generating fundamental positive sequence current, d-axis current and q-axis current in control system can be completely decoupled. When generating unbalance current and fluctuation current, only positive sequence component can be decoupled, and the other components can not, which does not affect the control effect. The control strategy of fundamental power module is shown in Fig. 3.

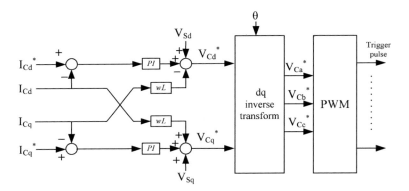

Fig. 3. Control strategy of fundamental power module.

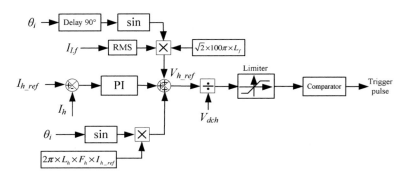

Fig. 4. Triangular wave comparison control strategy of harmonic power module.

3.4 Harmonic Power Module Control Strategy

Harmonic current control is to make the device output rapidly follow the harmonic demand current, the principle of which is similar to single-phase active power filter. To fix swiching frequency and simplify control, the output current of harmonic power module is controlled by triangular wave comparison, which is shown in Fig. 4. I_{h_ref} and I_h are output demand current and actual output current. I_{Lf} is the current on fundamental connection inductance. θ_i is the phase of I_{Lf} in every phase.

4 Controller Based on Series Real-Time EtherCat

EtherCat is an open real-time protocol for Ethernet communication, which has been widely used in the fields such as wind power generation, automatic production lines. Its terminal is similar to the central dispatch of railway, and EtherCat bus is similar to a high-speed train. The train can exchange signals with each station mutually, according to dispatcher's demands. Therefore, EtherCat has advantages of good real-time communication, no conflict and high tranfer rate.

The device proposed in this paper adopts a layered and distributed controller based on EtherCat. The main controller is in charge of human-computer interaction, information collection and distrubance calculation. The unit controller is in charge of module protection and PWM wave generation. EtherCat transfers signals and synchronizes time clock between the two types of controllers.

5 Experiments of 10kV/1MVA Industrial Prototype

Based on the topology, control strategy and EtherCat controller, a 10kV/1MVA controllable current quality disturbance industrial prototype has been created. This device is composed of 39 power modules. In each phase, there are twelve fundamental power modules with the swiching frequency 5Hz, and one harmonic power module with

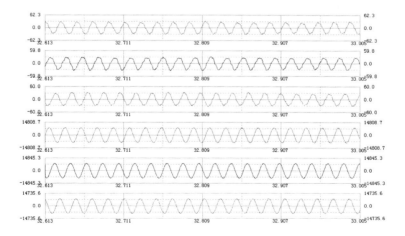

Fig. 5. Positive sequence current, 20A. The current leads the voltage 90°.

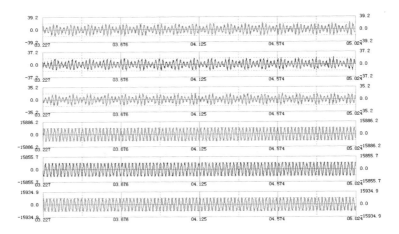

Fig. 6. Current fluctuation, 10Hz sine, 50%.

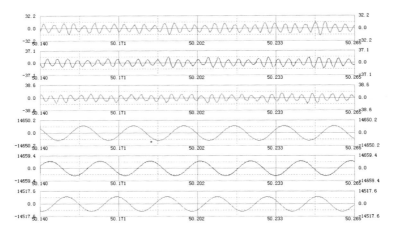

Fig. 7. Current harmonic, 5th.

the swiching frequency 2.5Hz. Expected effects have been achived. There are some of the results below. In each figure, the three waveforms on the top are the three-phase output current, and the three on the bottom are the three-phase system voltage.

6 Conclusion

This paper discusses the main circuit topology and control strategy of the high-power controllable current quality disturbance generator. Several types of disturbance experiment waveforms of the 10kV/1MVA controllable current quality disturbance generator are given to prove the effect. And the device has the features as follow.

- It adopts general controller based on series EtherCat. The controller has a high-speed communication rate, and its core control board has a good process performance. With the function of time clock synchronization, the device realizes distributed control of each power module. The controller is extendable, and can realize remote monitor and control. With great universal property, it can be applied in equipments containing half and fully devices, such as SVC, DSTATCOM, UPFC and so on.
- It adopts fundamental cascade H bridge modules and harmonic H bridge modules, realizing decoupling control of fundamental wave and harmonic wave, which is easy to accurately control harmonic current.
- It accesses the system without series transformer, eliminating the problems such as supersaturation, voltage distortion, voltage phase shifting, voltage drop, cost increasement, huge covering area by using series transformer.

References

1. Xiao, X., Xu, Y.: Power Quality Analysis and its Development. Power System Technology, 66–69 (2001) (in Chinese)
2. Poisson, O., Rioual, P., Meuneir, M.: Detection and Measurement of Power Quality Disturbances Using Wavelet Transform. IEEE Trans. Power Del., 1039–1042 (2000)
3. Lauttamus, P., Tuusa, H.: Simulated Electric Arc Furnace Voltage Flicker Mitigation with 3-level Current-controlled STATCOM. In: Proc. IEEE Applied Power Electronics Conf. Expo., pp. 1697–1701 (2008)
4. Yazdani, A., Crow, M.L., Guo, J.: An Improved Nonlinear Statcom Control for Electric Arc Furnace Voltage Flicker Mitigation. IEEE Trans. on Power Delivery, 2284–2290 (2009)
5. Ping, D., Kong, J., Chen, G.: The Research on Carrier Phase-shift SPWM and Circulating Currents in Parallel Converters. Power Inverter, 18–23 (2008) (in Chinese)
6. Jiang, X., Xiao, X., Yin, Z., Ma, Y.: Harmonic Analysis for Cascade Multi-level Converter based on Carrier-Shifted SPWM. Power Electronics, 57–59 (2005) (in Chinese)

Nonlinear Control of the Doubly Fed Induction Generator by Input-Output Linearizing Strategy

Guodong Chen[1,2], Luhua Zhang[2], Xu Cai[1], Wei Zhang[3], and Chengqiang Yin[4]

[1] Key Laboratory of Control of Power Transmission and Transformation,
Wind Power Research Center, School of Electronic Information and Electrical Engineering,
Shanghai Jiao Tong University, Minhang District, Shanghai 200240, China
[2] Technology Center, Shanghai Electric Power Transmission & Distribution Group,
No.270, Changning Rd, Shanghai, 200042, China
[3] Extra High Power Subcompany of Shandong Electric Power Corporation,
No.111, Wei Shi Rd, Jinan, 250021, China
[4] School of Automobile and Transportation Engineering Liaocheng University,
No.1, Hunan Rd, Shandong 252059, China

Abstract. With regard to a nonlinear system, the approximate linearized system is convenient to design the controller using method of linear systems, but it does not match to the nonlinear nature of the system. And that will result in degraded dynamic response. Vector control in doubly fed wind system aims for decoupled control of active and reactive powers and good dynamic performances. However, the active and reactive powers are only asymptotically decoupled. This paper addresses a nonlinear control algorithm based on the input-output linearizing and decoupling control strategy. The control strategy is tested by a 7.5 kW wind power generation test rig, and the experiment results validate that the input-output linearizing and decoupling control can provide improved dynamic responses and decoupled control of the DFIG in both steady and dynamic states.

Keywords: wind power; DFIG; decoupling control; input-output linearizing control.

1 Introduction

The doubly fed induction generator (DFIG) is a multiple-input multiple-output (MIMO) nonlinear system with strongly coupled variables. Generally the controller design theory is based heavily on the assumption that a linear model of a system will closely approximate the nonlinear behavior observed in reality[1]. A nonlinear controller would be able to deliver better results than the linear controller [2]. Once the rotor frequency ω_r becomes a state variable but a parameter of the DFIG system, the nonlinearity presented is the structural nonlinearity which is caused by products between the current components and the rotor frequency ω_r. The structural nonlinearity can only be mastered completely by nonlinear controllers [2-5].

This paper concentrates on design and implementation of DFIG control strategies based on the input-output linearizing and decoupling scheme to provide improved decoupled control and high-performance of dynamic response. Experiment results from a 7.5 kW wind power test rig, which involves the real time control system dSPACE acting as the host controller and SIEMENS 6RA70 DC master device performing as the wind turbine, are provided to demonstrate the feasibility of the proposed control strategy. Finally, conclusions are drawn.

2   The Input-Output Linearizing Control Law

Given the system:

$$\begin{cases} \dot{x} = f(x) + g \cdot u \\ y = h(x) \end{cases} \quad (1)$$

where x is state vector; u is the input; y is the output; f and g are smooth vector fields; h is smooth scalar function.

In order to obtain the input-output linearization of the multi-input multi-output system, the output y of the system is differentiated until the inputs appear.

$$\dot{y} = L_f h(x) + L_g h(x) u \quad (2)$$

Where $L_f h(x) = \partial h/\partial x \, f(x)$, $L_g h(x) = \partial h/\partial x \, g(x)$ represent Lie derivatives of $h(x)$ with respect to $f(x)$ and $g(x)$ respectively. If $L_g h(x) = 0$, then the input u does not appear and the output is differentiated respectively as

$$y^{(r)} = L_f^r h(x) + L_g L_f^{r-1} h(x) u. \quad (3)$$

Where r is the relative rank of y. If we perform the above procedure for each input y_i, we get a total of m equations in the above form, which can be written completely as

$$\begin{bmatrix} y_1^{(r)} & \cdots & y_m^{(r_m)} \end{bmatrix}^T = A(x) + E(x) \begin{bmatrix} u_1 & \cdots & u_m \end{bmatrix}^T \quad (4)$$

Where the $m \times m$ matrix $E(x)$ is defined as

$$E(x) = \begin{bmatrix} L_{g1} L_f^{r-1} h_1 & \cdots & \cdots & L_{gm} L_f^{r-1} h_1 \\ \cdots & \cdots & \cdots & \cdots \\ \cdots & \cdots & \cdots & \cdots \\ L_{g1} L_f^{r_m-1} h_m & \cdots & \cdots & L_{gm} L_f^{r_m-1} h_m \end{bmatrix}, \quad (5)$$

$$A(x) = \begin{bmatrix} L_f^r h_1 & \cdots & \cdots & L_f^{r_m} h_m \end{bmatrix}^T. \quad (6)$$

The matrix $E(x)$ is the decoupling matrix for the system. If $E(x)$ is nonsingular, then the original input u is controlled by the coordinate transformation:

$$u = -E^{-1}(x) A(x) + E^{-1}(x) v, \quad (7)$$

Where $v = \begin{bmatrix} v_1 & \cdots & \cdots & v_m \end{bmatrix}^T$.

Substituting (6) into (4) obtains a linear differential relation between the output y and the new input v

$$\begin{bmatrix} y_1^{(r)} & \cdots & y_m^{(r_m)} \end{bmatrix}^T = \begin{bmatrix} v_1 & \cdots & v_m \end{bmatrix}^T. \tag{8}$$

3 The Input-Output Linearizing Control of the DFIG

3.1 Mathematical Model of the DFIG

In the reference frame rotating at the synchronous angular speed, the stator and rotor voltage and flux vectors of a DFIG can be expressed as[6]

$$\begin{cases} u_s = R_s i_s + d\varphi_s/dt + j\omega_1 \varphi_s \\ u_r = R_r i_r + d\varphi_r/dt + j\omega_s \varphi_r \end{cases} \tag{9}$$

$$\begin{cases} \varphi_s = L_s i_s + L_m i_r \\ \varphi_r = L_m i_s + L_r i_r \end{cases}. \tag{10}$$

In the d-q reference frame

$$\begin{cases} u_{sd} = R_s i_{sd} + d\varphi_{sd}/dt - \omega_1 \varphi_{sq} \\ u_{sq} = R_s i_{sq} + d\varphi_{sq}/dt + \omega_1 \varphi_{sd} \\ u_{rd} = R_r i_{rd} + d\varphi_{rd}/dt - \omega_s \varphi_{rq} \\ u_{rq} = R_r i_{rq} + d\varphi_{rq}/dt + \omega_s \varphi_{rd} \end{cases} \tag{11}$$

$$\begin{cases} \varphi_{sd} = L_s i_{sd} + L_m i_{rd} \\ \varphi_{sq} = L_s i_{sq} + L_m i_{rq} \\ \varphi_{rd} = L_m i_{sd} + L_r i_{rd} \\ \varphi_{rq} = L_m i_{sq} + L_r i_{rq} \end{cases} \tag{12}$$

3.2 Rotor Side Controller Design

When the stator connects to a steady grid, the stator flux φ_s keeps constant. Therefore, $\varphi_s = u_s/\omega_1$, if the stator resistance is neglected.

In the stator flux field oriented frame

$$\begin{cases} \varphi_{sd} = |\varphi_s| = \varphi_s = U_s/\omega_1 \\ \varphi_{sq} = 0 \end{cases}. \tag{13}$$

Substituting (13) to (12) yields

$$\begin{cases} i_{sd} = \dfrac{1}{L_s}\varphi_s - \dfrac{L_m}{L_s} i_{rd} \\ i_{sq} = -\dfrac{L_m}{L_s} i_{rq} \end{cases}. \tag{14}$$

According (14) the direct and quadrature components of stator and rotor currents are linear dependent respectively, thus we chooses state vectors of the DFIG as follows

$$x = [x_1 \ x_2]^T = [i_{rd} \ i_{rq}]^T. \tag{15}$$

By substituting(12), (13) and (14), the following equations hold

$$\begin{cases} u_{rd} = R_r i_{rd} + \sigma L_r \dfrac{di_{rd}}{dt} - (\omega_1 - \omega_r)\sigma i_{rq} \\ u_{rq} = R_r i_{rq} + \sigma \dfrac{di_{rq}}{dt} + (\omega_1 - \omega_r)\sigma L_r i_{rd} \end{cases} \tag{16}$$

Where $\sigma = 1 - L_m^2/L_s L_r$

Arranging (16) in the form of (1)

$$\begin{cases} \dfrac{di_{rd}}{dt} = -\dfrac{R_r}{\sigma L_r} i_{rd} + \dfrac{\omega_1}{L_r} i_{rq} - \dfrac{\omega_r}{L_r} i_{rq} + \dfrac{u_{rd}}{\sigma L_r} \\ \dfrac{di_{rq}}{dt} = -\dfrac{R_r}{\sigma} i_{rq} - \omega_1 L_r i_{rd} + \omega_r L_r i_{rd} + \dfrac{u_{rq}}{\sigma} \end{cases} \tag{17}$$

Defining the input of the DFIG system

$$u = [u_1 \ u_2]^T = [u_{rd} \ u_{rq}]^T. \tag{18}$$

From(17), we have

$$\begin{cases} f_1(x) = -\dfrac{R_r}{\sigma L_r} i_{rd} + \dfrac{\omega_1}{L_r} i_{rq} - \dfrac{\omega_r}{L_r} i_{rq} \\ f_2(x) = -\dfrac{R_r}{\sigma} i_{rq} - \omega_1 L_r i_{rd} + \omega_r L_r i_{rd} \end{cases} \tag{19}$$

$$g = \begin{bmatrix} 1/\sigma L_r & 0 \\ 0 & 1/\sigma \end{bmatrix}. \tag{20}$$

Since the rotor side controller is set to decouple the active and reactive powers, the active and reactive powers of stator are selected as the output

$$y = \begin{bmatrix} y_1 \\ y_2 \end{bmatrix} = \begin{bmatrix} P_s \\ Q_s \end{bmatrix} = \begin{bmatrix} u_{sd} i_{sd} + u_{sq} i_{sq} \\ u_{sq} i_{sd} - u_{sd} i_{sq} \end{bmatrix}. \tag{21}$$

From (14) and (21)

$$\begin{cases} y_1 = \dfrac{\varphi_s}{L_s} u_{sd} - \dfrac{L_m}{L_s}(u_{sd} i_{rd} + u_{sq} i_{rq}) \\ y_2 = \dfrac{\varphi_s}{L_s} u_{sq} - \dfrac{L_m}{L_s}(u_{sq} i_{rd} - u_{sd} i_{rq}) \end{cases}. \tag{22}$$

Differentiating (22) until an input appears

$$\begin{cases} \dot{y}_1 = \dfrac{\dot{u}_{sd}}{L_s}(\varphi_s - L_m i_{rd}) - \dfrac{L_m}{L_s}\dot{u}_{sq}i_{rq} - \dfrac{L_m}{L_s}(u_{sd}f_1 + u_{sq}f_2) - \dfrac{L_m u_{sd}}{\sigma L_s L_r}u_{rd} - \dfrac{L_m u_{sq}}{\sigma L_s}u_{rq} \\ \dot{y}_2 = \dfrac{\dot{u}_{sq}}{L_s}(\varphi_s - L_m i_{rd}) + \dfrac{L_m}{L_s}\dot{u}_{sd}i_{rq} - \dfrac{L_m}{L_s}(u_{sq}f_1 - u_{sd}f_2) - \dfrac{L_m u_{sq}}{\sigma L_s L_r}u_{rd} + \dfrac{L_m u_{sd}}{\sigma L_s}u_{rq} \end{cases} \quad (23)$$

Rewriting (23) in the form of (4)

$$\begin{bmatrix} \dot{y}_1 \\ \dot{y}_2 \end{bmatrix} = A(x) + E(x)\begin{bmatrix} u_1 \\ u_2 \end{bmatrix}. \quad (24)$$

Where

$$A(x) = \begin{bmatrix} \dfrac{\dot{u}_{sd}}{L_s}(\varphi_s - L_m x_1) - \dfrac{L_m}{L_s}\dot{u}_{sq}x_2 - \dfrac{L_m}{L_s}(u_{sd}f_1 + u_{sq}f_2) \\ \dfrac{\dot{u}_{sq}}{L_s}(\varphi_s - L_m x_1) + \dfrac{L_m}{L_s}\dot{u}_{sd}x_2 - \dfrac{L_m}{L_s}(u_{sq}f_1 - u_{sd}f_2) \end{bmatrix}. \quad (25)$$

$$E(x) = \begin{bmatrix} -\dfrac{L_m u_{sd}}{\sigma L_s L_r} & -\dfrac{L_m u_{sq}}{\sigma L_s} \\ -\dfrac{L_m u_{sq}}{\sigma L_s L_r} & \dfrac{L_m u_{sd}}{\sigma L_s} \end{bmatrix}. \quad (26)$$

Since E(x) is nonsingular, the control scheme is given from (7) as

$$\begin{bmatrix} u_{rd} \\ u_{rq} \end{bmatrix} = E^{-1}(x)\left[-A(x) + \begin{bmatrix} v_1 \\ v_2 \end{bmatrix}\right]. \quad (27)$$

For a DFIG the demanded active power P_s^* is decided by the maximum power tracking scheme of the wind turbine according to the instant wind speed, and the demanded reactive power Q_s^* is set by the grid operator to support grid voltage. To follow the reference trajectory of P_s^* and Q_s^* with a certain dynamic, we use proportional-integral controller imposing to the linearized system stable poles for the desired performances. Hence, the new input v can be calculated by

$$\begin{bmatrix} v_1 \\ v_2 \end{bmatrix} = \begin{bmatrix} \dot{y}_1^* - k_{p1}e_1 - k_{i1}\int e_1 dt \\ \dot{y}_2^* - k_{p2}e_2 - k_{i2}\int e_2 dt \end{bmatrix}. \quad (28)$$

Where e_1 is the error between the demanded and the achieved active power, and e_2 is that of the reactive power.

$$\begin{cases} e_1 = y_1^* - y_1 = P_s^* - P_s \\ e_2 = y_2^* - y_2 = Q_s^* - Q_s \end{cases}. \quad (29)$$

We design the error dynamics to be convergent and stable by selecting the PI coefficients K_p, K_i to satisfy the polynomial of Hurwitz[7]

$$\begin{cases} \ddot{e}_1 + k_{p1}\dot{e}_1 + k_{i1}e_1 = 0 \\ \ddot{e}_2 + k_{p2}\dot{e}_2 + k_{i2}e_2 = 0 \end{cases}. \tag{30}$$

Equations (27) and (28) compose the controller of the rotor side converter. The structure of the controller is shown in Fig. 1.

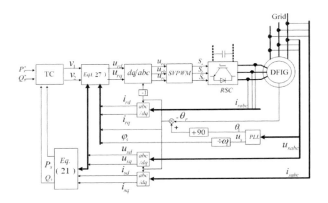

Fig. 1. Control diagram of DFIG using input-output linearizing control

4 Experiment Results

4.1 Responses to Step Change in Demanded Active Power

In Fig. 2 the active power of the DFIG experiences a step change from 3.75 kW to 7.5 kW, and the reactive power is maintained at 0. Fig. 2a shows that the stator phase voltage u_s and current i_s keep in phase and the current shifts in approximately one circle to 2 times the original amplitude. And at the start point of the step and recovers when i_s and the rotor current i_r increases in about one circle, which is illustrated in Fig. 2b.

Appling the vector control scheme in the test rig[8], results derived are presented in Fig. 2c and Fig. 2d. It can be gained that the response to step change in active power under vector control is slower than that under the input-output linearizing control scheme—it will take about three circles to reach the set point.

The achieved active and reactive powers P, Q in dynamic process under different control strategies are recorded in the dSPACE platform, shown in Fig. 3. In Fig. 3a the achieved reactive power is affected by the step change of active power and shifts in the dynamic process when using vector control strategy. That reveals coupled relations between the active and reactive powers in dynamic state.

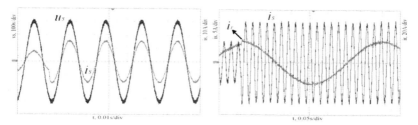

(a) Stator voltage and current-nonlinear control (b) Stator and rotor current-nonlinear control

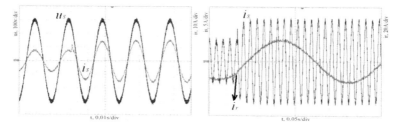

(c) Stator voltage and current-vector control (d) Stator and rotor current-vector control

Fig. 2. Responses to step change in the active power

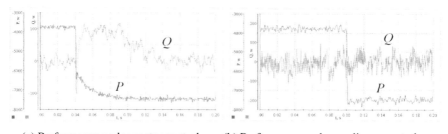

(a) Performance under vector control   (b) Performance under nonlinear control

Fig. 3. Responses of the active and reactive powers to step change in demanded active power

5 Conclusions

Nonlinear controllers are designed and proved to solve the structural nonlinear problems well. In this paper the input-output linearizing and decoupling control algorithm for the DFIG decoupling control is presented. The designed controller is implemented and tested in the laboratory on a 7.5 kW DFIG based wind turbine test rig. The experiment results from vector control and the input-output linearizing control are compared to validate that the input-output linearizing strategy can obtain decoupled control of the powers even in dynamic state and also responses much faster than that of the vector control.

Acknowledgement

The authors would like to acknowledge the Shanghai Government for supporting this research under the Project 10dz1203901 and 11QB1401500.

References

1. Hand, M.M., Balas, M.J.: Systematic controller design methodology for variable-speed wind. National Renewable Energy Laboratory, Golden, Colorado (2002)
2. Quang, N.P., Dittrich, J.-A.: Vector control of three-phase AC machines. Springer, Heidelberg (2008)
3. Xu, L., Yao, L., Sasse, C.: Grid integration of large DFIG-based wind farms using VSC transmission. IEEE Transactions on Power Systems 22(3), 976–984 (2007)
4. Byrnes, C.I., Isidori, A., Willems, J.C.: Passivity feedback equivalence and the global stabilization of minimum phase nonlinear systems. IEEE Transactions on Automatic Control 36(11), 1228–1240 (1991)
5. Benchaib, A., Rachid, A., Audrezet, E.: Sliding mode input-output linearization and field orientation for real-time control of induction motors. IEEE Transactions on Power Electronics 14(1), 3–13 (1999)
6. Krause, P.C.: Analysis of Electric Machinery. McGraw-Hill, New York (1986)
7. Boukas, T.K., Habetler, T.G.: High performance induction motor speed control using exact feedback linearization with state and state derivative feedback. IEEE Transactions on Power Electronics 19(4), 1022–1028 (2004)
8. Pena, R., Clare, J.C., Asher, G.M.: Doubly fed induction generator using back-to-back PWM converters and its application to variable-speed wind-energy generation. IEEE Proceedings: Electric Power Applications 143(3), 231–241 (1996)

Passive Location Method for Fixed Single Station Based on Azimuth and Frequency Difference Measuring

Tao Yu

51st Research Institute of CETC, Shanghai 201802, China
tyt0803@163.com

Abstract. On existing foundation of DF with single station, if the target is uniform motion alone straight line, the instantaneous path angle of target can be derived from the specific value of FDOA between adjacent detective nodes with making three continuous frequency measurements and after eliminating unknown velocity and wavelength. Then, the Doppler shift, wavelength, angle of advance and velocity can be solved successively. On this basis, provided approximately that incident wave of marker signal from target is regard as parallel beam, the functional relation between angular velocity and Doppler frequency difference can be obtained after making a deformation of differential operation for Doppler shift equation by introducing the concept of angle change rate. Thus, the calculating formula solving relative radial distance between target and fixed station can be determined. The simulation analysis shows that derived analytic formula are all correct.

Keywords: DF, Doppler shift, single station location, passive location, course angle, speed measure.

1 Introduction

Based on classical technique of DF and TOA location, fixed single-station can obtain the positional and motive information about the target by multipoint successive tracking [1]. In order to achieve the instantaneous locating for the target that is approximatively uniform motion, the location system need to simultaneously possess the functions measuring the angle as well as rate and the centrifugal acceleration [2, 3]. Theoretically, the positioning accuracy and rate of convergence can be increased by introducing the information about the angular velocity. Moreover, the location mode measuring phase change has the fast and accurate advantage [4-6]. But this is at the cost of increasing the complexity and difficulty in measuring. At the meantime, the measuring accuracy of angular velocity is generally required to mrad/s degree. In fact, such requirement had restricted the application. This is not only to put forward the higher requirement for system design, but also to increase the cost.

The passive location with fixed single-station which combines the beam direction from target and the measuring for Doppler shift is a current research subject. Not only can this method locate the place of moving target, but it also can estimate the velocity and flight path of moving target [7]. But, Doppler shift equation is a nonlinear equation which contains both location parameter and kinematical parameter. If we carry out directly the analysis of positioning and tracking making use of Doppler shift equation in rectangular coordinate system, the many unknown parameters may be introduced. For this reason, relative complex resolving method has to be adopted.

Different from existing method, the one presented in this paper have as following characteristic: (1) In the polar coordinate system, resolve directly the frequency and kinematical parameter of target based on two Doppler frequency difference equation between adjacent acquisition points. (2) According to the principle of angular rate, derive directly the calculating formula of distance between target and station on the basis that the course angle and speed have been determined.

2   Ratio of Frequency Difference

Provided that the target is uniform motion alone straight line in two-dimension plane, for the locating mode based on the azimuth angle-Doppler frequency difference with single station as depicted in fig.1, the station can obtain the azimuth angle and radiation frequency by tracking the target. The measurement for the azimuth angle is as bench mark in the north direction.

The Doppler shifts received by detective station in different moment are separately:

$$\lambda f_{di} = v_T \cos \beta_i \qquad (i = 1, 2, 3) \tag{1}$$

in where: λ is signal wavelength radiated by target; v_T moving speed of target; β_i angle of advance between radial distance and target flight direction.

According to the geometric relationship as shown in fig.1, substituting the relationship between interior angle and external angle $\beta_i = \alpha - \theta_i$ into Doppler formula yields:

$$\lambda f_{di} = v_T \cos(\alpha - \theta_i) \tag{2}$$

in where: α is instantaneous course angle of target; θ_i azimuth angle at the end of detective station.

By consequent measuring Doppler shift at least three times, we can obtain two equations of Doppler frequency difference:

$$\lambda(f_{d2} - f_{d1}) = v_T [\cos(\alpha - \theta_2) - \cos(\alpha - \theta_1)] \tag{3}$$

$$\lambda(f_{d3} - f_{d2}) = v_T [\cos(\alpha - \theta_3) - \cos(\alpha - \theta_2)] \tag{4}$$

After eliminating the moving speed and signal wavelength of target by using to ratio two equations, we have:

$$\frac{f_{d2} - f_{d1}}{f_{d3} - f_{d2}} = \frac{\cos(\alpha - \theta_2) - \cos(\alpha - \theta_1)}{\cos(\alpha - \theta_3) - \cos(\alpha - \theta_2)} \tag{5}$$

During practicable engineering design, Doppler frequency difference may be replaced by the measured value of radiation frequency:

$$\Delta f_i = f_{d(i+1)} - f_{di} = f_{t(i+1)} - f_{ti} \tag{6}$$

in where: Δf_i is frequency difference; f_{ti} the measured value of frequency radiated by target.

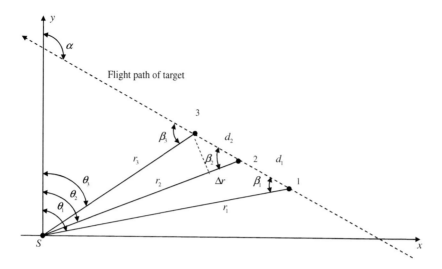

Fig. 1. Geometric model

3 Course Angle

The instantaneous course angle can be resolved from (5):

$$tg\alpha = \frac{p_f(\cos\theta_3 - \cos\theta_2) - (\cos\theta_2 - \cos\theta_1)}{(\sin\theta_2 - \sin\theta_1) - p_f(\sin\theta_3 - \sin\theta_2)} \qquad (7)$$

in where: $p_f = \dfrac{f_{d2} - f_{d1}}{f_{d3} - f_{d2}} = \dfrac{\Delta f_{d1}}{\Delta f_{d2}} = \dfrac{\Delta f_{t1}}{\Delta f_{t2}}$

The analog computation shows that the instantaneous course angle has ambiguity. In order to overcome the ambiguity, the locate system has to detect some times continuously during practicable application so that the traveling direction of target can be probably determined. The elementary calculating shows that the course angle may be calculated directly by Eq.(7) upper half-plane in two-dimensional space within $0° \leq \theta \leq 90°$ limits. But also, the angle of advance is:

$$\beta_i = \alpha - \theta_i \qquad (8)$$

Within $-90° \leq \theta \leq 0°$ limits, the angle of advance is:

$$\beta_i = -(\alpha - \theta_i) \qquad (9)$$

If the calculated value for angle of advance is plus, it indicates that the angle of advance and azimuth angle are on same side relative to radial ray. Otherwise, if the calculated value for angle of advance is negative, it indicates that the two angles are situated on the both sides of radial ray. Moreover, stipulating for: (1) the anticlockwise direction is plus for angle of advance; (2) the clockwise direction is plus for azimuth angle.

Provided that the flight speed of target $v = 100 \ m/s$, the flight distance $d = 1000 \ m$, the radial distance $r_1 = 100 \ km$, signal wavelength $\lambda = 0.15 \ m$, the fig.2 shows that the change curve of functional relationship between the angle of advance and course angle within $135^\circ \leq \beta_1 \leq 45^\circ$ limits at $\theta = 45^\circ$. At the same time, the analysis of relative error for course angle proves that the derived formula is all accurate.

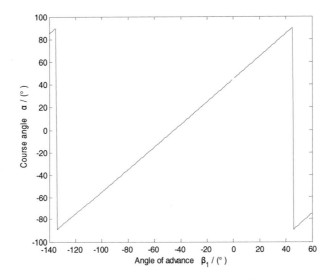

Fig. 2. Curve between the angle of advance and course angle

4 Velocity

As soon as the course angle is obtained, the angle of advance can be determined. Hence, the other parameter can be determined in proper order. As a transition in problem solver, firstly the ratio between speed and wavelength can be obtained from Eq.(3) or (4):

$$\frac{v_T}{\lambda} = \frac{\Delta f_{dij}}{\cos(\alpha - \theta_i) - \cos(\alpha - \theta_j)} \qquad (10)$$

As soon as this ratio is replaced into each Doppler shift equation, the Doppler shift value can be calculated:

$$f_{di} = \frac{v_T}{\lambda} \cos \beta_i = \frac{\Delta f_{dij} \cos(\alpha - \theta_i)}{\cos(\alpha - \theta_i) - \cos(\alpha - \theta_j)} \qquad (11)$$

Further, according to the measured value of frequency, the centre frequency of signal can be solved:

$$f_0 = f_t \pm f_d \tag{12}$$

Thus, the wavelength can be determined. Furthermore, again making use of the ratio between speed and wavelength, the target speed can be obtained from Doppler shift:

$$v_T = \frac{\Delta f_{dij}}{\cos(\alpha - \theta_i) - \cos(\alpha - \theta_j)} \lambda \tag{13}$$

Analog calculations verify that this formula is errorless. Fig.3 shows the curve of relative error of speed.

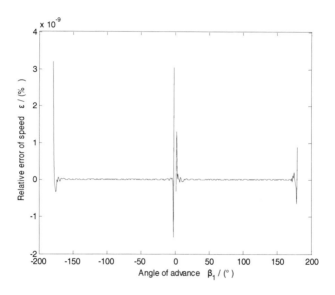

Fig. 3. Relative error of speed

5 Distance

As soon as both wavelength and Doppler shift is obtained, the radial distance of target can be directly resolved by angle rate. Firstly, by actualizing the differential modification for Doppler shift equation, the Doppler shift can be expressed by sinusoidal rate of the angle of advance:

$$\lambda f_{di} = \frac{v_T}{\omega_i} \frac{\partial \sin \beta_i}{\partial t} \tag{14}$$

in where: $\omega = v_{ti}/r_i$ is angular velocity; v_{ti} tangential velocity; r_i radial distance.

Analogously regarding the signal radiated from target as parallel wave, then according to the geometric relationship at dotted line closed to the radial distance r_2 as shown in fig.1, we has analogously:

$$\sin \beta_2 \approx \sqrt{d_2^2 - \Delta r^2}/d_2$$

in where: $\Delta r = r_2 - r_3$ is path length difference.

Also according to $\dot{r}_i = v_{ri} = \lambda f_{di}$, we can obtain:

$$\lambda f_{d1} = \frac{v_T}{\omega_2}\frac{\partial \sin \beta_2}{\partial t} = \frac{v_T}{\omega_2}\frac{\Delta r}{d_2}\frac{\lambda \Delta f_{d2}}{\sqrt{d_2^2 - \Delta r^2}} \quad (15)$$

in where: $\Delta f_{d2} = f_{d2} - f_{d3}$.

According to $ctg\beta_2 = \Delta r/\sqrt{d_2^2 - \Delta r^2}$ and $\lambda f_{d2} = v_T \cos \beta_2$, the functional relation between angular velocity and Doppler frequency difference can be obtained by arranging Eq.(15):

$$\omega_2 d_2 \sin \beta_2 = \lambda \Delta f_{d2} \quad (16)$$

Substituting $\omega_2 = v_{t2}/r_2$ and $d_2 = v_T \Delta t$ into the equation, the radial distance between fixed single station and target can be obtained:

$$r_2 = \frac{d_2 v_{t2} \sin \beta_2}{\lambda \Delta f_{d2}} = \frac{\Delta t \left[v_T^2 - (\lambda f_{d2})^2\right]}{\lambda \Delta f_{d2}} \quad (17)$$

in where: Δt is analogously the time interval which the station detects the target with consequent and fixed cycle.

The relative error curve of radial distance at $r_1 = 100$ km is shown in fig.4. within $0^\circ - 90^\circ$ limits. Obviously, the shorter the time period, the less the relative error.

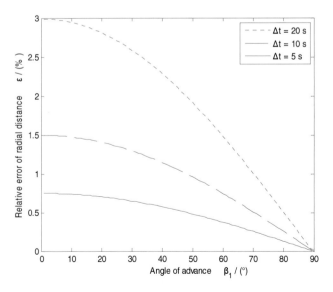

Fig. 4. Relative error of radial distance

6 Conclusion

The measuring for Doppler frequency difference can be directly changed into the one for measured frequency difference. And this is advantageous for engineering application. Not only the form measuring frequency difference makes detecting method simplified, but it is also more advantageous for increasing the measurement precision of the locate system. Even though the domestic technique of frequency measurement remain to increase further at the present time, measuring frequency difference should be a better method for measurement precision and should be easier to use in comparison to directly measuring Doppler shift.

Adopting the measurement mode of frequency difference is more favorable to passive sounding that can make locate system directly measure the kinematical parameter of target when the wavelength is unknown. And then, the wavelength can be solved based on the information of the azimuth and frequency difference. at the meantime, not only the directing ranging formula can be used for fixed single station, but also can be used for airborne and satellite-borne platform to carry out the passive and active ranging.

Strictly speaking, on condition that it is also difficult to improve the measurement accuracy both frequency and angle, the result derived in this paper is only some estimating formula about the speed and distance. But, the research meaning of this paper is further to provide the technical support for the applied analysis that have higher positioning accuracy and to improve the available tracing algorithm of passive location based on Doppler shift.

References

1. Sun, Z., Guo, F., Feng, D.: Passive location and tracking technology by single observer. National Defence Industry Press, Beijing (2008)
2. Wan, F., Ding, J., Tian, J.: Method of Fixed Single-site Passive Detection Location and Its Applications. Journal of Air Force Radar Academy 24(3), 160–162 (2010)
3. Wan, F., Ding, J.: Implementation of Single-site Passive Location by Using Multi-station Passive Location Systems. Journal of Air Force Radar Academy 24(2), 87–90 (2010)
4. Cheng, D., Li, X., Wan, S.: Single Passive Radar Location Based on Measurement Subset of Direction and Its Changing Rate. Modern Defence Technology 37(4), 132–136 (2009)
5. Liu, Y., Dou, X.: Research on Single Observer Passive Location Technology Based on Phase Difference Change Rate. Radio Engineering 40(6), 48–50 (2010)
6. Huang, D., Ding, M.: Introduction to Passive Location Using Phase Rate of Change. Modern Radar 29(8), 32–34, 51 (2007)
7. Ding, W., Ma, Y.: Research on Signal Observer Passive Location. Command Control & Simulation 30(1), 35–37 (2008)
8. Wan, F., Ding, J., Yu, C.: Fixed Single Observer Passive Detection and Location Using Spatial-Frequency Domain Information. Journal of Detection & Control 32(3), 91–95 (2010)

Online Predicting GPCR Functional Family Based on Multi-features Fusion

Wang Pu and Xiao Xuan

Dept of Machine and Electron, Jing-De-Zhen Ceramic Institute,
333403, Jing-De-Zhen, China
wp3751@163.com

Abstract. GPCRs are a superfamily of very important signal molecular receptors; there are many studies about the relationship between the structure and function of them. Because it is both time-consuming and laborious to acquire the functional information by experiment, this paper tries to resolve this problem by pattern recognition. The specific idea is to extract many sequence-derived features, design a predictor based on fuzzy K-nearest neighbor algorithm, and subject it to strict test, the overall success rate of 95.87% and 99.14% are achieved by jackknife and independent test. It indicates that the method of feature extracting and predictor design is successful, and this method can be used as an effective experiment added. For convenience, a user-friendly web server has been established at http://icpr.jci.edu.cn/wangpu/GPCRpred.

Keywords: GPCR; Features fusion; Predictor design; Web-server.

1 Introduction

G protein-coupled receptors (GPCRs), also known as seven-transmembrane domain receptors, 7TM receptors, heptahelical receptors, serpentine receptor, and G protein-linked receptors (GPLR), comprise a large protein family of transmembrane receptors that sense molecules outside the cell and activate inside signal transduction pathways and, ultimately, cellular responses [1]. G protein-coupled receptors are involved in many diseases, and are also the target of approximately 30% of all modern medicinal drugs [2].

Despite the lack of sequence homology between classes, all GPCRs share a common structure and mechanism of signal transduction. GPCRs can be grouped into 6 classes based on sequence homology and functional similarity: (1) Class A (Rhodopsin-like), (2) Class B (Secretin receptor family), (3) Class C (Metabotropic glutamate/pheromone), (4) Class D (Fungal mating pheromone receptors), (5) Class E (Cyclic AMP receptors) and (6) Class F (Frizzled/Smoothened) [3].

Facing the avalanche of protein sequence data generated in the Post-Genomic Age, it is highly desired to develop computational methods that can rapidly and effectively identify the functional families of GPCRs based on their primary sequences so as to provide useful information for classifying drug targets.

The simplest model used to represent a protein sample is its amino acid composition or AAC, which was widely used for predicting various protein attributes. However, the prediction quality might be considerably limited since all the sequence-order information would be lost accordingly. To avoid completely lose the sequence-order information, the pseudo amino acid composition or PseAAC was proposed [4]. The concept of PseAAC has provided a very flexible mathematical frame for incorporating various effects into the protein sample formulation, and hence been widely used to deal with many protein-related problems and sequence-related systems [5]. In this paper, we will apply multiple sequence-derived pseudo amino acid compositions (PseAAC). These features are capable of capturing information about amino acid composition, sequence order as well as various physicochemical properties of proteins. The resulting feature vectors are finally fed into a simple yet powerful classification algorithm, called fuzzy K nearest neighbor, to predict the functional types of GPCRs.

2 Dataset Construction

Protein sequences were collected from the G protein-coupled receptor database (GPCRDB release 10.0) at http://www.gpcr.org [6], which is a molecular class-specific information system that collects, combines, validates, and disseminates heterogeneous data on GPCRs. The original dataset contains 2,604 receptors, of which 1,884 belong to Class A, 309 to Class B, 206 to Class C, 65 to Class D, 10 to Class E, and 130 to Class F.

To avoid homology bias, a redundancy cutoff was imposed with the program CD-HIT [7] to winnow those sequences which have $\geq 60\%$ pairwise sequence identity to any other in a same subset. Finally we obtained a training dataset that contains 751 protein sequences with low homology, of which 526 belong to Class A, 82 to Class B, 77 to Class C, 34 to Class D, 4 to Class E, and 28 to Class F (Table 1). Then the filtered sequences make up the independent dataset for avoiding the predictor is overtrained.

Table 1. Breakdown of benchmark dataset.

Class	A	B	C	D	E	F	Overall
Training dataset	526	82	77	34	4	28	751
Independent dataset	1358	227	129	31	6	102	1358

3 Features Extraction and Fusion

A protein sequence $\mathbf{P}$ with L amino acid residues could be expressed as

$$\mathbf{P} = R_1 R_2 R_3 R_4 R_5 R_6 \cdots R_L \tag{1}$$

Then how to extract applicable features from the GPCR sequence for functional type prediction problem plays a key role in this study. In order to capture as much information of protein sequences as possible, a variety of statistic and physicochemical properties are used in the procedure of feature extraction.

3.1 Amino Acid Composition (AAC)

$$AAC = [f_1, f_2, \cdots, f_{20}]^T \quad (2)$$

where $f_i = n_i/L$, in which each $i(=1, 2, \cdots, 20)$ corresponds to a distinct amino acid and n_i is the number of amino acid i occurring in the protein sequence of length L, T is the transpose operator, same as follow.

3.2 Dipeptide Composition (DC)

To observe the interaction of the amino acid pairs with gaps, different orders dipeptides were generated as a conditional probability matrix of 20×20,

$$DC_d = \begin{bmatrix} P(A/A) & P(C/A) & \cdots & P(Y/A) \\ P(A/C) & P(C/C) & \cdots & P(Y/C) \\ \vdots & \vdots & \vdots & \vdots \\ P(A/Y) & P(C/Y) & \cdots & P(Y/Y) \end{bmatrix} \quad (3)$$

where $d = 1, 2$, or larger, and $P(a_1/a_2)$ indicates the probability that amino acid a_1 and a_2 with d gaps between the two residues appear at the same time.

Because the local residues couple strongly, here we choose $d = 1$ and $d = 2$, thus obtain 800 dipeptide compositions.

3.3 Complexity Factor (CF)

A protein sequence is actually a symbolic sequence for which the complexity measure factor can be used to reflect its sequence feature or pattern and has been successfully used in some protein attribute prediction [8]. Among the known measures of complexity, the Lempel-Ziv (LZ) [9] complexity reflects the order that is retained in the sequence, and hence was adopted in this study.

3.4 Fourier Spectrum Components (FSC)

Given a protein sequence $\mathbf{P}$, suppose $H(R_1)$ is the certain physicochemical property value of the 1st residue R_1, $H(R_2)$ that of the 2nd residue R_2, and so forth. In terms of these property values the protein sequence can be converted to a digit signal string $[H(R_1), H(R_2), \cdots, H(R_L)]$, for which we implement the discrete Fourier transform, obtaining the frequency domain values,

$$\{F_1, F_2, \cdots, F_L, \Phi_1, \Phi_2, \cdots, \Phi_L\} \quad (4)$$

where the amplitude component F_k and phase component Φ_k $(k=1,2,\cdots,L)$ can be calculated as below,

$$X[k] = \sum_{l=1}^{L} H(R_l) \exp\left[-j(\frac{2\pi l}{L})k\right] = F_k \exp(j\Phi_k) \tag{5}$$

where j represents the imaginary number.

The 2L Fourier spectrum numbers contain substantial information about the digit signal, and thereby can also be used to reflect characters of the sequence order of a protein. Furthermore, in the L phase components, the high frequency components are noisier and hence we only need to consider the 1st 10 phase components.

Then what physicochemical property values to choose? Here we adopt three kinds that have been used in the concept of Chou's pseudo amino acid composition [5], which have been used widely to predict various attributes of proteins, they are hydrophobicity, hydrophilicity and side-chain mass. Thus we can obtain 60 Fourier spectrum components.

3.5 Features Fusion

In summary, a comprehensive set of 881 features will be generated. These features are used to represent every protein sequence, however, different kinds of features are on different scales, thus the raw feature vector need to be standardized, and a protein sequence can be expressed by a standard vector as,

$$X = [x_1, x_2, \cdots, x_{881}]^T \tag{6}$$

where

$$x_k = \begin{cases} \dfrac{f_k}{\sum_{i=1}^{20} f_i + \sum_{j=1}^{841} w_j p_j}, & (1 \leq k \leq 20) \\[2ex] \dfrac{w_{(k-20)} p_{(k-20)}}{\sum_{i=1}^{20} f_i + \sum_{j=1}^{841} w_j p_j}, & (21 \leq k \leq 861) \end{cases} \tag{7}$$

where f_i $(i=1,2,\cdots,20)$ are the amino acid composition, $p_j(j=1,2,\cdots,841)$ are the pseudo amino acid composition, which could be dipeptide composition, complexity factor or Fourier spectrum components, and w_j are the weight factors, here we choose $w=20$ for DC, $w=10^{-3}$ for CF and $w=10^{-4}$ for FSC.

4 Classifier Design

Fuzzy K-NN classification method [10] is a special variation of the K-NN classification family. Instead of roughly assigning the label based on a voting from the nearest neighbors, it attempts to estimate the membership values that indicate how much degree the query sample belong to the classes concerned.

Suppose $\{\mathbf{P}_1, \mathbf{P}_1, \cdots, \mathbf{P}_N\}$ is a set of vectors representing N proteins in the training set which has been classified to M classes: $\{C_1, C_2, \cdots, C_M\}$, where C_i denotes the ith class. Thus, for a query protein $\mathbf{P}$, its fuzzy membership value for the ith class is given by:

$$\mu_i(\mathbf{P}) = \frac{\sum_{j=1}^{K} \mu_i(\mathbf{P}_j) d(\mathbf{P}, \mathbf{P}_j)^{-2/(\varphi-1)}}{\sum_{j=1}^{K} d(\mathbf{P}, \mathbf{P}_j)^{-2/(\varphi-1)}} \tag{8}$$

where K is the number of the nearest neighbors counted; $\mu_i(\mathbf{P}_j)$ is the fuzzy membership value of the protein $\mathbf{P}_j$ to the ith class (it is set to 1 if the real label of $\mathbf{P}_j$ is C_i; otherwise, 0); $d(\mathbf{P}, \mathbf{P}_j)$ is the distance between the query protein $\mathbf{P}$ and its jth nearest protein $\mathbf{P}_j$ in the training dataset; and φ (>1) is the fuzzy coefficient for determining how heavily the distance is weighted when calculating each nearest neighbor's contribution to the membership value. Various metrics can be chosen for $d(\mathbf{P}, \mathbf{P}_j)$ is Euclidean distance. After calculating all the memberships for a query protein, it is assigned to the class with which it has the highest membership value.

To provide an intuitive picture, a flowchart to show the process of how the classifier works is given in Fig. 1.

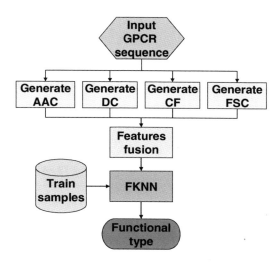

Fig. 1. Flowchart to show the process of how the classifier works

5 Results and Discussion

The leave-one-out (LOO) cross-validation is a rigorous and objective statistical test and is often used to examine the power of a new predictor. LOO involves using a single observation from the original sample as the validation data, and the remaining observations as the training data. This is repeated such that each observation in the sample is used once as the validation data.

The values of parameter φ and K used in (8) were determined by optimizing the overall LOO test success rate thru a 2-D grid search (Fig. 2). It was found that the highest overall LOO test success rate was obtained when $\varphi=1.21$ and $K=3$. With these optimized parameters, we further make independent test on the independent dataset.

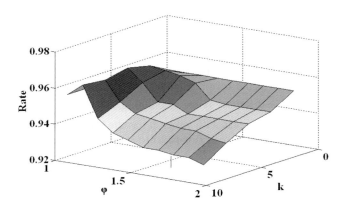

Fig. 2. Success rates by LOO test with different parameters.

Table 2. Prediction accuracies and sensitivitiess by LOO and independent test.

	Class A	Class B	Class C	Class D	Class E	Class F	Overall
LOO test	$\frac{516}{526}=98.10\%$	$\frac{75}{82}=91.46\%$	$\frac{74}{77}=96.10\%$	$\frac{28}{34}=82.35\%$	$\frac{4}{4}=100\%$	$\frac{23}{28}=82.14\%$	$\frac{720}{751}=95.87\%$
Independent test	$\frac{1358}{1358}=100\%$	$\frac{221}{227}=97.36\%$	$\frac{120}{129}=93.02\%$	$\frac{31}{31}=100\%$	$\frac{6}{6}=100\%$	$\frac{101}{102}=99.02\%$	$\frac{1837}{1853}=99.14\%$

The success rates by the LOO and independent test are given in Table 2, from which we can see that the overall success rate by the current approach is 95.95%, indicating that this approach can predict the functional types of GPCRs properly, or at least can play a complementary role to the existing method in identifying types of GPCRs. For the convenience of biologist and pharmacologist in using the predictor to screen and prioritize their research, the GPCRpred web-server has been designed by the technique of JSP. The web-sever is very simple to use, just input your query protein sequence in a frame and you will know the functional information.

Acknowledgments

The work in this research was supported by the grants from the National Natural Science Foundation of China (no. 60961003), the Key Project of Chinese Ministry of Education (no.210116), the Province National Natural Science Foundation of Jiangxi(no.2009GZS0064), the Department of Education of JiangXi Province (No. GJJ09271).

Referrences

1. Heuss, C., Ferber, U.: G-protein-independent signaling by G-protein-coupled receptors. Trends Neuro. Sci. 23, 469–475 (2000)
2. Overington, J.P., Al-Lazikani, B., Hopkins, A.L.: How many drug targets are there? Nat. Rev. Drug Discov. 5(12), 993–996 (2006)
3. Foord, S.M., Bonner, T.I., et al.: International Union of Pharmacology. XLVI. G protein-coupled receptor list. Pharmacol. Rev. 57(2), 279–288 (2005)
4. Chou, K.C.: Prediction of protein cellular attributes using pseudo-amino acid composition. Proteins 43, 246–255 (2001)
5. Chou, K.C.: Pseudo Amino Acid Composition and its Applications in Bioinformatics, Proteomics and System Biology. Current Proteomics 6, 262–274 (2009)
6. Horn, F., Weare, J., Beukers, M.W., et al.: GPCRDB: an information system for G protein-coupled receptors. Nucleic Acids Res. 26, 275–279 (1998)
7. Li, W., Godzik, A.: Cd-hit: a fast program for clustering and comparing large sets of protein or nucleotide sequences. Bioinformatics 22, 1658–1659 (2006)
8. Xiao, X., Shao, S.H., et al.: Using complexity measure factor to predict protein subcellular location. Amino Acids 28, 57–61
9. Gusev, V.D., Nemytikova, L.A., et al.: On the complexity measures of genetic sequences. Bioinformatics 15, 994–999 (1999)
10. Keller, J.M., Gray, M.R., et al.: A frezzy K-nearest neighbor algorithm. IEEE Transactions on Systems, Man, and Cybernetics 15, 580–585 (1985)

Information Architecture Based on Topic Maps

Guangzheng Li

Department of Marine Technology
Shandong Jiaotong University Maritime College
Weihai, China
ligz@sdjtu.edu.cn

Abstract. Aiming at the lack of information architecture is not fit to the demands of users, resulting in the feedback of a great deal of useless information, we present the new model information architecture based on topic maps. We define the framework, which includes four stages: Data mining, Information table, Visual structure and Views. Data mining is performed to obtain the topics and the associations between topics, and information table is defined to record them by using Distributed Hash Table. Visual structure includes information recommendation, information navigation and information retrieval, which main objective is to search information, which users really need. The results of visual structure will be displayed based on Topic map. A prototype system of visual information architecture has been implemented and applied to the massive information organization, management and service for education.

Keywords: information architecture; Topic Map; information system.

1 Introduction

With the rapid growth of knowledge resources and increasing individual demands of users, the problem of how to spend less effort finding information they need has been becoming more and more important. People can only appreciate what they can actually find, but the most information service systems are not fit to the demands of users, resulting in the feedback of a great deal of useless information. Information must be structured with both user and content in mind so people can successfully find what they are looking for as quickly and easily as possible. Information architecture focuses on designing effective navigation, organization, labeling, and search systems for websites and creating usable navigation [1].Effective information architectures enable people to quickly, easily and intuitively find the information, visualization is the one of choice. Information visualization is a research area that focuses on the use of visualization techniques to help people understand and analyze data. It presumes that "visual representations and interaction techniques take advantage of the human eye's broad bandwidth pathway into the mind to allow users to see, explore, and understand large amounts of information at once"[2]. Information visualization is about utilizing interactive graphics to present information and support interactions. In this paper we present a new model of information architecture based on Topic Map.

1.1 Information Architecture

Information architecture was first coined by Richard Saul Wurman in 1975. Wurman's initial definition of information architecture was "organising the patterns in data, making the complex clear" [3]. It was largely dormant until in 1996 by a couple of library scientists, Lou Rosenfeld and Peter Morville.They defined the role of the information architect as someone who clarifies the mission and vision for the site, balancing the needs of its sponsoring organization and the needs of its audiences in their book Information Architecture for the World Wide Web[1]. Information architecture has been used for a wide variety of topics. Chris Farnum defines the field of information architecture and the role of the information architect, and describes the principles and methodology of information architecture and provides a starting point for learning more and applying these ideas to information management [4]. Xiaoyan Wang analyses the relation between he component combination of information architecture and user experience and establishes the information architecture model on the basis of user experience, and discusses the implementation method [5]. Information architect has been a hot topic.

Various methods are used to capture and define information architecture. Some of the most common methods are: Sitemaps, Annotated page layouts, Content matrices, Page templates, and so on. Sitemaps [6] let users see all available content areas on one page, and gives them instant access to those site pages. Sitemaps can also help users find information on a cluttered site, providing a clean, simple view of the user interface and the available content, but Sitemaps are not necessarily indicative of the navigation structure. Page layout is the part of graphic design that defines page level navigation, content types and functional elements, annotations are used to provide guidance for the visual designers and developers who will use the page layouts to build the site [7]. Page templates can define the layout of common page elements, such as global navigation, content and local navigation. The development of topic map brings a new opportunity for information architecture.

1.2 Topic Map

Topic map(TM) is a new ISO standard (ISO/IEC 13250) [8] for describing knowledge structures and using them to improve the find ability of information. The Topic Maps family of ISO specifications is as follows [9]:

ISO/IEC 13250: Information Technology—Topic Maps
ISO/IEC 18048: Information Technology—TQML (Topic Maps Query Language)
ISO/IEC 19076: Information Technology—TMCL(Topic Maps Constraint Language)
ISO/IEC TR 29111: Information Technology—Topic Maps—Expressing Dublin Core Metadata using Topic Maps.

Topic Maps describe information structures and associating them with information resources. It absorbs the ideas contained in the semantic web, which establish a semantic web above the resource level, the semantic organization and joining between the physical resource entities and the abstract concepts are implemented. The structure of topic maps composed of Topics, Associations and Occurrences (TAO) [10], which is shown in Fig. 1.

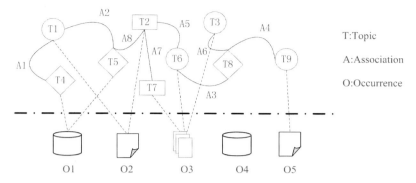

Fig. 1. The TAO Model of Topic Map

Topic (T): representing any concept, from people, countries, and organizations to software modules, individual files, and events.
Associations (A): representing hypergraph relationships between topics.
Occurrence (O): representing information resources relevant to a particular topic.
Topic Maps is increasingly used in information integration, knowledge management, e-learning, digital libraries, and as the foundation for Web-based information delivery solutions [11][12][13][14]. Topic map inherits the characteristics of knowledge organization methods such as index, term lists, thesauri, ontology, etc. Consequently, topic map adapts to information architecture.

- Information can be fully constructed based on different types of topics. Topic can be used to represent any specific entity or abstract concept in the real world, which has formal and morphological diversification. For example, topics can be expressed as knowledge elements (elements), knowledge unit, knowledge base and knowledge systems, etc. Therefore, the multi-granularity and multi-formals of information architecture can be realized based on topic map.
- All relevant knowledge can be system construction by the multiple associations between topics. Topic map reflects the associations between the knowledge and establishes a semantic web above the resource level. It absorbs the ideas contained in the semantic web, according with human intelligence realm to some extent.
- Graphic display based on TM is more perceivable, which can provide visual knowledge navigation mechanism and hence overcomes the shortcoming of linear display.

Topic maps are dubbed "the GPS of the information universe". Topic maps are also destined to provide powerful new ways of navigating large and interconnected corpora. It is more suitable for information architecture in a distributed environment.

2 Information Architecture Based on TM

Information architecture is the core technology to deal with the distributed information resources and provide high-quality knowledge services. In our framework of information architecture based on TM, we define four stages: Data mining, Information table, Visual structure and Views.

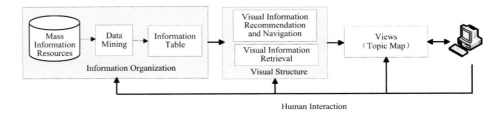

Fig. 2. The framework of information architecture based on TM

Data mining. Extraction of the elements of topic map is performed to obtain the topics and the associations between topics. Topics extraction is the scope of information extraction, and is regarded as the issue of sequence tagging. We use sequence data labeling technique to obtain the terminologies. But the association between topics is acquired based on semantic understanding, and is regarded as a classification problem. We use a classifier to determine which pairs of relational words are that we really need.

The following text clips are regarded as an example to show the association between topics.

> *Internet Layer* is part of *Internet Protocol Suite*
> *Internet Layer* includes *IP, ICMP, IGMP*, and more.
> *Internet Protocol (IP)* is the primary protocol that establishes the internet.
>
> The first major version of *IP*, now referred to as *Internet Protocol Version 4 (IPv4)* is the dominant protocol of the Internet.

The above text clips contain four pairs of partOf association between *Internet Layer* and *Internet Protocol Suite, IP* and *Internet Layer, ICMP* and *Internet Layer, IGMP* and *Internet Layer,* two pairs of synonymous association between *Internet Protocol* and *IP, Internet Protocol Version 4* and *IPv4* and a pairs of instanceOf association between *IP* and *IP4*.

Information table. Information table is defined to record the topics and associations by using Distributed Hash Table. The table mainly contains three attributes:

> *Information Table{TopicA, AssociationType,TopicB, Occurrence}*
> *TopicA*: a topic
> *AssociationType:* the association type
> *TopicB:* the topic is association with *TopicA*.
> *Occurrence:* the information resources relevant to *TopicA*.

For example, we established an information table about the above text clips, as shown in Table 1.

Table 1. A Information table of the text clips.

TopicA	AssociationType	TopicA	Occurrence
Internet Layer	partOf	Internet Protocol Suite	http://en.wikipedia.org/wiki/
IP	partOf	Internet Layer	http://en.wikipedia.org/wiki/
ICMP	partOf	Internet Layer	http://en.wikipedia.org/wiki/
IGMP	partOf	Internet Layer	http://en.wikipedia.org/wiki/
IP	synonymous	Internet Protocol	http://en.wikipedia.org/wiki/
IPv4	synonymous	Internet Protocol Version 4	http://en.wikipedia.org/wiki/
IPv4	instanceOf	IP	http://en.wikipedia.org/wiki/

Visual structure. The main objective of visual structure is to search information, which users really need. It includes information recommendation, information navigation and information retrieval.

- Information recommendation. How to accurately and effectively provide the information which the user needs is the core of the information recommendation research. We adopt a recommendation mechanism based on information item and user group which considers the effect of the user itself and its neighbors. First, the interesting node of users is set up, if the interesting node of user is matched in its interesting model, else, fuzzy match is to be carried out in user group in accordance with registration information. Second, the interesting nodes is sorted based on trend degree of interesting and will be returned to the user. Final, when user select the interesting node, return all the topics associated with the interesting node in a certain information radius by information architecture reasoning.

- Information navigation. Knowledge navigation extracts the most probable path as a new user's guide and this path could be learned from old user's a large number of interest paths. We present an information navigation algorithm, is shown as follows[15]:

    It is assumed that N users visited the topic map of the specific area. The topic node set is $T = \{T_i \mid 1 \le i \le p\}$, m is arbitrary user in N users, visited the Topic sequence is $\{N_{T_1}^m, N_{T_2}^m, ..., N_{T_L}^m\}, 1 < L < p$, navigation path is defined as follows:

$$\{T_1, T_2, ... Tu, ... T_p \mid T_u = T_i, if(T_i \mid \max P_u(T_i)\})1 \le i \le p\} \quad (1)$$

$P_u(T_i)$ represents the probability of every topic node appears in the navigation path at the u place.

- Information retrieval. Knowledge-based information retrieval is used. The users represented their own search requirement to the system, according to the users' requirement, through searching the information of the special field, and then the system outputs the information satisfying user's search requirement to the users. Information reasoning schemes are based on a fuzzy set theoretic framework. Concepts that describe domain topics and relationships between concepts are captured in an information base.

Views. The results of visual structure will be displayed based on Topic map. Based on the TM logical representation of information, the visual information map constructing tool is designed, it is free software coded in Java applet, to assist users in sharing, and navigating the domain knowledge [16].

3   Empirical Evaluation

We built the corpus of Computer Network, which includes 12017 topics, 2816 associations between topics and 5038domain-specific terms. We select a topic "TCP/IP protocol" as interesting node, when information radius is equal to 2, the Information architecture results are depicted in Fig.3.

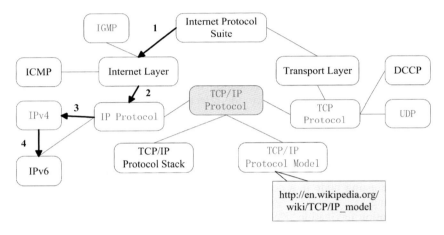

Fig. 3. The Information architecture results based on topic map

Topics and associations are represented in the information architecture results in which fillet rectangular node is regarded as a topic. The bold node is regarded as interesting node. Each edge is regarded as an association of topics. When user clicking the edge, it will display the association type. When clicking the nodes, it will display the occurrences which are associated with the topic. The bold lines in the topic map represent cognitive path tendency, the numbers on the bold line indicate the sequence of cognitive steps.

4   Conclusions

The proposed information architecture based on topic map provides us a means to organize, discovery and display information. Visual information structure based on TM not only achieves better the structure-based information reasoning results and provides users with intuitive access mechanisms for the required information. Information has been provided by a stereo knowledge map and hence overcomes the shortcoming of linear display. The ongoing work is information recommendation, information navigation and information retrieval can be carried out by computing cloud with huge computing ability and storage capacity distributed and parallel. We hope that the real visual information architecture system based on topic map will be widely deployed in the future.

Acknowledgments. This research is supported by the Key Research Programs of China Institute of Communication Education under Grant No. JJY1001-3.

References

1. Rosenfeld, L., Morville, P.: Information Architecture for the World Wide Web, 1st edn. O'Reilly, USA (1998)
2. Wurman, R.S., Bradford, P. (eds.): Information Architects. Graphis Press, Zurich (1996) ISBN:3-85709-458-3
3. Wurman, R.S.: Information Architects. Graphis Inc., New York (1997)
4. Farnum, C.: Information Architecture: Five Things Information Managers Need to Know. Information Management Journal 36(5), 33–40 (2002)
5. Wang, X.-y., Hu, C.-p.: Information Architecture Based on the User Experience. Information Science 08 (2006)
6. Site Map Usability Jakob Nielsen's Alertbox, August 12 (2008)
7. Barker, I.: What is information architecture Step Two Designs (2005), http://www.steptwo.com.au/papers/kmc_whatisinfoarch/index.html
8. ISO/IEC JTC 1/SC34 N323. (2002). Guide to the Topic Map Standards, International Organization for Standardization, http://www1.y12.doe.gov/capabilities/sgml/sc34/document/0323.htm
9. Pepper, S.: Topic Maps, 3rd edn. Encyclopedia of Library and Information Sciences (2010), doi:10.1081/E-ELIS3-120044331
10. Pepper, S.: The TAO of Topic Maps (2001), http://www.gca.org/papers/xmleurope2000/
11. Qiu, J., Yao, Y., Wang, Y., Wang, X.: Research of E-Government Knowledge Navigation System Based on XTM. In: Web Intelligence and Intelligent Agent Technology Workshops (WI-IAT 2006), pp. 586–589 (2006)
12. Olsevicova, K.: Application of Topic Maps in E-learning Environment. ACM SIGCSE Bulletin 37(3), 363 (2005)
13. Ouziri, M.: Semantic Integration of Web-Based Learning Resources: A Topic Maps Based Approach. In: Proceedings of the Sixth International Conference on Advanced Learning Technologies, pp. 875–879 (2006)
14. Stefan, S., Nastansky, L.: K-Discovery: Using Topic Maps to Identify Distributed Knowledge Structures in Groupware-Based Organizational Memories. In: Proceedings of the 35th Annual Hawaii International Conference on System Sciences (HICSS 2002), pp. 966–975 (2002)
15. Li, G., Lu, H., Ren, W.: Service-Oriented Knowledge Modeling on Intelligent Topic Map. In: Proceedings of the 1st IEEE International Conference on Information Science and Engineering (ICISE 2009), Nanjing, China, December 18-20, pp. 2394–2397 (2009)
16. Lu, H., Feng, B., Chen, X.: Visual Knowledge Structure Reasoning with Intelligent Topic Map. IEICE Transactions on Information and Systems E93-D(10), 2805–2812 (2010)
17. Lu, H., Feng, B., Chen, X.: Visual Knowledge Structure Reasoning with Intelligent Topic Map. IEICE Transactions on Information and Systems E93-D(10), 2805–2812 (2010)

Implementation of 2D-DCT Based on FPGA with Verilog HDL

Yunqing Ye and Shuying Cheng[*]

School of Physics and Information Engineering, and
Institute of Micro-Nano Devices & Solar Cells, Fuzhou University,
Fuzhou 350108, China
yeyunqing1986@163.com, sycheng@fzu.edu.cn

Abstract. Discrete Cosine Transform is widely used in image compression. This paper describes the FPGA implementation of a two dimensional (8×8) point Discrete Cosine Transform (8×8 point 2D-DCT) processor with Verilog HDL for application of image processing. The row-column decomposition algorithm and pipelining are used to produce the high quality circuit design with the max clock frequency of 318MHz when implemented in a Xinlinx VIRTEX- II PRO FPGA chip.

Keywords: 2D-DCT, FPGA, Verilog HDL.

1 Introduction

Discrete Cosine Transform (DCT), among various transforms, is the most popular and effective one in image and video compression. So far, it has been adopted by many international standards like JPEG, MPEG1, MPEG2, MPEG4, and H26x, etc. However, it is difficult to satisfy the requirement for real-time by software owing to its heavy quantity of computing. Therefore, a hardware method is adopted to satisfy the requirement for speed in many practical applications. Hardware implementation algorithms of 2D-DCT can be divided into two categories: row-column decomposition methods and non row-column decomposition methods [1-2]. 4096 times of multiplications and 4032 times of additions are required for the 8×8 point 2D-DCT calculation. The row-column decomposition method is usually adopted to reduce the computational complexity. In this paper, the row-column decomposition algorithm which changes 2D-DCT to two 1D-DCT (one dimensional point Discrete Cosine Transform) is used. In the design of 1D-DCT, direct multiplication operation has been replaced by pre-adder, partial multiplication based on look-up table, partial multiplication result shift and addition so as to reduce the hardware cost.

2 DCT Algorithms

In this section the 1D-DCT and 2D-DCT will be presented. For the 1×8 sequence $f(i)$, DCT can be defined as:

---

[*] Corresponding author.

$$F(k) = \frac{1}{2}C(k)[\sum_{i=0}^{7} f(i)\cos\frac{(2i+1)k\pi}{16}] \quad (1)$$

Where k = 0,1,2......7, C(0)=1/√2 (t=0), C(t)=1 (t≠0).

The equation for 2D-DCT of a (8×8) input sequence f(x,y) representing the image pixel values is shown below [3].

$$F(u,v) = \frac{1}{4}C(u)C(v)[\sum_{x=0}^{7}\sum_{y=0}^{7} f(x,y)\cos\frac{(2x+1)u\pi}{16}\cos\frac{(2y+1)v\pi}{16}] \quad (2)$$

Where u,v = 0,1,2......7; C(0)=1/√2 (t=0), C(t)=1 (t≠0).

3 Hardware Implementation of 2D-DCT

This section presents the implementation of the DCT on FPGA in detail.

3.1 Design Strategy

According to the definition of 2D-DCT above, the equation for 2D-DCT can be represented as following.

$$F(u,v) = \frac{1}{2}C(u)\sum_{x=0}^{7}[\frac{1}{2}C(v)\sum_{y=0}^{7} f(x,y)\cos\frac{(2y+1)v\pi}{16}]\cos\frac{(2x+1)u\pi}{16} \quad (3)$$

It can be introduced from (3) that 2D-DCT can be decomposed into two 1D-DCT, so the fast algorithm for computing 1D-DCT can be used for 2D-DCT. The matrix F=CfC^T can be used to rewrite (3), where C is on behalf of the 8×8 2D-DCT coefficient matrix [4]. It can be deduced from the matrix expression that the 2D-DCT calculation results can be obtained by the following four steps:

(1) calculating Y=Cf .
(2) transposing the matrix Y to get Y'.
(3) calculating F= CY'.
(4) transposing the matrix F to get the final result.

With row-column decomposition, the 8 point 1D-DCT is applied to each row of the input 8×8 pixel matrix, and a further 1D-DCT is applied to each column of 8×8 block of semi-transformed values. The final results of 2D-DCT can be achieved by transposing the 8×8 matrix which is obtained by 8 point 1D-DCT twice. The implementation architecture of 2D-DCT is shown in Fig.1. The 2D-DCT module includes a control module, data selection module, serial to parallel module, 1D-DCT module, parallel to serial module, memory matrix RAM module, bit interception module and transposing address generation module. In this paper, the 1D-DCT module is reused to meet the speed requirement and reduce hardware resource cost.

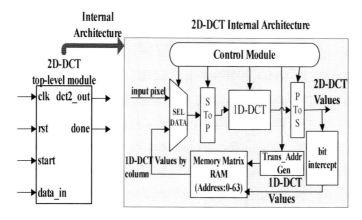

Fig. 1. The 2D-DCT implementation architecture

3.2 Verilog HDL Design of Some Modules

1) 1D-DCT module

Assuming that the input eight points are x0, x1, x2, x3, x4, x5, x6, x7, 1D-DCT computing equation can be simplified as (4) based on the characteristic of the symmetry and rotation for DCT coefficients.

$$\begin{aligned}
Y0 &= [(x0+x7)+(x1+x6)+(x2+x5)+(x3+x4)]*C4 = (s07341625)*C4 \\
Y1 &= (x0-x7)*C1+(x1-x6)*C3+(x2-x5)*C5+(x3-x4)*C7 \\
   &= f0\_7*C1+f1\_6*C3+f2\_5*C5+f3\_4*C7 \\
Y2 &= [(x0+x7)-(x3+x4)]*C2+[(x1+x6)-(x2+x5)]*C6 = (s07\_34)*C2+(s16\_25)*C6 \\
Y3 &= (x0-x7)*C3+(x1-x6)*(-C7)+(x2-x5)*(-C1)+(x3-x4)*(-C5) \\
   &= f0\_7*C3+f1\_6*(-C7)+f2\_5*(-C1)+f3\_4*(-C5) \\
Y4 &= [(x0+x7)+(x3+x4)-(x1+x6)-(x2+x5)]*C4 = (s0734\_1625)*C4 \\
Y5 &= (x0-x7)*C5+(x1-x6)*(-C1)+(x2-x5)*C7+(x3-x4)*C3 \\
   &= f0\_7*C5+f1\_6*(-C1)+f2\_5*C7+f3\_4*C3 \\
Y6 &= [(x0+x7)-(x3+x4)]*C6-[(x1+x6)-(x2+x5)]*C2 = (s07\_34)*C6+(s16\_25)*(-C2) \\
Y7 &= (x0-x7)*C7+(x1-x6)*(-C5)+(x2-x5)*C3+(x3-x4)*(-C1) \\
   &= f0\_7*C7+f1\_6*(-C5)+f2\_5*C3+f3\_4*(-C1)
\end{aligned} \quad (4)$$

Where $C1 = 1/2*COS(\pi/16)$, $C2 = 1/2*COS(2\pi/16)$, $C3 = 1/2*COS(3\pi/16)$, $C4 = 1/2*COS(4\pi/16)$, $C5 = 1/2*COS(5\pi/16)$, $C6 = 1/2*COS(6\pi/16)$ and $C7 = 1/2*COS(7\pi/16)$.

The pipeline architecture shown in Fig.2 is used to realize the 1D-DCT calculation described as (4).

The pipeline architecture shown in Fig.2 makes it possible to input the next eight data continuously so as to improve the processing speed.

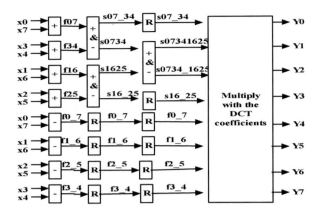

Fig. 2. The computing architecture figure of 1D-DCT

It can be found from (4) that the unit of each 1D-DCT expression is the multiplication of a pre-sum and a DCT coefficient (named as one_mult_block). As long as the one_mult_block module is achieved, the eight DCT values can be got by instantiating it. The implementation of the one_mult_block module will be mentioned below.

Since DCT coefficients are fractional, they have been expanded by 2^{14} times in order to overcome the difficulty in the implementation of fractional multiplication on FPGA. At the same time, the expanded DCT coefficients are extended to 15 bit wide and the fifteenth bit is the sign bit. The negative DCT coefficients in (4) are expressed in binary complement to participate in the computing. Therefore, we have implemented the multiplication of a pre-sum and a 15-bit DCT coefficient based on look-up table method shown in Fig.3. In order to simplify the graphical scale, the

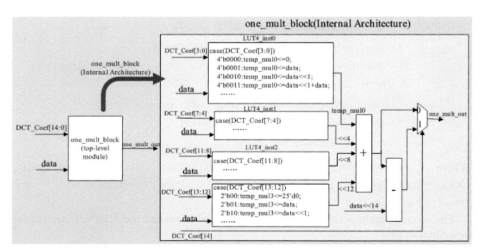

Fig. 3. The multiplication of a pre-sum and a 15-bit DCT coefficient based on look-up table method

expansion of the sign bit of the code marked in Fig.3 is omitted. It needs to extend the sum as wide as temp_mul * with the sign bit, otherwise, the results will be wrong if the input data is negative.

The 8 point DCT values can be obtained by instantiating one_mult_block module. Equation (4) shows that Y2 is equal to $(s07\_34)*C2+(s16\_25)*C6$. The method of calculating Y2 is showed in Fig.4.

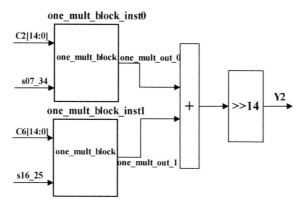

Fig. 4. The calculation process of Y2

2) Control module

The control module is implemented using a state machine. The state machine is divided into three states: idle state, row 1D-DCT state, and column 1D-DCT state. After the valid reset signal, the initial state is the idle state. When the valid pixel data come in, the state machine enters into row 1D-DCT state. After maintaining 64 clock cycles, it goes into column 1D-DCT state. While the column 1D-DCT state finished, the state machine recalls the idle state and sets the done signal meanwhile, then it just waits for the next 8×8 point 2D-DCT.

3) Trans_Addr_Gen module

This module achieves the goal of transposing the semi-transformed values by controlling the read and write addresses of the memory matrix. When the state machine is in the state of row 1D-DCT, the 1D-DCT results will be written to RAM by row address mode (0,1,2,3,4,5,6,7,8,9......).When the state machine is in the state of column 1D-DCT, the data sent to the data selection module will be read from RAM by column address mode (0,8,16,24,32,40,48,56,1,9......), and the corresponding 1D-DCT results will be written to RAM by column address mode. The generation method of the column address pattern is swapping the high three bits and low three bits of the row address pattern.

4 Implementation Results

4.1 Functional Simulation

The whole design is simulated and tested by a test bench using ModelSim and implemented in a Xinlinx VIRTEX- II PRO FPGA device.

The 8×8 data in Table 1 are used for test vectors. The simulation result is shown in Fig.5. The data in Table 2 are the 2D-DCT values of data in Table 1 got by the Matlab command (dct2).

It can be seen from Fig.5 that the results of 2D-DCT are very close to those in Table 2 which means the designed code can implement 8×8 2D-DCT correctly. The value of the

Table 1. 8×8 pixel matrix

170	153	153	153	170	153	153	153
170	153	170	187	170	153	170	153
170	153	170	170	204	187	221	204
187	153	187	187	204	187	238	221
170	187	170	153	187	187	170	153
187	170	187	153	204	170	170	153
204	187	153	153	136	136	136	136
205	187	153	153	153	153	153	119

Table 2. 2D-DCT values using matlab

1364	17	9	36	9	12	-13	36
24	-72	-21	-21	16	13	2	11
-87	71	10	6	8	5	-4	-24
-39	36	-14	4	-5	-27	12	2
4	-5	4	11	-8	-4	-2	-5
14	-20	28	-10	15	8	2	4
22	-28	-8	16	-3	-12	3	25
-29	11	8	10	-12	-21	-3	-18

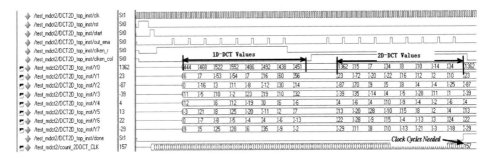

Fig. 5. Simulation results of 8×8 2D-DCT with Modelsim

```
Device utilization summary:
--------------------------------

Selected Device : 2vp30ff896-6

 Number of Slices:                    23  out of  13696     0%
 Number of Slice Flip Flops:          19  out of  27392     0%
 Number of 4 input LUTs:              44  out of  27392     0%
 Number of IOs:                       36
 Number of bonded IOBs:                4  out of    556     0%
 Number of GCLKs:                      1  out of     16     6%

---------------------------------------------------------------
===============================================================
Timing constraint: Default period analysis for Clock 'clk'
  Clock period: 3.135ns (frequency: 318.979MHz)
  Total number of paths / destination ports: 161 / 26
---------------------------------------------------------------
```

Fig. 6. The resource utilization and the maximum clock frequency

count_2DDCT_CLK register in Fig.5 reveals that it requires 157 clock cycles to complete the 2D-DCT operations in this design.

4.2   Implementation in FPGA Chip

The hardware description of this architecture for 2D-DCT implementation has been synthesized using ISE Xilinx 10.1 and mapped on the Xinlinx VIRTEX- II PRO (2vp30ff896-6) FPGA chip. The utilized resource and the maximum clock frequency are demonstrated in Fig.6.

5   Conclusion

This paper has proposed a architecture based on row-column decomposition for the computation of 2D-DCT. Verilog HDL was used to complete the code design of the top-level module and sub-modules in accordance with top-down process. The improvement is that direct multiplication operation has been replaced by shift and addition combined with pipelining which can save resources and improve the operation speed. The design implemented in the Xinlinx VIRTEX- II PRO (2vp30ff896-6) FPGA chip can complete 8×8 2D-DCT logic operations correctly at 318MHz clock frequency.

Acknowledgments

Financial support by the National Nature Sciences Funding of China (61076063) is gratefully acknowledged. The authors also wish to express their gratitude to funding from Fujian Provincial Department of Science & Technology, China (2009J01285) and Fuzhou Municipal Bureau of Science & Technology, China (2010-G-102).

References

1. Sun, C.C., Ruan, S.J., Heyne, B., Goetze, J.: Low-power and high-quality Cordic-based Loeffler DCT for signal processing. IET Circuits Devices Syst. 1(6), 453–461 (2007)
2. Khayatzadeh, A., Shahhoseini, H.S., Naderi, M.: Systolic Cordic Dct:An Effective Method For Computing 2D-Dct. In: Proceedings.Seventh International Symposium on Signal Processing and Its Applications, vol. 2, pp. 193–196 (2003)
3. Tumeo, A., Monchiero, M., Palermo, G., Ferrandi, F., Sciuto, D.: A Pipelined Fast 2D-DCT Accelerator for FPGA-based SoCs. In: IEEE Computer Society Annual Symposium on VLSI (ISVLSI 2007), pp. 331–336 (2007)
4. Atani, R.E., Baboli, M., Mirzakuchaki, S., Atani, S.E., Zamanlooy, B.: Design and Implementation of a 118 MHz 2D DCT Processor. In: IEEE Symposium on Industrial Electronics (ISIE 2008), pp. 1076–1081 (2008)

Impact of Grounding Performance of AMN to Conducted Disturbance Measurement

Xian Zhang[1], Qingdong Zou[2], Zhongyuan Zhou[2], and Liaolan Wu[1]

[1] Shanghai Entry-Exit Inspection and Quarantine Bureau of the People's Republic of China, No. 1208, Minsheng Road, Pudong New Area Shanghai, 200135
zhangxian@shciq.gov.cn
[2] Electromagnetic Compatibility Laboratory, Southeast University, Nanjing, China, 211189
qingdongz@gmail.com

Abstract. In this paper, the isolation and the voltage division factor changes of AMN are studied when AMN is grounded via various ground wires of different impedances, which is to simulate impacts brought by poor ground conditions on in-site test results of EUT. The result shows the isolation and the voltage division factor vary with changes of frequency and the length of the ground wire for AMN. In addition, the impedance of the ground wire will seriously affect the repeatability and the accuracy of the test results.

Keywords: artificial mains network (AMN), isolation, voltage division factor, test in site.

1 Introduction

As can be seen from Figure 1, in conducted disturbance measurement, conducted disturbance voltage from the equipment under test (EUT) is coupled to the EMI receiver via artificial mains network (AMN).

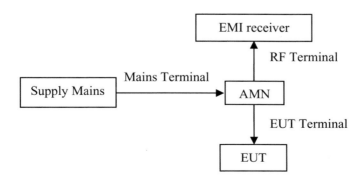

Fig. 1. Alternative arrangement for measuring conducted disturbance voltage

AMN, which is extraordinary important auxiliary equipment during conducted disturbance measurement, is required to provide a defined impedance at the terminal of EUT in a given radio frequency range (for example, 9 kHz~30 MHz), to couple the disturbance voltage from EUT to the measuring receiver and to isolate EUT from unwanted signals on supply mains at the same time.

The performance of AMN depends on the following key parameters: 1) AMN impedance of EUT terminal; 2) isolation between mains terminal and RF terminal; 3) voltage division factor (or insertion loss) between RF output port and EUT port of AMN. The impedance, the isolation and the voltage division factor of AMN are well-defined in CISPR 16-1-2[1]. If the absolute value of the voltage division factor is more than 0.5 dB[2], it should be accounted for as a modified final measurement value when calculating the EUT emission levels.

As conducted disturbance measurement carried out in a shielded room, which provides a good ground condition for AMN to work well, the difference of the isolation and the voltage division factor compared with its calibration is not huge. In addition, the isolation is generally over 40 dB, which satisfies test requirements. Therefore, the measured value of the voltage division factor can be added to the test results from EMI receiver as a final modified value.

In reality, some large scale machinery & electrical equipment can not be moved into the shielded room because of its bulk mass or heavy weight, or the shielded room's condition which can not ensure EUT work normally. In these situations, the measurement needs to be implemented in site, where the equipment needs to be operated.

In most cases, in-site conditions can not bear comparison with those in the shielded room, which has good ground conditions. So how will the isolation and the voltage division factor of AMN change? Will they meet the test requirements? The paper researches on these two questions and finds the trend how the isolation and the voltage division factor will change as AMN is grounded with various ground wires of different impedances.

2   Impedances of Ground Wires

Tested ground wires are copper wires of different length and cross sectional area, as can be seen from Table 1.

Table 1. Specifications of different ground wires

Ground wires	Length/cm	Cross sectional area/mm^2	Shape of cross section
1[#]copper strip	5	7.2	rectangle
1[#]copper wire	5	0.785	round
1[#]copper wire	20	0.785	round
1[#]copper wire	100	0.785	round

The impedances[3] of wires listed in Table 1 vary in the frequency range from 150 kHz to 30 MHz, which is described in Figure 2.

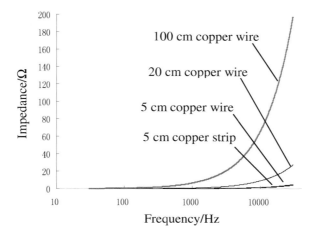

Fig. 2. Impedances of different wires vary with the frequencies

Conclusions can be drawn from Figure 2 as follows:

- At the same frequency, RF impedances of ground wires with the same material and cross sectional area are different if lengths are diverse. The impedance increases when the length enlarges.
- For the same ground wire, the skin depth will decrease, while the RF resistance, the inductive resistance and the impedance will increase with rising frequency.
- When the ground wire becomes longer, the RF resistance, the inductive resistance and the impedance will raise.
- The RF resistance trends to go up with bigger diameter of the ground wire. With the same cross sectional area of a wire, the equivalent diameter of a rectangle wire is less than a round wire. As a result, in the case of the same cross sectional area and length, the RF impedance of the rectangle is smaller in comparison with the round. In other words, the RF performance of the rectangle wire is better than the round, which is the reason why a flat wire is used for grounding in high frequency.

3 Isolation of AMN

The isolation of AMN is defined as the extent of isolating the mains terminal of AMN from the RF terminal when the EUT terminal is connected with a 50Ω load.

There are several steps to measure the isolation:

- The measurement of the isolation is carried out according to the Annex H of CISPR 16-1-2:2006。 The first step is to measure the output voltage U_1 when a signal source is connected with a 50Ω load. Specifically, the signal source is directly linked with the EMI receiver whose internal resistance is 50Ω, as can be seen from Figure 3.

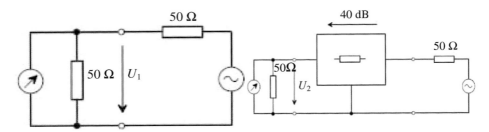

Fig. 3. Schematic diagram of measuring U_1 **Fig. 4.** Schematic diagram of measuring U_2

- The second step is to connect relevant EUT terminal (such as L-PE terminal or N-PE terminal) with a 50Ω matched load. At the same time, the output of signal source, which should be in accord with the first step, is connected to the corresponding mains terminal and reference ground. Meanwhile, measure the output voltage U_2 at RF terminal by EMI receiver. The schematic diagram is as Figure 4.
- The isolation can be calculated from the following formula:

$$F_D = U_1 - U_2 \tag{1}$$

U_1 is the reference voltage of signal source in dBμV; U_2 is the output voltage on RF terminal in dBμV. In order to find out the impact on the performance of AMN in various ground conditions, measure the isolation of AMN when it is grounded with the wires listed in Table 1. The actual layout is shown in Figure 5.
- Measure the isolation of AMN in the frequency range from 150 kHz to 30 MHz with a result reflected in Figure 6.

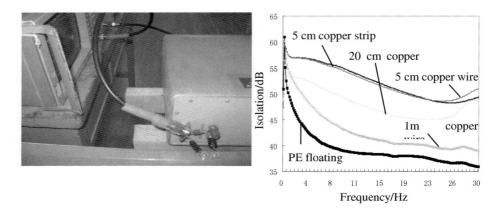

Fig. 5. Actual layout of the measurement   **Fig. 6.** The trend of the isolation

There are several conclusions that can be obtained from Figure 6.

- In the measured frequency range, the isolation is different whether AMN is grounded or not. When AMN is not grounded, the isolation is the least, the value of which is about 35 dB. For the 1m copper wire used as a ground wire, the isolation is less than 40 dB above 22 MHz, which does not meet the requirement of the 40 dB isolation. In other words, if AMN is not grounded, or if the length of the ground wire is too long, these will result in decreasing isolation. Then the disturbance from the supply mains may be coupled into the RF output of AMN, which will bring uncertainty in the measurement and impact on the repeatability and the accuracy of test results.
- As the frequency rises up, the isolation will decrease because of the increasing impedances of ground wires, which means that in high frequency situation the performance of AMN will degrade and the measurement result will be more easily affected by disturbances.
- When AMN is grounded via various ground wires, the isolation varies with impedances of ground wires. The smaller the impedance, the better the isolation. The isolating performance is the best when 5cm copper strip or 5cm copper wire is used to ground.
- The grounding performance of copper strip is a little better than copper wire in most frequencies as the impedance of the former is smaller than the latter for the same length and different cross sectional area.
- When AMN is grounded with the same cross sectional area but diverse length of copper wires (5cm, 20cm and 100cm accordingly), the isolation of 5cm copper wire is the biggest. With the increasing length of a ground wires, the impedance rises up while the isolation decreases. That is to say the isolation is bigger if the impedance of a ground wire is smaller, and the performance of AMN is better. Therefore, it had better to cut down the length of a ground wire to improve the performance of AMN.

4 Voltage Division Factor of AMN

The disturbance voltage at EUT terminal will produce a voltage drop from LRC components to RF terminal for measurement, which is characterized as voltage division factor. The value of it will directly affect the precision of measurement results. Voltage division factor can be tested by network analyzer or by signal generator and EMI receiver. The theory of these two methods is similar. Take the method using signal generator and EMI receiver for example as a schematic diagram shown in Figure 7.

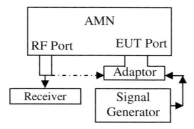

Fig. 7. Schematic diagram of measuring voltage division factor

Steps to carry out the measurement:

- A T type adapter should be connected to the EUT terminal. A signal produced by signal generator is fed in B terminal of the T type adapter, while EMI receiver is linked with a terminal of the adapter to receive the voltage signal, which is V_1.
- Make sure that the signal fed in B terminal is the same. A terminal is matched with a 50Ω load. Connect the EMI receiver to the RF terminal of AMN to get the voltage V_2.
- The voltage division factor can be obtained by the following equation:

$$F_V = V_1 - V_2 \tag{2}$$

where both V_1 and V_2 are in dBμV.

AMN is ground via these ground wires separately: 5cm copper wire, 20cm copper wire and 1m copper wire as listed in Table 1. In addition, measure the voltage division factor like the specific layouts shown in Figure 8. Finally, test results can be seen in Figure 9 and Figure 10.

Fig. 8. Specific layouts to measure voltage division factor

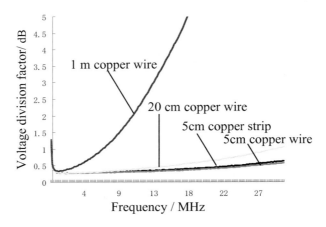

Fig. 9. Voltage division factor of various ground wires

From Figure 9, some useful conclusions can be summed up as follows:

- The value of voltage division factor varies greatly when AMN is grounded with different ground wires. Grounding performances of 5cm copper strip and 5 cm copper wire, values of which are both less than 1 dB, are quite similar and much better than 20cm copper wire and 1m copper wire. On the contrary, the performance of 1m copper wire is fairly bad as the voltage division factor is over 0.5 dB in the frequency range. During the measurement of disturbance voltage, the voltage division factor changes as the length of ground wires differs, which brings in diverse impact on measurement results. The longer the length of ground wires, the larger the voltage division factor. Therefore, it is better to reduce the length of the ground wire to weaken the coupling extent between EUT terminal and RF terminal.
- Voltage division factor increases with raising frequency. Those of 5cm copper strip, 5cm copper wire and 20cm copper wire change a little when the frequency increases. However, the effect of the 1m copper ground wire on the voltage division factor of AMN strengthens greatly with rising frequency. Namely, for longer length of a ground wire and higher frequency, the grounding impedance is larger and its influence on voltage division factor is greater.
- The voltage division factor of 1m ground wire trends as Figure 10 in accordance with frequency. It is shown that when the frequency goes up, the voltage division factor increases, which is about 20 dB in the frequency about 30 MHz. In other words, if AMN is grounded via a long wire, as both the grounding impedance and voltage division factor are quite large, the coupling performance between EUT terminal and RF terminal is not so good. In addition, voltage division factor is a function of frequency. Factors above will cause many difficulties in modifying measurement results.

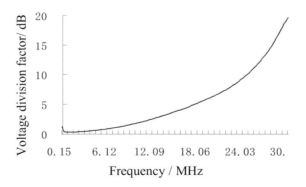

Fig. 10. Voltage division factor of 1m ground wire

5 Summary

To carry out conducted disturbance measurement for mains terminal of EUT, AMN shall be well electrically bonded to reference ground. Bad bond will result in raising ground impedance of AMN. Meanwhile, the isolating and coupling performance will become worse obviously with increasing length of ground wires and rising frequency. Or it will not satisfy the 40 dB isolating requirement defined in CISPR 16-1-2. In a word, it brings many difficulties in modifying measurement results, the accuracy of which is hard to guarantee.

For EUT which has to be measured in site, the surrounding electromagnetic field is very complicated. It is unable to provide a reference ground as good as in the shielded room. The increasing impedance of the ground wire will badly affect the isolating and coupling performance of AMN, which results that the repeatability of the measurement and the accuracy of the test results can not be guaranteed. As a result, it is better to use a voltage probe to carry out conducted disturbance voltage measurement when the equipment is in site.

References

1. CISPR 16-1-2:2006 Specification for radio disturbance and immunity measuring apparatus and methods Part 1-2: Radio disturbance and immunity measuring apparatus Ancillary equipment Conducted disturbances (2006)
2. ANSI C63.4-2003, American National Standard for Methods of Measurement of Radio-Noise Emissions from Low-Voltage Electrical and Electronic Equipment in the Range of 9 kHz to 40 GHz (2003)
3. Qiu, C., Zhao, D., Jiang, Q.: Structure Design Principle for Electronic Equipment (Revised Edition), pp. 316–317. Southeast University Press, Nanjing (2005)

Optimization of PID Controller Parameters Based on Improved Chaotic Algorithm

Hongru Li and Yong Zhu

College of Information Science and Engineering, Northeastern University,
Shenyang 110004, China

Abstract. The performance of PID controller depends on the combination of the control parameters. A kind of improved chaotic algorithm was put forward to optimize the PID controller parameters. In order to solve the shortcoming of low convergence rate of chaos optimization algorithm, the step acceleration method was introduced. And the improved chaotic algorithm combined the merits of both methods. Simulation results show that the improved chaotic algorithm is effective and the optimization PID controller parameters is superior to the PID controller parameters by conventional methods.

Keywords: chaotic algorithm, step acceleration method, PID, optimization.

1 Introduction

The conventional PID controller has been widely used in the most of modern industrial process control with its simple constructure. The control effect of a PID controller mainly depends on its optimal PID parameters. The traditional PID parameter setting method or Ziegler - Nichols parameter setting method is so complex that the best parameters can not be found easily and quickly, and the simulation result shows that there is strong oscillation or large overshoot [1]. Therefore, there appear some optimization methods which can adjust the PID parameters to the controlled object. According to the selected objective function, it is important to choose the proper optimization method. Assuming the objective function is a positive definite quadratic function, there are many methods such as the steepest descent method, Newton method, conjugate gradient method, DFP method and so on. They all have some boundednesses: the steepest descent method may have serrated phenomena and a low convergence speed; the Newton method cannot calculate the inverse matrix of Hesse matrix easily and so on. Along with the rapid development of computer technology, chaos optimization algorithm has become a new optimization algorithm and widely used in recent years [2]. The chaos optimization algorithm makes full use of some special dynamic characteristic, such as the initial value sensitivity, inherent randomness and ergodicity [3]. The chaos optimization algorithm can find the global optimal solution, but it takes plenty of search time. The step acceleration method can be used in local search, which has a small computation and fast optimization speed. The objective function of the two methods may not be continuous and differentiable, so they can be combined to have a large application scope. Preliminary optimal solution can be obtained by using chaos optimization algorithm, and then local search with step acceleration method improves the search speed and precision [4]. But the step acceleration method may fall into the local optimal solution. Therefore, some measures should be taken to improve the algorithm and make the solution more accurate.

2 The Improved Chaotic Algorithm

2.1 The Essential Theory

Combining chaotic optimization algorithm with step acceleration method can engender a new compound optimization algorithm. Firstly, find out the approximate range of the optimal solution by using the chaotic algorithm. If the searched solution is in the range of feasible solution, then local search with step acceleration method should be carried on to find out the best solution.

The main idea of chaotic algorithm is to linearly transform the variable to the chaotic variable which belongs to the range $X_k \in (0,1)$. Make full use of the ergodicity of chaotic system to find out another point X' which makes the value of the objective function lower. The logistic equation is as follows:

$$X_{k+1} = \eta * X_k * (1 - X_k) \ . \qquad (1)$$

When η equals 4, the logistic mapping is in chaos [5]. By using chaos characteristic of initial value sensitivity, there can be a lot of chaos value orbits depending on different initial value.

The step acceleration method consists of "detection search" and "mode moving". When the subprime solution X' is obtained, "detection search" use λ as the step length to find out a point making the value of objective function lower in the positive and negative direction. If the "detection search" is successful, the "mode moving" should be carried on in the X_{i+1}-X_i direction. The formula is as follows:

$$M_i = X_{k+1} + \xi * (X_{k+1} - X_k). \qquad (2)$$

ξ > 0, and ξ equals 1 generally. Thus, formula 2 can be transformed to formula 3:

$$M_i = 2 * X_{k+1} - X_k \ . \qquad (3)$$

Then, the point M_i is the new reference point. Carry on another "detection search" and "mode moving" continuously. If the "detection search" fails, the step length should shorten. Then carry on the "detection search" anew till the step length is less than the setting value μ [6]. But the step acceleration method can fall into the local optimal solution easily and frequently. Take two dimensional space for an example. Figure 1 is as follows:

Optimization of PID Controller Parameters Based on Improved Chaotic Algorithm

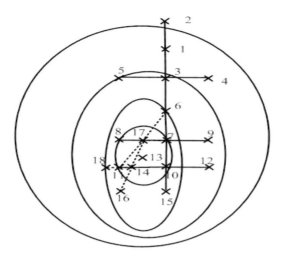

Fig. 1. Detection search and mode moving process

As shown in figure 1, the annular curve is the contour line of the objective function. Point 13 is the global optimal solution. Optimizing with the step acceleration method, the first path of detection search is point 1→point 2→point 3→point 4→point5, and the first optimal function is point 3. The value of objective function at point 3 is less than that of point 1, so the first detection search is successful. Then carry on the mode moving to the point 6. Carry on the second detection search with the new reference point 6. The second path of detection search is point 3→point 7→point 8→point 9, and the second optimal function is point 7. The value of objective function at point 7 is less than that of point 3, so the second detection search is successful. Then carry on the mode moving to the point 10. Carry on the third detection search with the new reference point 10. The third path of detection search is point 7→point 15→point 11→point 12. The value of objective function at the four points are all more than that of point 10, so the third detection search fails. Thus, the step length λ will lessen to 0.2 times. Now the range of detection search is a small neighborhood of point 10. Thus, the optimizing will fall into be a local optimal solution. In order to obtain a better solution, the step acceleration method should be improved, that is, the step length should change according to the latest detection search.

The main idea of improved step acceleration method is to make a through inquiry into the amplitude change of the basic point in the latest detection search. If the amplitude change is little in some direction, that means the new basic point is close to the global optimal solution in this direction. Thus, the step length should shorten in this direction. On the contrary, if the amplitude change is big, that means the new basic point is far from the global optimal solution in this direction. Thus, the step length should be unchanged in this direction. Based on the theory above, a boundary should be set which is called allowable changing rate δ. In normal circumstances, the parameter δ equals 1.2. If the amplitude change of a basic point is more than the allowable changing rate ($b_k^{(j-1)} / b_k^j > \delta$) in some direction, that means the new basic point is far from the global optimal solution in this direction. If the amplitude change of a basic point is less than the allowable changing rate ($b_k^{(j-1)}/b_k^j < \delta$ & $b_k^j/b_k^{(j-1)} < \delta$) in some direction, that means the new basic point is close to the global optimal solution in this direction. Thus, the step length should lessen.

According to the main idea of the improved step acceleration method, the search path can be described with the dotted line. The first search path of the improved step acceleration method is the same as that of unimproved. The first path of detection search is point 1→point 2→point 3→point 4→point5, and the first optimal function is point 3. The value of objective function at point 3 is less than that of point 1, so the first detection search is successful. Then carry on the mode moving to the point 6. Compared with the point 3, the amplitude change of point 6 is less than the allowable changing rate δ in the horizontal direction. According to the improved step length method, the step length of the second detection search will multiply a ratio β. In normal circumstances, the ratio β equals 0.5. Thus, the second path of detection search is point 3→point 7→point 17. The value of objective function at point 17 is less than that of point 6, so the second detection search is successful. Then carry on the mode moving to the point 11. Carry on the third detection search with the new reference point 11. The third path of detection search is point 8→point 16→point 18→point 14. The value of objective function at point 14 is less than that of point 11, so the third detection search is successful. Then carry on the mode moving to the point 16. But the value of objective function at point 16 is more than that of point 14. Thus, start the next detection search with the point 14. Compared with point 11, the amplitude change of point 14 is less than the allowable changing rate δ in the vertical direction. Thus, the step length of next detection search will multiply the ratio β. The procedures will continue in cycles. It will not come to a stop until the value of objective function changes too much. As can be seen form figure 1, the ultimate iterative value by the improved step acceleration method is closer to the global optimal solution than that of unimproved.

2.2 Realization of the Improved Algorithm

The realization procedures are as follows:

(1) Transform the to-be-optimized variable from its approximate range (a,b) to the range (0,1) by formula 4. Then, the parameter will be a chaotic variable.

$$X_i = \frac{X - a}{b - a} . \qquad (4)$$

Set the original value of the variable between 0 and 1. But 0.25, 0.5, 0.75 can not be set [7]. If the initial value is 0.25, 0.5 or 0.75, the orbit may be steady and the chaotic algorithm can not be used. Initialize the objective function J^*.

(2) Carry on the rough search with formula 1.

(3) Transform the chaotic variable to the original range with the formula 5.

$$X_i = X * (b - a) + a . \qquad (5)$$

(4) Calculate the value of objective function J. If $J < J^*$, then $J^* = J$. Set K = K+1.

(5) If K > num (num is the set number of optimization), the subprime solution X' can be obtained. Otherwise, go back to procedure (2).

(6) Carry on the refined search with the original value X' by the improved step acceleration method.

(7) If the value of objective function J^* does not change too much, the corresponding X_k^* can be seen as the optimal solution.

According to the procedures above, the more precise solution can be obtained with the improved algorithm than that of the unimproved one. The flow chart is as follows:

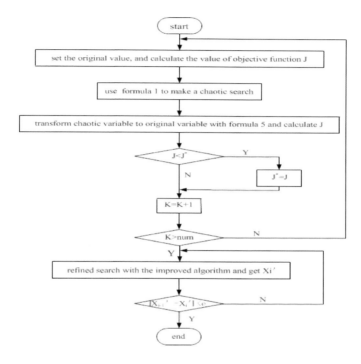

Fig. 2. The flow chat of the improved algorithm

3 Simulation Results

Set the transfer function of a system:

$$G(s) = \frac{K * \exp(-T_3 s + 1)}{(T_1 s + 1)(T_2 s + 1)} \quad . \quad (6)$$

K=1, T_1=T_2=0.435, T_3=0.17. Select ITAE Criterion as the objective function which is as follows:

$$J = \sum_{i=0}^{Ts} t_i * e(k) \quad . \quad (7)$$

Ts is the number of sample points. Thus, the problem can be transformed to find the best K_p, K_i, K_d which can make the value of objective function J least.

Based on improved step acceleration method, the structure schematic of optimization is as follows:

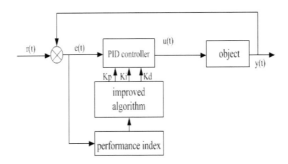

Fig. 3. The structure schematic of PID optimization

The output control law of PID controller is as follows:

$$u(k) = K*e(k) + K_i * \sum_{i=0}^{k} e(i) + K_d *[e(k)-e(k-1)] \:. \tag{8}$$

In the formula, $K_i=K_p*(T/T_i)$, $K_d=K_p*(T_d/T)$. Ti is the integral time constant. Td is the differential time constant. And T is the sample time constant.

In normal circumstances, the original value can be confirmed as follows:

$$K_p = \frac{T_1+T_2}{K*T_3}, T_i = T_1+T_2, T_d = \frac{T_1*T_2}{T_1+T_2} \:. \tag{9}$$

Set the parameter num properly, and the best optimal solution can be obtained. Compare the improved algorithm with the unimproved one, the optimizing result is as follows:

Table 1. Result of the two kinds of optimization

method	K_p	K_i	K_d	J
improved chaotic algorithm	2.636	2.227	0.443	0.671
unimproved chaotic algorithm	2.438	1.987	0.412	0.421

Set the parameters searched by the improved and unimproved algorithms. Simulate with the amplitude of 10, the response curves are as follows:

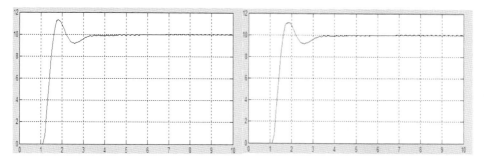

Fig. 4. Response curve with unimproved algorithm

Fig. 5. Response curve with improved algorithm

In figure 4, the overshoot of the system is 12.4%, while that of the system in figure 5 is 8.76%. That means the parameters searched with improved algorithm is closer to the global optimal solution than the unimproved one, which proves that the superiority and feasibility of the improved algorithm.

4   Conclusion

The paper introduces an advanced algorithm combined the chaotic algorithm with the improved step acceleration method. The simulation results show that the parameters searched with the improved algorithm is closer to the global optimal solution than the unimproved ones, which can decrease the overshoot. The convergence speed of the improved algorithm is fast, and it can be realized easily.

Acknowledgement

This work is supported by national natural science foundation of P. R. China Grant #61004083.

References

1. Lv, Q., Wu, Y.: Research Based on Chaotic PID Parameters Optimization Control Method to the Servo System. Journal of System Simulation 18, 750–752 (2006)
2. Kozic, S., Hasler, M.: Low-density Codes Based on Chaotic Systems for Simple Encoding. IEEE Transactions on Circuits and Systems I: Regular Papers 56, 405–415 (2009)
3. Wu, Y., Lv, Q., Tian, D.: Study on Adaptive Fuzzy PID Parameters Optimization Based on Chaos. Journal of System Simulation 20, 5224–5227 (2008)
4. Xu, H., Ma, M.: Application in the Nutrients System by Chaos Step Acceleration Method. Journal of Agriculture Mechanization Research 10, 168–170 (2007)
5. Duan, B., Sun, T., Li, Z., Mei, G., Zhang, G.: Improve Chaos GA-PID- Control in Digital Invert Power Supply. In: The 9th International Conference on Electronic Measurement and Instruments, Beijing, pp. 315–318 (2009)
6. Zhang, W., Xue, J.: Optimization Algorithm. NeuPress, Shenyang (2004)
7. Wu, T., Cheng, Y., Tan, J., Zhou, Y.: The Application of Chaos Genetic Algorithm in the PID Parameter Optimization. In: Proceedings of 2008 3rd International Conference on Intelligent System and Knowledge Engineering, pp. 230–234. Xiamen (2008)

Generalizations of the Second Mean Value Theorem for Integrals

Chen Hui-Ru[1] and Shang Chan-Juan[2]

[1] College of Mathematics and Information Science,
Huanggang Normal University, Huangzhou 438000, Hubei, China
[2] Institute of Education, Tibet Institute for Nationalities, Xianyang 712000, Shanxi, China

Abstract. The paper systematically summarizes such promotions as conclusion to open interval, to arbitrary point of the interval and to infinite interval. Meanwhile, the paper will further the exploration of new promotion on this basis.

Keywords: mathematical analysis, the second mean value theorem for integrals, open interval, infinite interval.

1 General Generalization

Theorem 1.1 [1]
(1) Suppose $g(x)$ is monotone decreasing and nonnegative on $[a,b]$, $f(x)$ is integrable on $[a,b]$, $n \in R$ and $n \geq 1$, then there exist $\xi \in [a,b]$, cause

$$\int_a^b f(x)g(x)dx = ng(b)\int_a^\xi f(x)dx ; \qquad (1.1)$$

(2) Suppose $g(x)$ is monotone increasing and nonnegative on $[a,b]$, $f(x)$ is integrable on $[a,b]$, $n \in R$ and $n \geq 1$ then there exist $\xi \in [a,b]$, cause

$$\int_a^b f(x)g(x)dx = ng(b)\int_\xi^b f(x)dx ; \qquad (1.2)$$

(3) Suppose $g(x)$ is monotone and nonnegative on $[a,b]$, $f(x)$ is integrable on $[a,b]$, $m,n \in R$ and $m \leq 1 \leq n$, then

(i) when $g(x)$ is monotone decreasing, there exist $\eta \in [a,b]$, cause

$$\int_a^b f(x)g(x)dx = ng(a)\int_a^\eta f(x)dx + mg(b)\int_\eta^b f(x)dx . \qquad (1.3)$$

(ii) when $g(x)$ is monotone increasing, there exist $\xi \in [a,b]$, cause

$$\int_a^b f(x)g(x)dx = mg(a)\int_a^\xi f(x)dx + ng(b)\int_\xi^b f(x)dx . \qquad (1.4)$$

The (1)(2) in the second mean value theorem for integrals are respectively special case of the (1)(2) in theorem 1.1 when $n=1$, although in contrast with the (3) in the second mean value theorem for integrals, the (3) in theorem 1.1 add this condition which is $g(x)$ is nonnegative on $[a,b]$, the conclusion is more meticulous than that of the second mean value theorem for integrals. Therefore, theorem 1.1 is the improvement and generalization.

2 Generalizing the Theorem to Open Interval

In the second mean value theorem for integrals, the ξ which meet the conclusion is limited in $[a,b]$, then if the condition remain unchanged, whether could the conclusion of theorem be sharpened to (a,b)? It turns out to be it can't, that is to say the ξ which meet the second mean value theorem for integrals can't be limited on the open interval (a,b), ξ might not be get in the open interval. The following two examples can explain

Case 2.1

$$f(x) = \sin x, \quad g(x) = \begin{cases} 0 & x = 2\pi \\ -1 & x \neq 2\pi \end{cases}$$

Only when $\xi = 0$ or 2π, the second mean value theorem for integrals can be true, while as $\xi \notin (0, 2\pi)$.

Case 2.2

In the interval $[a,b]$, $f(x) = 1$, $g(x) = \begin{cases} 1 & x \in [a,b] \\ 2 & x = b \end{cases}$, if there exist ξ cause

$$\int_a^b f(x)g(x)dx = g(a)\int_a^\xi f(x)dx + g(b)\int_\xi^b f(x)dx,$$

that is $b - a = g(a)(\xi - a) + 2(b - \xi) = 2b - a - \xi$, so there is $\xi = b$, and ξ is on the extreme point of $[a,b]$ from this point, we can think about that under what condition, we can get the ξ in the second mean value theorem for integrals in the open interval (a,b)? Searching for such condition is undonbtly meaningful.

Theorem 2.1 [2]

(1) Suppose $g(x)$ is incessant on $[a,b]$, $f(x)$ has continual derivative function $f'(x)$ on $[a,b]$, and $f'(x) \geq 0$, $f(a) \geq 0$ there must be $\xi \in [a,b]$, cause

$$\int_a^b f(x)g(x)dx = f(b)\int_\xi^b g(x)dx.$$

(2) $f'(x) \leq 0$, $f(b) \geq 0$ then there exist

$$\int_a^b f(x)g(x)dx = f(a)\int_a^\xi g(x)dx \ (a < \xi < b).$$

(3) If $f(a) \geq 0$ is canceled in (1), or $f(b) \geq 0$ is canceled in (2), the conclusion is

$$\int_a^b f(x)g(x)dx = f(a)\int_a^\xi g(x)dx + f(b)\int_\xi^b g(x)dx.$$

Although the condition in theorem 2.1 is very strong, the mid value point $\xi \in [a,b]$ in the conclusion can be sharpened to $\xi \in (a,b)$ [3].

Theorem 2.2 [4]

(1) If function $f(x)$ and $g(x)$ are bounded and integrable, function $g(x)$ were monotone in (a,b), $g(a+0) \neq g(b-0)$, then there exist $\xi \in (a,b)$, cause

$$\int_a^b f(x)g(x)dx = g(a+0)\int_a^\xi f(x)dx + g(b-0)\int_\xi^b f(x)dx,$$

(2) Suppose function $f(x)$ and $g(x)$ are bounded and integrable on $[a,b]$, function $g(x)$ is monotone increasing and nonnegative in (a,b), $g(a+0) \neq g(b-0)$, than there exist $\xi \in (a,b)$, cause

$$\int_a^b f(x)g(x)dx = g(b-0)\int_\xi^b f(x)dx.$$

From theorem 2.2 we can know that, under weaker condition of the second mean value theorem for integrals, if we add $g(a) \neq g(b)$, the mid value point $\xi \in [a,b]$ can be sharpened as $\xi \in (a,b)$.

The ξ in theorem 2.2 might not be only, let's look at the example below

Case 2.3

Suppose $f(x) = \sin x$, $g(x) = x$, $x \in [0, 2\pi]$. It's easy to know that, when $\xi = \dfrac{\pi}{2}$ or $\dfrac{3\pi}{2} \in (0, 2\pi)$, all have

$$\int_0^{2\pi} f(x)g(x)dx = g(0+0)\int_0^\xi f(x)dx + g(2\pi - 0)\int_\xi^{2\pi} f(x)dx,$$

$$\int_0^{2\pi} f(x)g(x)dx = g(2\pi - 0)\int_\xi^{2\pi} f(x)dx.$$

Now the problem is, under what condition the ξ is only? Then let's talk about the sufficient condition for ξ be the only one in the open interval (a,b).

Theorem 2.3 [4]
(1) If function $f(x)$ and $g(x)$ meet the condition of 2.2(1) in $[a,b]$, and $f(x)$ is identically positive (or negative), then there exist only $\xi \in (a,b)$, cause

$$\int_a^b f(x)g(x)dx = g(a+0)\int_a^\xi f(x)dx + g(b-0)\int_\xi^b f(x)dx.$$

(2) If function $f(x)$ and $g(x)$ meet the condition of 2.2(2) in $[a,b]$, and $f(x)$ is identically positive (or negative), then there exist only $\xi \in (a,b)$, cause

$$\int_a^b f(x)g(x)dx = g(b-0)\int_\xi^b f(x)dx.$$

3 Generalizing the Theorem to Any Point on the Interval

The conclusion of the second mean value theorem for integrals just exist one point, obviously it has limit, we hope that the conclusion can be true on any point in the interval, thus we give the conclusion below

Theorem 3.1 [5, 6]
Suppose function $f(x)$ is integrable on the closed interval $[a,b]$.

(i) If function $g(x)$ is (strictly monotore) decreasing on the closed interval $[a,b]$, and $g(x) \geq 0$, $f(x) > 0$ then for arbitrary point $\xi \in [a,b]$, there must be two different points $\alpha, \beta \in [a,b]$, meet the condition $\alpha < \xi < \beta$, cause

$$\int_\alpha^\beta f(x)g(x)dx = g(\alpha)\int_\alpha^\xi f(x)dx.$$

(ii) If function $g(x)$ is (strictly monotone) increasing on the close interval $[a,b]$, and $g(x) \geq 0, f(x) > 0$, then for arbitrary point $\eta \in [a,b]$, there must be two different points $\alpha, \beta \in [a,b]$, $\alpha < \eta < \beta$, cause

$$\int_\alpha^\beta f(x)g(x)dx = g(\beta)\int_\eta^\beta f(x)dx.$$

(iii) If function $g(x)$ is (strictly monotone) on the closed interval $[a,b]$, $f(x) > 0$ ($f(x) < 0$), then for arbitrary point $\xi \in [a,b]$,, there must be two different points $\alpha, \beta \in [a,b]$ and $\alpha < \xi < \beta$ cause

$$\int_\alpha^\beta f(x)g(x)dx = g(\alpha)\int_\alpha^\xi f(x)dx + g(\beta)\int_\xi^\beta f(x)dx$$

In theorem 3.1, after adding some condition, we generalize he conclusion to arbitrary point in the interval, but the condition is too sharp, then we get the same result under weaker condition.

Theorem 3.2 [7, 8]
Suppose function $f(x)$, $g(x)$ are integrable and identically positive (or negative) in the closed interval, to arbitrary $\xi \in (a,b)$

(i) If function $g(x)$ is locally strictly monotone decreasing (or increasing) at ξ, then there exist $\alpha, \beta \in [a,b]$, and $\alpha < \xi < \beta$ cause

$$\int_\alpha^\beta f(x)g(x)dx = g(\alpha)\int_\alpha^\xi f(x)dx.$$

(ii) If function $g(x)$ is locally strictly monotone decreasing (or increasing) at ξ, then there exist $\alpha, \beta \in [a,b]$, and $\alpha < \xi < \beta$ cause

$$\int_\alpha^\beta f(x)g(x)dx = g(\beta)\int_\eta^\beta f(x)dx.$$

We can also use the means above to improve the third form of the second mean value theorem for integrals to:

(iii) Suppose function $f(x)$ is integrable on the closed interval $[a,b]$, $f(x)$ is identically positive (or negative), to arbitrary point $\xi \in (a,b)$, if $g(x)$ is locally strictly monotone at ξ, then there exist $\alpha, \beta \in [a,b]$, meet the condition $\alpha < \xi < \beta$, cause

$$\int_\alpha^\beta f(x)g(x)dx = g(\alpha)\int_\alpha^\xi f(x)dx + g(\beta)\int_\xi^\beta f(x)dx.$$

4 Generalizing the Theorem o Infinite Interval

In the second mean value theorem for integrals, although finite invertal $[a,b]$ can't be sharpened to (a,b) freely, it can be generalived to $[a,+\infty)$ or $(-\infty,b]$ or $(-\infty,+\infty)$, Next I will give specific explain and proof.

Theorem 4.1[9]
Suppose $g(x)$ is monotone bounded on $[a,+\infty)$, $f(x)$ is integrable on $[a,+\infty)$, and $f(x)$ has no except $+\infty$, then there exist $\xi \in [a,+\infty)$, cause

$$\int_a^{+\infty} f(x)g(x)dx = g(a)\int_a^\xi f(x)dx + g(+\infty)\int_\xi^{+\infty} f(x)dx \text{ Here}$$

$$g(+\infty) = \lim_{x \to \infty} g(x).$$

Theorem 4.2

Suppose $g(x)$ is monotone bounded on $(-\infty,+\infty)$, $f(x)$ is integrable, and $f(x)$ has except $-\infty$ and $+\infty$, then there exist $\xi \in (-\infty,+\infty)$, cause

$$\int_{-\infty}^{+\infty} f(x)g(x)dx = g(-\infty)\int_{-\infty}^{\xi} f(x)dx + g(+\infty)\int_{\xi}^{+\infty} f(x)dx,$$

Here $g(+\infty) = g(-\infty) = \lim_{x \to \infty} g(x)$.

References

1. Zhang, Q.-z.: Generalization of the Mean Value Theorem of Integrals. Journal of Shangqiu Teachers College 24(6), 120–122 (2008)
2. Department of Mathematics of Jilin University Mathematics, Mathematics Analysis, pp. 205–207. People's education press, Beijing (1979)
3. Jin, Y.-g.: On the calculus Mean Theorem. Journal of Chongqing Normal University (Natural science edition) 19(3), 83–85 (2004)
4. Yu, H.-l.: Journal of Shandong Normal University (Natural Science) 19(3), 83–85 (2004)
5. Li, K.-d.: Converse Proposition of the second mean Value Theorem for Integrals. Huanghuai Journal 10(1), 67–69 (1994)
6. Feng, M.-q.: An improvement on the Mean Value Theorem of Integral. Journal of Beijing Institute of Machinery 22(4), 40–43 (2007)
7. Lou, M.-z.: A Remark of the second mean Value Theorem for Integrals. Huanghuai Journal 10(3), 65–66 (1994)
8. Yu, L.-f.: Understanding of the Calculus Mean Value Theorem. Ningbo City College of Vocational Technology 22(2), 24–29 (2006)
9. Zhu, B., Wang, L.: Some generalizations and Applications of the second mean Value Theorem for Integrals. Henan University of Technology 20(12), 49–50 (2005)

Research into Progressiveness of Intermediate Point and Convergence Rate of the Second Mean Value Theorem for Integrals

Chen Hui-Ru and He Chun-Ling

College of Mathematics and Information Science, Huanggang Normal University, Huangzhou 438000, Hubei, China

Abstract. The intermediate of the second mean value theorem of integrals have been studied from two aspects: the general position of intermediate in some sector can be confirmed, which provides a new method for approximation, through asymptotic studies on intermediate of the second mean value theorem of integrals. For another, This paper attempts to further explore the convergence rate of intermediate.

Keywords: the second mean value theorem for integrals, intermediate point, progressiveness, convergence rate.

The mean value theorems in the mathematic analysis are existence theorems, which can be applied to the intermediate point in a certain interval of a specific equation. These theorems only affirm the existence of intermediate point, but they do not Point out a method to find it. By our study into progressiveness of intermediate point the general positron of it in a certain internal can be ascertained, which provide a new approach to approximate calculation.

1 The Case of Interval Length Approaching Zero

In this case progressiveness of intermediate point of mean value theorem is: when $x \to a$, that is, the interval length approaches zero, the intermediate point ξ is in limit position of interval (a, x). Here are the three forms of the seamed mean value theorem for integrals:

1 Suppose $g(t)$ is continuous, monotone; nonincraeasing and non-negative in $[a, x]$, besides, $g(x) \neq g(a) \neq 0$ and $f(t)$ is integrable in $[a, x]$, then there at least exists a ξ in (a, x), which makes

$$\int_a^x f(t)g(t)dt = g(a)\int_a^\xi f(t)dt .\qquad(1.1)$$

2 Suppose $g(t)$ is continuous, nonotone, nondecreasing and nonnegative in $[a,x]$, besides, $g(x) \neq g(a)$ and $f(t)$ is integrable, then there a least exists a ξ in (a,x), which makes

$$\int_a^x f(t)g(t)dt = g(x)\int_\xi^x f(t)dt. \tag{1.2}$$

3 Suppose $g(t)$ is monotone in $[a,x]$ and $g(x) \neq g(a)$, besides, $f(t)$ is integrable in $[a,x]$, the there at least exists a ξ which makes

$$\int_a^x f(t)g(t)dt = g(a)\int_a^\xi f(t)dt + g(x)\int_\xi^x f(t)dt. \tag{1.3}$$

According to the three forms of the second mean value theorem for integrals above, the results of Theorem 1.1, Theorem 1.2 and Theorem 1.3 can be got as follows:

Theorem 1.1. Under the condition Form 1 of the second mean value theorem for integrals, let $\lim_{t \to a} f(t)/(t-a)^\alpha = A$, the intermediate point ξ in Formula (3.1) $\in (a,x)$ has a progressive estimation formula

$$\lim_{x \to a}[\frac{\xi-a}{x-a}]^{a+1} = 1 \text{ 且}$$

$$\left|[\frac{\xi-a}{x-a}]^{a+1} - 1\right| \leq \frac{\alpha+1}{|A|}[\frac{1}{|f(a)|}\omega(p', x-a) + \omega(\varphi, x-a)].$$

in which A is a nonzero constant and α is a real number and $\alpha \geq 0$, besides, $\omega(p', x-a)$ and $\omega(\varphi, x-a)$ are the continuous models of single-sided pointwise of $[\int_a^x f(t)g(t)dt/(x-a)^\alpha]'$ and $\int_a^x f(t)dt/(x-a)^{\alpha+1}$ when they are at point a.

If the condition of $\lim_{x \to a} g(x) = B$ is added to Theorem 1.1, then in Formula (1.1) the intermediate point $\xi \in (a,x)$ has a progressive estimation formula

$$\lim_{x \to \infty}\frac{\xi-a}{x-a} = [\frac{B}{g(a)}]^{1/(\alpha+1)}$$ in which A, B are constants and $A \neq 0$ and $\alpha \geq 0$ [2].

Theorem 1.2. Under the condition of Form 2 of the second mean value theorem for integrals, let $\lim_{t \to a} g(t)/(t-a)^\alpha = b$ and $\lim_{t \to a} f(t)/(t-a)^\beta = c$, then the intermediate point ξ in Formula (1,2) $\in (a,x)$ has a progressive estimation a formula

$$\lim_{x \to a} \frac{\xi - a}{x - a} = [\frac{\alpha}{\alpha + \beta + 1}]^{1/(\beta+1)},$$

in which b, c are nonzero constants and α, β are real numbers, besides, $\alpha \geq 0$ and $\beta \geq 0$.

Theorem 1.3. Under the condition of Form 3 of the second mean value theorem for integrals, let $\lim_{t \to a}[g(t) - g(a)]/(t-a)^\alpha = A$ and $\lim_{t \to a}[f(t)/(t-a)^\beta] = B$, then in Formula (1,3) the intermediate point $\xi \in (a, x)$ has a progressive estimation formula

$$\lim_{x \to a} \frac{\xi - a}{x - a} = [\frac{\alpha}{\alpha + \beta + 1}]^{1/(\beta+1)}.$$

In which A, B are nonzero constants and α, β are real numbers, besides, $\alpha \geq 0$ and $\beta \geq 0$.

If change $\lim_{t \to a}[g(t) - g(a)]/(t-a)^\alpha = A$ into $\lim_{t \to a} g'(t)/(t-a)^\alpha = A \neq 0$, the conclusion should be changed into accordingly $\lim_{x \to a} \frac{\xi - a}{x - a} = [\frac{\alpha + 1}{\alpha + \beta + 2}]^{1/(\beta+1)}$ [2]. The conclusion show that in the second mean value theorem for integrals when the internal length of $[a, b]$ is short enough then the intermediate point can be regarded as $a + [\frac{\alpha + 1}{\alpha + \beta + 2}]^{\frac{1}{\beta+1}}(b-a)$ approximately. But when it is replaced, how about the error? The theorems below give a satisfactory answer.

Theorem 1.4[2]. Suppose functions $f(t)$ and $g(t)$ satisfy those conditions

(1) $f(t)$ is monotone and differentiable in $[a, x]$, and

$$\lim_{t \to a} g'(t)/(t-a)^\alpha = A \neq 0, \quad (\alpha \geq 0);$$

(2) $g(t)$ is continuous in $[a, x]$, and $\lim_{t \to a}[f(t)/(t-a)^\beta] = B$, $(\beta \geq 0)$;

(3) $\xi = a + [\frac{\alpha + 1}{\alpha + \beta + 2}]^{\frac{1}{\beta+1}}(x-a)$.

Let $R(x) = \int_a^x f(t)g(t)dt - g(a)\int_a^\xi f(t)dt - g(b)\int_\xi^x f(t)dt$, then

$$R(x) = o((x-a)^{\alpha+\beta+2}), (x \to a).$$

Theorem 1.3 can be derived from Theorem 1.2. Suppose $g(t)$ is monotone increasing in $[a,x]$ and $G(t) = g(a) - g(t)$, then $g(t)$ is monotone nondecreasing and nonnegative, besides, $g(t) \neq g(a) = 0$ and Formula 1.3 is equal to

$$\int_a^x G(t)g(t)dt = G(x)\int_\xi^x f(t)dt,$$

Therefore $G(t)$ and $f(t)$ satisfy the conditions of Theorem 1.2. And then theorem 1.3 can be derived from Theorem 1.2, which is a similar to the way of using Theorem 1.2 and Theorem 1.3 to estimate convengence rate of intermediate point with the help of Theorem 1.1.

In text [3] the results of Theorem 1.5, 1.6 and 1.7 are as follows:

Theorem 1.5. Suppose $f(t)$ is a integrable function and $g(t)$ is a nonnegative monotone decreasing function in $[a,x]$, and $g(a) > 0$ and $g(x) = 0$, $g(a) = \cdots = g^{(m-1)}(a) = 0$, $g^{(m)}(a) \neq 0$, $f(t)$ satisfies $f(a) = f'(a) = \cdots f^{(n-1)}(a)$, $f^{(n-1)}(a) \neq 0$, then the intermediate point ξ ascertained by the second mean value Formula (1.1) for integrals satisfies

$$\lim_{x \to a} \frac{\xi - a}{x - a} = \sqrt[n+1]{\frac{m}{m+n+1}}.$$

Theorem 1.6. Suppose $f(t)$ is a integrable function and $g(t)$ is a nonnegative monotone increasing function in $[a,x]$, and $g'(a) = \cdots = g^{(m-1)}(a) = 0$, $g^{(m)}(a) \neq 0$, $f(t)$ satisfies $f(a) = f'(a) = \cdots f^{(n-1)}(a) = 0$, $f^{(n)}(a) \neq 0$. Then the intermediate point ξ ascertained by the second mean value Formula (1.2) for integrals satisfies

$$\lim_{x \to a} \frac{\xi - a}{x - a} = \sqrt[n+1]{\frac{m}{m+n+1}}.$$

Theorem 1.7. suppose $f(t)$ is a integrable function and $g(t)$ is a monotone function in $[a,x]$, $g'(a) = \cdots = g^{(m-1)}(a) = 0$, $g^{(m)}(a) \neq 0$, $f(t)$ satisfies $f(a) = f'(a) = \cdots f^{(n-1)}(a) = 0$, $f^{(n)}(a) \neq 0$, then the intermediate point ascertained by the second mean value Formula (1.3) for integrals satisfies

$$\lim_{x \to a} \frac{\xi - a}{x - a} = \sqrt[n+1]{\frac{m}{m+n+1}}.$$

2 The Case of Interval Length Tending to Infinity

Discuss about the limit position of the intermediate point ξ in interval $[a,+\infty)$ when $x \to +\infty$. Therefore it needs talking about the second mean value theorem for integrals in interval $[a,+\infty)$ which has been discussed as above and the conditions of function will be emphasized. Here are the three forms of the second mean value theorem for integrals:

1. Suppose function $g(x)$ is integrable, monotone docnoasing and nonnegative in $[a,+\infty)$, besides, $g(x) \neq g(a) \neq 0$ and $f(x)$ is integrable in $[a,+\infty)$, then $\forall x \in (a,+\infty)$, $\exists \xi \in (a,x)$, cause

$$\int_a^x f(t)g(t)dt = g(a)\int_a^\xi f(t)dt. \tag{2.1}$$

2. Suppose function $g(x)$ is integrable, monotone incensing and nonnegative in $[a,+\infty)$, besides, $g(x) \neq g(a)$ and $f(x)$ is integrable in $[a,+\infty)$, then $\forall x \in (a,+\infty)$, $\exists \xi \in (a,x)$, cause

$$\int_a^x f(t)g(t)dt = g(x)\int_\xi^x f(t)dt. \tag{2.2}$$

3 Suppose $g(x)$ is integrable and monotone in $[a,+\infty)$, besides $g(x) \neq g(a)$ and $f(x)$ is integrable in $[a,+\infty)$, then $\forall x \in (a,+\infty)$, $\exists \xi \in (a,x)$, cause

$$\int_a^x f(t)g(t)dt = g(a)\int_a^\xi f(t)dt + g(x)\int_\xi^x f(t)dt \tag{2.3}$$

About the three forms of the second mean value theorem for integrals text [4] covers the results of Theorem 2.1, 2.2 and 2.3 as follows:

Theorem 2.1. Under the condition of Form 1 of the second mean value theorem for integrals, additionally suppose $\lim_{x\to\infty} g(x) = A$, $\lim_{x\to\infty} x^\alpha f(x) = B$. Then the intermediate point ξ in Formula (2.1) $\in (a,x)$ has a progressive estimation formula

$$\lim_{x\to\infty} \frac{\xi - a}{x - a} = [\frac{A}{g(a)}]^{1/(1-\alpha)}.$$

in which A, B are nonzero constants and α is a constant, $\alpha < 1$.

Theorem 2.2. Under the condition of Form 2 of the second mean value theorem for integrals, additionally suppose that $\lim_{x\to\infty} x^\alpha g(x) = A$, $\lim_{x\to\infty} x^\beta f(x) = B$. Then the intermediate point ξ in Formula (2.2) $\in (a,x)$ has a progressive estimation formula

$$\lim_{x \to a} \frac{\xi - a}{x - a} = [\frac{\alpha}{\alpha + \beta - 1}]^{1/(1-\beta)}.$$

in which A, B are nonzero constants, α and β are real numbers, besides, $\alpha < 0$, $\beta < 1$.

Theorem 2.3. Under the condition of Form 3 of the second mean value theorem for integrals, additionally suppose that $\lim\limits_{x \to \infty} x^\alpha g(x) = A$, $\lim\limits_{x \to \infty} x^\beta f(x) = B$, Then the intermediate point ξ in Formula (2,3) $\in (a, x)$ has a progressive estimation formula

$$\lim_{x \to a} \frac{\xi - a}{x - a} = [\frac{\alpha}{\alpha + \beta - 1}]^{1/(1-\beta)}$$

in which A, B are nonzero constants, α and β are real numbers, besides, $\alpha \geq 0$, $\beta \geq 0$.

Morever by using Taylor Expansion some new conclusions can be drawn like Theorem 2.4, 2.5 and 2.6 which are to be presented in text [5].

Theorem 2.4. Suppose functions $f(t)$ and $g(t)$ satisfy these conditions:

(1) $g(t)$ is monotone decreasing and nonnegative in $[a, x]$ and $\lim\limits_{t \to +\infty} g(t) = K$;

(2) $f(t)$ has a continuous derivative of order $m+1$ and $f^i(a) = 0$, $i = 0,1,2,\cdots,m-1$, $f^m(a) \neq 0$;

(3) $\lim\limits_{x \to +\infty} \frac{f(x)}{(x-a)^{m+1}} = B \neq 0$;

Then the intermediate point ξ of Formula (2.1) satisfies

$$\lim_{x \to +\infty} \frac{\xi - a}{x - a} = [\frac{K}{g(a)}]^{\frac{1}{m+2}}.$$

Theorem 2.5. Suppose functions $f(t)$, $g(t)$ satisfy there conditions:

(1) $g(t)$ is monotone increasing in $[a, x]$ and has a continuous derivative of order $n+1$

(2) $f(t)$ has a continuous derivative of order $m+1$ and $f^i(a) = 0$, $i = 0,1,2,\cdots,m-1$, $f^m(a) \neq 0$;

(3) $\lim\limits_{x \to +\infty} \frac{g(x)}{(x-a)^{n+1}} = A$, $\lim\limits_{x \to +\infty} \frac{f(x)}{(x-a)^{m+1}} = B$, and $AB \neq 0$;

Then the intermediate point ξ in Formula (2.2) satisfies

$$\lim_{x \to +\infty} \frac{\xi - a}{x - a} = [\frac{n+1}{m+n+3}]^{\frac{1}{m+2}}$$

Theorem 2.6. Suppose functions $f(t)$, $g(t)$ satisfy these conditions:

(1) $g(t)$ has a continuous derivative of order $n+1$ in $[a, x]$ and is monotone in $[a, x]$. $f(t)$ has a continuous derivative of order $m+1$ and m, n are natural hombres;

(2) $g^i(a) = 0$, $i = 1, 2, \cdots, n-1$, $g^n(a) \neq 0$, $f^j(a) = 0$, $j = 0, 1, 2, \cdots, m-1$, $f^m(a) \neq 0$;

(3) $\lim_{x \to +\infty} \frac{g(x)}{(x-a)^{n+1}} = A$, $\lim_{x \to +\infty} \frac{f(x)}{(x-a)^{m+1}} = B$, and $AB \neq 0$;

Then the intermediate point ξ in Formula (2.3) satisfies

$$\lim_{x \to +\infty} \frac{\xi - a}{x - a} = [\frac{n+1}{m+n+3}]^{\frac{1}{m+2}}.$$

References

1. Zhang, S.-y., Liu, C.-f.: The new Development of Research on the Asymptotic Behavior of the "Medial Point" of the Theorem of mean Reaction(I). Academic From of Nandu 20(6), 13–20 (2000)
2. Chen, J.-b., Mei, C.-l.: The Appoachability for "Medial Point" of the Second Integral Mean Value Theroem and Its Error Estimate. Journal of Lishui Teachers College 26(2), 6–9 (2004)
3. Yu, X.-m.: The Theorem of "Meddle Point" of the Second Integral Mean Value. Journal of Shangluo Teachers College 16(2), 13–14 (2002)
4. Zhang, S.-y.: The new Develpoment of Research on the Asymptotic Behavior of the "Medial Point" of the Theorem of mean Reaction. Journal of Xuchang Teachers College 19(2), 22–26 (2000)
5. Yu, Y.: Asymptotic Properties for the "Middle Point" of the second Mean Value Theorems of Integral 26(3), 85–88 (2009)

Flooding-Based Resource Locating in Peer-to-Peer Networks

Jin Bo

Science and Technology Department, Shunde Polytechnic, Foshan 528333, China
wlmingyue@163.com

Abstract. In peer-to-peer (P2P) networks, flooding-based resource discovery is very popular. With flooding, a node searching for a resource contacts its neighbors in the system, which in turn contact their own neighbors and so on until a node that shares the requested resource is located. Flooding process assumes no knowledge about the network topology or the resource distribution thus offering an attractive method for resource discovery in dynamically P2P systems. In this paper, according to the previous researches in flooding-based P2P networks, we first classify the existing flooding-based resource location technologies into some subcategories, and then compare the search efficiency and analyze some typical systems in each subcategory. The above comparison and analysis will become the good guidance to design new resource locating algorithm in P2P networks.

Keywords: Resource locating, flooding, search efficiency, peer-to-peer.

1 Introduction

Peer-to-Peer (P2P) networks have emerged as new Internet computing mode during the past few years. Such networks often involve thousands or millions of live nodes. To provide a good resource location service in such large-scale networks, we hope the search scheme can give high quality service, such as high query success rate and short response time, to the end users, and bring low overhead to the network.

Napster [1] pioneered the idea of peer-to-peer file sharing, in such network, a dedicated central server maintains indices of the files that are published by all the active nodes in the network. When a node wants to download files, it first sends requests for files to the server. The server searches for matches in its index, and return a list of nodes that hold the matching file. The node then opens direct connections with one or more of the nodes that hold the requested file, and downloads it. Because all the nodes in such network resort to the central serve to retrieve the interesting files, the central serve may become bottleneck of the network.

These centralized networks have been replaced by new decentralized networks such as flooding-based discovery that distributes both the download and search capabilities. These networks establish an overlay network of nodes. In this paper, we consider decentralized resource discovery mechanisms based on flooding. With flooding, a node that wants a particular resource contacts its neighbors in the systems, which in turn contact their own neighbors until a node that provides the requested resource is reached. Flooding enables resource discovery without directories or knowledge of the specific topology of the system, thus, offering an attractive mechanism for resource discovery in dynamically evolving networks. For example, Gnutella [2], a popular peer-to-peer file sharing system, utilizes some form of flooding-based discovery.

The rest of this paper is organized as follows: Section 2 describes the related work. We overview the flooding-based resource locating technologies in p2p networks in Section 3. Section 4, 5 and 6 analyze the three type searching methods p2p systems, and we conclude this paper in Section 7.

2 Related Work

There are some researchers that survey several resource sharing technologies in P2P networks. Stephanos *et al.* in [3] survey the peer-to-peer content distribution technologies. In this survey, they propose a framework for analyzing peer-to-peer content distribution technologies. their approach focuses on nonfunctional characteristics such as security, scalability, performance, fairness, and resource management potential, and examines the way in which these characteristics are reflected in—and affected by—the architectural design decisions adopted by current peer-to-peer systems. They also study the current peer-to-peer systems and infrastructure technologies in terms of their distributed object location and routing mechanisms, their approach to content replication, caching and migration, their support for encryption, access control, authentication and identity, anonymity, deniability, accountability and reputation, and their use of resource trading and management schemes.

Researchers in [4] provide a survey of major searching techniques in peer-to-peer networks. They first introduce the concept of P2P networks and the methods for classifying different P2P networks. Next, they discuss various searching techniques in unstructured P2P systems, strictly structured P2P systems, and loosely structured P2P systems. The strengths and weaknesses of these techniques are highlighted. Searching in unstructured P2Ps covers both blind search schemes and informed search schemes. Blind searches include iterative deepening, *k*-walker random walk, modified random BFS, and two-level *k*-walker random walk. Informed searches include local indices, directed BFS, intelligent search, routing indices, attenuated bloom filter, adaptive probabilistic search, and dominating set based search. The discussion of searching in strictly structured P2Ps focuses on hierarchical Distributed Hash Table (DHT) P2Ps and non-DHT P2Ps. Searching in non-hierarchical DHT P2Ps is briefly overviewed. The presentation of the hierarchical DHT P2Ps pays more attention to Kelips [5] and Coral [6], whereas that of searching in non-DHT P2Ps focuses on SkipNet [7] and TerraDir [8]. The description of searching in loosely structured P2Ps focuses on Freenet [9]. They conclude this chapter by summarizing open problems in searching the P2P networks.

Similarly, some other researchers in [10] study hybrid search schemes for unstructured peer-to-peer networks. They quantify performance in terms of number of hits, network overhead, and response time. Their schemes combine flooding and random walks, look ahead and replication. They consider both regular topologies and topologies with supernodes. They introduce a general search scheme, of which flooding and random walks are special instances, and show how to use locally maintained network information to improve the performance of searching. Their main findings include: (1) A small number of supernodes in an otherwise regular topology can offer sharp savings in the performance of search, both in the case of search by flooding and search by random walk, particularly when it is combined with 1-step replication. They quantify, analytically and experimentally, that the reason of these savings is that the search is biased towards nodes that yield more information. (2) There is a generalization of search, of which flooding and random walk are special instances, which may take further advantage of locally maintained network information, and yield better performance than both flooding and random walk in clustered topologies. The method determines edge criticality and is reminiscent of fundamental heuristics from the area of approximation algorithms.

The authors in [11] provide a quantitative comparison of full text keyword search in structured and unstructured P2P systems. They examine three techniques (and optimizations to those techniques) proposed in the literature: using a DHT along with inverted lists and Bloom filters; using a super-peer network; and using a random walk over an unstructured network. They use real web documents and user queries to measure the cost for both document publishing and query processing, in terms of bandwidth and response time. Their results show that all three techniques use roughly the same bandwidth to process queries (with the super-peer technique having a slight edge). The structured network provides the best response time (30 percent better than a super-peer network), but has a high cost of document publishing, using six times as much bandwidth as the super-peer system. The random walk technique requires no publishing, but has a very long response time unless multiple random walks operate in parallel.

3 Overview

In an overlay network where each node has several neighbors and maintains a local cache. When a node A needs a particular type of resource r, it initially searches its own cache. If it finds the resource there, it extracts the corresponding contact information and the search ends. If resource r is not found in the local cache, A sends a message querying all or a subset of its neighbors, which in turn propagate the message to their neighbors, and so on.

To avoid overwhelming the network with search requests, search is limited to a maximum number of steps, t (similar to the Time To Live, TTL) parameter in many network protocols). In particular, the search message contains a counter field initialized to t, any intermediate node that receives the message first decrements the counter by 1. If the counter value is not 0, the node proceeds as normal; otherwise, the node does not contact its neighbors and sends a positive (negative) response to the inquiring node if r is found (not found) in its cache.

When the search ends, the inquiring node A will either have the contact information for resource r or just a set of negative responses. In the latter case, node A assumes that a node offering the resource cannot be found. Note that we make no assumption about network connectivity. Disconnectedness may indeed occur because the network is dynamic. This is quite possible in P2P systems.

In this paper, we will consider three different search strategies based on what subset of its neighbors each node contacts, namely, the flooding, teeming and random paths strategies.

4 Pure Flooding

With pure flooding, node A that searches for a resource r checks its cache, and if the resource is not found there, A contacts all its neighbors. In turn, A's neighbors check their caches and if the resource is not found locally, they propagate the search message to all their neighbors. The procedure ends when either the resource is found or a maximum of t steps is reached. The scheme, in essence, broadcasts the inquiring message.

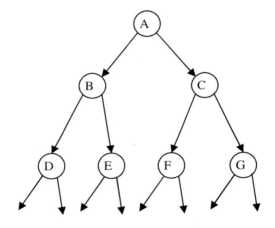

Fig. 1. Pure Flooding

As the search progresses, a d-ary tree is unfolded rooted at the inquiring node A. An example is shown in Fig. 1. Note that the term "tree" is not accurate in graph-theoretic terms since a node may be contacted by two or more other nodes, but we will use it here as it helps to visualize the situation. This search tree has (at most) d^j different nodes at the jth level, $j \geq 0$, which means that at the jth step of the search algorithm, there will be (at most) d^j different nodes contacted.

In the original Gnutella architecture [2], it uses the flooding mechanism to distribute Ping and Query messages: each Gnutella node forwards the received messages to all of its neighbors. The response messages received are routed back along the opposite path through which the original request arrived. To limit the spread of messages through the network, each message header contains a TTL field. At each hop, the value of this field is decremented, and when it reaches zero, the message is dropped.

In [12], Yang and Garcia-Molina borrowed the idea of iterative deepening from artificial intelligence and used it in P2P searching. This method is also called expanding ring. In this technique, the querying node periodically issues a sequence of BFS searches with increasing depth limits $D_1 < D_2 < \ldots < D_i$. The query is terminated when the query result is satisfied or when the maximum depth limit D_0 has been reached. In the latter case, the query result may not be satisfied. All nodes use the same sequence of depth limits called policy P and the same time period W between two consecutive BFS searches.

5 Teeming

To reduce the number of messages, a variation of flooding called teeming [13], probabilistic flooding [14], or random BFS [15] are proposed. At each step, if the resource is not found in the local cache of a node, the node propagates the inquiring message only to a random subset of its neighbors. We denote by ε the fixed probability of selecting a particular neighbor. In contrast with flooding, the search tree is not a d-ary one any more (Fig. 2). A node in the search tree may have between 0 to d children, $d\varepsilon$ being the average case. Flooding can be seen as a special case of teeming for which $\varepsilon = 1$.

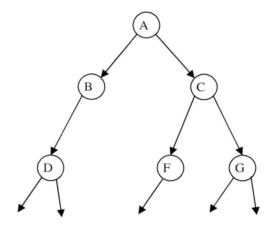

Fig. 2. Teeming

In the k-walker random walk algorithm [18], k walkers are deployed by the querying node. That is, the querying node forwards k copies of the query message to k randomly selected neighbors. Each query message takes its own random walk. Each walker periodically "talks" with the querying node to decide whether that walker should terminate. Nodes can also use soft states to forward different walkers for the same query to different neighbors. k-walker random walk algorithm attempts to reduce the routing delay. On average, the total number of nodes reached by k random walkers in H hops is the same as the number of nodes reached by one walker in kH hops. Therefore, the routing delay is expected to be k times smaller.

A similar scheme is the two-level random walk [16]. In this scheme, the querying node deploys k_1 random walkers with the TTL being l_1. When the TTL l_1 expires, each walker forges k_2 random walkers with the TTL being l_2. All nodes on the walkers' paths process the query. Given the same number of walkers, this scheme generates less duplicate messages but has longer searching delays than the k-walker random walk.

Another approach is called the modified random BFS, and it was proposed in [17]. The querying node forwards the query to a randomly selected subset of its neighbors. On receiving a query message, each neighbor forwards the query to a randomly selected subset of its neighbors 7 (excluding the querying node). This procedure continues until the query stop condition is satisfied. No comparison to the k-walker random walk was given in [17]. It is expected that this approach visits more nodes and has a higher query success rate than the k-walker random walk.

6 Random Paths

Although, depending on ε, teeming can reduce the overall number of messages, both teeming and flooding suffer from an exponential number of messages. One approach to eliminate this drawback is performing a random path or random walker [13], [18] search as follows: Each node contacts only one of its neighbors (randomly). The search space formed ends up being a single random path in the overlay network. This scheme propagates one single message along the path and the inquiring node will be expecting one single answer.

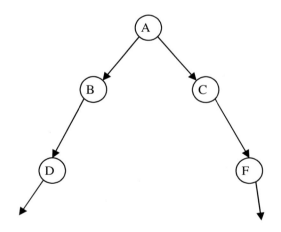

Fig. 3. Teeming

This scheme is generalized as follows: The root node (i.e., the inquiring node A) constructs $p \geq 1$ random paths. In particular, if r is not in its cache, A asks p out of its d neighbors (not just one of them). All the other intermediate) nodes construct a simple path as above, by asking exactly one of their neighbors. This way, we end up with p different paths unfolding concurrently (Fig. 3). The search algorithm produces fewer messages than flooding or teeming but needs more steps to locate a resource.

Authors in [19] proposed a novel resource locating method in open multi-agent systems. In such algorithm, each agent maintains a limited size local cache in which it keeps information about k different resources, that is, for each of k resources, it stores the contact information of one agent that provides it. This creates a directed network of caches. The general mechanism for locating a resource is as follows. The agent that requires a resource first looks at its cache. If no contact information for the resource is found in the cache, the agent selects other agent(s) from its cache, contacts them and inquires their local cache for the resource. This procedure continues until either the resource is located or a maximum number of steps is reached.

7 Conclusions

In this paper, we consider the general problem of discovering resources in a distributed environment. In particular, we study the performance of a number of flooding-based approaches to this problem. According to the previous researches in flooding-based P2P networks, we first classify the existing flooding-based searching technologies into some subcategories in P2P networks, and then compare the search efficiency and analyze some typical systems in each subcategory. The comparison and analysis can be good guidance to design new resource locating algorithm in P2P networks.

References

1. Napster (2003), http://www.napster.com
2. Gnutella (2003), http://rfc-gnutella.sourceforge.net/developer/index.html
3. Stephanos androutsellis-theotokis, and Diomidis spinellis. A survey of peer-to-peer content distribution technologies. ACM Computing Surveys 36(4), 335–371 (2004)
4. Li, X., Wu, J.: Searching techniques in peer-to-peer networks. In: Wu, J. (ed.) Handbook of Theoretical and Algorithmic Aspects of Sensor, Ad Hoc Wireless, and Peer-to-Peer Networks. CRC Press, Boca Raton (2005)
5. Gupta, I., Birman, K., Linga, P., Demers, A., Van Renesse, R.: Kelips: Building an efficient and stable P2P DHT through increased memory and background overhead. In: Kaashoek, M.F., Stoica, I. (eds.) IPTPS 2003. LNCS, vol. 2735. Springer, Heidelberg (2003)
6. Freedman, M., Mazieres, D.: Sloppy hashing and self-organizing clusters. In: Kaashoek, M.F., Stoica, I. (eds.) IPTPS 2003. LNCS, vol. 2735. Springer, Heidelberg (2003)
7. Harvey, N.J.A., Jones, M.B., Saroiu, S., Theimer, M., Wolman, A.: Skipnet: A scalable overlay network with practical locality properties. In: Proceedings of USITS, Seattle, WA, pp. 113–126 (March 2003)
8. Silaghi, B., Bhattacharjee, B., Keleher, P.: Query routing in the TerraDir distributed directory. In: Proceedings of SPIE ITCOM 2002, Boston, MA (August 2002)

9. Clarke, I., Sandberg, O., Wiley, B., Hong, T.W.: A distributed anonymous information storage and retrieval system. In: Proceedings of the ICSI Workshop on Design Issues in Anonymity and Unobservability, Berkeley, California (June 2000), http://freenet.sourceforge.net
10. Gkantsidis, C., Mihail, M., Saberi, A.: Hybrid search schemes for unstructured peer-to-peer networks. In: Proceedings of Infocom (March 2005)
11. Yong Yang, M.R., Dunlap, R., Cooper, B.F.: Performance of Full Text Search in Structured and Unstructured Peer-to-Peer Systems. In: IEEE INFOCOM (April 2006)
12. Yang, B., Garcia-Molina, H.: Improving search in peer-to-peer networks. In: Proc. of the 22nd IEEE International Conference on Distributed Computing, IEEE ICDCS 2002 (2002)
13. Dimakopoulos, V.V., Pitoura, E.: Performance Analysis of Distributed Search in Open Agent Systems. In: Proc. Int'l Parallel and Distributed Processing Symp, IPDPS 2003 (2003)
14. Banaei-Kashani, F., Shahabi, C.: Criticality-Based Analysis and Design of Unstructured Peer-to-Peer Networks as Complex Systems. In: Proc. GP2PC 2003, Third Int'l Workshop Global and Peer-to-Peer Computing (2003)
15. Kalogeraki, V., Gunopulos, D., Zeinalipour-Yazti, D.: A Local Search Mechanism for Peer-to-Peer Networks. In: Proc. 11th ACM Conf. Information and Knowledge Management, CIKM (2002)
16. Jawhar, I., Wu, J.: A two-level random walk search protocol for peer-to-peer networks. In: Proc. of the 8th World Multi-Conference on Systemics, Cybernetics and Informatics (2004)
17. Kalogeraki, V., Gunopulos, D., Zeinalipour-yazti, D.: A local search mecha nism for peer-to-peer networks. In: Proc. of the 11th ACM Conference on Information and Knowledge Management, ACM CIKM 2002 (2002)
18. Lv, Q., Cao, P., Cohen, E., Li, K., Shenker, S.: Search and Replication in Unstructured Peer-to-Peer Networks. In: Proc. 16th ACM Int'l Conf. supercomputing (ICS 2002), pp. 84–95 (2002)
19. Leontiadis, E., Dimakopoulos, V.V., Pitoura, E.: Cache Updates in a Peer-to-Peer Network of Mobile Agents. In: Proceedings of the Fourth International Conference on Peer-to-Peer Computing, P2P (2004)

A Kind of Service Discovery Method Based on Petroleum Engineering Semantic Ontology

Lixin Ren[1] and Li Ren[2]

[1] MOE Key Laboratory of Petroleum Engineering in China University of Petroleum, Beijing 102249, China
[2] Land operating area, PetroChina Jidong Oilfield Company, Tangshan 063004, China
`lixin_ren@163.com`

Abstract. In this paper, we introduce a kind of service discovery method based on petroleum engineering semantic matching. We propose an OWL (Web Ontology Language) based ontology for service description, and for supporting logic-based service matchmaking. We design a petroleum engineering semantic matching algorithm according to the ontology. The algorithm takes the petroleum engineering semantic meaning of service into account. It improves the veracity of petroleum engineering service matching, and paves the way for fuzzy matching and petroleum engineering service composition.

Keywords: Petroleum engineering, semantic, OWL, ontology, service.

1 Introduction

For petroleum engineering, users often require seamlessness to get the service. Service discovery based on semantics of this strategy will meet user requirements [1]. Service discovery is the important research field in universal environment [2]. In order to match user expectations, service discovery need consider the functional and non-functional aspects. Non-functional aspects include QoS and User Preferences etc. Meet the functional matching services may not meet the performance requirements of user expectations. But how to match functional expectations are the most basic requirements of functional expectations, matching algorithm in high-performance functions is based on non-functional factors on the results of matches. Semantic web service description language OWL-S does not describe the QoS ontology [3]-[5]. In this paper, we introduce a kind of service discovery method based on petroleum engineering semantic matching. We propose an OWL (Web Ontology Language) based ontology for service description, and for supporting logic-based service matchmaking. We design a petroleum engineering semantic matching algorithm according to the ontology.

The overall structure of the ontology of petroleum engineering is shown in Figure 1. Space and Time are semantic description of the service locations and service time, Person is used to describe the user's personal information, is mainly used for service matching. The ontology model of petroleum engineering simplifies a number of other features such as security, user privacy, etc.

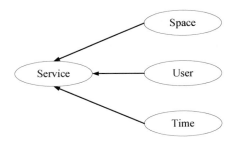

Fig. 1. Basic structure of the ontology

2 The Ontology Description

2.1 Location Description

Description of spatial information services possible methods includes: An Internet address, description of services offered by the network, URI Uniform Resource Locator used to describe, such as http ://www. sun code. cn / services/ Weather Report. Real-life address, description of services, from life, such as hospitals, schools, location, and so on. A coordinate system can be used for local area or a small environment, the location and orientation in the description of the equipment.

In the description of spatial information, the first need to define the namespace description.

```
<rdf:RDF
    xmlns =http://www.titucs/pc/space#
    xmlns:owl =http://www.w3.org/2002/07/owl#
    xmlns:rdf =http://www.w3.org/1999/02/22-rdf-syntax-ns#
    xmlns:rdfs=http://www.w3.org/2000/01/rdf-schema#
    xmlns:xsd ="http"//www.w3.org/2001/XMLSchema#">
```

Where, xmlns = http://www.suncode.cn/sapce #, define the default namespace, that is, when the document appears when the label is not prefix the namespace reference. xmlns: xsd = http://www.w3.org/2000/10/ XML Schema #,defined in the document to appear as a prefix xsd namespace entries, this is the introduction of XML Schema to define vocabulary. xmlns: owl = http://www.w3.org/2002/07/owl #, defines the term there should be looking for owl prefix namespace. OWL is a common statement in the document to add OWL vocabularies, such as owl: Object Property.

Class constraints include setting the definition of the class type ID, to determine the logic between class and class relations, and so on. Complete class definition is as follows:

```
<owl:Class rdf:ID="Space"/>
<owl:Class rdf:ID="Address">
    <rdfs:subClassOf rdf:resource="#Space">
</owl:Class>
<owl:Class rdf:ID="Coordinate">
    <rdfs:subClassOf rdf:resource="#Space">
</owl:Class>
<owl:Class rdf:ID="NetworkAddress">
    <rdfs:subClassOf rdf:resource="#Address">
</owl:Class>
<owl:Class rdf:ID="PhysicalAddress">
    <rdfs:subClassOf rdf:resource="#Address">
</owl:Class>
```

2.2 Time Description

Time ontology namespace using xmlns = http://www.suncode.cn/space #. Time Ontology property includes: startTime (Used to describe the properties of the service start time). endTime (Used to describe the properties of the service end time). That the relationship between class attributes, such as before, after, withen, equals, etc.

For Example, attributes can be defined as follows:

```
<owl:ObjectProperty rdf:ID="after">
    <rdfs:domain rdf:resource="#Afternoon">
    <rdfs:range rdf=resource="#Morning">
</owl:ObjectProperty>
```

2.3 User Information Description

User information can be divided into two categories, one for the common information such as user name, phone number, email address, etc.; another feature for the user information, such as personality, body shape, hobbies and so on.

User ontology namespace using xmlns=http://www.suncode.cn/userInfo#. Defined as follows:

```
<owl:Class rdf:ID="User">
  <rdfs:subClassOf>
    <owl:Restriction>
      <owl:onProperty rdf:resource="#Name"/>
      <owl:cardinality rdf:datatype="http://www.w3.org/2001/XMLSchema#nonNegativeInteger">1</owl:cardinality>
    </owl:Restriction>
  </rdfs:subClassOf>
  <rdfs:subClassOf>
    <owl:Restriction>
      <owl:onProperty rdf:resource="##gender"/>
      <owl:cardinality rdf:datatype="http://www/w3/org/2001/XMLSchema#nonNegativeInteger">1</owl:cardinality>
    </owl:Restriction>
  </rdfs:subClassOf>
</owl:Class>
```

2.4 Service Description

The service can be divided into five categories, including ComunicationService (mobile phone), EntertainmentServcie (TV), ComputingPerpheral, HouseholdService, UserDefinedService.

Ontology described in the service definition, at first you need to introduce Space, Time and Person ontology defined, as follows:

```
<owl:imports rdf:resource= "http://www.tjtucs.cn/pc/space.owl"/>
<owl:imports rdf:resource= "http://www.tjtucs.cn/pc/user.owl"/>
< owl:imports rdf:resource= "http://www.tjtucs.cn/pc/time.owl"/>
```

Service ontology namespace is xmlns=http://www.suncode.cn/service#。 The following illustrates how to defined service ontology to describe the specific.

```
<TV:Service rdf:about="http://www.suncode.cn/TV/movie">
<TV:hasProvider rdf:resource="#Channel2"/>
<TV:hasStartTime rdf:resource="#Channel2StartTime"/>
<TV:hasEndTime rdf:resource="#Channel2EndTime"/>
<TV:hasDescription&xsd:string>
This movie is about sports.
</TV:hasDescription>
</TV:Service>
```

This example describes one television service, which is individual of class TV. From this description, the inference engine may be knows the TV service start time, end time, description, service providers and other information, Inference engine can infer the same time the service is a subclass of Entertain Service.

3 Semantic Process Algorithm

3.1 Balance Genealogy Measure Algorithm

Through the field of information analysis and comparison of various algorithms, integrated semantic Web service matching has above characteristics. Semantic matching algorithm is based on BGM (Balance Genealogy Measure) algorithm. Set matching algorithm on the premise that the concept to be matched leaf nodes and are defined in the concept of spanning tree [6]: A concept set to the lowest common parent node, each node in the concept of set paths constitute this set of concepts of the "spanning tree". BGM algorithm main idea is that if there is more than the first tree in a leaf node and a node in the second tree is the best match, then reduce the duplication of the similarity matching. BGM checks interval parameter p [0, 1], is used to control the repetition rate of decline in the similarity matching [7].

3.2 Matching

In order to meet the requirements of seamless service based on the above algorithm by adding a QoS matching. Considering the advantages and disadvantages of a variety of QoS models, based OWL ontology language, in this paper, OWL-S service ontology to extend, define a semantic QoS model-OWL-QoS ontology. Ontology of this model includes QoSProfile, QoSAtribute, QoSMetric and QoSValue four classes.

(1) QoSProfile. Shown in Figure 2, which corresponds to all the web services CommonQosAttribute need to have the Qos parameters, such as price, response time, reliability, reputation and so on..
(2) QoSAtribute. QosAttribute is used to describe all of the QoS parameters of the superclass, as shown in Figure 3. Qos parameters can be said of such measure, monotonicity, the share of the entire QoSProfile weight and so on.

Qos parameter value used to indicate the measure, shown in Figure 10. The main types of numeric (Numerical value) and language type (Linguistical value) two. Numerical model used to describe integers, real numbers that the Qos parameters. Language is used to describe the type of values that can not be measured with Qos parameters. These are mostly adverbs of degree, like the fast, fast, slow, very slow and the like.

Both from the semantic and numerical analysis of the characteristics of Qos parameters, combined with general semantic similarity model is proposed for Qos comprehensive information matching similarity model, the specific algorithm see [8]. This model can solve the different concepts of information as described in the matching of heterogeneous Qos.

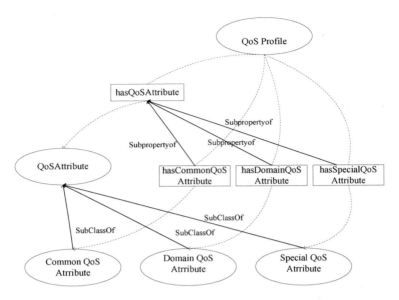

Fig. 2. QoSProfile

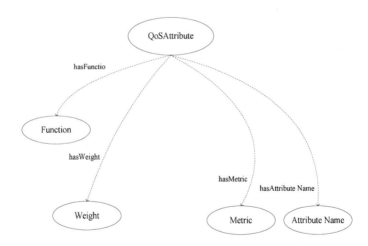

Fig. 3. QoSAtribute

4 Test

We have done many experiments and tests [9]-[11]. Now here we give some examples of them.

Program A: Based on previous similarity calculation service matching algorithms and filters to study the output sequence of the three services that most closely matches the performance indicators.

Program B: Set the threshold on the matching process (take 0.7), filtered through the threshold, the output sequence.

Program C: Adding QoS to use the proposed service discovery after matching algorithm, matching the highest QoS services.

From three matching system to assess the performance of Web services, including system response time and precision and recall. System response time which means the system from a user submits a request to return to the time sequence matching service; precision refers to the matching system to find out the results and the results given by experts in the field of coincidence degree.

Three programs examine the system response time is shown in Figure 4.

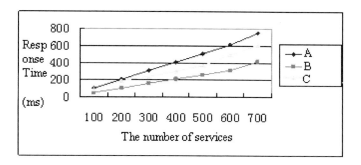

Fig. 4. Response Time

5   Conclusion

For petroleum engineering, we propose an OWL (Web Ontology Language)-based ontology for service description, and for supporting logic-based service matchmaking. Besides service description, service matchmaking is another important problem in service discovery. We introduced a semantic matching algorithm according to the ontology. The algorithm takes the semantic meaning of service into account, enable interact between user and device or between device and device. It improves the veracity of service matching, and paves the way for fuzzy matching and service composition. The experiment and tests can be seen that semantic based service discovery model has a faster response time, and precision, and can output to the user the best service to meet the requirements.

References

1. Bakhouya, M., Gaber, J., Koukam, A.: Service discovery and composition in ubiquitous computing. In: Proceedings of 2004 International Conference on Information and Communication Technologies: From Theory to Applications, pp. 489–490 (2004)

2. Dogac, A., Kabak, Y., Laleci, G.B.: Enriching ebXML registries with OWL ontologies for efficient service discovery. In: Proceedings of 14th International Workshop on Research Issues on Data Engineering: Web Services for e-Commerce and e-Government Applications, pp. 69–76 (2004)
3. Masuoka, R., Labrou, Y., Parsia, B., Sirin, E.: Ontology-enabled pervasive computing applications. Intelligent Systems 18, 68–72 (2003)
4. Zuo, z., Zhou, m.: Web Ontology Language OWL and its description logic foundation. In: Proceedings of the Fourth International Conference on Parallel and Distributed Computing, Applications and Technologies, PDCAT 2003, pp. 157–160 (2003)
5. Kopena, J., Regli, W.C.: DAMLJessKB: a tool for reasoning with the Semantic Web. Intelligent Systems 18(3), 74–77
6. Helal, S.: Standards for service discovery and delivery. IEEE Pervasive Computing 1(3), 95–100 (2002)
7. Zhang, D.G.: Web-Based Seamless Migration for Task-oriented Nomadic Service. International Journal of Distance Education Technology (JDET) 4(3), 108–115 (2006)
8. Weiser, M.: The computer for the twenty-first century. Scientific American 265(3), 94–104 (1991)
9. Wang, X.H., Zhang, D.Q., Gu, T., Pung, H.K.: Ontology based context modeling and reasoning using OWL. In: Proceedings of the Second IEEE Annual Conference on Pervasive Computing and Communications Workshops, pp. 18–22 (2004)
10. Zhang, D.G.: A kind of new decision fusion method based on sensor evidence. Journal of information and Computational Science 5(1), 171–178 (2008)
11. Zhang, D.G.: A Kind of new approach of context-aware computing for ubiquitous application. International Journal of modeling, Identification & Control 8(1), 10–17 (2009)

Study of Coal Gas Outburst Prediction Based on FNN

Yanli Chai

School of Information and Electrical Engineering
China University of Mining & Technology
Xuzhou, Jiangsu, China

Abstract. This paper made detail research on fuzzy logic system and neural network, and set up dynamic fuzzy neutal network (FNN) prediction model for serious coal and gas accident. The model can express qualitative knowledge, and also has self-learning function and ability to deal with qualitative knowledge. The paper verified the feasibility of the model by simulation experiment and realized prediction of serious coal and gas accident.

Keywords: FNN, coal and gas accident, prediction.

1 Introduction

Coal and gas outburst is a spontaneous ejection of gas and coal from the internal coal wall to mining space. Its generated huge dynamic effect seriously threatens the safety production of coal mine, and is one of the most severe natural accidents in coal mine. The coal and gas flow from the outburst is instantly full of the whole tunnel and damages facilities and suffocates people. It even would cause gas explosion and coal dust outburst, which result in more serious loss of life and economic. High gassy coal mines and outburst mines account for a significant portion of coal mines in our country. As the mining depth getting deeper and deeper, instant release energy of gas also gets stronger. Coal and gas outburst accidents continue to show an upward tendency. Coal and gas outburst majorly happens in coal seam tunnel heading, raise advance, and cross-cut uncovering coal. Some coal mines also have coal and gas outburst in coal face. And more than 90% of coal dust has explosion hazard.

Coal and gas outburst is very complicated. There is no accurate influence factor. And the correlation between outburst accident and outburst factors is also not sure. The mechanism of coal and gas outburst is not defined and still stays in different hypothesis stage. The risk prediction of coal seam outburst actually is assortment prediction problem of prophase factor of coal and gas outburst to later forecast object. [1] Fuzzy neural network (FNN) is the combination of fuzzy system and neural network, which brings together the advantages of neural network and fuzzy system. It obtains association, recognition, self-adaptation and fuzzy information processing in the same time. The combination of fuzzy system's method in processing uncertainty and the connection structure and learning method of ANN, endows fuzzy neural network with features like fuzzy expression, connection learning and distributed information processing. On the one hand, with the help of mature fuzzy theory and technology, it objectively realizes the correct expression and process of uncertain information and inaccurate relationship in outburst prediction. On the other hand, the information mapping of neural network can capture the correlation of outburst influence factor and outburst accident from outburst historical data, thus to realize the correct prediction of coal and gas.

2 Structure and Features of Fuzzy Neural Network (FNN)

The key concept of fuzzy logic was first raised by Zadah from 1972 to 1974. The core of fuzzy control is fuzzy control rule base. Rule base is the collection of some uncertainty reasoning rules. And fuzzy logic reasoning is one of the major methods for uncertainty reasoning.

Fuzzy reasoning system is also called system based on fuzzy rules. As Figure 1 shows, fuzzy reasoning system mainly includes five function blocks: rule base, data base, reasoning mechanism, fuzzy input interface and defuzzification interface. Rule base includes many fuzzy IF-THEN rules. Data base define the membership function of individual fuzzy set in fuzzy rules. Reasoning mechanism makes logic operation of fuzzy rules. Fuzzy input transforms the clear input to fuzzy cell matching with output. Defuzzification transforms the fuzzy calculated result to clear output.

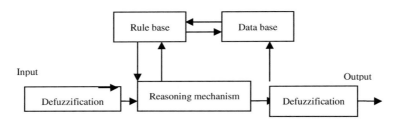

Fig. 1. Fuzzy Reasoning Systems

General rule base and data base known collectively as knowledge base. Process in the execution of fuzzy IF-THEN by fuzzy reasoning is as below:

1) Obfuscation, which is to compare input variable and membership function in the initial part and obtain the marked membership value (compatible degrees) of each language.
2) Process 6 input value fuzzily and obtain active right of each rule by calculation (special T- Norm operator, is usually multiplied or minimized) of the respective fuzzy membership function.
3) Generate effective result of each rule (fuzzy or clear).
4) Defuzzification, which is to superpose all effective result to generate a specific output.

The majority of fuzzy neural network adopts multilayer forward network structure. Considering the specific relater rules between influence factor and accident of coal and gas outburst, this paper chooses five-layer forward type of fuzzy neural network, which includes input layer, obfuscation layer, fuzzy relation mapping layer, defuzzication layer, and output layer.

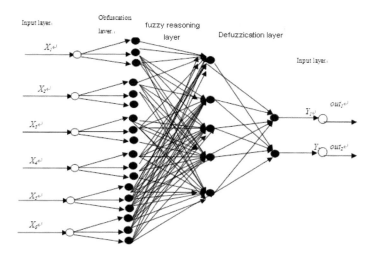

Fig. 2. Standard Structure of D-FNN

First layer: Input layer. Input layer directly send the input value $x = [x_1, x_2, \cdots, x_n]$ to next layer from input node like human factor (x1), gas pressure (x2), gas diffuse initial velocity (x3), coal damage type (x4), thickness of each layer of soft coal (x5), and surrounding rock air permeability (x6).

Second layer: Obfuscation layer. Each node represents a language variable value. Its function is to calculate individual input component and transform to membership function μ_i^j of fuzzy set.

Third layer: Rule antecedent layer. Each node of the layer represents a fuzzy rule. Its function is to match the antecedent of fuzzy rule a_i, and calculate the fitness ai of each rule according to formula (1)

$$a_i = \mu_{A'}(x_1)\mu_{A'}(x_2)\mu_{A_3'}(x_3)\mu_{A_3'}(x_4)\mu_{A_3'}(x_5)\mu_{A_3'}(x_6) \tag{1}$$

Fourth layer: Rule consequent conclusion layer. It defines the conclusion of rule node. Each generated rule is corresponding to the output $\overline{a_i}$ generated by input. Its node number is m, the same as that of the third layer, and the connection weight is 1. To realize normalization calculation, it is:

$$\overline{a_i} = \frac{a_i}{\sum_{i=1}^{m} a_i}, i = 1,2,\cdots,m \qquad (2)$$

Fifty layer: Output layer. All rule nodes was connected to output node through connection wire. It can realize clear calculation based on formula (3) and complete center average and defuzzication operation, which is

$$y_i = \sum_{j=1}^{m} \omega_{ij} \overline{\alpha_j}, i = 1,2,\cdots,r \qquad (3)$$

Connection weight w_i^j equals to the central value of membership function of y_i's jth language value. The above formula can also be written in vector form:

$$y = \omega \overline{\alpha} \qquad (4)$$

In them,

$$y = \begin{bmatrix} y_1 \\ y_2 \\ \vdots \\ y_r \end{bmatrix}, \omega = \begin{bmatrix} w_{11} & \omega_{12} & \cdots & w_{1m} \\ w_{21} & \omega_{22} & \cdots & w_{2m} \\ \vdots & \vdots & \vdots & \vdots \\ w_{r1} & w_{r2} & \cdots & w_{rm} \end{bmatrix}, \overline{\alpha} = \begin{bmatrix} \overline{\alpha_1} \\ \overline{\alpha_2} \\ \vdots \\ \overline{\alpha_m} \end{bmatrix}$$

After the steps, we complete the learning process (training process) of the whole fuzzy neural network system according to the input and output sample data of coal and gas outburst.

3 Study in Coal and Gas Outburst Prediction Based on D-FNN

3.1 Steps of Model of Gas Outburst Prediction Based on D-FNN Are as Below

a. Process property information to obtain fuzzy membership function and complete obfuscation.
b. Train fuzzy neural network and construct standard fuzzy neural network structure with 6 input and 2 output based on existing model.
c. There are still many redundant connection and nodes in the network after successful training. We can decrease the number of network connection and nodes according to a certain network reduction algorithm. This helps the acquisition of rules that follow.

3.2 D-FNN Algorithm

The learning algorithm based on gradient descent firstly describes the system structure, and then confirm all the free parameters of the system based on sample collection. The D-FNN system based on model Mamdani has below form:

$$y = f(x) = \frac{\sum_{i=1}^{M} y^{-t} \left\{ \prod_{t=1}^{n} \exp\left[-\left(\frac{x_1 - \overline{x_1}}{\sigma_1^l} \right)^2 \right] \right\}}{\sum_{i=1}^{M} \left\{ \prod_{t=1}^{n} \exp\left[-\left(\frac{x_1 - \overline{x_1}}{\sigma_1^l} \right)^2 \right] \right\}} \quad (5)$$

$$y = f(x) = \frac{\sum_{l=1}^{M} \overline{y}^{-t} \left\{ \prod_{t=1}^{n} \exp\left[-\left(\frac{x_1 - \overline{x_1}'}{\sigma_1^l} \right)^2 \right] \right\}}{\sum_{l=1}^{M} \left\{ \prod_{t=1}^{n} \exp\left[-\left(\frac{x_1 - \overline{x_1}'}{\sigma_1^l} \right)^2 \right] \right\}} \quad 6)$$

M is confirmed rule number, x1 is input, y is output. All of them are free parameter. The above system can be seen as feedforward network system. So the task is changed to train the parameter of feedforward network with gradient descent method. Set objective function as:

$$e^p = \frac{1}{2} [f(x_0^p) - y_0^p]^2 \quad (7)$$

In them: $(x_0^p, y_0^p), p = 1, 2, \cdots, N$ is the input-output pair of sample collection,

According to gradient descent method, we can get:

$$\begin{cases} \overline{y}'(q+1) = \overline{y}'(q) - \alpha \frac{\partial e}{\partial \overline{y}'} \bigg|_q \\ \overline{x}'(q+1) = \overline{x}'(q) - \alpha \frac{\partial e}{\partial \overline{x}'} \bigg|_q \\ \sigma_1'(q+1) = \sigma_1'(q) - \alpha \frac{\partial e}{\partial \overline{\sigma_1}'} \bigg|_q \end{cases} \quad (8)$$

In them, $q = 0, 1, 2, \cdots, \alpha$, α is learning step length. According to chain derivation rule of composite function, we can get:

$$\begin{cases} \overline{y}^l(q+1) = \overline{y}^l(q) - \alpha \dfrac{f-y}{b} z^l \\ \overline{x}_1^l(q+1) = \overline{x}_1^l(q) - \alpha(f-y)\dfrac{\overline{y}^l(q)-f}{b} \\ \sigma_1^l(q+1) = \sigma_1^l(q) - \alpha(f-y)\dfrac{\overline{y}^l(q)-f}{b} z^l \dfrac{2\left(x_{01}^p - \overline{x}_1^l(q)\right)^2}{\left(\sigma_1^l(q)\right)^2} \end{cases} \quad (9)$$

In them:

$$z^l = \prod_{t=1}^{n} \exp\left[-\left(\dfrac{x_1 - \overline{x}_1^l}{\sigma_1^l}\right)^2\right] \quad (10)$$

$$b = \sum_{l=1}^{M} z^l \quad (11)$$

$$a = \sum_{l=1}^{M} \overline{y}^l z^l \quad (12)$$

In them, $f(x) = a/b$.

d. Combine the rule antecedent and consequent, we can get the rule form we need:

$$\lim_{k \to x} d(k+1) = \dfrac{\lim_{k \to x} d(k) + Q_2}{1 + Q_2 + b} \quad (13)$$

Get

$$d(\infty) = \dfrac{Q_2}{Q_2 + b} \quad (14)$$

4　Results Analysis of Simulation

We can clearly see the change of gas concentration gradient from Figure 2. We can analyze and recognize in advance in the early stage of the upward gas concentration, to realize early prevention of gas outburst.

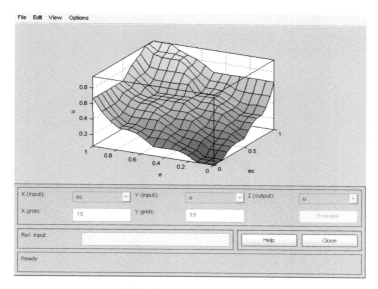

Fig. 2. Prediction surface

5 Conclusions

The writer firstly summarizes the current research situation of neural network and the shortage of fuzzy controller itself, and then makes analysis research of gas concentration data with fuzzy neural network. At last, he realizes the prediction of gas outburst through simulation and achieves the objective of early warning of gas accident. This paper relies on national nature science fund project—research of formation mechanism and management method of serious gas accident in coal mine production. It has very important theory and practical meaning and realizes outstanding warning of coal and gas outburst. This paper provides theoretical base and technical support for administrators' further decision making, and provides strong scientific bases for disaster emergency, disaster rescue and reconstruction work in our country.

References

1. Xin, H., Xu, L., Ji, H., Yang, G.: Present Situation and Problem analysis on Prediction Technology of Coal and Gas Outburst. Shanxi Coking Coal Science & Technology 9(9), 8–10 (2009)
2. Zhao, S., Ji, H., Sheng, Z.: Analysis on the Situation of Prediction Technology of Coal and Gas Outburst. Shanxi Coking Coal Science & Technology 10(10), 24–29 (2009)
3. Hao, J., Yuan, C.: FNN for predicting coal and gas outburst. Journal of Coal Science and Engineering 12(24), 624–627 (1999)
4. Sun, J.: The causes and lessons of "2·22" gas explosion disaster at Tunlan Coal Mine. Journal of Coal Science and Engineering 1(35), 72–74 (2010)
5. Chen, H., Qi, H., Wang, O., Long, R.: The Research on the Structural Equation Model of Affecting Factors of Deliberate Violation in Coalmine Fatal Accidents in China. Systems Engineering—Theory & Practice 8(8), 127–136 (2007)

Study of Coal and Gas Prediction Based on Improvement of PSO and ANN

Yanli Chai

School of Information and Electrical Engineering
China University of Mining & Technology
Xuzhou, Jiangsu,China
cyl_cumt@163.com

Abstract. Since BP algorithm based on gradient descent easily fall into local extreme value and has insurmountable shortages like slow convergence speed, we should combine PSO (Particle Swarm Optimization) and ANN (Artificial Neural Network). We bring PSO to the optimization connection weights of ANN and apply to the prediction of serious coal and gas accident. The simulation result proves that the algorithm has fast convergence speed and is easy to realize.

Keywords: PSO, ANN, prediction, coal and gas accident.

1 Introduction

High gassy coal mines and outburst mines account for a significant portion of coal mines in our country. As the mining depth getting deeper and deeper, instant release energy of gas also gets stronger. Coal and gas outburst accidents continue to show an upward tendency. Coal and gas outburst majorly happens in coal seam tunnel heading, raise advance, and cross-cut uncovering coal. Some coal mines also have coal and gas outburst in coal face. And more than 90% of coal dust has explosion hazard. Considering the current serious situation of coal mine production and safety management problem, this paper take the mine safety monitoring data as research object. Through the improvement of PSO and establishment of mapping between particle swarm dimension space and ANN connection weights, it realizes the prediction of coal and gas outburst.

2 Basic PSO and Parameters

PSO (Particle Swarm Optimization) is a representative colony intelligent method motivated by social behavior of bird group movement behavior. Bird group usually suddenly change their fly direction, scatter or gather during their flight. Their behavior is usually hard to predict, but the entire group always maintain consistency, and most suitable distance between each individual. While using PSO to solve optimization problem, the solution of the problem corresponds to a "particle" or "agent" of search box. Every particle has its own position, speed and fitness values. In each interaction, particle updates itself through following two "extreme value": one is the best solution that particle itself finds out, called individual extreme value (Pbest); the other one is the optimized solution found from the entire group, called global extreme value (Gbest).

PSO do not have crossover and variation operation. It is simple with fast running speed. Since information sharing would generate evolutionary advantage thoughts in biological group, PSO make use of it to find the optimized solution through the collaboration between the individuals.

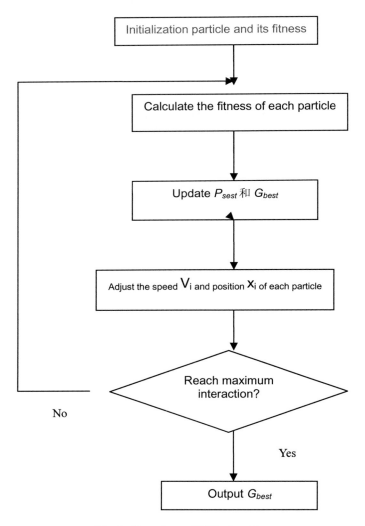

Fig. 1. Flow chart of PSO algorithm

Only when the particle in current position has better fitness values than it in the experienced best position p_i, the best position of the particle can be replaced by the current position. Thus, PSO, in a certain sense, also includes a certain form of "option" mechanism. The evolution interactive of PSO is just a self-adaptive process. It is a kind of self-adaptive between speed and position, but not only a replacement process of speed and position. Obviously, it makes full use of the information change of evolution process in evolution interactive. Particle can fly in a relatively ideal direction thought by group experience. Hypothetically, in D dimension search space, there are m particles in a swarm. And they fly in a certain speed. The i particle is a vector of D dimension $x_i = (x_{i1}, x_{i2}, \cdots, x_{id})^T$, $i = 1, 2, \cdots, m$, so the position of the i particle in D dimension search space is x_i. We can substitute x_i into a objective function to calculate its fitness value and then measure the good and the bad of x_i according to the fitness value. The speed of particle i is $v_i = (v_{i1}, v_{i2}, \cdots, v_{id})$, and $1 \leq i \leq m, 1 \leq d \leq D$; the optimized value experienced by particle i is $p_i = (p_{i1}, p_{i2}, \cdots, p_{id})$; the optimized value of all particles found in the entire group is $p_g = (p_{g1}, p_{g2}, \cdots, p_{gd})$; the position and speed of particle change according to the below equation:

$$v_{id}^{k+1} = \omega v_{id}^k + c_1 randd_1(P_{id}^K - x_{id}^K) + c_2 randd_2(P_{gd}^K - x_{id}^K)$$
$$x_{id}^{K+1} = x_{id}^K + v_{id}^{k+1}$$

ω: Inertia weight

$c_1 = c_2 = 2$, is learning factor

$rand_1, rand_2$ is pseudorandom number equally distributed in interval $[0,1]$

v_{max}: maximum speed

3 BP Network (Back Propagation)

BP algorithm is a search algorithm based on gradient method, a supervised learning algorithm. Actually it transforms the input/output problem of a group of sample to a nonlinear optimization problem. Through negative gradient descent algorithm, it is a learning method, solving weights problem by interaction operation. The difference of output vector and expected vector was minimized as much as possible through the repeatedly adjustment training of network weights and deviation. The training is over when the square error of network output layer is smaller than the designed error, and then we can save the network weights and deviation. Provided BP network is three-layer network, and input neuron numbered i, output neuron numbered k as fig. 2 shows:

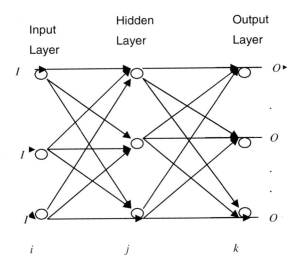

Fig. 2. BP of ANN schematic diagram

BP algorithm strictly follows the error gradient descent principle to adjust weights. After entering into flat region, even thought output vector has big difference with input vector, the training can only proceed slowly at the cost of increasing interactions because small error gradient minimize weights adjustment. However, PSO never follows error gradient descent principle. It makes its "descendant" rightly inherit more information from ancestor by its self renewal, so it can find out the optimized solution in a shorter time. PSO updates the best position of every current particle and the best position of the entire group to achieve a global optimization effect at last.

4   Study of Algorithm of Particle Swarm Optimization Based on ANN

PSO, as an emerging evolutionary algorithm, has features of fast convergence speed, high robustness, and strong global searching power. It does not need the help of characteristic information (like descent) of problem itself. The combination of PSO and ANN and the optimization of ANN connection weights by PSO can better overcome the problem of BP ANN. It not only can exert the generalization ability of ANN, but also raise the convergence speed and learning ability of ANN.

The use of PSO in ANN optimization mainly includes two aspects: one is network learning (also called network training), which is connection weights between each layer of optimized network; second is topological structure of optimized network.

The essence of BP network based on PSO is to map the weights and threshold value of PSO to particle and optimize these parameters through the continual updates of particles' speed and position, thus to achieve the network training objective. The key two points of learning algorithm in weights optimization are as below:

(1) Establish mapping between dimension space and ANN connection weights of PSO particle. Dimension component of each particle in particle swarm corresponds to a connection weight in ANN. That is to say the number of connection weights in ANN equals the number of dimensions of each particle in PSO learning algorithm.

(2) Use square error of ANN as the adaptive function of PSO and minimize the square error of network by the powerful searching capacity of PSO algorithm. Based on the above analysis, the learning algorithm can be designed as below: code all connection weights between neurons to real vector to represent the individual of the group; generate randomly the vector group and then interacts according to the original steps of algorithm. Network training process based on PSO and BP is as below:

Step 1. Confirm the BP network structure and parameter based on PSO

Confirm network input neuron I according to the input vector length of sample; confirm output neuron O according to the output vector of sample; confirm hidden layer neuron H and particle group scale N according to experience; set initial and final inertia weight as ω_{start} and ω_{end}, learning factor as c_1 and c_2, network training maximum interactions as G_{max}, initial network as ω_{ij}, and v_{jk}, θ_j, ψ_k as random number in (0, 1).

Step 2. Establish mapping relation between particles and parameters needs optimization in PSO

Use one-dimensional matrix to show the parameters needs optimization in a three-layer BP network.

Step 3. Calculate fitness function

We still need to establish a fitness function to evaluate the individual quality of each particle in the group. In the training network, the error square between network actual output value and expected output value, and the mean error square can be shown by below formula:

$$SSE = \sum_{p=1}^{P}\sum_{k=1}^{0}(y_k - c_k)^2$$

$$MSE = \frac{1}{p}\sum_{p=1}^{P}\sum_{k=1}^{0}(y_k - c_k)^2$$

Step 4. Update individual optimal value and global optimal value

Compare the current fitness value of individual particle with that of last generation, and update the individual extreme value if the current fitness value is better than that of last generation; compare the optimal current fitness value of individual particle with that of last generation, and update the global extreme value if the optimal current fitness value is better than that of last generation.

Step 5. Updates of speed and position

Update speed and position according to speed and position formula and restrict speed

Step 6. End the algorithm and output the optimal network

End the algorithm according to termination conditions, such as if it meets maximum interactions, or if training error smaller than design value; output optimal solution result. Flowchart of algorithm is shown as below:

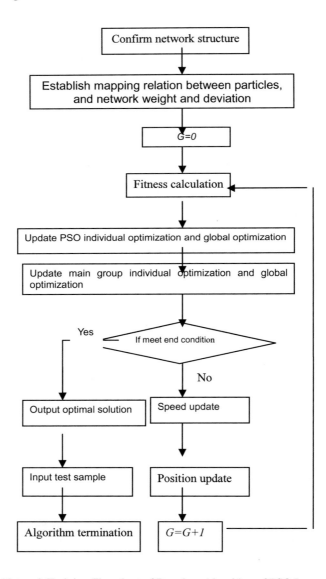

Fig. 3. Network Training Flowchart of Based on Algorithm of PSO Improvement

5 Simulation and Results Analysis

We choose 6 outstanding index coal face gas pressure P, gas disperse initial speed Δp, coal's Sturdiness coefficient, coal's damage type and outburst overall index D and K, as input parameter with relative analysis of impact factor. Output neuron can be divide into four categories according to gas density and outburst strength: danger zone without outburst (000), danger zone with outburst (001), threaten zone with outburst (010) and serious threaten zone with outburst (100). Simulation result can be seen from Fig. 3-Fig.4. The network error is smaller than 0.001. Undersize learning rate will cause overlength of BP network training time. The over adjustment of weight in training will excite the function to saturation, thus almost stop the adjustment of network weight and stop training completely consequently. After improve the BP network with PSO, the training error result quickly achieve accuracy requirements.

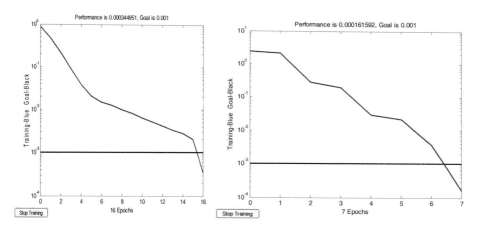

Fig. 3. BP network training error curve normalizing

Fig. 4. BP network training error curve after normalizing

6 Conclusions

The happening intrinsic mechanism of coal and gas outburst is complicated, and the relative rules between influence factor and incidents of outburst are not precise and clear. With PSO and ANN, this paper objectively realizes the correct expression and dispose of inaccurate information and relationship in outburst. The adoption of PSO algorithm to improve BP network, effectively overcomes the shortages of traditional BP algorithm, such as slow convergence speed and easily fell into local minimum value, etc. And the generation ability of BP network also has big improvements. It develops fast due to it fast convergence speed, fewer designed parameters and easy realization. It transforms from qualitative prediction to quantitative prediction, and raises the scientificity and accuracy of short term prediction of coal and gas outburst. The application provides theoretical base and technical support for administrator's further decision making, and provides strong scientific bases for disaster emergency, disaster rescue and reconstruction work in our country.

Acknowledgment

This research work was supported by the National Natural Science Foundation of China, No. 50534050.

References

1. 30 case analysis of MATLAB ANN, 1st edn. Beihang University Press (April 2004)
2. Yubo, T.: Mix ANN technology. Science Press (June 2009)
3. Prahalad, C.C.K., Ramaswamy, V.: The future of competition: co-creating gunique value with customers. Harvard Business School Press (2004)
4. Jizhen, Liao, H., Wu, Q.: PSO and its application, 1st edn. Science Press (January 2009)
5. Yang, C., Zhao, W.: Research of BP ANN in Coal and Gas Outburst Prediction. Journal of Lanzhou Jiaotong University (December 2009)
6. Hao, J., Yuan, C.: FNN for predicting coal and gas outburst. Journal of Coal Science and Engineering (December 1999)
7. Wang, L.: Tian Shuicheng, On the Construction of Intelligent Gas System. Science and Technology Review 22(25) (2007)
8. Hao, J.: Application of ANN for coal and gas outburst prediction, pp. 189–195. Xi'an Jiaotong University Press, Xi'an (1998)
9. Nan, C., Peng, X.: Prediction of Coal and Gas Outburst Area based on pattern recognition. Journal of Northeastern University, Natural Science (September 2004)
10. Wang, W., Cai, L.: Study of Determining Bayesian Network Topology Structures. Mini-Micro Systems 23(4), 435–437 (2002)
11. Wu, C., Zeng, Y., Zhang, Z.: Application of Auto-adapting Neural Networks in Forecasting Gas Content with Geological Factors. Journal of Liaoning Technical University (5), 609–612 (2003)
12. Nie, B., He, X.: Present Situation and Progress Trend of Prediction Technology of Coal and Gas Outburst. China Safety Science Journal 13(3), 40–43 (2003)
13. Hu, Q.: Research on the Technology of Protecting Efficiently Coal and Gas from Gushing. Journal of Huainan Institute of Technology 22(4), 11–14 (2002)

A New Web-Based Method for Petroleum Media Seamless Migration

Jiaxu Liu, Yuanzheng Wang, and Haoyang Zhu

Land operating area, PetroChina Jidong Oilfield Company, Tangshan 063004, China
liu_jiaxu@163.com

Abstract. In this paper, we propose and design a new kind of Web-based method of petroleum media seamless migration. The method can ensure the petroleum media seamless migration fast, smoothly and correctly. The implemented method will better solve the migration of petroleum media on the application of mobile work. Tests have shown that our suggested method is effective.

Keywords: Petroleum, Web, media, migration.

1 Introduction

In petroleum engineering, people have put new requirements for network applications. The Internet of Things [1]-[3] as a new network is onto history's stage, which in the past in China is only known as the sensor network. The Internet of Things is based on the Internet, making the items "talk" through existing means of modern information technologies, as agreed in the agreement. And the main means of modern information technologies include RFID, infrared sensors, global positioning system and laser scanner. The Internet of Things can be understood to have the three layers, which are perception layer, network layer and application layer. And we regard the perception layer as human skin and facial features used to identify objects and collect information. We see the network layer as the human brain and central nervous system, responsible for the information transmission and processing. And the application layer is seen as the social role for the combination of industry to achieve a wide range of intelligent.

Taking the characteristics of the Internet of Things and the Internet into consideration, we believe that realizing petroleum media seamless migration strategy for applications layer of the Internet of Things in different architectures is feasible and a certain significance[4]-[6]. So, we do some researches about the media seamless migration on the "high level" of the Internet of Things. The success of the study will give people work and life a huge impact. Information that the things in the world have can be accessed at any time and in any corner. And after a series of operations, they can be displayed in a larger device or PC, playing the rest parts of audio and video to a universal degree based on Web.

In order to better achieve the seamless migration strategy for the petroleum audio and video on the application layer based on Web, transmission speed, transmission quality and migration failure must be guaranteed. In order to ensure that the petroleum media are seamlessly, quickly and accurately migrated to the new environment where the media can be initiative on display, such as the music continuing to play and the movie continuing to play, we did the following related designs.

2 Design of Petroleum Media Migration Strategy

2.1 Design of the File Size

Send smaller files to the server through the transmission module, then the breakpoint information to be going to reach, then by the control module to control playback of audio and video. Smaller files transport strategy based on the information of location is described as follows.

First, find the first position after cutting. First of all, for the audio and video files, according to a certain size of the cutting, as well as both the first location and the middle location of the smaller files given the corresponding flag bit. And then, when the client pauses the player, the pause position will be recorded. According to the position, do some searches on the smaller files. When the corresponding smaller file is found, we give the following two strategies. Firstly, if the position is forward for the file, the cut will once again be given. Secondly, if the record is the back of the file, we will ignore the smaller and send the next file directly; thirdly, send the small file and the subsequent files.

Second, find the location and then cut. First of all, according to the client side getting the location of the file, position location; then, call the cutting function to cut the file according to the size pre-set; finally, sent the smaller files.

The algorithms of Cutting modules is as follows. The cutting modules algorithm: Get the size of original file; Set size of splitting the file; Determine the reasonableness of the size; Store the name and the file number; Create separate files; store the original file data; Get the current path of the program, noting the file name not included here; Create a folder to store the segmented files.

2.2 Design of Sending the Key Frame of Petroleum Audio and Video

We design the algorithm to determine how many key frames are there in the smaller file based on the above algorithm mentioned and give the flag position to the key frame [6]-[7]. So you can send the key frames firstly. And this way will prevent the phenomenon from occurring, such as the intermittent and the stagnation of audio and video. Under the good network condition, send more aid frames to enhance the playback quality; and if the network is bad, only some of the key frames are transmitted in order to achieve playback.

The brief description of the strategy is as follows. Firstly, send small videos to the server; secondly, make the use of the search module to search the key frame. After getting the key frames, give the special flag, and then point them with the pointers. When the command of migration is sent by the client, the server will do the following steps. Get the corresponding smaller files, and get key frames, send key frame firstly. Then, after specified client receives the audio and video playback breakthrough point, the self-organization module control the play module to play. It will prevent the intermittent and stagnation of audio and video from occurring, improve the effect and ensure the QoS. The key frames selection strategy is as follows.

Using color histogram as features of image to construct the "frame" matrix. The lens with n frames is defined as an n-dimensional feature vector. its n-dimensional vector is set as follows.

$$F_i = (f_{i1}, f_{i2}, \ldots, f_{in}) \qquad (1)$$

Therefore, the distance between two adjacent frames can be expressed as follows.

$$D_{ij} = \sum_{kn=1}^{N} |h_i(k) - h_j(k)| \qquad (2)$$

$h_j(k)$ is N quantization color histogram of k frame. So the frame matrix can be expressed as the distance between two adjacent frames. To further enhance the accuracy of extracting key frame, an adaptive threshold is used to quantify the frames matrix. Suppose the distance of two adjacent frames is suited to the Gaussian distribution, the adaptive threshold can be defined as $T = \mu + \alpha\sigma$. The probability density function defined is as follows.

$$f(x) = \frac{1}{\sqrt{2\pi}\sigma} \exp(-\frac{(x-\mu)^2}{2\sigma^2}) \qquad (3)$$

So the "frame" matrix is defined as follows.

$$D_{ij} = \begin{cases} 0, & M_{ij} < T \\ 1, & M_{ij} > T \end{cases} \qquad (4)$$

In order to better extract key frames, the problem can be modeled as an optimal solution to global problems. We use the genetic algorithm to realize it.

The Sure Stream technology [8]-[9] can create different files depending on the connection speeds, which uses the mechanism called client/server to detect changes in broadband. However, the method can set the size of the buffer only depending on the bandwidth when the connection occurs. So the strategy is not suitable for the changing network.

2.3 Design Based on RTP / RTCP under JMF Architecture

The RTCP reports are used to detect the trend of the changing network and take advantage of the trend to achieve the intelligence flow mechanism based on dynamic detection of changes in broadband.

Get feedback information by RR control package, that is to say which read the report from receipt for statistical analysis. Three performances from RTCP are used to detect the current network congestion.

First, the transmission delay jitter estimate: denoted by Ji, it is the estimated average deviation of the arrival time of two adjacent packets. Supposed the arrival time is relatively stable, the value is zero. Otherwise the value is quite large. The value will be reflected from the zero state to the non-zero state, to the zero state, which is such a circular process of change.

Second, the delay differences under the situation of continuous data packet transmission, denoted by M. The formula is defined as follows.

$$P(i+1,i) = (Receive_i - Send_i) - (Receive_{i+1} - Send_{i+1}) \quad (5)$$

Receive$_j$ and Send$_j$ separately represent the jth packet sent and received RTP timestamp. And the value represents delay of two adjacent packets. When the value declines, the available bandwidth of network is larger, or smaller.

We use the GetControls method from SessionManager to get BufferControl interface. Since the launch of a streaming media session will occupy a certain Buffer resources on the system resources, and also because the JMF supports RTP/TRCP protocol, so you can access the information contained in RTCP packets, which can change the size of the Buffer of the corresponding Session according to network conditions, and then adapt them to fit current network status, and ultimately realize the intelligent flow mechanism.

2.4 Design Based on QoS of TCP

The control of QoS on the client side can be utilized with the sender. Add the timestamp and serial number to the header of package from the sender. In the above paragraph, the use of RTP header has been explained, which is related to the serial number and timestamp. In fact, the purpose of setting the serial number is to determine whether the packets are lost, setting the time stamp to estimate delay and delay jitter. As the network latency is constantly changing, so we use the average delay. The average delay formula is as follows.

$$D_{new} = \lambda D_{cu} + (1-\lambda) D_{old} \quad (6)$$

D_{new} represents delay of a new round, λ on behalf of a trade-off factor for the regulation of the size of delay. It is easy to see that the size of the value reflects the sensitivity of the network. In order to better reflect the network changes, its value should be set to 0.68. To more accurately check the network congestion, we introduce the idea of packet loss interval, during which it is the number of transmitted packets. The number of transmitted packets is Larger; the network is better. Otherwise, the network congestion is serious and need to be improved. Periodically, the receiver will send the results checked to the sender. Furthermore, the sender can control the sending rate. Of course, some packets lost need to resent. Thus, if the client is unable to timely receive data sent by the client, it will suppose that network is crowded; and it will make the appropriate regulation.

3 Implementation of Petroleum Media Migration Method

3.1 Play Module in the Web-Based Architecture

Through Play module is based web player plug-in. The advantages of the plug are omitted here. In use, we add the appropriate parameters according to the principle of need. Control the player to play by calling the corresponding interface function. It needs to be embedded in Web pages <body> tag. In request module, we use the Ajax technique.

In push module, we refer to the implementation of Comet [10]-[11] application. In the Comet applications, we give the following modules implemented: doPost () module and onEvent () module. The doPost () module is to get the main parameters for the intercept and wake onEvent () function module.

We describe the data flow just like this: When the client opens the "home", the self-organization module gets context information and migrates by the control module and sends context information by sending module. The receiver module captures the information. After the information is analyzed, it will wakes up event handling module and pushes the information to the specified other client by the push module .Use a hidden frame to achieve the push module. In the end, context information is completed for the migration and self-organizing module control play module to play streaming media.

3.2 Petroleum Streaming Media Migration

The interface of player module used in JMF [12]-[13] is the Player. And the object will regard audio and video stream as input of data, then, send the audio and video streaming to the appropriate device, such as audio, screen. The approach is very similar to the situation. The CD player reads the songs in CD recording, then its signal to the speaker. Corresponding algorithm is omitted here.

In order to enhance the display effect, we have added some unnecessary features, such as recycling. In the following presentation, we will omit some non-essential features to achieve, only giving the algorithm of the key function such as the receiving and migrating function. In the B/S, the modules which need to be implemented are congestion control module, transfer size module, Socket transfer module and so on. The key algorithms of the seamless migration strategy under B/S are as follows.

Migration module algorithm is as follows: Convert the type of location of the breakpoint type to CString type; Convert integer into CString; Create Socket, cut files, transfer file.

Receiver module is as follows: Receive data; Exchange between values you type CString variables; Schedule thread.

Thread module algorithm is as follows: Define CActiveMovie3 type pointer variable; Create socket; Receive data; Combine the smaller files; Resolve the string and respectively assigned; Set the file needs to be played; Play the file.

In order to achieve more features, you can also add other functions and properties.

4 Test

In the B/S structure, take the audio and video for example, we can describe the seamless migration process of audio and video like this. Click the Hyperlink, then, self-organizing modules organize play module to play streaming media. By clicking the Pause button, self-organization module controls control module to access the context information, then, by clicking on the transfer button, the audio and video streaming media will be displayed at the breakpoint on the specified computer or specify a scene perceived in the environment. The significance of doing this is that the device is in another place far away or the information can not be directly accessed. Of course, you can directly gather the URL of data source and the breakpoint of data source. The situation is taken for example as follows.

Audio and video data streams under P2P architecture can be described in the following situations. Supervisors concern about the video stream or audio stream got from the camera or the listener. When they see or hear the suspect, they would redirect the video stream or the audio to other supervisors. It will help improve the detection speed. Because of the space constraining, it will achieve that all members in every place can watch or listen to the suspected place. It facilitates a breakthrough in the case. Similarly, you could exchange instant messages, to the effect that you could watch the chat and exchange instant results at the same time.

In order to produce visual effects, we create the following logos: increasing the rendering of web player module, setting Location tag at the bottom and the corresponding text in text box. All these show the accuracy and unity of the information after migration. Rendering is achieved by adding the appropriate parameters of playback module. And location tag is implemented by the DOM technology, and the value of text box achieved by setting the label property.

In the P2P architecture, the migration process can be described just like this. Get video stream or audio stream by sensors. Then the data is migrated by the Internet and need some cloud computing in the data center and be stored. When the stream is visited, the stream is guided to the client. When the client needs to migrate the steam, the stream is guided to the other client, and the stream continues to play. The significance of the kind of migration strategy is to focus on necessity and importance of audio and video. And quality information point is transferred each other, facilitating other people who is in need to watch the stream.

The test results are just like Figure 1 and 2.

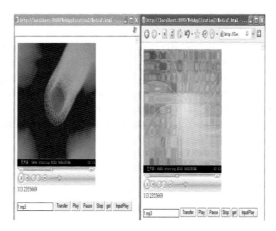

Fig. 1. Media Player Audio Transition Graph

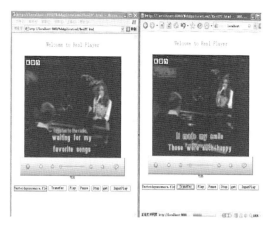

Fig. 2. RealPlayer Video Transition Graph

5 Conclusion

In order to support petroleum engineering, we propose and implement seamless migration strategies under Web-based architecture for the application layer of the Internet of Things, taking the characteristics of the Internet of Things into consideration. In the B/S architecture, we make good use of Comet server push philosophy, better enhancing the data timely and fast. And the Ajax technology is used to realizing the asynchronous request. We can achieve the migration of files the browser supports.

References

1. An, B.: Application of RFID and Internet of Things in Monitoring and Anti-counterfeiting for Products. In: Proc of International Seminar on Business and Information Management (2008)
2. WOLFW. Cyber-physical Systems. Computer (2009)
3. JMF Wikipedia [EB/OL] (2010), http://baike.baidu.com/view/209561.htm
4. Hu, K.-m.: RTP/RTCP Transport model design of Streaming Media based on JMF. Sichuan Institute of Technology, Natural Science (2010)
5. Liu, X.-j., Liu, B., Yu, D., et al.: B/S and C/S hybrid architecture model based on Ajax technology. Computer Applications (2009)
6. Lin, J., Sed Igh, S., Miller, A.: A general framework for quantitative modeling of dependability in Cyber-Physical Systems: a proposal for doctoral research. In: Proc of 33rd Annual IEEE International Computer Software and Applications Conference (2009)
7. Zhang, D.G.: Web-Based Seamless Migration for Task-oriented Nomadic Service. International Journal of Distance Education Technology (JDET) 4(3), 108–115 (2006)
8. Zhang, D.-g.: Seamless migration strategy of Pervasive computing. Control and Decision 20(1) (January 2005)
9. Sha, L.: Cyber-Physical Systems: A New Frontier. In: 2008 IEEE International Conference on Sensor Networks, Ubiquitous and Trustworthy Computing, SUTC 2008 (June 2008)

10. Zhang, D.G.: A kind of new decision fusion method based on sensor evidence. Journal of information and Computational Science 5(1), 171–178 (2008)
11. Zhang, D.G.: A Kind of new approach of context-aware computing for ubiquitous application. International Journal of modeling, Identification & Control 8(1), 10–17 (2009)
12. DOLINRA. Deploying "Internet of things". In: International Symposium on Applications and the Internet (2006)
13. Georgoulas, D.: Intelligent Mobile Agent Middleware for Wireless Sensor Networks: A Real Time Application Case Study. In: The Fourth Advanced International Conference on Telecommunications 2007 (2007)

The Summary of Low-Voltage Power Distribution Switch Fault Arc Detection and Protection Technology

Aihua Dong[1], Qiongfang Yu[1], Yangmei Dong[2], and Liang Li[1]

[1] School of Electrical Engineering & Automation, Henan Polytechnic University,
Jiaozuo City, Henan Province, China
[2] Document Micro Duplication Center, National Library of China, Beijing, China
{dah,yuqf,liliang74}@hpu.edu.cn, dym0207@163.com

Abstract. Aimed at the situations of arc occurring frequently and serious harm of the power distribution switch, the paper analyzes arc and its detection technology developing process, emphatically discusses the arc main detection methods and their principles and looks into the future of the improving accuracy of arc detection and distinction. In the end the paper points out that arc early warning will be the arc main research direction in future.

Keywords: low-voltage, distribution switch, detection methods, summary.

1 Introduction

In the low-voltage power supply system, low voltage distribution switch play the control role of on and off power lines, when the charged line failure, also has the role of cut off the fault and protect the safe operation of power grid. But arc is a frequently phenomenon in a low-voltage power distribution switch. And arc is divided into two types of non-fault and fault in a low-voltage power distribution switch. When the power distribution switch opens, stops inductive load, both ends of contacts also produce arc, but under the action of arc extinction device the arc very quick extinguishment, cannot cause the fault. This kind of arc caused by switch normal work is called non-fault arc. The contacts and junctions of distribution switch, the factors of oxidation, loose, overload and pressure spring's elastic aging would lead to contacts' bad contacts, contact resistance increases, prone to partial discharge or spark discharge, if it is not overhauled processed in time, the serious arc discharge can appear, namely forms the arc. Besides, the overheating because of long time overload would accelerate the generatrix surface's aging in distribution switch cabinet, and because of the instability of the load, current's instability, the generatrix connections temperature fluctuated, the metal objects' thermal expansion and contraction is easy to make bus connections loose, cause the poor contact, partial discharge, then develop into arc fault. This kind of arc caused by not good contact of switch contact, junction and wire attachment point in the power distribution switch cabinet is called the fault arc. According to statistics [1], the electrical fire caused by distribution switchgear arc fault occupied a large proportion of fire, and has the trend to continue increase. Every year, the distribution switchgear burned by arc short circuit up to thousands in national average, and half of the total accidents caused by arc, and arc fault has the most serious degree of damage [2]. According to the accident statistics by BGFE [3], 25% of the current incident is arc fault accident, 34% of current accident casualties occurred in switchgears. Arc fault has great energy, high temperature, severe damage, only 0.5A arc current is sufficient to cause electrical fires, and 2A ~ 10A arc current can produce 2000 ℃ ~ 4000 ℃ partial high temperature[4]. At the same time, high temperature produces by arc fault also can melt some certain substances, the generated steam contains large amounts of harmful substances, which threaten the around people's life security. For the more, if the fault arc caused in low voltage distribution switch can not be eliminated in time, it would developed into line fault and the vicious accident of power system instability and large area power outage would also been caused [5], more secondary damage been caused further. Fault arc can also radiate a huge light, heat and other energy which can cause explosion and fire, resulting in personal injury and severe economic and property losses. Therefore the technique summery on fault arc detection and protection of low-voltage power distribution switch, description and analysis the existing all kinds of fault arc detection methods and utilizing the good detection technology and learning experiences, the in-depth study on effective detection method of fault arc will be promoted, to realize the prevent or reduce the occurrence of arc fault and increase the safety and reliability of distribution line.

2 Fault Arc Detection Methods

For more and more accidents caused by fault arc, there is more and more research for the fault arc. From the early 90s of last century, the arc fault's physical properties, mechanism and detection method are studied at home and abroad [6-14], some valuable results have been gained, many fault arc detection methods have been proposed.

2.1 Voltage Detection Method

When the arc comes into being, the circuit's voltage and current will be changed. The arc fault current waveform is approximately sinusoidal, but the arc voltage waveform is more complicated. The marked feature of the arc voltage in zero regions is: the time constant θ is very small in arc zero regions, arc voltage change rate is greater, and reach the largest at the time of current crossing the zero. This feature can be used in the identification of the arc voltage waveform [15-16]. The principle of the voltage detection method is showed in Fig. 1.

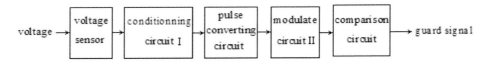

Fig. 1. Arc voltage method schematic diagram

First, the circuit voltage which contain the arc message is detected with voltage sensor, conditioned by conditioning circuit I, which including transformation, amplification, filtering and other links, the voltage signal treated by conditioning circuit I is transported to pulse converter, then the voltage bust information is changed into pulse signal. This pulse signal is again treated by conditioning circuit II, such as second shaping, conversion, etc. the conditioning circuit II is mainly composed by one-shot trigger and integrator. The pulse signal crossed through one-shot trigger is changed into the pulse signal which the frequency changed with pulse signal and its amplitude and width is constant. The pulse signal is used as the comparator input signal after be integral transformed by integrator, when the input signal's amplitude exceed the comparator's set value, the comparator output the protection signal.

This method need to detect the arc voltage signal, to extract the arc voltage's drop message. But arc voltage signal's detection could be interfered by network load and other factors, so misjudgment is prone to occurred.

2.2 Arc Light Detection Method

The arc light is the most obvious and the fastest changed physical parameter when the arc occurring. It can realize the protection by the arc light detection and according to the arc light strength. The arc light detection system is one of the ideal protection programs for middle and low voltage distribution switch, mainly realizing though the detection of arc light intensity. Fig.2 is the ABB Company's Arc Guard System schematic diagram, it consist three main parts: arc protection, current detection unit and arc light detector. The arc protection is the core part, it has two input signal channel: the one is current signal channel, the other one is arc light signal channel. Detected by current transformer CT, treated by current detection unit, the arc current signal is transferred to arc protection through channel 1. Detected by detector, the arc light is transferred to arc protection through channel 2 after it be amplified and filtered. The arc protection receive the arc current signal and fault arc signal, convert the two signals to digital by A/D converter, and after the digital processing, logical comparing and determining, when the both current signal and the arc light signal reach to the set value, sent out the arc fault protection signal, control the tripping mechanism of power distribution switch, to cut off fault arc line.

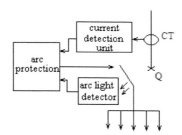

Fig. 2. ABB arc protection Schematic diagram

There are too many interference sources in the environment, so a single arc light criterion is difficult to realize the reliability of arc fault protection, but it can be used as one of the main criterion for arc fault protection system. This kind of fault arc protection system with both arc light signal and the current signal detection can improve the reliability of the system.

2.3 Ultrasonic Detection Method [17]

Arc sound is also an obvious physical phenomenon that generated in the process of arc producing, so the arc sound can be used as one of the main criterion for the arc fault alarming. The typical structure of ultrasonic detection system is showed in Fig.3. The system is constructed by different types of piezoelectric sensors, pre-amplifier, filter, post-amplifier and digital acquisition circuit. Finally the detection data is input to computer to deal with.

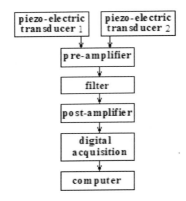

Fig. 3. Ultrasonic detection system diagram

Piezoelectric sensors is ultrasonic transducer, it can change the ultrasonic signal into electrical signal. Ultrasonic transducer is the system's key component. The piezoelectric sensors with the reliable performance and the frequency range from 40 ~200 kHz can be selected. Its core components is the imports coating piezoelectric, its frequency error is small, external structure uses multilayer impedance matching and excellent lining material, has good vibration damping properties, and has the good time and temperature stability, high sensitivity. The preamplifier uses instrumentation amplifier, its main features are high sensitivity, high input impedance, low output impedance, not only can amplify the signal, but also can filter the DC component. The filter circuit is composed by low-pass filter and high-pass filter. There still has some noise at the process of the arc fault sound detection, such as low-frequency mechanical vibration (under 20 kHz), high frequency electromagnetic radiation interference and high-frequency noise by signal itself, so the signal must be low pass and high pass filtering. Because the detected ultrasonic signal has different attenuation according to distance, for different detection distance and arc sound intensity, the post-amplifier circuit is set adjustable magnification from 10 to 100 times. For the specific circumstances, the sampling frequency, sampling points and sampling time of digital acquisition circuit can be adjusted; signal waveforms and collected data can be saved. The protection of fault arc can be achieved by analysis and processing the collected data by computer.

2.4 Ultraviolet Detection Method [18]

We can see from power distribution switch arc discharge mechanism that with the molecular excitation, ionization, recombination, charge exchange, electron attachment and the incidence of radiation, the different discharge stages produce the luminescence phenomenon with different spectral. In the discharge spectrum, contains a lot of ultraviolet light. So, the discharge intensity can be judged by detect the UV radiation intensity in the distribution switch [19], then can judge whether there has arc in distribution switch. The past UV detection method mostly based on the full UV band limited by the manufacturing process and testing equipment, easily be interfered by sunlight, only be used in the specific weather conditions, application inconvenience. With the development of detection technique, blind UV detection method appeared. The UV wavelength range is 10~400 nm, sunlight is also contain ultraviolet, the ultraviolet wavelength less than 280nm were almost all absorbed by the ozone in the atmosphere, the ultraviolet which its wavelength range from 315 to 400nm mostly can be transmitted through the atmosphere, the wavelength range from 280 to 315nm is only 2%, usually the ultraviolet which its wavelength less than 280nm is called sun blind zone. Most wavelength range of ultraviolet produced by arc discharge is 280~400 nm, also a small part of the wavelength range from 230 to 280nm, so the detection of this part of ultraviolet can be used as criterion to judge the fault arc production.

The UV sensor working principle is based on photoemission effect of metal and electronic fan flow theory. After the voltage applied between UV sensor's anode and cathode, the electric field between the poles established. When the UV injects into the cathode surface and the incident light energy is greater than the cathode surface escaping power, photoemission is generated, photoelectron is escaped. Photoelectron accelerate to anode under the influence of the electric field, produce ionization with gas particles, New electronic produced by ionization collision with other particles, rapidly form a large current between the anode and cathode and discharge, this phenomenon is called electron fan flow. According to Townsend discharge theory, in the uniform electric field, Fan flow discharge current depends on the photocurrent density, electric field strength, tube gas pressure and the gas type [20]. The typical UV sensor's channel working wavelength is from 185 to 260nm in the blind date such as R2868 produced by HAMAMATSU Company; it can effectively avoid the interference of sunlight and has higher sensitivity [21].

UV sensor's drive circuit is showed in Figure.4. The drive circuit's working voltage is 300~350V DC voltage. This high voltage can be realized by DC/DC high voltage DC modules, its the input voltage is 5V DC voltage, the 300~350V voltage can be gained through adjust the input resistance. This drive circuit's working principle is: when there has not ultraviolet, the sensor is in high-impedance state, its output voltage is zero, the Schmitt Trigger 74HC14 input is low voltage, output is high voltage, that means drive circuit output high voltage, LED light off; when distribution switch generate arc discharge, UV sensor receives UV from arc, the sensor is conduction state, capacitance C1 charge to capacitance C2, through resistance R2 and UV sensor, and generate instantaneous current on resistance R3, Schmitt Trigger 74HC14 input is high voltage, then output a low voltage, LED light on. After the UV sensor conducting, the power of capacitance C1 gradually reduce, voltage gradually decreasing, the input voltage of UV sensor also gradually reduce at this time, when the voltage is lower than its sustaining voltage, sensors stop conduction, then external power source charge to capacitor C1, sensor input voltage gradually increase again, when reach the start voltage, if receive UV, UV sensor conduct again. The use of Schmitt Trigger 74HC14 chip is to shape the output pulse, to gain the stable output waveform pulse.

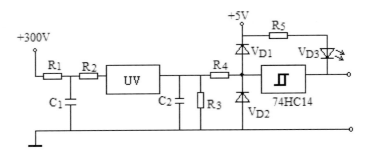

Fig. 4. Drive circuit of UV sensor

3   Arc Fault Pre-warning Will Be the Trend in the Future

The great harm of arc has been the consensus, researchers carried out lot of research work to the arc fault detection and the protection method, gained encouraging results, carried out a lot of arc fault identification method, summed up in two categories as followed. Firstly, single-criterion method: including voltage/current detection methods, pressure detection methods, arc detection methods, electromagnetic detection methods, arc sound detection methods, infrared temperature detection methods, ultrasonic detection methods. Single-criterion method needs to detect some signal only. For the detection signal reaches the threshold value, the device sends the protection signal. This method has the advantages of less detection, single to identify, but has a disadvantage of high err-detection rate. Secondly, Multi-criteria method: including double criterion methods such as arc and current, arc sound and current, arc and temperature; tri-criterion methods such as sound-light-pressure criterion method, light-pressure-current criterion method, and four criterion methods such as sound-light-pressure-current criterion method. More than two signals be detected in multi-criteria method, and the information fusion technology be used, which can improve the reliability of fault arc identification. Practice shows that the there has the clear chi-chi sound and intermittent flash at the period of distribution switchgear's discharge before the arc comes into being, and at the same time, the released power gathered near or around the space of discharge point, form the electromagnetic waves and spread in all directions, so there must has the obvious generation of light, sound, electricity and magnetic, but the pressure generated in this time is very small, the duration time is short, so is difficult to capture. After analysis, the signal of light, round, electric and magnetic is selected as arc fault's alarming signal to detect is feasible.

Fault arc generation process can be divided into two stages [22-23]: the first stage is partial discharge and spark discharge before the arc burning. The phenomenon of electrical pulse, gas resultant, ultrasound, electromagnetic radiation, light, local overheating would occur at this stage; the second stage is fault arc burning, the arc current and significant arc light would occur at this stage. The arc's research is mainly focus on the arc fault alarm at present, that is the failure has occurred, to find it through the signal detection, then send the alarm signal, so this is a passive arc protection method, the practice proved that this method's alarm accuracy is not high, on the other hand, because protection has not high speed action, it is difficult to quickly cut off the fault arc. The best arc fault protective method is to prevent the occurrence of arc, which is timely detect the hidden arc fault and forecast or warning the arc fault. The sound and light that indicated discharge would occur during the charge before the arc fault's occurrence, with the generation of electricity and magnetic. Therefore deeply research the physical characteristics of light, electric, sound, magnetic, reveal the intrinsic link with the arc occurred, find the effective method of filtering and noise reduction; study multi-information fusion algorithm, through analysis the existing information fusion method, build mathematical model of arc fault warning based on the multi-information fusion, to improve the reliability of the fault information's identification, timely and accurately send out the warning of arc fault. The forecast of arc fault changed the traditional passive protection mode, realized the early prevention of arc fault, the distribution switchgear with hidden faults be repaired or replaced timely, eliminate the arc fault before the accident occurrence, would play an important role for reduce or avoid power outages and electrical fires. So the arc fault's early forecast has the important social and economic benefits, has the very broad prospects for development.

Acknowledgments

This paper contents have been funded by scientific and technological project of Henan Province (the project serial number: 0624460012), expresses the heartfelt acknowledgments; A large number of papers have been referenced in the process of paper's writing, thanks to the authors of all the references by this opportunity.

References

1. Du, Y., Gu, N.: Present situation and accidents analysis of distribution switchgear in domestic power systems. Power System Technology 26, 70–76 (2002)
2. Liu, X.: Low-voltage distribution cabinet technological innovation and development. Building Electricity 3, 29–31 (2006)
3. Lan, H., Zhang, R.: Current research and development trends on faults arc detection method in switch cabinet. High Voltage Engineering 34, 496–499 (2008)
4. Gregory, G.D., Wong, K., Dvorak, R.F.: More about arc fault circuit interrupters. IEEE Transactions on Industry Applications 40, 1006–1011 (2004)
5. Tian, G.: The protection based arc detecting to internal fault of LV/MV switchgear and its application. Electric Engineering 10, 62–64 (2004)

6. Klaus, D.W., Balnaves, D.: Internal fault s in distribution switchgear—where are we now and where are we going. In: International Conference on Trends in Distribution Switchgear, London, U K (1998)
7. Land, H.B., Eddins, C.L., Gauthier, L.R.: Design of a sensor to predict arcing fault s in nuclear switchgear. IEEE Transactions on Nuclear Science 50, 1161–1165 (2003)
8. Bartlett, E.J., Vaughan, M., Moore, P.J.: Investigation into electro-magnetic emissions from power system arcs. In: EMC York 1999: Conference Publication. IEEE, York (1999)
9. Mamis, M.S., Meral, M.E.: State-space modeling and analysis of fault arcs. Electric Power Systems Research 76, 46–51 (2005)
10. Sidhu, T.S., Singh, G., Sachdev, M.S.: Arcing fault detection using artificial neural networks. Neurocomputing 23, 225–241 (1998)
11. Sidhu, T.S.: Protection of power system apparatus against arcing faults. In: International Conference on Global Connectivity in Energy, Computer, Communication and Control, York, USA (1998)
12. Sidhu, T.S., Sachdev, M.S., Sagoo, G.S.: Detection and location of low-level arcing fault s in metal-clad electrical apparatus. In: Developments in Power System Protection: Conference Publication No. 479. IEEE, London (2001)
13. Sidhu, T.S., Sagoo, G.S., Sachdev, M.S.: Multi-sensor secondary device for detection of low-level arcing faults in metal-clad MCC switchgear panel. IEEE Transactions on Power Delivery 17, 129–134 (2002)
14. Cai, B., Chen, D., Wu, R.: Online detecting and protection system for internal faults arc in switchgear. Transactions of China Electrotechnical Society 20, 83–87 (2005)
15. Fang, Z., Qiu, Y., Wang, h.: Simulation of the discharge process of dielectric barrier discharge. High Voltage Engineering 32, 62–65 (2006)
16. Kang, J.: Exact digital simulation and calculation of fault arcs. Relay 30, 14–16 (2002)
17. Zhang, Y., Yang, Y., Yang, C., Wang, b.: The study on Ultrasonic measure in partial discharge of electric equipment. Piezoelectrics & Acoustooptics 32, 414–416 (2010)
18. Weng, J., He, W., Li, C.: On-line discharge monitoring system for high-voltage power equipment. Electric Power Automation Equipment 30, 135–138 (2010)
19. He, W., Chen, T., Liu, X., et al.: On-line monitoring system of faulty insulator based on non-touching UV pulse method. Automation of Electric Power Systems 30, 69–70 (2006)
20. Wang, J., He, W., Li, J., et al.: Non-contact electroscope for Ultra-high-voltage appliances based on Ultra-violet pulse detecting. Transactions of China Electrotechnical Society 23, 138–140 (2008)
21. Zhou, L., Yang, H., He, W., et al.: Study of on-line ultraviolet detector for HV insulator. Chinese Journal of Scientific Instrument 28, 109–111 (2007)
22. Yang, J., Zhang, R., Du, J.: Study on application of arcing protective system faults based on multiple information fusion. High Voltage Apparatus 43, 194–196 (2007)
23. Land, H.B., Eddins, C.L., Gauthier, L.R., et al.: Design of a Sensor to Predict Awing Faults in Nuclear Switchgear. IEEE Trans. on Nuclear Science (2003)

Multi-disciplinary Modeling and Robust Control of Electric Power Steering Systems

Ailian Zhang[1], Shujian Chen[2], and Huipeng Chen[3]

[1] School of Mechanical Engineering and Automation,
Wuhan Textile University, Wuhan 430073, China
[2] Wuhan Polytechnic Institute, Wuhan 430074, China
[3] HangZhou DianZi University, HangZhou 310018, China
zalchuchu@sina.com, chensj_611@sina.com, hpchen@hdu.edu.cn

Abstract. Multi-disciplinary modeling and analysis were carried out for an electric power steering (EPS) system. Model-based design method was proposed and the EPS control system was designed by this means. A kind of backstepping robust controlling method was presented based on the analysis of S/T curves of singular values. The shortage of one-ring EPS control system was discussed and the Two Closed Loop EPS control system was proposed which inner ring controller was designed using single neuron PID algorithm and outer ring controller using the new proposed robust controlling method. The results showed that the method is concise and effective, and the controller has excellent robustness and stability.

Keywords: Electric power steering, Multi-domain, H∞ control, System Simulation, Robust Control.

1 Introduction

The nonlinearity is enhanced in Electric Power Steering (EPS) systems with the usage of motor supporting devices, and more, the uncertain changes of driving road conditions, and the unpredictable variations of the inherent characteristics due to the wear and aging of some components in driving motion, such cases make it difficult to control the EPS system in real situations. The robust control method is needed to design the EPS system for achieving the given performance requirements in these uncertain circumstances. The control system is a key part of an EPS and used to meet the stability, tracking and anti-jamming. Traditional control system using single-loop control strategies is simple and easy to implement, but with some defects such as poor control accuracy, low efficiency, and high control order. This paper presented a new double closed loop control strategy. The inner loop is the electrical current loop of the motor using single neuron adaptive PID algorithm [4] to reach no force difference and fast response. The outer ring is the position loop using the improved H∞ robust control algorithm [5] to meet the tracking accuracy of torque and the requirements of steering feeling. The new control strategy does not need to re-construct the original system, in addition the robustness of the system is improved obviously, the design of the controller is predigested, the order of the controller is reduced.

The principle of the steering column-type Multi-disciplinary EPS system is shown in Figure 1. According to the rotation numbers and the driver's turning intentions, the torque sensor will detect the steering torque when there is some deviation in the expectation angle between the input shaft and output shaft. ECU attains the power torque by computing and sends instructions to the power motor according to the steering torque, the motor speed and the power status. The torque generated by the motor and amplified through the worm body is provided to the steering column and drives the steering column.

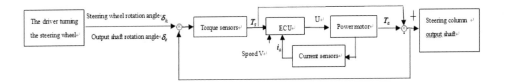

Fig. 1. The run principle of EPS multi-domain system

2   Multi-disciplinary Modeling for EPS Systems

2.1   Dynamics Models of EPS Systems

The EPS system is divided into three parts: mechanical part, electric power part and control part. According to Newton's theorem and Kirchhoff's law, the dynamic equations and the electrical characteristics equations of all parts can be obtained. The mechanical steering component includes the steering column, the torque sensor, the output shaft, the steering resistance and the pinion, its dynamics equations are as follows:

$$I_h \ddot{\delta}_h + D_h \dot{\delta}_h + T_s = T_h + f(\delta_h, \dot{\delta}_h) \tag{1}$$

$$T_s = D_{ts}(\dot{\delta}_h - \dot{\delta}_e) + K_s(\delta_h - \delta_e) \tag{2}$$

$$I_e \ddot{\delta}_e + D_e \dot{\delta}_e = T_e - T_r + f(\delta_e, \dot{\delta}_e) \tag{3}$$

$$T_e = T_a + T_s \tag{4}$$

$$F_r = K_r x_r \tag{5}$$

$$m_r \ddot{x}_r + b_r \dot{x}_r = \frac{T_r}{r_p} - F_r + f(x_r, \dot{x}_r) + F_d \tag{6}$$

The electric power component includes the power motor and the reduction gear, the straight-line power mode is used to realize the controller. The relationship of the target current with the torque sensor is: $i_{ar} = K_a(V)(T_s - 1)$, $K_a(V)$ is power factor. The equations of the dynamics and electrical characteristics are as follows:

$$I_m \ddot{\delta}_m + D_m \dot{\delta}_m = T_m - T_g + f(\delta_m, \dot{\delta}_m) \tag{7}$$

$$T_a = i_{ms} T_g \tag{8}$$

$$u_a = L \frac{di_a}{dt} + R i_a + K_e \dot{\delta}_m \tag{9}$$

$$\delta_m = i_{ms} \delta_e \tag{10}$$

H-bridge field-effect transistor is used in the control hardware, which controls the electric motor with the bipolar mode of operation using PWM method; the robust control of the double closed-loop control strategy is used in the software part.

2.2 Control Models of EPS Systems

A control block diagram with physical meanings is obtained by analyzing the EPS system dynamics model, it is as Figure 2.

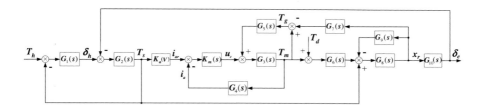

Fig. 2. EPS control block model diagram

The details of the transfer functions and their specific expression in the figure 2 are as the Table 1.

A good power is the key of EPS power control, and it keeps a smooth and comfortable steering road feel for driver while achieving the turning portability. The uncertain changes of driving and roads conditions, and the unpredictable variations of the inherent characteristics due to the wear and aging of some components in driving motion, such cases change the EPS dynamic characteristics. In order to meet the performance requirements of the system under these uncertainties cases, the above EPS system needs to be improved with the robust control method to enhance the robustness and adaptability.

Table 1. The transfer functions of EPS system dynamics models

Dynamics model of the steering wheel: $G_1(s) = 1/(I_h s^2 + D_h s)$	Current sensor model: $G_4(s) = 1/K_t$
Torque sensor dynamics model: $G_2(s) = D_s s + k_s$	Motor reducer model: $G_6(s) = i_{ms}$
Motor dynamics model: $G_3(s) = \dfrac{K_t(I_m s + D_m)}{K_c[(Ls+R)(I_m s + D_m) + K_e K_t]}$	Steering spring resistance model: $G_9(s) = K_r r_p$
	Inner loop motor controller: $K_m(s)$
Transfer function of calculating the motor electromotive force: $G_5(s) = K_e/(I_m s + D_m)$	$M_r = I_e/r_p^2 + m_r$
Transfer function of calculating the motor load: $G_7 = i_{ms}(I_m s^2 + D_m s)/r_p$	$D_r = D_e/r_p^2 + b_r$
Pinion rack model (motor rigidity): $G_8(s) = 1/(M_{rm} s^2 + D_{rm} s)$	$M_{rm} = M_r r_p + i_{ms}^2 I_m / r_p$
Transfer function of the rack displacement to the output shaft rotation: $G_{10}(s) = 1/r_p$	$D_{rm} = D_r r_p + i_{ms}^2 D_m / r_p$

3 Robust Control System of the Electric Power Steering

A new control structure for the EPS system is proposed to solve some problems in the conventional control structure, it is a double closed loop control structure. The new control structure diagram of the EPS system is shown in the Figure 3. The original structure needs not to change, only to add a new controller $K_E(s)$ outside the inner ring of the motor.

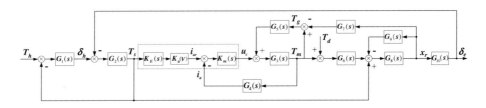

Fig. 3. The new control structure of EPS system

3.1 Inner Single Neuron PID Controller

The single neuron adaptive PID controller structure is shown in the Figure 4. The input of converter is $y_r(k)$ and $y(k)$ in the figure; the output of converter is the state variables $x_1(k), x_2(k), x_3(k)$ which are needed when neurons learn the control. Here $x_1(k) = y_r(k) - y(k) = e(k)$ $x_2(k) = \Delta e(k)$, $x_3(k) = e(k) - 2e(k-1) + e(k-2)$. $z(k) = x_1(k) = y_r(k) - y(k) = e(k)$ is the capability indicator or progressive signal. The K is ratio factor of the neurons. If K is greater than zero, the neurons generate the control signals by relating search, it is $u(k) = u(k-1) + K\sum_{i=1}^{3} w_i(k)x_i(k)$.

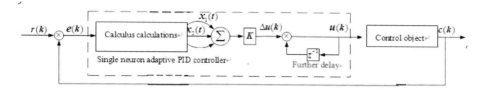

Fig. 4. Single neuron adaptive PID controller diagram

The single neuron adaptive PID controller achieves the adaptive and self-learning function by adjusting the weighting factors. The modified Hebb[7] learning algorithm is adopted to adjust the weighting factors here.

3.2 Outer Loop Robust Controller

The EPS control system must have a good robustness because of the stringent requirement of the steering system security. In this paper, the H∞ robust control algorithm is adopted for the design of the controller. Because the weighting function determines largely the controlling capability, designers often calculate and test a lot by their experience to obtain the desired weight function. For this reason, the paper proposed a backstepping loop shaping method (ECLGS). The method starts from the actual project, it inversely deduces the controller K according to the bandwidth frequency, high frequency asymptote and the largest singular value structure sensitivity function T. Thus, it ignores the weight function. According to the relevance of S and T, it determines indirectly the shape of the sensitivity function S, which ensures the robustness of the system performance.

The bandwidth frequency of the closed-loop system is ω_b, to construct the $T(s)$ easily, it makes the angular frequency equal the bandwidth frequency approximately here, because the relationship of the time constant T_c and ω_c is $\omega_c = 1/T_c$, there is $\omega_b \approx \omega_c = 1/T_c$; the frequency of the high frequency asymptote is $-20n$ dB, the " n " must be a integer here, it is usually the integer from 1 to 3; to ensure that the system tracks the target value without static error, the largest singular value selected is the integer 1, so complement sensitivity function $T(s)$ can be constructed as follows:

$$T(s) = \frac{1}{(T_c s + 1)^n} = \frac{G(s)K(s)}{1 + G(s)K(s)} \qquad (11)$$

The ideal S/T curve controller in the system can be obtained by the formula(11):

$$K(s) = \omega_b^n / [(s+\omega_b)^n - \omega_b^n]G(s) \qquad (12)$$

According to the Figure 3 and the Figure 4, the EPS multi-disciplinary transfer function under the new controlling structure is as follows:

$$G(s) = \frac{G_2(s)G_6(s)G_8(s)G_{10}(s)}{G_2(s)G_8(s)G_{10}(s)+(1+G_1(s)G_2(s))(1+G_8(s)G_9(s))} \qquad (13)$$

The bandwidth of the system determines the system response speed, according to performance requirement of the EPS system requirements, it makes the angular frequency $150 rad/s$ equal the system bandwidth approximately, it is $\omega_b = 150 rad/s$; to suppress effects of sensor noise and other uncertainties on the control precision and to guarantee the system robustness, the high-frequency asymptote slope rate is took as -60dB/dec, at present "n" is 3. like this, the singular value curve of the T is structured to the inertial system frequency characteristic curve whose largest singular value is the three steps of one. According to the equation (14), the overall controller of the ECLGS can be obtained as following equation (16) shows.

$$K(s) = \frac{3375000}{s(s^2+450s+67500)G(s)} \qquad (14)$$

According to the design of the inner ring, it gets the system response time of the inner loop, it is $T_{in} < 0.001s$. But According to the unit step response simulation which is input by the system steering wheel torque without the outer ring, we can obtain the response time T_{out} of the system outer ring, it is $T_{out} > 1.57s$, because $T_{out} \gg T_{in}$, the inner closed loop transfer function can be approximated as part of all-pass [8], it is $G_m(s) \approx 0.05$, at this time, the inner loop is noneffective on the design of the outer loop controller which can work independently. According to the stability theory of the control system [9], greater the gains of the open loop of the System, easier the closed- loop loses stability. If the closed-loop system can be stable on the maximum open loop gains, then the closed-loop system must be stable when the open loop gain is less than the maximum. Therefore, it only needs to consider the maximum power gains to ensure the stability of closed-loop system on other power gains for the design of EPS controller. In the paper, the EPS system will have the greatest power gain $K_a(V) = 4.29$ under the zero speed, the outer loop controller is $K_E(s) = K(s)/[K_a(V)G_m(s)] = 4.622K(s)$, it is as follows:

$$K_E(s) = \frac{15599250(0.00001819s^4+0.005903s^3+2.192s^2+139.622s+513.131)}{s^2(0.00099s^4+2.9378s^3+1683.6188s^2+391093.3125s+33429206.25)} \qquad (15)$$

4 Simulation and Result Analysis

In order to achieve the multi-domain system simulation for the EPS, the Simulink model shown in the Figure 3 is built In the Matlab / Simulink, the fixed-step 0.001s is used in simulation. In the case of the maximum power gain at zero speed (the system is the most unstable), the simulation is respectively used to the EPS system in the time domain and the frequency domain (S / T curve).

The controller obtained by ECLGS algorithm is a 6th order controller. Under the action of the controller, the frequency domain curve of the sensitivity function S and the system closed-loop function T is obtained respectively as the S / T curve in the Figure 5 (a). By the graph we can know that the slope to close the door in the closed loop system is -60dB/dec, the multitude margin is 69.85 and the Phase margin is 19.09° when the System sensitivity function gain in the low frequency is 0.2%, so the system has the good anti-jamming performance in the main work frequency range (the low frequency range or within $150 rad/s$) and the good noise suppression in the high frequency range, as well as the good robust stability.

The unit step response of the hand power to the steering wheel is simulated respectively in case of using the outer controller and not using the controller, we can gain the unit step response curve of the hand power to the steering wheel as the Figure 5 (b). By the curve in the figure we can know that the System step response settling time is 0.1482 seconds without controlling, and there is the 24.4% overshoot. When the ECLGS controller is used, the system step response settling time is reduced to 0.0355 seconds, there is no overshoot. Those show that the ECLGS controller can provide the system a good response performance if the robust is stable.

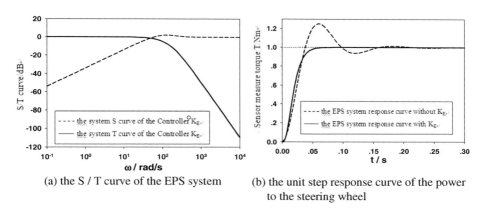

(a) the S / T curve of the EPS system

(b) the unit step response curve of the power to the steering wheel

Fig. 5. The S / T curve of the EPS system and the step response curve of the steering wheel practicing hand

5 Conclusion

This paper presented a backstepping-type H∞ robust control algorithm. It avoids a large number of repeated calculations for choosing the weighting function in the classical H∞ method. The multi-disciplinary model is established after the deep analysis of the EPS system. The model-based design method was proposed and the EPS control system is designed. There are some improvements to the traditional EPS single-loop control structure, a double closed-loop EPS control structure is put forward, the single neuron PID algorithm was adopted for the inner loop controller design and a new robust control algorithm was proposed for the outer loop controller design. The new control structure improved significantly the robustness of the system without reconstruction, and simplified the controller design and reduced the order of the controller. The frequency-domain S / T curves simulated shows that the EPS system has a good noise suppression, interference and robust stability under the action of the controller. The simulation of the hand steering wheel input to the system shows that the system under the controller action has a good time-domain response performance. The results showed that the controller designed in the paper can meet fully the robust control requirements of the EPS systems.

References

1. Lin, Y., Shi, G.: Electric power steering technology status and trend. Highway and Transportation Science and Technology 18(3), 79–82 (2001)
2. Tang, X., Yu, J.: On feedback scheduling of output feedback model-based net worked control systems. Control and Decision 24(1), 141–144 (2009)
3. Wu, Y., Jiang, Z., Chen, L.: Simulation optimization research of the multi-domain modelling based on the Modelica language. System Simulation Academic Journal 21(12), 3748–3752 (2009)
4. Norgaard, N., Ravn, O., Poulsen, N.K., et al.: Neural networks for modeling and control of dynamic systems. Springer, London (2000)
5. Kim, J.H., Song, J.B.: Control logic for an electric power steering system using assist motor. Mechatronics 12(3), 447–459 (2002)
6. Kohno, T., Takeuchi, S., Momiyama, M., et al.: Development of electric power steering (EPS) system with. H∞ control [J]. SAE Paper, 2000-01-0813
7. Rubaai, A., Kotaru, R.: Neural net-based robust controller design for Brushless DC motor Drives. IEEE Trans. on System, Man and Cybernetics 29(3), 460–474 (1999)
8. Luo, X., Shu, L., Li, G.: The all-pass system function analysis of the Bending Discrete Fourier Transform. Communication Technology 42(7), 282–284 (2009)
9. D'Azzo, J.J., Houpis, C.H., Sheldon, S.N.: Linear Control System Analysis and Design, 5th edn. Taylor & Francis, London (2003)

A Novel Analog Circuit Design and Test of a Triangular Membership Function*

Weiwei Shan[**], Yinchao Lu, Huafang Sun, and Junyin Liu

National ASIC System and Research Engineering Center, Southeast University
210096 Nanjing, China
wwshan@seu.edu.cn, flyinmo@163.com, se_filly@seu.edu.cn,
ljyseu@163.com

Abstract. In this paper, a novel analog design of a triangular membership function circuit is presented with a compact structure for hardware implementations of fuzzy-neural systems. This circuit is consisted of two simple transconductors and a current mode minimization circuit, which generates a linear triangular curve by getting the minimum part of two properly set transconductors. The characteristics (width, height, slope and position) are easy to adjust. This triangular MFC was realized in 0.6μm CMOS technology and tested with an output signal of a high precision and a high speed of 20MHz, showing that it is suitable for real-time applications.

Keywords: fuzzy logic; triangular membership function; analog circuit.

1 Introduction

Since fuzzy logic theory being introduced by L. A. Zadeh in 1965, it has been used extensively in many areas such as process control, signal processing, system identification and so on. Not only advanced fuzzy software methods have been studied, but also many hardware implementations have been reported. Fuzzy hardware implementations have gained wide attention for their advantages of high speed for real-time applications.

For hardware implementations, there are usually three kinds of design methodology, digital [1], analog [2] [3] [6] and mixed signal approach [4] [5], whereas most of the mixed-signal FLCs use analog circuit structures with digital memory. Digital design provides precise performance and programmability whereas requiring extra A/D and D/A converters to interface sensors and actuators respectively. On the other hand, analog approach excels in chip size and power consumption comparing to its digital counterpart. Therefore, analog and mixed-signal implementation methods are the most studied. In analog or mixed-signal implementations, the fuzzy functional blocks work in either current-mode or voltage-mode. Current-mode circuits have demonstrated many advantages such as simple circuitry, low power consumption and

---

[*] This work was sponsored by the National Scientific Foundation of China (Grant No. 61006029) and Jiangsu Scientific Foundation (Grant No. BK2010165).
[**] Corresponding author.

high speed [4]. However, in current-mode circuits, voltage input is preferred because the interface signal in practical applications is usually a voltage signal, or else a voltage to current converter is needed to get a current input signal.

Since Membership Function Circuit (MFC) is the first operational block in a fuzzy system, its characteristic should be considered carefully. Usually, several MFCs are needed in a fuzzy logic controller (FLC) according to the fuzzy partitions of the input variables, therefore, membership function circuit with compact structure is preferred in order to save the chip area. Many designs of MFC have been published, however, many of them are designed with simple structure but low accuracy, whereas some others provide a precise Membership Function (MF) but with relatively complex structure. In this paper, an alternative way is presented, which provide a circuit with relatively high accuracy and speed under a compact structure.

Based on two source-coupled differential pair transconductors, an analog circuit of Triangular MF is designed by getting the minimum part of two output currents of transconductors. The HSPICE simulation results as well as the test results of the fabricated circuit (under 0.6um CMOS mixed-signal technology provided by CSMC) show that the proposed circuit is able to generate a triangle wave with high precision and frequency response.

2 Triangular Membership Function Circuit

2.1 Transconductor Circuit

Since an ideal triangle curve as (1) is difficult to be realized directly by analog circuits, an approximation is necessary in designing a triangular MFC. Here x_0 is the center of the triangle curve, and $x_2 - x_1$ is its width.

$$\mu(x) = \begin{cases} -\dfrac{|x - x_0|}{x_2 - x_0} + 1, & x_1 < x < x_2 \\ 0, & \text{else} \end{cases} \quad (1)$$

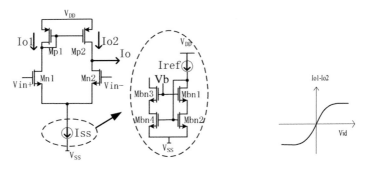

Fig. 1. (a) Source-coupled differential pair transconductor (b) Io1 - Io2

In this paper, a novel triangular MFC is designed by taking advantage of the linear part of the nonlinear characteristic of the transconductor as shown in Fig. 1(a) to approximate the ideal triangle curve.

This source-coupled differential pair transconductor is the most widely used one with a differential input signal. The difference of Io1 and Io2 is nonlinear as depicted in equation (2), however, it is linear around the area of $v_{id}=0$, which is shown clearly in Fig. 1(b).

$$I_{o1}-I_{o2} = \begin{cases} \sqrt{2I_{ss}\beta}v_{id}\sqrt{1-\frac{\beta}{2I_{ss}}v_{id}^2}, & \text{when } |v_{id}| = |v_{in+} - v_{in-}| \leq \sqrt{\frac{I_{ss}}{\beta}}, \beta = \frac{1}{2}\mu_n C_{ox}\frac{W}{L} \\ I_{ss}\,\text{sgn}(v_{id}) & \text{others} \end{cases} \quad (2)$$

2.2 Triangular Membership Function Circuit

Two such transconductors with a minimization circuit can generate a triangular MF by getting the minimum of outputs of the two properly set transconductors, where each output current rises or falls with opposite orientation. The building block is depicted in Fig. 2.

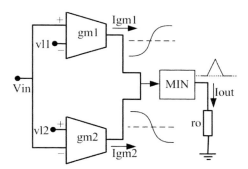

Fig. 2. Building block of triangular MFC

Reference voltages Vl1 and Vl2 define the points where output signal starts falling to –Iss (for gm2) or rising to Iss (for gm1) respectively in each transconductor, because the input signal is connected to the positive input port of gm1 and the negative input port of gm2. By setting Vl1 < Vl2 and keeping the difference of Vl1 and Vl2 as the width of the MF, a triangular MF can be derived which resembles a standard triangular curve, which is shown as the dc simulation result of Fig.3. Here Igm1 and Igm2 are the input currents of the minimization circuit coming from the outputs of the two transconductors with negative parts converted to zero. From this dc analyses, it can be noted that this MFC generates a precise triangular curve.

Since the triangular MFC utilizes the linear part of the transconductors. Thus according to equation (2), the transconductors are properly set to increase the linearity by adopting a large Iss and small W/L ratios of differential pairs.

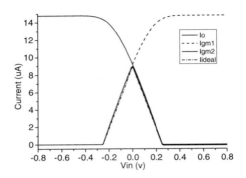

Fig. 3. DC output current. (a) Triangular MF output current

For the minimization circuit, a current-mode bounded-difference minimization circuit [3] can automatically convert the negative current of the transconductor to zero. The minimum operation is formulated using simultaneous bounded-difference equations as (3), where $\ominus$ is a bounded-difference operator defined by equation (4) [3].

$$MIN(x,y) = x \ominus (x \ominus y) \tag{3}$$

$$x \ominus y = \begin{cases} x-y & \text{when } x \geq y \\ 0 & \text{others} \end{cases} \tag{4}$$

A modified version of a two-input minimization circuit is depicted in Fig.4(a), where the input current copies of Ix and Iy are generated by cascode current mirrors in order to reduce the errors caused by current mirrors. This minimization circuit is also suitable for the minimization operation in fuzzy inference.

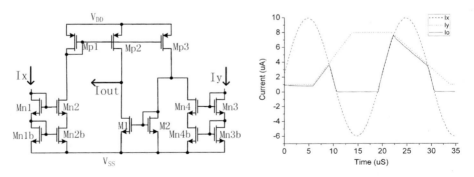

Fig. 4. (a) Current mode minimization circuit.   (b) Its transient response

The transient response of current-mode minimization is shown as Fig. 4(b). This minimization circuit follows the minimum input current when both currents are positive, or else its output current will be zero.

3 Test Results

This circuit was designed and fabricated in a 0.6μm standard CMOS technology, with its micro-photo shown as in Fig. 5. It occupied a small area of 237um*130um=0.0308 mm².

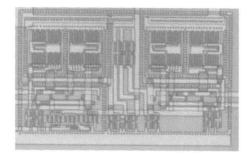

Fig. 5. Microphoto of the fabricated MFC.

The transient response of the triangular membership function circuit is shown in Fig. 6 (a), where the dashed line is the 10 kHz input signal whose peak value is from -1V to 1V; the solid line is the output voltage triangular signal when connecting a resistor to the ground to convert the output current to a voltage signal. The test result is shown in Fig. 6 (b), where the lower line is the input signal, and the upper line is the output signal. It can be seen that the fabricated circuit yields the same output result as that of simulated one.

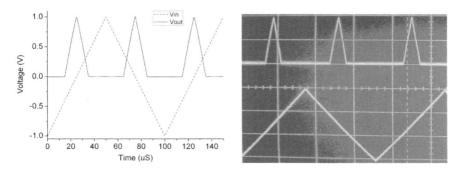

Fig. 6. Transient response of the triangular MFC (a) Simulation result (b) Test result.

The location of the triangular MF can be altered by tuning the reference voltage of Vl1 and Vl2 while fixing the difference of them, as shown in Fig. 7. The slope of the triangular MFC is electrically tunable by tuning the value of Iss of the transconductor, shown in Fig. 8(a). The width of the triangle is tuned by fixing one of the reference voltage of Vl1 and Vl2, and change the other voltage, as shown in Fig. 8 (b) by fixing Vl2=0.25v while setting Vl1=-0.2v, -0.25v and -0.3v respectively.

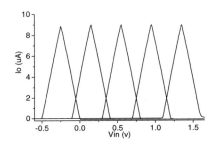

Fig. 7. Triangular MF output current with different centers

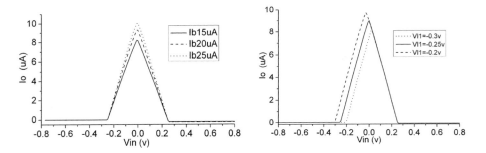

Fig. 8. Output current with (a) different slopes and peak values (b) different widths.

Table 1. Main features of analog Triangular Membership Function Circuit.

Triangular MFC	[2]	[5]	[7]	This paper
Voltage supply	±5v	±1.5v	6-7v	±2.5v
Max delay	70ns	--	--	50ns
Current/power (a)	--	0.3mW	0.14mW	200uA
Precision	high	high	low	high
Transistor count (b)	25	28	12	21
CMOS Technology	2um	0.5um	2.4um	0.6um

(a) Static power consumption
(b) Not include the transistors generating the bias current.

The maximum delay of the triangular MFC is 50ns, corresponding to a 20MHz fuzzification speed, which is fast enough for real-time control applications. When a positive step input voltage from 0v to 1v is applied to the triangular MFC with a load capacitor of 20fF, the fall time for the output voltage to settle to 10% of the steady state is 32ns. On the other hand, when a negative step input voltage is applied, the rise time for the output voltage to settle to 90% of the steady state is 50 ns.

Table 1 shows main features of the proposed Triangular Membership Function Circuit comparing to some other analog circuits. We can see that the proposed Triangular MFC exhibits a characteristic in high precision, low delay time and a compact circuitry.

4 Conclusions

A CMOS analog implementation of a triangular membership function for fuzzy systems was designed, fabricated and tested in this paper. This circuit is constructed of a compact structure with two differential pair transconductors and a minimization circuit. It works in current-mode with voltage input and current output, which makes a good interface to practical applications. Its behaviors are verified by HSPICE simulation as well as experimental results which show that it is suitable to be integrated into fuzzy logic controllers for real time applications.

References

1. Jin, W., Jin, D., Zhang, X.: VLSI design and implementation of a fuzzy logic controller for engine idle speed. In: Proceedings of 7th International Conference on Solid-State and Integrated Circuits Technology, pp. 2067–2070 (2004)
2. Tokmarci, M., Alci, M., Kilic, R.: A simple CMOS-Based Membership Function Circuit. Analog Integrated Circuits and Signal Processing 32, 83–87 (2002)
3. Sasaki, M., Inoue, T., Shirai, Y., Ueno, F.: Fuzzy Multiple-input Maximum and Minimum Circuits in Current Mode and Their Analyses Using Bounded-Difference Equations. IEEE Trans. Computers 39(6), 768–774 (1990)
4. Chen, C.Y., Hsieh, Y.T., Liu, B.D.: Circuit implementation of linguistic-hedge fuzzy logic controller in current-mode approach. IEEE Trans. Fuzzy Systems 11(5), 624–646 (2003)
5. Kachare, M., Ramírez-Angulo, J., Carvajal, R.G., López-Martín, A.J.: New Low-Voltage Fully-Programmable CMOS Triangular/Trapezoidal Function Generator Circuit. IEEE Trans. Circuits I 52(30), 2033–2042 (2005)
6. Dualibe, C.: Design of Analog fuzzy logic Controller in CMOS Techonology-Implementation, test and Application, pp. 109–117. Kluwer Acdemic, Dordrecht (2003)
7. Samman, F.A., Sadjad, R.S.: Analog Mos Circuit Design for Reconfigurable Fuzzy Logic Controller. Circuits and Systems 2, 28–31 (2002)

Novel Electronically Tunable Mixed-Mode Biquad Filter

Sajai Vir Singh[1,*], Sudhanshu Maheshwari[2], and Durg Singh Chauhan[3]

[1] Department of Electronics and Communications, Jaypee Institute of Information Technology, sector-128, Noida, -201304 (India)
sajaivir@rediffmail.com
[2] Department of Electronics Engineering, Z. H. College of Engineering and Technology, Aligarh Muslim University, Aligarh-202002 (India)
[3] Uttarakhand Technical University, Dehradun-248001 (India)

Abstract. This paper presents a novel electronically tunable mixed-mode biquad filter. The proposed filter employs only three multi-output current controlled current conveyor trans-conductance amplifiers (MO-CCCCTAs) and two grounded capacitors. With the current as an input, the proposed filter can realize low pass (LP), band pass (BP), high pass (HP), band reject (BR) and all pass (AP) in current-mode and LP, BP and BR responses in trans-impedance-mode. When the voltage acts as an input, the proposed filter can realize LP, BP, HP, BR and AP responses in trans-admittance-mode and LP, BP and BR in voltage-mode. Moreover, the filter also offers the feature of quality factor control of independent of pole frequency through only single bias current. PSPICE simulation results are included for verification of the proposed circuit.

Keywords: CCCCTA, mixed-mode, tunable filter, biquad.

1 Introduction

Because of its wide applications in communications, measurements and instrumentations, there has been a great attention on the design and study of analog filter using different current-mode active elements [1-23]. Depending on the type of signals at their inputs and outputs, analog filter can be classified as current-mode [2-7], voltage-mode [8-11], trans-admittance-mode [12] and trans-impedance-mode filter [13-14]. The trans-impedance-mode and trans-admittance-mode filters act as a bridge transferring from voltage-mode to current-mode and vice versa, respectively. In some analog signal processing applications it may be desirable to have active filters with input currents and/or voltages and output currents and/or voltages that are the mixed-mode filter. Mixed-mode filter circuits [15-22] can be classified either as multiple-input type or single input type. However, most of these structures are multiple-input type [15-18], which can realize one standard filter function in one mode at a time. In some applications, however, simultaneous outputs of many different filter functions may be required and a very little work has been done in the domain of single input-type mixed-mode filters [19-22]. The mixed-mode filter structure reported in [19] employs four CFOAs, nine resistors and

---

[*] Corresponding author.

two grounded capacitors and can realize all the standard filter functions in each mode but not from the same configuration. Moreover, this reported circuit needs switches to realize BR or AP response and could not employ orthogonal tunability ω_0 and Q. The filter circuit reported in ref. [20] uses single FDCCII, three resistors and two grounded capacitors and realizes all standard filter functions in voltage and trans-impedance-mode but realizes only two filter functions (BP and HP) in current and trans-admittance-mode. The another valuable mixed mode filter circuit in ref. [21] uses five MCCCIIs, two grounded capacitors and realizes LP, BP and HP in all the four modes. It also provides orthogonal tunability of ω_0 and Q. However, all the reported filter circuits [19-21] use excessive number of active and/ or passive components and suffer from unavailability of input voltage at high input impedance in voltage and trans-admittance-mode. Recently reported mixed-mode filter [22] employs three CCCCTAs and two grounded capacitors. The filter [22] can realize LP, BP and HP responses in all the four modes from the same configuration. The filter can also realize BR and AP responses in current and trans-admittance-mode, with interconnection of relevant output currents. Moreover, it provide the feature of availability of input voltage at high input impedance in voltage and trans-admittance-mode but it suffers from the feature of Q control of independent of ω_0 through single bias current and uses more hardware in term of number of transistors. In this paper a novel electronically tunable mixed-mode biquad filter is presented. The proposed mixed-mode filter employs only three MO-CCCCTAs and two grounded capacitors. The proposed filter can realize LP, BP, HP, BR and AP responses in current-mode and trans-admittance-mode. It can also realize LP, BP and BR in trans-impedance-mode and voltage-mode. The mixed-mode filter exhibits low active and passive sensitivities. The performances of proposed circuit are illustrated by PSPICE simulations.

2 CCCCTA Descriptions and Filter Design

CCCCTA is relatively new active element [23] and has received considerable attention as current mode tunable active element. The current-voltage properties of MO-CCCCTA can be expressed by the following equations

$$V_{Xi} = V_{Yi} + I_{Xi}R_{Xi}, I_{Zi} = I_{Xi}, I_{\pm O} = \pm g_{mi}V_{Zi} \qquad (1)$$

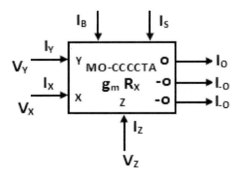

Fig. 1. MO-CCCCTA Symbol

where R_{xi} and g_{mi} are the parasitic resistance at x terminal and transconductance of the ith CCCCTA, respectively. R_{xi} and g_{mi} depend upon the biasing currents I_{Bi} and I_{Si} of the CCCCTA, respectively. The schematic symbol of MO-CCCCTA is illustrated in Fig.1. For BJT model of CCCCTA [23] shown in Fig.2, R_{xi} and g_{mi} can be expressed as

$$R_{Xi} = \frac{V_T}{2I_{Bi}} \text{ and } g_{mi} = \frac{I_{Si}}{2V_T} \qquad (2)$$

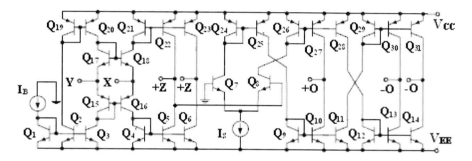

Fig. 2. Internal Topology of CCCCTA

The proposed mixed-mode second order filter circuit as shown in Fig. 3 uses three MO-CCCCTAs and two grounded capacitors. By routine analysis of the circuit in Fig.3, the following current and voltage responses are obtained.

$$I_1 = \frac{-(C_1C_2R_{X2}s^2 + g_{m2})R_{X1}I_{in} + sg_{m1}R_{X2}C_2V_{in}}{D(s)}, I_2 = \frac{g_{m1}R_{X2}C_2s(R_{X3}I_{in} + V_{in})}{D(s)} \qquad (3)$$

$$I_3 = \frac{-g_{m2}(R_{X3}I_{in} + V_{in})}{D(s)}, I_4 = \frac{(C_1C_2R_{X2}s^2 + g_{m2})(R_{X3}I_{in} + V_{in})}{D(s)} \qquad (4)$$

$$I_5 = \frac{-g_{m1}R_{X2}C_2s(R_{X3}I_{in} + V_{in})}{D(s)}, I_6 = \frac{g_{m2}(R_{X3}I_{in} + V_{in})}{D(s)} \qquad (5)$$

$$V_1 = \frac{sC_2R_{X2}(R_{X3}I_{in} + V_{in})}{D(s)}, V_2 = \frac{(R_{X3}I_{in} + V_{in})}{D(s)} \qquad (6)$$

$$V_3 = \frac{R_{X1}(C_1C_2R_{X2}s^2 + g_{m2})(R_{X3}I_{in} + V_{in})}{D(s)} \qquad (7)$$

where $D(s) = C_1C_2R_{X1}R_{X2}s^2 + sg_{m1}R_{X3}R_{X2}C_2 + g_{m2}R_{X1}$ (8)

From the above equations we can see that various mixed-mode filter functions, with input current or voltage and output current or/and voltage, can be obtained.

Case I. With $V_{in}=0$ and I_{in} as input signal, the following filter responses in current as well as trans-impedance-mode are simultaneously obtained.

(a). An inverted BR from I_1,
(b). A non-inverted BP from I_2,
(c). An inverted LP from I_3,
(d). A non-inverted BR from I_4,
(e). An inverted BP from I_5,
(f). A non-inverted LP from I_6,
(g). A non-inverted HP from $I_3 + I_4$,
(h). An inverted AP from I_1+I_2,
(i). A non-inverted BP at V_1,
(j). A non-inverted LP at V_2,
(k). A non-inverted BR at V_3.

Case II. With $I_{in}=0$ and V_{in} as input signal, the following filter responses in voltage as well as trans-admittance-mode are simultaneously obtained.

(a). A non-inverted BP from I_1 and I_2,
(b). An inverted LP from I_3,
(c). A non-inverted BR from I_4,
(d). An inverted BP from I_5,
(e). A non-inverted LP from I_6,
(f). A non-inverted HP from $I_3 + I_4$,
(g). A non-inverted AP (if $R_{X1}=R_{X3}$) from $I_4 + I_5$,
(h). A non-inverted BP at V_1,
(i). A non-inverted LP at V_2,
(j). A non-inverted BR at V_3.

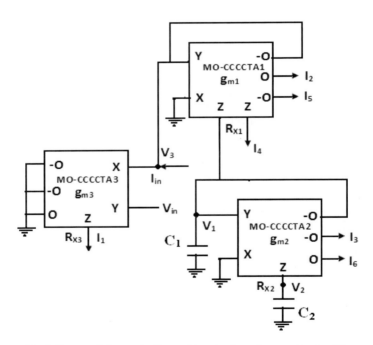

Fig. 3. Proposed electronically tunable mixed-mode second order filter

From above description it is clear that the proposed structure can be viewed as mixed-mode filter. The pole frequency (ω_0), the quality factor (Q) and Bandwidth (BW) ω_0/Q of the proposed filter in each mode can be expressed as

$$\omega_0 = \left(\frac{g_{m2}}{R_{X2}C_1C_2}\right)^{\frac{1}{2}}, Q = \frac{R_{X1}}{g_{m1}R_{X3}}\left(\frac{C_1 g_{m2}}{C_2 R_{X2}}\right)^{\frac{1}{2}}, \text{and } BW = \frac{\omega_0}{Q} = \frac{g_{m1}R_{X3}}{C_1 R_{X1}} \quad (9)$$

Substituting intrinsic resistances as depicted in eqn. (2), it yields

$$\omega_0 = \frac{1}{V_T}\left(\frac{I_{B2}I_{S2}}{C_1 C_2}\right)^{\frac{1}{2}}, Q = \frac{2}{I_{S1}}\frac{I_{B3}}{I_{B1}}\left(\frac{C_1}{C_2}I_{S2}I_{B2}\right)^{\frac{1}{2}} \quad (10)$$

From eqn. (10), we see that the Q can be adjusted independently from ω_0 by varying I_{B1} or I_{B3} or both. It is also clear that high value of Q can be obtained by adjusting the ratio of I_{B3} and I_{B1}. By maintaining $I_{B2}= I_{S2}= I_{S1}$, it can be remarked that the pole frequency can be electronically adjusted by I_{B2} and I_{S2} without affecting the quality factor.

Taking the non-idealities of CCCCTA into account, the relationship of the terminal voltages and currents can be rewritten as follow.

$$V_{Xi} = \beta_i V_{Yi} + I_{Xi}R_{Xi}, I_{Zi} = \alpha_i I_{Xi}, I_O = \gamma_{pi}g_{mi}V_{Zi}, I_{-O} = -\gamma_{ni}g_{mi}V_{Zi} \quad (11)$$

Where β_i, α_i, γ_{pi} and γ_{ni} are transferred ratios of the ith CCCCTA which deviate from 'unity' by the transfer errors. In the case of non-ideal and reanalyzing the proposed filter in Fig. 3, denominator of the response in each case become

$$D'(s) = C_1 C_2 R_{X1} R_{X2} s^2 + \alpha_1 \beta_1 \gamma_{n1} R_{X2} C_2 g_{m1} R_{X3} s + \alpha_2 \beta_2 \gamma_{n2} g_{m2} R_{X1} \quad (12)$$

In this case, the ω_0 and Q are changed to

$$\omega_0 = \left(\frac{\alpha_2 \beta_2 \gamma_{n2} g_{m2}}{C_1 C_2 R_{X2}}\right)^{\frac{1}{2}}, Q = \frac{R_{X1}}{\alpha_1 \beta_1 \gamma_{n1} g_{m1} R_{X3}}\left(\frac{\gamma_{n2}\alpha_2\beta_2 C_1 g_{m2}}{C_2 R_{X2}}\right)^{\frac{1}{2}} \quad (13)$$

The active and passive sensitivities of the proposed circuit in Fig.3 are low and, can be found as

$$S^{\omega_0}_{C_1,C_2,R_{X2}} = -\frac{1}{2}, S^{\omega_0}_{\gamma_{n2},\alpha_2,\beta_2,g_{m2}} = \frac{1}{2} \quad (14)$$

$$S^{Q}_{C_2,R_{X2}} = -\frac{1}{2}, S^{Q}_{\beta_2,\alpha_2,C_1,\gamma_{n2},g_{m2}} = \frac{1}{2}, S^{Q}_{R_{X1}} = 1, S^{Q}_{R_{X3},\alpha_1,\beta_1,\gamma_{n1},g_{m1}} = -1 \quad (15)$$

3 Simulation Results

The proposed mixed-mode filter was verified through PSPICE simulations. In simulation, the MO-CCCCTA was realized using BJT model as shown in Fig.2, with the transistor model of HFA3096 mixed transistors arrays [7] and was biased with ±2.5V DC power supplies. The circuit was designed for Q=1 and $f_0=\omega_0/2\pi=2.43$MHz. The active and passive components were chosen as $I_{B1}=I_{B2}=I_{B3}=29.5\mu A$, $I_{S1}=I_{S2}=I_{S3}=120\mu A$ and $C_1=C_2=0.15$nf. Fig.4 shows the simulated gain responses of the LP, HP, BP, BR and AP of current and trans-admittance-mode filter of Fig.3. Fig.5 shows the simulated gain responses of the LP, BP and BR of and trans-impedance-mode and voltage-mode filter of Fig.3. The simulation results show the simulated pole frequency as 2.4MHz that is ~1.2% in error with the theoretical value. Fig.6 (a) shows the magnitude responses of BR function for different values of I_{B3}, by keeping $I_{B1}=I_{B2}=29.5\mu A$, $I_{S1}=I_{S2}=I_{S3}=120\mu A$, and $C_1=C_2=0.15$nf. The quality factor was found to vary as 1, 2, 4, 6.2 and 12.4 by keeping constant pole frequency as 2.37MHz for five values of I_{B3} as 30μA, 60μA, 120μA, 180μA and 360μA, respectively, which shows that the quality factor of the response can be electronically adjusted by input bias current I_{B3} without affecting pole frequency.

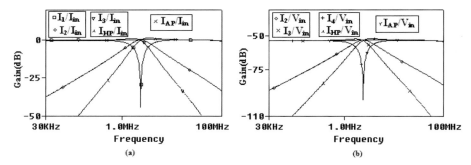

Fig. 4. Simulated gain responses of LP, BP, HP, BR and AP of the circuit in Fig.3 (a) in current-mode (b) in trans-admittance-mode

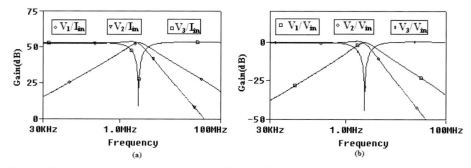

Fig. 5. Simulated gain responses of LP, BP and BR of the circuit in Fig.3 (a) in trans-impedance-mode (b) in voltage-mode

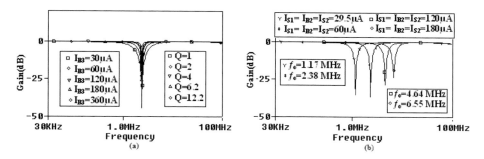

Fig. 6. (a) BR responses in current-mode, for different value of I_{B3} (b) BR responses in current-mode, for different value of $I_{B2}=I_{S1}=I_{S2}$

Fig.6 (b) shows magnitude responses of current-mode BR function where I_{B2}, I_{S2} and I_{S1} were equally set and changed for several values, by keeping its ratio to be constant for constant Q(=2). Other parameters were chosen as $I_{B1}=I_{B3}=29.5\mu A$, $I_{S2}=I_{S3}=120\mu A$, and $C_1=C_2=0.15nf$. The pole frequency was found to vary as 1.17MHz, 2.38MHz, 4.64MHz and 6.55MHz for four values of $I_{B2}=I_{S2}=I_{S1}$ as 29.5μA, 60μA, 120μA and 180μA, respectively, which shows that pole frequency can be electronically adjusted without affecting the quality factor.

4 Conclusion

In this paper an electronically tunable mixed-mode biquad filter using only three MO-CCCCTAs and two grounded capacitors has been proposed. The proposed filter offers the following attractive features: (i) realization of LP, BP, and BR responses in voltage-mode, current-mode, trans-admittance-mode and trans-impedance-mode, from the same configuration, (ii) ability to realize HP and AP in the current-mode and trans-admittance-mode from the same topology too, (iii) availability of input voltage at high input impedance in voltage and trans-admittance-mode, (iv) use of only grounded capacitors which makes the structure less sensitive to parasitic and easier to integrate, (v) low sensitivity figures. (vi) Q control of independent of ω_0 through only single bias current, (vii) easily obtained high Q-value filter by adjusting the ratio of two bias currents of the MO-CCCCTAs.

References

1. Soliman, A.M.: Current Conveyor Filters Classification and Review. Microelectronics J. 29, 133–149 (1998)
2. Soliman, A.M.: New Current-Mode Filters Using Current Conveyors. Int'l J. Electronics and Communications (AEÜ) 51, 275–278 (1997)
3. Khan, I.A., Zaidi, M.H.: Multifunction Translinear-C Current Mode Filter. Int'l J. Electronics 87, 1047–1051 (2000)
4. Maheshwari, S.: Current-Mode Filters with High Output Impedance and Employing Only Grounded Components. WSEAS Trans. on Electronics 5, 238–243 (2008)

5. Tsukutani, T., Sumi, Y., Yabuki, N.: Versatile Current-Mode Biquadratic Circuit Using Only Plus Type CCCIIs and Grounded Capacitors. Int'l J. Electronics 94, 147–1156 (2007)
6. Keskin, A.Ü., Biolek, D., Hancioglu, E., Biolkova: Current-Mode KHN Filter Employing Current Differencing Transconductance Amplifiers. Int'l J. Electronics and Communications (AEÜ) 60, 443–446 (2006)
7. Senani, R., Singh, V.K., Singh, A.K., Bhaskar, D.R.: Novel Electronically Controllable Current Mode Universal Biquad Filter. IEICE Electronics Express 1(14), 410–415 (2004)
8. Maheshwari, S.: High Performance Voltage-Mode Multifunction Filter with Minimum Component Counts. WSEAS Transactions on Electronics 5, 244–249 (2008)
9. Praveen, T., Ahmed, M.T., Khan, I.A.: A Canonical Voltage-Mode Universal CCCII-C Filter. J. Active and Passive Electronics Devices 4, 7–12 (2009)
10. Chang, C.M., Tu, S.H.: Universal Voltage-Mode Filter with Four Inputs and One Output Using Two CCII+s. Int'l J. Electronics 86, 305–309 (1999)
11. Menaei, S., Yuce, E.: All-Grounded Passive Elements Voltage-Mode DVCC-Based Universal Filters. Circuits Syst. Signal Process. 29(2), 295–309 (2008)
12. Singh, S.V., Maheshwari, S., Mohan, J., Chauhan, D.S.: An Electronically Tunable SIMO Biquad Filter Using CCCCTA. In: Contemporary Computing. CCIS, vol. 40, pp. 544–554 (2009)
13. Soliman, A.M.: Mixed-Mode Biquad Circuits. Microelectronics J. 27, 591–596 (1996)
14. Minaei, S., Topcu, G., Cicekoglu, O.: Low Input Impedance Trans-impedance Type Multifunction Filter Using Only Active Elements. Int'l J. Electronics 92, 385–392 (2005)
15. Abuelma'atti, M.T.: A Novel Mixed-Mode Current-Controlled Current Conveyor-Based Filter. Active Passive Electronics Components 26, 185–191 (2003)
16. Abuelma'atti, M.T., Bentrica, A., Al-Shahrani, S.M.: A Novel Mixed-Mode Current-Conveyor-Based Filter. Int'l J. Electronics 91, 91–97 (2004)
17. Yuce, E.: Fully Integrable Mixed-Mode Universal Biquad with Specific Application of the CFOA. Int'l J. Electronics and Communications (AEÜ) 64, 304–309 (2010)
18. Chen, H.P., Liao, Y.Z., Lee, W.T.: Tunable Mixed-Mode OTA-C Universal Filter. Analog Integrated Circuits and Signal Processing 58, 135–141 (2009)
19. Singh, V.K., Singh, A.K., Bhaskar, D.R., Senani, R.: Novel Mixed-Mode Universal Biquad Configuration. IEICE Electronics Express 2, 548–553 (2005)
20. Lee, C.N., Chang, C.M.: Single FDCCII-Based Mixed-Mode Biquad Filter with Eight Outputs. Int'l J. Electronics and Communications (AEÜ) 63(2), 736–742 (2009)
21. Zhijun, L.: Mixed-Mode Universal Filter Using MCCCII. Int'l J. Electronics and Communications (AEÜ) 63(2), 1072–1075 (2009)
22. Maheshwari, S., Singh, S.V., Chauhan, D.S.: Electronically Tunable Low Voltage Mixed-Mode Universal Biquad Filter. IET Circuits, Devices and Systems (2011) (accepted), doi:10.1049/iet-cds.2010.0061
23. Siripruchyanun, M., Jaikla, W.: Current Controlled Current Conveyor Transconductance Amplifier (CCCCTA): A Building Block for Analog Signal Processing. Electrical Engineering 90, 443–453 (2008)

Localization Algorithm Based on RSSI Error Analysis

Liu SunDong[1], Chen SanFeng[1], Chen WanMing[2], and Tang Fei[1]

[1] Shenzhen institute of information Technology, Guangdong, Shenzhen, China
[2] Department of Automation, University of Science and Technology of China, Hefei, China

Abstract. This article proposes a new localization algorithm for wireless sensor network (hereinafter referred to as WSN), mainly used for handling the problems that WSN meets in actual geographical environment and made the algorithm more adoptable to actual application environment. First, this article analyses the two factors that generate error in distance measurement applying RSSI in actual environment. Then on the basis of this it proposes RSSI signal theoretical model's online updating method and the way of applying two network nodes to realize localization, in order to reduce RSSI error' interference to localization precision. At last the simulation test verifies the feasibility of the algorithm.

Keywords: wireless sensor, network (WSN), network node, RSSI.

1 Introduction

In WSN, node localization is one of the key technologies. Not only in field of WSN application such as environmental monitoring, target tracing but also in the internal agreements of WSN such as location-based topology structure control and routing algorithm, they all need to acquire the information on node localization. Therefore, it is significant to study the localization algorithm of node.

Documents [Cesare A. Giovanni V. 2006; Ol K. et al. 2006; Chen W. 2006; Flammini A. et al. 2006; Mikko K. et al. 2006] have undertaken a series studies on RSSI and RSSI-based localization algorithm for wireless sensor networks. However they do not provide corresponding strategies to both the online updating for the parameter of RSSI signal transformation theoretical model and the RSSI error caused by obstacles. In the actual environment, if no online updating for the parameter of RSSI signal transformation theoretical model, the model may be not suitable for the environment applied in this network, furthermore, it results in failure equivalence between the intensity of received signal in RSSI and actual distance within nodes; if no corresponding measures for the RSSI error caused by obstacles, the signal attenuation would be great once there is obstacle leading to a large deviation between converted distance and the actual one. Therefore, if RSSI plays the role of localization for WSN nodes, online updating of model parameter and error treatment for obstacle would be inevitably vital issues.

This paper will focus on the actual geography environment in which WSN locates and analyze the big error of nodes' distance (such as incorrect model parameter of signal transformation, error caused by small scale obstacle) generated through

RSSI—intensity attenuation of received signal within nodes. Then the article will propose methods to reduce those errors and put forward the localization algorithm.

2 Localization Algorithm Based on RSSI Error Analysis (ERSSI)

There are many reasons for the error generated in distance measurement applying RSSI. The article only takes two key factors into consideration: 1. Error generated by theoretical model or experience model which measures the distance based on received signal's attenuation intensity 2. Error generated by small obstacle.

Aiming at the two aspects of error source analysis mentioned above, the article puts forward localization algorithm based on RSSI error analysis (ERSSI) in purpose of reducing the error's influence mentioned above.

1. For error generated by theoretical model, the method of partial TDOA to assist RSSI parameter calibration will be adopted. The procedure is described as follows:

Install transmitting and receiving device of ultrasonic signal on part of nodes, those nodes are called ultrasonic nodes for convenience. In the reason that this device will add extra hardware resources to nodes, only few nodes will install this kind of device, to realize the updating to the parameter of RSSI theoretical model under the online condition. If the whole WSN environment is approximately alike, it only needs to set a minimum of three ultrasonic nodes in this network. Those three ultrasonic nodes are neighbors and could communicate with each other, and also receive ultrasonic signals from neighboring nodes. Ultrasonic Node A transmits both electromagnetic signal and ultrasonic signal at the same time. When Ultrasonic Node B receives the electromagnetic signal, it opens the ultrasonic receiver to receive ultrasonic signal, and record the time interval between two received signals. The distance between two nodes can be calculated by multiplying the interval by ultrasonic velocity. The process above is similar to localization algorithm of TDOA [Niculescu D. Nath B. 2001]. Error will generate in acquiring distance with the method mentioned above but compared with error in RSSI distance measurement, its error is relatively smaller which could be approximately regarded as the actual distance between nodes. Apply this distance information to update RSSI theoretical model formula (1) and then transmit the updated parameter information to other nodes in the same region in network. It realizes the online updating for RSSI theoretical parameter by this method.

The whole network only needs to set a minimum of three ultrasonic nodes, therefore, compared with former network nodes, the cost of hardware does not increase a lot after the setting of ultrasonic nodes. The const of ultrasonic nodes newly added can totally omit when the whole network is consist of thousands or millions of common nodes.

2. For error caused by small obstacle between nodes, the procedure is described as follows:

When obstacle widens the measured distance between nodes, for Node A under determined, it is the extremely small probability that this distance would become the minimum or second minimum value of the distance between Node A and all its neighboring nodes. In other words, there is a large probability that this measured distance could not be the minimum or second minimum value. If so, using the two nodes nearest and second-nearest Node A to localize Node A, then the localization of

Node A will not be influenced by obstacle; once the small probability became true, i.e. the distance widened by obstacle was the minimum of second minimum in all distances, the influence on localization would not be a great one because this is the minimum or second minimum distance which is a really small value. Furthermore, the influence of error could be tried to eliminate during refinement—the last step of localization.

As shown in Figure 1, Node A is the node under determined; it acquires the distances from Node B, C, D and E with method of RSSI distance measurement. Sort these distances in ascending order, which is RAC, RAB, RAD, RAE respectively, and choose the smallest and second smallest RAC, RAB to localize Node A. When the position of Node B and C are known, according to analytic geometry, the position of Node A could only be position A or A1 illustrated in figure, and A and A1 are symmetric with respect to Line LBC, are mirror images of one another. Hence, Node A has two possible positions according the nearest Nodes B and C. And if RAC, RAB has no error or very small error, then one out of these two positions must have 0 error in the actual position of A or tiny error in it. Because the measured distance error and the length of distance increase in an unequal proportion, that is, the longer the distance is, the bigger distance error. In this reason the error in measured distance of the nearest points B and C is relatively very small. If so, the error between then one out of these two positions and the actual position must be tiny. By this means, the influence on other nodes' bigger errors for Node A could be eliminated under certain conditions. After that another third point D is used to determine A's specific position. Calculate the distance from Node D to the two possible positions respectively, and compare the two distances with RAD, then the one nearer RAD is the localization position of Node A. Since the third node D is just used to the comparison of these two positions, as long as there is not a big error in RAD, Node D will not influence A's actual localization.

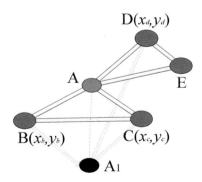

Fig. 1. Illustration of the Localization Method

The above method tries to limit the error factor for Node A localization onto the Node B and C which is very near Node A and have relatively smaller error in measured distance; then to limit the error in choice of specific position onto the third nearest point D, in order to reduce the influence of other nodes behind obstacles and with bigger errors as much as possible.

Suppose the coordinates of Node B, C and D in Figure 2 are acquired, which are (x_b, y_b), (x_c, y_c) and (x_d, y_d) respectively, Node A is the node under determined and its coordinate is unknown, as (x_{Blind}, y_{Blind}). The distance from Node A to Node B, C, D are acquired through the method of online updating for RSSI distance measurement mentioned above, which are $R_{Blind,b}$, $R_{Blind,c}$, $R_{Blind,d}$ respectively, and $R_{Blind,b} < R_{Blind,d}$, $R_{Blind,c} < R_{Blind,d}$.

The calculation of node localization is as follows:

$$\begin{cases} (x_{Blind} - x_b)^2 + (y_{Blind} - y_b)^2 = R^2_{Blind,b} \\ (x_{Blind} - x_c)^2 + (y_{Blind} - y_c)^2 = R^2_{Blind,c} \end{cases} \quad (1)$$

The formula (1) can find the two values of Node A's coordinate, which are correspondent to A and A1 in Figure 2, i.e. $(x_{Blind}, y_{Blind}) = (x_a, y_a)$ or $(x_{Blind}, y_{Blind}) = (x_{a1}, y_{a1})$.

Then use Node D's coordinate and distance between it and Node A to determine the final coordinate of Node A. After calculating, we can find:

$$\left| (x_d - x_a)^2 + (y_d - y_a)^2 - R^2_{Blind,d} \right| < \left| (x_d - x_{a1})^2 + (y_d - y_{a1})^2 - R^2_{Blind,d} \right| \quad (2)$$

Distance between coordinate position A and Node D is more closer than the measured distance, therefore choose (x_a, y_a) as the coordinate of Node A.

After finishing the above localization, the method of circulating refinement could be applied to reduce localization error according to demand of practical application. There are various means of circulating refinement, document [Savarese C. et al. 2002] proposes refinement algorithm.

The method mentioned above, can effectively reduce the error generated by theoretical model for RSSI distance measurement in actual environment and the error of measurement and localization caused by small obstacle.

3 Experiment

Evenly distribute 200 nodes on a regular plane in random, set the transmitting signal radius of node as 2 units, and regard the RSSI signal that it possibly receives from outside of 2 units as noise, take it out of consideration.

In Figure 2, the horizontal axis represents the influential intensity of the size of obstacle to measured distance. Its values represent that the existence of obstacle makes the node's measured distance (1+k) times larger than actual one. The vertical axis represents the localization error of node. The figure shows that the larger the size of obstacle is, increasing the node localization error is. The figure also reflects that this algorithm has a better localization precision.

Analyzing reasons for the above: the data for localizing every node in this chapter is measured distance information provided by nearest neighboring nodes, and then if the distance value is smaller, there is less possibility of obstacle existence during distance measurement; even obstacle exists but its influence is tiny. Since this algorithm just uses 2+1 nearest points from the node, for the large values in distance measurement caused by obstacle, the localization algorithm in this article does not use these data; therefore the localization effect is less influenced by small obstacle. However, the localization error of this algorithm will increase with the increasing size of obstacle.

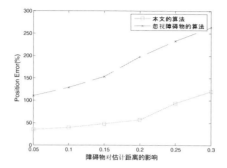

Fig. 2. Relationship Between Localization Error and the Size of Obstacle

In Figure 3, once the size of obstacle is fixed, the horizontal axis represents the proportion of numbers of obstacle, in another word, the proportion of numbers of distance measurement information influenced by obstacle to the total numbers. The vertical axis represents localization error of node. The figure shows that larger proportion of obstacle numbers is, larger localization error of node becomes.

Analyzing reasons for the above: this algorithm just uses 2+1 nearest points from the node, and uses data influenced by obstacle as lest as possible. However, with the increasing numbers of distance measurement information influenced by obstacle, the operating process of this algorithm inevitably uses part of data with lager error, leading to a lower localization precision than that of Figure 3; what's more, its localization error will increase faster with the increasing numbers of information influenced by obstacle.

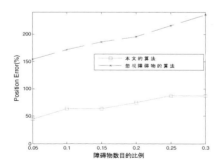

Fig. 3. Relationship Between Localization Error and Numbers of Obstacle

4 Conclusion

Focusing on actual geography environment of WSN application, this article mainly analyzes the error possibly generated within the application of RSSI distance measurement, and proposes a localization updating algorithm responding to the reducing of these errors. This algorithm has remarkable effect on small size obstacle.

Acknowledgment

This research was support by the Natural Science Foundation of GuangDong Province (9151802904000014).

References

1. Alippi, C., Vanini, G.: A RSSI-based and calibrated centralized localization technique for Wireless Sensor Networks. In: Proceedings of the Fourth Annual IEEE International Conference on Pervasive Computing and Communications Workshops, PERCOMW 2006 (2006)
2. Chen, W., Mei, T., Meng, M.Q.-H., et al.: A localization algorithm based on discrete imprecision range measurement in wireless sensor networks. In: Proc. of the 2006 IEEE International Conference on Information Acquisition (ICIA 2006), Weihai, Shandong, China, pp. 644–648 (August 2006)
3. Flammini, A., Marioli, D., Mazzoleni, G., et al.: Received Signal Strength Characterization for Wireless Sensor Networking. In: IMTC 2006 Instrumentation and Measurement Technology Conference, Sorrento, Italy, April 24-27, pp. 207–211 (2006)
4. Kohvakka, M., Suhonen, J., et al.: Transmission power based path loss metering for wireless sensor networks. In: The 17th Annual IEEE International Symposium on Personal, Indoor and Mobile Radio Communications, PIMRC 2006 (2006)

Moving Target Detection Based on Image Block Reconstruction

Guo Sen[1] and Liao Jiang[2]

[1] Shenzhen Institute of Information Technology, Shenzhen, China
[2] Lu Tong High Technology Company, Nanchang, China

Abstract. Background reconstruction is one key issue in computer vision research. The existing algorithms can not satisfy the requirements of real-time and robustness contemporarily. In this paper, a novel background reconstruction algorithm is presented. At first, image is partitioned into several sub-blocks rationally, these sub-blocks belong to background or no can be judged by its motion energy calculated by frame difference. Background updating can be carried out according to the analysis of variation of motion energy of four corners of image, At last, this algorithm is applied in one human target detection, which shows that this algorithm can achieve better result.

Keywords: frame difference; block reconstruction; background reconstruction.

1 Introduction

Moving target detection is one of the key issues in image understanding. The method most frequently applied in moving target detection is background subtraction, which means it realizes the difference between image and background, then removes background to obtain target foreground. It is widely applied in fields such as video surveillance, human-computer interaction, traffic assistant and robot vision. Background subtraction applied in moving target detection needs to base on the premise of precise background reconstruction. In practical application, changes of illumination and shadow in circumstance and slight shift of camera position all can vary the background. Therefore, there should be self-adaptive character for background reconstruction algorithm to adapt new circumstance.

The most frequently used background reconstruction algorithm is time-averaged background image. It sums and then averages the image sequence in a time span to obtain a frame of approximate background image. This kind of method is low in complexity and good at real-time, but it would easily mix moving target in foreground into background image which produces mixture effect and influences the separating effect of moving target. In recent years, many researchers devote to developing different background models in hope of reducing the influence of dynamic scene changing on motion segmentation. For instance, Haritaoglu in his document [1] made use of minimum, maximum strength and maximum value of time difference to conduct statistical modeling for every pixel in a scene, and updated background periodically. In Mckenna's document [2], adaptive background model featuring in the combination of pixel color and gradient information is used to cope with shadows and unreliable color cues. In Karmann and Brandt, Kilge's document [3], adaptive background model based on Kalman Flitering is applied to cope with time variation of weather and light. In their document [4], Stauffer and Grimson makes use of the adaptive mixture Gaussian background model and updates it by online estimation which reliably copes with the problems of illumination changes and disorder motion in background. The methods mentioned above can achieve satisfied effect but complicated in real-time, which could not be applied in occasions having a higher real-time demanding, such as video surveillance.

This paper proposes a moving target detection algorithm based on block target reconstruction. At first, the image will be rationally segmented into several sub-blocks under certain constraint conditions, and sub-block energy characters are acquired through method of frame difference. According to this, the sub-blocks are discriminated into background sub-blocks and foreground moving target sub-blocks. After that, the background estimation of the scene could be reached according to background sub-block set. In the period of background updating, illumination change in the scene is judged according to values of consecutive frame difference for four sub-blocks of image in the corners (corner block); accordingly, decision is made for the necessity of background reconstruction. The above algorithm is applied in a surveillance scene for one human target which proves the good real-time and robustness of this method.

2 Image Region Segmentation

This paper applies image sub-block to represent the image sub-region mentioned above in way of image blocking. The judgment for segmenting sub-block in size and the sub-block as common background sub-region between frames are closely related to scene condition, scene target and background definition. This paper takes a human-target surveillance scene as example to illustrate the method and principle of image region segmentation.

While segmenting the scene, scene image $F_k(x,y)$ and difference scene image $D_k(x,y)$ will be divided into $M \times N$ sub-blocks $\{F_k(j)\}_{j=1}^{M \times N}$, $\{D_k(j)\}_{j=1}^{M \times N}$ averagely in M rows and N columns from left to right, and from top to bottom. As illustrated by Figure 1.

B_{11}	B_{12}	..	..	B_{1N}
..	..	..	..	..
..	..	..	..	..
B_{M1}	B_{M1}	..	..	B_{MN}

Fig. 1. Scene Segmentation

This paper adopts frame difference method to detect moving target's pixel in order to determine whether the sub-block is background sub-block. The experiment shows that the detecting result of frame difference method would generate internal holes within target. Therefore the sub-block should not be too small to lead to one or several sub-blocks in the internal hole of target. Furthermore, it results in misjudging these sub-blocks as background blocks. However, choice of sub-block is not the bigger the better. If the sub-block is bigger, the condition of undistorted reconstruction is more difficult to satisfy. Hence the rational choice of sub-block size is one key to effect of this algorithm.

The above description of scene model shows that the sizes of scene targets are almost the same. The experiment proves that the sub-block in size of one-fourth average target area generally will not be covered into the holes. Taking the two constraints about size of sub-block into consideration, this paper sets the width and height value of sub-block between one eighth and one fourth of target average width and height. Choose a integer corresponding to the image size to segment and it would satisfy the Formula (1):

$$\frac{1}{8}H < H_b < \frac{1}{4}H \text{ and } \frac{1}{8}W < W_b < \frac{1}{4}W$$
$$H\%H_b = 0 \text{ and } H\%H_b = 0$$
(1)

In the formula, H and W represent the height value and width value of frame image. H_b and W_b are the average height value and width value of sub-block, "%" represents complementation.

3 Background Updating Algorithm

In the target detecting system based on background difference method, when significant changes leads to the necessity of reconstructing background, the system has to stop temporarily and wait for the result of background reconstruction to restart its work. This algorithm is low in complexity so that it can reconstruct the background in a short time.

The pseudo-code of background updating algorithm is as follows:

Algorithm: background updating
Parameters: T—number of frames limited in a short time; L_j—reconstruction sign of sub-block j; $F_k(j)$—sub-block j's image in Frame k; $B(j)$—background sub-block j;
Procedure:
1 Module parameter initializes (set all reconstruction signs of sub-blocks are 0)
2 Circulation from the next frame of the current one to frame T
   2a calculate every sub-block $D_k(j)$ of consecutive frame difference;
   2b circulation (j) from the first sub-block to sub-block $M \times N$
       if $L_j = 0$ and $F_k(j) \subset B_{k,k+1}(x, y)$ then $L_j = 1$;
   2c if all $L_j = 1$, then exit the circulation;
3 If exists j let $L_j = 0$, then linear estimation $B(j)$;
4 Module ends, outputs reconstruction result $B(j)$;

The threshold in algorithm is chosen according to application. Based on the analysis above, energy threshold, fourth-order statistics threshold has larger discrimination tolerance so that they are easier to determine. In the scene studied by this paper, the process of background subtraction is rather short, it generally does not exist sub-block which needs to estimate.

In actual reconstruction process, it could experience fierce changes of illumination; the construction should be stop under such situation and restart until the end of fierce changes.

The timing of background reconstruction is related to conditions of background changes, and most of changes are caused by illumination change in the circumstance. If the background needs to be reconstructed because of illumination changes, the variation of which always exceeds a certain degree. This paper chooses sub-blocks in the four corners of the image (corner block) and uses their values of consecutive frame difference to judge illumination changes, which shows in Figure 2.

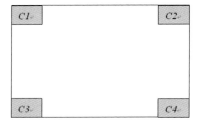

Fig. 2. Illustration of Four Corner Blocks for Judging Illumination Changes in Circumstance

Generally speaking, the regions of four corner blocks would not be influenced by target motion at the same time. Therefore, their changes could be utilized for the analysis of illumination changes in the scene. In the study of this paper, it chooses the average value of corner block pixel changes as the variation of corner block grayscale C_i ($i=1...4$), through setting a threshold C to determine whether the illumination changes has exceeded a certain degree and the background needs to be reconstructed.

4 Experiment and Analysis

In order to detect the background reconstruction method proposed by this paper, it chooses sequence images under different illumination conditions to conduct the experiment. The result shows in Figure 3 and Figure 4.

Hardware equipment: Host Pentium 4, CPU clock speed 3.0 G, Memory 1G;
Developing environment: Visual C++6.0.

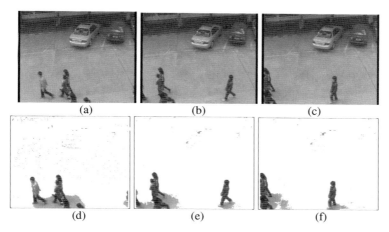

Fig. 3. Moving Target Detection Result of Scene One

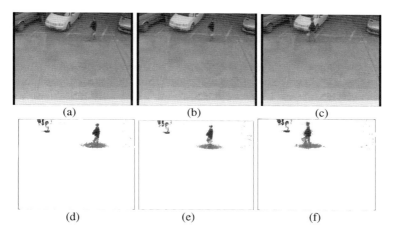

Fig. 4. Moving Target Detection Result of Scene Two

This paper proposes a fast and robust background reconstruction method, not consuming as much calculation time as method of updating model based on complicated background. This method adopts simple algorithm to reconstruct background step by step. Compared with other complicated algorithms, the method in this paper achieves better background reconstruction effect under the condition of speed satisfying.

5 Conclusion

This paper puts forward background difference method for target detection, and proposes a fast background reconstruction method based on block segmentation strategy according to illumination changes in the scene and determining rules for background sub-block. Compared with current algorithms, this algorithm has real-time and robustness. This paper proves the tenability of the algorithm with an experiment of live video surveillance.

Acknowledgment

This research was support by the Natural Science Foundation of GuangDong Province (10151802904000013), Science and Technology Planning Project of shenzhen city (JC200903180648A) and (SY200806300267A).

References

1. Haritaoglu, I., Harwood, D., Davis, L.W.: Real-time Surveillance of People and Their Activities. Proceedings of IEEE Transaction Pattern Analysis and Machine Intelligence 22(8), 809–830 (2000)
2. McKenna, S.: Tracking Groups of People. Computer Vision and Image Understanding 80(1), 42–56 (2000)
3. Kilger, M.: A Shadow Handler in a Video Based Real Time Traffic Monitoring System. In: Proceedings of IEEE Workshop on Applications of Computer Vision, Palm Springs, pp. 54–60 (1992)
4. Stauffer, C., Grimson, W.: Adaptive Background Mixture Models for Real Time Tracking. In: Proceedings of IEEE Conference on Computer Vision and Pattern Recognition, vol. 2, pp. 246–252 (1999)
5. Hou, Z., Han, C.: Background Reconstruction Algorithm Based on Pixel Grayscale Classification. Journal of Software 16(09), 1568–1576 (2005)
6. Feng, C., Qi, F., Chen, M.: Background Difference Method of Multiple Distribution Model in Long-term Video Surveillance System. Journal of Infrared and Millimeter 21(1), 59–63 (2002)
7. Yu, S., Xiao, D., Zhou, J., Jiang, G.: Adaptive Background Subtraction Algorithm. Journal of Chinese Computer Systems 24(7), 1331–1334 (2003)

Explore Pharmacosystematics Based on Wavelet Denoising and OPLS/O2PLS-DA

Zhuo Wang, Jianfeng Xu, and Ran Hu

School of Software, Nanchang University, Nanchang, China
ncofzw@gmail.com

Abstract. The aim to analyze the different pharmaceuticals have different or similar effect, the paper put forward explore pharmacosystematics based on wavelet denoising and bidirectional orthogonalization Partial Least Squares-Discriminant Analysis (OPLS/O2PLS-DA). The first to data preprocessing using wavelet denoising(WDS), the second to classify and train the sample of pharmaceuticals based on OPLS/O2PLS-DA,the third to discriminate the pharmaceutical which unknown type. The result indicate that the different pharmaceuticals have different or similar effect ,the new pharmaceutical may classify based on effect using OPLS/O2PLS-DA. The method was proved to be feasible and effective after tested with 8 kinds of pharmaceuticals experimental data.

Keywords: pharmaceuticals; wavelet denoising; partial least squares analysis.

1 Introduction

Metabolomics [1-4] is a new discipline of analyzing metabolites qualitatively and quantitatively, metabolomics as a branch of science concerned with the study of systems biology in clinic diagnosis and pharmaceutical.

Data Reducing dimensions include derivative,multiplicative signal correction,standard normal variate,exponentially weighted moving average, wavelet denoise spectral,orthogonal signal correction,rough sets,fuzzy,and so on.

Wavelet denoising is a technical success to reseach and apply in data preprocessing of images, physics, signal, and other sciences such as [5-8].

OPLS/O2PLS-DA,widely used in chemistry, sociology, economics chemistry, biology, clinical ,pharmaceuticals, etc.

In the paper, explore pharmacosystematics based on wavelet denoising and OPLS/O2PLS-DA.the result indicate the method can classfy and differentiate the different pharmaceuticals have different or similar effect.

2 The Basis Theory of Wavelet Denoising

Wavelet denoising (http://www.umetrics.com/) differs from wavelet denoise spectral in that compression is done variable wise and not observation wise.after the removal of the wavelet coefficients,the X variables are transformed bace to the original domain.after transforming all the variables to the wavelet domain,the denoising is done by performing the inverse wavelet transform with either the seleceted coefficients from the selected details.all other coefficients are set to 0.

3 Bidirectional Orthogonalization Partial Least Squares-Discriminant Analysis

Partial Least Squares (PLS) (http://www.umetrics.com/), as in multiple linear regressions, the main purpose of partial least squares regression is to build a linear model.

The PLS [9] model accomplishing these objectives can be expressed as:

$$X = TP + E \tag{1}$$

$$Y = UQ + F \tag{2}$$

$$U = T + B \tag{3}$$

$$B = (T^T T)^{-1} T^T U \tag{4}$$

Where X is a matrix of X-variables, Y is a matrix of Y-variables, T is a matrix of scores that summarizes the X-variables, U is a matrix of scores that summarizes the Y-variables, E is a matrix of residuals, the deviations between the original values and the projections. F is a matrix of residuals, the deviations between the original values and the projections is the inner relation. P is a matrix of loading showing the influence of the variables, Q is a matrix of loading showing the influence of the Y-variables. B is the inner relation.

Partial Least Squares-Discriminant Analysis (PLS-DA) [10-15] attempts to derive latent variables, which maximize the covariance between the measured variable(s) X and the response variable(s) Y; the 'dummy' indicator vector (matrix for more than two groups) .As a supervised method, PLS-DA makes use of the classificatory information of sample belongings in the feature extraction.

R2X and R2Y are the fraction of the sum of square of the entire X's and Y's explained by the current latent variable (LV) of PLS-DA, and represent the variance of X and Y variables, respectively.

Orthogonal Partial Least Square Analysis (http://www.umetrics.com/) (OPLS) [15-17] developed a modification of PLS, which is designed to handle variation in X that is orthogonal to Y. O2PLS is a generalization of OPLS [18-19].O2PLS is bidirectional to Y and Y to X, in other words, X and Y can be used to predict each other. The O2PLS model l can be written as, for model l of X:

$$X = T_p P'_p + T_o P'_o + E \tag{5}$$

For model l of Y:

$$Y = U_P Q'_P + U_O Q'_O + F \qquad (6)$$

Where a linear relationship exists between T_P and U_P. the score vectors in T_P and T_O are mutually orthogonal. T_O : matrix of scores that summarizes the X variation orthogonal to Y, U_O : matrix of scores that summarizes the Y variation orthogonal to X, P_0 expresses the importance of the variables in approximating X variation orthogonal to Y, in the selected component, Q_O expresses the importance of the variables in approximating Y variation orthogonal to X, in the selected component.

4 Explore Pharmaceuticals Based on Bidirectional Orthogonalization Partial Least Square Discriminate Analysis after Wavelet Denoising

In this experiment, the first class blank group include serial number 1-10 sample space; the second class include serial number 11-40 sample space, which 3 type pharmaceuticals have the same effect; the third class include serial number 51-78 sample space, which 3 type pharmaceuticals have the same effect, the forth class include serial number 41-50 sample space, which 1 type pharmaceutical,the fifth class include serial number 79-87 sample space, which 1 type pharmaceutical.class 1-3 as train sets,class 4-5 as test sets.

In this work, Data preprocessing using WDS,and keep coefficients account for 99.50% of the variance as fig 1. The principal component of OPLS/O2PLS-DA for train sets such as fig 2,Comp[2]P in M2.R2Y(cum) equal to 0.891222, M2.Q2(cum) equal to 0.862971, The principal component of OPLS/O2PLS-DA for all sets such as fig 3,Comp[4]P in M4.R2Y(cum) equal to 0.844811, M4.Q2(cum) equal to 0.755719.The pharmaceutical which train sets such fig 4 and figure 6 indicate that the 11-40 sample space of 3 type pharmaceuticals have same effect are in class 2, number 51-78 sample space of 3 type pharmaceuticals have same effect are in class 3. The pharmaceutical which all sets such figure 7 indicate that the 11-40 sample space of 3 type pharmaceuticals have same effect are in class 2,number 41-50 sample space of 1 type pharmaceutical in class 4,the class 2 and class 4 have overlapping show they have same effet. the number 51-78 sample space of 3 type pharmaceuticals have same effect are in class 3 ,number 79-87 sample space of 1 type pharmaceutical in class 5,the class 3 and class 5 have overlapping show they have same effet. fig 5 is plan picture indicate the class 3 and class 5 have overlapping show they have same effet, the class 2 and class 4 result undesirability.

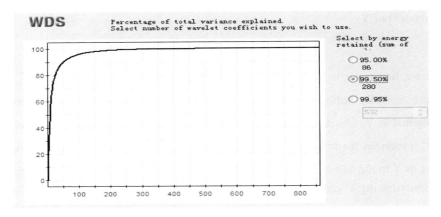

Fig. 1. Keep coefficients account for 99.50% of the variance

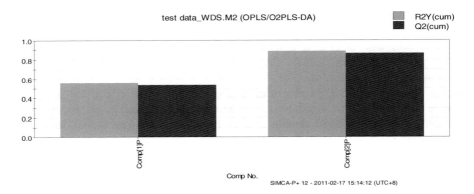

Fig. 2. Principal component of train set

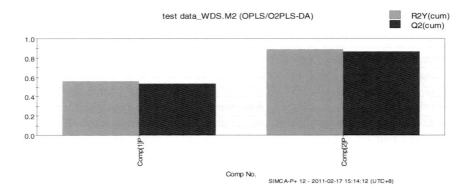

Fig. 3. Principal component of all set

Explore Pharmacosystematics Based on Wavelet Denoising and OPLS/O2PLS-DA   759

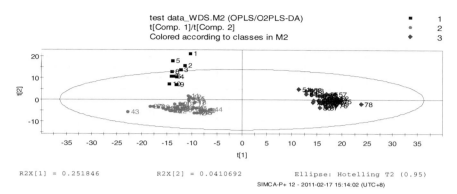

Note: t [1] and t [2] is a matrix of scores that summarizes the X-variables

Fig. 4. The plan of classification result for train sets

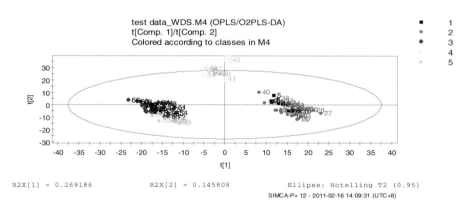

Note: t [1] and t [2] is a matrix of scores that summarizes the X-variables

Fig. 5. The plan of classification result for all sets

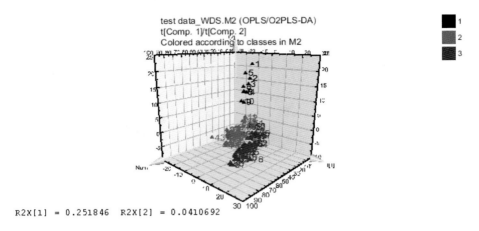

Fig. 6. The three-dimension of classification result for train sets

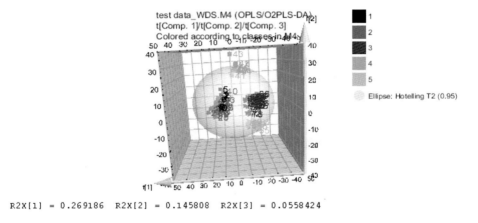

Fig. 7. The three-dimension of classification result for all sets

5 Conclusion

In the paper, explore pharmacosystematics based on wavelet denoising and bidirectional orthogonalization Partial Least Squares-Discriminant Analysis (OPLS/ O2PLS-DA).The result indicate that the different pharmaceuticals have different or similar effect,The method was proved to be feasible and effective after tested with 8 kinds of pharmaceuticals experimental data.

References

1. Holmes, E.: "Metabonomics": understanding the metabolic responses of living systems to path physiological stimul via multivariate statistical analysis of biological NMR spectroscopic data [J]. Xenobiotica 29, 1181–1189 (1999)
2. Nicholson, J.K., Holmes, E., Lindon, J.C., Wilson, I.D.: The challenges of modeling mammalian biocomplexity. Nature Biotechnology 22(10), 1268–1274 (2004)
3. Jia, L., Chen, J., Yin, P.: Serum metabonomics study of chronic renal failure by ultra performance liquid chromatography coupled with Q-TOF mass spectrometry. Metabolomics 4, 183–189 (2008)
4. Zhou, J., Xu, B., Huang, J., Jia, X., et al.: 1H NMR-based metabonomic and pattern recognition analysis for detection of oralsquamous cell carcinoma. Clinica Chimica Acta 401, 8–13 (2009)
5. Zou, L., Tao, C., Zhang, X., Zhou, R.: Estimation of event related potentials using wavelet denoising based method. In: Zhang, L., Lu, B.-L., Kwok, J. (eds.) ISNN 2010. LNCS, vol. 6064, pp. 400–407. Springer, Heidelberg (2010)
6. Yang, R., Ren, M.: Wavelet denoising using principal component analysis. Expert Systems with Applications 38, 1073–1076 (2011)
7. Lin, J.-L., Liu, J.Y.-C., Li, C.-W., et al.: Motor shaft misalignment detection using multiscale entropy with wavelet denoising. Expert Systems with Applications 37, 7200–7204 (2010)
8. Luisier, F., Vonesch, C., Blu, T., Unser, M.: Fast interscale wavelet denoising of Poisson-corrupted images. Signal Processing 90, 415–427 (2010)
9. Guo, C.: Computational Chemistry. Chemical Industry Press, BeiJing (2004)
10. Ma, C., Wang, H., Lu, X., et al.: Terpenoid metabolic profiling analysis of transgenic Artemisia annua L. by comprehensive two-dimensional gas chromatography time-of-flight mass spectrometry. Metabolomics 5, 497–506 (2009)
11. Römisch, U., Jäger, H., Capron, X., et al.: Characterization and determination of the geographical origin of wines. Part III: multivariate discrimination and classification methods. Eur. Food Res. Technol. 230, 31–45 (2009)
12. Liang, X., Zhang, X., Dai, W., et al.: A combined HPLC-PDA and HPLC-MS method for quantitative and qualitative analysis of 10 major constituents in the traditional Chinese medicine Zuo Gui Wan. Journal of Pharmaceutical and Biomedical Analysis 499, 31–936 (2009)
13. Williams, P., Geladi, P., Fox, G., et al.: Maize kernel hardness classification by near infrared (NIR) hyperspectral imaging and multivariate data analysis. Analytica Chimica Acta 653, 121–130 (2009)
14. Nie, B., Du, J., Xu, G.: Classification and Discrimination for Traditional Chinese medicine' Nature Based on OSC-OPLS/O2PLS-DA. In: IFITA 2010 Proceedings, pp. 354–357 (2010)
15. Trygg, J., Wold, S.: J. Chemometrics 16, 119–128 (2002)
16. Eriksson, L., Johansson, E., Kettaneh-wold, N., Trygg, J., Wikstrom, M., Wold, S.: Multi- and Megavariate data analysis, Part II. In: Method Extensions and Advanced Applications, ch. 23, Umetrics Academy (2005)
17. Trygg, J.: Prediction and spectral profile estimation in multivariate calibration. Journal of Chemometrics 18, 166–172 (2004)
18. Trygg, J.: O2-PLS for Qulitative and Quantitative Analysis in Multivariate Calivration. Journal of Chemometrics 16, 283–293 (2002)
19. Trygg, J., Wold, S.: O2-PLS,a Two-Block(X-Y) Laten variable regression(LVR) Method with an integral OSC Filter. Journal of Chemometrics 17, 53–64 (2003)

The Study on Pharmacosystematics Based on WOSC-PLS-DA

Zhuo Wang, Jianfeng Xu, and Ran Hu

School of Software, Nanchang University, Nanchang, China
ncofzw@gmail.com

Abstract. The aim to study the different pharmaceuticals have different or similar effect, the paper put forward study on pharmacosystematics based on wavelet compression, orthogonal signal correction and Partial Least Squares-Discriminant Analysis (WOSC-PLS-DA). The first to data preprocessing using wavelet compression(WCS) and orthogonal signal correction(OSC) ,the second to train the sample of pharmaceuticals based on WOSC-PLS-DA,the third to classify the pharmaceutical which unknown type. The result indicate that the different pharmaceuticals have different or similar effect ,the new pharmaceutical may classify based on effect using WOSC-PLS-DA.

Keywords: pharmaceuticals; wavelet compression;orthogonal signal correction; partial least squares analysis.

1 Introduction

Metabolomics [1-4] is a new discipline of analyzing metabolites qualitatively and quantitatively, metabolomics as a branch of science concerned with the study of systems biology in clinic diagnosis and pharmaceutical.

Data Reducing dimensions include derivative,multiplicative signal correction,standard normal variate,exponentially weighted moving average, wavelet denoise spectral,orthogonal signal correction,rough sets,fuzzy,and so on.

Wavelet compression is a technical success to reseach and apply in data preprocessing of chemistry, physics, biology, and other sciences such as [5-8].

The orthogonal signal correction (OSC) is a power method which dimension reduction and noise reduction. The method removes variation in the data matrix between samples that is not correlated with the Y-vector [9-10].the resulting dataset was filtered to allow pattern recognition be absorbed in the variation correlated to characteristic of interest within the sample population, which improves the productivity and separation power of pattern recognition methods [11].

Partial Least Squares (PLS), as in multiple linear regression, the main purpose of partial least squares regression is to build a linear model.

2 The Basis Theory of Wavelet Compression

Wavelet compression may compress the data to reduce the numbet of observations while keeping all the main information.Wavelet transformation (http://www.umetrics.com/) is a linear transformation,similar to the fourier transform.

Wavelet compressing the signal using variance, to calculte the variance spectrum of the coefficient matrix set in the wavelet domain,the positions of a selected number of the largest variance coefficients are located,and those columns are extracted from the wavelet coefficient matrix into a compressed data matrix.Save the original positions of the extracted coefficients which used both in future compression of spectra,and in the reconstruction of loadings,or scores etc.a new project is created with the compressde dataset.

3 Orthogonal Signal Correction

The OSC (http://www.umetrics.com/). algorithm is similar to the nonlinear iterative partial least squares (NIPALS) algorithm [11-12], which removes only so much of X as is unrelated(orthogonal) to Y OSC commonly used in PCA and PLS, and can be summarized as follows[10]:

Step 1: Specifies a starting "loading" or "correction" vector.p'in the first dimension this can be a row of 1's, or the loading of the first principal component of X; in the second dimension, p' can be the average spectrum, or the second principal components loading etc.

Step 2: Calculates a "score vector", $t = XP/P'P$ in the ordinary "NIPALS"way.

Step 3: Orthogonalizes t to Y: $t_{new} = (1 - Y(Y'Y)^{-1}Y')t$

Step 4: Calculates a weight vector, $w = (X'X)^{-}t_{new}$

Step 5: Calculates a new loading vector $P' = t_{new}'X/(t_{new}'t_{new})$

Step 6: Subtracts the "correction" from X, to give the "filtered" X.Continue with the nest "component", then another one, etc., until satisfaction. $X_{new} = X - t_{new}P'$

OSC effectively removes information not correlated to the target parameter and substantially decreases the number of latent variables required to construct calibration models. Consequently, OSC filtered data give a much simpler calibration model with fewer latent variables, and thus the interpretation of the model becomes easier [12].

4 Partial Least Squares-Discriminant Analysis

Partial Least Squares-Discriminant Analysis (PLS-DA) [13-18] attempts to derive latent variables, which maximize the covariance between the measured variable(s) X and the response variable(s) Y; the 'dummy' indicator vector (matrix for more than two groups) .As a supervised method, PLS-DA makes use of the classificatory information of sample belongings in the feature extraction.

R2X and R2Y are the fraction of the sum of square of the entire X's and Y's explained by the current latent variable (LV) of PLS-DA, and represent the variance of X and Y variables, respectively.

5 Study on Pharmacosystematics Based On WOSC-PLS-DA

In this experiment, the first class blank group include serial number 1-10 sample space; the second class include serial number 11-40 sample space, which 3 type pharmaceuticals have the same effect; the third class include serial number 51-78 sample space, which 3 type pharmaceuticals have the same effect, the forth class include serial number 41-50 sample space, which 1 type pharmaceutical,the fifth class include serial number 79-87 sample space, which 1 type pharmaceutical.class 1-3 as train sets,class 4-5 as test sets.

In this work, Data preprocessing using WOSC,and keep coefficients account for 99.50% of the variance as figure 1. The principal component of PLS-DA for train sets such as figure 2,Comp[2]P in M7.R2Y(cum) equal to 0.838761, M7.Q2(cum) equal to 0.825216, The principal component of PLS-DA for all sets such as figure 3,Comp[3]P in M4.R2Y(cum) equal to 0.547671, M4.Q2(cum) equal to 0.488123.The pharmaceutical which train sets such figure 4 and figure 6 indicate that the 11-40 sample space of 3 type pharmaceuticals have same effect are in class 2, number 51-78 sample space of 3 type pharmaceuticals have same effect are in class 3. The pharmaceutical which all sets such figure 7 indicate that the 11-40 sample space of 3 type pharmaceuticals have same effect are in class 2, except number 43 is the outlier, number 41-50 sample space of 1 type pharmaceutical in class 4,the class 2 and class 4 have overlapping show they have same effet. the number 51-78 sample space of 3 type pharmaceuticals have same effect are in class 3 ,number 79-87 sample space of 1 type pharmaceutical in class 5,the class 3 and class 5 have overlapping show they have same effet. fig 5 is plan picture indicate the class 3 and class 5 have overlapping show they have same effet, the class 2 and class 4 result undesirability.

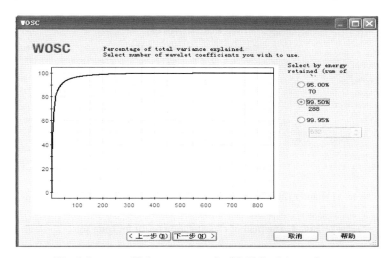

Fig. 1. keep coefficients account for 99.50% of the variance

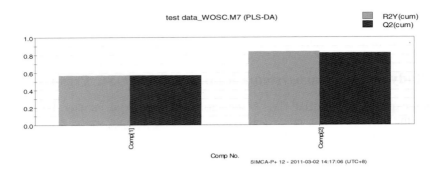

Fig. 2. Principal component of train set

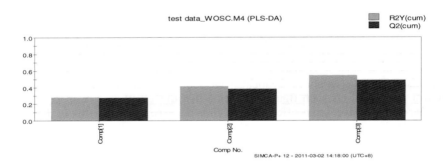

Fig. 3. Principal component of all set

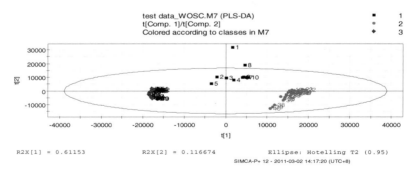

Note: t [1] and t [2] is a matrix of scores that summarizes the X-variables

Fig. 4. The plan of classification result for train sets

The Study on Pharmacosystematics Based on WOSC-PLS-DA 767

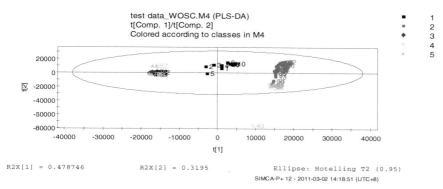

Note: t [1] and t [2] is a matrix of scores that summarizes the X-variables

Fig. 5. The plan of classification result for all sets

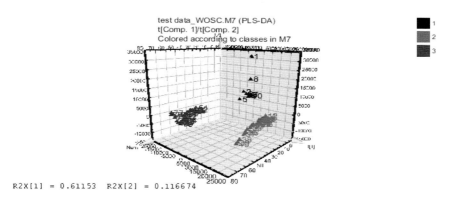

Fig. 6. The three-dimension of classification result for train sets

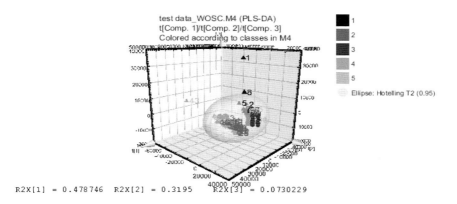

Fig. 7. The three-dimension of classification result for all sets

6 Conclusion

In the paper, put forward study on pharmacosystematics based on wavelet compression, orthogonal signal correction and Partial Least Squares-Discriminant Analysis (WOSC-PLS-DA).The result indicate that the different pharmaceuticals have different or similar effect,The method was proved to be feasible and effective after tested with 8 kinds of pharmaceuticals experimental data.

References

1. Smith, T.F., Waterman, M.S.: Identification of Common Molecular Subsequences. J. Mol. Biol. 147, 195–197 (1981)
2. Holmes, E.: "Metabonomics": understanding the metabolic responses of living systems to path physiological stimul via multivariate statistical analysis of biological NMR spectroscopic data. Xenobiotica 29, 1181–1189 (1999)
3. Nicholson, J.K., Holmes, E., Lindon, J.C., Wilson, I.D.: The challenges of modeling mammalian biocomplexity. Nature Biotechnology 22(10), 1268–1274 (2004)
4. Jia, L., Chen, J., Yin, P.: Serum metabonomics study of chronic renal failure by ultra performance liquid chromatography coupled with Q-TOF mass spectrometry. Metabolomics 4, 183–189 (2008)
5. Zhou, J., Xu, B., Huang, J., Jia, X., et al.: 1H NMR-based metabonomic and pattern recognition analysis for detection of oralsquamous cell carcinoma. Clinica Chimica Acta 401, 8–13 (2009)
6. Zen, X.: The theory of wavelet analysis and its application data mining. University of Electronic Science and Technology of China (2006)
7. Jahnke, T., Udrescu, T.: Solving chemical master equations by adaptive wavelet compression. Journal of Computational Physics 229, 5724–5741 (2010)
8. Pan, G.-F., Yang, H.-Z., Kong, J.: Application of spectroscopy technique to water quality analysis based on wavelet data compression. Journal of Infrared and Millimeter Waves 29(5), 397–400 (2010)
9. Dechevsky, L.T., Gundersen, J., Grip, N.: Wavelet compression, data fitting and approximation based on adaptive composition of lorentz-type thresholding and besov-type non-threshold shrinkage. In: Lirkov, I., Margenov, S., Waśniewski, J. (eds.) LSSC 2009. LNCS, vol. 5910, pp. 738–746. Springer, Heidelberg (2010)
10. Holmes, E., Tsang, T.M., Jeffrey, T., Huang, J., et al.: Metabolec profiling of CSF:Evidence that early intervention may impact on disease progression and outcome in schizophrenia. Pls Medicine 3, 1420–1428 (2006)
11. Wold, S., Antti, H., Lindgren, F., Ohman, J.: Orthogonal signal correction of near-infrared spectra. Chemometrics intelligent lab systems 44, 175–185 (1998)
12. Brindle, J.T., Antti, H., Holmes, E., Tranter, G., Nicholson, J.K., et al.: Rapid and noninvasive diagnosis of the presence and severity of coronary heart disease using 1H-NMR-based metabonomics. Nat. Med. 8, 1439–1444 (2002); Clerk Maxwell, J.: A Treatise on Electricity and Magnetism, 3rd ed., vol. 2., pp.68–73. Clarendon, Oxford (1892)
13. Kim, K., Lee, J.-M., Lee, I.-B.: A novel multivariate regression approach based on kernel partial least squares with orthogonal signal correction. Chemometrics and Intelligent Laboratory Systems 79, 22–30 (2005)

14. Ma, C., Wang, H., Lu, X., et al.: Terpenoid metabolic profiling analysis of transgenic Artemisia annua L. by comprehensive two-dimensional gas chromatography time-of-flight mass spectrometry. Metabolomics 5, 497–506 (2009)
15. Römisch, U., Jäger, H., Capron, X., et al.: Characterization and determination of the geographical origin of wines. Part III: multivariate discrimination and classification methods. Eur. Food Res. Technol. 230, 31–45 (2009)
16. Liang, X., Zhang, X., Dai, W., et al.: A combined HPLC-PDA and HPLC-MS method for quantitative and qualitative analysis of 10 major constituents in the traditional Chinese medicine Zuo Gui Wan. Journal of Pharmaceutical and Biomedical Analysis 49, 931–936 (2009)
17. Williams, P., Geladi, P., Fox, G., et al.: Maize kernel hardness classification by near infrared (NIR) hyperspectral imaging and multivariate data analysis. Analytica Chimica Acta 653, 121–130 (2009)
18. Nie, B., Du, J., Xu, G.: Classification and Discrimination for Traditional Chinese medicine' Nature Based on OSC-OPLS/O2PLS-DA. In: IFITA 2010 Proceedings, pp. 354–357 (2010)
19. Trygg, J., Wold, S.: J. Chemometrics 16, 119–128 (2002)

Research on Sleep-Monitoring Platform Based on Embedded System

Kaisheng Zhang and MingXing Gao

College of Electrical Engineering of Shaanxi University of Science and Technology,
Xi'an, 710021, P.R. China
zhangkaisheng@sust.edu.cn, gmx0023@163.com

Abstract. It is a most important study that monitoring and caring the sleep. A sleep-monitoring platform based on embedded system was designed and analyzed in this paper. It was designed using embedded microprocessor as hardware of system monitoring and utilizing wireless communication technology to complete the data of pulse transmission, software flow chart of the system was designed at the same time. The platform will not only provide a powerful tool for in-depth analysis the generation mechanism and characteristics of variation of pulse wave signal, but also provide a simple, convenient and reliable way to reveal the mysteries of cardiovascular disease diagnosis and establish a non-invasive detection for scientific and accurate cardiovascular parameters.

Keywords: Sleep; Monitoring; Platform; Research; Embedded Microprocessor.

1 Introduction

The prevalence of sleep disorders has become an important factor to jeopardize public health[1]. SAS (Sleep Apnea Syndrome) is one of the most common sleep disorders which affect around one out of every 100 people. The disease is divided into obstructive sleep apnea syndrome (OSAS) and central sleep apnea syndrome (CSAS). In the United States, Studies suggest that about two million people suffer from SAS, approximately 30 to 60-year-old adults, 1/4 male and 1/10 female have clinical manifestations of the disease[2]. With the advances in the diagnosis and treatment, awareness and monitoring of these disorders have a greater improved. Monitoring technology has developed earlier in developed countries, especially began in the 20th century, 30 years. Nearly 20 years, with the rapid development of biomedical measurement technology, sensor technology, communication technology and computer technology, the way of care gradually toward the development of systematic monitoring and network and the scope of care is also gradually expanding[3]. Various experiments and studies have shown that the pulse wave contains more wealth information. It will make the monitoring technology for sleep apnea based on the pulse wave becomes a hot spot. Recent wireless communications has enabled the design of low-cost, miniature, lightweight. Thus, the way of wireless remote monitoring at home will become an important monitoring means.

In existing technology, a way to get the data of pulse signals is researched without affect sleep. Domestically, the most of studies on sleep is also though calculating parameters of the cardiovascular system and then analyzed the quality of sleep by analyzing the pulse signal. Ting Li who get pulse signals by using the photoelectric pulse sensor[4]. Jingyu Guo who get human pulse signals without bring physical load by using fingertip pulse sensor[5]. The means of communication of signals from the traditional wired into a wireless, so as to realize the remote monitoring. For the interactive interface of host computer, Kunliang Xu achieve the pulse during sleep using the software platform based on the LabVIEW (Laboratory Virtual instrument Engineering Workbench), which makes the sleep care flexible and efficient[6]. Therefore, the sleep-monitoring platform was designed to complete acquisition, transmission, analysis and display of pulse signal. It will make the means of sleep monitoring become intelligent, miniaturized and portable constantly.

With the development of CPU hardware of embedded system from the 8 to 32-bit and the rapid development of embedded software system environment. High-end 32-bit CPU development environment is mature and a 32-bit embedded processors dominate in microcontroller gradually replaced the original 8-bit[7]. ARM processor is prevalent in the global, 32-bit RISC embedded processor of ARM has already started to become a high-end embedded application and design mainstream. Therefore, the combination of embedded technology, virtual instrument technology and wireless communication technology will be an important means of monitoring technology in the modern sleep-monitoring fields[8].

2 The Design Principle of System

Using pulse sensor to get parameters of the cardiovascular system and evaluate sleep quality can reduce the physical and mental load of sleeper. Pulse as a physiological signal, is the external reflection of physiological important information such as heart and blood vessels. The heart beat makes blood spread out along the arteries then forms Pulse. The relaxation and contraction of human blood, which dissemination along the arterial wall in the heart of a cycle will produce pulse wave. The detecting the pulse not only provide a physical reference information for blood pressure, blood flow measurements and other physiological testing but also the pulse wave itself can also give the value information of many diagnosis.

The real-time monitoring of sleep in this paper is realized by means of wireless communication. This system uses embedded processor LPC2103 as microcontroller and utilizes the virtual machine to complete the data collection, storage and waveform display.

Embedded system is a special computer system based on computer technology. The software and hardware can be tailored to meet the system demand which need strict requirements for functionality, reliability, cost, size and power consumption of the system. It mainly consists of embedded microprocessors, peripheral hardware devices, embedded operating system and user application software. The embedded system, which is operated embedded in the main equipment, is used to achieve control, monitoring and management functions of other equipment. Wireless communication technology uses the electromagnetic waves to transmit information without physical media, and it has a certain anti-jamming capability utilizing digital signal coding transmission [9]. 2.4GHz public frequency band is used for a short-range wireless communications without applying specific band will be good in open and rich in resources available. The LabVIEW of NI (National Instruments) Company is a kind of graphical programming software which is designed for instrument control, data acquisition, analysis and expression. It will enhance the capability for users building their own instrument system with efficient and economical hardware equipment in standard computer. Thus, combined the LABVIEW and general data acquisition equipment will design a virtual instrument which can make the biomedical signal analysis and various ECG monitoring flexible and efficient.

The System used 32-bit high-speed processor ARM7 as hardware platform. The core chip is LPC2103 of the PHILIPS Company's ARM7TDMI-S. Pulse sensor will use the infrared pulse sensor HKG-07, it can detect the micro-fingertip corresponding changes occurred in blood volume because of heart beat. The shape of the sensor is clip or finger-style, so it is used without affecting sleep. The wireless communication module uses the 2.4GHz module JF24C, which integrates a high frequency shift keying (GFSK) transceiver circuit features to complete the high-speed data transmission capabilities which realized by the small size of it.

The characteristic value of pulse wave K is defined by the professor Zhichang Luo[10] at Beijing University of Technology, the formula about K is as follows:

$$K = (P_m - P_d)/(P_s - P_d) \quad (1)$$

$$P_m = \frac{1}{T} \times \int_0^T P(t)dt \quad (2)$$

P_m is the mean arterial pressure (MAP), which is equivalent to the average pulse pressure value $P(t)$ in a cardiac cycle. P_s is systolic blood pressure and P_d is diastolic blood pressure. In practice, you can use the peak, valley and the average amplitude of the pulse wave in a cycle instead of P_s, P_d and P_m of formula (1).

The value of peak, valley, and the average amplitude of pulse wave in the System will be measured and recorded as P_s, P_d and P_m values. After the pulse characteristic value K is calculated, then obtain the part of the cardiovascular parameters. It will provide important reference information for the diagnosis of diseases and assessment of sleep quality.

3 System Desing

The main sleep monitoring means in this paper is transmitting the pulse signals which acquired by pulse sensor to the PC by wireless transmission module, then though programming to realize the real-time data collection, storage and waveform display using LabVIEW. It includes 5 modules, data acquisition, signal conditioning, wireless communication module, data communication and monitoring platform interface. The overall system block diagram is shown in fig 1.

The introduction of each module is as follows:

a) Data acquisition
Using the infrared pulse sensor acquire the body pulse signal without interfering with sleep.

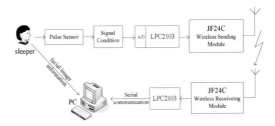

Fig. 1. The overall system block diagram

b) Signal conditioning
Complete amplification, filtering, and A/D for pulse signals.
c) Wireless communication module
Complete the design for wireless module interface and serial communication of LPC2103 and JF24C to ensure the reliability of wireless transmission of data. The Control Center LPC2103 complete pulse wave acquisition and conversion and send the digital pulse signals through the wireless transmitter module. Wireless receiver module receives digital pulse signals into the LPC2103 though serial port and finally to the host computer.
d) Data communication.
Using the LabVIEW achieve serial communication.
e) Design of monitoring platform interface.
When hardware transferred the received data to your computer, using virtual instruments achieve real-time data collection, storage and pulse waveform display. In the test platform, facial image information during sleep is captured via computer's camera and also the pulse signal can be displayed simultaneously.

The system uses PHILIPS Company's ARM7TDMI-S LPC2103 as microcontroller. It uses 1.8V and 3.3V dual power supply, the maximum operating frequency of 70MHz, integrated 8-channel 10-bit internal A/D converter, the conversion time 2.44us. Wireless communication module using the module 2.4G JF24C. The module integrates a feature of high frequency shift keying (GFSK) transceiver circuit to achieve the small size of chip realizing special capabilities of high-speed data transmission.

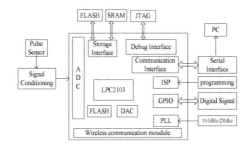

Fig. 2. The hardware design diagram of system

2.4GHZ public frequency band is used for a short-range wireless communications without applying specific band will be good in open, rich in resources available. The hardware design diagram of system is shown in fig 2.

The software design of this system includes the design of embedded systems programming and design of PC programming.

It mainly completes the data acquisition, A/D, and wireless transceiver. The software flow chart is shown in figure3:

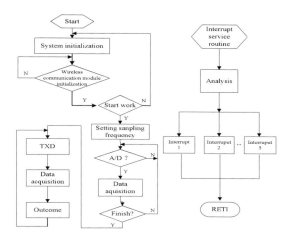

Fig. 3. System software flow chart

The pulse signals will be transmitted to the host computer by the protocol of serial port. The data storage, display and analysis of the pulse wave on the computer by the LABVIEW programming though serial communications. The software functions structure diagram on PC machine is show in fig 4.

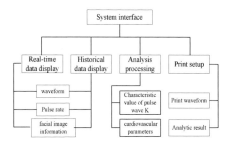

Fig. 4. The software functions structure diagram on PC machine.

4 Conclusion

As people increase in work pressure and the faster pace of life, sleep disorders have been prevalent in the crowd. The survey found that several major sleep disorders including apnea syndrome, rapid eye movement (REM) cycle behavior disorder, periodic limb behavior disorders and so on. It was reported that about 10% ~20% of adults suffer from varying degrees of insomnia and 4%~6% of the population suffer from apnea syndrome. People have poor sleep quality will affect the efficiency of work and study in the short term, as a long run will induce many diseases. Therefore, sleep is an important and indispensable component of a healthy life. Thus, the monitoring and analysis of sleep quality will have become an important part of assessment person's health status, diseases prevention. It will make the research of sleep has a great significance.

Acknowledgment

This project was supported by the Graduate Innovation Fund of Shaanxi University of Science and Technology.

References

1. Pandi-Perumal, S.R., Verster, J.C., Kayumov, L., Lowe, A.D., Santana, M.G., Pires, M.L.N., Tufik, S., Mello, M.T.: Sleep disorders, sleepiness and traffic safety:a public health menace. Brazilian Journal of Medical and Biological Research 39, 863–871 (2006)
2. Angius, G., Lraffo: A Sleep Apnoea Keeper in a Wearable Device for Continouous Detection and Screening during Daily Life. Computers in Cardiology 35, 433–436 (2008)
3. Milenkovic, A.: Wireless sensor networks for personal health monitoring: Issues and an implementation. Computer Communications 29, 2521–2533 (2006)
4. Li, T., Li, Y.: A non-invasive detection of pulse wave analysis system. Journal of Biomedical Engineering 25, 1059–1062 (2008)
5. Guo, J., Xue, F.: Finger on the pulse wave detecting based on VB System. Henan University (Natural Science) 29, 91–93 (2008)
6. Xu, K., Du. Program, H.: design of biomedical signal data acquisition based on LabVIEW. Yunnan University (Natural Science) 28, 110–112 (2006)
7. Lee, L.-N.: On the status and development of embedded systems. Changchun University of Technology (Natural Science) 25, 74–75 (2004)
8. Rotariu, C., Costin, H.N., Puscoci, S., Andruseac, G., Costin, C.C.: An Embedded Wireless Module for Telemonitoring. In: International Conference on Advancements of Medicine and Health Care through Technology, vol. 23, pp. 103–106 (2007)
9. Chang, L., Zhang, S., Yang, Y.: Information of pulse waveform characteristics. Beijing University of Technology 22, 71–79 (1996)
10. Wu, W.H., Bui, A.A.T., Batalin, M.A., Au, L.K., Binney, J.D., Kaiser, W.J.: MEDIC: Medical embedded device for individualized care. Artificial Intelligence in Medicine 42, 137–152 (2008)

Study on Sleep Quality Control Based on Embedded

Kaisheng Zhang[*] and Zhen Li

College of Electrical Engineering of Shaanxi University of Science and Technology,
Xi'an, 710021, P.R. China
zhangkaisheng@sust.edu.cn, liz_2010@126.com

Abstract. The importance of research on embedded sleep quality control is analyzed, at the same time, system monitoring hardware and system software flow chart are designed by utilizing embedded microprocessor. It fuses the cognitive science, the biological cybernetics and the information theory, utilizes and draws lessons from the theory and method of cognitive behavioral science as well as the new achievement of modern microelectronic technology and computer technology. An objective, quantitative, non-interference system of sleep quality control system, which is much close to natural sleep patterns of human, is constructed according to the character of the transition process of human sleep and normal sleep characteristics, The main characteristic of the study lies in how to improve people's sleep quality, in addition, on the basis of studying the transitional stage of sleep, highlight analyzing the mechanism of sleep to control and guide people's sleep.

Keywords: Sleep Quality; Control; Style; Study; Embedded.

1 Introduction

Sleep is an important physiological phenomenon for human, about one-third of a person's lifetime is spent in sleep, and the whole body including the central nervous system achieves recovery and rest during sleep. The reconstruction and regeneration of neurons and relevant organizations need the rest of body, and only sleep state is able to provide this kind of rest. Human can eliminate fatigue, restore the body's vigor; protect the brain, promote growth; develop immunity from disease, stimulate airframe rehabilitation; delay senility, promote longevity; keep mental health and assist skin care through moderate sleep. Sleep is an important step in body's recovery and integration together with consolidation of memory, sleep is also an important and indispensable component of health, and it is essential for healthy life to sleep. Situation of sleep is one of the most important indexes evaluating Human health. Sleep deprivation directly influences people's works and lives, and if this continues, some diseases may arise. Sleep disorder may lead to many kinds of illnesses, and has direct relation with many common diseases.

As quick life rhythm, pressures of daily life and lack of exercise, human sleep quality decrease, this may produce many negative consequences such as body immunity force drops, resistance to illness lower and so on. An investigation shows that

---

[*] Corresponding author.

considerable crowd suffers from sleep disorders and sleep-related diseases, seriously affecting the body's function of organs such as cardial vessel and muscles and doing harm to health of many people.

Difficulty in falling sleep is one of the most common phenomenon in insomnia, "uncontrolled cognitive activities " cause people suffering from varying degrees of insomnia during the transitional stage from awake to sleeping.. In the research of recent years, there are evidences that cognitive activities can be stopped though Psychological control based on cognitive science. As the reason of difficulty in falling sleep mostly is that continuous thoughts cannot be depressed, How to monitor people's sleep effectively, control "thoughts" in sleep to improve sleep quality is worthy of special attention.

2 Research Content

A. Whole scheme

Those harmful thoughts, which are uncontrolled by individual volition, will be suppressed due to external stimulation that can make attention be shifted. Therefore, stimulation can be provided continuously by embedded processor according to the sleep status reflected by pulse signal, in order to make the source of these distracting thoughts continuously be shifted, until the finally inactivation. According to this basic cognition principle, the reaction and pulse signal are obtained after providing appropriate sound stimulation to the user to monitor the sleep state, and then to regulate the law of sound stimulation under the current sleep state of the user to improve sleep quality. The Implementation structure diagram of the system is shown in fig 1.

The main contents involved are: the pulse signal and cognitive behavioral response signal of human body are collected by pulse sensor, as well as signal conditioning circuit related is designed; the data of pulse waveform is analyzed; the stimulus identity is controlled and stimulation is generated by using stimulator: scheme of stimulation is adjusted combining the user's response and behavior with change of pulse signal to shorten sleep latency and improve sleep quality efficiently.

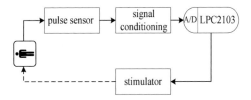

Fig. 1. System Structure Diagram

B. Signal processing of the sensor

The HKG-07A series infrared pulse sensor is applied in the design, this model detect corresponding changes of fingertip occurred in blood volume, which caused by the beating of the heart, using an infrared rays.

 The signals from the sensor have the characteristics of small amplitude, strong interference, high impedance and low frequency, so it is easy to bring in outside interference during the collecting procedure, for instance, power-line interference, baseline drift caused by human activities, high-frequency interference and human respiratory movement, as a result, the obtained signal is often accompanied by severe background noise. In order to meet the requirements, sensor signals should be processed by signal conditioning circuit.

The amplifier circuit is designed as two-stage amplification in order to get better anti-interference. The weak signals from the sensor should be amplified, and then through a 40Hz low-pass filter and a 50Hz notch filter, finally, analog to digital conversion is completed by analog to digital converter at a sample frequency of 5 KHz. Signal conditioning circuit is composed by one-stage amplifier, circuit for rejection of baseline drift, 50Hz notch filter circuit, low-pass filter and secondary amplifier circuit. Signal conditioning circuit diagram is shown in Fig 2.

One-stage amplifier applies differential structure to improve common mode rejection ratio. AD620 is adopt differential amplifier for its high common mode rejection, therefore, static and power-line interference can be restrained effectively. Rejection circuit of baseline drift is mainly composed of operational amplifier U2 and U4 by using the principle of negative feedback. To apply rejection circuit of baseline drift can not only eliminate statics effectively, but also reject the baseline drift caused by human body cold, tension and breathing trembling. The principle is that the output of operational amplifier U2 charges C3 though R3. If the level of negative side input drift or wave, the charging result will make the positive side appear a potential which is equal to changes in baseline level of negative side input signal, then a more stable output can be obtained by subtraction of the two sides. In order to reduce interference power frequency, the parameters of notch filter should be set up as:

$R5 = R6 = 2R7 = 9.6K\Omega$ $C4 = C5 = 2C6 = 0.33\mu F$, the frequency of the notch filter is calculated as:

$$f = \frac{1}{2\pi RC} \approx 50H_z$$

The result can satisfy the request.

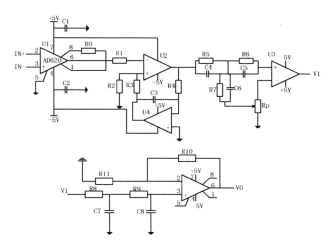

Fig. 2. Software Control Frame Chart

C.The hardware design

32-bit high-speed ARM7 processor is selected as hardware platform in the article. The core chip we use is ARM7TDMI-S (LPC2103) for PHILIPS Corp. It has some outstanding advantages, for example, adopting double power supply of 3.3v and 1.8v, the maximum operating frequency can reach 70MHz, and combining 8K byte internal static RAM with 32K byte on-chip flash program memory. 10 bit building A / D converter of LPC2103 provides 8 channel analogue inputs, and multiple 32-bit and 16-bit timer can be used to realize PWM outputs. As the price of LPC2103 is cheap, and has Low Power consumption as well as small package, it can fully meet the requirements of the system design. The embedded system takes LPC2103 as a core, including other circuit designs such as necessary power, reset, clock and so on, the Hardware Design is shown in fig 3.

To apply HKG-07A series infrared pulse sensor, the sensor investigates microvascular volume changes of fingertips due to the beating of the heart using infrared ray,and then export pulse signal which synchronize in pulse beating after amplification, conditioning and shaping, at the same time, in the context of sound stimulation , micro-signal generated by the reaction of finger can also be measured by the sensor.; the ARM embedded system adopts LPC2103 to measure and record the user's pulse information and behavioral responses information, and deal with the information by real-time processing, pattern recognition and confirmation of the signal, then produces sound stimulation as predefined property.

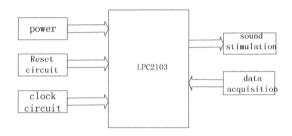

Fig. 3. Hardware Design Block Diagram System Structure Diagram

D. Software Design

The software system uses ADS1.2 as its development environment and wrote in C language. ADS is abbreviation to ARM Developer Suite, ADS1.2 can completely provide a complete development environment on Windows interface and is easy to use C language in development. ADS1.2 supports all series of ARM microcontroller before ARM10 and has the feature of high compiling efficiency and excellent system base function.

The system program design develops in the method of modularization, and the main program is mainly include initialization of system varies modules as well as transfer and harmonize of subroutine. Software control block diagram is shown in Fig 4.

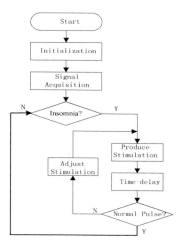

Fig. 4. Software Control Frame Chart

3 Technical Difficulties and Characteristics of the Study

In recent years, with the development of informationization, Intelligence and networking, the technique of embedded system has access to expansive space and abundant opportunity. The embedded system is endowed with those characteristics such as tiny kernel, high specificity, and good real-time property and so on. Therefore, it has gained extensive application in many fields and explosive growth. With the advance in science, the pulse can be displayed and recorded accurately through the sensor by making use of the microprocessor. Analyzing the pulse of the object can possibly acquire an accurate the physical condition of the object measured as quickly as possible.

Technical difficulties of this study lie in collecting and processing pulse signal of human body, as well as controlling properties of stimulation; Some Problems that should be said attention to in the study of the project: On the one hand, the pulse signal has characteristics of low frequency, small amplitude and easy to introduce interference, so we should adopt high sensitivity and test repeatedly to design reasonable conditioning circuit for amplification, filtering and other treatment. On the other hand, only the sound of appropriate loudness can be very good for depressing harmful thought effectively. If the sound stimulation is too weak, user's attention can not be diverted very well to achieve the purpose of suppression of harmful thinking activities, while excessive sound stimulation will become a new source of interference of natural sleep. Therefore, we need to combine the changes of user's pulse signal with his cognitive behavior, then test repeatedly to select appropriate sound as stimulation. The main feature of this article is to discuss the problem of how to improve human sleep quality, and emphasize how to control and guide people's sleep by analyzing the mechanism of sleep on the basis of studying the transitional stage of sleep.

4 Conclusion

Sleep disorders is the main reason for formation and development of many diseases, difficulty in falling sleep, the reason causing which mostly is continuous thoughts cannot be depressed, is a kind of common phenomenon in insomnia, therefore, how to monitor people's sleep effectively, control "thoughts" in the sleep to improve sleep quality is worthy of special attention. As electronic technology, computer technology and Network Technology develop rapidly, electronic measurement technology has been applied in different fields widely, on these bases, it is possible to improve sleep quality though controlling and guiding human sleep relying on microprocessor. The sleep mechanism of human is a bit complicated, and varies from person to person, the related process can not be described in accurate mathematical models, however, the mechanism of poor sleep quality can be found out through a great deal of test, then computer technique can be utilized to improve people's sleep quality. The study of sleep mechanism and sleep quality control is based on interdisciplinary interference and Infiltration, and is a perfect combination of computer technology, bio-medicine and signal processing technology. In addition, it has far-reaching significance of improving the life quality and guaranteeing people's physical and mental health. research of sleep has a great significance.

Acknowledgment

This project was supported by the Graduate Innovation Fund of Shaanxi University of Science and Technology.

References

1. Hoshiyama, M., Hoshiyama, A.: Heart Rate Variability Associated with Rapid Eye Movements during Sleep. Computers in Cardiology 34, 689–692 (2007)
2. Wang, R.Y., et al.: Sleep Monitoring Control System. Sleep Monitoring Control System, 119–121 (2006)
3. Chung, G.S., Choi, B.H., Jeong, D.-U., Park, K.S.: Noninvasive Heart Rate Variability Analysis Using Loadcell-Installed Bed During Sleep. In: Proceedings of the 29th Annual International Conference of the IEEE EMBS Cité Internationale, pp. 2357–2360 (2007)
4. Yu, M.S., et al.: The New Model for Sleep Medicine Monitors. Chinese Medical Equipment Information, 4–6 (2003)
5. Angius, G., Raffo, L.: A Sleep apnoea keeper in a wearable device for continous detection and screening during daily life. Computers in Cardiology 35, 433–436 (2008)
6. Zhang, D.B.: The Principle, Design and Application of Embedded Systems, pp. 448–449. Machinery Industry Press, Beijing (2005)

7. Zhang, Y.: Sleep Monitoring System. Engineering Master's Degree thesis of South China University of Technology, pp. 27–28 (2005)
8. Hoshiyama, M., Hoshiyama, A.: Heart Rate Variability Associated with Rapid Eye Movements during Sleep. Computers in Cardiology 34, 689–692 (2007)
9. Douet, Z.Z., et al.: The New Development and Challenges of Embedded System Design. In: The Application of Microcontroller and Embedded Systems, pp. 15–18 (2004)
10. Vernon, M.K., Dugar, A., Revicki, D.: Measurement of non-restorative sleep in insomnia: A review of the literature. Sleep Medicine Reviews 14, 205–212 (2010)

Study on Embedded Sleep-Monitoring Alarm System

Kaisheng Zhang and Wenbo Ma

College of Electrical Engineering of Shaanxi University of Science and Technology,
Xi'an, 710021, P.R. China
zhangkaisheng@sust.edu.cn, aaa19860830@163.com

Abstract. The importance of embedded sleep-monitoring alarm system is analyzed. The system of monitoring hardware and software flowchart are designed under the control of embedded microprocessor. The contradictions between increasing demands for health care and insufficient number paramedic are alleviated in this paper. It can relieve mental pressure and lighten physical burden of the sleeping patients in the familiar and comfortable home environment and real-time monitoring of physical health. The system can reduce serious harm of acute disease, because timely treatment of patient is provided if get in touch with the first aid center timely and correctly.

Keywords: Sleep; Monitoring; Alarm; Research; Embedded.

1 Introduction

Human physiological function can bring a series of change in the sleep process, such as tactile hypoesthesia, shallow breath, heart rate is slowing down, carbon dioxide combining power in the blood is increasing, sensitivity of respiratory center to carbon dioxide is weakening, pulmonary ventilation is reducing, etc. Normal people can quickly adapt to these changes, while patients and elderly people have certain harm, so sudden illness most occurred at night.

In sudden illness, cardiovascular disease is the number one enemy of human health. Statistics from the World Health Organization showed that each year about 175,000,000 people died of cardiovascular disease, which accounting for over one third in all the cause of death, and this trend is continuing to expand. The data shows that 71% of heart attack at home or in the workplace, 60% to 70% of people died outside the hospital due to the loss rescue time (50% to 75% of them died at home, 8% to12% die at their post, and 6% died in public places). The population of china accounts for about 22% of the population of the world, but medical resources accounted for only 2% of the world, and limited medical resources mostly concentrated in the big hospital of city. It is difficult to implementation of real-time monitoring physical condition in daily life, because most of the medical equipments are expensive and complicated to operate, which are just limited to use in the hospital ward. If the sudden illness could not be known in time, it will delay the best rescue time. Embedded Sleep-Monitoring Alarm System is proposed based on the above reasons, it can monitoring and alarm patients of heart disease, cerebral hemorrhage and other sudden illness, especially for elderly and mobility person. It gains valuable time to save the life of patients.

2 System Design

The system can alarm and monitor sudden illness under the control of embedded microprocessor, the pulse sensor detects pulse signals during sleep condition.

Pulse signal is a weak low-frequency signal, which frequency range is 0.2 Hz to 45Hz and amplitude range is 0 mV to 20mV. It is vulnerable to affected by baseline drift, human activities, frequency interference and other factors in pulse signal acquisition process. In order to meet the requirements of later work, collected pulse signals need signal processing.

Acquisition and processing: Amplifying and filtering and A / D conversion of the collected pulse signals.

Threshold setting: Pre-set abnormal thresholds and emergency thresholds.

Disease diagnosis: Physical condition is automatic diagnosed according to the collected pulse signals and setting thresholds.

Wireless transmission: Wireless transmission abnormal data and alarm signal to monitoring alarm module.

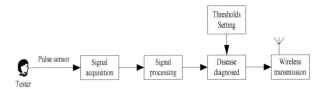

Fig. 1. Acquisition and Processing Module

The monitoring and alarm system immediately takes appropriate measures when abnormal or emergency physical condition is diagnosed. Beyond the abnormal threshold, send SMS and store data. Reach the emergency threshold, telephone alarm and automatic open the door.

Send SMS: Abnormal physical conditions of testers are sent to the doctor and family members by SMS.

Store data: Store abnormal pulse signals for better diagnosis.

Telephone Alarm: Prerecord of home address and patient conditions are automatically telephone the 120 first-aid center.

Automatic open the door: Open the door automatically by wireless for rescuers to enter the room easily.

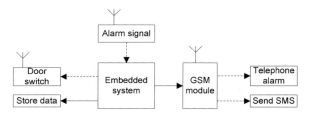

Fig. 2. Monitoring and alarm module

The major function of the main program is initialization modules of embedded sleep-monitoring alarm system, which makes every module into a normal working state and coordinates workflow in order to achieve real-time monitoring and intelligent alarm functions. Main program flow chart is shown in Fig 3.

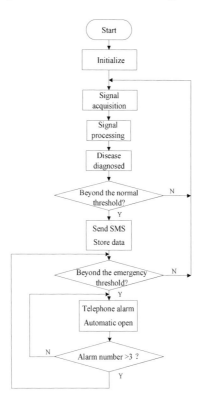

Fig. 3. Main program flow chart

3 Technical Difficulties and Characteristics Research

Theoretical analysis of the model and comparison of clinical detection between the health and the patients with cardiovascular disease in different ages testify that the changes in physiology and pathology of cardiovascular can lead to corresponding changes of pulse wave map feature and area. K is the important physiological index to medical diagnostic and examination of cardiovascular disease, which can indicate characteristics value change of pulse wave regularly and sensitively. HRV (Heart Rate Variability) can diagnose cardiac arrest, bradycardia, tachycardia and other cardiovascular diseases. K (pulse wave characteristics value) and HRV are judgment indicators.

A.Diagnostic index K
K (pulse wave characteristics value) based on changes of pulse wave map area is defined as:

While
$$K = \frac{P_m - P_d}{P_s - P_d}$$

$$P_m = \frac{1}{T}\int_0^T P(t)dt$$

Pm is MAP (mean arterial pressure), which is equal to average value of pulse pressure P (t) in a cardiac cycle. Ps is SBP (systolic blood pressure) and Pd is DBP (diastolic blood pressure).

Research shows that the pulse signal of K value can reveal health and disease important information, which is theoretical foundation of K value to select pulse cycle wave. Flowchart of K value corresponding wave chose of pulse signal is shown in fig 4.

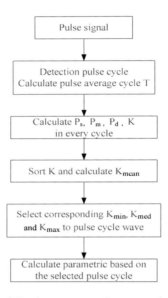

Fig. 4. Flowchart of K value corresponding wave chose of pulse signal

Take a random pulse signal as an illustration, K value corresponding periodic wave is chosen to select appropriate wave. Table□ is computing result of K value, TP5 is corresponding Kmax, TP1 is corresponding Kmed, and TP9 is corresponding Kmin.

Draw a practical conclusion after repeated experiments of similar example above: Main characteristic parameters of average pulse cycle signal and pulse characteristics value Kmed corresponding periodic wave are relatively close. In practical applications, Kmed corresponding pulse wave pulse signal instead of average pulse cycle wave to reduce calculation and complexity.

Due to characteristic information reduced to just one characteristic index K, which is simple, explicit and regular in physiological changes. The value of the characteristic test is convenient, the results of detection repeatability and stability are good, and the blood flow parameter for clinical cardiovascular nondestructive detection provides a simple method. K can be used as an important cardiovascular clinical examination physiological index.

B. Diagnostic index HRV

Pulse frequency is the same to heart rate, which frequency under the impact of age and gender. HRV (Heart Rate Variability) can be displayed and record by sensor and microprocessor accurately. Physiology condition of measurement tester can be accurate acquired through contrasting pulse of the normal person to the pulse of the measurement tester. If meet the risk can send an alarm to the related person quickly. HRV is a potentially simple, non-invasive diagnostic screening tool for sudden illness.

C. Technical difficulties

The technical difficulties of the study are how to process the collected pulse signals; how to extract related disease information from pulse waves; how to take appropriate measures for individual different cases; how to normal work in undisturbed sleep situations of testers; how to reduce the false alarm probability.

The measures should be taken in this study as follows : Set the abnormal threshold and the emergency threshold for different age, gender, physique, disease of monitoring object based on the existing medical research or follow doctor's orders. Select small size, low power consumption, high sensitivity, and strong anti-jamming capability pulse sensor to reduce power consumption, size and weight of the system. Choose comfort and solid configuration and wear method to long wearing, which is designed as close cute intimate finger stall style. Signals are transmitted by wireless communication technology to the devices on bedside, even patients rolled over in the bed which not affect monitoring. Automatically identify false alarm of physiological signal acquisition device falls off, which acquisition pulse signal by pulse sensors and measure body temperature by temperature sensors at the same time. Mild abnormal physical condition is sent to the doctor and family members by SMS and store abnormal data. Reach the emergency condition, immediately telephone the 120 emergency center and automatic open the door for rescuers to enter the room easily. The alert system is divided into two levels, which ensure timely warning and avoid frequent moves. It can more accurately determine physical condition in sleep to reduce the false alarm probability.

4 Conclusion

Embedded Sleep-Monitoring Alarm System can real-time monitor and intelligent alarm expediently and uninterruptedly for the potential patients, which reduce the risks when no one nursing and decrease medical costs. The system can automatically and immediately open the door to inform the rescuer and the emergency center which nearest closest to the patient once they have sudden illness. The system can monitor and alarm patients of heart disease, cerebral hemorrhage and other sudden illness, especially for elderly and mobility person. It gains valuable time to save the life of patients.

Acknowledgment

This project was supported by the Graduate Innovation Fund of Shaanxi University of Science and Technology.

References

1. Spivey, G.: A randomized trial of home telemonitoring in a typical elderly heart failure population in North West London: results of the Home-HF study. European Journal of Heart Failure 11, 319–325 (2009)
2. Kim, C.-G.: Wireless Sensor Networks for Home Health Care. Journal of Systems Architecture, 161–176 (2008)
3. Kassebaum, J.A.: Sensor Networks for Medical Care. In: EMBS Annual International Conference (August 2006)
4. Xiu, D., Spencer, J.: Recognition and Quantification of Sleep Apnea by Analysis of Heart Rate Variability Parameters. Journal of Computational Physics (May 2007)
5. Brunelli, C.: An Embedded Wireless Module for Telemonitoring. Advancements of Medicine and Health Care, 1143–1150 (2007)
6. Hundt, C.: Sensors and sensing in biotelemetry and home care applications. Electronic Notes in Theoretical Computer Science 29, 363–368 (2007)
7. Sbeyti, H.: Design and Simulation of Portable Telemedicine System for High Risk Cardiac Patients. In: Biological and Medical Sciences, pp. 199–202 (2006)
8. Leong, M.P., Jin, C.T.: SMS-Based Platform for Cardiovascular Tele-Monitoring. EURASIP Journal on Applied Signal Processing 7, 629–633 (2003)
9. Takahashi, P.Y., Hanson, G.J.: A randomized controlled trial of telemonitoring in older adults with multiple chronic conditions: the Tele-ERA study. BMC Health Services Research 10, 255–262 (2010)
10. Nahas, M.: The Current State of Telemonitoring: A Comment on the Literature. Telemedicine and e-Health, 344–354 (2005)
11. Benussi, L.: Wireless sensor networks for personal health monitoring: Issues and an implementation. Nuclear Instruments and Methods in Physics Research, 98–99 (2007)
12. Penzel, T., Bunde, A.: Comparison of Detrended Fluctuation Analysis and Spectral Analysis for Heart Rate Variability in Sleep and Sleep Apnea. IEEE Transactions on Biomedical Engineering (October 2005)
13. Schmitt, D.T., Stein, P.K., Ivanov, P.C.: Performance Evaluation of a Wireless Body Area Sensor Network for Remote Patient Monitoring. IEEE Transactions on Biomedical Engineering (May 2009)
14. Haque, M.A., Hasan, M.K.: Research on the Meter's Pulse Signal Processing Based on Sleep-Monitoring Alarm System. Medical Engineering & Physics (February 2001)
15. Munzenberger, R., Slomka, F., Hofmann, R.: Stratification Pattern of Static and Scale-Invariant Dynamic Measures of Heartbeat Fluctuations Across Sleep Stages in Young and Elderly. Microelectronics Journal (September 2003)

CUDA Framework for Turbulence Flame Simulation

Wei Wei[1] and Yanqiong Huang[2]

[1] Henan University of Technology, 450000, Zhengzhou, China
hust_wade@yahoo.com.cn
[2] University of Exeter, EX44BL, Exeter, UK
yh269@exeter.ac.uk

Abstract. This paper proposed linear B-spline weights for flame simulation based on turbulence, which works fine and has less calculation. The turbulence movement of flame described by function was added into flame simulation which increased the authenticity of flame. For real-time applications, this paper presented a strategy to implement flame simulation with CUDA on GPU, which achieved a speed up to 2.5 times the previous implementation.

Keywords: turbulence; linear; CUDA.

1 Introduction

Reeves [1] proposed particle systems approach for fire simulation and other irregular objects in 1983. Perlin [16] developed the Perlin noise in 1985, which widely used in turbulence. Dynamic fluid equation based on grid was proposed by Ferziger [2] as a method of flame simulation in computer graphics, which was complex and included large amount of calculation. The real-time simulation of flame was difficult to realize and the calculation based on gird was difficult to guarantee stability. Perlin adopted the definition of solid texture and added noise to achieve flame animation [3]. Ebert described a new technique which efficiently combined volume rendering and scan line [4]. Scott and his colleagues [5] presented a technique to animate amorphous materials such as fire, smoke and dust in real-time on graphics hardware with dedicated texture memory. Perry proposed the flame spread model in 1994 [6]. On this basis, Wang Jizhou [8] presented a survey on the development of flame simulation in computer animation, with a detail introduction to the classification of the works as well as different kinds of methods employed in the field. Nguyen and his colleagues [9] simulated flame by hydrodynamic equations in 2002, which simulated the turbulence movement of flame more accurately. Wavelet Noise [18] was developed as a replacement for Perlin Noise [16]. The noise is guaranteed to exist only over a narrow spectral band, which makes more sophisticated filtering possible. Rasmussen et al. [19] used the method of Stam and Fiume [7] to break up artifacts when emulating a full 3D simulation using 2D slices, and Lamorlette and Foster [20] used it to add detail to a procedural flame modeling system. Neyret [21] used the same multi-scale intuition as Kolmogorov to animate textures.

2 Flame Modeling

Flame can be seen as the procession of fuel reacts with oxygen, which releases light and heat. The movement of flame can be described into laminar and turbulent flow. As was recently demonstrated for graphics [17], a calculus identity can be used to construct a divergence free vector field by taking the curl of a scalar field. We can respectively construct 2D and 3D vector fields using the noise function ω:

$$W_{2D}(x) = (\frac{\partial \omega}{\partial y}, -\frac{\partial \omega}{\partial x})$$
$$W_{3D}(x) = (\frac{\partial \omega_1}{\partial y} - \frac{\partial \omega_2}{\partial z}, \frac{\partial \omega_3}{\partial z} - \frac{\partial \omega_1}{\partial x}, \frac{\partial \omega_2}{\partial x} - \frac{\partial \omega_3}{\partial y}) \quad (1)$$

We primarily use the 3D case. The 3D case requires three different noise tiles, which we have denoted $\omega 1$, $\omega 2$ and $\omega 3$, but in practice we use offsets into the same noise tile. The derivatives can be evaluated directly because ω uses B-spline interpolation.

$$[\frac{t^2}{2}, \frac{1}{2}+t(1-t), \frac{(1-t)^2}{2}] . \quad (2)$$

Instead of the usual quadratic B-spline weights, this paper proposed linear B-spline weights which works fine and has less calculation:

$$[2t, |1-2t|, 2|1-t|] . \quad (3)$$

The statistical properties of these particles based on equations of fluid and were more stable in the numerical solution. The following equation defined the changes of particles.

$$dX^{(i)} = U^{(i)}dt . \quad (4)$$

$$dU^{(i)} = \frac{3}{4}C_0 <\omega> U^{(i)}(t) - <U> dt + \sqrt{C_0 k <\omega>} dW \quad (5)$$

Equation (4-5) defined the position X of the particle according to the velocity U, and velocity calculation based on the simplified Langevin model [10]. Symbol <> represented the average. Constant C_0 was a standard value of turbulent motion, and k was the disturbance kinetic energy, and dW was the derivative of W(x). The turbulent movement of the flame can be modeled combined with particle system, which overcame the reality limitations of particle system for flame.

3 Rendering with CUDA

Zhao Chunxia [14] proposed a flame model based on particle system, which discussed particle's attributes in detail and emphasized the color variation and dynamic wavering of the flame. Li Jianming [15] based on physical model simulation for the calculation of the flame, high complexity and difficult problem in real-time simulation, proposed a fluid-based model and real-time GPU-accelerated simulation of the flame, but frequently data interaction between CPU and GPU affected the efficiency, and real-time rendering problem was not been fundamentally improved.

In order to ensure real-time rendering, we implemented the parallel version of our technique using the NVIDIA CUDA [NVIDIA. 2009] language, which allows us to use the graphics processor without using shading languages. In the context of CUDA, the CPU plays the role of the Host, which controls the graphics processor and calls Device. It sends data, calls the Device to execute some functions, and then copies back its results.

Each graphics processor of an NVIDIA graphics card is divided into several multiprocessors. NVIDIA CUDA [NVIDIA. 2009] divides the processing in blocks, where each block is divided in several threads. Each block of threads is mapped to one multiprocessor of the graphics processor. When the CPU calls the Device to execute a function, it needs to inform how the work will be divided in blocks and threads. Maximum performance is achieved when we maximize the use of blocks and threads for a given graphics processor.

Each of the multiprocessors is a group of simple processors that share a set of registers and some memory, which is the shared memory space. The shared memory size is very small; usually 16KB or 32KB on graphics cards running on 1.3 compute capable device, but it is as fast as the registers. The communication between two multiprocessors must be done through the Device memory, which is quite slow if compared to the shared memory. There is also the Constant Cache and Texture Cache memory, which has better access times than the Device memory, but it is read-only for the Device. Before the execution of the code in the Device, CPU must send the data to its Device memory to be processed later. The memory copy from the Host (CPU) memory to the Device memory is a quite slow process, and should be minimized. Besides, the NVIDIA CUDA Programming Guide [22] says that one single call to the memory copy function with a lot of data is much more efficient than several calls to the same function with a few bytes. The performance of application can be improved by making good use of these restrictions of CUDA. To avoid several memory transactions between the Host (CPU) and the Device, all attributes in contiguous memory areas was stored together, and treat it like an array. At the position k stored an attribute of the particle P_k (Fig. 1). Proceeding in this way, several unnecessary copies were avoided, which improved the overall performance. Device memory can be allocated on a linear, but also can be assigned for the CUDA array form. CUDA memory can be 1 dimensional, 2 dimensional and 3D (2.0 version). Memory types included unsigned8, 16 or 32-bit int, 16-bit (only driver API can do) float, 32-bit float. This allocated memory can only process through kernel function in CUDA.

In order to verify that our parallel implementation can be executed faster than the sequential one, a couple of tests were accomplished. All the tests were executed in an Intel X5450 3.00GHz, NVIDIA Quadro FX 3700 graphics card.

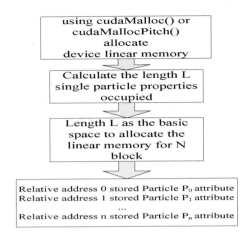

Fig. 1. Data structure used on GPU

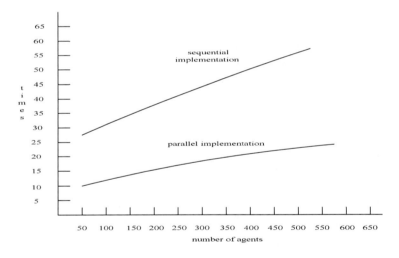

Fig. 2. Speed up achieved using the parallel

The graphic in Fig. 2 shows the speed up achieved using the parallel implementation over the sequential version of the technique. As we can see, test showed the parallel version was above more than twice faster than the sequential one (exactly the lowest point in the graphic is at 2.8 times). Besides that, sequential version with the agent increases, the time increased faster than parallel version.

In addition, according to the NVIDIA CUDA Programming Guide [NVIDIA. 2009], the graphics processor cannot handle all the data in a parallel way. The division of the work in blocks of threads lets the graphics processor scheduler run some blocks of thread while others wait for execution. Because of this, the computation of 256 local maps in a parallel way does not give a speed up of 256 times.

To explain what the cause of the graphics peaks is, the NVIDIA CUDA Programming Guide says that each algorithm implemented with CUDA has an optimal point, in which the amount of blocks and threads uses the most possible number of resources available in the graphics processor simultaneously.

This paper presented a strategy to implement flame simulation on GPU under CUDA Framework, which constitutes a great advantage when compared to the traditional method. We implemented a parallel version of this algorithm using the NVIDIA CUDA [NVIDIA. 2009] language, which allows us to use the graphics processor avoiding the use of shading languages. The parallelism was explored, reducing the amount of memory transactions between CPU and GPU. Our result shown that the GPU implementation improves up to 2.5 times the sequential CPU version.

As future work, the exploration of this method used on parallel architectures and explored the use of other shading languages. It would be interesting to compare the possible improvements in performance using other languages.

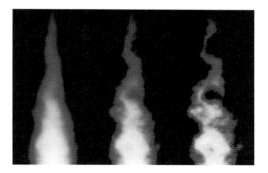

Fig. 3. Real-time flame simulation

The properties of each particle changed in terms of the turbulence movement every frame rendering. Frame rate was around 60 FPS, which ensured smooth and real-time effects (Fig. 3).

4 Conclusions

This paper presented linear B-spline weights for flame simulation based on turbulence, which works fine and has less calculation. The turbulence movement of flame described by function was added into flame simulation which increased the authenticity of flame. For real-time applications, this paper presented a strategy to implement flame simulation with CUDA on GPU, which achieved a speed up to 2.5 times the sequential implementation. In further studies, smoke can be joined in flame simulation.

References

1. Reeves, W.T.: Particle systems-a technique fur modeling a class of fuzzy objects. Computer Graphics (S0097-8930) 17(3), 359–376 (1983)
2. Ferziger, J.H., Peric, M.: Computational Methods for Fluid Dynamics, 3rd edn. Springer Press, Heidelberg
3. Perlin, K.: An image synthesizer. ACM Computer Graphics 19(3), 287–296 (1985)
4. Ebert, D.S., Richard, E.P.: Rendering and animation of gaseous phenomena by combining fast volume and scan line A-buffer techniques. ACM Computer Graphics 24(4), 357–366 (1990)
5. Scott, A.K., Crawfis, R.A., Reid, W.: Fast Animation of Amorphous and Gaseous Phenomena. In: Volume Graphics 1999, Swansea, Wales, pp. 333–346 (1999)
6. Perry, C.H., Picard, R.: Synthesizing flames and their spread. In: Siggraph 1994. Technical Sketches Notes, US (1994)
7. Stam, J., Eugene, F.: Depicting fire and other gaseous phenomena using diffusion processes. In: Computer Graphics Proceedings, Annual Conference Series, ACM SIGGRAPH, pp. 129–136. ACM Press, Los Angeles (1995)
8. Wang, J., Gu, Y.: Flame Simulation Method of Review. Journal of Image and Graphics 12(11), 1961–1970 (2007) (in Chinese)
9. Quang, N.D., Ronald, F., Wann, J.H.: Physically based modeling and animation of fire. ACM Transactions Graphics (S0730-0301) 21(3), 721–728 (2002)
10. Pope, S.B.: Turbulent Flows. Cambridge University Press, Cambridge (2000)
11. Subramaniam, S., Pope, S.B.: A mixing model for turbulent reactive flows based on Euclidean minimum spanning trees. Combustion and Flame 115(4), 487–514 (1998)
12. Lamorlette, A., Foster, N.: Structural modeling of natural flames. In: Proceedings of ACM SIGGRAPH 2002, pp. 729–735 (2002)
13. Adabala, N., Manohar, S.: Modeling and rendering of gaseous phenomena using particle maps. Journal of Visualization and Computer Animation 11, 279–293 (2000)
14. Chunxia, Z., Yan, Z., Shouyi, Z.: Three-dimensional particle-based fire simulation systems approach. Computer Engineering and Applications 28 (2004) (in Chinese)
15. Li, J., Wu, Y., Chi, Z., He, R.: GPU-based fluid model and the flame acceleration real-time simulation. Journal of System Simulation 19 (2007) (in Chinese)
16. Perlin, K.: An image synthesizer. In: Proceedings of ACM SIGGRAPH, pp. 287–296 (1985)
17. Bridson, R., Hourihan, J., Nordenstam, M.: Curl-noise for procedural fluid flow. In: Proceedings of ACM SIGGRAPH (2007)
18. Cook, R., Derose, T.: Wavelet noise. In: Proceedings of ACM SIGGRAPH (2005)
19. Rasmussen, N., Nguyen, D.Q., Geiger, W., Fedkiw, R.: Smoke simulation for large scale phenomena. In: Proceedings of ACM SIGGRAPH (2003)
20. Lamorlette, A., Foster, N.: Structural modeling of flames for a production environment. In: Proceedings of ACM SIGGRAPH (2002)
21. Neyret, F.: Advected textures. In: ACM SIGGRAPH/EG Symposium on Computer Animation, SCA (2003)
22. http://developer.download.nvidia.com/compute/cuda/3_0/toolkit/docs/NVIDIA_CUDA_ProgrammingGuide.pdf

Multi-agent Crowd Collision and Path Following Simulation

Wei Wei[1] and Yanqiong Huang[2]

[1] Henan University of Technology, 450000, Zhengzhou, China
hust_wade@yahoo.com.cn
[2] University of Exeter, EX44BL, Exeter, UK
yh269@exeter.ac.uk

Abstract. This paper aimed for research agent crowd collision and behavior simulation under emergency. The agent crowd collision model which used particles to represent agent body was established by mathematical physics and was more precise compared to the existing method. The key-frame animation was optimized which used to achieve the simulation of multi-agent behavior. System was realized on PC platform and the results were tested.

Keywords: 3D real-time rendering; virtual fire; skeleton animation.

1 Introduction

The wide use of crowd simulation in games, entertainment, medical, architectural and safety applications, has led to it becoming a prevalent area of research. One of the main obstacles to interactive rendering of crowds lies in the computation needed to give them a credible behavior. Multi-agent systems are used to model a network of loosely coupled dynamic units, often called agents. Based on the pioneering work on distributed behavior models by Reynolds [11], the study of multi-agent simulation has grown tremendously over the last two decades. Many simulation algorithms have been developed based on simple models and local rules. Besides computer graphics, multi-agent systems are widely used to model the dynamics of crowds, robots and swarms in traffic engineering, virtual environments, control theory, and sensor networks.

2 Crowd Collision Simulation Methods

The collision detection test is used to make each agent aware of the surrounding environment; it is essential for tasks such as path planning and obstacle avoidance. There are many techniques to detect interference between geometric objects [1]. Many of them use hierarchical data structures, for example, hierarchical bounding boxes [2, 3], sphere trees [4], BSP trees [5] and Octrees [6]. However, the majority of these approaches try to solve the harder problem of exact interference between complex objects. For this reason, they tend to be much more precise than what is needed to simulate crowd. This paper proposed a more precise method which enhances the reality and gets small error.

2.1 Set Up Agent Crowd Collision Behavior Models

A particle can be thought of as an agent without size or orientation, which has a position, velocity, and applied forces. This paper proposed the crowd collision of a rigid agent body, X (t) and V (t) denote the position and velocity, respectively, of the center of mass of the rigid agent body. The linear momentum [10] of the agent was

$$P(t) = mV(t) .\tag{1}$$

Newton's second law of motion stated that the rate of change of linear momentum was equal to the applied force:

$$\dot{P}(t) = F(t) .\tag{2}$$

Where m was the mass of the body and F (t) was the applied force on the object. The equations of motion pertaining to position and linear momentum were:

$$\dot{X} = m^{-1}P, \dot{P} = F .\tag{3}$$

Matrix R (t) represented orientation vector of the agent. In the coordinate system of the rigid body, b denoted the time-independent position of a point relative to the origin (the center of the mass). The world coordinate of the point was:

$$Y(t) = X(t) + R(t)b .\tag{4}$$

For two equal mass rigid agents, the collision between them can be calculated and realized (Fig. 1). Once all these quantities were known, the numerical particle was ready to be rendered. In particle lives by each frame, the following four tasks must be completed:

1) Particle source generates new particles;
2) Update particle attributes;
3) Delete dead particles;
4) Rendering particles.

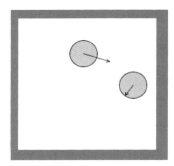

Fig. 1. Collision of Two Rigid Agents.

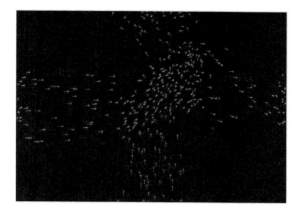

Fig. 2. Multi-agent Crowd Cosillion Behavior Simulation.

3 Movement Control of Three-dimensional

3.1 Movement Control of Three-Dimensional

Methods of realizing agent motion control mainly including kinematic method (forward and reverse kinematics method) [7], dynamic method, movement captured method, the key frame method as well as skeletal animation method [8]. Using former two methods to simulate agent motion was rather difficult, lack of motion detail and movement fidelity and was weak at real-time performance. Motion Captured method made use of motion captured equipment to directly collect the data of real agent motion through optical or magnetic sensors, and then, according to the recorded data, rendered the agent body animation, yet this approach required specialized hardware and were expensive [9]. Skeleton animation technology used a series of bones to promote skin to move. Its advantage lies in the fact that a line segment represented of the bone joints, and all the complex motions were defied by the very small amount changes of the line segment position.

3.2 Key Frame Animation

The standard implementation of keyframe animation involves interpolating positions, orientations and scales as separate channels [10].

The *keyframe Controller* class stores the position, orientations, and scales separately. A sequence of keyframe positions is (t_i, P_i) for $0 \leq i < n$. The time is assumed to be ordered, $t_0 < t_1 < \ldots < t_{n-1}$. A normalized time is computed,

$$u = \frac{t - t_i}{t_{i+1} - t_i} . \tag{5}$$

and the position keys are interpolated by

$$P = (1-u)P_i + uP_{i+1} . \tag{6}$$

A sequence of keyframe orientations is (t_i, q_i) for $0 \leq i < n$, where q_i is a unit quaternion that represents a rotation. The *KeyframeController* class allows you to specify a common set of times if that is what your animation uses.

The normalized time is computed using (7). The orientation keys are interpolated using the *slerp* function for quaternions,

$$q = slerp(t; q_i \bullet q_{i+1}) = \frac{q_i \sin((1-t)\theta_i) + q_{i+1} \sin(t\theta_i)}{\sin \theta_i} . \tag{7}$$

Where θ_i is the angle between q_i and q_{i+1} and $q_i \bullet q_{i+1} = \cos \theta_i$.

A sequence of scales is (t_i, σ_i) for $0 \leq i < n$. The natural manner for scales is geometric rather than algebraic:

$$\sigma = \sigma_i^{1-u} \sigma_{i+1}^u . \tag{8}$$

For normalized time $u \in [0,1]$. σ is actually linear in the logarithm of scale:

$$\log(\sigma) = (1-u)\log(\sigma_i) + u\log(\sigma_{i+1}) . \tag{9}$$

However, this paper proposed a linear interpolation of scale which works fine and has less calculation:

$$\sigma = (1-u)\sigma_i + u\sigma_{i+1} . \tag{10}$$

4 Implementation

System was tested on PC platform, by the environment of which was intel Core2 1.66GHZ, ATI Radeon X1300 graphics cards (Fig. 3).

Fig. 3. Simulated agents escape behavior.

It was still able to reach a higher frame rate by gradually increasing the number of characters in the scene as shown, when the number of characters in scene increases to 220, the frame rate still can reach 80 which guaranteed smooth screen and with the increasing agent number, the change of FPS becoming more and more stable (Fig. 4).

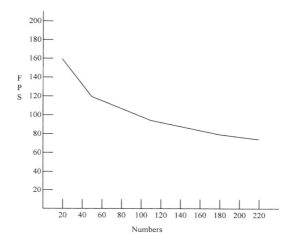

Fig. 4. Test results.

5 Conclusion

Crowd visualization was a vast research topic and our current research tried to improve the results of the existing system on several fronts. The agent crowd collision model which used particles to represent agent body was established by mathematical physics and was more precise compared to the existing method. The key-frame animation was optimized which used to achieve the simulation of multi-agent behavior.

References

1. Lin, M., Gottschalk, S.: Collision detection between geometric models: A survey. In: Proceedings of IMA Conference on Mathematics of Surfaces (1998)
2. Cohen, J., Lin, M., Manocha, D., Ponamgi, M.: I-COLLIDE: An Interactive and Exact Collision Detection System for Large-Scaled Environments. In: Proceedings of ACM Interactive 3D Graphics Conference, pp. 189–196 (1995)
3. Stefan Gottschalk, M.L., Manocha, D.: Obb-tree.: A hierarchical structure for rapid interference detection. In: SIGGRAPH 1996 Conference Proceedings, pp. 171–180 (1996)
4. Hubbard, P.M.: Collision detection for interactive graphics applications. IEEE Transactions on Visualization and Computer Graphics 1(3), 218–230 (1995) ISSN1077-2626
5. Naylor, B.F.: Binary space partitioning trees as an alternative representation of polytopes. Computer-Aided Design 22(4), 138–148 (1990)

6. Samet, H.: The Design and Analysis of Spatial Data Structures. Series in Computer Science reprinted with corrections edition. Addison-Wesley, Reading (1990)
7. He, K., Yuming, J.: Based on inverse kinematics of the virtual human walking. System Simulation 16(6), 1343–1345 (2004)
8. Seron, F.J., Rodriguez, R., Cerezo, E., Pina, A.: Adding Support for High-Level Skeletal Animation. IEEE Transactions on Visualization and Computer Graphics [C]. [S.l.]: [s.n.] (2002)
9. Boulic, R., Becheiraz, P., Emering, L., Thalmann, D.: Integration of Motion Control Techniques for Virtual human and Avatar Real-time Animation. In: Proc. of VRST 1997, pp. 111–118. ACM press, New York (1997)
10. Eberly, D.H.: 3D Game Engine Design, 1st edn. Morgan Kaufmann, San Francisco (2000), Book & CD-ROM
11. Reynolds, C.W.: Flocks, herds and schools: A distributed behavioral model. In: SIGGRAPH 1987: Proceedings of the 14th Annual Conference on Computer Graphics and Interactive Techniques, pp. 25–34. ACM, New York (1987)

Study on a New Type Protective Method of Arc Grounding in Power System

Wenjin Dai and Cunjian Tian

Department of Electrical and Automatic Engineering
Nanchang University
Nanchang, 330031 China
`dwj480620@yahoo.com.cn, 447848006@qq.com`

Abstract. Arc grounding is a hidden trouble that threatens electric power system and distribution networks. At present, it is frequent to use the methods of the grounding scheme by arc suppression coil or small resistance, but certain shortcomings still exist in this way. This paper presents a new protective method, which is based on microcomputer. Its principle and a numerical algorithm are presented also. With this method, we can eliminate arc fast and make the system operate stably.

Keywords: arc grounding, new protective method microcomputer-based, numerical algorithm, arc suppression coil.

1 Introduction

Along with the electrical network scale rapid expansion, the electrical network accident also frequently occurs. But in 10~35KV system,it mainly for single-phase earth fault.The experience indicated,when this kind of voltage rank electrical network develops the certain scale, the interior over-voltage,specially the electrical network has the single-phase intermittent arc light earth and the ferromagnetic resonance which produces under the special condition has become a big threat to the safe operating of this kind of electrical network equipment. And it is most serious of over voltage by single phase arc light earth. In order to solve this kind of problem, at present, there are two more universal methods, namely: neutral point after arc suppression coil earth protection, as well as neutral point after small resistance earth protection. In addition,recently two new methods,which are XHB(the protection plan of combination of arc suppression and over-voltage) and arc suppression by Microcomputer,have appeared.This paper gives another kind of new protection plan and has given the analysis and the introduction to its principle and algorithm.

2 Algorithm Derivation and Principle of Arc Protection

2.1 Situation of a Line Supplied by One Active Networks

Considering a three-phase arcing fault on an overhead line,presented in Fig.1.At the fault point, the fault is fed with currents i and i_a, namely: $i_a = i + i_2$, which is the arc current.In Fig.1.,the voltage across the arc is designated by $u_a(t)$ .When the system is in the normal steady state,the current across the line is designated by i_L.

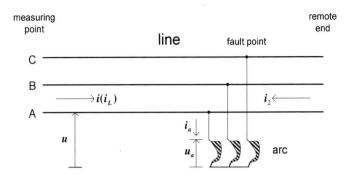

Fig. 1. Three-phase arcing fault on overhead line

The circuit from Fig.1 is equivalent single-phase circuit depicted in Fig.2. By applying the Kirchoff's law, we can obtain the following differential equation from Fig.2.

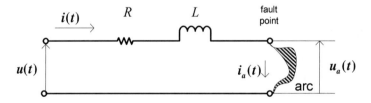

Fig. 2. Equivalent single-phase circuit of the faulted line

$$u = Ri + L\frac{di}{dt} + u_a + \xi \qquad (1)$$

where ξ is zero-mean Gaussian noise.The random measuring errors and the errors in the problem modelling are modelled as a Gaussian noise.The equation (1) present the arc voltage.we assumed the arc voltage to be a square wave form,which is in the same direction with the arc current i_a .And it can be expressed as follows:

$$u_a(t) = U_a \operatorname{sgn}(i_a) \qquad (2)$$

By substituting (2) into (1), we can obtain the following:

$$u = Ri + L\frac{di}{dt} - U_a \operatorname{sgn}(i_a) + \xi \qquad (3)$$

The exact are current was measured by the relay. And the relay deplay the current was i. Therefor:

$$\text{sgn}(i_a) = \text{sgn}(i) \tag{4}$$

2.2 The Situation of a Line Supplied by Two Active Networks

Two active networks are presented in Fig.3. We neglect the arc voltage and analyse the flow of currents depicted in Fig.3. We assume the fault distance is ℓ.

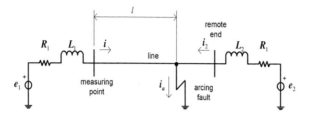

Fig. 3. Single-line equivalent circuit of the faulted overhead line supplied by two active networks

The circuit from Fig.3 is equivalent by the circuit depicted in Fig.4, in which voltage u_r is the voltage at the normal line operation. By applying the superposition principle, the circuit from Fig.4 is equivalent by the sum of two circuits presented in Figs.5 and 6.

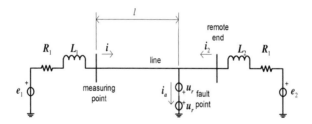

Fig. 4. Equivalent circuit derived from the circuit presented in Fig.3

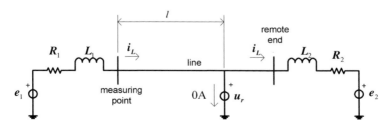

Fig. 5. Normal line operating state

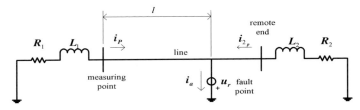

Fig. 6. Faulty state

In Fig.6, all the currents present the short circuit currents distributed through the network. By analyzing the circuit from Fig.6, we can concluded that the phase of i_a is the same as the phase of i_p, therefore there is the following:

$$\text{sgn}(i_a) = \text{sgn}(i_p) \tag{5}$$

By the superposition principle, the fault current is the sum of currents presented in Figs.5 and 6, we can obtain the following:

$$i = i_L + i_P \tag{6}$$

By Eqs.(4) and (5),one obtains the following important conclusion:
$\text{sgn}(i_a) = \text{sgn}(i - i_L)$.

Thus, the mathematical model describing the transients in the circuit from Fig.1 can be now represented as follows:

$$u = Ri + L\frac{di}{dt} + U_a \, \text{sgn}(i - i_L) + \xi \tag{7}$$

In Eqs.(7),the rate of change of current(di/dt) can be used the numerical calculation, namely.

$$\frac{di(t)}{dt} \approx \frac{i_{n+1} - i_{n-1}}{2T} \tag{8}$$

The line voltage and currents can be uniformly sampled with the sampling frequency $f_s = 1/T$. We sample N voltage, N+2 fault current and pre-fault current, which can be used as an input to the algorithm.So we obtains the following:

$$\begin{bmatrix} u_1 \\ u_2 \\ \cdot \\ \cdot \\ \cdot \\ u_N \end{bmatrix} = \begin{bmatrix} i_1 & i_2 - i_0 & \text{sgn}(i_1 - i_{L1}) \\ i_2 & i_3 - i_1 & \text{sgn}(i_2 - i_{L2}) \\ \cdot & \cdot & \cdot \\ \cdot & \cdot & \cdot \\ \cdot & \cdot & \cdot \\ i_N & i_{N+1} - i_{N-1} & \text{sgn}(i_N - i_{LN}) \end{bmatrix} \begin{bmatrix} R \\ L_e \\ U_a \end{bmatrix} + \begin{bmatrix} \xi_1 \\ \xi_2 \\ \cdot \\ \cdot \\ \cdot \\ \xi_N \end{bmatrix} \tag{9}$$

Where $L_e = \dfrac{L}{2T}$, Eqs.(9) can be expressed in the following matrix form:

$$V = Ax + \xi \tag{10}$$

where $x = [R \ \ L_e \ \ U_a]^T$, $u = [u_1, \ ..., \ u_N]^T$, A is an N*3 coefficient matrix and $\xi = [\xi_1, \ ..., \ \xi_N]^T$. By minimising the random noise and using the least error square method, we can obtain the following equation:

$$x = (A^T A)^{-1} A^T v \tag{11}$$

According to U_a, it can be concluded which kind of fault occurred. If $U_a > 0$, the fault is transient with arc. Otherwise, if $U_a = 0$, the fault is permanent without arc. So, we use Microcomputer to sample line voltage and current. Then, we compute U_a according to Eqs.(9). If $U_a > 0$, we can know the fault is transient with arc. So we trip the power transmission line connecting source by using microcomputer.

3 Simulation of the Fault of Arc Models and without Arc Based on MATLAB

3.1 Mathematics Equation of Arc Models

3.1.1 Mayr Arc Model

The Mayr arc model is ifered based on three basic principles of the thermal equilibrium, hotly Inertia, as well as hotly drifts. The Mayr arc model is defined as

$$\frac{1}{g}\frac{dg}{dt} = \frac{d \ln g}{dt} = \frac{1}{\tau}(\frac{ui}{p} - 1) \tag{12}$$

where, g the conductance of arc
u the voltage across the arc
i the current across the arc
τ arc time-constant
p arc radiation power

The physics significance of the mayr arc model is very clear. Namely, the arc temperature will elevate, the heat will drift to strengthen and arc conductance g will have the increase, when arc power ui is bigger than radiation power p. Because of hotly Inertia or arc time-constant τ, arc temperature will elevate or arc conductance g will have little increase. The Mathematics equation of Arc Models is suitable for little current, including zero area arc process.

3.1.2 Cassie Arc Model

The Cassie arc model is defined as

$$\frac{1}{g}\frac{dg}{dt} = \frac{d\ln g}{dt} = \frac{1}{\tau}(\frac{u^2}{u_c^2}-1) \tag{13}$$

Where u_c constant arc voltage, which is the reference voltage in the method of the IEC arc transient state restores the voltage (TRV) and equal to the TRV peak value. Other physical quantities meanings is the same as Eqs.(11).

The typical arc model also has other some types, like Habedank arc model, Modified Mayr arc model, Schavem aker arc model,Schw arz arc model and KEMA arc model and so on.

3.2 Simulation of the Fault of Arc Model Based on Mayr Arc Model

According to Eqs.(11), we can obtain the graph in Fig.7 by using Matlab software.

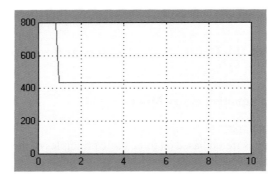

Fig.7. The wave for the fault of arc model based on Mayr arc model

On other hand, the wave for the fault of permanent without arc simulating in Matlable software is as following:

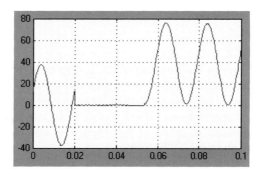

Fig. 8. Wave for the fault of permanent without arc

According to Figs.7 and 8, we know the voltage of the fault is not zero when the fault is transient with arc, or is zero when the fault is permanent without arc.

4 Conclusion

We may fast the extinguishing of arc, causing the electrical power system stable movement by using microcomputer to the arc light earth protection. Thus, we reduce the harm which brings by arc.

References

1. Zuo, Q., Jin, Z.: The research for the protection of electrical power system arc light earth. Electric Power Science and Project (4), 31–33 (2002)
2. Bao, L., Jin, Z.: The installment of arc suppression over voltage protection Based on PLC. Automated Technology and Application 23(1), 33–35 (2004)
3. Terzija, V.V., Radojevi, Z.M.: Numerical algorithm for medium voltage overhead lines protection and adaptive autoreclosure. Electrical Engineering 29, 95–99 (2003)
4. Shao, P., Qing, Y., Jing, L.: Simulation of arc model Based on MATLAB. Electrical Power System and its Automated Journal 17(5), 64–66 (2005)

The Development of Economizing-Energy Illumination System

Wenjin Dai and Lingmin Liang

Department of Electrical and Automatic Engineering
Nanchang University, Jiangxi, China
123095572@qq.com

Abstract. The illumination control system designed in this article is a new economizing-energy control system. The system can preferably adapt to the characteristics of the HID lamps. It can reduce the interference to electric power system, prolong the life of fired lamps and save energy. It is much reliable and has energy saving effect. Moreover, the system has little noise. It can be applied to all kinds of illumination sets.

Keywords: Economizing Energy, Illumination System, Microprocessor Control.

1 Introduction

Illumination field is a developing field today. Now, the study in the international illumination field concentrates mostly on the two aspects. One is energy saving, and the other is the improvement on the illumination quality. The aspect dealt with in this article is the energy saving aspect. Compared with incandescent lamp, fluorescent lamp and high-voltage mercury lamp, the high-pressure sodium lamp has higher lighting efficiency and more complex shine process.

At present, manual control, time control and optical control are widely adopted in the existing illumination systems. All the systems are opened and closed according to the given time and the daylight time. The systems, which use program control, can also only utilize the switching and multiplex control according the given time. As a result, the actual illumination luminance is decreased greatly. In an addition, we must erect one or more lines. In practice, this system cannot solve the problem that the lamps have shorter lifetime and higher power consumption.

In our opinion, the fundamental way is to vary the lamp voltage according to our needs. Therefore, two problems must be solved. The first is to reduce the voltage to the rating range, and the second is to keep the voltage stable.

2 General Design

There are many ways to vary the voltage. The easiest way is that one appropriated resistance is to be set in the power supply loop. By this way, the load voltage can be decreased. But the resistance itself has much power consumption. So this way is inapplicable. The system has higher cost and bigger volume, and system cannot realize the auto-regulation voltage when the power supply is changed. Therefore, the way making use of reactor or resistance to decrease the voltage cannot be adopted.

On the other hand, it cannot assure the power supply voltage is in the rating range. The high-pressure sodium lamp is a sensible element, and it has negative resistance character. So the supplied voltage cannot be over low or high. When the voltage drop over 5 percent, the lamp will black out and influence illumination. To avoid this, we must keep the power supply voltage in the rating range. If we use thyristor control, although the output wave shape is not sine wave, it has not affect on the natural illumination. It has many distinct advantages. It is small weight, can easily be installed and automatically control. After many experiments, we design one system controlled by thyristor.

3 System Design

3.1 Hardware

We use the microprocessor as the core of the illumination system. The system has simple hardware and many functions. The system is very simple, its cost is quite low and it works stably. We select 8031 microprocessor of Intel. The microprocessor has 128 byte RAM, 32 pieces of I/O lines, two 16 bits timers.

The schematic circuit of the system is shown in Fig.1.

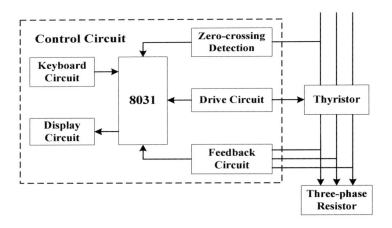

Fig. 1. System schematic circuit

(1) The Zero-crossing Detection Circuit

In the circuit, the zero-crossing time is needed to provide the exact zero-crossing pulse for the voltage regulating circuit controlled by microprocessor. Then the trigger impulse signal can keep synchronization strictly with thyristor circuit. Fig. 2 is the wave form of AC power supply detector. The photo electricity coupling parts can restrain noise intervene. The response time is very short.

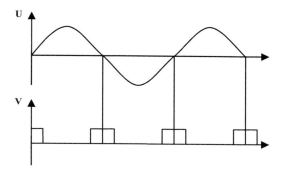

Fig. 2. The wave form of AC power supply detector

(2) The Display and the Keyboard Circuit

The microprocessor drives the numeral tube (common cathode) and the keyboard. The chip 8155 has 256-byte RAM memory capacity, two programmable 8bits parallel ports, which named PA and PB, and one 6 bits parallel port named PC. The chip can join directly the microprocessor MCS-51 port without any additional hardware. The display is adopted the scan mode. The system displays the system time, the working state and voltage value at intervals. The inner timer of the microprocessor can generate the timing clock. The software carries out the function of counting. The keyboard can set the on-off time, the step-down value, the start-up time, the start intervals, the system time and so on. We adopt DS12887AH as the clock chip, which has lithium batteries, 114byte RAM. In the case of abrupt power down, the chip can protect the memory information and make the clocks continue to work. The chip cannot be influenced by the external power supply.

(3) The Drive Circuit

After being insulated and amplified, the output impulse of the microprocessor can join the controlled poles of thyristor to control the start-up. The main circuit of thyristor is commonly high-pressure, but the microprocessor control circuit is low-pressure. To assure the equipment safe and reliable run, the start-up impulse caused by the microprocessor I/O lines can only give to the thyristor circuit after insulated, which can also decrease the noise from the electric network.

3.2 Software

Fig. 3 is the flow chart.

The zero-crossing signal is the input signal of external interruption 0. In the case of the interruption permission, 8031 responds the interruption and comes into the interruption service routine. T0 is used to control the output impulse of the microprocessor, and T1 is used to control the display of the numeral tube. The value of AC phase angle is converted into the initial value of the timer. When the value of the timer is overflowed, one of the pins named P1.0~P1.5 of the chip 8031 outputs one positive impulse. After being insulated and amplified, the impulse is to start up the thyristor. The initial value will determine the start-up time of the thyristor. After the initial value based on the deviation being changed, the microprocessor can make the output voltage equal to the given voltage. Thus the microprocessor has the function of voltage stabilization.

There are three tables in the program, which deals with the keyboard. The first table and the second one are the time lists. The third table is the list of the state and the output voltage. Considering that the three-phase supply will fluctuate when converted from 360 degree to 1 degree, we deal with the angle in a small degree. The time will be delayed as converted from analog to digital. We have taken steps, such as detecting not only time, just permitting small fluctuation and little delaying time, and so on, to decrease this error.

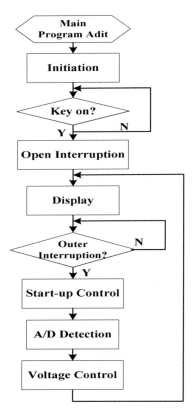

Fig. 3. The flow chart

The compulsive start-up procession may damage the lamps. We have laid out one subroutine to deal with the start-up procession of the lamps. To assure the security and the reliability of the system, we have considered the case such as power down.

4 Conclusion

The system has original design principle and simple hardware and good economizing-energy. The program is written in assembly language. After being applied the new technology, the lamps and the lanterns have less actual power consumption than rating power consumption. The system has evident economizing-energy effect, which our experiments have proved. The system has achieved the goal of economizing energy and prolonging the working life of the lamps and the lanterns. Although the average voltage of the lamps and the lanterns just drop down in a small degree, the power consumption will fall obviously, the instantaneous lightness of the lamps and the lanterns will not be changed obviously. Moreover, the visual brightness of the eyes does not be changed obviously because of the persistence of vision. In practice, the system can economize 20 percent of the energy. The system performance is reliable. It can also be applied to other lighting facilities, such as fluorescent lamp, incandescent lamp and so on.

References

1. Severns, R.: dv/dt Effects in MOSFETs and BJT Switches. In: IEEE PESC 1991 Record, pp. 256–258 (1991)
2. Bose, B.K.: Evaluation of modern power semiconductor devices and future trends of converters. IEEE Trans. and Appl. (1992)
3. Hobson, L., Hinchlitte: High frequency Switching of Power. MOSFETs. In: PCIM, pp. 70–77 (1988)
4. Scott, W., Hamachi, G., Mayo, R., Ousterhout, J.: Changes to Magic in Version4, Computer Science Division, Electrical Engineering and Computer Sciences, University of California (1986)
5. Nishimura, H.: Journal of the Illuminating Engineering Society, 70–76 (Summer 1998)
6. Frchnrich, H.-J.: Journal of the Illuminating Engineering Society, 131–140 (Summer 1998)
7. Intel Embedded Controller Application Handbook (1989)
8. Intel Corporation, Microcontroller Handbook (1984)

The Membrane Activity's Design and Evolution about acE Service Flow*

Xiaona Xia[**], Baoxiang Cao, and Jiguo Yu

School of Computer Science, Qufu Normal University, Rizhao Shandong, 276826, China
xiaxn@sina.com

Abstract. Starting from semantic composition of services, acE service flow as architecture researching point, this paper has designed and argued dynamic flow evolution mechanism. First of all, the architecture service topology concept is raised, to achieve dynamic composition of participation granularities and adaptive evolution for flow, the corresponding model expression and flow composition rule' collection is given; Then, equivalent service logic and flow evolution relation is organized, π-calculus mechanism is extended, and is to complete opening flexible architecture requirement and autonomous migration process driven by membrane activity. Finally, the decision-making and planning mechanism about acE service flow is finished analyzed, and furthermore, the validity and feasibility of this researching project is verified.

Keywords: acE Service Flow; Membrane Activity; π-calculus Extension; Bisimulation Service; Semantics; Composite Service.

1 Introduction

The dynamic evolution of the existing service composition cover migration, evolution time, composite mode, evolution operation and classification etc[1-8], at the same time, this issue has done pioneering work related to: dynamic composite service topology supported by architecture-centric decision-making, formal integration and flexible achievement of service flow, and self-adaptive evolution routing of internetware etc. About architecture-centric mechanism, semantic logic and service composition, the design and implementation of evolution routing, relying on Natural Science Foundation of Shandong Province "Architecture-centric Topology Research and self-adaptive Evolution Routing Implementation based on Agent Internetware" and the Key Science-Technology Development Project of Shandong Province "PLM Component Library Building", this issue takes a portrait of self-adaptive architecture-centric

---

[*] This paper is supported by National Natural Science Foundation of China (No.60072014), Natural Science Foundation of Shandong Province of China (No.ZR2009GM009), the Key Science-Technology Development Project of Shandong Province of China(No. 2009GG10001014) and Promotional Foundation (2005BS01016) for Middle-aged or Young Scientists of Shandong Province, SRI of SPED(J07WH05), DRF and UF(3XJ200903, XJ0609)of QFNU.
[**] Corresponding author.

evolution's research about service composition, and aims to solve the overall deployment of business process in architecture-centric environment and self-reflect of local granularities.

Using calculus composition adapts to describe dynamic business logic of π-calculus[9], the dynamic calculus is expanded based on acE service flow, and to achieve opening flexible platform and goal-controlled composite service migration process. Finally, the engine decision-making and planning structure will be analyzed and argued during acE service flow implementation and dynamic service composing evolution, to further verify the validity and feasibility of this research project.

2 Related Work

Reference[1] focuses on dynamic evolution of composite service to ensure the gotten architecture goal, it presents a reasonable evolution operating set, further, it gives availability-oriented composite service evolution method, and designs an online service instance migration algorithm, then achieves the service implementation engine to support dynamic evolution and effective credible intention implementation pre-condition for service composition reliability.

Reference[6] thoroughly discusses evolution accuracy of service composition process, proposes verification topology logic about process structure's liveness and boundedness, flexibly plans and selects goal-oriented specific component service. Through the concept and solution of change regions, it achieves some specific running instance's migration of some specific flow structure.

Through membrane activities, Reference[9, 10] expands a kind of Mπ-calculus to express location concept for process interaction and message transferring, corresponding operation semantics is defined to structure synchronization, and in-depth analyzes behavior bisimulation, turns out that the weak transaction equivalence is a bisimulation relationship, that point of view supports dynamic rule-driven relationship for main architecture decision-making adjustment in service compositions.

Reference[11] expands π-calculus for transaction bisimulation relationship, in order to describe natural semantic of transaction in compensable relationship and not to add any new operation calculus, it introduces "Transaction Membrane" to express transaction scope, and achieves communication among processes in different scopes, the whole process is Membrane Activity process. From the semantic point of view, membrane activity can flexibly describe the constant changing process with different transaction positions, that is associated with π-calculus process and more friendly in multi-layer semantics.

3 Service Composition Evolution Sequence of acE Service Flow

Architecture-centric concept and service sequence evolution composed by membrane is the key technology to acE service flow. About describing service composition and sequence evolution, optimal workflow net is very appropriate[14], extended to acES-WFN.

3.1 acES-WFN

acES-WFN=$(P, SM, SFTh, i, o, SM_0)$, if is to meet architecture-centric adjustment and orthogonal process thread, iif:
- Service sequence membrane meets OISR;
- For every SM starting from initial service membrane SM_0, there exists a gotten sequence from SM to end membrane SM_e, formal as:

$\forall SM(SM_0 \xrightarrow{*} SM) \Rightarrow (SM \xrightarrow{*} SM_e)$.

- No matter what service membrane process, if the process can reach SM_e, the initial node is only SM_0, formal as:

$\forall SM(M_0 \xrightarrow{*} SM \wedge SM \geq M_e) \Rightarrow (SM = SM_e)$.

- P in aces-WFN can have idle SM_i, but not have dead SM_d, expressed as: $\forall r \in acR$ (acR is Architecture-centric decision-making rules, "*" stands for rule sub-sets of some acR), $\exists SM, SM'$, meet $s.t.SM_0 \xrightarrow{*} SM \xrightarrow{r} SM'$

- Service sequence thread $SFTh ::= (P_1, P_2, ..., P_n) \times (SM_0, ..., SM_i, ..., SM_e)$

- SM is the extended by place, i and o is the basic identity of place, that represent initial and final node, that is, $\forall SM$, there is a initial place $^{\bullet}i = \phi$ and a final place $o^{\bullet} = \phi$, $SM ::= (i, p_1, p_2, ...p_i, ...o) \times (t_1, t_2, ..., t_i, ...t_n)$, and $t_i \in acR$.

acES-WFN is extended by $S-WFN$ [13], its operation logic meet Replacement Operation and Addition Operation etc[1], according to the different character of addition actions, Addition Operation is divided into Sequence add($S-WFN_1 \rightarrow S-WFN_2$), Parallel_add($S-WFN_1 \parallel S-WFN_2$) and Choice_add($S-WFN_1 + S-WFN_2$), their specific definitions, extending processes and corresponding formal sequence structures' description, is given by Reference[1], by the operation and adjustment, it can keep the rationality and stability of a $S-WFN$.

3.2 SM Service Sequence Transferring

$S-WFN \xrightarrow{Extends} acE-WFN$, During meeting these basic condition of acE-WFN, to accept acR deployment and achieve orthogonal $SFTh$ needs routing transferring of membranes.

The rational workflow process of acES-WFN=$(P, SM, SFTh, i, o, SM_0)$ embodies, the routing transferring of SM sequence from SM_0 to SM_e is a transferring sequence of acES-WFN evolution, it is named transferring trace, expressed as smP, $(SM_1, SM_2, ..., SM_x) \times acR \xrightarrow{Evoluted} smP$. smP is adjusted by acR and recorded in amP_List.

The routing transferring of smP meets:

- $smP = SM_0 SM_1 ... SM_e \in amP\_List$, SM_m and SM_n are any two membranes of smP, meet $0 \leq m < n$, $m+1 \leq n \leq e$, SM_m is the predecessor membrane of SM_n, SM_n is the successor membrane of SM_m, denoted by $SM_m \prec_{smP} SM_n$.

- $\forall smP \in amP\_List$, if SM_m and SM_n all appear in smP, meet $SM_m \prec_{smP} SM_n$, SM_m and SM_n are the predecessor sand successor relationship, denoted by $SM_m \triangleright_{smP\_List} SM_n$.

- $\exists smP, smP' \in amP\_List$, if $SM_m \prec_{smP} SM_n$, $SM_n \prec_{smP'} SM_m$, then SM_m and SM_n occur concurrent sequence implementation relationship, that is: $SM_m \parallel_{smP\_List} SM_n$.

- $\forall smP \in smP\_List$, if $SM_m \in smP$, $SM_n \in smP$, 但$SM_m \prec_{smP} SM_n$ and $SM_n \prec_{smP'} SM_m$ all don't meet, SM_m and SM_n exist predecessor and successor relationship, it is the "Isolated membrane activity", about amP_List, SM_m and SM_n are only participation and selection relationship, that achieves orthogonal membrane, denoted by $SM_n \propto_{smP\_List} SM_n$.

4 The Flexible Transferring of Service Instance

To ensure the effectiveness of evolved instance, the reasonable inputting and outing data of acES_ $EvoSelection$ () is the selection and composition of process about membrane, the algorithm is to achieve the gotten membrane states for architecture-centric goals, that is the effective transferring of state tracing.

acES-WFN^0 = $(P^0, SM^0, SFTh^0, i^0, o^0, SM_0^0)$ and acES-WFN^e = $(P^e, SM^e, SFTh^e, i^e, o^e, SM_0^e)$ are two orthogonal evolving and evolved service flow, SM^0 is the initial membrane instance driven by acR in acES-WFN^0, they finish the OISR OF acE membranes, and finish the effective transferring of $SM_0^0 \xrightarrow{*} SM_0^e$, iif:

(1) SM^e is the effective gotten membrane activity.

(2) There is the OISR from SM_0^e to SM^e in acES-WFN^e.

(3) For every acES-WFN, the transferring of the entire membrane states can reach SM^e, and finish the service sequence from P^0 to P^e, transferred tracing can not repeat the implemented transferring.

The first condition achieves membrane in P meets reachable and terminative; The second one ensures the membrane of corresponding OISR in P can be implemented during transferring; the third means that finished P does not have idle membrane to waste the resource of amP after transferring is finished.

Flexible transferring must achieve the reachable and terminative process of membrane states, its implementation is:

$SM\_GottenTrans()$

{ *Input* :

acES-WFN$^0 = (P^0, SM^0, SFTh^0, i^0, o^0, SM_0^0), P, acR, comP, comSFTh$

Input : acES-WFN$^e = (P^e, SM^e, SFTh^e, i^e, o^e, SM_0^e)$

Output : SM^e

$TransPath(acES-WFN^e)$ // Get service routing of acES-WFN^e, and achieve the effective set of $SFTh$ and P.

$Set\_P' = \phi$

$TransPath(acES-WFN^e) \xrightarrow{Split\_P} Set\_P$ // Capture service sequence set.

$TransPath(acES-WFN^e) \xrightarrow{Split\_STh} Set\_STh$ //Capture orthogonal thread set.

for $\forall\ p \in Set\_P$ // p is the basic service granularity in P.

{ $_{acR}P = P \setminus (SM^e \setminus SM^0)$ // It is the composite process of evolving transferring about membrane.

$_{acR}Set\_P' = Set\_P' \cup P$ }// $comP$ is achieved by a cycle operation.

$comP = comP \setminus (SM^0 \setminus SM^e)$ // It is the adding and deleting process of finished evolving transferring for membrane.

for $\forall\ SFTh \in Set\_SFTh$// $SFTh$ is the basic service sequence thread.

{ $_{acR}SFTh = SFTh \setminus (SFTh^e \setminus SFTh^0)$ // It is the composite process about corresponding service threads.

$Copy(SFTh', SFTh)$

$_{acR}Set\_SFTh' = Set\_SFTh' \cup SFTh$

$Orthogonal(_{acR}Set\_SFTh')$ }// Evolution thread is achieved by cycle operation.

$comSFTh = comSFTh \setminus (SFTh^0 \setminus SFTh^e)$ //It expresses the adding and deleting process of orthogonal threads.

}

Evolving transferring is as in Fig.1, Fig.2 is the evolving process of orthogonal thread. Fig.1 and Fig.2 express the process and result of algorithm $SM\_GottenTrans()$.

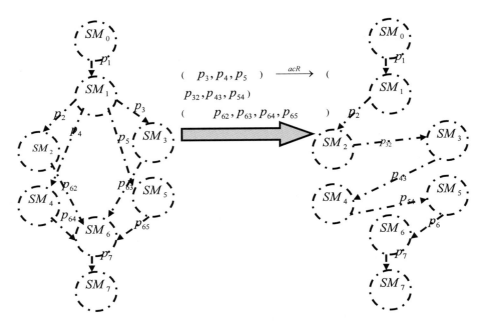

Fig. 1. Selection and Reaching Process of Membrane Evolution

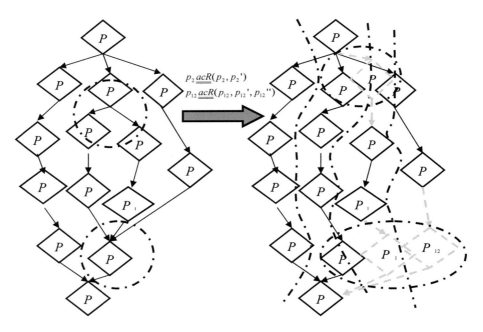

Fig. 2. Evolving Process of Orthogonal Thread

5 The Platform's Building for Dynamic Evolving Service Composition

Based on the above analysis and argumentation, the service's selection and composition platform is built to support dynamic evolving of acE service flow, self-adaptive service-oriented architecture adjustment is the control center, shown in Fig.3.

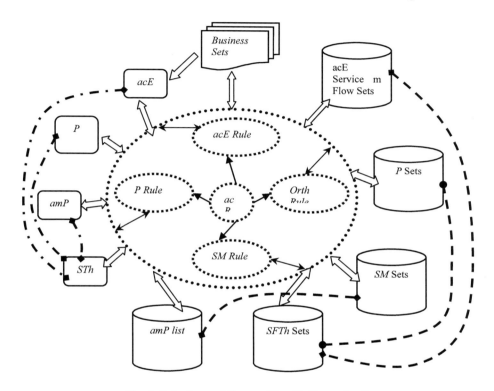

Fig. 3. Selection and Composition Platform of acE

Evolving operation set is ensured to be the basic participation granularity for evolution strategy, the platform configuration is the formal storage management, it is to achieve the center rule "Cycle structure that acR is the core adjustment, the corresponding structure includes service sequence rule(P Rule), acE composition selection rule(acE Rule), Orthogonal thread implementation rule(Orth Rule) and service Membrane Rule(SM Rule),this is the "decision-making logic mechanism" from business requirement to finishing architecture goal of acE service flow. Starting from Business set, "decision-making logic mechanism" achieves acE service flow sets, P Sets, SM Sets, SFTh Sets and amP list. The intention "Cycle" in "decision-making logic mechanism" completes flow structure's composition and separation, effectively controls flow granularities in different stages, achieves selection and composition of service process, these logical internal relationships and flow process, decision-making, sequence process, all meet semantic granularity's description.

References

1. Zeng, J., Sun, H.-L., Liu, X.-D., et al.: Dynamic Evolution Mechanism for Trustworthy Software Based on Service Composition. Journal of Software 21(2), 261–276 (2010)
2. Song, W., Ma, X.-X., Lv, J.: Instance Migration in Dynamic Evolution of Web Service Compositions. Chinese Journal of Computers 32(9), 1816–1830 (2009)
3. von de Aalst, W.M.P., Jablonski, S.: Dealing with workflow change: Identification of issues and solutions. Int'l Journal of Computer System Science & Engineering 15(5), 267–276 (2000)
4. Papazoglou, M.P.: The challenges of service evolution. In: Bellahsene, Z., Leonard, M. (eds.) Proc. of the 20th Int'l Conf. on Advanced Information Systems Engineering, pp. 1–15. Springer, Heidelberg (2008)
5. von de Aalst, W.M.P., Basten, T., Verbeek, H.M.W., Verkoulen, P.A.C., Voorhoeve, M.: Adaptive workflow-On the interplay between flexibility and support. In: Filipe, J., Cordeiro, J. (eds.) Proc. of the 1st Int'l Conf. on Enterprise Information Systems, pp. 353–360 (1999)
6. Rinderle, S., Reichert, M., Dadam, P.: Correctness criteria for dynamic changes in workflow systems-A survey. Data & Knowledge Engineering 50(1), 9–34 (2004)
7. Andrikopoulos, V., Benbernou, S., Papazoglou, M.P.: Managing the evolution of service specifications. In: Bellahsene, Z., Leonard, M. (eds.) Proc. of the 20th Int'l Conf. on Advanced Information System Engineering, pp. 359–374. Springer, Heidelberg (2008)
8. Ryu, S.H., Casati, F., Skogsrud, H., Benatallah, B., Saint-Paul, R.: Supporting the dynamic evolution of Web service protocols in service-oriented architectures. ACM Trans. On the Web 2(2), 1–46 (2008)
9. Milner, R., Parrow, J., Walker, D.J.: A calculus of mobile process. Part I/II[J]. Journal of Information and Computation 100(1), 1–77 (1992)
10. Yuan, M., Huang, Z., Cao, Z., et al.: An Extended π-Calculus and Its Transactional Bisimulation. Journal of Computer Research and Development 47(3), 541–548 (2010)
11. Wang, J.-S., Li, Z.-J., Li, M.-J.: Compose Semantic Web Services with Description Logics. Journal of Software 19(4), 957–970 (2008)
12. Bukhres, O., Elmagarmid, A., Kuhn, E.: Implementation of the flex transaction model. IEEE Data Engineering 16(2), 28–32 (1992)
13. Zhang, A., Nodine, M., Bhargava, B.: Global scheduling for flexible transactions in heterogeneous distributed database systems. IEEE Trans. on Knowledge Data Engineering 13(3), 439–450 (2001)
14. von de Aalst, W.M.P., von Hee, K.: Workflow Management Models, Methods, and Systems. The MIT Press, Cambridge (2002)

Design of Intelligent Manufacturing Resource Planning III under Computer Integrated Manufacturing System

Baoan Hu and Min Wu

School of Science and Technology, Nanchang University,
Nanchang, 330029, China
85303178@qq.com, 515038795@qq.com

Abstract. In this paper, the study of MRPIII under CIMS is discussed based on practice. The whole design idea of the system is introduced in detail. Its function and the network design are involved. At the end, the whole system is presented.

Keywords: Computer Integrated Manufacturing, Intelligent Manufacturing Resoure Planning (MRPIII), Design system.

1 Introduction

Computer Integrated Manufacturing System (CIMS) is a high technology in manufacture project. It gathers many subjects, such as Systems Engineering, Management Engineering,Computer Science and Modern Mechanical Manufacturing, accordingly, it forms a integrated enterprise network controlled by computer. This network involves market analysis,production decision, product design, process planning, product manufacturing and the selling. It combines the information management system with the control system of the enterprise in a whole. With the development of the modern science and technology, CIMS has become the focus of the research of the manufacturing in the world nowadays.

Manufacturing Resource Planning (MRPII) is a closed loop production management system centers on Material Requirements Planning (MRP). It manages and controls all the manufacturing resource of the enterprises, such as manpower, equipment, materiel and capital. It extends market forecast, production schedule, material requirements, capacity requirements, stock control and workshop control to the vendition of the product, which is the end of the whole production course. When the changing market needs the adjusting of plan, we can use the simulating method to make rapid decisions and managements for all kinds of resource of the enterprises for the purpose of the best management effect. Actually, MRPII is a effective management system based on the information technology and the theory of Management. As a extensive resource coordination system, it stands for a new kind of idea of production management and is a new manner of production organization. Based on MRPII , Intelligent Manufacturing Resource Planning (MRPIII) combines the intelligentized technology, such as Expert System (ES) and Just In Time Technology (JIT), with MRPII and forms a new system (MRPIII=MRPII+ES+JIT). It can implement intelligentized production management and orient the production process of the enterprises.

2 Design of MRPIII

2.1 General Design Idea

Based on the request of the production management and the implementation of the enterprise CIMS , MRPIII is put forward. It can meet the requirements of the intelligentized decision of the enterprise manager, ensure the validity of the production decision of the enterprise,improve the production management of the enterprise and supply the enterprise with a exact, fast, convenient and effective software for the management of the resources. At the same time, it supplies exact and perfect data for the enterprise, making it convenient for the uniform implementation of the management and control of information for the total enterprise.

The whole framework of MRPIII is shown in Fig.1:

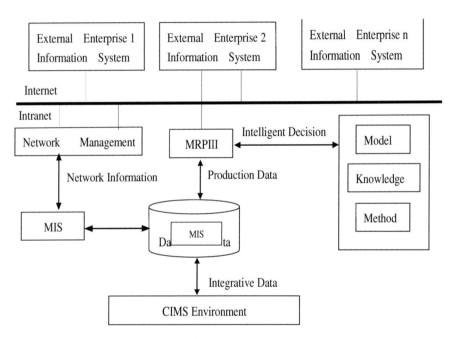

Fig. 1. The Whole Framework of MRPIII

MRPIII is located in basic CIMS environment. It can make intelligentized capacity analysis about production process within certain resources restriction, regulate the balance of capacity, support the intelligent decision of load, offer the inquiry about purchasing plan and the intelligentized choice for providers. It can also regulate the production schedule, offer the inquiry into the information about the rate of production progress and analyze the risk of production.

In the CIMS environment that MRPIII is located, the information system of each external enterprise adds the communion between the network management system and the data information(consists mostly of the rate of production progress and purchasing information)of other external exterprises. Internet and Intranet connect evry enterprise network management system or directly from with each other, which realizes the communications between the MRPIII of enterprises and external information system. With a certain communication protocol, the network management system manages communication process, deals with communication conflict and makes the communication known for users and each information system. At the same time, MRPIII obtains basic correlative data from the MIS of the enterprise, with which it can offer the information support and the function support for the decision-maker of the enterprise . The corresponding model storage, knowledge storage and method storage offer intelligence support for production decision.

Database manages and controls internal information that is from MIS, MRPIII and the like, and external information that is from the the information system of external enterprise. It can supply the whole CIMS with integrated data information.

2.2 System Function Design

MRPIII introduced in this paper is a computer-aided enterprise management system that takes planning control as its mainbody. It emphasizes the logic of the combination of MRP and JIT, which is just the emphasis point of this paper. According to the system implementation demand , this system can be divided into five sub-modules on the principle of system optimization :
(1) basic information management
(2) implementation information inquiry
(3) order bargaining decision
(4) material requirements planning
(5) purchasing bargaining decision

Detailed function framework is shown in Fig.2 :

In Fig.2, the sub-modules of material requirement planning includes: master production schedule, material requirement planning, balance of production capacity, adjustment of material requirement planning and inquiry about MRP report. This system covers every product and management department inside the enterprise and it carries out automatic and uniform management. It provides memory, analysis, updating, inquiry, optimization, statistics, report and printing for the information of the enterprise. The information can be production schedule, production process management, raw material purchasing, stock management and the selling. This system can run reliably and has less maintain problem. There are some uniform standard interfaces between sub-systems,which make it easy to expand the functions.

2.3 Application Network Design

2.3.1 Network System Design Demand

Taking the actuality of ordinary enterprises into account, we put forward the following demands for network design for the purpose of meeting every above request about the construction of MRPIII:

1.Supporting the network computer pattern of the combination of client /server and Intranet;

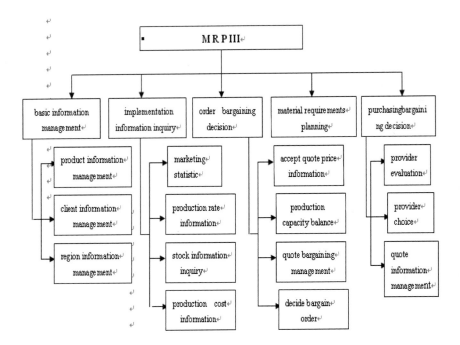

Fig. 2. The Framework of Whole Function of MRPIII

2. Regarding TCP/IP as common communication protocol, and at the same time, supporting NETBEUI and IPX/SPX;
3. Supplying each function department of the enterprise with information service, sharing network resource, such as compute, memory, peripheral equipment, software and data. With Web browser, every department can inquire about the documents promulgated by the leader of the factory and release his own information, and the leader can inquire about the information of every department too;
4. Realizing concentrative network management, ensuring the flexible and safe running of the network, and supporting the dynamic construction of the multi-departmental project group.

2.3.2 Network Integrated Wire Layout

The system of wire layout is the basal establishment and underlaying circumstance of the construction of the total network. Whether the construction is good or not will directly influence the quality of communication and the stability and reliability of the performance of the total enterprise network.. Therefore, the system of wire layout should conform to the need of the information network development and take the computer network into a whole and long consideration. The structure wire layout should be adopted. We adopt the system of compositive wire layout of AMP Co. in America as our basal wire layout that supports various network communication.

2.3.3 Network Hardware Configuration Scheme

According to the demand of the above design idea and the enterprise development on the basis of the design standard about high ratio of performance to price, we present the following solution about the network hardware.

Concrete configuration is as follows:

The whole system adopts ethernet network as backbone network.
(1) Server : HP series, for its good ratio of performance to price; HPLC3PIII550, D8594A, 128M, 9.1G hard disk;
(2) HUB : Dlink916, 10/100M,16XRJ45;
(3) Desktop computer network card : 530TX, 10/100M, RJ45, PCI network card;
(4) System of wire layout : twisted-pair of AMP Co.;
(5) Client (terminal) : PIII-500, 64M compatible computer.

The configuration of the system software :
Network operating system : WINDOWS NT SERVER4.0 ;
Desktop operating system : Windows98 ;
Database Management System : SQL Server7.0 ;
Front end developing language : PowerBuilder 6.0 ;

3 Conclusion

In the present market, many clients require the enterprises to provide a variety of products, accept a small quantity of order and allow the change of the standard or the performance of the product, but the traditional enterprise can't meet these demands. In this paper, we present a manufacturing resource management system centers on MRPIII, which can optimize the design of the production schedule, supervise the implementation of the production schedule, carry out scientific stock management and make it convenient to realize the uniform implementation of the information management and control for the total enterprise. As is proved in practice, the implementation of MRPIII can lead to the following results: the cost decreases by about 5%, the production cycle condenses by about15% and the stock falls by 15-20% in the economic batch production.

References

1. Meger, A.D.: The integration of information system in manufacturing. OMEGA (3), 25–27 (1987)
2. Bolwinjin, P.T., Boorsma, J.: Flexible manufacturing integrated technological innovation. Robotics and Computer Integrated Manufacturing (2), 89–99 (1986)
3. Loos, P., Allweyer, T.: Application of production planning and scheduling in the process industries. Computers in Industry (36), 199–208 (1998)

The Design of Improved Duffing Chaotic Circuit Used for High-Frequency Weak Signal Detection

Wenjing Hu, Zhizhen Liu, and Zhihui Li

School of Electrical Engineering of Shandong University, Jinan, China, 250061
hwj@sdu.edu.cn

Abstract. As the existing Duffing chaotic circuit is only fit for low-frequency weak signal detection, this paper proposes an improved Duffing chaotic detection circuit, which overcomes the problem of instant saturation caused by high magnification and makes the Duffing chaotic circuit suitable for high-frequency weak signal detection. Adjusting the parameters of the circuit appropriately, the circuit can realize the weak signal detection ranged from low to high frequency. The performance of this circuit to detect high-frequency weak signal has been studied by simulation of Multisim. The simulation result shows that the circuit is very sensitive to the initial value. When a slight increase in the amplitude of the sinusoidal signal, the phase transition from the chaotic state to the large-scale periodic state takes place at once, so it can be used to detect high-frequency weak signal. In addition, we find that the circuit is made up of two low-pass filters in series. the amplitude-frequency characteristic curve of the circuit is obtained by simulation of Multisim,, and it verifies the performance that the circuit can also filter high-frequency noise.

Keywords: Chaos, Duffing chaotic circuit, High-frequency noise, Weak signal detection.

1 Introduction

At present it is very popular to use the phase transition of Duffing oscillator to detect weak signal in the field of weak signal detection. For example, many researches attempt to apply this method to metal detection, vibration measurement, diagnosis, exploration seismology and so on. Many theoretical studies have shown that this method to detect weak sinusoidal signal in noisy background is better than some traditional detection method both in accuracy and signal recognition, and related research materials become more and more[1]. However, these materials are mostly limited to Matlab simulation, not yet study the implementation of the Duffing oscillator circuit deeply. A few of EWB simulation circuits are only fit for low -frequency (ω is about 1rad/s) signal detection. For the circuit design of high-frequency signal detection, high magnification will cause the instant saturation of the circuits, and make the circuits not work. All of these will limit the application of this method[2].

In order to overcome the problems mentioned above, we propose an improved Duffing chaotic circuit, which avoids the instant saturation of the circuits. If adjusting the circuit parameters appropriately, the circuit can realize the weak sinusoidal signal detection ranged from low to high frequency. Through the simulation by Multisim, we make intensive study about the critical threshold of the circuits and its performance of weak signal detection under high-frequency. Through the analysis of the circuit structure, we also find that the circuit is made up of two low-pass filters in series.

2 Duffing Oscillator and Its Circuit Implementation

2.1 The Theory of Weak Signal Detection by Duffing Chaotic System

We discuss the Duffing equation which is usually used in chaotic weak signal detection. Its state equations are expressed as:

$$\begin{cases} \dfrac{dx}{dt} = y \\ \dfrac{dy}{dt} = \sqrt{2}r\sin(\omega t) - 0.5y + x^3 - x^5 \end{cases} \quad (1)$$

ω is the frequency of the to-be-detected signal. The detection principle of the system is that the weak sinusoidal signal can make the system change from chaotic state to large-scale periodic state even in the noisy background. According to the Minikev-method, The stable and unstable manifolds in Poincare map of equation (1) will intersect only in the case of low-frequency, so the chaotic solution may appears. In order to detect higher frequency signals, we make a time-scale transformation, and get new dynamic equations as follows

$$\begin{cases} \dfrac{dx}{dt} = \omega y \\ \dfrac{dy}{dt} = \omega(\sqrt{2}r\sin(\omega t) - 0.5y + x^3 - x^5) \end{cases} \quad (2)$$

This system still has the critical phase transition from the chaotic state to the large-scale periodic state even in high-frequency. In the new phase space $(x\text{-}y)$, the phase velocity of x,y also increases ω times, so we can change the ω in equation (2) to adapt to different frequency detection. In order to make the circuit easily, we rewrite the equation (2) in integral form

$$\begin{cases} x = \int \omega y \, dt + C_1 = \omega \int y \, dt + C_1 \\ y = \int \omega(\sqrt{2}r\sin(\omega t) - 0.5y + x^3 - x^5) dt + C_2 \\ = \omega \int (\sqrt{2}r\sin(\omega t) - 0.5y + x^3 - x^5) dt + C_2 \end{cases} \quad (3)$$

C_1 and C_2 are integral constants. If $x(0)=0$, $y(0)=0$, then C_1 and C_2 are zero. In order to realize equation (3) by circuits, Integrators must be used. When equation (3) is realized by circuits, if equation (3) is multiplied by ω ($\omega=10^6$rad/s here)at first, then integrate, to meet the function of the equation (3), magnification of the amplifier must be $k=10^6$, while the DC bias power source of the circuit is only -15v and +15v, the circuit is saturated instantly. To solve this problem, we design the operation circuit to finish integration first and then multiplication. Thus this method can avoid the instantaneous saturation of the circuit, make the circuit work properly., and make high-frequency signal detection possible. This is the innovation of this paper, as we know, no related literatures proposed it. The existing EWB simulation circuits are mostly designed based on equation (1) only fit for low-frequency signal detection even though the parameters of these circuits are adjusted, and they are only made by virtual, not actual devices[3].

2.2 The Design of Duffing Chaotic Circuit

Equation (3) can be realized by sinusoidal voltage sources, analog operational amplifiers, analog multipliers, resistors and capacitors. The Duffing chaotic circuit designed here can detect weak signals ranged from low to high frequency.

In order to see the operation relationship of equation (3) clearly, first, we give the operation block diagram shown as in Fig. 1. R_1C_1 is the time constant of the inverting differential integrator made up of operational amplifier. R_8C_2 is the time constant of the inverting integrator. R_1C_1 and R_8C_2 are corresponding to elements in the following Fig. 3.

The integrator is usually made up of operational amplifier shown as in Fig. 2, the relationship between the input and output voltages is

$$U_0 = \frac{-1}{RC}\int U_i dt$$

The output voltage U_o is proportional to the integration of the input voltage U_i, the scale factor is $-1/(RC)$. Therefore,, considering the scale factor, R_1C_1, R_8C_2 are introduced in the operational block diagram.

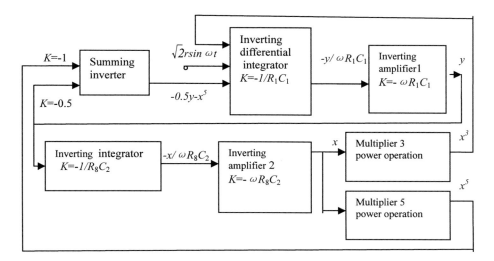

Fig. 1. Operational block diagram

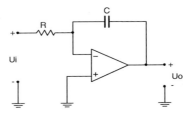

Fig. 2. Analog integrator

Fig. 3 is the Duffing oscillator electronic circuit, whose components are all actual devices. Operational amplifiers are 3554AM. The advantage of this circuit is that, by adjusting R_1C_1, R_8C_2 properly, the circuit can detect both low and high frequency signals

Here we take $\omega=10^6 \text{rad/s}$ (f=159155hz) as an example to introduce the circuit design. From Fig. 1, we know that the magnification of inverting amplifier 1 is

$$K_1 = -R_1 C_1 \omega.$$

Taking into account the performance of the operational amplifier, the closed-loop magnification of single-stage operational amplifier ordinarily is less than 100, here suppose $k_1=100$, then

$$R_1 C_1 = k_1/\omega = 100/1000000 = 10^{-4}.$$

Then follow the design method of the differential integrator[4]. Select proper R_1 and C_1, here

$$R_1 = 1\text{K}\Omega \; ; \; C_1 = 100\text{nF}.$$

So we complete parameter design of the inverting differential integrator in Fig. 1. The Magnification k_2 of the inverting amplifier 2 can be selected as 10, so we get

$$R_8 C_2 = k_2/\omega = 10/1000000 = 10^{-5.}$$

Then select appropriate R_8 and C_2, here

$$R_8 = 1\text{K}\Omega \; ; \; C_2 = 10\text{nF}.$$

So we complete parameter design of the inverting integrator in Fig. 1.

If ω is taken other values, the adjusting method is similar to the method above. By adjusting R_1C_1, R_8C_2 properly, we can detect signals of different frequencies. Other circuit components in Fig. 3 are selected, here $R_1=R_2=R_3=R_4=R_6=R_8=R_{10}=1\text{K}\Omega$, $R_7=100\text{K}\Omega$; $R_{11}=10\text{K}\Omega$; $R_{12}=20\text{ K}\Omega$; $R_{13}=R_{14}=10\text{ K}\Omega$; $R_5=R_9=10\text{K}\Omega$. R_5 and R_9 are DC negative feedback resistor used to avoid integral drift. DC bias supply: $V_2=V_4=V_6=V_8=V_{10}=-15\text{V}$; $V_3=V_5=V_7=V_9=V_{11}=15\text{V}$.

3 Circuit Simulation and Performance Analysis

3.1 Detection Performance Based on Initial Sensitivity

We conducted a simulation of the circuit in Fig. 3 by Multisim (Electronics Workbench). First set the system absolute-error-limit to $10^{-12,}$ relative-error-limit to 0.001, and the source frequency f to 159155hz. Since the source voltage in Multisim is RMS, for convenience, we choose RMS[5]. Combined circuit in Fig. 3 with operational block diagram Fig. 1, we can get node ① voltage u_1 is x and node ② voltage u_2 is y in equation (2). Connect A channel of the oscilloscope to node ①, B channel of the oscilloscope to node ②, Timing Diagrams of x, y or two-dimensional phase plane trajectories can be observed. In addition, the input signal source V1 is

$$u = \sqrt{2}r \sin \omega t$$

When r is different value from 0 to 1, timing diagram and phase orbits given by Multisim are similar to that given by Matlab. When the RMS of the source voltage is r=0.52556171v, the system performs as the critical state of chaos, shown as in Fig. 4(a). If the RMS of the signal increases only 0.00000001v to r=0.52556172v, the system changes to the large-scale periodic state from chaotic state, shown as in Fig. 4(b). So we know that the Duffing circuit is still very sensitive to the initial conditions, qualified with the prerequisites to detect weak sinusoidal signals in high-frequency. Simulation results also show that Duffing circuit has its unique features different from Matlab simulation results. Analysis is as follows:

(1) The critical threshold from the chaotic state to the large-scale periodic state is different from the value given by Matlab, there are two reasons:

1) The Numerical Methods used in Multisim are different from that in Matlab. Because of the initial value sensitivity of chaotic systems, different algorithms result in different orbits.

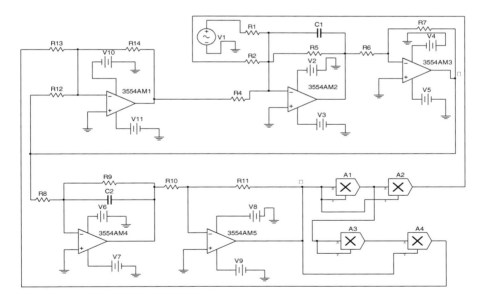

Fig. 3. Duffing chaotic circuit

2) Matlab simulation of Duffing equations is strict Numerical simulation, while the Duffing circuit is not strict Duffing equations. Because the circuit uses the actual devices of real models. Especially the operational amplifiers are real, which have the input offset voltage V_{os}, input offset current I_{os}, and input bias current I_{bs}. All of them have effects on the critical threshold. The critical threshold (amplitude) of the circuit is 0.74325649, which is a little bigger than the critical threshold 0.73111185 in Mtalab[6]. The main reason is that the open loop amplification of operational amplifiers is not infinity, resulting that the closed- loop gain of operational amplifiers slightly is smaller than the designed value, so the amplitude of the source voltage should be a little bigger to make the circuit into large-scale periodic state.

(2) The circuit has a transition process at the initial response stage, which is shown in Fig. 5. The response stage is very short, lasting about 50μs, similar to the periodic state. This process is shown in two-dimensional phase plane trajectory diagram as Fig. 4(a). The ring outside of the phase diagram is the transition process. As r increases, the ring gradually increases. At the critical state of chaos, it is very clear and is the biggest one of all.

3.2 The Study of the Performance of Dufffing Circuit in Filtering Out High-Frequency Noise

Through Analysis of the circuit structure in Fig. 3, we know that the circuit is made up of inverting integrator and inverting differential integrator. Since integrators themselves are

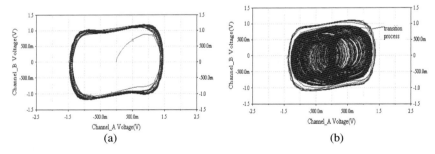

Fig. 4. The Chaotic state and the large-scale periodic state of the Dufffing circuit, (a) r=0.52556171 the chaotic critical state; (b) r=0.52556172 the large-scale periodic state

low-pass filters, so the circuit is made up of two low-pass filters in series. We can say that the Duffing chaotic circuit also has the function to filter high-frequency noise. In order to prove this function, we use the powerful simulation of Multisim and get the amplitude-frequency characteristic of the system after AC simulation of the circuit, shown as in Fig. 6. From Fig. 6 we know that the 3dB bandwidth B_0 is about 0~200kHz, the corresponding upper cut-off frequency is about 200kHz. When $f > f_0$, each 10 times increase in the frequency, the gain will be declined by about 25dB. This shows that the circuit has strong attenuation in high-frequency noise. In addition, for the second-order low-pass filter, its equivalent noise bandwidth B_e is about 0~1.22 f_0

and almost 0~244 kHz. The range of B_e is very close to the range of B_0 also shows that, in the total noise energy through the system, the energy of the high-frequency above the cutoff frequency takes a little proportion[7]. On the other hand, this phenomenon proves that the circuit has a strong inhibitory effect on high-frequency noise.

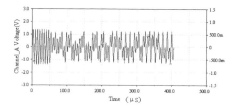

Fig. 5. Initial response of the voltage u_1

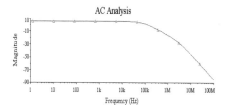

Fig. 6. Amplitude-frequency characteristic of the Duffing circuit

4 Conclusion

This paper discusses the realization of the Duffing circuit in high-frequency signal detection, overcomes the instantaneous saturation problem caused by high magnification, improves the Duffing chaotic circuit not only to fit for low-frequency signal detection but also fit for high-frequency signal detection. If the performance of operational amplifiers is good enough, this circuit can realize the broadband signal detection from low to high frequency by adjusting its parameters.

Through Multisim simulation of the circuit, we study the performance of the Duffing circuit. Simulation results show that: at high-frequency, the system still has sensitivity to the initial value and the critical phase transition from the chaotic state to the large-scale periodic state, so it can be used for the detection of high-frequency weak signals. Through the analysis of the circuit structure, we find that the circuit is made up of two low-pass filters in series. Therefore, the Duffing chaotic circuit can also filter part of the high-frequency noise. By the amplitude-frequency characteristic curve of the circuit, we prove that the circuit can filter high-frequency noise. This phenomenon is difficult to find by Matlab and other numerical simulations.

References

1. Zhang, S., Li, Y., et al.: A New Method for Detecting Line Spectrum of Ship-radiated Noise Using Duffing Oscillator. Chinese Science Bulletin 52(14), 1906–1912 (2007)
2. Wang, Y., Xiao, Z., et al.: Simulation and Experimental Study on The Chaos Circuit of Duffing Oscillator. Journal of Circuits and Systems 13(1), 133–135 (2008)
3. Gao, J., Yang, H.: The analysis of Operational Amplifiers Application, pp. 87–89. Beijing Institute of Technology Press, Beijing (1989)
4. Li, Z., Luo, L.: Basis of Electronic Technology and Its Applications, pp. 51–53. Higher Education Press, Beijing (2003)
5. Xiong, W., Hou, C., et al.: Mutisim7 Circuit Design and Simulation Application, pp. 71–72. Shanghai Jiaotong University Press, Shanghai (2005)
6. Hu, W., Liu, Z.: Study of Metal Detection Based on Chaotic Theory. In: 8th World Congress on Intelligent Control and Automation, pp. 2309–2314. Shandong University, Jinan (2010)
7. Gao, J.: Detection of Weak Signals, pp. 41–45. Tsing Hua University Press, Beijing (2004)

A Method of Setting Radio Beacon's Frequency Based on MCU

Yongliang Zhang, Jun Xue, Wenhua Zhao, and Xinbing Fang

China Satellite Maritime Tracking and Controlling Department,
Jiangyin, Jiangsu Province, 214431, China
zylelet@gmail.com

Abstract. The paper proposed a method of designing radio beacon by using Micro-programmed Control Unit (MCU) for convenience and agility of setting the beacon's frequency. It uses the MCU, which has Universal Serial Bus (USB) function controller inside, to communicate with Personal Computer (PC) by USB interface. The software of PC and MCU are designed and the beacon's frequency may be set by PC online or by MCU offline. The tests confirm that the design is correct and reliable and meets the needs of real applications.

Keywords: radio beacon; C8051F320; USB communication; frequency setting.

1 Introduction

Radio beacon is an important part of the inspection system for radio equipment technical status. Generally, it is true to use different beacons for different radio equipment and use different beacons for different frequencies of radio equipment, which results of high cost of equipping beacons and heavy workload of managing them. In recent years, a multi-frequency beacon was developed and the above problems are solved. But the firmware of the beacon has to be refreshed when new frequency is set. The operation is complex.

The paper proposes a new design method of setting radio beacon's frequency by using Micro-programmed Control Unit (MCU). With Universal Serial Bus (USB) function controller inside, the MCU communicates with the host (Personal Computer, PC) through USB interface. The software of the host and the MCU are designed. Then it is implemental that the host sets the frequency of a radio beacon online and the MCU does it offline. Thus, when setting a new frequency of radio beacon, an operator just runs the user-friendly software on the host. It is convenience and needs no more study.

2 Silicon Labs' C8051F320

Silicon Labs' C8051F320 [1] are used in the radio beacons, which are fully integrated mixed-signal System-on-a-chip MCUs. The block diagram of C8051F320 is shown in Fig. 1. The main concerned features are

- High-speed pipelined 8051-compatible micro-controller core (up to 25MIPS)
- In-system, full speed, non-intrusive debug interface (on-chip)
- USB function controller with 8 flexible endpoint pipes, integrated transceiver, and 1kB FIFO RAM
- Supply Voltage Regular (5V-to-3V)
- Precision programmable 12MHz internal oscillator and 4x clock multiplier
- High accuracy programmable 12MHz internal oscillator and on-chip clock multiplier
- In-system programmable 16kB Flash
- 2304 bytes internal RAM (256+1k+1k USB FIFO);
- On-chip Power-On Reset, VDD Monitor and Missing Clock Detector

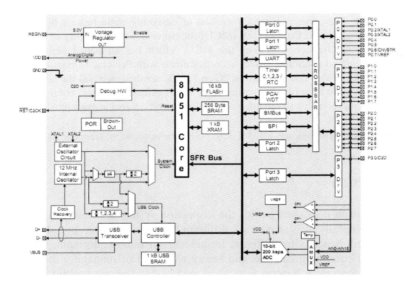

Fig. 1. C8051F320 block diagram

C8051F320 devices are truly stand-alone System-on-a-chip solutions. The Flash memory also has the ability to reprogram the system and can be used for non-volatile data storage. It is implemental to field upgrade the firmware of 8051. User software has complete control of all peripherals and you can turn off any or all of the peripherals to conserve power. The on-chip Silicon Labs 2-Wire (C2) Development Interface allows non-intrusive (uses no on-chip resources), full speed, in-circuit debugging using the production MCU installed in the final application.

On-chip USB Controller is a USB 2.0 compliant Full or Low Speed function with integrated transceiver and endpoint FIFO RAM. It is one of key features of C8051F320 devices and makes it easy and quick to develop USB communication applications.

3 System Design of Setting Frequency

The system hardware of setting beacon's frequency consists of a host (PC) and a set of MCU circuit, as shown in Fig. 2. The host calculates the value of frequency control registers (FCRs) according to the input frequency and sends the data to MCU through USB interface. The MCU receives the data and writes them into FCRs, then outputs the data signal to frequency generator (such as Frequency Synthesizer). Thus setting frequency online is realized. At the same time, the MCU writes the data into Flash. Therefore the MCU may read the data from Flash, write them into FCRs and output the data signal when the radio beacon is offline. Thus setting frequency offline is realized.

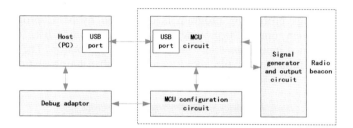

Fig. 2. System hardware of setting beacon's frequency

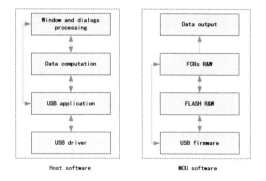

Fig. 3. System software of setting beacon's frequency

The system software of setting beacon's frequency consists of the host software and the MCU software, as shown in Fig. 3. The software are developed by using C language in Microsoft VC ++ 6.0 and Silicon Laboratories IDE V3.84.

According to function, the host software are assigned as window and dialogs processing, data computation, USB applications and USB driver. Window and dialogs processing program generates a graphical user interface, responses controls' operation and displays running status and About dialogs. Data computation program calculates FCRs' values by using input frequency. USB application program communicates with the MCU. The MCU software are assigned as USB firmware, FCRs reading and writing (R&W), Flash reading and writing (R&W), and data output. USB Firmware communicates with the host. Data output program writes FCRs' data to the predefined ports and displays beacon's running status. USB application program and USB firmware use some USBXpress API functions. And USB driver are USBXpress Driver.

4 Software Design of the System

The design of the system software focuses on the programs of window and dialogs processing, data computation, MCU application and USB communication.

4.1 Window and Dialogs Processing Program

Window and dialogs processing program is developed by MFC programming, as shown in Fig. 4. when running. It provides USB device selection list, "update device list" button, frequency input box, "Set" button and "Close" button. The program also provides response functions of the above controls with running status dialogs and About dialog.

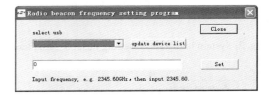

Fig. 4. Graphics User Interface of host software

4.2 Data Computation Program

Data computation program calculates the values of FCRs by using input frequency. For example, using Silicon Labs' Frequency Synthesizer Si4136 [2] as frequency generator, the calculating equation is

$$f_{out} = (2N/R) \bullet f_{ref} \qquad (1)$$

where f_{out} is the signal's frequency of beacon's output, f_{ref} is the reference frequency, N and R are the divided parameters needed to calculate.

Si4136 needs to identify nine FCRs' values based on N and R. Therefore, the data computation program has to determine N and R. While f_{ref} depends on oscillator selection, for convenience of computation, R is fixed and N is calculated by

$$N = 0.5 R f_{out} / f_{ref} \qquad (2)$$

4.3 MCU Application Program

The MCU receives the host's data and sets beacon's frequency online or offline. When setting the frequency online, the MCU writes the data into FCRs and Flash in USB interrupt service program (ISP). When setting the frequency offline, The MCU is powered on, self-loaded, and initiated, then reads the data from the Flash and writes them into FCRs. The main program's and the ISP's routines are shown in Fig. 5.

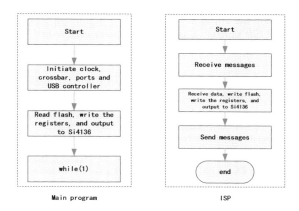

Fig. 5. MCU application program

4.4 USB Communication Program

USB communication program are developed by using Silicon Labs' USBXpress Development Kit [3]. It consists of three components, namely USB firmware, USB application and USB driver.

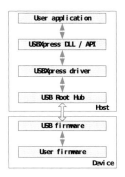

Fig. 6. USBXpress data flow

The Silicon Laboratories USBXpress Development Kit provides a complete host and device software solution for interfacing Silicon Laboratories C8051F32x, C8051F34x, and CP210x devices to the Universal Serial Bus (USB). No USB protocol or host device driver expertise is required. Instead, a simple, high-level Application Program Interface (API) for both the host software and device firmware is used to provide complete USB connectivity. The USBXpress Development Kit includes Windows device drivers, Windows device driver installer, host interface function library (host API) provided in the form of a Windows Dynamic Link Library (DLL), and device firmware interface function library (C8051F32x and C8051F34x devices only). The included device drivers and installation files support MS Windows 2000/XP/Server 2003. USBXpress data flow is shown in Fig. 6.

It is quick to develop the applications of USB communication between the host and the MCU by using USBXpress. The API functions called in USB application and USB firmware are shown in Fig. 7, which are described in details in Reference [3].

SI_GetNumDevices()
 //Returns the number of devices connected
 SI_GetProductString()
 //Returns a descriptor for a device
 SI_Open()//Opens a device and returns a handle
 SI_Close()//Cancels pending IO and closes a device
 SI_Read() //Reads a block of data from a device
 SI_Write() //Writes a block of data to a device
 SI_FlushBuffers()//Flushes the TX and RX buffers
 SI_SetTimeouts()
 //Sets read and write block timeouts
 SI_GetTimeouts()
 //Gets read and write block timeouts
 SI_CheckRXQueue()
 //Returns the number of bytes in RX queue
 API functions called in USB application
	USB_Clock_Start() //Initializes the USB clock
	 USB_Init()//Enables the USB interface
	 Block_Write()//Writes a buffer of data to the host
	 Block_Read()//Reads a buffer of data from the host
	 Get_Interrupt_Source()
	 //Indicates the reason for an API interrupt
	 USB_Int_Enable() //Enables the API interrupts
	 USB_Int_Disable() //Disables API interrupts
	 USB_Disable() //Disables the USB interface
	 USB_Suspend() //Suspend the USB interrupts
	 USB_Get_Library_Version()
	 //Returns USBXpress firmware library version
	
	 API functions called in USB firmware

Fig. 7. API functions in USBXpress Development Kit

It is necessary to add SiUSBXp.h, SiUSBXp.dll and SiUSBXp.lib files provided by USBXpress to the host software project when designing USB application program. After installed the USB driver, the host software allow the user to observe and change the status of C8051F320 device I/O peripherals. When the device interfaces with the host through USB ports, USB application begins to enumerate the devices and shows them in device list. Then the specified device is selected and data are transferred between the host and the device.

USB application program is called in the response function of "Set" button. After the button is clicked, at first the variable corresponding to input frequency edit box is updated, then data computation program is called to calculate the values of FCRs, finally USB application program is called to transfer the data between the host and the MCU.

It is necessary to add C8051F320_defs.h and USB_API.h files provided by USBXpress to the MCU software project when designing USB firmware program. After USB clock and interface are initialized, USB firmware program is called in USB ISP to transfer the data between the host and the MCU.

5 Tests and Conclusion

The hardware and software of radio frequency (RF) beacons are developed by using the method. The running status of the host software is shown in Fig. 8. while the output signal of the beacon is shown in Fig. 9. The beacon accesses the host's USB port and is shown in device list, namely "xb01". Input frequency is 2280.80MHz and the signal with the frequency is observed by the spectrum analyzer as soon as the "Set" button is clicked. While the beacon is removed from the host and restarted, the same signal is also observed. And with long time observation the frequency of the signal is stable.

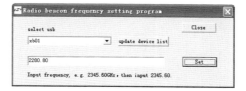

Fig. 8. Running status of the host software

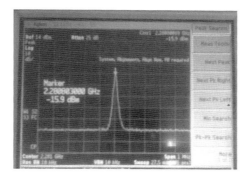

Fig. 9. Output signal of the radio beacon

More tests of setting various frequencies and observing the output signals of the beacon within allowed frequencies have also been done. The results of the tests are expected and confirm the correctness and reliability of the design method. Some other parameters of radio beacons are also able to be set with convenience and agility to meet the needs of practical applications.

References

1. C8051F320/1: Full Speed USB, 16 k ISP FLASH MCU Family. Silicon Labs (2009)
2. Si4136/Si4146: ISM RF Synthesizer with Integrated VCOs for Wireless Communications. Silicon Labs (2010)
3. USBXpress programmer's guide. Silicon Labs, AN169 (2011)

Study on Modeling and Simulation of UPFC

Tingjian Zhong, Zunnan Min, Raobin, and Huangwei

Jiang Xi Vocational &Technical College of Electricity, Nanchang, China
Nanchang, 330031 China
jxdlztj@163.com

Abstract. As the most representative and multiplex device of Flexible A.C. TransmissionSystem (FACTS), the Unified Power Flow Controller (UPFC) is one of the researchfocuses in FACTS recently. UPFC has the characteristic functions as follows: fast anddynamical adjusting the parameters of electricity transmission system, such asvoltage, impedance, phase angle, real power and reactive power, expanding thecapacity of electricity transmission, improving the stability of power system and optimizing the operation of power system.

This paper analyses the basic principle of UPFC and its controlling function and completes its parallel part mathematical modeling by considering the dynamic effect of capacitor. The model of UPFC parallel part in this paper uses a PID controller and its control system adopts a d-q decoupled strategy. With MATLAB, this paper realizes the power flow analysis and power system voltage stability computation. The simulation results of the power flow indicate that UPFC has strong control capability of power flow and voltage. The computation of the latter proves that UPFC can improve the system stability.

Keywords: FACTS, UPFC, Matlab Simulation, Power Flow Regulate.

1 Introduction

The flexibility alternating current transmission system (FACTS), proposed by famous electric power expert Hingorani from American electric research institute in 1986. FACTS is the defined by IEEE, which is the alternating current transmission system using the electric power electron converter and other static electric power electric power compensation controller. FACTS technology will achieve continuous regulation control of power system voltage, parameters(such as line impedance), phase angle and power flow by combined the power electronic technology with modern control technology. FACTS has changed the last low adjustment and control and optimization technology without high precise. As a result, three major electrical parameter of the power flow distribution in power system can be affected.

Doctor Gyugyi.t from USA west room science and technology centre first has brought forward the concept of Unified Power Flow Controller in 1991, which can abbreviate to UPFC. UPFC is a FACTS family, the most complex and most attractive of a compensator, which has synthesized a number of FACTS devices flexible control means, and the most powerful FACTS device, is a value of a control system in the flexible AC transmission technology. UPFC is a new power flow control device, which is composed by static synchronous compensator (STATCOM) and the series compensation of the static synchronous series compensator (SSSC). UPFC plays a role in series compensation, shunt compensation and phase shifting regulation and so on. It also can achieve the accurate adjustment of line active and reactive power and improve transmission capacity and system oscillation damping.

Currently UPFC research work concentrates on four areas: UPFC mathematical model, UPFC control strategy, UPFC electric power electron converter and UPFC system application. But in this paper the former two areas will be mainly studied.

2 UPFC Basic Structure and Principle Analysis

2.1 Basic Working Principle of Unified Power Flow Controller (UPFC) System

UPFC basic structure theory and phase diagram are shown in Figure 2.1.

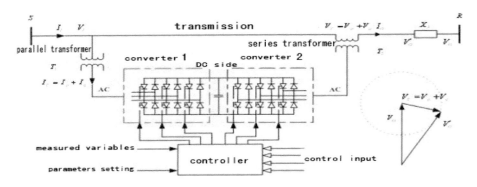

Fig. 2.1. UPFC basic structure theory and phase diagram

It can be seen from Figure 2.1 that UPFC consists of two voltage source type power converter, the DC link connecting two power converter, two transformers and control system. The power converter 1 is parallel connection to power system through the transformer T_1 (the parallel converter part), the power converter 2 is serial access in power system through the transformer T_2 (the series converter section), usually the transformer 1 is called parallel converter, transformer 2 is called series converter.

In the UPFC, the power transmission system parameters are achieved through the series converter injecting voltage into power transmission, in other word, the output of series converter transformer T2 injects voltage $\dot{V}_T$ into the power system, the voltage amplitude and phase are available change ($0 \leq \arg(\dot{V}_T) \leq 2\pi, 0 \leq |\dot{V}_T| \leq V_{T\max}$), this voltage is superimposed to the UPFC series access point voltage V_O on the system, which can synthesize the voltage V'_O with variable gain and phase, the phasor diagram is shown in Figure 2.1. Because of its controllability, the power from the sending side to acceptance side can be controlled. According to the phasor diagram in the Figure 2.1, the transmission power can be calculated such as type (2-1)

$$P_r = \frac{V_o' V_r}{X_l} \sin \delta \qquad Q_r = \frac{V_o' V_r}{X_l}(1 - \cos \delta) \qquad (2\text{-}1)$$

2.2 Unified Power Flow Controller (UPFC) Model in the Steady-State Model

The two converters of UPFC are used voltage-type converter, if only considering fundamental component, we use a controlled current source I_{pq} to be equivalent parallel converter section, using a controlled voltage source Vpq equivalent series converter part, so the steady-state model of UPFC can be obtained (as shown in Figure 2.2). In Figure 2.4, the steady-state model of UPFC, the sent voltage is expressed by $V_S \angle \delta_S$, the acceptance by $V_R \angle \delta_R$.

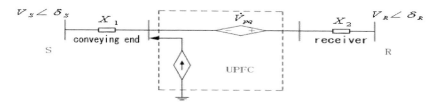

Fig. 2.2. The steady-state model of UPFC

2.3 Unified Power Flow Controller (UPFC) Control Functions

As part of UPFC by the parallel and series components, its control features, including parallel and serial control function control function, discussed separately below.

2.3.1 Parallel Compensation Control

According to the steady-state model of UPFC, parallel converter part is equivalent with a controlled current source, it is divided into active component and reactive component. The active component is used to maintain of a stable DC voltage, provide power support to series converter and ensure its functionality. And reactive component can be used as parallel reactive power compensation,static dynamic reactive power compensation to maintain the node voltage dynamic stability. When the parallel converter to work independently, it is a static compensator (STATCOM), if you do not consider the loss, the equivalent current source for the reactive current.

2.3.2 Series Compensation Control

To simplify the function analysis, it is assumed that UPFC is installed in the middle of the power transmission line, the terminal voltage of s ending side S and acceptance side R are respectively $\dot{V}_S$ and $\dot{V}_R$, voltage virtual value is the equation $|\dot{V}_S|=|\dot{V}_R|=V$, the phase angle difference between them is δ. According to the steady-state model of UPFC it is assumed parallel converter section to provide sufficient power to support, and only part of the role of the series converter. Therefore, a controlled voltage source can be equivalent series converter part, as shown in Figure 2.3. The UPFC series compensation can be achieved through controlling the phase and amplitude of controlled voltage source.

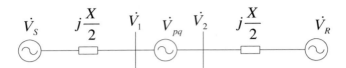

Fig. 2.3. Series converter equivalent circuit

2.3.3 Power Flow Control

With UPFC single-machine infinite power system equivalent circuit is shown in Figure 2.4. In order to facilitate the steady flow analysis, the UPFC parallel side and series side, respectively controlled current source and controlled voltage source to be equivalent. $V_1 \angle \delta_1$ is the bus voltage and phase angle when UPFC access system, $V_2 \angle \delta_2$ is the export voltage and angle after the line crosses UPFC, $V_R \angle \delta$ for for the line end of the department infinite bus system voltage and phase angle, i_d and i_q respectively for the equivalent current source of parallel side inverter. Under the d-q axis components, $V_{pq} \angle \delta_{pq}$ for the series inverter side of the equivalent voltage source of voltage and phase angle; I_{ld} and I_{lq} respectively for the series current; X_T and X_L are the equivalent reactance of transformer and line; generation machine sub-transient electric potential E' is constant, X_d and X_q for the generators d-q-axis sub-transient reactance.

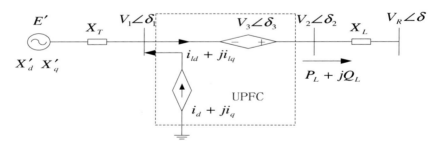

Fig. 2.4. UPFC equivalent circuit

In the design and operation of UPFC it is important to note that the system may change the running tide of UPFC controllable region, which may lead to the trend of UPFC can not achieve control objectives. Especially when subjected to large disturbance, the system state change dramatically, it may happen, we should pay attention to take appropriate actions to resume UPFC normal operation as soon as possible.

3 UPFC Parallel Part Model and Controller Design

3.1 Mathematical Model of Parallel Converter

There are two functions for parallel converter, which consists of maintaining access to end node voltage of UPFC and DC bus voltage stability. The controller structure varied, both based on fuzzy adaptive control, optimal control, neural network intelligent controller, also based on the linear theory of traditional PI decoupling controller. According to the normal operation of power system UPFC are permitted access to certain end node voltage fluctuations (usually 5%) of the characteristics of the UPFC control access to end node voltage using the voltage sag compensation with current control.

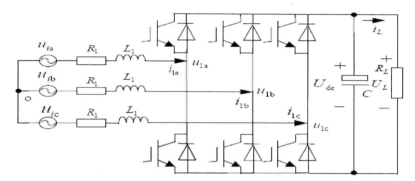

Fig. 3.1. UPFC circuit model of parallel converter

UPFC shunt converter circuit model is shown in Figure 3.1. Main circuit switching devices use IGBT and anti-parallel diode, and as an ideal switch, the switch function S_n ($n = a, b, c$ respectively three-phase) reflect, u_{ia}, u_{ib} and u_{ic} represent the three-phase output voltage of PWM inverter, L_1 and R_1 represent the filter inductors and parallel transformer equivalent inductance and resistance, C said inverter DC capacitor, R and L are on behalf of load effects by the series converter and switching losses caused by the two converters.

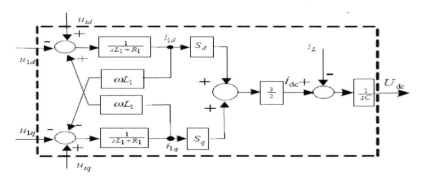

Fig. 3.2. The mathematical model of the parallel converter under d-q coordinate system

3.2 Parallel Converter Controller Design

This paper uses a belt loop decoupling control strategy in order to obtain good dynamic response performance and a better power flow control capability. A modified double loop PI decoupling control system is proposed as shown in Figure 3.4 based on the traditional PI decoupling control. Control system consists of current controller, capacitor voltage controller, current-feedback controller, as well as voltage sag controller with the current compensation. Figure 3.3 shows that, d-axis current controller $G_{id(s)}$ and the q-axis current controller Giq (s) are using PI control, and their signal path entirely similar, it will be recorded as they are unified. Close to the actual situation, consider the PWM inverter delay and feedback channel filtering, d-q current control system is shown in Figure 3.4.

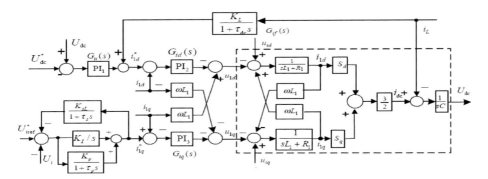

Fig. 3.3. Loop decoupling control system

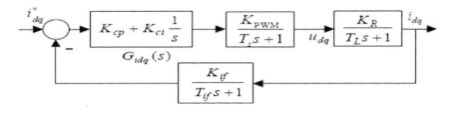

Fig. 3.4. Current controller

The capacitor voltage Udc controller is shown in Figure 3.5, where $W_{ci(s)}$ is the current loop controller to make the d-axis current i_d have little change in the transient process. So using the first order inertia approximately takes the place of second-order system, $W_{ci(s)}$ can be expressed as $1/(2T_{sf}s+1)$. Voltage controller open-loop transfer function is shown in the equation 3-1.

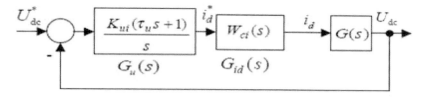

Fig. 3.5. Capacitor voltage Udc controller

$$W_{ov}(s) = \frac{K_{ui}K_0(\tau_u s+1)(1-T_2 s)}{s(1+T_P s)(2T_{sf} s+1)} \tag{3-1}$$

4 UPFC Simulation Study

4.1 UPFC Used in the Three-Phase Power System Simulation

UPFC simulation model is shown in Figure 4.1. Model mainly is composed by the main circuit model of UPFC module, UPFC shunt side control module and the measurement system module. There are other parallel side of the UPFC control module output voltage calculation module, UPFC DC voltage calculation module, and the main parameters of the signal power flow summary of the module.

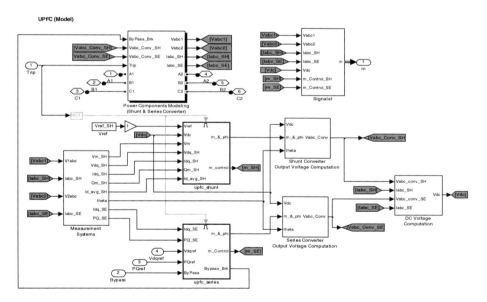

Fig. 4.1. UPFC simulation model

In this article according to the simulation model, the unified power flow controller (UPFC) in the power transmission system is simulated, operating characteristics of UPFC in this study. Using the two power plants as both ends of the power supply, transmission line is connected through a double circuit line and two transmission line distributed parameters and in place connected to the load 200MVA. UPFC is installed in the two loops between the distributed parameter transmission line.

By a double-circuit 230kV transmission line connected to two power stations, issued a total of 1500 MW, through two transformers Tr1 and Tr2, gave an 200MW of load and capacity of the receiving end of a 15000MVA, the acceptance of terminus of three-phase RL impedance with power equivalent. In this model, UPFC installed in line L2 to the right, used to control the 500kv bus B3 of active and reactive, while controlling the voltage bus B_UPFC.

4.2 UPFC on the System Power Flow Control and Impact on Power

First, the UPFC model will be established to set the flow control mode. Then according to the above model of the transmission system, transmission system, Powergui provided by a simulink initializes the load and motor function, can set the initial grid current parameters, and using Powergui, calculates the steady-state current grid nodes RMS voltage.

When the UPFC is not put into operation, by the steady-state current and voltage RMS Powergui can calculate the power plant 2 has 899MW of power to issue 2, sent through the transformer, the remaining 101MW to flow throughout the network ring. 99MVA transformer 2 will therefore bear the overload.

UPFC input through the Bypass module scheduled start time to 3 seconds, from Powergui calculate the flow through the bus B3 of the active initial steady-state value of 587MW, by Pref module reference value is set Pref = 5.87pu with the active flow through the bus B3 initial steady-state value equal to the first 5 seconds time to set the active reference value Pref = 6.87pu. Meanwhile, the default module Qref as Qref =-0.27pu, And the system bus B3 initial flow through the reactive equivalent. The simulation results obtained are as follows:

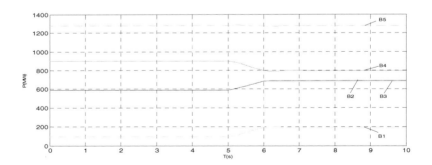

Fig. 4.2. Bus B1, B2, B3, B4, B5 active curve

Figure 4.2 is the flow through the bus B1, B2, B3, B4, B5 of the active curve. From the figure it is clear that the first time 3 seconds into the system without UPFC significant impact to the system. This is because at this time UPFC device Pref module settings and B3 of the active flow through the same bus. Therefore, the first 3 seconds time UPFC active power control settings, consistent with the system, so there is no significant change in system operation.

System operation to 5th seconds, active power flow through the bus B3 steady rise to a stable value in the first 6 seconds. This is because the Pref module will set the value into 6.87pu, UPFC started to adjust the line active power. Meanwhile, the figure also shows the active power flow through bus B4 starts to steadily decline since the fifth seconds and to achieve a stable value in the first 6 seconds time. This value can be calculated by Powrgui 796MW, it flows through the transformer active power down to a rating of 2 less than 800MW, thus avoiding the transformer 2 is in overload status. Figure 4.2can also be seen from the flow through the active bus B5 remained basically unchanged. UPFC active power line started at the adjusted time of 1 second after the system reached the preset value of stable operation, which explains the UPFC model can be fast, accurate and stable flow of the system to adjust.

In addition, Figure B2 and B3 stream of the active bus curves almost coincide. This is because, before the UPFC input, the difference of B2 and B3 between active lines L1 and L2 only is the line loss, relative to the flow through bus B2 and B3 of the active is very small, calculated by the difference between Powergui only to 2MW, so this time the two curves almost coincide. Similarly, when the UPFC input, the calculated difference between the two is only 3MW, so little overlap between the two curves remain. This also shows that UPFC active power consumption of the device itself is small, less than 1MW, compared to flow through the active transmission line is very small.

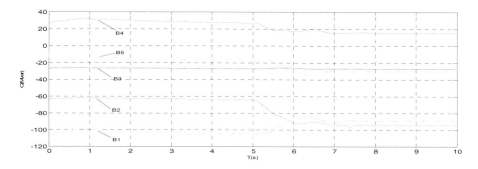

Fig. 4.3. Bus B1, B2, B3, B4, B5 reactive curve

Figure 4.3 is the flow through the bus B1, B2, B3, B4, B5 reactive power curve. From the figure we can see that flow through the bus B3 reactive power has remained stable. Because presetting Qref is - 0.27pu in Qref module and system initial reactive power flow through the bus B3 are equal, flow through the bus B3 to maintain the same initial value. Several other buses achieve a stable after 2 to 3 seconds adjustment and reactive power has little change before and after UPFC input to ensure the stability of the system voltage, which plays an important role.

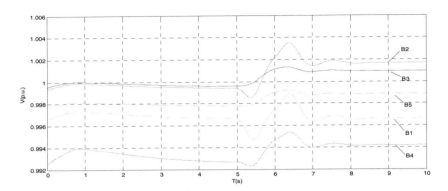

Fig. 4.4. Bus B1, B2, B3, B4, B5-voltage curve

Figure 4.4is the bus B1, B2, B3, B4, B5 voltage curve. As can be seen from the chart, the change range of the bus voltage stable value is very small before and after UPFC input, less than 0.1%; there is not more than 0.6% of range change only in the adjustment process of UPFC. UPFC input voltage stabilize from 3 to 4 seconds.

5 Conclusion

In the UPFC power flow control mode, We can see from the simulation results that UPFC power flow control functions would adjust fast, accurately and steadily the system flow to effectively alleviate the trend of transmission system congestion, without adding new transmission lines and new electrical equipment cases it can make full use of existing transmission circuits and electrical equipment to increase the transmission capacity. Meanwhile, UPFC put into operation, a very small impact on the system voltage, system voltage to maintain stability. Therefore, UPFC operation of the transmission line active and reactive current are well controlled, and this is an important application of UPFC device characteristics, the characteristics of the stable operation of the transmission system are very important.

References

1. Hingorani, N.G., Gyugyi, L.: Understanding FACTS, Concepts and Technology of Flexible AC Transmission Systems. IEEE Press and John Wiley&Sons, Inc., New York (2000)
2. Nabavi Niaki, A., Iravani, M.R.: Steady-state and dynamic models of unified power flow controller(UPFC) for power system studies. IEEE Transaction on Power Systems 11(4), 1937–1943 (1996)
3. Stefanov, P.C., Stankovic, A.M.: Modeling of UPFC operation under unbalanced conditions with dynamic phasors. IEEE Transaction on Power Systems 17(2), 395–403 (2002)
4. Fujita, H., Watanabe, Y., Akagi, H.: Transient analysis of a unified power flow controller and its application to design of the DC-link capacitor. IEEE Transaction on Power Electronics 16(5), 735740 (2001)

5. Makombe, T., Jenkins, N.: Investigation of a unified power flow controller. IEE Proc-Gener Transm. Distrib. 146(4), 400408 (1999)
6. Dash, P.K., Mishra, S., Panda, C.: A radial basis function neural network controller for UPFC. IEEE Transaction on Power Systems 15(4), 1293–1299 (2000)
7. Kang, Y.L., Shrestha, G.B., Lie, T.T.: Application of an NLPID controller on a UPFC to improve transient stability of a power system. IEE Proc-Gener Transm. Distrib. 148(6), 523529 (2001)
8. Garcfa-Gonzalez, P., Garcia-Cerrada, A.: Control system for a PWM-Based STATCOM. IEEE Transaction on Power Dlivery 15(4), 12521257 (2000)
9. Xu, L., Agelidis, V.G., Acha, E.: Development considerations of DSP-controlled PWM VSC-based STATCOM. IEE Proc-Electr Power Appl. 148(5), 449–455 (2001)

The Research Design of the Pressure-Tight Palette for Oil Paintings

Ruilin Lin

Department of Commercial Design, Chienkuo Technology University,
No. 1, Chieh Shou N. Rd., Changhua City 500, Taiwan
linrl2002@gmail.com

Abstract. The product developed by this study is an innovatively designed pressure-tight palette for oil paintings. Near the opening of the box of the product, there is a ring-shaped and raised wedging edge, with a ring-shaped supporting edge under it and a lid. By the two opposite edges at the bottom of the lid are the two breaches extended to the proper spots at the bottom on the contiguous sides, forming the supporting points. When the lid of the box is closed, pressure can be imposed on the lid, and air will be exhausted through the raised wedging edge and the two breaches, to create the pressure-tight effect. To open the lid, force can be imposed on it. The depending point is the fulcrum while the breach is pressed toward the supporting edge and air can get in through the opening between the raised wedging edge and the lid when the other breach comes off the supporting edge. Then the lid can be easily opened. This way, colors can be well kept in the palette for future use.

Keywords: pressure-tight, oil painting, palette.

1 Design Concept

Currently in the market, the pressure-tight technology is mostly used in window design. Pressure-tight windows can block noises from outside or specially designed to sustain indoor temperature while it is too cold outside. Also sometimes this technology is applied on car windows to solve rain drops problems. With light and soft designs, this technology helps to extend ranges of product effects. For example, mist problem can be solved easily and conveniently. Furthermore, after the pressure-tight technology is applied in designing, costs can be saved, problems can be solved rapidly and efficiently, and expected ideals can be reached [2] [5] [6].

Discussions in studies of oil painting related issues are mostly regarding exploration of painting styles, usage of painting skills, and recreation of masterpieces. Of course, there have been oil painting simulations, teaching, and designing using Autodesk Drawing Software or other 3D computer software [1] [3] [4]. However, there have been very few studies or inventions related to research and development of tools for oil paintings. Thus, the innovatively designed product of this study is a pressure-tight palette for oil paintings, featuring well preservation of colors and convenience to open. Oil painters can simply close their palettes while painting or resting, instead of spending a lot of efforts going through minute and complicated process to clean up like they did before. This way, efforts can be saved, novices may be more interested in oil painting, and more artists may devote themselves into the field of oil painting.

2 Theoretical Foundations

The traditional palettes for oil paintings which people generally use are made of two block-shaped plates. The two plates are combined by a pivot component, so that they can be either folded together or spread up becoming one single plate. By fixing the two plates with a fixing component, it would be difficult to fold the two plates using external forces. One can hole the palette by putting his thumb through the hole on the plate.

To put away a traditional palette which comprises two plates, it has to be folded. And colors for oil painting are rather expensive. If they are not shoveled into color bottles, they may dry out and wasted for they cannot be used anymore. To avoid this kind of wasting, painters usually have to shovel colors back into color bottles one by one and kept them there. However, different colors may be mixed during the process, and when they are needed next time, painters have to put those them back on palettes again. This complex process often causes a lot of inconvenience.

Because of these known problems of traditional palettes, it is all consumers' wish that a more practical palette can be invented. And this has also become the development goal and direction for related manufacturers. Therefore, based on the researcher's years of experiences in researching and developing related products, considering the above-mentioned goal, the inventor went through detailed design and evaluation and finally developed this practical and innovative product.

3 Content

Traditional palettes are for color mixing only. When users are done with their paintings, if colors are not put back to color bottles, they may dry out. For the purpose of preserving colors so that they can be used again, users must spend time and efforts shovel colors into color bottles and take them out again next time when they are needed again. This complex process always causes user a lot of inconvenience. This is the technical problem of traditional palettes which needs to be solved.

The innovatively designed product of this study is a pressure-tight palette for oil paintings. Near the opening of the box of the product, there is a ring-shaped and raised wedging edge, with a ring-shaped supporting edge under it and a lid. By the two opposite edges at the bottom of the lid are the two breaches extended to the proper spots at the bottom on the contiguous sides, forming the supporting points. When the lid of the box is closed, pressure can be imposed on the lid, and air will be exhausted through the raised wedging edge and the two breaches, to create the pressure-tight effect. To open the lid, force can be imposed on the lid. The depending point is the fulcrum while the breach is pressed toward the supporting edge and air can get in through the opening between the raised wedging edge and the lid when the other breach comes off the supporting edge. Then the lid can be easily opened. These are the features of the technology developed by this study to solve the problem.

The innovatively designed product of this study helps to preserve colors properly for reuse through the pressure-tight effect created by the structure with a box and a lid. With this structure, users can open their pressure-tight palette for oil paintings in a short time. The technology, methods, and effects of this innovatively designed product are described later with illustrations to help readers understand the goal, structure, and features with deeper insight.

4 Mechanical Application

The innovatively designed product is a pressure-tight. It comprises a lid and a box. By the two opposite edges at the bottom of the lid are the first and second breaches extended to the proper spots at the bottom on the contiguous sides, forming the supporting points. And above the first breach on the top of the lid there is a pressing point. On the box near the opening there is a ring-shaped and raised wedging edge, with a ring-shaped supporting edge under it. Between the raised wedging edge and the supporting edge, there is a trench. The contact surface on the box with the first breach is called the first closing area, while the contact surface on the box with the second breach is called the second closing area. The second closing area is a little bit higher than the first closing area. From the side view, there is a slope called the first slope. Near the first closing area, there is the second slope, which is at different height from the first slope. And on the top of the second closing area, there is a concave edge. With this edge, when the lid is closed, force can be imposed on the pressing point. Then the supporting point becomes a fulcrum for the lid while the lid gets closer to the first and second slopes. The first breach gets closer to the supporting edge. Air goes through the second breach, leaving the supporting edge, and enters through the gap between the raised wedging edge and the lid. Also, the concave structure by the top edge of the second closing area allows air to enter the box rapidly, so that the lid can be opened easily.

The lid and the box are made of metal plates or plastic ones with elasticity. When the lid is closed, the air inside can be emitted through the gaps between the raised wedging edge and the first and second breaches to create the pressure-tight effect by imposing force on the lid. On one side of the box, a ring-shaped component with a

(1) Lid	(2) Box	(4) Ring-shaped component
(41) Hole	(10) First breach	(11) Supporting point
(13) Second breach	(15) Pressing point	(21) Raised wedging edge
(23) Supporting edge	(25) Trench	(27) First closing area
(271) First slope	(28) Second closing area	(281) Second slope
(282) Concave edge	(30) Fixing component	(50) Block-shaped plates
(51) Pivot component	(52) Fixing component	(53) Hole

Fig. 1. Mechanical Component Diagram

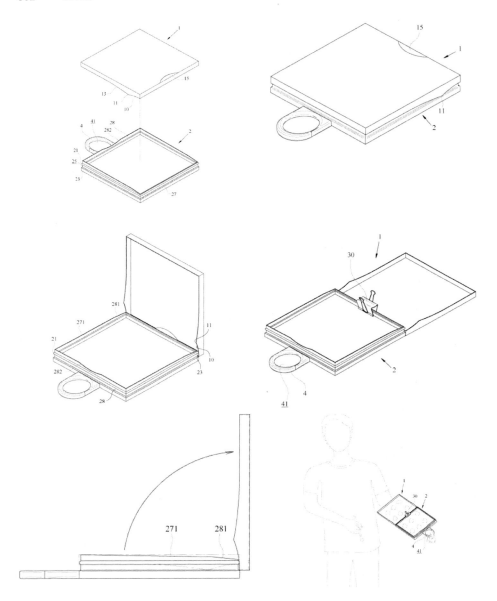

Fig. 2. Pressure-tight palette for oil painting

hole can be installed. When the lid of the box is opened, a fixing component can be used to fix the lid and the box so that they form one plane. When using the palette, user's thumb should go through the hole of the ring-shaped component so that his arm can support the whole pressure-tight palette for oil paintings to save some efforts (figure 2). This is the specific detailed description of the implementation of the technology applied to this innovatively designed product.

5 Contribution

The contribution of the mechanical design of this innovatively designed product is as below:

This study proposes an innovatively designed palette product which improves color preservation with more convenience. In the aspect of the structure, by the two opposite edges at the bottom of the lid are the two breaches extended to the proper spots at the bottom on the contiguous sides, forming the supporting points. On the box near the opening there is a ring-shaped and raised wedging edge, with a ring-shaped supporting edge under it. Between the raised wedging edge and the supporting edge, there is a trench. The contact surface on the box with the first breach is called the first closing area, while the contact surface on the box with the second breach is called the second closing area. The second closing area is a little bit higher than the first closing area. On the top of the second closing area, there is a concave edge. The second slope is at different height from the first slope. And above the first breach on the top of the lid there is a pressing point. The lid and the box are made of metal plates or plastic ones with elasticity. The lid and the box are made of metal plates or plastic ones with elasticity. After the lid is opened, the box and the lid are fixed with a fixing component. On one side of the box, a ring-shaped component with a hole can be installed, so that painters can easily hold their palettes with their thumbs through the holes. Of course, this design not only helps users to open and put away their palettes, but also solves the trouble of colors being mixed in the process of shoveling colors back to color bottles, while users can conveniently reuse those colors to reduce wasting

The pressure tight theorem is applied to the innovatively designed palette for oil paintings in this study. With the designed structure, this innovative product helps to overcome many difficulties which may be encountered in the process of creation with traditional palettes for oil paintings. Generally speaking, the advantages of this innovative product include:

(1) Conceptual Innovation
The research and development of the innovative product in this study is from the perspective of painters. The design is improved according to the difficulties which may be encountered while using traditional palettes. Solutions with innovative concept are proposed to solve these problems. The result of this study is beyond traditional concept and thoughts. It is worthy to promote it.

(2) Convenience for putting away
The pressure-tight theorem is applied to the design of the palette with a simple structure, so that oil painters can save the time spent on cleaning their palettes and get some rest instead. All they have to do is to close their palettes. Putting their palettes away can be done easily with fewer complex steps.

(3) Being Energy Saving and Eco-Friendly
Colors for oil paintings are usually more expensive than general watercolors or poster colors. When not being used, colors may dry out and can therefore not be able to be

used anymore, causing waste. With the press-tight design of this study, colors for oil paintings can be preserved for a longer period of time. When painters need to rest, they don't have to shovel their colors back to color bottles or throw them away. All they need to do is to fold their palettes. The problem of colors being mixed can be avoided and waste can be reduced.

(4) Market Expansion

There are issues of inconvenience and waste while using traditional palettes for oil paintings. The innovatively designed product of this study, a pressure-tight palette for oil paintings, benefits a lot of users who love oil painting creation. The perspective of market expansion is very good.

References

1. Ribeiro, C., Bruno, P., Carvalho, A.C., et al.: Magnetic memory of oil paintings. Journal of Applied Physics 102(7) (2007), doi: 10.1063/1.2786072, 074912 - 074912-3
2. Li, J., Bai, M., Xie, N., Li, H.: Numerical analysis magnetic airtight installation for rotary forced-air cooler. In: IEEE 10th International Conference on Computer-Aided Industrial Design & Conceptual Design, pp. 730–733 (2009)
3. Liu, J.: Oil painting structure which based on autodesk drawing software. In: 2010 2nd International Conference on Information Engineering and Computer Science (ICIECS), pp. 1–4 (2010), doi:10.1109/ICIECS.2010.5677759
4. Liu, J., Qian, X.: Drawing program of Oil Painting which based upon computer. In: 2010 2nd International Workshop on Database Technology and Applications (DBTA), pp. 1–4 (2010), doi:10.1109/DBTA.2010.5659099
5. Ling, W., Jin, H.: Research on energy-saving technology of window surrounding for the sandwich insulation walls of rural housings in cold regions of China. In: 5th International Conference on Responsive Manufacturing-Green Manufacturing (ICRM 2010), pp. 111–115 (2010)
6. He, X.M., Liu, W.J.: Car rain protection airproof performace test method based on ultrasonic. In: 2010 International Conference on Electrical and Control Engineering (ICECE), pp. 565–568 (2010)

Product Innovative Development of the Automatic Protection Facility Product for Pool Drain Covers

Ruilin Lin

Department of Commercial Design, Chienkuo Technology University,
No. 1, Chieh Shou N. Rd., Changhua City 500, Taiwan
linrl2002@gmail.com

Abstract. This study aims to develop an innovative product, an automatic protection facility for pool drain covers, which can be implemented on drains of swimming pool filtration systems. The protection device includes: a support bracket, a blocking device, and a touch device. The support bracket is used to fix the base installed to the drains with several through holes. The blocking device is folded when the through holes are connected to the drains and is unfolded when the through holes are disconnected from the drains. The touch device controls the state of the blocking device (folded, unfolded). With this product, when a swimmer is accidently dragged down by the pool drain, actively or passively touching the touch device triggers the blocking device to unfold and block the suction from the pump, so that the swimmer may not get hurt or drowned.

Keywords: pool drain cover, automatic protection facility, product innovative development.

1 Design Concept

During summer, parents usually take their children to swimming pools for water activities. However, many accidents may be caused by drain systems. For example, a child's buttocks may be sucked due to the powerful suction from a filtration system, causing his buttocks swelling. Or a child may be sucked deep into a pumping tube of a pool filtration system and drowned. These accidents occur because the surface of drains is flat. Once drains are covered by someone or something, very strong suction may therefore caused may suck other objects into pumping tubes, or in some cases, swimmers may be sucked by the pumping holes and cannot escape. Usually there are safety pistons or safety lids installed for pool drains. But accidents still occur for sometimes they are not perfectly placed or installed.

The purpose of the production innovative design in this study is to develop an automatic protection facility which can efficiently prevent swimmers from being sucked in by the pumping holes, or avoiding accidents due to misplaced or loose safety pistons or safety lids. It can be installed by pool drains of filtration systems. In other words, it is a protection facility used to prevent accidents such as swimmers being sucked in by pumping holes.

2 Theoretical Foundations

In the aspect of related literature, a study used the swimming pool construction of the Tayih Landis Hotel in Tainan as an example to prove that swimming pools for high hotel buildings or congregate housings should be rectangle-shaped and of 150m^2~250m^2 in area and 1~1.2m in depth. Bottoms of pools should be near-flat but slightly oblique. Overflow shape should be eliminable slopes. For overflowed water recycling, the Finnish style and shock-proof, light, compound style stainless tiles should be used. Sterilization systems of chlorine water with ozone, and prioritized the heat-alternated heating systems are first to be considered and included in designs. In the aspect of construction, electric welding procedure and sequence should be implemented for stainless pool bodies, RC-made pool bodies should be made of light and high-intensity water-tight concrete and highly-tensile wieldable steel. And construction for pool body integration, water-proof treatment, arrangement of pillars, periphery fixing, and equipment should all follow the principles of being safe, comfortable, convenient, and economic [3]. Of course, some scholars believe that although indoor swimming pools are not influenced by weather, terrain, and location, good swimming environments should still be provided, so that swimmers can exercise in comfortable and good environments. Thus, the gray relational analysis is applied to analyze the cost considerations of swimming pool environments and consumers' satisfaction. The result of this study shows that the important influential factors include: (1) water temperature, (2) number of swimmers, (3) water capacity, (4) citric acid, (5) bleach, (6) enzyme clarificant, (7) chlorine neutralizer, and (8) accumulated dosing time by a dosing machine [2].

Moreover, there is also a study regarding elementary school students' perception of swimming coaches' leading styles. The result shows that the students had the highest level of perception of the coaches' training and teaching behaviors and the lowest level of perception of their autocratic behaviors. The students from different grades with different swimming time and levels had significantly different perception of the coaches' leading styles. The coaches' leading styles are positively and significantly related to the students' satisfaction. The prediction power of the students' perception of the coaches' social support behaviors, training and teaching behaviors, and returning and rewarding behaviors was significant, with 44.8% of variation being explained [4]. Furthermore, there is a study regarding the current status and differences of the swimming abilities of the students from elementary schools in Taipei City and Taipei County in Taiwan. The result shows that most of the students believed swimming classes help to improve their swimming abilities (57.5%). 79.1% of the students could swim the freestyle stroke, 43% the breaststroke, 48.1% the backstroke, and 13.7% the butterfly stoke. The 4 swimming abilities of the students were significantly different. Their swimming abilities can be improved by (1) improving swimming pool water quality, (2) improving swimming pool equipment, and (3) hiring good swimming coaches [5]. In addition, some scholar applied bio control engineering operations to explore issues such as the relationship between central nervous system and muscle, helping human muscle to function completely under normal circumstances [1].

3 Content

The innovative product developed by this study is an automatic protection facility for pool drain covers, which includes a support bracket which fixes the base installed by the drains with several through holes, a blocking device installed on the bottom side of the support bracket, and a touch device which controls the state of the blocking device (folded, unfolded). There are several through holes on the support bracket. The blocking device is folded when the through holes are connected to the drains and is unfolded when the through holes are disconnected from the drains. Normally, the blocking device is folded, so that water can go into drains via the through holes on the drain covers and the support bracket. When a swimmer is sucked to the drain covers due to the suction of the pump, he can touch the touch device to unfold the blocking device and block the suction, thus avoiding an accident.

The blocking device of the automatic protection facility for pool drain covers consists of several prop stands and flexible connecting pieces which connect 2 prop stands. In addition, the touch device is installed on the top side of the support bracket to be placed inside the through holes of drain covers. Therefore, if a swimmer is sucked to the drain cover due to the suction from the pump, some part of his body may touch the touch device or he may directly put his finger through a through hole and touch the touch device inside to unfold the blocking device and block the suction from the pump for the purpose of protection.

4 Mechanical Application

The innovative protection facility for pool drain covers developed by this study should be installed under the drain covers with several through holes of the swimming pool filtration systems. The protection facility consists of a support bracket, a blocking device, and a touch device. Support bracket: It is made of metal, joint-less, and round-shaped with a base. The body part is extended from the center of the base toward outside, while the support part is extended vertically toward the body part. The top side of the base faces the pool and the bottom side faces the drains. There are several through holes on both sides. To install this facility, the base part of the support bracket should be fixed to the pool drains with through holes under the pool drain covers which also come with several through holes. And the through holes of the support bracket should match those of the drain covers. Pool water can go through the through holes of the drain covers and the support bracket to the through holes of the pool drain.

Blocking device: It is installed on the bottom side of the support bracket. It is folded when the through holes are connected to the drains and is unfolded when the through holes are disconnected from the drains. It is composed of several prop stands and several flexible connecting pieces which connect two prop stands. The prop stands are connected to the body of the support bracket away from the base. The blocking device can be folded or unfolded. When the blocking device is folded, the prop stands are against the support bracket with the connecting pieces closed between the prop stands so that the through holes on the support bracket are connected to the drains. When the blocking device is unfolded, the prop stands are perpendicular to the support bracket with one end of the stands against the support bracket, with the connecting pieces open between the prop stands so that the through holes on the support bracket are blocked from the drains. Touching device: It is installed on the top side of the support bracket. A part of it is inside the through holes of the drain covers. Touching the touch device makes the status of the blocking device to change from "folded" to "unfolded". Furthermore, there can be more than one touch devices installed on the top side of the support bracket. They can be arranged where they can be reached through the through holes, increasing the chances of sucked swimmers touching one of the devices.

(10) Automatic protection facility for pool drain covers	(20) Support bracket	(21) Base
(23) Body	(25) Supporting part	(211) Through hole
(30) Blocking device	(31) Prop stand	(33) Flexible connecting piece
(40) Touch device	(50) Pumping hole	(51) Hole wall
(53) Drain cover	(531) Through hole	(60) Pool water

Fig. 1. Mechanical Component Diagram

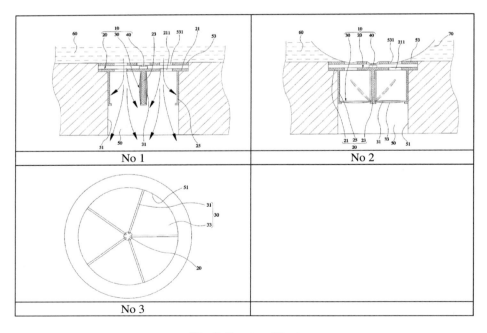

Fig. 2. Structure Chart

In practice, under normal circumstances, the blocking device is folded. When the pump starts working, the water inside the pool goes into the drain through the through holes of the drain covers and the through holes of the support bracket due to the suction power. This is the process of water change. If during this process, a swimmer is sucked to the pumping holes due to the suction from the pump, part of the swimmer may be sucked into the pumping holes and touch the touch device. Then the blocking device is unfolded. The prop stands expanded toward the body of the support bracket until they are vertical to and against it, causing the flexible connecting pieces to expand and block the suction from the pump. This way the swimmer can get away from the drain and an accident is avoided. Otherwise, the swimmer may also stick his finger through one of the through holes of the drain covers to reach the touch device and unfold the blocking device. In addition, when the blocking device is unfolded, the prop stands can be supported by the support bracket, so that they may not be deformed easily due to water pressure. In sum, the automatic protection facility developed by this study with a blocking device which can be unfolded to block the through holes of drain covers and stop the suction caused by the pump, successfully avoiding accidents of swimmers being sucked by pumping holes [Figure 1, 2].

5 Contribution

This study develops an innovative automatic protection facility for pool drain covers to prevent swimmers from being sucked by pumping holes, not able to escape, drowned, and leaving their families to grieve. Overall, the advantages of this innovative products are listed below:

(1) Humane Care
The innovative product of this study is designed from users' angle by considering dangerous situations which may be encountered while swimming. The friendly design comes from the attitude of humane care to provide safety and help swimmers to escape immediately after being sucked. This is especially helpful for children without big strength. The design consideration which helps to save lives shows human-based caring element in the design.

(2) Conceptual Innovation
This innovative product is developed by throwing away traditional and old concepts and ideas, using brand new technologies and materials to solve problems swimmers (especially young ones) may encounter while swimming. The achievement of this study makes it worthy to promote the product.

(3) Low Cost
The structural design in this study is improved according to the requirements of the automatic protection facility for pool drain covers. Although the originative value is high, the cost is low. Related business owners can apply this product to reduce chances of accidents.

(4) Market Expansion
The traditional automatic protection facility for pool drain covers often fail to stop accidents due to loosened pistons or safety lids. This study improves the design and

solves the current problems. The structure design which helps to save lives allows swimmers to escape in a very dangerous situation and prevents accidents from happening. Thus, the future prospective of this innovative product in the market is very promising.

References

1. Avella, D.A.: Decomposition of EMG patterns as combinations of time-varying muscle synergies. In: First International IEEE EMBS Conference on Neural Engineering, Conference Proceedings, pp. 55–58 (2003), doi:10.1109/CNE.2003.1196754
2. Hsiang, H.C.: Using the Grey Relational Theory to Study Water Quality Management–A Case Study for the Indoor Swimming Pool of Taipei Rapid Transit Corporation, Department of Mechanical Engineering, Tatung University, Master dissertation (2004)
3. Jon, I.G.: The Research on the Establishment of the Natatorium in the High-rise Construction-To Take the Natatorium of Tayih Landis Hotel in Tainan City As Example, Department of Architecture, National Cheng Kung University, Master dissertation (2004)
4. Wang, T.Y.: Satisfaction Level of Learning Determined by the Leading Style of the Elementary School Swimming Coach, Department of Leisure and Recreation Management, Asia University, Master dissertation (2010)
5. Chen, W.L.: The Study of Swimming Capacity among The Elementary School Students in Taipei County, Department of Recreation and Leisure Industry Management, National Taiwan Sport University, Master dissertation (2010)

A Dynamic Optimization Modeling for Anode Baking Flame Path Temperature

Xiao Bin Li[1,*], Leilei Cui[1], Naijie Xia[1],
Jianhua Wang[1], and Haiyan Sun[2]

[1] School of Electrical and Electronic Engineering, Shanghai Institute of Technology, 201418,
Shanghai P.R. China
`xbli@sit.edu.cn`
[2] School of Ecological Technology and Engineering, Shanghai Institute of Technology, 201418,
Shanghai, P.R. China

Abstract. Anode baking flame path temperature is an uncertainties multivariable control system. Baking flame path temperature accurately control is very important to realize to save energy, lower energy consumption and reduce pollutants discharge. In order to control baking flame path temperature accurately, the second order plus time delay control model is identified and optimized. This model based on the data acquired from the anode baking furnace scene, which is depend on APSO (Adaptive Particle Swarm Optimization). The model simulation and practice results show that the proposed model is efficient and effective.

Keywords: Dynamic optimization, anode baking, flame path temperature, modeling.

1 Introduction

With aluminum electrolytic technology development, large-scale production of environmental protection, energy saving, high efficiency has become an inevitable trend. Aluminum electrolysis production is composed of alumina conveying, carbon (anode baking), electrolysis and casting. In this process carbon provides electrolyzer with anode, whose performances influence aluminum electrolysis's quality. For example, in aluminum production costs, anode carbon accounted for 13% of the total cost, while t he power consumption accounted for 34% which is directed related with anode quality. In conclusion the anode quality is the main factor in energy saving for electrolytic aluminum production.

The research of anode baking began in 1980s, Furman A proposed the anode baking's mathematical simulation[1], The American applied for the patent of" Method and apparatus for producing uniformly baked anodes"[2]. R.T. BUI simulated the dynamics of the anode baking ring furnace[3]. The domestic research in anode baking began in 1990s. The research includes the anode baking burning frame, anode baking flame

---

* This work is partially supported by NSFC No.51075258, Science Foundation of SIT (YJ2011-33) and (YJ2011-22).

path's construction and the control of the heave oil supplement temperature .anode baking process, baking gas temperature and anode baking temperature and so on[4]. There is a large gap between the developed country and the domestic technology, which focus on the current efficiency, the high DC consumption, the huge anode consumption and the big anode effect coefficient and so on.

Due to the anode quality, it is important to improve aluminum electrolytic current efficiency and reduce electrolytic energy consumption. During the baking, accurate flame path temperature control for anode block quality plays a decisive role. It is the basis of baking high-quality anode[5].

Therefore, the study collected a large number of anodes baking process flue temperature data, built its optimal control model based on APSO method. Flue temperature from the model and the actual output are compared to verify the validity of the model, this optimal control model create preconditions for precise control of flame path temperature and energy saving in aluminum electrolysis.

2 Establishment of Baking Flue Temperature Model

2.1 The Choice of Model Structure

The relationship of anode baking flue temperature and pulse electromagnetic valve current is as follows:

f (flue temperature) = the current of pulse electromagnetic valve

Through the field data of anode baking flue temperature and pulse electromagnetic valve current, the second-order delay system model is built. The equation is as follows:

$$y(k) = a_1 y(k-1) + a_2 y(k-2) + b_1 u(k-1-d) + b_2 u(k-2-d) \tag{1}$$

where d =delay factor
Then write it in the form of ARMAX:

$$A(q)y(t) = q^{-d}B(q)u(t) + C(q)e(t) \tag{2}$$

where n= the order of the model

$$A(q) = 1 - a_1 q^{-1} - a_2 q^{-2} - \ldots - a q^{-m}, B(q) = b_0 + b_1 q^{-1} + \ldots + b_n q^{-m}$$

where $C(q)$ is the input parameter of noise characteristic, whose order is 1, $d = \tau/Ts$ is delay factor, Ts is sample time.

2.2 Identification of Model Parameters

There are many Model parameter identification methods, common used are least square and maximum likelihood estimation method. While in the process of system identification, two conditions are necessary. Firstly, objective functions are continuous and derivable. Secondly, gradient information is used for local search. Because of the advantages of identification and optimization, APSO algorithm is chosen to build the anode baking flame path temperature model.

An adaptive particle swarm optimization is random global optimization algorithm based on population evolution. No special requirements are needed in the form of the optimization objective function. The system parameters identification based on PSO is essentially a parameters optimization process in the solution space. The unknown parameters are optimized to fit the measured model data well. APSO algorithm for model parameter identification is given in Fig. 1. The detailed steps of identification and optimization is shown in reference [6].

(1) Parameters Coding and Initialization: The population particles and their velocity are coded in real number. Each particle is described on two dimensions. Suppose n is the population size. After initializing the population, a random matrix emerged including each particle' position and their velocity. We take an optimization objective function.

$$J = \min\left(\sum_{i=1}^{m}(y(i) - \hat{y}(i))^2\right) \quad (3)$$

where m is the number of identification sample data. $\hat{y}$ is the output of the identified model. y is the actual process output.

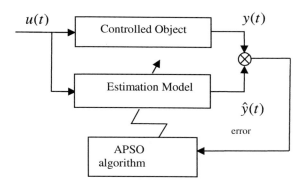

Fig. 1. System Identification Diagram

The fitness function f is defined as

$$f = 1/(J+1) \quad (4)$$

The denominator of the fitness function $J+1$ is designed to prevent calculating overflow when the value of objective function tends to 0. Each particle' initial position is $pbest$ position and each individual particle' fitness is calculated. Hence, the fitness of the initial global position $gbest$ is the best of all the individual position.

(2) Adaptive Adjustment of Inertia Weigh: inertia weigh w is given by

$$w = w_{Max} - iter \times \frac{w_{Max} - w_{Min}}{iter_{Max}} \quad (5)$$

where w_{Max} is maximum inertia weigh; w_{Min} is minimum inertia weigh; *iter* is current generation number; $iter_{Max}$ is total generation number.

(3) Adaptive variation global extremum *gbest* : global fitness variance of the particle swarm σ^2 is set to the evaluation criteria. When σ^2 is less than the given value, it means that the particle is associated with a local optimum. Hence, the mutation probability p_m is calculated as follows:

$$p_m = \begin{cases} k, \sigma^2 < \sigma_d^2 \\ 0, others \end{cases} \quad (6)$$

where k is the random number in the interval $[0.1, 0.3]$. The value of σ_d^2 is according to the actual situation, which is much less than the maximum of σ_d^2.

A section random perturbation operator is designed to fulfill the mutation. operation of *gbest*. The generation number is described as $iter_1 = 0.5 \times iterMax$. where $gbest_k$ is denoted the k th dimension of the *gbest*, η is the random variable which obeyed normal distribution in the interval(0, 1). Thus,

$$gbest_k = gbest_k \times (1 + \alpha \cdot \eta) \quad (7)$$

where α =0.5, when $iter \leq iter_1$, α =0.1, when $iter > iter_1$.

(4) The Particle Velocity Updating: The individual velocity is updated according to the following equation:

$$v(k+1) = w \cdot v(k) + c1 \cdot rand() \cdot (pbest(k) - x(k)) + c_2 \cdot rand() \cdot (gbest(k) - x(k)) \quad (8)$$

where $v(k)$ is the k th generation velocity, $x(k)$ is the k th generation particle' current position. , $rand()$ is the random number between 0 and 1, c_1 and c_2 represent the "self cognition", $c_1 = c_2 = 2$, w is inertia weigh.

During generations, each particle's maximum velocity is bounded to v_{Max} ; each particle's minimum velocity is bounded to v_{Min}.

(5) The Particle Position Updating: The individual position is updated according to the following equation:

$$x(k+1) = x(k) + v(k+1) \quad (9)$$

During generations, each particle's position is bounded in a certain interval.

(6) Evaluation of Each Particle: Calculate the particles fitness after generation. If its fitness is better than the fitness of *pbest*, the new position will be set to *pbest*. If the optimum particle fitness in the swarm is better than the fitness of *gbest*, the new position will be set to *gbest*.

(7) If the end conditions are met, the global extreme value *gbest* is the required optimum, the algorithm finish. Otherwise, jump to step (2) to continue the generation.

3 Simulations and Analysis

3.1 Data Acquisition

There are nine flame paths, eight workpiece rooms in the anode baking furnace, which is produced by HuaLu Aluminum Company of Aluminum Corporation of China Ltd. There are three layers in each workpiece room, And each layer could place seven anode carbon blocks. So each furnace can place a total of 3×7×8×18=3024 blocks, each furnace production cycle is 28 hours. There are altogether 7 carbon blocks, with each same size of 703×5240×5180 and 8 blocks, with each same size of 10394×5240×5184.

Table 1. Flame path temperature (℃)

	1号	2号	3号	4号	5号	6号	7号	8号	9号
1	815	869	835	831	813	810	853	866	849
2	867	869	872	869	869	927	869	869	869
3	878	880	924	879	879	935	878	879	878
4	896	895	894	896	896	928	894	895	895
5	907	908	909	910	911	933	907	910	910
6	917	918	916	919	920	934	915	917	918
7	924	923	924	923	926	940	922	925	924
8	929	932	932	931	932	940	934	930	931
9	938	939	939	939	936	944	937	937	940
10	947	946	946	946	947	945	947	949	946
11	955	953	954	952	956	954	953	953	953
12	960	959	959	961	960	961	961	962	960
13	969	970	969	970	970	968	970	970	970
14	975	979	973	976	975	976	974	974	975
15	980	983	967	982	981	982	982	982	982
16	988	991	983	990	990	988	989	989	989
17	996	998	1000	997	997	998	996	998	987
18	1005	1003	1005	1004	1005	1002	1003	1003	1004
19	1009	1014	1018	1016	1018	1014	1018	1015	1016
20	1020	1027	1025	1021	1019	1020	1018	1020	1020

$$G(s) = \begin{bmatrix} g11 & g12 & 0 & 0 & 0 & 0 & 0 & 0 & 0 \\ g21 & g22 & g23 & 0 & 0 & 0 & 0 & 0 & 0 \\ 0 & g32 & g33 & g34 & 0 & 0 & 0 & 0 & 0 \\ 0 & 0 & g43 & g44 & g45 & 0 & 0 & 0 & 0 \\ 0 & 0 & 0 & g54 & g55 & g56 & 0 & 0 & 0 \\ 0 & 0 & 0 & 0 & g65 & g66 & g67 & 0 & 0 \\ 0 & 0 & 0 & 0 & 0 & g76 & g77 & g78 & 0 \\ 0 & 0 & 0 & 0 & 0 & 0 & g87 & g88 & g89 \\ 0 & 0 & 0 & 0 & 0 & 0 & 0 & g98 & g99 \end{bmatrix} \quad (10)$$

There are 3 flame shelves, 2 gas-excluded shelves, 1 temperature-measured and pressure-measured shelf, 1 zero-pressure shelf, 1 blower, 2 cooling shelf and some industry field Ethernet control system .9 flame path are coupled mutually .Only considering the coupling between the adjacent flame path, 25 flame path models are built. The data are collected in Table 1 and Table 2, the flame path models equations are as follows(10).

Table 2. Pulse electromagnet valve current (mA)

	1号	2号	3号	4号	5号	6号	7号	8号	9号
1	8.81	4.00	4.83	9.00	9.68	10.4	6.62	5.57	6.61
2	10.7	8.08	13.7	9.20	11.9	11.0	10.5	8.50	9.62
3	10.4	8.77	13.1	10.3	12.5	6.52	12.2	7.85	9.42
4	10.2	8.49	5.48	9.71	12.7	4.00	11.0	8.20	9.73
5	9.41	6.53	7.48	8.19	10.0	4.00	10.5	9.34	8.01
6	9.38	6.95	7.80	8.34	10.8	4.00	9.40	8.06	8.75
7	9.91	6.69	7.80	8.88	11.1	4.00	10.4	7.63	7.86
8	9.75	6.71	7.67	8.55	11.1	4.00	9.87	8.54	8.83
9	9.46	6.68	7.74	8.07	10.9	4.00	10.3	7.07	8.04
10	10.3	6.97	8.42	8.12	10.8	4.97	11.0	7.97	8.51
11	9.83	7.40	8.79	8.21	12.1	6.12	11.8	9.31	8.12
12	9.60	7.22	8.53	8.40	10.3	8.01	10.9	7.73	8.73
13	10.6	6.76	8.63	8.65	11.4	9.04	10.9	8.74	8.42
14	10.2	7.20	9.31	9.27	11.6	9.09	10.9	8.43	9.14
15	10.2	7.10	11.8	9.21	12.7	9.17	10.5	8.31	9.23
16	10.6	7.08	15.3	9.62	11.6	8.75	11.2	9.02	9.45
17	10.5	7.72	16.8	9.66	11.6	10.2	12.1	8.63	9.91
18	10.9	7.61	13.8	11.7	12.1	8.21	12.5	11.1	9.37
19	11.8	8.35	12.6	10.4	12.6	10.3	11.9	11.4	9.34
20	10.7	9.44	13.2	10.7	11.8	12.1	11.9	11.6	9.93

3.2 Simulation Results and Analysis

APSO identification steps are as mentioned above. The anode baking temperature identification model parameters are as follows:

$y_{11}(k)=0.5012y_{11}(k-1)+0.1816y_{11}(k-2)+0.01211u_1(k-1-d)-0.002306u_1(k-2-d)$, d=13 ;

$y_{12}(k)=0.4328y_{12}(k-1)+0.2153y_{12}(k-2)+0.003312u_2(k-1-d)-0.002133u_2(k-2-d)$, d=9 ;

$y_{21}(k)=0.3876y_{21}(k-1)+0.1223y_{21}(k-2)+0.01325u_1(k-1-d)-0.0156u_1(k-2-d)$, d=11 ;

$y_{22}(k)=0.6654y_{22}(k-1)+0.1567y_{22}(k-2)+0.01257u_2(k-1-d)-0.0188u_2(k-2-d)$, d=12 ;

$y_{23}(k)=0.5236y_{23}(k-1)+0.1321y_{23}(k-2)+0.001098u_3(k-1-d)-0.06657u_3(k-2-d)$, d=11 ;

$y_{32}(k)=0.5011y_{32}(k-1)+0.2087y_{32}(k-2)+0.006675u_2(k-1-d)-0.003512u_2(k-2-d)$, d=10 ;

$y_{33}(k)=0.5123y_{33}(k-1)+0.2133y_{33}(k-2)+0.004212u_3(k-1-d)-0.002875u_3(k-2-d)$, d=10 ;

$y_{34}(k)=0.7765y_{34}(k-1)+0.1538y_{34}(k-2)+0.005212u_4(k-1-d)-0.00406u_4(k-2-d)$, d=12 ;

$y_{43}(k)=0.4833y_{43}(k-1)+0.1675y_{43}(k-2)+0.007765u_3(k-1-d)-0.001445u_3(k-2-d)$, d=11 ;

$y_{44}(k)=0.6788y_{44}(k-1)+0.1854y_{44}(k-2)+0.00213u_4(k-1-d)-0.0004439u_4(k-2-d)$, d=11 ;

$y_{45}(k)=0.5321y_{45}(k-1)+0.1033y_{45}(k-2)+0.003214u_5(k-1-d)-0.002833u_5(k-2-d)$, d=10 ;

$y_{54}(k)=0.6754y_{54}(k-1)+0.1025y_{54}(k-2)+0.005567u_4(k-1-d)-0.002368u_4(k-2-d)$, d=11 ;

$y_{55}(k)=0.5087y_{55}(k-1)+0.1368y_{55}(k-2)+0.00335u_5(k-1-d)-0.001037u_5(k-2-d)$, d=10 ;

$y_{56}(k)=0.5126y_{56}(k-1)+0.2237y_{56}(k-2)+0.0076u_6(k-1-d)-0.002175u_6(k-2-d)$, d=10 ;

$y_{65}(k)=0.5512y_{65}(k-1)+0.1897y_{65}(k-2)+0.006754u_5(k-1-d)-0.001879u_5(k-2-d)$, d=10 ;

$y_{66}(k)=0.5873y_{66}(k-1)+0.2135y_{66}(k-2)+0.007566u_6(k-1-d)-0.001567u_6(k-2-d)$, d=10 ;

$y_{67}(k)=0.5132y_{67}(k-1)+0.1897y_{67}(k-2)+0.00569u_7(k-1-d)-0.004563u_7(k-2-d)$, d=11 ;

$y_{76}(k)=0.5563y_{76}(k-1)+0.1538y_{76}(k-2)+0.05032u_6(k-1-d)-0.0003356u_6(k-2-d)$, d=13 ;

$y_{77}(k)=0.5869y_{77}(k-1)+0.1857y_{77}(k-2)+0.02566u_7(k-1-d)-0.001833u_7(k-2-d)$, d=11 ;

$y_{78}(k)=0.5547y_{78}(k-1)+0.1899y_{78}(k-2)+0.01578u_8(k-1-d)-0.00504u_8(k-2-d)$, d=10 ;

$y_{87}(k)=0.5166y_{87}(k-1)+0.2387y_{87}(k-2)+0.07689u_7(k-1-d)-0.002356u_7(k-2-d)$, d=11 ;

$y_{88}(k)=0.5128y_{88}(k-1)+0.2876y_{88}(k-2)+0.02236u_8(k-1-d)-0.003518u_8(k-2-d)$, d=10 ;

$y_{89}(k)=0.6093y_{89}(k-1)+0.1876y_{89}(k-2)+0.008765u_9(k-1-d)-0.002356u_9(k-2-d)$, d=10 ;

$y_{98}(k)=0.6087y_{98}(k-1)+0.1056y_{98}(k-2)+0.03512u_8(k-1-d)-0.002234u_8(k-2-d)$, d=12 ;

$y_{99}(k)=0.5123y_{99}(k-1)+0.1067y_{99}(k-2)+0.02135u_9(k-1-d)-0.002679u_9(k-2-d)$, d=13。

Compared the anode baking flame path temperature model output and the actual temperature output, each flame path error is shown in Figure 2.

From Fig. 2, it can be concluded that the anode baking flame path temperature output from the APSO algorithm and the actual output fit well. Errors are maintained in ± 1°C. This model can achieve the precise control of the anode baking flame path temperature.

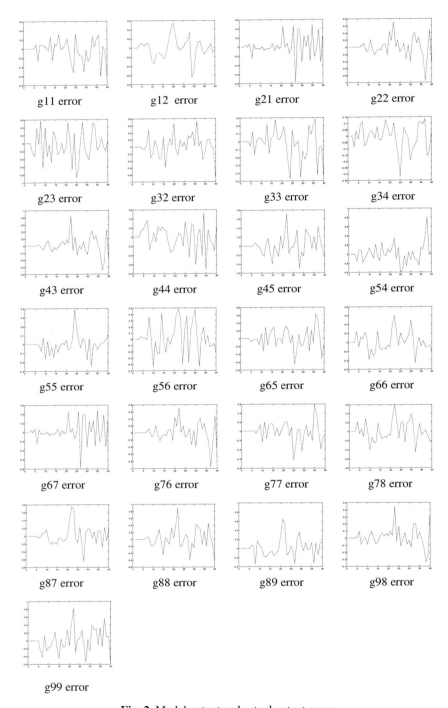

Fig. 2. Model output and actual output error

4 Conclusion

1) Transfer function model is used to describe the relationship between input and output. The anode baking temperature can be well represented through a second-order lag model.
2) The anode baking flame path temperature control model can be well identified and optimized through APSO algorithm.
3) The anode flame path temperature model building makes it possible to control the anode baking flame path temperature precisely and achieve energy saving for aluminum electrolysis.

References

1. Furman, A.: A mathematical model simulation at anode baking furnace. Light Metals, 545–549 (1980)
2. Benton, et al.: Method and apparatus for producing uniformly baked anodes, United States Patent 4, 354,828 (1982)
3. Thibault, M.A.: Simulation the dynamics of the anode baking ring furnace. Light Metals,1144–1148 (1985)
4. Li, X.-b., Wu, Y.-x., Kou-De-min: Modeling and Multivariable Decoupling Control for Anode Baking Flue Gas Temperature. Control Engineering 16-2, 153–157 (2009)
5. Kocaefe, Y.S., Demedde, E., Kocaefe, D.: A 3d mathematical model for the horizontal anode baking frnace. Light Metals, 529–534 (1996)
6. Wu, Y.-x., Li, X.-b., Sun, H.-y.: Parameter Identification and Optimization Based on Adaptive Particle Swarm Optimization Algorithm. Science Technology and Engineering 8, 3777–3782 (2008)

An Improved Ant Colony System for Assembly Sequence Planning Based on Connector Concept

Hwai-En Tseng

Department of Industrial Engineering and Management
National Chin-Yi University of Technology
35, Lane215, Section 1 Chung-Shan Road, Taichung City 41101, Taiwan
hwai_en@seed.net.tw

Abstract. Assembly sequence planning (ASP) needs to take into subtle consideration certain constraint factors such as geometric data from CAD, tools and fixtures so as to work out a specific assembly sequence. Guided genetic algorithms proposed by Tseng (2006) overcame the restrictions of traditional algorithms through a new evolution procedure. In this study, ant colony system (ACS) algorithms are adopted to solve the ASP problem. By comparing the result of guided genetic algorithms with that of ACS, it has been found that ACS can effectively provide better quality solutions to the ASP problem.

Keywords: genetic algorithms, ant colony system, connectors, adjacency list.

1 Introduction

In the modern manufacturing environment, a well-organized assembly sequence planning (ASP) can have potential benefits for the application such as error reduction and the cost down of total production [1]. During the process of ASP, a planner needs to consider the constraint factors such as geometric data from CAD, the combination property of parts, tools to organize the suitable assembly sequence. In short, the purpose of ASP problems is to find out the optimal/near-optimal assembly sequence under the premise that the precedence graph is known.

In the past, many researchers have applied traditional heuristic and graph searching methods to solve the ASP problem [2]. It is time-consuming to use such methods when the number of constraints is larger. To cope with such complicated problems, many researchers try other approaches such as genetic algorithms (GAs) to find suitable assembly sequences in a more efficient way [3].

Since Dorigo and Gambardella (1997) applied ant colony system (ACS) algorithms successfully in the Traveling Salesman Problem (TSP) [4], the algorithms have been getting more and more popular. In this study, the author attempted to improve the ACS algorithms and apply them to ASP problems. The model is operated by connector-based environment. That is to say, connectors function as assembly elements in product description and serve as concept product building blocks in the design stage. Therefore, more engineering features can be included [3].

2 Connector Related Information

In this study, four properties of connectors are taken into consideration: combination, assembly tools, assembly directions, and precedence relationships.

(1) Combination property
As shown in Table 1, product components can be assigned to specific connectors according to these four combination categories.

Table 1. Classification of connector types

Type		Code	Example
Fixed fastener	Disassembled	FD	Screw, bolted joint, key, spline, wedge
	Not disassembled	FND	Pressing fits, riveted joints, welding
Movable fastener	Disassembled	MD	Snap ring, bearing, spring
	Not disassembled	MND	Races and ball-bearing balls

(1) FD, fixed fastener disassembled, (2) FND, fixed fastener not disassembled, (3) MD, movable fastener disassembled, (4) MND, movable fastener not disassembled.

(2) Assembly tools
According to the degree of difficulty, there are four types of assembly tools (Table 2).

Table 2. Classification of assembly tools.

Level	Force magnitude	Tool name	Assembly operation
T_1	None	Hand	No tools are needed; i.e., the assembly is manual.
T_2	Small	Work-bench, handgun, screw-driver, spanner, pliers	Use a simple hand tool to assemble, no strict interference occurs between components.
T_3	Medium	Screw driver, spanner, racket spanner	Use simple hand tool to assemble; other tools are needed to support the assembly work.
T_4	Large	Hacksaw, heavy sledge-hammer, crusher, torsional twister, chassis	Use a special tool to assemble the product; the operation may cause a destructive result.

(3) Assembly direction
In the product assembly task, the direction property (+x, -x, +y, -y, +z, and –z) will affect the execution of assembly action.

(4) Precedence relationship
The precedence relationship of connectors depends on the engineering features or geometric information of connectors. For instance, $C_1 \rightarrow C_2$ means that Connector C_1 has a higher priority in assembly sequence than Connector C_2.

Figure 1(a) illustrates the pen example and Figure 1(b) shows the precedence graph. The engineering information, including combination types, assembly tools and assembly directions of connectors, is listed in Table 3. Take the connector composed of the ink cartridge and the tip as an example. Its combination property belongs to the fixed disassembled (FND) category; the assembly direction is X; and a hand vice (T3) should be used as the assembly tool.

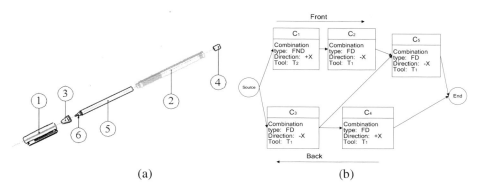

Fig. 1. (a) Graph of parts, (b) connector-based precedence graph for the pen example.

Table 3. Connector information for a pen.

	Connector name	Combination type	direction	Tool	Component owned by connector
C_1	Interference fit	FND	x	T2	5,6
C_2	Insert	FD	-x	T1	3,6
C_3	Screw	FD	-x	T1	2,3
C_4	Snap fit	FD	x	T1	1,2
C_5	Turn	FD	-x	T1	2,4

(1) FD, fixed fastener disassembled, (2) FND, fixed fastener not disassembled, (3) MD, movable fastener disassembled, (4) MND, movable fastener not disassembled.

In this study, the final score calculation of the ant's path are based upon the similarity of the connector's engineering data. In Formula (1), $SS_{(i,j)}$ represents the element in the ith row and jth column of the matrix SS, namely, the engineering information similarity between Connector i and Connector j. The value of similarity $S_{(i,j)}$ can be calculated by Formula (1):

$$SS_{(i,j)} = W_c \times C_{(i,j)} + W_t \times T_{(i,j)} + W_d \times D_{(i,j)} \tag{1}$$

where
W_c: weight of the combination property;
W_t: weight of the assembly tool property;
W_d: weight of the assembly direction property;
$SS_{(i,j)}$: similarity of engineering information between Connector i and Connector j. If i=j, then $S_{(i,j)}$=0; i, j = 1,2,3.........; n is the number of connectors.
$C_{(i,j)}$: when the combination types of connectors are the same, $C_{(i,j)}$=1; otherwise, $C_{(i,j)}$=0.
$D_{(i,j)}$: when the assembly directions of connectors are the same, $D_{(i,j)}$=1; otherwise, $D_{(i,j)}$=0.
$T_{(i,j)}$: when the assembly tools of connectors are the same, $T_{(i,j)}$=1; otherwise, $T_{(i,j)}$=0.

The calculation of weights in Formula (1) starts with the ranking of the relative importance of engineering properties. The weight for each engineering property can be obtained from Formula (2) (Barron and Barrett, 1996):

$$W_r = \frac{e+1-r}{\sum_{l=1}^{e} l} = \frac{2(e+1-r)}{e(e+1)} \tag{2}$$

where
e: the number of engineering properties; r: the rank of the engineering properties.

Connector	C_1	C_2	C_3	C_4	C_5
C_1	—	0	0	0.1666	0
C_2	0	—	1	0.8333	1
C_3	0	1	—	0.8333	1
C_4	0.1666	0.8333	0.8333	—	0.8333
C_5	0	1	1	0.8333	—

Fig. 2. Similarity matrix SS for engineering information.

In this study, the objective function is based upon the similarity of engineering data between connectors. As shown in Figure 2, based on the similarity matrix of connector's engineering information, SS, the sum of the similarity of engineering information between adjacent connectors can be calculated by Formula (3):
where

$$F = \sum_{h=1}^{n-1} SS_{h,h+1}^{A} \qquad (3)$$

$SS_{h,h+1}^{A}$: Similarity of engineering data between Connector h and Connector h+1 in an assembly sequence; h = 1,2,3……n; n is the number of connectors in the assembly sequence.

If the assembly sequence for a pen connector is 1→2→3→4→5, then the value of the objective function can be obtained from the similarity matrix SS through Formula (3) and Figure 2: F=$SS^A_{1,2}$+$SS^A_{2,3}$+$SS^A_{3,4}$+$SS^A_{4,5}$ =2.8333.

In the precedence graphs, the edges linked by each node are defined as the degrees, which can be further divided into outdegree and indegree. They are defined as follows:
Outdegree: For a vertex C_i, its outdegree is the number of directed edges that emanate from that vertex C_i. If the precedence relationship of C_i and C_j is C_i →C_j; the number of C_j is the outdegree of C_i; $C_j \neq C_i$.
Indegree: For a vertex C_i, its indegree is the number of directed edges that lead to that vertex C_i. Suppose the precedence order will be C_j→C_i; C_j indicates the number of C_i 's indegree; $C_j \neq C_i$.

In this study, the outdegree and indegree are used to construct the the successor lists (SL) and predecessor lists (PL). SL represents the lists where successors in each node in the forward searching, from the *Source* to the *Sink*, are recorded while PL represents the lists in which predecessors in each node in the backward searching, from the *Sink* to the *Source*, are taken down. The SL and PL store for each connector of its successors and predecessors (Figure 3).

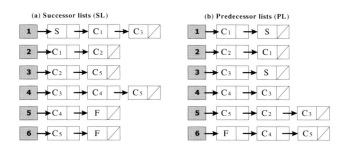

Fig. 3. Data storage for pen case (a) Successor lists (SL) (b) Predecessor lists (PL)

3 Improved Ant Colony System for Assembly Sequence Planning

The definitions of the terms used in this study are listed below:
t: number of iterations.   m: number of ants.
n: number of connectors. τ_0: the initial pheromone.
q: q is a random number uniformly distributed, $0 \leq q \leq 1$.
q_0: a parameter which determines the relative importance of exploitation versus exploration, $0 \leq q_0 \leq 1$.
ρ: the rate of pheromone evaporation, $0 \leq \rho \leq 1$.

α: the importance of pheromone, $\alpha > 0$. β: the importance similarity, $\beta > 0$.
$\tau_{(i,j)}(t)$: the current pheromone intensity on Connector i and Connector j.
$SS_{(i,j)}$: the similarity between Connector i and Connector j.
best_path: a matrix to record the overall optimal solution.
$\Delta \tau_{(i,j)}(t)$: the reciprocal of the optimal solution in m solutions.
path$_{(i,j)}(t)$: a matrix to record the times the ant takes a path from Connector i to Connector j. If the ant moves from C_i to C_j, then add 1 to *path*$_{(i,j)}$.
The improved ACS algorithms are shown in Figure 4. Formula (4), is the exploitation mechanism. When $q \leq q_0$, the ant will make their choices upon the basis of pheromone ($\tau_{(i,j)}$) and similarity $SS_{(i,j)}$. The second one, Formula (5), is the exploration mechanism. (Step 9).

In order to reduce the differences of pheromone among connectors, which will sometimes lead to the local optimal solution, a local update of the pheromone is conducted after each iteration. The evaporation of pheromone will not process until all of the ants complete all of the connectors. Formula (6) shows how to calculate the local update (Step 11). In ACS algorithms, Step 12 deals with the global update and Formula (7) denotes the calculation of global update.

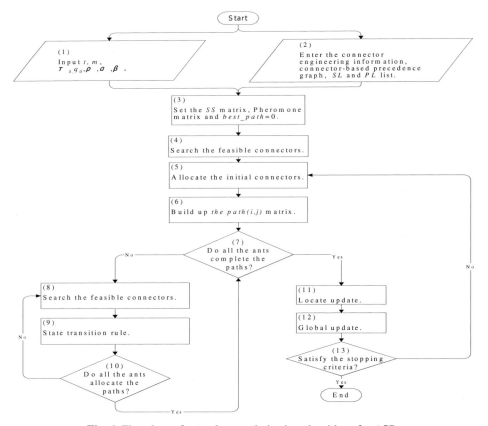

Fig. 4. Flowchart of ant colony optimization algorithms for ASP

$$J = \begin{cases} \arg\max_{g \in M_k(i)} \left\{ [\tau_{(i,g)}]^\alpha \times [SS_{(i,g)}]^\beta \right\} & \text{if } q \leq q_0 \\ \dfrac{[\tau_{(i,j)}]^\alpha \times [SS_{(i,j)}]^\beta}{\sum_{g \in M_k(i)} [\tau_{(i,g)}]^\alpha \times [SS_{(i,g)}]^\beta} & \text{if } q > q_0 \end{cases} \quad (4)$$

$$(5)$$

$$\tau_{(i,j)} \leftarrow (1-\rho) \times \tau_{(i,j)} + path_{(i,j)} \times \rho \times \tau_0 \tag{6}$$

$$\tau_{(i,j)} \leftarrow \tau_{(i,j)}(t) + \rho \times \Delta\tau_{(i,j)} \tag{7}$$

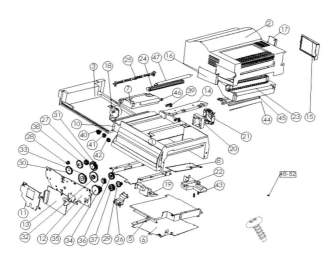

Fig. 5. Laser-printer parts. Parts numbered 48-92 are screws:

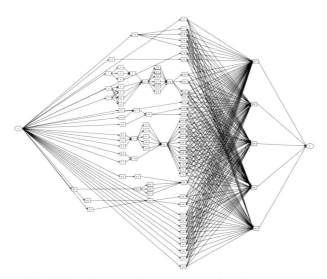

Fig. 6. Precedence graph of connectors for a laser printer.

4 Practical Examples

In this study, Borland C++ 6.0 was used for the programming. A printer were used to verify the feasibility of ACS algorithms. The tests were conducted on an Intel Pentium 4, with a 2.4-GHz CPU and 512MB RAM. Finally, the results were compared with those of the GAs proposed by Tseng (2006).

92 parts compose the printer (Figure 5) and they are combined into 91 connectors. Figure 6 shows the precedence graph of the printer connectors. In this case, the number of the ants in ACS algorithm is set to be 50; the initial pheromone 0.005; the pheromone evaporation rate 0.6; α 0.5; β 5; q_0 0.9. For GAs, the parent population size is 100; the crossover rate 75%; the mutation rate, 50%. In terms of the engineering data, the combination type, assembly tools, and assembly direction are equally important. Similarly, ten trials were conducted to compare the results of GAs under a 30-second termination limitation. The comparison of the results for these two algorithms is shown in Table 4. The results in Table 4 demonstrate that, either in the average objective value or the maximum objective value, the ACS performs better than Guided-Gas proposed by Tseng [3].

Table 4. Comparison between ACS and Guided-Gas

Method	Average value	Max value
ACS	81.6332	82
Guided-GAs	76.8669	77.667

5 Conclusions

In this study, the authors attempted to solve the ASP problem by ant colony system. Compared with the results of Tseng's Guided-GAs [3], the ACS algorithms are better in the quality of solutions. In the future, the automatic precedence graph in CAD systems is an issue worthy of in-depth exploration.

References

1. Otto, K., Wood, K.: Product Design-Techniques in Reverse Engineering and New Product Development. Prentice-Hall, London (2001)
2. Gottipolu, R.B., Ghosh, K.: Representation and selection of assembly sequences in computer-aided assembly process planning. International Journal of Production Research 35, 3447–3465 (1997)
3. Tseng, H.E.: Guided genetic algorithms for solving larger constraint assembly problem. International Journal Production Research 44(3), 601–625 (2006)
4. Dorigo, M., Gambardella, L.M.: Ant Colony System: A cooperative leaning approach to the traveling salesman problem. IEEE Transactions in Evolutionary Computation 1(1), 53–66 (1997)

Study on Hysteresis Current Control and Its Applications in Power Electronics[*]

Ping Qian[1] and Yong Zhang[2]

[1] Shanghai Institute of Technology, Shanghai 210094, China
[2] Anhui University of Technology, Maanshan 243002, China
zhangy422@126.com

Abstract. The paper, which is based on some related materials, presented a survey of hysteresis current control in power and electronics. The applications contain hysteresis current controlled dual-buck half bridge inverter, hystersis current method for active power filter based on voltage space vector and hystersis-band current tracking control of grid-connected inverter.

Keywords: Hysteresis current, Voltage space vector, Three-level hystersis control.

1 Introduction

The actual current signal is compared with the given current signal of the inverter by hysteresis current control.If the actual current signal exceeds the given current signal a certain range, we can change the switching state of the inverter to control the change of the actual current signal in order to track the given current signal.

Hysteresis current control has a series of advantages such as quick response, internal current limiting capacity and stability. Based on the above advantages, hysteresis current control is widely used in power inverter, AC drive and active power filter and so on.

2 Hysteresis Current Controlled Dual Bridge Inverter[1]

Improving the switching frequency and achieving high efficiency are the major topics and development directions of the study on the inverter. Dual buck half bridge inverter (DBI) provides a reliable way to solve the two topics which are often mutually contradictory. The existence of discontinuous conduction mode operation state requires the bias of inductor current for DBI implement with linear controllers like ramp comparison SPWM controllers. A novel operation scheme for DBI and a hysteresis current controlled dual buck half bridge inverter(HCDBI) are proposed. Experiment Verifies the high efficiency and good dynamic performance can be obtained by using this inverter.

---

[*] Project supported by the Shanghai Committee of Science and Technology, China(Grant No. 08530511500).

2.1 Proposal of Dual Half Bridge Inverter(DBI)

N.R.Zargari gave the circuit of dual buck half bridge inverter [2] (see Fig.1).

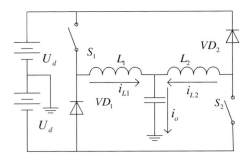

Fig. 1. BUCK inverter circuit

Because of the existence of the inductor L1 and L2, DBI solves the problem of the passing of the traditional half bridge inverter and it doesn't need to set dead time. Besides, body diodes of the power diodes don't work and continuous current flows through the diode on the same bridge arm, so the optimization of the power diode could be decoupled.

2.2 Hysteresis Controlled Inverter

The methods in the literature[3,4] are based on ramp comparison SPWM controllers. In order to make DBI work normally, it is necessary to set a minimum bias current which brings lots of conduction losses, so it is possible to use hysteresis current control to eliminate voltage distortion which is caused by bias current and discontinuous current and make DBI work in the mode which doesn't have bias current(see Fig.2).

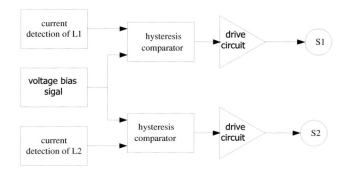

Fig. 2. Principle of current of HCDBI

Its working principle is that voltage external loop achieve stable voltage control and current internal loop controls two filtering inductors' current separately. When the inductors of the inverter output forward current, the circuit of S1,VD1,L1 and C works. Hysteresis current controllers limit the high-frequency ripple of the i_{L1} in the width of ring set by modulating the filtering output. S2,VD2 doesn't work and the current of the L2 is 0. Otherwise, S1,VD1,L1 doesn't work and S2,VD2,L2 work. Hysteresis current control has a strong power of self-adaptation. It can rectify the switching frequency automatic and reduce the width of the discontinuous current.

3 Hysteresis Current Control Method for Active Power Filter Based on Voltage Space Vector

The performance of active power filter depends on the methods of current controlling in the circuit of current tracking control largely. Based on the analysis of voltage space vector, a novel Hysteresis current control method is proposed in literature[5].This method gives the best suitable switching of voltage vector using the space distribution of current error vector and reference voltage vector, and then controls the current error below the hysteresis width. Voltage space vector is introduced into the control system to eliminate the phase interference, and it can be realized simply without complicated vector transform.. The proposed method has quick current response, and can limit the current error and the switching frequency with good current tracking performance.

Equivalent circuit of typical active power filter based on voltage source inverter is as in Fig.3 where actual switches are instead by ideal switches.

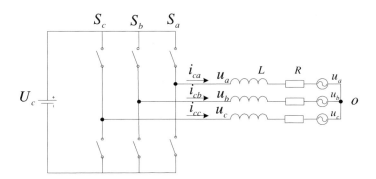

Fig. 3. Equivalent circuit diagram of APF

The interaction between the APF output voltage and switching function is as in table.1 (U_c is as the voltage base).

Table 1. Switch modes and output voltages of APF

S$_{abc}$	u$_a$	u$_b$	u$_c$	U$_k$
000	0	0	0	U$_0$
100	2/3	-1/3	-1/3	U$_1$
110	1/3	1/3	-2/3	U$_2$
010	-1/3	2/3	-1/3	U$_3$
011	-2/3	1/3	1/3	U$_4$
001	-1/3	-1/3	2/3	U$_5$
101	1/3	-2/3	1/3	U$_6$
111	0	0	0	U$_7$

Space vector and stationary orthogonal coordinate $\alpha - \beta$ are introduced to eliminate the effects of phases. When α-axis and a-axis coincide, the function of these two coordinates is as follows:

$$f = f_\alpha + jf_\beta = \frac{2}{3}(f_a + f_b e^{j2\pi/3} + f_c e^{j4\pi/3}). \qquad (1)$$

From the Fig.6, vector formula of output voltage of APF is

$$u = Ldi_c / dt + Ri_c + U. \qquad (2)$$

When output current is instruction current vector i_c^*,

$$u^* = Ldi_c^* / dt + Ri_c^* + U. \qquad (3)$$

In the formula (3), u^* is APF's output reference voltage vector which corresponds to instruction current vector i_c^*. From the formula (2) and formula (3), formula (4) can be derived as follows:

$$Ld\Delta i / dt = U_{AB} |\Delta i|. \qquad (4)$$

From the formula (4), the right APF's output reference voltage vector U_K (k=0,...7) can be selected to control the conversion rate $d\Delta i / dt$ of current deviation vector Δi in order to control Δi.

APF's output reference voltage vector U_K (k=0,...7) which correspond to 8 modes of the switches make vector space be divided into 6 triangle regions and the area of u^* can also be divided into 6 triangle regions as I -VI in the Fig.4(a). In order to make the

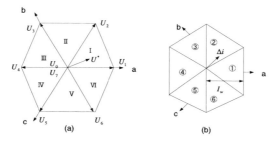

Fig. 4. Region division of u^* 、 Δi

distinguishing of the positive electrode and the negative electrode uncomplicated, Δi is divided into 6 triangle regions as in the Fig.4(b).The width of hysteresis is I_w.

From the analysis of the formula (4), once reference voltage vector u^* and current deviation vector Δi are fixed, the locations of the space region of vectors are fixed. In order to control current tracing, a right voltage vector U_k is selected to make the conversion rate $d\Delta i / dt$ of current deviation vector and current deviation vector Δi in the opposite direction and the $|\Delta i|$ is limited in a certain width of hysteresis. Thus the current tracking control can be conducted.

4   Hysteresis-Band Current Tracking Control of Grid-Connected Inverter [6]

Grid-connected inverter and its control technologies attract more and more attention in the word today when energy shortage is serious. The key of grid-connected inverter is the control technology of grid-connected current [7,8]. Its control target is to make grid-connected current track the frequency and phases of grid voltage and the change of grid's given capacity. Furthermore, the total distortion should be low in order to reduce the harmonic's effects on the grid.

Three-level hysteresis control which has outstanding advantages is one of control methods of PWM inverter. It has freewheeling state in addition to energy input and energy feedback. U_{AB} is unipolar modulation in the half output period and current ripple is smaller than two-level modulation when there is the same switching frequency [9,10].

4.1   Grid-Connected Inverter System

Single-phase PWM grid-connected inverter is as in Fig.5. DC source can be solar, wind and other renewable energy power generation equipment.

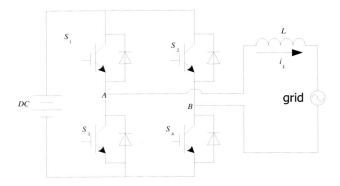

Fig. 5. Grid-connected single-phase PWM inverter

The control system which uses hysteresis control strategy is as in Fig.6. Sync signal which is detected by part of phase detection and the given amplitude of grid-connected current are delivered to sine wave generator and generate reference current signal i_{ref} which has the same phase and the same frequency with grid voltage. Then the feedback signal i_L of grid-connected current is compared with i_{ref} and the bias is controlled to generate switching control signals.

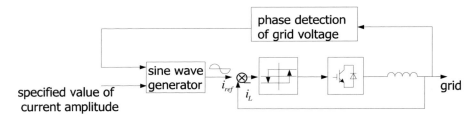

Fig. 6. The control schematic of grid-connected inverter

4.2 Principle of Three-Level Hysteresis Control

As in Fig.7, the bias of the reference current and the output feedback current is compared to hysteresis –band threshold H. When output current is in positive half cycle and the bias is above hysteresis –band upper threshold, S1,S3 or S2,S4 conduct, U_{AB} =0, inductor current decreases. So the bias will decrease. When the bias is below hysteresis –band lower threshold, S1,S4 conduct, $U_{AB} = U_d$, and inductor current rises. So the bias will decrease. When output current is in negative half cycle, there is also experiencing a similar situation.

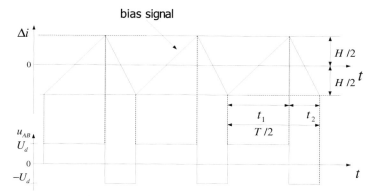

Fig. 7. Three-level hysteresis control method

5 Conclusion

As control algorithms of modern power electronic technology is more and more advanced, hysteresis current have a series of advantages such as quick response, strong internal current limiting and strong stability. Hysteresis control technology based on advanced control algorithms will be applied and developed in more widely occasions.

References

1. Feng, H., Jun, L.: Hysteresis current controled dual buck half bridge inverter. Transactions of China Electrotechnical Society 19(8), 73–77 (2004)
2. Zargari, N.R.: Two switch high performance current regulated DC/AC converter module. IEEE Transaction on Industry Applications 31(3), 583–589 (1995)
3. Peng, F.Z., Mckeever, J.W., Adams, D.J.: Cascade multilevel inverters for utility applications. In: IECON 1997, vol. 2, pp. 9–14, 437–442 (1997)
4. Hanson, D.J.: A transmission SVC for national grid company PLC incorporating a Mvar STATCOM. IEE Colloquium(Digest) 500(23), 1–8 (1998)
5. Guo, Z.-y., Zhou, Y.-q., Liu, H.-c., Hu, K., Zeng, X.-d.: A novel Hysteresis current control method for active power filter based on voltage space vector. Journal Chinese Electrical Electrical Engineering Science 27(1), 112–117 (2007)
6. Gu, H.-r., Yang, Z.-l., Wu, W.-y.: Research on hysteresis-band current tracking control of grid-connected inverter. Journal Chinese Electrical Electrical Engineering Science 26(9), 108–112 (2006)
7. Zhang, X., Zhang, C., Cao, R.: Study on nonlinear control of FV parallel feed inverter. Acta Energiae Solaris Sincica 23(6), 770–773 (2002)
8. Kjaer, S.B., Pedersen, J.K., Blaabjerg, F.: A review of single-phase grid-connected inverters for photovoltaic modules. IEEE Transactions on Industry Applications 41(5), 1292–1306 (2005)
9. Sun, C., Bi, Z.-j., Wei, G.-h.: Modeling and simulation of a three-phase four-leg inverter based on a novel decoupled control technique. Journal Chinese Electrical Electrical Engineering Science 24(1), 124–130 (2004)
10. Baker, D.M., Agelidis, V.G., Nayer, C.V.: Comparison of tri-level and bi-level current controlled grid-connect single-phase full-bridge inverters. In: IEEE ISIE, Guimaraes, Portugal, pp. 463–467 (1997)

Design of Pulse Charger for Lead-Acid Battery

Ping Qian and Maopai Guo

Shanghai Institute of Technology, Shanghai
No.100, Haiquan Road, Fengxian District, Shanghai, China 200235
qping@sit.edu.cn

Abstract. In this paper, conventional means for lead-acid battery charging is briefly introduced, and the polarization phenomenon lasting in the charge process is further analyzed. Aiming to curtail this effect and improve charging efficiency, the circuit based on pulse is designed. It is proved that the charger can improve the charge performance significantly.

Keywords: Sealed lead-acid battery, Polarization, Pulse charger.

1 Introduction

Due to its reliability, low expense and stability, lead-acid battery have been widely used in different areas such as communication station, railroad system and national defense occasion. Most electronic equipment nowadays adopts lead-acid battery as back-up power. In addition, the application of lead-acid battery in emergency lamp, car and yacht are increasing rapidly. However, as a result of improper charging, batteries always have a shorter life time than expected. It indicates that the conventional charging technology is not agreeable for lead-acid battery.

In face of the facts, A charger based on pulse mode is proposed, it charges the battery using pulse way, which is capable to improve the charging efficiency and prolong battery life.

2 Characteristics of Battery Charging

There are different way to charge the lead-acid battery including constant current, constant voltage, two phase, fast charging and balance charging. It is proved that, for the battery nominal as 2V, when the charging voltage is below 2.32-2.35V, no matter the charging current is, the gassing is slight. [1] Only when the voltage reach at 2.35-2.40V, the gassing effect turns to obvious, when the voltage goes above 2.40V, also called gassing voltage point, the gassing becomes fierce. When the charging voltage higher than the gassing voltage, the actual effect of the charging current is for producing gas not for charging the battery, which proved a harm to the battery and wasting of electricity. [2]

In order to shorten the charging process and improve charging efficiency, The best way is to control the charging voltage under the gassing voltage point. The energy stored into the battery from the charging start until at the gassing point is determined by charging current. Hence, for the fast charging purpose, the current value should be set higher. However, the higher current inclines to bring about the polarization effect, which prevent the energy transferring to the battery. To keep the charging current as high value, the polarization effect should be eliminated.

The instantaneous discharge of the battery during the charging phase is proved as a effective method to clear up the polarization. In the discharge duration, the redundant electrons accumulated on the two electrodes move towards the contrary direction of the charging current, so the redundant electrons accumulated on the electrodes reduces rapidly and the polarization effect goes down. Meanwhile, the positive and negative ions in the electrolyte also move towards the contrary direction of the charging current. This play a role of mixing the electrolyte, which can effectively control the polarization caused of density difference. In addition, a part of heat energy generated from ohm polarization is transferred to the load, which can prevent the battery from overheat. [3]

3 Pulse Charger

Based on the principal, In order to cut down the polarization influence and improve charging efficiency. The pulse charger is designed in this paper. Thecharger operates as the sequence of charging-pause-discharging-pause-charging, illustrated as the following figure.

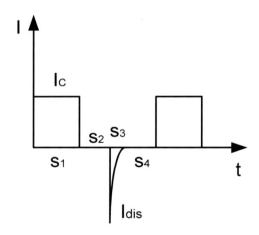

Fig. 1. Curve of current

Terminal voltage of the battery will increase during the charging process. In order to control the current and the voltage. The whole charging process isdivided as three current stages of 1.65A, 0.75A and 0.5A. When the batteryvoltage is low, the current is set as 1.65A and the battery voltage is measuredconstantly, when the voltage reach to 14.4V namely the voltage at the gassingpoint, then decease the current as the next level. [4]

4 Main Circuit

The charger is designed for the battery of 12V 12AH. For the small-sized power system. Flyback converter is a good choice, it is simple in structure and safe in electric, refer to the figure as below, the output is designed as 15V 30W.

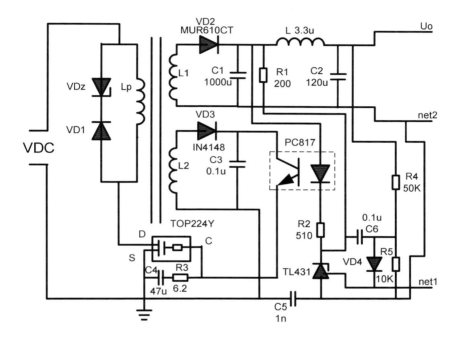

Fig. 2. Main circuit

We choose TOP224Y as the switch device, it's the typical three terminal component of TOPSwitch- II series. The pins specified as C(control pin), S(source pin) and D(drain pin).The internal MOSFET can withstand high voltage to 700V. The AC/DC conversion efficiency near to 90%. It combines the switch element and the control circuit in a single chip, with the function of auto-reset, over-heat protection and over current protection. Since the high integration, the device is compact in size. The flyback converter using this component features fewer external device, small size and high reliability. The IC functions as follow, at the fixed switch frequency 100KHZ, the duty cycle is regulated by the feedback current I_C, thus to realize the control of the constant output voltage. For example, when the output voltage goes below the required value, under the function of optical coupler, the feedback current I_C goes down, D increase, and V_0 goes up, thus stabilize the output voltage as the required value. [5]

5 Battery Set

The battery set consists of battery, operational amplifier and series of mosfet. See the figure as below.

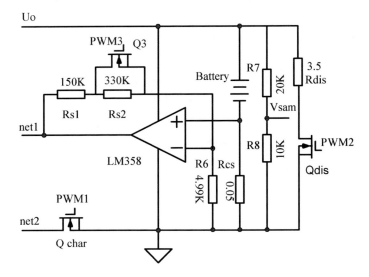

Fig. 3. Battery set

Q_{char} is placed for charging, once turned on, the battery is in charging state. Q_{dis} for discharge purpose, if switched on, the battery turns to discharge state. When Q_{char} and Q_{dis} are both switched off, the battery is at the standing state. During the charging stage, the voltage dropped on the sense resistor R_{CS} is measured and placed on the non-inverting terminal of the operational amplifier, and the voltage will be amplified as

$(R_{S1}+R_{S2})$ /R6 times. Since the voltage of the TL431 reference pin is 2.5V, thus by setting the value of $(R_{S1}+R_{S2})$ can control the charging current value and realize constant current charge. The equivalent value of R_{S2} could be adjusted by changing the output of PWM3.

During the charging stage (Q_{char} is on), the control function is dominated by the output of the operational amplifier, When Q_{char} turns off, the output is regulated by the feedback voltage through R4,R5 and stabilized at 15V.

The charger operates as the sequence of 2S charge - 20mS pause - 10mS discharge - 25mS pause and the cycle is repeated afterwards until the battery is fully charged.

6 MCU Circuit

The following shows the MCU unit, the 5V power is obtained by the TL431. MCU realize measurement and control functions.

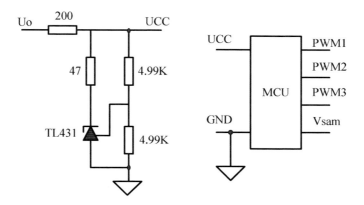

Fig. 4. MCU circuit

The MCU is selected as STC12C5410AD, which having 4 routes of PWM, 8 channels of A/D and various of programmable I/O port. It is easy to fulfill our requirements through these resources.

During the charging process, the battery voltage is measured by the A/D channel, three routes of PWM signal are sent out to the control pin of the switch device, thus realize pulse charge.

7 Conclusions

Due to the adverse effect of polarization, It is compulsory to put high voltage on the battery terminals for fully charging when using the conventional chargers. The high voltage generally leads to water loss, excessive charging, or even heat uncontrolled, which is more likely to cause the battery capacity loss and life shortening. However, the pulse chargers are capable of clearing up the slight sulfuration arised from charge shortage or occasionally no charging. The negative pulse of 10mS could remarkably weaken the polarization voltage, so the battery is likely to accept large current under low voltage. As the result, the time spent in charging the battery to 90% of the capacity is greatly reduced and the life time of the battery is extended accordingly.

References

1. Chan, H.L.: A new battery model for use with battery energy storage systems and electric vehicles power systems. In: Power Engineering Society Winter Meeting 2000, vol. 1, pp. 470–475. IEEE, Los Alamitos (2000)
2. Zhao, Z.: Photovoltaic power and its application. Chinese science press, Beijing (2005)
3. Du, C.: The characteristics of VRLA charge and discharge. Telecom power technologies (2006)
4. Chen, J.: Study on the multi-stage constant current Charging method for small-sized VRLA batteries. Chinese LABAT Man (2004)
5. Sha, Z.: Application of single-chip SMPS. China Machine Press, Beijing (2009)

Hot-Line XLPE Cable Insulation Monitoring Based on Quick Positive and Negative DC Superposition Method[*]

Jianbao Liu[1,2], Zhonglin Yang[2], Liangfen Xiao[2], Leping Bu[2], and Xinzhi Wang[2]

[1] Huazhong University of Science and Technology/Electrical and Electronic Colloge,
Wu Han, China
[2] Navy Engineering University/School of Electrical and Information Engineering,
Wu Han, China
ljbtt@hotmail.comUT

Abstract. According to the feature of DC superposition theory which is applied widely in monitoring the insulation condition of power cable with cross-linked polyethylene insulation (XLPE), such as high precision but time-consuming, so a quick positive and negative DC superposition monitoring method is presented in this paper. In which the Positive and negative DC voltages are injected to detection circuit respectively, and the steady voltages of the two conditions mentioned above are calculated from the transient voltages of sampling resistance. Considering the algebraic average of positive and negative steady voltages as finally steady voltage, so the value of insulation resistance for XLPE power cable is obtained. The proposed method not only has avoided infection in measuring precision resulted from water treeing, but also taken into account distributing capacitance; the monitoring time has been decreased. WINBOND W78E58B has been used in signal processing and the experiment results have affirmed the availability.

Keywords: Insulation monitoring; Quick positive and negative DC superposition detection; XLPE Power cable; Hot-line monitoring.

1 Introduction

With the automatic degree of electrical equipment rising, the reliability and continuity requirements of electricity supply are also increasing. Protecting power cables well-insulated is an important part of ensuring reliability of the power system. Traditional offline methods of detecting the aging cable insulation works in the power cut condition; so they can not meet the actual demand of application, especially when the continuity of power supply is exigent. While hot-line monitoring methods induct data processor into measurement systems, and the insulation resistances of power cables are detected on line in the work situation, so the detected data information at running state of power system is more reasonable.

---

[*] National Natural Science Foundation of China supports the project in this paper. Contact number is: 50737004.

Currently, for power cables with cross-linked polyethylene insulation (XLPE) which is widely used in power system, hot-line insulation monitoring methods are DC superposition theory[1], DC component method [2], Dielectric loss tangent method [3], Low-frequency method [4], partial-discharge method [5] and so on. It has been proved that the DC superposition theory has the feature of higher precision and the measurement results can reflect the actual situation of insulation. However, it is time-consuming to measure insulation resistance exactly at the presence of distributed capacitance on the power cables and water treeing rectification effect [6-8].

So in this paper, a method for monitoring XLPE cable insulation base on quick positive and negative DC superposition method is presented, and XLPE power cable $R_x C_x I_d$ mathematical equivalent model is constructed. Since the distributing capacitance between conductor of power cable and grand has been considered, the measuring time of insulation resister has been decreased. Furthermore, the water treeing generating from aging of power cable has been accounted in power cable mathematical equivalent model, which is expressed by I_d, so error resulted from disturbing of water treeing has been counteracted through positive and negative DC superposition.

2   DC Superposition Theory

Schematic diagram of DC superposition theory is shown in Fig.1. DC voltage is put on XLPE power cable by the way of inductor L, neutral point of three-phase voltage transformer and power distribution equipments. There are AC voltage and DC voltage existing at the same time between the ektexine of power cable and its inner conductor of electricity. The voltage transformer is used to separate DC voltage from AC voltage, to avoid high AC voltage reacting on DC voltage. Parallel capacitor C has been used to filter out AC components, so that there is direct current I only flowing through the measuring device M.

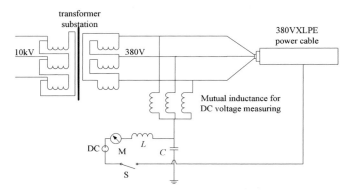

Fig. 1. Schematic diagram of DC superposition theory

Equivalent circuit of DC superposition theory is shown in Fig.2. R_m is the inner resistance of the detection devices; R_x is the equivalent insulation resistance for XLPE cable.

When the switch S in Fig.1 is closed, the leakage current I on power cable can be measured by meter M, so the insulation resistance value of power cable can be obtained through formulary E/I (E is the DC power supply voltage). It can be seen that DC superposition theory has the ability of measuring the insulation resistance on the condition that the power cable shows the characteristic of resistance. But owing to the water treeing rectification effect, the characteristic of current source is existent. Therefore positive and negative DC superposition method has been developed.

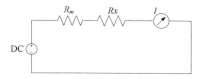

Fig. 2. Equivalent circuit of DC superposition method

Positive and negative DC superposition method brings under subjection to DC superposition theory. Its rule is that making use of the difference of direct currents which are generated from the positive and negative DC-EMF in power cable for data processing, avoiding the impact of one-way stray interference effects such as the water treeing rectification, but not considering the impact of distributed capacitance to the ground of the cable. Furthermore this hot-line monitoring method uses high-radio filter with large inductance, large bulk capacitor without source, to filter the interference of high frequency capacitor current of the ground loop, so the time constant decided by the ground distributed capacitor, filter capacitor, inductor and sense resistance, etc. is too large, resulting in a longer measurement time and not conducive to hot-line monitoring. DC current component charges slowly as the time variety and maybe result the current polarity reversal. So the more serious that the power cable water treeing aged, the time of hot-line monitoring will be longer, and the error caused from measurement will be greater.

3   Quick Monitoring Method for XLPE Cable Insulation

The corresponding equivalent model of quick positive and negative DC superposition method presented in this paper is a Parallel model integrated the insulation resistance R_x, distributed capacitance C_x which exists between the power cable and the ground ,and the equivalent current source I_d caused by water treeing (i.e. $R_xC_xI_d$ Equivalent model), shown in Fig.3. This method eliminates the water treeing and other one-way interferences which cause I_d through superposing positive and negative DC power supplying orderly, and uses quick detection method to reduce DC power splice time, quickly and accurately draws the value of cable insulation resistance.

Fig.4 shows the equivalent circuit of quick positive and negative DC superposition method, measurement process is as follows:

First, push the switch S to position 1, monitoring loop is composed by the sampling resistance R_0, the current limiting resistance R_1, the DC Superposition power supply E, and power cable $R_xC_xI_d$ equivalent circuit. R_m is inner resistance of the DC source.

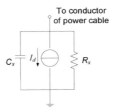

Fig. 3. Simple equivalent model of XLPE power cable

When the above circuit switches on in the time t_1, sample the voltage $u_0(t)$ on R_0 with the same time interval ΔT, get three voltage values: $u_0(t_1)$, $u_0(t_1+\Delta T)$, $u_0(t_1+2\Delta T)$, and then calculate the transient voltage values of corresponding distributed capacitance $u_x(t_1)$, $u_x(t_1+\Delta T)$, $u_x(t_1+2\Delta T)$ based on the formulary as follows:

$$\begin{cases} u_x(t_1) = E - u_0(t_1) \cdot (R_0 + R_1 + R_m)/R_0 \\ u_x(t_1 + \Delta T) = E - u_0(t_1 + \Delta T) \cdot (R_0 + R_1 + R_m)/R_0 \\ u_x(t_1 + 2\Delta T) = E - u_0(t_1 + 2\Delta T) \cdot (R_0 + R_1 + R_m)/R_0 \end{cases} \quad (1)$$

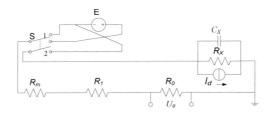

Fig. 4. Equivalent circuit of proposed method

When the monitoring circuit connects to the grid, the voltage $u_x(t)$ of the distribution capacitor changes exponentially as follows:

$$u_x(t) = U_{xsz}\left(1 - e^{-t/\tau_1}\right) \quad (2)$$

Among them, U_{xsz} is the steady-state voltage of distribution capacitors when positive DC power is applied, τ_1 is the time constant for the transition process.

Put the transient voltage values of distribution capacitance calculated in (**1**) into (**2**), obtain formulary (**3**):

$$\begin{cases} u_x(t_1) = U_{xsz} \cdot \left(1 - e^{-t_1/\tau_1}\right) \\ u_x(t_1 + \Delta T) = U_{xsz} \cdot \left(1 - e^{-(t_1 + \Delta T)/\tau_1}\right) \\ u_x(t_1 + 2\Delta T) = U_{xsz} \cdot \left(1 - e^{-(t_1 + 2\Delta T)/\tau_1}\right) \end{cases} \quad (3)$$

Solve equations (3), obtain U_{xsz} Value:

$$U_{xsz} = \frac{u_x^2(t_1+\Delta T) - u_x(t_1) \cdot u_x(t_1+2\Delta T)}{2u_x(t_1+\Delta T) - u_x(t_1) - u_x(t_1+\Delta T)} \quad (4)$$

So the voltage of insulation resistance on power cable in case of positive DC power supply is get.

The first step of measurement is completed, then push the switch S to position 2, the negative DC source E is applied. Noticeable, the cable distribution capacitor has initial voltage U_c in opposite directions to superimposed DC source E, the voltage $u_x(t)$ of the distributed capacitor changes based upon the exponential rule:

$$u_x(t) = U_{xsf} + (U_c - U_{xsf}) \cdot e^{-t/\tau_2} \quad (5)$$

Among them, U_{xsf} is steady-state voltage of the distribution capacitor when negative DC source is injected; τ_2 is the time constant for the transition process.

From the time t_2, the voltages $u_0(t)$ of resistance R_0 are sampled with the same time interval ΔT, three voltage values: $u_0(t_2)$, $u_0(t_2+\Delta T)$, $u_0(t_2+2\Delta T)$ are get, and the corresponding instantaneous voltage values of the distributed capacitance $u_x(t_2)$, $u_x(t_2+\Delta T)$, $u_x(t_2+2\Delta T)$ are calculated from equations (**6**):

$$\begin{cases} u_x(t_2) = -E + u_0(t_2) \cdot (R_0 + R_1 + R_m)/R_0 \\ u_x(t_2+\Delta T) = -E + u_0(t_2+\Delta T) \cdot (R_0 + R_1 + R_m)/R_0 \\ u_x(t_2+2\Delta T) = -E + u_0(t_2+2\Delta T) \cdot (R_0 + R_1 + R_m)/R_0 \end{cases} \quad (6)$$

Put the transient voltage values of the distributed capacitor obtained from (**6**) into (**5**), voltage values of the distributed capacitance are obtained as followed:

$$\begin{cases} u_x(t_2) = U_{xsf} + (U_c - U_{xsf}) e^{-t_2/\tau_2} \\ u_x(t_2+\Delta T) = U_{xsf} + (U_c - U_{xsf}) e^{-(t_2+\Delta T)/\tau_2} \\ u_x(t_2+2\Delta T) = U_{xsf} + (U_c - U_{xsf}) e^{-(t_2+2\Delta T)/\tau_2} \end{cases} \quad (7)$$

Solution of above equations, obtain the steady-state voltage U_{xsf} of distribution capacitors when the negative DC power is injected:

$$U_{xsf} = \frac{u_x(t_2) u_x(t_2+2\Delta T) - u_x^2(t_2+\Delta T)}{u_x(t_2) + u_x(t_2+2\Delta T) - 2u_x(t_2+\Delta T)} \quad (8)$$

According to Fig.4, the direction of equivalent current source I_d is contrary when positive or negative DC power supply is injected in the monitoring circuit, so the impact on the voltage of sampling resistance is equal on the value and opposite on the direction, and therefore, by using algebraic mean value of $U_{xs\cdot z}$ and $U_{xs\cdot f}$, the branch steady-state voltage U_{xs} is obtained from (**9**) below:

$$U_{xs} = (U_{xsz} + U_{xsf})/2 \quad (9)$$

Thus, by using quick positive and negative DC superposition method, I_d caused by the stray interference and other has been eliminated; the insulation resistances R_x of XLPE power cable can be calculated according to the equivalent circuit in Fig.4:

$$R_x = \frac{U_{xs} \cdot (R_0 + R_1 + R_m)}{E - U_{xs}} \tag{10}$$

The presented quick insulation monitoring method for hot-line XLPE cable based on positive and negative DC superposition method, using $R_xC_xI_d$ cable equivalent model, considers the problems of XLPE power cable insulation monitoring comprehensively. From any time that the detection device access to the loop, by collecting the voltage signal of the sampling resistance with the same interval three times respectively on the process of positive and negative DC power superimposed, the insulation resistance value of power cable has been worked out. In this way, the time of insulation monitoring has been reduced significantly and the accuracy of measurement is satisfying. On the other hand, because the process of the DC current flowing through the voltage transformer is transitory, the problems of transformer saturation will not be caused. So there is no zero sequence voltage in power supply circuit, which maybe arouse relay to act mistakenly. Therefore the proposed method is suitable to monitor insulation of XLPE power cable applied in the high voltage transmission power system.

4 Experimental Analyses

This paper has accomplished experimental analysis of the proposed quick positive and negative DC superposition method in Laboratory. WINBOND W78E58B processor has been used to deal with the digital signal in the insulation monitoring device that makes up of seven modules, including module for digital signals analysis and processing, module for A/D conversion, module for man-machine conversation, and module for sampling and filtering, module of DC resource and module for safeguard. Fig.5 and fig.6 show the faceplate and interior configuration of the manufactured insulation monitoring device based on the method which has been proposed in this paper.

Fig. 5. Faceplate of monitoring device   **Fig. 6.** Interior configuration of monitoring device

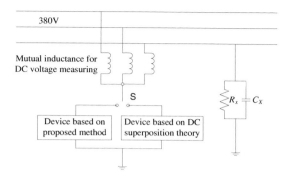

Fig. 7. Test circuit of monitoring device

In order to validate performance of the proposed insulation monitoring method, experimental analyses has been done in this paper. Experimental circuit is shown in Fig.7. Monitoring devices based on the proposed method and DC superposition theory have been connected in parallel to the grid by a transformer neutral point, the isolation transformer output 380VAC to construct AC power grid environment, and the resistance R_x and capacitance C_x simulate the insulation resistance and distributed capacitance of XLPE power cable.

In the case of using the same insulation resistance R_x, setting different values of distribution capacitance, the result of two devices is presented in Table 1. The insulation monitoring device using DC superposition method shows the test results in real time, and the measured data is updated with time continuously.

Table 1. Different testing data between two devices

C_x(mF)	R_x(MΩ)	quick positive and negative DC superposition method		DC superposition theory R_x (MΩ)		
		R_x(MΩ)	Measurement time (s)	$t = 8s$	$t = 16s$	$t = 24s$
0		569.58	7	551.85	551.85	551.85
0.15		574.25	9.5	515.26	539.47	539.47
0.45	557.63	569.11	12	361.48	481.54	529.56
1.5		576.14	10.5	147.62	248.55	348.15
6.8		580.49	11.5	57.64	86.57	108.56

As can be seen from the Tab.1, if the distributed capacitance is not set (C_x=0), two monitoring devices have similar results and the measurement time is also similar. When the distributed capacitance increases, the measurement time of quick positive and negative DC superposition method does not change significantly. While, the waiting time for calculating the final results of the device using DC superposition theory increases remarkably.

Setting several Values of distributed capacitance C_x to compare the infection of distributed capacitance on measurement precision of insulation resistance, the results are shown in Tab.2.

Table 2. Measurement results with different C_x

Initialization Of $R_x(M\Omega)$	$C_x=0\mu F$	$C_x=0.15\mu F$	$C_x=0.45\mu F$	$C_x=1.5\mu F$	$C_x=6.8\mu F$
2.59	2.61	2.61	2.62	2.61	2.61
6.04	6.15	6.14	6.14	6.16	6.15
9.07	9.29	9.28	9.28	9.27	9.28
9.97	9.99	10.04	10.02	10.02	10.01
20.58	19.95	20.21	20.00	20.16	20.24
57.59	58.12	58.54	59.29	59.10	58.84
82.52	83.21	82.84	82.88	83.45	82.71
153.22	152.78	152.11	151.84	152.44	151.89
250.58	248.68	243.87	245.35	246.27	244.37
364.01	365.54	366.24	369.25	359.84	361.26
557.63	569.58	574.25	569.11	543.14	580.49
842.57	838.12	848.45	824.5	858.54	845.81
988.97	1005.7	1015.4	1009.8	1006.2	998.43

From the tab.2, it can be concluded that the absolute measurement error of insulation resistance is about 0.7%~2.4% when the cable insulation problems in the serious range with R_x <10MΩ; the absolute measurement error is about 0.2%~2.9% when the cable insulation problems in the moderate range with $10M\Omega < R_x < 100M\Omega$; and the absolute measurement error is about 0.3%~4.1% when the cable insulation problems in the mild range with $100M\Omega < R_x < 1000M\Omega$.

Accordingly, the manufactured insulation monitoring device based on the quick positive and negative DC superposition method can ensure the accurate measurement of cable insulation resistance in despite of serious, moderate and mild state. The metrical data of insulation resistance can reflect correctly the state of cable insulation, which can be taken for reference for operators.

5 Conclusions

Based on the physical characteristics of XLPE power cable, this paper has constructed $R_xC_xI_d$ equivalent mathematical model, in which the factors of distributed capacitance and water treeing have been considered sufficiently.

Considering that the process of charging in distributed capacitance is time-consuming, this paper present quick positive and negative DC superposition method. By calculating steady voltage through transient voltages of sampling resistance, the value of insulation resistance can be obtained in short time. And the influence of current resource I_d caused by water treeing has been eliminated through superposing positive and negative DC power supplying orderly.

A set of insulation monitoring device based on proposed quick positive and negative DC superposition method has been manufactured; experimental results show that the presented method in this paper has the advantages of accuracy and timesaving.

References

1. Soma, K., Aihara, M., Kataoka, Y.: Diagnostic Method for Power Cable Insulation. IEEE Trans. 21(6), 1027–1032 (1986)
2. Oonishih, Uranof, Mochizukit, et al.: Development of new diagnostic method for hot-line XLPE cable with water trees. IEEE Trans. on Power Delivery 2(1), 1–7 (1987)

3. Yagiy, Tanakah, Kimurah: Study on Diagnostic Method for Water Treed XLPE Cable by Loss Current Measurement. In: Conference on Electrical Insulation and Dielectric Phenomena (CEIDP), Atlanta, USA, pp. 653–656 (1998)
4. Akayama, T.N.: On-line Cable Monitor Developed in Japan. IEEE Trans. on Power Delivery 6(4), 1359–1365 (1991)
5. Tian, Y., Lewin, P.L., Davies, A.E., et al.: Partial Discharge Detection in High Voltage Cable Using VHF Capacitive Coupler and Screen Interruption Techniques. IEEE ISK 20(1), 26–31 (2002)
6. Kim, C., Jang, J., et al.: Finite element analysis of electric field distribution in water treed XLPE cable insulation (1): The influence of geometrical configuration of water electrode for accelerated water treeing test. Polymer Testing 26, 482–488 (2007)
7. Chen, G., Tham, C.H.: Electrical treeing characteristics in XLPE power cable insulation in frequency range between 20 and 500 Hz. IEEE Trans. On Dielectrics and Electrical Insulation 16(1), 179–188 (2009)
8. Kim, C., Jin, Z., et al.: Investigation on water treeing behaviors of thermally aged XLPE cable insulation. Polymer Degradation and Stability 92, 537–544 (2007)

Study on the Propagation Characteristics of Electromagnetic Waves in Horizontally Inhomogeneous Environment

Lujun Wang[1], Yanyi Yuan[1], and Mingyong Zhu[2]

[1] Naval Academy of Armament, 100161, Beijing, P.R. China
[2] College of Marine, Northwestern Polytechnical University, 710072, Xi'an, P.R. China

Abstract. Evaporation duct is a typical structure of atmospheric refractivity, which can trap electromagnetic waves and change the propagation characteristics. Evaporation duct has significant influence on the radar and communication system operating on the sea. Most study on the propagation characteristics of electromagnetic waves assumed that the evaporation duct was range-independent. Much less effort has been devoted to the range-dependent cases. The purpose of this paper is to study on the characteristics of electromagnetic waves in horizontally inhomogeneous environment by simulation based on the parabolic equation model. Some numerical calculations were compared the experimental results.

Keywords: Evaporation duct, electromagnetic waves, communication system.

1 Introduction

Evaporation duct (ED) is an anomalous structure in the ocean atmosphere [1-11], which is associated with the rapid decrease of humidity above the ocean surface. The propagation of radio waves in evaporation duct is very different from the standard atmospheric environment. It has significant influences on the microwave communication system and radar system over the sea surface.

Much work has been done about evaporation duct, such as the generating mechanism, statistical features and the propagation characteristics. Several methods are available for modeling the wave propagation in evaporation duct, such as parabolic equation (PE) method, ray optics, waveguide mode analysis and the hybrid method [8-10]. Many researches usually assume that the evaporation duct is horizontally homogeneous. But in fact, both the height and strength of evaporation duct vary obviously in different range because of the complex meteorological factors, such as air temperature, air humidity, wind speed, wind direction, and sea surface temperature. In this paper, parabolic equation method is used to analyze the evaporation duct propagation characteristics in horizontally inhomogeneous environment. Furthermore, the experimental results of measuring the microwave path loss in evaporation duct are analyzed.

2 PE Model for Calculating Electromagnetic Propagation

2.1 PE Model

Starting with Maxwell's equations and assuming e^{jwt} time dependence, we can derive,

$$\nabla \times \nabla \times \vec{H} - \frac{\nabla \varepsilon}{\varepsilon} \times \nabla \times \vec{H} - \omega^2 \mu \varepsilon \vec{H} = 0 \qquad (1)$$

where ε is the permittivity in the region $r > a$ (a is earth radius) and it is a function of geometrical position r, θ, and ϕ. The magnetic permeability, μ, is assumed to be constant and ω is the angular frequency. Equation (1) will be reduced to an initial value problem where the source is included via the appropriate initial condition. As suggested by Fock, the following substitution is made:

$$\vec{H}_\phi = \frac{u(r,\theta)}{r\varepsilon(r,\theta)\sqrt{\sin\theta}} e^{ika\theta} \qquad (2)$$

where $k = \omega\sqrt{\mu\varepsilon(a,0)}$ and $\varepsilon(a,0)$ is the permittivity just above the earth's surface. The resulting equation is then transformed to a rectangular coordinate system with origin located on the surface directly below the source using $z = r - a$ and $x = r\theta \approx a\theta$, where z is the altitude and x is the distance along the surface. Finally, after making the parabolic approximation $|\partial_x^2 u| \ll k|\partial_x u|$, we obtain

$$\partial_z^2 u + 2ik\partial_x u + k^2(n^2 - 1 + 2z/a) = 0 \qquad (3)$$

where n is refractive index and z is the height from the earth surface.

2.2 Split-Step Fourier Method

Equation (3) along with the appropriate boundary conditions at $z = 0$ and infinity is an initial value problem that will allow a marching-type numerical solution. The solution is obtained from one range to the next using the Fourier split-step algorithm. This method is based on the fact that the Fourier transform of (3) has a simple solution at $x + \delta x$ in terms of the solution at x provided the refractive index. The split-step approach was chosen because it is straightforward to implement on a computer using a fast Fourier transform (FFT) routine, and it's stable and accurate for values of δx that are small relative to scales of the horizontal variation of ε, but that may be large compared to a wavelength.

The FFT is defined as

$$U(x,p) = F[u(x)] \equiv \int_{-\infty}^{\infty} u(x,z)e^{-ipz} dz \qquad (4)$$

where $p = k\sin\theta$ and θ is the angle from the horizontal. Then the split-step solution at $x + \delta x$ is

$$u(x_0 + \Delta x, z) = \exp[i(k/2)(n'^2 - 1)\Delta x] \cdot \\ F^{-1}\{U(x,p)\exp[-i(p^2\Delta x/2k)]\} \qquad (5)$$

In (5), n' is a "modified" refractive index, given by $n'^2 = n^2 - 2z/a$, which contains the factor accounting for a spherical earth. When n' is a constant, the equation (5) is an exact solution of (3) for a flat-earth problem with uniform refractive index. When considering the horizontally inhomogeneous case that n' varies with x and z, substitution of (5) into (3) results in terms that do not satisfy the parabolic wave equation. These terms, which give an indication of the error associated with a single range step, are expressed as

$$error = [2i\partial_z n' \partial_z - 2k\partial_x n' + \frac{i}{n'}(\partial_z n')^2 + i\partial_z^2 n'] \cdot \qquad (6)$$
$$(n'k\Delta x)u + (\partial_z n')^2 (n'k\Delta x)^2 u$$

As shown in (6), the accumulative error depends on the gradients of n' as well as on δx. When the "modified" refractive index varies slowly with height and range, the accumulate error comes from the δx. In practice, the error introduced by this procedure is found to be negligible provided δx is small enough for the split-step solution to converge.

It is noted that the FFT method needs the boundary conditions and the initial electromagnetic field, which were discussed in [8-10].

3 Simulations and Experiments

Based on the PE method, the propagation loss (L) from the source to any node can be calculated. The distribution of L reflects the influences of the ED environment on the propagation characteristics. In this section, simulation results and experimental observations will be discussed.

3.1 Simulation Conditions

The frequency of a microwave signal source was assumed to be 13 GHz. The wavelength λ is 2.3 cm. The transmitter height was 3 m. Horizontal polarization was used, and the elevation angle was 0 degree. The calculation height was 50 m with spacing of 0.02 m, and the range was 150 km with spacing of 0.4m. For PE method, the smaller step sizes in range and height are required for converging. Based on a lot of simulations, it is found that the following step sizes are adequate: about $\lambda/3$ in height grid, and about 10λ for range-dependent or 200λ for range-independent environment in range grid. When horizontally inhomogeneous conditions are modelled, different refractivity profiles are input at several ranges. The PE program performs linear interpolations in both range and height at the intermediate calculation positions. It's difficult to obtain the refractivity profiles at many ranges in an experiment about evaporation duct. In the following examples, two refractivity profiles at the source spot and the receiving spot are used. Although the refractivity profiles considered in these examples are relatively simple, it can predict propagation characteristics in horizontally inhomogeneous ED environment. In constructing the ED profiles, neutral stability condition is assumed and the M-value (modified refraction profile) at any height, h, is calculated as

$$M(h) = M_0 + 0.125h - 0.125\delta \ln(\frac{h+z_0}{z_0}) \qquad (7)$$

where M_0 is the M-unit value at the surface and δ is the ED height, $z_0 = 1.5 \times 10^{-4}$.

3.2 Simulation Results and Analysis

Figure 1 displays two propagation loss diagrams for different ED environment. Fig. 1(a) shows the propagation losses in dB for 15 m → 10 m ED environment (the evaporation duct height in source and receiver are 15 m and 10 m, respectively), and Fig. 1(b) shows the propagation losses for 15 m → 20 m ED environment. From Fig.1 it can be seen that the duct has more obvious effect on the propagation losses. The microwave energy extends along the sea surface and the signal is stronger in this region than that in higher region. Electromagnetic waves are trapped partly in the ED. But the signal intensity is different for these two range-dependent cases. The trapping capability of 15 m → 20 m ED environment is larger than that of 15 m → 10 m ED environment.

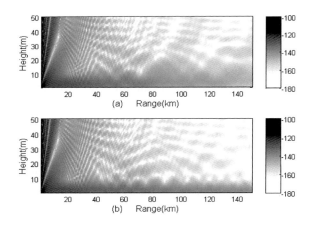

Fig. 1. Propagation loss for two range-dependent environments. (a)15 m → 10 m; (b) 15 m → 20 m.

Figure 2 shows the propagation losses versus range at height of 3 m. Compared with the horizontally homogeneous evaporation duct of 15 m, the propagation loss at height of 3m and at range of 150 km is less 1 dB in 15 m → 20 m ED environment. However, the propagation loss in 15 m → 10 m ED environment is about 8 dB higher than that of the evaporation duct with 15-m height. The difference is so obvious that it will influence the performance of a microwave communication system.

Figure 3 shows the propagation losses versus height at range of 150 km. When the receiving height is lower than 4 m, Fig.3 has similar features with Fig.2.

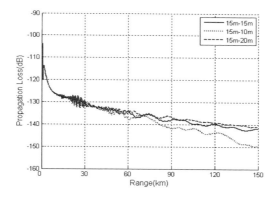

Fig. 2. Propagation loss versus range at a height of 3 m.

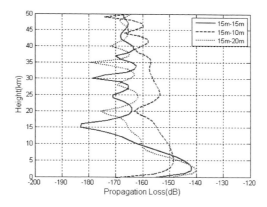

Fig. 3. Propagation losses versus height at a range of 150 km

Assuming that the evaporation duct height in source position is 15 m, the propagation losses at 3-m height and 150-km range were calculated with the ED height changing from 5 m to 25 m at the receiving position. The results are displayed in Fig.4. When the ED heights at the receiving position are less than that of the source position, the propagation losses are bigger. When the ED heights at the receiving spot are higher than 15 m, the propagation losses are smaller.

The impacts on the electromagnetic waves of horizontally inhomogeneous environment are different for the diverse ED cases. Table 1 shows the propagation losses at 3-m height and 150-km range in different ED environments. In the environment with lower ED height, 13 m for example, the propagation loss in horizontally inhomogeneous environment is 19.2 dB higher than that of the horizontally homogeneous environment. However, it is only 2.3 dB difference for environment with higher ED height. So the horizontally inhomogeneous environment has more significant influence on the microwave propagation in the environment with lower ED height.

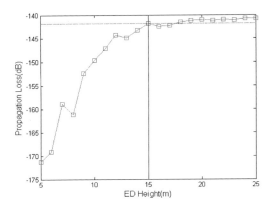

Fig. 4. Propagation losses versus the ED height at receiving point.

Table 1. Propagation losses at 3-m height and 150-km range.

Cases	ED environment	Propagation loss (dB)
1	13 m → 13 m	140.8063
2	13 m → 7 m	159.9944
3	20 m → 20 m	140.1598
4	20 m → 13 m	142.5128

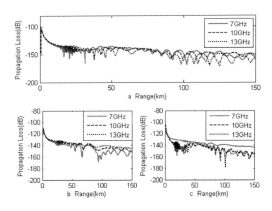

Fig. 5. Propagation losses at height of 3m with three evaporation duct environments. (a)13 m → 13 m; (b)13 m → 7 m; (c)13 m → 20 m.

The horizontally inhomogeneous environment has different influence on electromagnetic propagation at different frequencies. The microwave signals with 7 GHz、10 GHz、13 GHz were simulated. Fig.5 shows the propagation losses at height of 3 m for three frequencies and three evaporation duct environments. In the horizontally homogeneous ED environment (case (a)), the propagation losses at three frequencies have small differences, but the complex interfering structures are different at three frequencies. In 13 m → 7 m ED environment, which makes against the microwave propagation, the propagation loss of lower frequencies is bigger than that of higher frequencies. The main reason is that the environment with lower evaporation duct height can not trap microwave with lower frequency.

In 13 m → 20 m ED environment, which is suitable for the wave propagation, the propagation loss at low frequency is smaller because of lower absorption loss.

3.3 Analysis of the Experimental Results

Australia researchers founded a microwave communication system to transfer data from the Great Barrier Reef to the Australian mainland [11]. The link range was about 80 km. They found that the actual propagation loss appeared to be about 20-30 dB larger than expected from the predicted result. There is not complete explanation for this result. They just suspected that the system may have been miss-aligned during the experiment. From the simulation results of this paper, the possible reason of this phenomenon is that the evaporation duct environment is horizontally inhomogeneous.

During an experiment in the South China Sea, the authors' group found a similar phenomenon. The communication range was about 120 km and the frequency was 13 GHz. We also found that the measured propagation loss at height of 3 m, about 165±5 dB, was larger than the predicted values, 150 dB. It is known that the communication system was installed and aligned exactly. So the main reason of the problem was the horizontally inhomogeneous ED environment, which affected the propagation characteristics. Because of the limited experiment situations, the horizontally inhomogeneous assumption will be confirmed in future.

4 Conclusions

Some simulation results for modeling electromagnetic propagation in the evaporation duct using PE method has been demonstrated. The influences of the horizontally inhomogeneous ED environment on propagation losses were analysed. Several conclusions were obtained:

(a) The effect of the microwave propagation from the low ED to the high ED is better than the opposite conditions.
(b) The low ED height environment has more obvious horizontally inhomogeneous effectiveness on microwave propagation.
(c) The horizontally inhomogeneous environment has more impact on the higher frequency microwave.

While the conclusions looks very promising, some work need to be done to fully confirm the conclusions. It's expected that the conclusions can give some help for designing a microwave communication system operating over the sea surface.

Acknowledgments

This work was supported by the science and technology development project of Shaanxi province (No.2010KJXX-02), and the foundation of state key lab on acoustics (No. SKLOA201002).

References

1. Anderson, K.D.: 94-GHz Propagation in the Evaporation Duct. IEEE Trans. Antennas Propag. 38, 746–753 (1990)
2. Anderson, K.D.: Radar Detection of Low-Altitude Targets in a Maritime Environment. IEEE Trans. Antennas Propag. 43, 609–613 (1995)
3. Yang, K.D., Ma, Y.L., Shi, Y.: Temp-spatial Distributions of Evaporation Duct for the West Pacific Ocean. Acta Physica Sinica 58, 663–674 (2009)
4. Lin, F.-j., Liu, C.-g., et al.: Statistical Analysis of Marine Atmospheric Duct. Chinese Journal of Radio Science 20, 64–68 (2005)
5. Dai, F.-s.: The Refractivity Models in the Marine Atmospheric Surface Layer and Their Applications in the Evaporation Duct Analysis. Chinese Journal of Radio Science 13, 280–286 (1998)
6. Liu, C.-g., Huang, J.-y., et al.: Characteristics of the Lower Atmospheric Duct in China. Journal of Xidian University 29, 119–122 (2002)
7. Yao, S.-y.: The Analysis on Characteristics of atmospheric Duct and It's Effects on the Propagation of electromagnetic waves. Acta Meteorologica Sinica 58, 605–616 (2000)
8. Kuttler, J.R., Dockery, G.D.: Theoretical Description of the Parabolic Approximation/Fourier Split-Step Method of Representing Electromagnetic Propagation in the Troposphere. Radio Science, 381–393 (1991)
9. Marcus, S.: A Model to Calculate EM Fields in Troposphere Duct Environments at Frequencies through SHF. Radio Sci. 17, 895–901 (1982)
10. Barrios, E.: Parabolic Equation Modeling in Horizontally Inhomogeneous Enviroments. IEEE Trans. Antennas and propagation. 40, 791–797 (1992)
11. Woods, G.S., Ruxton, A., Huddlestone-Holmes, C., Gigan, G.: High-Capacity, Long-Range, Over Ocean Microwave Link Using the Evaporation Duct. IEEE J. Oceanic Eng. 34, 323–330 (2009)

Multimedia Sevice Innovation Design of Graffiti in Modern Commercial Design

Xiaoyan Wang[1], Zhanxi Zhao[1], and Yantao Zhong[2]

[1] College of Mechanical and Electrical Engineering, Hohai University,
Jinling North Road.200, 213022 Changzhou, China
[2] Department of Arts, Changzhou University,
Gehu Road.1, 213164 ChangZhou, China
wxy826@126.com

Abstract. Graffiti has become a new form of culture with free creation in modern commercial design, and has paid attention to the fashion show and expression of primitive beauty, has more feeling and fun sense in design expression. The graffiti multimedia sevrvice innovative design is mainly manifested in the initial impetus graffiti from street culture in the edge of the wandering into commercial art outside the mainstream consumer, to realize business marriage of graffiti and popular clothing, be permeated with vigor, endowed modern commercial design with fresh modern style, also graffiti understands to relax advertising with concise and optional way, improve public the art sense. The innovative design of Graffiti annotates a foothold new graphics design performance from implies formfrom multidimensional and the space.

Keywords: multimedia, innovation, street, marriage.

1 Service-Orinetation of Fashion

In this era of fashionable change heavy and complicated, the trend of the restoring ancient ways has came. The commercial design follows basic concepts and framework of graffiti culture, in the meantime, will be mixed popular element of universal, the trend of continuously show newness has been unstoppably become fashionable pronoun. Perhaps is tired of mainstream design circle prudent middle-class taste flashy for years, many new business designers are trying to search inspiration from the weird and the wild in the graffiti art. The graffiti agitation in Modern commercial design, will make the computer, painter, ps as creator's new weapons and equipment, display themselves through the Internet than metope and shuttle of the metro wider stage. More important is that former street graffiti artists has jumped as new prominent designer of identity obvious ascending design company. Their join inject booster for the gradually insipid machinery of design circle, will freely spraying graffiti of free creation culture with brand-new form in every corner of design.

Pop and fashion is indispensable design elements in modern commercial design, "vogue" with "advanced" and "modern" meaning, the fashion language in business design filled with fun feeling and hinted effection. Commodities as constantly changed consumer is always in a generation with a generation alternate, a wave with a wave push billow, popular fashion in the modern fast-food melt, the fast food change consumption age regardless of how much or not being value has its existing significance and value. The commercial design formed specific fashion culture characteristics because of its commercial market and cultural market changes driving. Philosopher said that everything is impossible in addition to change. The fashion lasting change of business design is reflected that. Graffiti in the application of commercial design usually accompanied by strong fashion meaning, because combination of graffiti and commercial design itself is that design concept satisfies people psychological guidance of innovation and chang, whether lifestyle or advertising forms of commodity must need to change as time passes. Since long rational modern design has dominated the commodity vision, with the design thought of various thoughts run, social incentives and rich life, the consumption of goods more begins to pay attention to culture consumption and art enjoyment. This fashion is different from previous fashion such as pop musical, general fashion does not need to comply with the principle of beauty, weird and chaos can be fashionable and popular performance morpheme, but graffiti of commercial design is different, because graffiti itself is a kind of art forms, it performances more dynamic and fashion in the unique design. The graffiti commercial design is not advocated sentimental affectation, exaggeration noisy, pays attention to show and expression of the original beautiful. Fashion likes fire, also likes the wind and cloud, locating in a fashion pace of scurry for himself than let it stay and pursues fashion as used to create fashion, moving in the static, change again in a constant is long idea of graffiti transformation in commercial design.

2 The Tangible Interpretation of Street Culture

The late 1990s, street culture enter young Chinese perspective with the tide of favotite Korea and Japan, surging a hip-hop upsurge, in the 21st century, street culture spread deeply to chinese cities, for people of new generation, street culture has become the symbol of type cool and fashion, it has also become a part of their lives. Street culture has five important elements of hip-hop: apparel, street dancing, Hip - Hop DJ, singing white (MC) and graffiti. The average person must be the most easy to remember that size large, loose optional hip-hop clothing, and with style hip-hop, because these draw close to the common people's life, easily accept and learn. Hip-hop culture went wherever incited many young people fanaticism. Such as one of the typical representative street culture is hip-hop culture: because hip-hop appear on the street, so called hip-hop, and with a strong participation, presentation and competition. With hip-hop young people make hip-hop display public individual character, express aggressive attitude of life, they emphasize concept that is "doing their own, enjoy life, and have courage to challenge". Another of the typical representative street culture is street basketball culture: street basketball in China also is the new things, but its vitality is very strong, is a kind of art of rhythm. Basketball of impacting wall, spine under the sun, drippy sweat constitute a youth segment of painting full strong vitality. It is a movement, is also a kind of performance, a kind of fashion.

The "hip-hop fashion, dance graffiti" baseball cap graffiti contest in June-July 2008 MLB was held (Fig. 1 shows an example). During the campaign, the numerous people of loving sports, pursuiting fashion, loving baseball, advocating unique, thoroughly showed about their display design talent and painting talent in the platform of MLB providing. At the beginning of the epidemic of street culture, sharp businessmen have caught the underlying great commercial value. As matchs of street basketball, extreme sports, hip-hop contests successively have performed in the major cities, street culture once again raised the upsurge of fashion consumption. And in the meanwhile of "hip-hop clan" frantically chasing "hip-hop movement", businessmen worried about competing for their brand, struggling naming matchs. Graffiti as an important representative of street culture, easier to be physicale, make public individual character, graffiti and hip-hop, rap etc street culture suppling each other, thus performanting. Street culture walks into the mass consumer culture, drive the graffiti of the unmainstream art wandering on the edge entering the group of commercial consumption, and also the rise of commercial graffiti as a tangible image better interpretating the invisible connotation of street culture.

Fig. 1. "Hip-Hop culture, dance graffiti" baseball cap graffiti contest

3 Business Marriage of Popular Clothing

The street is not the only export of graffiti, with young consumer culture growing, many patterns of graffiti designer appeared in commodities such as garments and shoes, graffiti elements have integrated into our life with its vigor and the characteristics of the passion, become to be everywhere! In recent years, graffiti began gradually to go into the fashion design field, enter into the commercial market. Fashion designers outside the prescribed series made graffiti as a series of part of the fashion design to advocat personality, early swing in fashionable bound. Cross use

Graffiti art, clothing, shoes, leather, watches, perfumes... Many young brands involved, many young people unabled to bear design on own T-shirt, and began to get acceptance of mainstream brands. Graffiti has become the passive to be active, become things via window, magazines, advertising, catwalks to consume. Angels, devils, god, palace, ancient times, sex, life, love, peace and war, dissatisfaction and rebel, homosexuality, and twisted psychological... designers used them with fashion design in the form of graffiti. The original just one naked thing, once combining them with art, they will need a soul to tell a story. Modern graffiti has endowed with fresh modern style, such as Japanese manga style, American street style, and as blank style of prints, and as picture painted style of Taiwan, more and more clothing brand use graffiti, scenery, character to daub, up even drawing personally many flowery unruly design on clothes, to express blooming youth.

Fig. 2. Vivian Westwood's spring-summer 2007 series

British punk godmother Vivian·westwood's 2007 spring-summer series (Fig. 2 shows an example), she's never law-abiding spirit developed acme once again through the graffiti art. This conference as colorful as the graffiti type background, displayed colorful personality fashion. The models' hairs with dying various exciting color, of course, fashion colour also was not drab, purple, brown, olive green, red, gems blue, black and white, incarnadine, she adorned with the child draws type graffiti design and greatly small bowknot (Fig. 3 shows an example), it makes people see the stylist still retaining childlike side. Vivian·west wood's designs cater to favorite the the 1980s' young, especially London youth "punk", "teddy GeEr", making the west wood's clothing into a world influence. Although her designs didn't become the governor of Paris fashion design, also failed to form trend, but her influence was mainly in concept, her design concept not only greatly impacted traditional fashion, and represents the radical young generation. And the idea is to coincide with the idea of graffiti art, integration of graffiti design and business is not without aiming, but closely contaction can be organically integrated, common expression creating spark, burst out fearless attitude of multicultural, open-minded, new generation.

Fig. 3. Vivian Westwood's spring-summer 2007 series

4 Amorous Feelings Penetration of Fashion Advertising

Modern advertisements have become a spectacular mass culture landscape of industrial society, as the most intense speak style in diversified era, with the help of the media spreading advantage, advertising words possess strong penetration, position, even influence compulsion, direct role in one's subconscious and unconscious level, not only preset public consumption idea, also produce great impact on modern existence state.

The connection of Graffiti and advertising was not accidental, the origin of graffiti is because of hoping more people to grasp, understand, and it is the same as the origin of advertising. Graffiti mainly exists in print advertising and wall advertisement in advertising (Fig. 4 shows an example). Plane advertising through the artistic inspiration paints pattern of rich highlight to adequately overstep advertising theme, and adequately attract eyeball. Graffiti advertising, straightforward, dynamic, is an important means of looking young people gathered. Abandoning the common imaging of advertising often used or expression of computer synthesis technique, graffiti advertising chooses concise and optional way, especially suit needs of fashion young people. With the constantly tighter and expensive in AD position of CG, making wall as advertising media, is also the trend. From the Angle of utilization the wall advantage is very big, and expenses will be economical compared with other aspects. On the other hand, for environmental protection and artistic rendering is also a very good way, at the same time of propaganda also can improve public art sense. Graffiti art going out from the noise of the city, agitated, unrest and mixed and disorderly, constituting a whole state with passers-by and streets, contacting with urban population closely, reflecting a kind of optional, edge state. Sometimes it's hard to say what graffiti image itself is, those graphics or symbols making inner exchanges with seeing them, everyone can have their own understanding, but no matter what, it is hoped that life can be obtained vitality and vigor from the mainstream express.

Fig. 4. Flat advertising and wall advertising

5 Epilogue

Graffiti in the expression of modern commercial design not only should reflect good graphic visual effect, more is to feel one kind of emptiness, a three-dimensional space, a sense of the leap space of impact. We should not only think design from the angle of formal beauty, more need from the standpoint of multidimensional and space to annotate a new graphics implies. Ignoring the dimensional existence, and constantly pursuing plane visual effect, will only result as unreal novels such as castles in the air.

Commercial graffiti has strong symbolic and implicated, symbol is used for a simply and accurately expressing complex idea, to say about the meaning of the object with the related things. It is not only a symbol, the more with its special spirit and emotional characteristics. Visual space creation of Commercial graphic design, mainly executing on the basis of people's spatial perception and dimensional feeling understanding. It contains space concept of a number of people in the sport's subjective modelling, complex and novelty, showing fime of human thought. With the development of society and changes of people's aesthetic concept, the commercial art design of flat-screen already cannot deduce deeper connotation. Only multidimensional and space design form can meet the aesthetic ideology of modern people. Releasing human nature, pursuiting vogue, feeling space changes, enjoying the nature, this has become a very strong modern people's psychological needs, people began to seek more graphic forms to meet their demands.

References

1. Zhong, Y.: From Insulation to Conductor: Creative Thinking Development of Graffiti. Decoration 213, 98–99 (2011)
2. Fu, L., Wang, A.: The Modern Ultra Graph Creativity Chart Standard. Henan Arts Publishing House, Zhengzhou (1995)
3. Hu, H., Zheng, H.: Decorative Pattern Design, pp. 5–6. Wuhan University of Technology Publishing House, Wuhan (2005)
4. Zhu, X., Li, B.: Art Design Discussion. JiangXi Art Publishing House, NanChang (2002)
5. Wang, L.: Arts Morphology, pp. 23–24. Southwest Normal University Publishing House, Chongqing (2004)

Using Bypass Coupling Substrate Integrated Circular Cavity (SICC) Resonator to Improve SIW Filter Upper Stopband Performance

Boren Zheng[1,2], Zhiqin Zhao[1], and Youxin Lv[1]

[1] School of Electronic Engineering, University of Electronic Science and Technology of China, Chengdu, Sichuan 610054, China
[2] School of Information Science and Engineering, Chongqing Jiaotong University, Chongqing, 400074, China
borenzh@gmail.com

Abstract. In this paper, a novel method of using bypass coupling SICC resonator to generate transmission zeros (TZs) in filter stopband to improve upper stopband attenuation is presented. A SIW quasi-elliptic function filter with bypass coupling SICC resonators is designed and fabricated to validate the method. The results show the method is effective to improve the filter stopband performance.

Keywords: Filters, Substrate integrated circular cavity (SICC), Bypass coupling, Transmission zeros (TZs).

1 Introduction

In microwave and millimeter-wave communication systems, require of filter with stringent selective, low insertion and potential integration into the circuit become more and more intense. The waveguide filters are widely used because of their high Q value, high power capability and outstanding selective. However, they are bulky, heavy and not suited for integration. Substrate integrated waveguide (SIW) filter provides a low-profile, low-cost, possible integration and low-weight scheme while maintaining high performance, which is satisfied with the needs perfectly, but the discontinuities in the post wall of the SIW cavity resonator are needed to generate the appropriate coupling for a small number of resonators in the filter. These discontinuities lead to poor stopband attenuation with an increasing amount of power carried across the coupling sections. Moreover substrate dissipation causes a loss to the SIW resonator Q value, the Q descending make the filter selective to deteriorate. SIW filter low stopband performance can be got by the cut-off frequency of waveguide easily, where has more difficulties in upper stopband. Many slightly different approaches were applied to improve the stopband performance. The defected ground structure, conventional step impedance resonator and the E-plane discontinuities are well known as the approach [1]-[3], which are difficult to realize for SIW filters implemented on a single-layer substrate. Multiple transmission zeros generated by cross-coupling or nonphysical cross-coupling of higher order modes be used to improve the stopband attenuation are published in [4]-[5]. Nevertheless, the transmission zeros cannot be far away from the desired passband due to the limitation of the physical structure, and the insertion loss will rise in passband with separate paths for energy flow. Other methods are presented recently, in [6], it proposed using zigzag meandered topology to introduce extra cross-couplings transmission zeros in stopband to get a better fitting of the electrical response. Using optimal design of the angle between the input and output ports of SICC filter to engender transmission zeros in stopband is offered in [7].

In this paper, a novel method to improve filter upper stopband performance using bypass coupling SICC resonator is presented. The bypass coupling SICC resonator is introduced in SIW filter though an inverter to engender transmission zeros in the filter stopband. The characteristics of the inverter with SICC resonator in a practical SIW filter are analyzed. This structure of the inverter with SICC resonator brings the bypass coupling of the SICC resonator in the SIW and does not affect the SIW filter passband. Transmission zero can be set up at arbitrary place in filter stopband through the SICC to enhance the filter stopband attenuation. A k-band quasi-elliptic function SIW filter with outstanding stopband performance is designed and fabricated to validate the method. The filter has a passband about 2.5GHz form 17GHz to 19.5GHz and excellent stopband attenuation over a frequency range of 23.1–26 GHz better than 45 dB. The results show the technique is useful to improve the filter stopband performance and is simple to implement in design.

2 Characteristics of Inverter with Bypass Coupling SICC

An inverter prototype with bypass coupling resonators is shown in Fig.1 (a), it consists of one passband resonator as the L_1, C_1 and two bypass bandstop resonators as the L_2, C_2 and L_3, C_3 respectively, the $J_{n-1,n}$ and $J_{n,n+1}$ are the filter inverters. Two stopband resonators are coupled in the circuit though inverter to generate transmission zeros in stopband. The admittance viewed from the input/output (not include the $J_{n-1,n}$, and $J_{n,n+1}$) is given by

$$B(\omega) = jb_1\left(\frac{\omega}{\omega_{01}} - \frac{\omega_{01}}{\omega}\right) + \frac{J_1^2}{jb_2\left(\frac{\omega}{\omega_{02}} - \frac{\omega_{02}}{\omega}\right)} + \frac{J_2^2}{jb_3\left(\frac{\omega}{\omega_{03}} - \frac{\omega_{03}}{\omega}\right)} \quad (1)$$

where $\quad b_i = \omega_{0i}C_i \quad \omega_{01} = 1/\sqrt{L_iC_i} \quad i = 1,2,3.$

The inverter parameters J_1 and J_2 are assumed to be frequency independent, two transmission zeros at $\omega_{z1} = \omega_{02}, \omega_{z2} = \omega_{03}$ may be produced as described in equation (1), [8]. To satisfy the transmission zeros place in stopband be controlled individually and not be interacted with the filter passband, the J must undertaking as a constant in filter passband. The J with bypass coupling resonator is frequency dependence in a practical filter. The actual J parameter of an inverter with bypass coupling SICC resonator can be calculated by the equation (2) though the scattering matrix of the inverter [9],

$$J = \sqrt{\frac{1+|S_{11}(\omega)|}{1-|S_{11}(\omega)|}} \quad (2)$$

where the $S_{11}(\omega)$ is the reflection coefficient of the inverter with a bypass coupling SICC resonator at the frequency ω. As shown in Fig.1(b), the inverter with bypass coupling SICC resonator should produce a TZ at the ω_3 (where the ω_3 is the resonant frequency of the resonator and the J approaches infinity), and the J is frequency independence as constant-approximation in the passband($\omega_1 \leq \omega \leq \omega_2$). The bypass resonator is constructed with SICC resonator, and the SICC height and radius meet with $h < 2.1R$, owing to the TM_{010} mode is selected as the operating mode in the SICC only [10], the SICC resonant frequency as transmission zero is completely controlled by the SICC radius and height, and the J parameter of the inverter will be frequency independent in the filter passband.

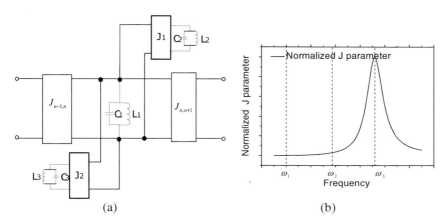

Fig. 1. Characteristics of inverter with bypass coupling resonator (a) Inverter prototype, (b) Frequency vs. the J of the inverter with SICC

SICC resonator is shown in Fig.2 (a). Dielectric filled structure of the conventional cylindrical waveguide resonators can be readily implemented in substrate by replacing the vertical metallic walls by closely spaced via posts and covering the bottom and top metal layers. The current lines of *TM* mode waveguide are along the waveguide. Because via-holes are along current lines, *TM* mode is selected as the operating mode in SICC uniquely. The resonant frequency (unloaded) of SICC can be computed by equantion(3), where μ_r and ε_r are relative permeability and permittivity of substrate respectively, c is the speed of light in free space, p'_{mn} and p_{mn} are the nth roots of the mth Bessel function of the first kind and its derivative , R is the SICC radius; h is the length of the SICC along the z-axis. The *TM₀₁₀* is the dominant mode

$$f_{mnl} = \begin{cases} \dfrac{c}{2\pi\sqrt{\mu_r \varepsilon_r}} \sqrt{\left(\dfrac{p'_{nm}}{R}\right)^2 + \left(\dfrac{l\pi}{h}\right)^2} & TE_{mnl} \quad \text{mode} \\ \dfrac{c}{2\pi\sqrt{\mu_r \varepsilon_r}} \sqrt{\left(\dfrac{p_{nm}}{R}\right)^2 + \left(\dfrac{l\pi}{h}\right)^2} & TM_{mnl} \quad \text{mode} \end{cases} \quad (3)$$

in one SICC resonator provided $h<2.1R$ [10], the resonant frequency of the SICC is determined by the height and radius and the spurious resonant frequency is suppressed by the dominant mode. Furthermore, the TM_{010} mode existing in the SICC resonator is circular symmetrical, the inverter location has not rigorous demand.

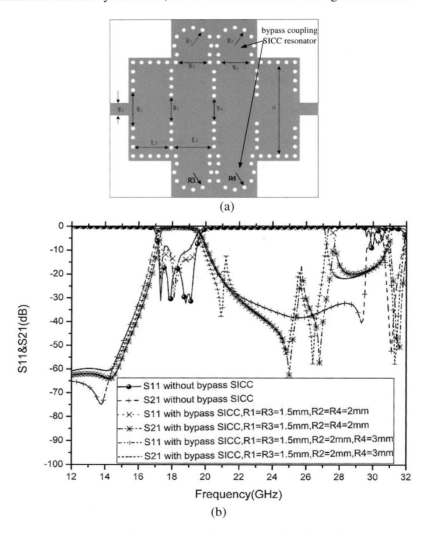

Fig. 2. (a) Proposed quasi-elliptic function filter structure (b) Simulation results of the SIW filter with different bypass coupling SICC resonators

According to the method as depicted in Fig.1 (a), Fig.2 (a) shows proposed quasi-elliptic function filter structure of a fourth-order oversized TE$_{101}$/TE$_{301}$ SIW filter. The setting bypass coupling SICC resonators are to improve upper stopband action. The details and the dimensions of the filter will present in section 3.Using the ansoft HFSS software, a simulated S parameter comparison result of the SIW filter with the different bypass coupling resonators shows in Fig.2 (b). It is seen that two transmission zeros have been produced in the stopband at 25.1GHz and 27GHz by the two different dimension bypass coupling SICC resonators with $R_1=R_3=1.5mm$, $R_2=R_4=2mm$, three transmission zeros have been generated at 21.3GHz, 25.1GHz and 26.9GHz by the three different dimension bypass coupling SICC resonators with $R_1=R_3=1.5mm, R_2=2mm, R_4=3mm$. The SIW filter passband has a litter change but the stopband attenuation augmented with the additional bypass SICC. The stopband attenuation is elevated with the rising number of same dimension resonators as depicting in Fig.2 (b) at 25.1GHz, one SICC with $R=2mm$ is 55dB and two SICCs with similar radius is 65dB. The equation (1) has illuminated the phenomenon, if the $\omega_{02} = \omega_{03}$, the admittance will be double or more at the frequency ω_{03}.

3 Design of SIW Filter with Bypass Coupling SICC

Using the oversized TE$_{101}$/TE$_{301}$ coupling scheme, the oversized SIW's dimension is to be excited high mode has been described in [4]. Because the inverter with bypass coupling SICC does not affect the filter passband, the bypass coupling SICC resonators are sited in the proposed quasi-elliptic function filter directly as depicted in Fig.2 (a). The final results of the filter dimensions are optimized by the ansoft HFSS as follows: metallized via-hole diameter is 0.5 mm, the space between via-holes is 1mm, $W_1=1.28mm$, $W_2=3.87mm$, $W_3=2.76mm$, $W_4=2.45mm$, $W_6=2.81mm$, $W_5=3.12mm$, $a=10.5mm$, $L_1=4.52mm$, $L_2=4.03mm$, $R_1=2.12mm$, $R_2=1.45mm$, $R_3=2.12mm$, $R_4=1.45mm$.The filter was fabricated on a single-layer Rogers R04003 substrate ($\varepsilon_r = 3.55$, $\sigma = 0.0027$) with a height of 0.508 mm by standard printed circuit board (PCB) processes. The photograph of the fabricated filter is shown in Fig.3 (a).

Fig.3 (b) shows the simulation and measurement results of the designed filter. It is observed that the transmission zeros in the stopband are produced by the bypass coupling SICC resonator and are slightly lower than simulation due to the fabrication accuracy. The filter 3dB passband is about 2.5GHz (from 17GHz to 19.5GHz). The insertion loss is -1.6 dB at 17.5 GHz and -1.2 dB at 19 GHz that includes the insertion loss of two SMA connectors. Removing the connector losses -0.6dB, the complete filter (including the input/output microstrip) has an insertion loss level of -1 dB in passband and the measured passband ripple (17.3-19.3GHz) is lower than 0.4 dB. This filter has a characteristic of a small return loss lower than -13 dB in the passband and an excellent selectivity performance in the stopband. The upper stopband attenuation over a frequency range of 23.1–26 GHz is better than 45 dB because the two transmission zeros are settled in there. The method of engendering transmission zeros in stopband though bypass coupling SICC resonator to improve the stopband performance is validated and it can be employed in other filter design easily.

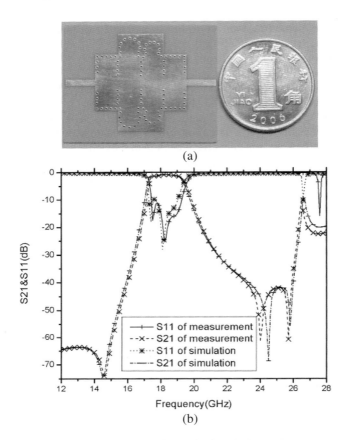

Fig. 3. (a) Photograph of the fabricated filter, (b) Simulation and measurement results of the proposed filter

4 Conclusion

A novel method of using bypass coupling SICC resonator to introduce transmission zeros in SIW filter stopband is proposed, design approach of generating the transmission zeros in stopband to improve stopband attenuation is presented. A SIW quasi-elliptic function filter with bypass coupling SICC resonators is designed and fabricated to validate the method. The measured results show a good meeting with the designed data and the method is valid to improve filter upper stopband selectivity. The filter has a 3dB passband about 2.5GHz from17GHz to 19.5GHz and excellent stopband attenuation over a frequency range of 23.1–26 GHz better than 45 dB. Using SICC as the bypass coupling resonator, its resonant frequency can be entirely controlled by radius and height owing to the TM_{010} mode is selected as only the operating mode while the SICC height h and the radius R meet with $h < 2.1R$. Due to the operating mode is exclusivity in SICC, to place transmission zeros in the desired location is flexibly by using the inverter bypass coupling. The technique is effective to improve the filter stopband performance and is simple to implement in the other filters.

References

1. Improved Upper Stopband Performance Using Defected Ground Structure. IEEE Microw. Wireless Compon. Lett., 20(6) (June 2010)
2. Morelli, M., Hunter, I., Parry, R., Postoyalko, V.: Stopband performance improvement of rectangular waveguide filters using stepped impedance resonators. IEEE Trans. Microw. Theory Tech. 50(7), 1654–1664 (2002)
3. Budimir, D.: Optimized E-plane bandpass filters with improved stopband performance. IEEE Trans. Microw. Theory Tech. 45(2), 212–219 (1997)
4. Chen, X.-P., Wu, K., IEEE, Drolet, D.: Substrate Integrated Waveguide Filter With Improved Stopband Performance for Satellite Ground Terminal. IEEE Trans. Microw. Theory Tech. 57(3) (March 2009)
5. Lenoir, P., Bila, S., Seyfert, F., Baillargeat, D., Verdeyme, S.: Synthesis and design of asymmetrical dual-band bandpass filters based on equivalent network simplification. IEEE Trans. Microw. Theory Tech. 54(7), 3090–3097 (2006)
6. Mira, F., et al.: Design of ultra-wideband substrate integrated waveguide (SIW) filters in zigzag topology. IEEE Microw. Wireless Compon. Lett., 19(5) (May 2009)
7. Tang, H.J., et al.: Optimal design of compact millimetre-wave SIW circular cavity filters. 15th Electron. Lett. 41(19) (September 2005)
8. Chen, X.-P., Wu, K., et al.: Dual-Band and Triple-Band Substrate Integrated Waveguide Filters With Chebyshev and Quasi-Elliptic Responses. IEEE Trans. Microw. Theory Tech. 55(12) (December 2007)
9. Liang, J.F., Liang, X.P., Zaki, K.A., Atia, A.E.: Dual-mode dielectric or air-filled rectangular waveguide filters. IEEE Trans. Microw. Theory Tech. 42 (June 1994)
10. Wu, d., Fan, y., He, z.: Vertical Transition and Power Divider Using Substrate Integrated Circular Cavity. IEEE Microw. Wireless Compon. Lett. 19(6) (June 2009)

Lossless Robust Data Hiding Scheme Based on Histogram Shifting

Qun-Ting Yang[1], Tie-Gang Gao[2], and Li Fan[1]

[1] College of Information Technical Science, Nankai University, Tianjin, 300071, China
patrickyoung@mail.nankai.edu.cn
[2] College of Software, Nankai University, Tianjin, 300071, China
gaotiegang@nankai.edu.cn

Abstract. This paper presents a lossless robust data hiding scheme based on histogram shifting. The proposed scheme divides a cover image into non-overlapping blocks and then calculates the arithmetic difference of each block with a private key. A threshold is set to decide the actual payload, and another threshold is adaptively determined to conceal the presence of hiding data. Experimental results show that the original cover image can be recovered without any distortion after the hidden data have been extracted if the stego image has not been altered, and the hidden data can still be extracted while the image quality is satisfactory. Meanwhile, compared with previous works, the scheme has higher payload and better image quality.

Keywords: lossless data hiding; histogram shifting; trace of secret data.

1 Introduction

As an important way to protect secret information, data hiding technique has been extensively studied in the past. Data hiding hides secret information into a cover object to create a stego medium with little distortion. In general, data hiding includes digital watermarking and steganography. Watermarking is used for copyright protection, authentication, and transaction tracking, etc., while steganography undetectably alters a cover object to conceal a secret message.

Several data hiding techniques about embedding data in images have been proposed. Wu and Tsai [1] introduced a steganography based on pixel-value differencing. In 2003, Tian [2] proposed a reversible data embedding a scheme with high payload and good imperceptibility. The above mentioned methods are fragile to any distortion. In order to be robust against common signal processing operations such as JPEG compression, Solanki et al [3] proposed a data hiding scheme in images based on information theoretic analyses to hide large volumes of data with low perceptual degradation.

In general, embedding secret data into a cover image often leads to the degradation of embedded image and the original image can't be recovered forever. However, in some applications, such as military image systems, medical diagnosis, and art works, it is critical for an original image to be reversed after the hidden data are extracted. This technique, which is also called reversible or distortion-free data hiding, has been employed. The first reversible data hiding scheme was introduced in 1997[4]. Then plenty of reversible data hiding schemes were proposed [5], [6].

The rest of this paper is organized as follows. Previous related works are reviewed in Section 2. The proposed scheme is presented in Section 3. Experimental results are given in Section 4. Finally, conclusions are drawn.

2 Related Works

Based on Ni et al's [7] scheme, Zeng [8] et al proposed a scheme by introducing a new embedding mechanism, which enhances the capacity and make the algorithm more robust. In this scheme, the cover image is divided into a number of non-overlapping blocks firstly. Based on patchwork theory, each arithmetic difference $\alpha^{(k)}$ for a block equals zero approximately and the distribution of occurrences number of α is shown in Fig.1(a). Next, Zeng introduced two thresholds T and G, and explore extra space to embed data. The image blocks with arithmetic difference $\alpha \in [-T,T]$ are used to embed information data. If a bit 0 is to be embedded, this block remains intact; and if a bit 1 is to be embedded, data was embedded by shifting the arithmetic difference β_2. The thresholds G influences the robustness. The image histogram after embedding is showed in Fig.1(b).

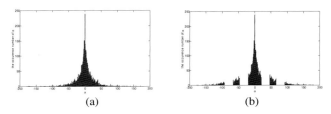

(a)   (b)

Fig. 1. The image distribution of α. (a) The distribution of α after exploring extra space. (b) The distribution of α after data embedding.

Despite claiming good imperceptibility by these data hiding schemes, it inevitably leaves obvious detectable traces to attacker. In general, in a normal image without secret data, the occurrences number of α decreases in a smooth fashion macroscopically. However, to a stego image generated by Zeng's scheme, the occurrences number of α is distinctly dissimilar to that of normal image, as shown in Fig. 1(b). Based on the distribution, the hiding bits and hiding capacity could be extracted easily. This defect is found in other schemes, such as Ni et al's [7], Vleeschouwer et al's [9], and Kim et al's [10] schemes. Note that in previous schemes mentioned above, to get better imperceptibility, larger image blocks are used to get a less modification of each pixel, which correspondingly decrease the hiding payload.

The proposed scheme improves these schemes through a new embedding mechanism. With a private key, the trace of secret data is concealed. To further enhance the security of this scheme, the threshold used to expand extra space is revised adaptively. Meanwhile, the proposed scheme has higher payload and better image quality compared with these schemes mentioned above.

3 The Proposed Scheme

3.1 Statistical Features Used to Embed Data

The pixel grayscale values in a local image block are often highly correlated and have spatial redundancy. Within an image block, the grayscale value of a selected pixel equals the average of values of pixels around approximatively. This statistical feature exists in other tested images. The difference α between a given pixel and its ambient pixels is calculated as formula 1, where, b is the grayscale value of selected pixel, M is the N members set made up with ambient pixels, and v_i is grayscale value of the ambient pixel i. Since most values of α are very close to zero, the hiding data could be embedded by shifting α of each image block.

$$\alpha = b - \frac{1}{N}(\sum_{i \in M} v_i) \tag{1}$$

3.2 Data Embedding Algorithm

Let image I with size of $M \times N$ be the cover image and I_m be the stego image. The secret data ESD is a binary sequence encrypted by an encryption system with private key $key1$. The data embedding procedure is summarized as follows.

Step 1: Divide I is into $m \times n$ blocks and get $Q = \lfloor M/m \rfloor \times \lfloor N/n \rfloor$ blocks. With a private key key_2, a random integer sequence with length of Q ($Q_i \in [1, m \times n]$) is generated, where Q_i is the selected pixel within an image block.

Step 2: Scan the cover image and calculate the arithmetic difference α of each image block. The histogram of α for a fixed pixel p at the right is shown in Fig. 2(a). To conceal the information hiding trace, pixel $p = Q_i$ in ith image block is selected in a predefined order. It is noticed that the histogram of α has similar feature, as shown in Fig. 2(b). The block with $\alpha \in [-T,T]$ is utilized to embed secret data by shifting the selected pixel p within each block. The shifting rule is given by:

$$R_k(i,j) = \begin{cases} S_k(i,j) + (T + 2 \times G) & \alpha \in (T, +\infty) \\ S_k(i,j) + (T + G) & \alpha \in (0,T] \text{ and } ESD(m) = 1 \\ S_k(i,j) - (T + G) & \alpha \in [-T, 0] \text{ and } ESD(m) = 1 \\ S_k(i,j) - (T + 2 \times G) & \alpha \in (-\infty, T) \\ S_k(i,j) & \alpha \in [-T, T] \text{ and } ESD(m) = 0 \end{cases} \tag{2}$$

where, $S_k(i,j)$ is the pixel value of selected pixel p in the kth image block, $R_k(i,j)$ is the pixel value of p after shifting, and $ESD(m)$ is the corresponding embedding bit. The distribution of α after shifting is shown in Fig. 2(c). The range $\alpha \in [-T,T]$ is called the inner zone, used to embed bit 0; the range

$\alpha \in [T+G, 2\times T+G]$ and $\alpha \in [-(2\times T+G), -(T+G)]$ is called the outer zone, used to embed bit 1. Because of the randomization of ϱ, the unsmooth fashion generated by shifting will be covered by the total histogram distribution. Without key_2, the histogram of α according to the same selected pixel p is shown in Fig.2 (d). It noticed that the data hiding trace is concealed by the private key key_2. After the histogram shifting, the stego image I_m is generated.

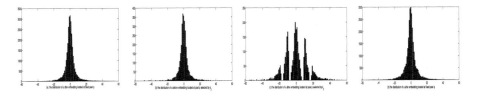

Fig. 2. The distribution of α located at fixed pixel and selected pixel generated by k_0 respectively

The image block with $\alpha \in [-T,T]$ are utilized to embed secret data, the larger T is, the more secret information could be embedded. The threshold G is utilized to separate the inner zone and outer zone. The larger G is, the more robust the scheme proposed is. However, the visual quality will be reduced as the value of T and G increase, meanwhile too large space between the inner zone and outer zone may reveal the presence of hidden data. In the practice, the value of G is determined adaptively by slope of two adjacent points on a fitting curve. After embedding, the values of α could form a set $X = \{x \mid x = \alpha_i, -255 \le i \le 255\}$, and the corresponding occurrence numbers of α could form a set $Y = \{y \mid y = occur\_num(x)\}$. The dominating elements of X (such as $x = \alpha i (-20 \le \alpha i \le 20)$) and its corresponding element in Y are utilized as data set to fit a smooth curve using the spline fitting. From the curve after fitting, every two adjacent points could be used to calculate the slope K_i by $(y_i - x_j)/(\alpha_i - \alpha_j)$. If adjacent K_i and K_j are both larger than the threshold β_1 (e.g. 0) in the positive axis or they are both less than β_2 (e.g. 0) in the negative axis, the threshold G should be revised by $G = G - 1$. Fig. 3 is an example of adjacent slopes K_i and K_j are both larger than β_1. The bar is the distribution of arithmetic difference and the curve is the smooth curve of fitting.

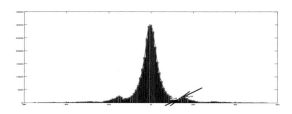

Fig. 3. The adjacent slopes k_i and k_j are both larger β_1

In the embedding procedure, there is possibility to cause an overflow/underflow by shifting α as mentioned above. For a grayscale image, the permitted pixel range is $[0, 255]$. If the selected pixel value of the cover image block is larger than $255 - S$ and a number S needs to be added to the grayscale value of this pixel, overflow will occur. Similarly, if the selected pixel value is lower than S and S needs to be subtracted from the grayscale value of this pixel, underflow will occur. To prevent overflow and underflow, if the overflow occurs, the value of selected pixel is altered to $255 - S$; and if the underflow occurs, the value of selected pixel is altered to S. Meanwhile, the overhead information made up with the coordinates and original value of these pixels can be compressed losslessly and transmitted to the receiver. Fortunately, the possibility to cause an overflow or underflow is very slim.

3.3 Data Extraction and Original Image Recovery Algorithm

Step 1: The stego image I'_m is divided into non-overlapping image blocks size of $m \times n$ in the same predefined order. With private key key_2, the selected pixel p of each image block was gotten. Then calculate the arithmetic difference α of the image block.

Step 2: The extraction and recovery procedures are carried on simultaneously. To each image block, one bit secret data, denoted by F''_k, could be extracted as follows.

$$F''_k = \begin{cases} 1 & -T \leq \alpha \leq T \\ 0 & \alpha \in [T, T+2\times G] \text{ or } \alpha \in [-(T+2\times G), -T] \\ no\ data & others \end{cases} \quad (3)$$

The recovery procedure is shown as follows.

$$R''_k(i,j) = \begin{cases} R'_k(i,j) - (T+2\times G) & \alpha \in (T+2\times G, +\infty) \\ R'_k(i,j) + (T+G) & \alpha \in (T+G, 2\times T+G] \\ R'_k(i,j) - (T+G) & \alpha \in [-(2*T+G), -(T+G)] \\ R'_k(i,j) + (T+2\times G) & \alpha \in (-\infty, -(T+2\times G)] \\ R'_k(i,j) & \alpha \in [-T, T) \end{cases} \quad (4)$$

Step 3: If the overhead information exists, the including pixel value and coordinates are utilized to revise the recovery image. Finally, the extracted secret data bits are concatenated and decrypted by the same decryption system.

4 Experimental Results

Six images called "Artwork", "Goldhill", "Lena", "Vessel", "Pepper" and "Plane" are used as cover images to evaluate the performance of the proposed scheme, shown in Fig 4. The sizes of these cover images are all 512×512 pixels.

Fig. 4. Test images: (a) Artwork. (b) Goldhill. (c) Lena. (d) Vessel. (e) Pepper. (f) Plane

Thanks to the private key K_0 and revision of threshold, the trace of hiding secret data is concealed. Take images "Lena" and "Pepper" as examples, the distribution of histogram without private key K_0 are shown in Fig. 5(a) and Fig. 5(b). From the histogram, it is difficult to find the trace of secret data and extract them.

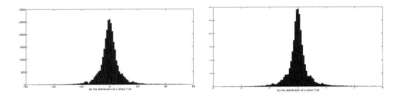

Fig. 5. Lena: the histogram of α : (a) Lena (b) Pepper

In this set of experiments, if T is set to α_{max}, the proposed scheme will get its maximum payload 28900 bits. The embedding ratios of actual payload to maximum payload about six test images are shown in Fig. 6. Where x and y represent the value of T and the ratio of actual payload to maximum payload respectively.

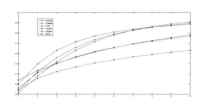

Fig. 6. The embedding ratios of actual payload to maximum payload with different T

The PSNR (peak signal noise ratio) between the original cover image C and the stego image S both sized $M \times N$ is shown as follows.

$$PSNR = 10 \times \log_{10}(\frac{255^2 \times M \times N}{\sum_{i=1}^{M}\sum_{j=1}^{N}(C(i,j)-S(i,j))^2}) \qquad (5)$$

As for the test images sized 512×512 in the experiments, the PSNR could be calculated as follows.

$$PSNR = 10 \times \log 10 \left(\frac{255^2 \times M \times N}{(M/3 \times N/3) \times (P_2 \times P_1 (T+G)^2 + (T+2 \times G)^2 * (1-P_1))} \right) = \log 10 \left(\frac{9 \times 255^2}{P' + P''} \right) \quad (6)$$

where, P_1 is the possibility of $\alpha \in [-T,T]$ in an image block and P_2 is the possibility of one secret bit was embedded in inner zone. Hence P' and P'' is the possibility of α shifting by $T+G$ and $T+2 \times G$, respectively. The value of P_2 is 0.5 if secret data is random sequence. In practical applications, if bit 1 shares a larger number in the secret data, bit 1 is embedded in the inner zone; otherwise bit 0 is embedded in the inner zone. Besides, if overhead information exists, the PSNR will be larger. Note that in these experiments, only image "Goldhill" needs little overhead information.

The comparison results between Zeng et al's scheme and proposed scheme about image quality and actual payload on test images are tabulated in Table 1. Note that the parameter in Zeng's scheme is selected to obtain maximum payload and in the proposed scheme the threshold T is set to 6.

Table 1. The performance comparison between Zeng's algorithm and the proposed scheme

	Zeng's scheme		Proposed scheme	
	PSNR	Actual payload (bits)	PSNR	Actual payload (bits)
Artwork	38.54	4096	41.30	19031
Gold hill	31.82	4096	40.76	19031
Lena	38.60	4096	41.78	22821
Vessel	38.56	4096	41.11	18989
Pepper	37.26	4096	42.00	22574
Palne	38.60	4096	41.20	23384

In robustness testing, all the stego images are compressed by LuraWave SmartCompress 2.2. The BER (bit error rate) confronted with JPEG compression with different quality factor is shown in Fig.7. The average BER is below 2% when the JPEG quality factor is 90 and the BER reach nearly 20% when the JPEG quality factor is 80. Hence, the hidden data can be extracted if the stego image goes through JPEG compression to some extent.

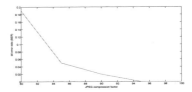

Fig. 7. The average BER of test images under different JPEG quality factor

5 Conclusion

A lossless robust data hiding scheme is proposed. Based on the histogram of arithmetic difference of each block, a threshold is set to decide the actual payload, and another threshold is adaptively determined to conceal the presence of hiding data. The original cover image can be restored completely after the hidden data have been extracted correctly if the stego image remains intact. Meanwhile, the hidden data are robust against non-malicious attacks such as JPEG compression to some extent.

References

1. Wu, D.C., Tsai, W.-H.: A steganographic method for images by pixel-value differencing. Pattern Recognition Letters 24, 1613–1626 (2003)
2. Tian, J.: Reversible data embedding using a difference expansion. IEEE Transactions on Circuits and Systems for Video Technology 13(8), 831–841 (2003)
3. Solanki, K., Jacobsen, N., Madhow, U., Manjunath, B.S., Chandrasekaran, S.: Robust image-adaptive data hiding using erasure and error correction. IEEE Trans. Image Process. 13(12), 1627–1639 (2004)
4. Barton, J.M.: Method and Apparatus for Embedding Authentication Information within Digital Data, US Patent 5 (1997)
5. Chang, C.C., Kieu, T.D.: A reversible data hiding scheme using complementary embedding strategy. Information Sciences (Article in Press)
6. Lee, C.C., Wu, H.-C., Tsai, C.-S., Chu, Y.-P.: Adaptive lossless steganographic scheme with centralized difference expansion. Pattern Recognition 41, 2097–2106 (2008)
7. Ni, Z., Shi, Y.Q., Ansari, N., Su, W., Sun, Q., Lin, X.: Robust Lossless Image Data Hiding Designed for Semi-Fragile Image Authentication. IEEE Transactions On Circuits and Systems for Video Technology 18(4) (April 2008)
8. Zeng, X.-T., Ping, L.-D., Pan, X.-Z.: A loss less robust data hiding scheme. Pattern Recognition 43, 1656–1667 (2010)
9. De Vleeschouwer, C., Delaigle, J.F., Macq, B.: Circular interpretation of bijective transformations in lossless watermarking for media asset management. IEEE Tran. Multimedia 5, 97–105 (2003)
10. Kim, K.-S., Lee, M.-J., Suh, Y.-H., Lee, H.-K.: Robust lossless data hiding based on block gravity center for selective authentication. In: IEEE International Conference on Multimedia and Expo (ICME), Cancun, Mexico (2009)

Open Fault Diagnose for SPWM Inverter Based on Wavelet Packet Decomposition

Zhonglin Yang[1,2], Jianbao Liu[2], and Hua Ouyang[2]

[1] Wuhan University / School of Electrical Engineering,
Wu Han, China
blueduny@sina.com
[2] Navy University of Engineering /School of Electrical and Information Engineering,
Wu Han, China
ljbtt@hotmail.com

Abstract. By introducing the double Fourier transform of the switch function, the frequency spectrums of DC side current of inverter in normal, open fault of one transistor and open fault of two transistors in one phase are analyzed. The RMS value of the wavelet packet decomposition coefficients of DC side current was calculated, and could be the symptom of open fault. The experimental results demonstrated that the proposed approach is feasible.

Keywords: Inverter, fault diagnose, wavelet packet decomposition, DC side current.

1 Introduction

The induction motor system supplied by voltage source inverter is used widely in industry filed. The reliability of inverter directly affects the normal operation of the system. In order to improve the system's reliability, the fault tolerant techniques are often applied to the power conversion unit [1]-[3]. The premise of fault tolerant techniques is the faults' effective detection, diagnosis, localization and isolation, for which a number of methods of inverter's fault diagnosis have been proposed. The method based on the current vector monitoring of time domain is proposed in [4]. The open circuit fault diagnosis methods based on the analysis of the current-vector trajectory and of the instantaneous frequency in faulty mode are proposed in [5]. The method based on the monitoring of the average current Park vector is proposed in [6]. The method of rapid open-circuit fault diagnosis based on the inverter's voltage model is proposed in [7]. The intelligent control techniques are applied in the inverter's fault diagnosis [8, 9]. These methods above-mentioned regard the inverter's output voltages as the detectives signals, and require three sensors to detect the signals, so the size of diagnostic system is larger. In this paper, a new inverter's diagnostic method which regards the DC side current as the detective signals is proposed. It only need a current sensor and reduces the size of the diagnostic system. The proposed method is verified by experimental results.

2 Analysis of Inverter DC Side Current's Spectrum

2.1 Inverter DC Side Current in Normal

SPWM voltage source inverter is widely used, therefore, this paper focus on such inverters. Fig.1 is the SPWM voltage source inverter main circuit topology [10].

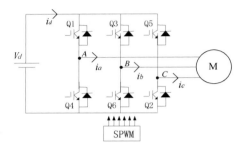

Fig. 1. Main circuit of SPWM voltage source inverter

The operation of the inverter can be illustrated by three binary signals k_i, which are the so-called switching functions. Logic "1" means that the corresponding ac phase is connected to the positive terminal of the dc-side. Logic "0" means that the ac phase is connected to the zero terminal of the dc-side [11]. The relationship between inverter DC input current and output current yields

$$i_d = k_a i_a + k_b i_b + k_c i_c \tag{1}$$

The switching functions are confirmed by the carrier signal and modulated signal. It be assumed that the initial phase of the carrier signal is 0 and the initial phases of the modulated signals are 0, -2π/3, 2π/3. With the double Fourier transformation, the switching functions can be expressed as follows [12]:

$$\begin{cases} k_a = \dfrac{1}{2} + \dfrac{M}{2}\sin(\omega_0 t) + \dfrac{2}{\pi}\sum_{m=1}^{\infty}\sum_{n=-\infty}^{\infty}\dfrac{1}{m}J_n\left(mM\dfrac{\pi}{2}\right)\cdot\sin\left[(m+n)\dfrac{\pi}{2}\right]\sin(m\omega_c t + n\omega_0 t) \\ k_b = \dfrac{1}{2} + \dfrac{M}{2}\sin(\omega_0 t - \dfrac{2\pi}{3}) + \dfrac{2}{\pi}\sum_{m=1}^{\infty}\sum_{n=-\infty}^{\infty}\dfrac{1}{m}J_n\left(mM\dfrac{\pi}{2}\right)\cdot\sin\left[(m+n)\dfrac{\pi}{2}\right]\sin\left[m\omega_c t + n\left(\omega_0 t - \dfrac{2\pi}{3}\right)\right] \\ k_c = \dfrac{1}{2} + \dfrac{M}{2}\sin(\omega_0 t + \dfrac{2\pi}{3}) + \dfrac{2}{\pi}\sum_{m=1}^{\infty}\sum_{n=-\infty}^{\infty}\dfrac{1}{m}J_n\left(mM\dfrac{\pi}{2}\right)\cdot\sin\left[(m+n)\dfrac{\pi}{2}\right]\sin\left[m\omega_c t + n\left(\omega_0 t + \dfrac{2\pi}{3}\right)\right] \end{cases} \tag{2}$$

Where M is the modulation index, ω_0 is the angular frequency of modulated signal, ω_c is the angular frequency of carrier signal, J_n is the n-order Bessel function.

When the inverter runs in normal, the inverter output currents are symmetrical three-phase sinusoidal signals, that is

$$\begin{cases} i_a = I\sin(\omega_0 t + \delta) \\ i_b = I\sin(\omega_0 t + \delta - 2\pi/3) \\ i_c = I\sin(\omega_0 t + \delta + 2\pi/3) \end{cases} \tag{3}$$

Where δ is the output current's phase angle relative to the output voltage, which is confirmed by the load impedance. Inserting (2) and (3) into (1) gives the DC side current, as expressed in

$$i_d = \frac{3}{4}MI\cos(\delta) + \frac{2I}{\pi}\sum_{m=1}^{\infty}\sum_{n=-\infty}^{\infty}\frac{1}{m}J_n(mM\frac{\pi}{2})\sin\left[(m+n)\frac{\pi}{2}\right]$$

$$\cdot\begin{cases}\sin(m\omega_c t + n\omega_0 t)\sin(\omega_0 t + \delta) \\ +\sin[m\omega_c t + n(\omega_0 t - 2\pi/3)]\sin(\omega_0 t + \delta - 2\pi/3) \\ +\sin[m\omega_c t + n(\omega_0 t + 2\pi/3)]\sin(\omega_0 t + \delta + 2\pi/3)\end{cases} \quad (4)$$

The first part of i_d is the DC component, the second part is the harmonic frequency signal of carrier signal and their edge webbing signals. In the condition $\omega_c \gg \omega_0$, only the low frequency components are considered during the inverter fault diagnosis. Therefore, the inverter DC side current in normal contains only a low frequency part of the DC component.

2.2 Inverter DC Side Current When One Transistor Open

When one transistor is inoperative, the bridge arm will not properly conduct. The low frequency part of the DC-side current contains not only the DC component, and contains other low-frequency Components. Taking Q1 open for example, the inverter DC side current is analyzed. Since the transistor Q1 open-circuit fault occurs, the upper part of the arm can not turn on. The relationship between inverter DC input current and output current yields

$$i_{d1} = k_b i_{b1} + k_c i_{c1} \quad (5)$$

Since the phase voltages are balanced with sinusoidal PWM modulation before and after the fault, the phase currents will be balanced sinusoidal with dc offset after the fault[10], namely:

$$\begin{cases}i_{a1} = -I_0 + i_{ah}(t) \\ i_{b1} = I_0/2 + i_{bh}(t) \\ i_{c1} = I_0/2 + i_{ch}(t)\end{cases} \quad (6)$$

Where $i_{ah}(t), i_{bh}(t), i_{ch}(t)$ respectively express the a, b, c-phase output current except the DC component, which contains the harmonics of the modulation signal. Inserting (2) and (6) into (5) gives the DC side current, as expressed in

$$i_{d1} = \frac{I_0}{2} - \frac{MI_0}{4}\sin(\omega_0 t) + \frac{i_{bh}(t) + i_{ch}(t)}{2} + \frac{M}{2}[\sin(\omega_0 t - 2\pi/3)i_{bh}(t) + \sin(\omega_0 t + 2\pi/3)i_{ch}(t)]$$

$$+ \frac{2}{\pi}\sum_{m=1}^{\infty}\sum_{n=-\infty}^{\infty}\frac{1}{m}J_n\left(nM\frac{\pi}{2}\right)\sin\left[(m+n)\frac{\pi}{2}\right]\begin{cases}\sin\left[m\omega_c t + n(\omega_0 t - 2\pi/3)\right](I_0/2 + i_{bh}(t)) \\ +\sin\left[m\omega_c t + n(\omega_0 t + 2\pi/3)\right](I_0/2 + i_{ch}(t))\end{cases} \quad (7)$$

From the Eq (7) we can find: When a transistor open-circuit fault occurs, the low frequency component of the inverter DC side current contains not only of the DC component, also contains the modulated signal and its harmonic components which do not appear when the inverter runs in normal.

2.3 Inverter DC Side Current When One Phase Open

When the open-circuit fault of one phase occurs, there is no output of the phase current. So the three-phase inverter is actually equivalent to a single-phase inverter. In order to facilitate the analysis and description, assuming open-circuit fault occurs in a phase (i.e. $Q1$, $Q4$ open), the inverter DC side current is analyzed. The relationship between inverter DC input current and output current yields

$$i_{d2} = k_b i_{b2} + k_c i_{c2} \tag{8}$$

$$\begin{cases} i_{b2} = I'\sin(\omega_0 t + \delta') \\ i_{c2} = -I'\sin(\omega_0 t + \delta') \end{cases} \tag{9}$$

Where δ' is the initial phase of output current relative to the a-phase modulated signal. Inserting (2) and (9) into (8) gives the DC side current, as expressed in

$$i_d' = \frac{\sqrt{3}}{4} MI'[\sin(2\omega_0 t + \delta) - \sin(\delta)] \\ + \frac{2I'}{\pi}\sum_{m=1}^{\infty}\sum_{n=-\infty}^{\infty}\frac{1}{m}J_n\left(mM\frac{\pi}{2}\right)\sin\left[(m+n)\frac{\pi}{2}\right]\begin{cases}\cos[m\omega_c t + n(\omega_0 t - 2\pi/3)]\cos(\omega_0 t + \delta) \\ -\cos[m\omega_c t + n(\omega_0 t + 2\pi/3)]\cos(\omega_0 t - \delta)\end{cases} \tag{10}$$

From the above equation we can find: When the open-circuit fault of one phase occurs, the DC side current contains the second harmonic of the modulated signal which do not appear when the inverter runs in normal. Therefore, the second harmonic of the modulated signal is the fault feature components of inverter's one phase open-circuit fault.

3 Inverter Open-Circuit Fault Diagnose Based on Wavelet Packet Decomposition

On the base of the wavelet's two-scale equation, w_n is commanded to meet the following recursive two-scale equations

$$\begin{cases} w_{2n}(t) = \sqrt{2}\sum_k h_k w_n(2t-k) \\ w_{2n+1}(t) = \sqrt{2}\sum_k g_k w_n(2t-k) \end{cases} \tag{11}$$

The muster of functions $\{w_n(t)\}_{n\in Z}$ is called as the wavelet packet generated by scaling functions[12]. Wavelet packet transform successfully resolves the wavelet's problem of "low resolving power in high-frequency band". It divides the frequency band more detailed, shown in Figure 2

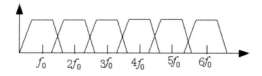

Fig. 2. The frequency division of the wavelet package transformation

According to above analysis, we know that the low frequency component of the inverter DC side current contains the modulated signal when a transistor open-circuit fault occurs and contains the second harmonic of the modulated signal when the open-circuit fault of one phase occurs. However, during the operation of the inverter, the small offset of modulation signal's frequency may appear. The direct detection would be difficult to achieve good results. At this time, we can decompose the DC current with the wavelet packet and figure out the RMS value of the wavelet packet decomposition coefficients of corresponding node, namely

$$x_{rms}(j,n) = \sqrt{\sum_{k=1}^{N} d_{j,n}^2(k)} \quad (12)$$

Where j is the number of layers for the signal decomposition, n is the frequency parameter of the wavelet packet decomposition (node number), $x_{rms}(j,n)$ is the RMS value of the wavelet packet decomposition coefficients, N is the length of the corresponding sequence of nodes.

With wavelet packet decomposition, the different frequency components of signal are decomposed to the appropriate frequency band. The RMS value of the coefficient mainly depends on the magnitude and distribution of harmonic components, which reflects the characteristics of the signal. When the open-circuit faults occur, the signal's energy corresponding to the same frequency band will vary in large range and the RMS value of the wavelet packet decomposition's coefficients will change significantly. Therefore, the RMS value of the wavelet packet decomposition's coefficients can be regarded as the characteristics of the inverter open-circuit fault indicators.

4 Experiment and Result Analysis

The experimental circuit is shown in Figure 1. The inverter is followed with a asynchronous induction motor. In the conditions of normal operation, one transistor open-circuit fault and one phase open-circuit fault, the DC side currents were collected, and were passed through a low-pass filter which the cut-off frequency is taken to be 500Hz. After filtering the high frequency components, the 4 layers wavelet packet decomposition was used. The RMS values of the wavelet packet decomposition coefficients corresponding to the fault node were calculated and shown in Table 1.

Table 1. The RMS value of the coefficient of wavelet package decomposing(j=4).

	The node corresponding to the DC component	The node corresponding to the modulated signal	The node corresponding to the second harmonics of modulated signal
The RMS value in normal	4.5885	0.0893	0.1096
The RMS value when one transistor open	6.0178	4.1198	1.9950
The RMS value when one phase open	5.4992	0.5209	6.8317

The data in Table 1 show that when the inverter is normal, the RMS values corresponding to the modulation signal and it's second harmonic are very small, which indicates that the DC side current contains only the DC component in the low-frequency band; when one transistor open-circuit fault occurs, the RMS values corresponding to the modulation signal and it's second harmonic have a larger growth, which indicates that the DC side current contains the modulation signal and it's second harmonic; when one phase open-circuit fault occurs, the RMS value corresponding to the second harmonic of modulation signal has a greater growth, while the RMS value corresponding to modulated signal is still small, which indicates that the DC side current contains not only the DC component but also the second harmonic of modulated signal in the low-frequency band.

5 Conclusion

The inverters are widely used in various industrial fields. When the inverters can not run in normal, the serious consequences will occur. Therefore, the study of inverter's fault diagnosis method to improve the reliability of inverters has the important economic and practical significance. In this paper, the frequency spectrums of DC side current of inverter in normal, open fault of one transistor and open fault of two transistors in one phase are analyzed and the fault characteristic frequencies are found.. During the operation, the frequency offset of modulated signal may appear which makes again the inverter's fault diagnosis. The method based on the wavelet packet decomposition is present. This method only needs a current sensor, and reduces the size of the diagnostic system. The experiential results also show that the method can overcome the adverse effects of modulated signal offset; the diagnostic result is accurate and reliable.

References

1. Welchko, B.A., Lipo, T.A., Jahns, T.M., et al.: Fault tolerant three-phase AC motor drive topologies: a comparison of features, cost, and limitations. IEEE Transactions on Power Electronics 19(4), 1108–1116 (2004)
2. An, Q.-t., Sun, L., Zhao, K., et al.: Diagnosis Method for Inverter Open-circuit Fault Based on Switching Function Model. Proceedings of the CSEE 30(6), 1–6 (2010) (in Chinese)
3. Bolognani, S., Zordan, M., Zigliotto, M.: Experimental fault-tolerant control of a PMSM drives. IEEE Transactions on Industrial Electronics 47(5), 1134–1141 (2000)
4. Smith, K.S., Ran, L., Penman, J.: Real-time detection of intermittent misfiring in a voltage-fed PWM inverter induction-motor drive. IEEE Transactions on Industrial Electronics 44(4), 468–476 (1997)
5. Peuget, R., Courtine, S., Rognon, J.P.: Fault detection and isolation on a PWM inverter by knowledge-based model. IEEE Transactions on Industry Applications 34(6), 1318–1326 (1998)
6. Mendes, A.M.S., Marques, A.J.: Voltage source inverter fault diagnosis in variable speed AC drives by the average current Park's vector approach. In: IEMDC 1999 Conference, Washington, USA, pp. 704–706 (1999)

7. Ribeiro, R.L.A., Jacobina, C.B., Silva, E.R.C., Silva, E.R.C., et al.: Fault detection of open-switch damage in voltage-fed PWM motor drive systems. IEEE Transactions on Power Electronics 18(2), 587–593 (2003)
8. Khanniche, M.S.: Wavelet-fuzzy-based algorithm for condition monitoring of voltage source inverter. Electronics Letter 40(4), 267–268 (2004)
9. Martins, J.F.: Unsupervised neural-network-based algorithm for an on-line diagnosis of three-phase induction motor stator fault. IEEE Transactions on Industrial Electronics 54(1), 259–264 (2007)
10. Kastha, D., Bose, B.K.: Investigation of Fault Modes of Voltage-fed Inverter System for Induction Motor Drive. IEEE Transactions on Industry Applications 30(4), 1028–1038 (1994)
11. Jiang, Y., Ekstrom, A.: Genetal analysis of harmonic transfer though converters. IEEE Transactions on Power Electronics 12(2), 287–292 (1997)
12. Jin, M.: Approaches for modeling and forecasting the conduction interfere of the electric power and electronic system. Naval University of Egineering, Wuhan (2006) (in Chinese)
13. Tao, X., Qi, W.: Application of Multiscale Principal Component Analysis Based on Wavelet Packet in Sensor Fault Diagnosis. Proceedings of the CSEE 27(9), 28–31 (2007) (in Chinese)

The Method for Determination the Freshness of Pork Based on Edgedetection*

Tianhua Chen, Suxia Xing, and Jingxian Li

College of Information Engineering, Beijing Technology and Business University
Beijing, People's Republic of China.100048
cth188@sina.com

Abstract. According to the biochemical mechanism of the pork Metamorphic process, made e some sample slice during the pork corruption time, got images through the optical microscope equipment application of Canny algorithm is proposed to detect cell edges, calculated the number of intact cells per unit area on the slice, found relationship between the number of intact cells and the freshness of pork, the results had corresponding relationship with the national standard TVB-N values, the study presents a new and simple identification method for determination of pork freshness.

Keywords: Pork freshness; Cell edge detection; Canny operator; Non-destructive testing; Food safety.

1 Introduction

As people's living standards improved, pork has become an important meat among urban and rural residents in China. Traditional pork freshness detection methods include sensory testing, physical and chemical testing, microbiological testing, instrument analysis[1]. Sensory test based on pork color, structure, odor, goods with a comprehensive judgments demanding on the ability of inspectors, it generally takes a long time after the system of training and practice can perform testing. Testing need special equipment and devices, the detection process is complex, test time is relatively long. Microbiological testing and analysis methods are similar to the physical and chemical analysis, all require different types of ancillary equipment, testing cycle length[2].Therefore, there is an urgent need for a fast, effective, non-destructive testing fresh pork scientific detection methods, we propose a detection based on Canny operator of pork fat in the metamorphic process of evaluating the number of intact cells of fresh pork rapid non-destructive testing methods.

---

* Fund Projects: Beijing Municipal Education Commission Science and Technology Innovation Platform - Signal Detection and Intelligent Information Processing Platform (201151).

2 Acquisition of Cell Image

In this paper, the image edge detection method was used to analyze the sample, so the first step is collecting slice pork samples images. Because the sample served in glass, in order to avoid glass reflection and refraction of light influence the role of image capture, using the end of light illumination on the sample.

2.1 Image Acquisition System

Image acquisition system is a CCD-based optical system, the main room by the light, light source, CCD camera, image acquisition card and PC machine. Room light and light images provide detection of pork slices as environment; camera take images, capture the image sized 512 × 512 pixels, by the frame grabber digitized into the computer. The freshness healthy animals meat is usually sterile, the pork corruption is usually caused by too long storage time, in a certain environment, temperature and humidity, due to the organization of microbial enzymes and bacteria on the surface[3]. The process from full fresh pork to the depth of corruption, corruption in pork fat during the first decomposition of deterioration, which contains water, proteins and lipids and other nutrients in the decomposition process and to provide the conditions for the growth of bacteria. Pork and corruption during the first gathering of bacteria is the adipose tissue area[4]. This selection of pork adipose tissue as a sample, the process of corruption in the sample each time point, remove the part of the adipose tissue samples to determine fresh point of the samples at that time.

The experiment samples slaughtered by the formal sold in Beijing market were passed the inspection without water and frozen pork from the health inspection and quarantine authorities, select the number of meat extract as samples, used the new blade sterilized with alcohol slaughtered in Extraction of adipose tissue into 12 samples with length, width and thickness are 2×1×0.2cm, and samples placed at 25℃ under natural corruption, made images each hour.

2.2 Sample Image Capture

The samples were removed from the constant temperature environment on the stage in the microscope, adjust the loading platform to the specified location. The microscope magnification was 100 times, collected the sample images each hour in the same location. The corruption of the meat is usually caused by external microbial contamination, from the surface of the meat to the deep spread along the connective tissue, so that by the secretion of collagenase hydrolysis of collagen connective tissue mucin, while producing gas, and finally broken down into amino acids, water, Carbon dioxide and ammonia, creating offensive smell.

The surface of microorganisms get into inner layer of the meat along blood vessels, and deep into the muscle tissue, in a certain range of the corruption, the propagation of microorganisms is still limited to the gap between the cells, when the depth of corruption occurs when part of micro-organisms into the muscle fibers[5].during the meat Corruption, the protein produced by the decomposition of water, sugar, carbon dioxide and other substances to provide a large population of bacteria living environment, when it reaches the depth of corruption reached a peak when the rate of bacterial growth and reproduction rate began to slow down.

3 Cell Image Treatment

Since the propagation of bacteria to the host for nutrition, with the deepening of the level of corruption, placing the sample under a microscope, you can clearly observe the different time periods, pork adipose tissue cells gradually deformed, become blurred. Pork and corruption, the protein structure is damaged, the cell wall bending, duplication, reduce the number of intact cells[6].Therefore,evaluation of edge detection can be used for fresh pork, through digital image processing methods to determine the number of statistical integrity of fresh pork, not only can improve the calculation speed and classification accuracy for detecting the pork freshness provides a new Methods.

3.1 Slice Image Processing

Sample slice image processing include histogram equalization and color space conversion. Histogram-based image enhancement method is based on the theory of probability and statistics, so through changing the shape of the histogram to achieve the effect of enhancing image contrast. Histogram equalization is to balance information with images obtained by CCD, correct the image histogram, compress image gray level with less pixels, draw the part with more pixels, increase the dynamic range of pixel gray value, enhance the contrast of the image, so the original section image details are not clear will become more balanced clear after the treatment. The images collected by CCD is based RGB color model, according to other processing needs, the images need to be converted to value color based on HSI. Slice image processing algorithm flow shown in Figure 1.

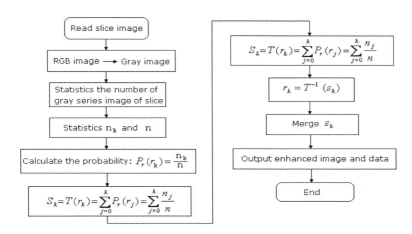

Fig. 1. Sample images of the algorithm process

3.2 Slice Image Processing

Through CCD imaging device observe the fat tissue, the magnification is 100 times, the optical zoom can clearly see the structure of fat cells. The captured image with size of 512×512 pixel,can be seen fat cells of fresh adipose tissue is approxim-ately circular or oval through the observation, but with the corruption of the sample, the fat cells and tissues are constantly changing. Fig.2(a), (b), (c), (d) samples were corrupted 0 hours, 3 hours, 9 hours and 12 hours depth of the corruption of the image.

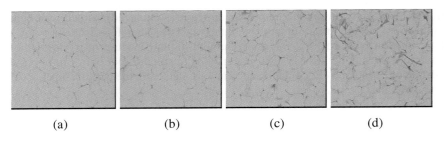

Fig. 2. Morphological changes of the corruption processing

0 hours. Fig.2 (a) is fresh meat (0 hours) cell slice image, 0 hours fat cells of fresh samples are full and uniform, linear cell wall can be seen clearly and easily distinguish the cell number.

3 hours. Fig.2 (b) are cells slice images after the 3 hours, through 3 hours after the oxidation corruption, it is clear that cell membranes become curved, not smooth, though still able to see cell wall clearly, but a small number of cells and adjacent cells wall occur overlap, and another small part of the cells wall had rupture phenomenon, not a complete fat cells yet.

9 hours. Fig .2 (c) is the cell slice image after 9 hours, the sample after 9 hours after corruption can be clearly seen overlap among a large number of cell walls, the phenomenon of increasing cell wall adhesion, only a small amount of intact cells. 12 hours. Fig. 2 (d) the cells in the slice images of the depth of corruption, had the deep corruption. In the figure, cannot observe full fat cells, the cell wall overlap and bend, has been hard to tell the complete cell structure.

3.3 Cell Contour Extraction

Canny operator using contour extraction of fat cells. Canny operator is a better edge detection characteristics and are widely used edge detection operator, has a high signal to noise ratio and detection accuracy, is a high performance step-type edge detection operator[7]. The superior detection characteristics in the following areas:

(1) good edge detection performance. Canny operator can guarantee the successful detection of cell edges, not to non-edge points as edge detection;

(2) edge detection has high signal to noise ratio. Canny operator can not only successfully detected cell edge, the edge of the weak signal a better ability to respond, the operator of the output signal to noise ratio is high;

(3) edge of the monitor give the accurate position. Canny edge operator was used to detect the edge point and the actual position with high accuracy;

(4) good single-edge characteristics[8]. Canny operator has a better detection of edge points only once capacity. Canny edge detection operator is shown in Figure3.

The Method for Determination the Freshness of Pork Based on Edgedetection 957

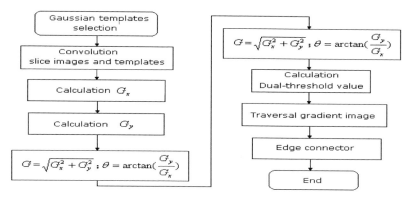

Fig. 3. Canny cell edge algorithm operator map

Canny operator is applied every 1 hour set of all 12 samples collected from the edge extracted image slices, placing 0 hours, 6 hours, 9 hours and 12 hours later, one set of samples of the cell edge detection results in Figure 4.

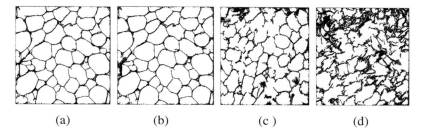

(a)　　　　　　(b)　　　　　　(c)　　　　　　(d)

Fig. 4. Cell edge extraction results

4 The Experimental Results

Analysis cell's size and shape on 12 different samples, the number of fat cells were statistic different adipose different time. The experimental data shown in Table 1:

Table 1. Statistics of intact cells of pork

	0 h	3 h	9 h	12 h		0 h	3 h	9 h	12 h
No1	36	35	21	12	No7	31	31	20	11
No2	33	33	26	13	No8	34	33	25	13
No3	35	35	22	10	No9	33	33	21	11
No4	28	27	18	9	No10	30	29	20	8
No5	30	29	20	11	No11	27	27	19	10
No6	35	35	19	10	No12	29	28	18	10

Experiments show that with the longer placement time, the oxidation level of corruption has deepened, the number of pork tissue intact cells decreased. Just slaughtered fresh pork, the full number of cells per unit area than all the 25 samples of fresh pork; through 3 hours later, the number of intact cells between 15 and 24, when the sample has a certain degree of significant deterioration, color and taste of the more obvious changes occurred in the samples of fresh pork for the times; experienced more than 12 hours, the level of corruption is very deep, can smell the distinct odor, careful observation, can be observed visually significant organizational change, the number of intact cells is less than 14, the sample of pork for corruption, experimental test results with national standards based on TVB-N test results.

5 Conclusion

This study collected from Beijing passed the test on the general market a total of 12 pork samples, the samples were placed in 25 ℃ environment, after 0 hours, 3 hours, 12 hours of constant after the number of intact cells detected with traditional national standard TVB-N data control. The results show that: For the pork, the full number of cells per unit area is greater than 25, detected as fresh pork; intact cells at between 15 and 24, the detection of fresh pork for the times; intact cells is less than 14, detected as corrupt pork. In this paper, Canny operator based on the number of intact cells calculated fresh pork TVB-N detection methods and the results of national testing standards compliance, is a fast, convenient, non-destructive testing method for fresh pork.

References

1. Cai, J., Wan, X., Chen, Q.: Feasibility Study for the Use of Near-Infrared Spectroscopy in the Quantitative Analysis of TVB-N Content in Pork. Acta Optica Sinica 29(10), 2808–2812 (2009)
2. Cheng, F., Liao, Y.-t., Ma, J.-w.: Research development on nondestructive determina-tion of pork quality. Journal of Zhejiang University(Agriculture and Life Sciences) (2), 199–206 (2010)
3. Canny, A.: A Computational Approach to Edge Detection. IEEE Trans. on PAMI 8(6), 679–698 (2006)
4. Guo, P., Qu, S., Chen, Y., Fu, Y., Yu, R.: Study on Intellectual Detection Techniques to Freshness of Pork. Transactions of the Chinese Society for Agricultural Machinery (8), 78–81 (2006)
5. Geng, A.-q., Zheng, G.-f., Qian, H.: A review of methods for the determination of freshness of meat. Meat Industry (12), 37–39 (2006)
6. Fortin, A., Robertson, W.M., Tong, A.K.W.: The eating quality of Canadian pork and it s relationship with intramuscular fat. Meat Science 69(2), 297–305 (2005)
7. Yu, R.-x., Guo, P.-y.: Pork Freshness Detection Based on Photo-electric Micro-Technique. Journal of Beijing Technology and Business University(Natural Science Edition) 2, 69–73, 78–81 (2008)
8. Feng, Q., Yu, S.-l., Huang, X.-q.: ZHANG Wei.A Novel Edge Detection Algorithm for Nucleated Cell. Journal of Image and Graphics 14(10), 2004–2009 (2009)

Remote Monitoring and Control System of Solar Street Lamps Based on ZigBee Wireless Sensor Network and GPRS

Lian Yongsheng, Lin Peijie, and Cheng Shuying[*]

Institute of Micro/Nano Devices and Solar Cells,
School of Physics & Information Engineering, Fuzhou University
Fuzhou 350108, P.R. China
`chengsying@yahoo.com.cn`

Abstract. In order to avoid the shortcomings and deficiencies of solar street lamp systems at present market, in this paper, a new kind of solar street lamps monitoring & control system is designed. By combining ZigBee technology with MPPT algorithm in the design of solar street lamp controller, the wireless monitoring network can be constituted by solar lamps themselves, and the conversion efficiency of solar energy is improved. The working state of solar lamps distributed in one area can be monitored and controlled by the remote monitoring center through GPRS and Internet. The system consists of three main parts: solar street lamp controller, ZigBee wireless network, remote monitoring center. By designing the system hardware and software, and testing the experimental result, it is proven that the system is highly reliable, cost-effective, low power consumption, fast establishing and has a broad application prospect.

Keywords: ZigBee network; MPPT; solar street lamp controller; monitoring and control system.

1 Introduction

With the worsening of environment and the intensifying of world energy crisis, solar energy as a renewable energy is becoming more and more important. Solar street lamps have attracted much attention around the world [1]. But most of the solar street lamps at present market are isolated units without remote monitoring function. They have some shortcomings such as failure to monitor the real-time working state of solar lamps, failure to give an alarm for repairing when some solar lamps are out of order, failure to achieve Maximum Power Point Tracking (MPPT) of solar panels and control charging and discharging of batteries reasonably. These deficiencies limit conversion efficiency, working reliability and large-scale promotion of solar street lamps. Therefore, how to make full use of solar energy and achieve centralization management effectively and conveniently has become the urgent problem to be solved in solar street lamps [2].

ZigBee is a low-power, short-distance wireless network standard. Based on IEEE 802.15.4 wireless protocol, it generally consists of a large number of distributed nodes

---

[*] Corresponding author.

which organize themselves into a multi-hop wireless network [3]. Each node has one or more sensors, embedded processors and low-power radios. As ZigBee technology represents a wireless sensor network which is highly reliable, cost-effective, low power consumption, programmable and fast establishing, it has attracted ongoing research attention. It is very suitable for civilian application areas, including lighting control, environment monitoring, traffic control and industrial automation, etc [4].

In view of all these factors, this paper presents a new solar street lamps remote monitoring and control system based on ZigBee wireless sensor network and GPRS. In addition to the system architecture design and implementation, a novel kind of solar street lamp controller combined MPPT algorithm with ZigBee wireless networking is developed. With such controller, any solar street lamp installed in one area can flexibly join in the wireless monitoring network established by ZigBee coordinator. This designed controller can not only improve the conversion efficiency of solar energy with MPPT, but also provide an easy way to monitor and control the working state of solar street lamps, because the wireless network is constituted by solar lamps themselves without installing additional devices. Through the collaboration of ZigBee network, GPRS and Internet, users on Internet can easily access the real-time data of solar lamps distributed in a large area and the remote monitoring center can be freely located on any remote PC.

2 System Hardware Design

Monitoring and control system of solar street lamps based on Wireless Sensor Network (WSN), as illustrated in Fig. 1, could be divided into three main parts: solar street lamp nodes, ZigBee wireless network and remote monitoring center. Among the solar street lamp nodes, one will be as ZigBee coordinator responsible for establishing network, others can be configured to be routers or end devices according to the actual distribution of solar street lamps.

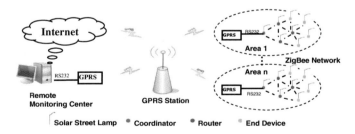

Fig. 1. System Architecture

A great number of solar street lamp nodes, distributed in one area, dynamically constitute a wireless monitoring network, in which, each node not only can collect parameters as voltage, current, and power from the solar panels, batteries and LED lamps, but also can control states of batteries and memorize all the collected parameters, and route them to coordinator. Data from solar street lamp nodes is transferred to remote monitoring center by GPRS and Internet.

By designing a visual management software for monitoring and controlling, the monitoring center can analyze and process the parameters of solar street lamps, give an alarm at the emergencies e.g. over-charging or over-discharging of batteries, the sudden change of solar lamps, and make decision in preventing from wrong state and correcting it. By sending control commands to ZigBee network, the working status of the system such as the brightness of LED lamps can also be adjusted by monitoring center according to the weather and the capacities of the batteries. The whole monitoring system of solar lamps owns the following characteristics: large capacity of network, flexible disposition, high reliability, fast self-establishing, low cost, low power consumption, and no influence on the natural environment.

2.1 Hardware Overview of Solar Lamp Node

The MCU used in the solar lamp node is CC2430 which is chipcon's leading SoC chip for ZigBee and highly suitable for systems with ultra low power consumption. The hardware diagram of each solar lamp node is illustrated in Fig. 2. It is composed of solar panel, battery, LED lamp and solar lamp controller. All the modules are indirectly powered by the solar panel. Specific circuit design will be introduced in the following section.

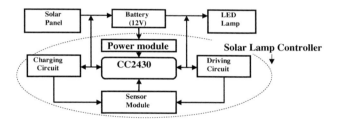

Fig. 2. Structure of Solar Lamp Node

As the core unit of solar lamp node, the designed solar lamp controller has the basic functions as following: (1) Ensuring that the battery is charged by the solar panel with MPPT to get as much energy as possible. (2) LED lamp will be automatically turned on when the night comes and it will be turned off in the daytime. (3) Collecting voltage, current and power from solar panel, battery and LED lamp for monitoring. (4) Providing protection for solar panel, battery and LED lamp. (5) Setting up wireless monitoring network based on ZigBee protocol.

2.2 Peripheral Design of ZigBee Wireless Module

To realize both Wireless Networking and MPPT by using only one MCU, the peripheral modules of the CC2430 is designed based on the basic application circuit according to the Data Sheet [5]. Fig. 3 shows the connection of CC2430 with the peripheral modules, including Solar Lamp Controller, GPRS Module (only for ZigBee coordinator), Sensor Module and Power Module.

Two internal timers in CC2430 are used to generate two PWM waveforms for driving the MOSFETs of solar lamp controller. Four ADC channels are used to perform conversions of the parameters collected by the sensor module, and save the results in memory (through DMA). A RS232 interface circuit is used to transmit the data to GPRS module by UART cable. As a 12V lead-acid battery is used in the solar lamp system, 5V and 3.3V voltages are respectively provided to the sensor module and the wireless communication module by the power module composed of LM7805 and AMS1117.

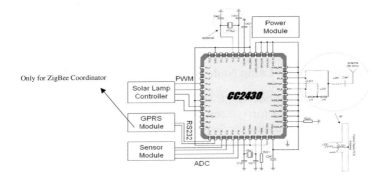

Fig. 3. Design of CC2430 Peripherals

2.3 Charging Circuit with Voltage and Current Sensor

With the change of sunshine intensity and solar cell junction temperature, output voltage and current of the solar panel will change nonlinearly. Therefore, there is a unique point of maximum output power of the solar panel in specific working conditions. In order to get as much electricity as possible under the same sunshine and temperature, MPPT should be considered when designing the charging circuit of battery illustrated in Fig. 4.

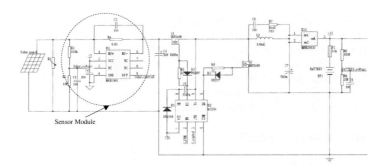

Fig. 4. Charging Circuit of the Battery

The core part of the Charging Circuit is a BUCK converter whose equivalent resistance and output voltage can be raised or reduced by adjusting the Duty Ratio of PWM to drive the MOSFET (Q1). The output power of the solar panel will reach the

Maximum Power Point (MPP) when the equivalent resistance of BUCK is equal to the internal resistance of the solar panel. The real-time values of output voltage and current of the solar panel will be collected by the sensor module, and transmitted to the ADC of CC2430 for conversions. According to the changing trend of the output power, CC2430 will adjust the Duty Ratio of PWM to achieve MPPT. In addition, the voltage value of battery must be acquired to judge whether it is over-charged or not.

2.4 LED Driving Circuit with Current Sensor

Fig. 5 shows the driving circuit of LED lamp which is responsible for adjusting the brightness of LED lamp according to the weather and the capacities of the batteries. Here another BUCK converter is adopted. By adjusting the Duty Ratio of PWM to drive the MOSFET (Q3), the discharging of the batteries and the brightness of LED lamp can be controlled effectively. The real-time current of LED lamp will be collected by the sensor module composed of a 0.03Ω sampling resistance and MAX4080. All the parameters collected from driving circuit and charging circuit will be sent to monitoring center.

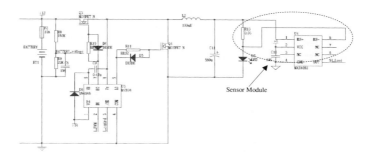

Fig. 5. Driving Circuit of LED Lamp

3 Design of System Software

The system software consists of two main parts: the software of processor CC2430, which is responsible for MPPT, controlling LED lamps, establishing wireless network, receiving control commands and sending the parameters of solar lamps; software of remote monitoring center, which is designed to be a visual management interface for monitoring and controlling.

3.1 Software of Processor CC2430

The software of processor CC2430 is developed on the basis of the ZigBee protocol (Z-Stack). In order to meet the specific application needs, some related functions in the application layer and network layer are modified. MPPT algorithm and controlling program of LED lamp are integrated into the ZigBee stack, so the CC2430 can realize MPPT and wireless communication by time sharing management. The flow diagram of processor CC2430 is illustrated in Fig. 6.

As for the realization of the solar panel MPPT, there are many kinds of methods. Main methods include Constant Voltage Method, Short-Current Pulse Method, Open Voltage Method, Perturb and Observe Method, Incremental Conductance Method, Artificial Neural Network Method, Temperature Method, etc. And each method has a variety of algorithms [6] [7]. Here Perturb and Observe Method is used. CC2430 changes the Duty Ratio of PWM to vary the output voltage and current of the solar panel, then adjusts the changing direction (PWM +blank or PWM-blank), according to real-time comparison between the preceding and later value of output power until the Maximum Power Point is found.

The self-organization and data transmission of the wireless sensor network are based on Z-Stack. The ZigBee network consists of a coordinator, several routers and end devices. Coordinator is responsible for building and maintaining the network, and undertaking the data transfer mission between the ZigBee network (solar street lamp nodes) and the remote monitoring center. As shown in Fig. 6, every solar lamp node firstly initializes its hardware and application variables. Coordinator establishes a new network, and other nodes join in the network by scanning the channels. Then the program will step into application layer to control the charging and discharging of solar street lamp. After having collected all the sensor data, it will process the collected data and form a data packet, then send data packet to the remote monitoring center by managing the communication of ZigBee network, GPRS and Internet. And the control commands received by coordinator from remote monitoring center will be retransmitted to every solar lamp node and dealt with.

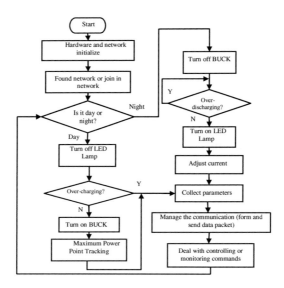

Fig. 6. Flow Diagram of Processor CC2430

3.2 Software of Remote Monitoring Center

Microsoft Visual C++ 6.0 is used to build a visual interface for remote monitoring center. And this monitoring software can be installed and launched on any Windows-based operating system. So the remote monitoring center can be located on any PC connected to the Internet.

The general management interface of the software is presented in Fig. 8. It can be divided into two main parts: the monitoring part is responsible for receiving and showing the messages of solar street lamps; the controlling part sends control commands to the ZigBee network. By setting some parameters such as physical address and lamp number, the software can set up a full peer-to-peer communication between the monitoring center and any solar lamp in one area.

Client/Server model based on TCP/IP protocol is applied as the communication model. Once the monitoring software is launched, the IP and Port Number of remote monitoring center (Server of TCP) are sent to the ZigBee coordinator (Client of TCP) by GPRS module with short message service (SMS). Then the coordinator can make communication connection with remote monitoring center by GPRS network and Internet according to the received IP and Port Number.

4 System Testing

The designed remote monitoring and control system of solar street lamps was tested on the campus playground of Fuzhou University. In the experimental test, ten solar lamps distributed on the playground constituted a small wireless monitoring network (separation distance: 70~120m), and the monitoring software was installed in the computer of our laboratory as the remote monitoring center. One GPRS module was connected to the serial port of PC for sending IP and Port Number, the other was connected to the ZigBee coordinator. One of the designed solar lamp systems and its controller was photographed and shown in Fig. 8.

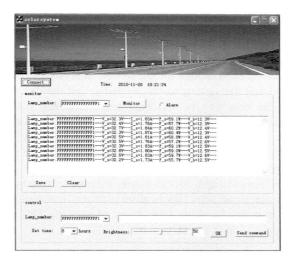

Fig. 7. The interface of Remote Monitoring Center

By setting some parameters, a communication between the monitoring center and ZigBee network was set up. Then the expected functions of monitoring and control were achieved successfully. The monitoring center can choose the communication mode peer-to-peer or one-to-many. The messages from solar lamps are listed in specific window of the interface, shown in Fig. 7. The status of solar lamps can be adjusted by sending control commands. In addition, the data can be stored in PC for further analysis.

Fig. 9 illustrates the daily changing curve of some parameters (Vs, Is, Ps: output voltage, current and power of solar panel) of one solar lamp monitored by the system, which was recorded once per hour between 7AM and 5PM. It is perceptible that the maximum value of output power of solar panel occurs between 11AM~2PM. It can be seen from Fig. 9 that the battery is charged by the solar panel with MPPT, and the MPP changes with the real-time change of the sunshine intensity. Coordinating with the monitoring system, the remote monitoring center has realized accurate and efficient remote and real-time detection on the parameters of solar street lamps.

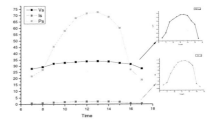

Fig. 8. The designed Solar Lamp System and Controller

Fig. 9. The daily changing curve of parameters monitored by the system

5 Conclusion

Based on ZigBee and GPRS, a monitoring and control system of solar street lamps is designed and implemented. Through the theoretical analysis and experimental test, it has been confirmed that the system works normally, and the designed software is so user-friendly that the data can be analyzed and monitored from any solar lamp distributed in one area. The application of ZigBee Wireless Sensor Network provides a good solution to the shortcomings of solar street lamp systems at present market. This system has many advantages such as high reliability, simple structure, low-cost, low power consumption, reasonable rates, etc. It will promote the popularization and application of solar street lamps. In the future, the whole system will be improved constantly. Meanwhile, as an actual application platform, the system can be expanded easily to different domains by using other types of sensors. This monitoring system has a very broad application prospect.

References

1. Shigehiro, N., Keiji, M., Hiroki, M., Kengo, I., Kayoko, O.: Solar street light. Shapu Giho/Sharp Technical Journal (93), 54–58
2. Lee, J.D., Nam, K.Y., Jeong, S.H., Choi, S.B., Ryoo, H.S., Kim, D.K.: Development of Zigbee based Street Light. In: Power Systems Conference and Exposition, PSCE 2006. IEEE Press, Los Alamitos (2006)
3. ZigBee Alliance, ZIGBEE SPECIFICATION (December 2006)
4. Jiang, P., Xia, H., Wu, K.: Design of Water Environment Data Monitoring Node Based on ZigBee Technology. In: International Conference on Computational Intelligence and Software Engineering, December 11-13, pp. 1–4 (2009)
5. CC2430 chip datasheet, Chipon Product from Texas Instrument, `http://www.ti.com`
6. Hohm, D.P., Ropp, M.E.: Comparative Study of Maximum Power Point Tracking Algorithms Using an Experimental, Programmable, Maximum Power Point Tracking Test Bed. In: Proc. Photovoltaic Specialist Conference, pp. 1699–1702 (2000)
7. Esram, T., Chapman, P.L.: Comparison of Photovoltaic Array Maximum Power Point Tracking Techniques. IEEE Trans. Energy Conv. 22(2), 439–449 (2007)

Non-line of Sight Error Mitigation in UWB Ranging Systems Using Information Fusion

Xiangyuan Jiang* and Huanshui Zhang

School of Control Science and Engineering,
Shandong University, Jingshi Road 73, Jinan, 250061, P.R. China
shiangyaujiang@gmail.com, hszhang@sdu.edu.cn

Abstract. An information fusion(IF) smoother based on biased Kalman filtering(BKF) and maximum likelihood estimation(MLE) is proposed for ranging error mitigation with both time of arrival(TOA) and received signal strength(RSS) measurement data in IEEE 802.15.4a protocol to improve the ranging accuracy. In this study, the line of sight(LOS) and non-line of sight(NLOS) condition in ultra-wide band(UWB) sensor network is identified by a joint hypothesis test with residual variance of TOA measurement and maximum likelihood ratio of RSS measurement. Then an information fusion smoother is proposed to accurately estimate the ranging measurement corrupted by NLOS error. Simulation results show that proposed hybrid TOA-RSS information fusion approach indicates a performance improvement compared with the usual TOA-only method.

Keywords: NLOS mitigation, Biased Kalman filter, Maximum likelihood estimation, Information fusion, UWB ranging system.

1 Introduction

Potentially providing accurate ranging, the UWB systems with NLOS propagation errors may still encounter severe degradation of position accuracy [1]. So the NLOS mitigation is absolutely necessary in UWB location and tracking system. The methods in coping with detailed applications differ in a variety of situations, and can be divided into three categories.

In the first category, NLOS mitigation is usually introduced to achieve a more precise TOA estimation from the received multi-path signal. Wu et al [2] proposed NLOS error mitigation method based on the signal propagation path loss model.

In the second category, since some location systems focus on the estimation of a stationary or slow moving node(MN), of which the motion dynamics may not be a major concern in the problem. To guarantee successful identification of NLOS anchor node(AN) and NLOS mitigation, effective location estimation

---

* This work is supported by the National Natural Science Foundation for Distinguished Young Scholars of China(No.60825304), and the National Basic Research Development Program of China(973 Program)(No.2009cb320600).

algorithms such as least square method [3], linear programming [4] and maximum likelihood method [5] is essential.

In the last category, the system is interest in the location and system dynamics of targets in motion or MNs. Kalman filtering [6] and particle filtering [7] fit the functionality. Such method could processing the time series of range measurements at each AN in time, nevertheless, they rely on the model and noise statistic seriously.

In this paper, a NLOS identification and error mitigation scheme for UWB systems based on state information fusion is proposed. The process of NLOS identification and mitigation use both TOA and RSS measurement data, which could get more information compared with the traditional TOA-only method mentioned above. Since RSS is already available in the receivers (for network quality indicator, power control and etc.), the algorithm requires little or no additional hardware, no additional transmissions and low computation complexity. Investigation of the effectiveness of the information fusion scheme in mitigating errors during the LOS-to-NLOS and NLOS-to-LOS transitions is especially focused.

2 System Model

As mandatory ranging protocol of IEEE 802.15.4a, the process of TW-TOA [8] ranging with RSS measurement could be considered as a multi-sensor multi-scale sampling system. Then a multi-scale sampling model, as shown in Fig. 1, is employed to describe the ranging protocol.

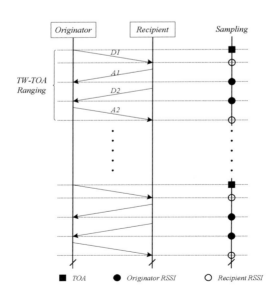

Fig. 1. The TOA and RSS samplings in TW-TOA ranging procedure.

First, the originator, which usually is the AN, sends a range request packet $D1$ and the recipient, which usually is the MN, replies with an acknowledgment $A1$. The recipient also transmits a time-stamp packet, $D2$, following the transmission of $A1$. Finally, the originator sends an acknowledgement, $A2$, for the time stamp.

Information fusion is introduced to mix the estimation results of biased kalman filter and MLE according to different sampling rate to get a more accurate range estimation.

Assuming that the range measurement corresponding to TOA data between AN and MN at time i can be modeled as

$$r(i) = d_A(i) + n(i) + N(i) \tag{1}$$

where $d_A(i)$ is the true range, $n(i)$ is the measurement noise as a zero-mean additive Gaussian random variable with standard deviation σ_m, and $N(i)$ is the NLOS error component in the received range signal. Assume that T_0 is the arrival time of the first path in the first cluster. The arrival time T_0 corresponds to the positive NLOS error component $N(i)$ at the time instant i and $N(i) = T_0 c$, where c is the speed of light. For the LOS cases, let $T_0 = 0$. The arrival time T_0 for the NLOS cases can be modeled as an exponential distribution

$$P(T_0) = \Lambda \exp[-\Lambda(T_0)] \tag{2}$$

where $\Lambda(1/ns)$ is the cluster arrival rate, whose statistical value in different environment could be found in Table 1 [?].

Due to large-scale fading, the normalized path loss has been modeled as a log-normal random variable

$$s(j) = 10\beta \log(d_S(j)/d_0) + R \tag{3}$$

where β is the path loss exponent, d_0 a reference distance, usually $d_0 = 1$, and R is a zero-mean Gaussian random variable δ with different variances in LOS and NLOS scenarios.

Table 1. The parameter of path loss and multi-path fading in UWB channel model

Parameter	Residential Environments		Outdoor Environments	
	LOS	NLOS	LOS	NLOS
β	1.79	2.70	1.76	2.50
δ (dB)	2.75	4.10	0.83	2.00
Λ (1/ns)	0.047	0.120	0.0048	0.0243

3 Proposed Information Fusion Algorithm

3.1 Adaptive Measurement Noise Variance Update for TOA Measurement

From the range measurement model in (1), we can define a state vector by: $X(i) = [d_A(i) \ \dot{d}_A(i)]^T$ where $\dot{d}_A(i)$ expresses the velocity of the MN according to AN. and dynamics of the state equation can be expressed as

$$X(i+1) = \phi X(i) + \Gamma W(i) \tag{4}$$

The state transition and noise transition matrices are written respectively as

$$\phi = \begin{bmatrix} 1 & T_s \\ 0 & 1 \end{bmatrix}, \Gamma = \begin{bmatrix} T_s^2/2 \\ T_s \end{bmatrix}$$

T_s is the sample period, and $W(i)$, dependent on the acceleration of the MN, is the process driving noise. Here, we assume $W(i)$ as a Gaussian random process with zero mean and covariance σ_ω^2. The measurements are represented as

$$Z(i) = \varphi X(i) + V(i) \tag{5}$$

where $Z(i)$ is the TOA measurement data comes from the according AN, $\varphi = [1 \ 0]$ is the measurement matrix, and $V(i)$ is the measurement noise with covariance σ_v^2.

Let $\hat{\sigma}_m$ be the calculated standard deviation of the measured TOA range data $r(i)$ at time i. The calculation repeats for every sliding block of C data samples. The standard deviation was calculated as

$$\hat{\sigma}_m(i) = \sqrt{\frac{1}{C} \sum_{k=i-(C-1)}^{i} (r(k) - \hat{d}(k))^2} \tag{6}$$

where $\hat{d}(k)$ is the estimated TOA in the information fusion estimation.

To mitigate the positive NLOS error component in TOA measurement, the ranging diagonal element of measurement noise covariance matrix is adjusted according to the decision from the LOS/NLOS hypothesis test in a Kalman filter. The adjustment was written as

$$\hat{\sigma}_v^2(i) = \begin{cases} \hat{\sigma}_m^2(i-1) + [\varepsilon(i)/\beta]^2, & \text{If } \varepsilon > 0 \text{ and NLOS detected} \\ \sigma_m^2, & \text{otherwise} \end{cases} \tag{7}$$

Where $\beta > 1$ is an experimentally chosen scaling factor, and ε is the innovation of kalman filter whose iterative operations will be summarized from (16) to (22). When the NLOS status is detected and the innovation term $\varepsilon > 0$ is obtained, the standard deviation $\hat{\sigma}_m$ can be used as the estimated standard deviation of the measurement noise.

3.2 Likelihood Estimation and CRLB for RSS Measurement

The TOA and RSS sampling rate are different, and their proportion is M:1. During TW-TOA ranging, there is $M = 4$.

$$s(i) = \sum_{j=(i-1)M-1}^{iM} s(j)/M \qquad (8)$$

From the range measurement model in (3), the dependence of the RSS measurement on d_S is explicitly noted. The likelihood function can be succinctly employed as

$$f(s(i)|d_S(i)) = (2\pi\delta^2)^{-1/2} \exp\left(-\frac{(s(i) - 10\beta \log_{10}(d_S(i)))^2}{2\delta^2}\right) \qquad (9)$$

With differentiating once produced, the Maximum likelihood estimation for d_S could be obtained

$$\hat{d}_S(i) = 10^{\left[\frac{s(i)}{10\beta}\right]} \qquad (10)$$

where $s(i)$ shown in (8) could bee seen as a complete statistic, based on which we could get a mean variance unbiased estimation of the distance. Because the shadow fading of path loss model is a white Gaussian noise, the Cramer-Rao lower bound could yield

$$P_S(i) = \frac{[log(10)\hat{d}_S(i)\delta]^2}{M[10s(i)]^2} \qquad (11)$$

$P_S(i)$ is the lower band of ranging covariance with RSS measurement, it could demonstrate the efficiency of MLE to some extent, and is necessary during state fusion in the following.

3.3 NLOS Identification

Although the NLOS processing with information fusion is focused on time series of measurement data, the NLOS mitigation is also usually accompanied with the functions of NLOS identification.

According to the decision from the TOA LOS/NLOS hypothesis test, the adjustment was written as

$$\begin{aligned} H_0 &: \hat{\sigma}_m < \gamma\sigma_m \text{ LOS detected} \\ H_1 &: \hat{\sigma}_m \geq \gamma\sigma_m \text{ NLOS detected} \end{aligned} \qquad (12)$$

Here, the channel state $H = H_0$ corresponds to LOS propagation and $H = H_1$ corresponds to NLOS propagation, where the scaling parameters β_L, β_N and δ_L, δ_N correspond to the path loss parameters under LOS and NLOS conditions respectively as in Table 1. γ is chosen experimentally to reduce the probability error.

The RSS LOS/NLOS hypothesis test

$$\begin{cases} f(s(i)|\hat{d}, H_0) = (2\pi\delta_L^2)^{-1/2} \exp\left(-\frac{(s(i) - 10\beta_L \log_{10}\hat{d})^2}{2\delta_L^2}\right) & \text{LOS detected} \\ f(s(i)|\hat{d}, H_1) = (2\pi\delta_N^2)^{-1/2} \exp\left(-\frac{(s(i) - 10\beta_N \log_{10}\hat{d})^2}{2\delta_N^2}\right) & \text{NLOS detected} \end{cases} \quad (13)$$

For each of the given estimates, we determine the conditional probabilities. The formulation of the joint channel state identification is as follows

$$Q = \frac{f(s(i)|\hat{d}, H_0)\gamma\sigma_m}{f(s(i)|\hat{d}, H_1)\hat{\sigma}_m(i)} \quad (14)$$

where $\hat{d}$ is the estimation of information fusion smoother in instant $i - 1$.

3.4 Smooth Range Estimation Using Information Fusion

The proposed information fusion structure is illustrated in Fig. 2, which consists of four major parts, NLOS identification, filtering, MLE and combination.

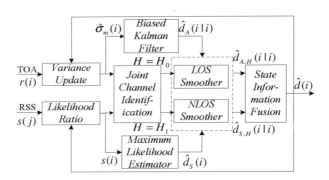

Fig. 2. The architecture of proposed information fusion smoother.

Summary of Information Fusion Smoother

NLOS identification

$$\begin{array}{l} H_0 : LOS \; detected \quad \text{If } Q < 1 \\ H_1 : NLOS \; detected \; \text{If } Q \geq 1 \end{array} \quad (15)$$

The iterative operations of the Kalman filter can be summarized as follows

$$\hat{X}(i|i-1) = \phi \hat{X}(i-1|i-1) \quad (16)$$
$$\varepsilon(i) = Z(i) - \varphi \hat{X}(i|i-1) \quad (17)$$

$$\hat{\sigma}_v^2(i) = \begin{cases} \hat{\sigma}_m^2(i-1) + (\varepsilon(i)/\beta)^2, & \text{If } \varepsilon(i) > 0 \text{ and NLOS detected} \\ \sigma_m^2, & \text{otherwise} \end{cases} \quad (18)$$

$$P(i|i-1) = \phi \hat{P}(i-1|i-1)\phi^T + \Gamma \sigma_\omega^2 \Gamma \quad (19)$$

$$K(i) = P(i|i-1)\phi^T[\varphi P(i|i-1)\varphi^T + \sigma_v^2]^{-1} \quad (20)$$

$$\hat{X}(i|i) = \hat{X}(i|i-1) + K(i)\varepsilon \quad (21)$$

$$P(i|i) = P(i|i-1) - K(i)\varphi P(i|i-1) \quad (22)$$

Maximum likelihood estimation

$$\hat{d}_S(i) = \begin{cases} 10^{\left[\frac{s(i)}{10\beta_L}\right]} & H = H_0 \\ 10^{\left[\frac{s(i)}{10\beta_N}\right]} & H = H_1 \end{cases} \quad (23)$$

CRLB

$$P_S(i) = \begin{cases} \dfrac{[log(10)\hat{d}_S(i)\delta_L]^2}{M[10s(i)]^2} & H = H_0 \\ \dfrac{[log(10)\hat{d}_S(i)\delta_N]^2}{M[10s(i)]^2} & H = H_1 \end{cases} \quad (24)$$

Combination

$$\hat{P}_{1,1}(i|i) = \left[P_{1,1}(i|i)^{-1} + P_S(i)^{-1}\right]^{-1} \quad (25)$$

$$\hat{d}(i) = \frac{1}{\hat{P}_{1,1}(i|i)}[P_{1,1}(i|i)^{-1}\hat{d}_A(i|i) + P_S(i)^{-1}\hat{d}_S(i)] \quad (26)$$

where $P_{1,1}(i|i)$ the covariance matrix of $d_A(i)$.

4 Simulation Result

Simulation results are provided to demonstrate the validity of the proposed ranging estimator. We assume that the MS has a steady velocity of $1.5m/s$ and moves in a straight line. The sample length is 150, and the sample interval is equal to $0.1s$.

The LOS or NLOS of BS is changed for each 50 samples in an alternate way. In each simulation case, 10 Monte Carlo runs are performed with the same parameters. values $\sigma_\omega^2 = 0.15m/s^2$, $\beta = 5$, $\sigma_m = 0.1m$, $\Lambda = 0.047$ and $\gamma = 3$ are used in the simulations.

Fig. 3 compares the ranging estimation between the proposed information fusion smoother and the biased Kalman smoother in [6]. It can be seen that the proposed smoother effectively mitigates the NLOS error components and successfully reduce the identification errors during the NLOS-to-LOS channel status transition.

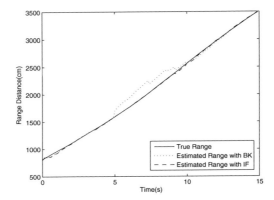

Fig. 3. Comparison of range estimation between the proposed IF and BKF in [6]

5 Conclusions and Future Works

Simulation results demonstrate that the performance of the proposed information fusion smoother approaches higher range accuracy. The priori knowledge of RSS measurement noise covariance is necessary for information fusion smoother. For future works, we can construct more accurate measurement models.

References

1. Guvenc, I., Chia-Chin, C.: A Survey on TOA Based Wireless Localization and NLOS Mitigation Techniques. Communications Surveys Tutorials 11(3), 107–124 (2009)
2. Wu, S., Ma, Y., Zhang, Q.: NLOS Error Mitigation for UWB Ranging in Dense Multipath Environments. In: Wireless Communications and Networking Conference, Kowloon, pp. 1565–1570 (2007)
3. Guvenc, I., Chong, C.C., Watanabe, F.: NLOS Identification and Mitigation for UWB Localization Systems. In: Wireless Communications and Networking Conference, Kowloon, pp. 1571–1576 (2007)
4. Venkatesh, S., Buehrer, R.: NLOS Mitigation Using Linear Programming in Ultrawideband Location-aware Networks. IEEE Transactions on Vehicular Technology 56(5), 3182–3198 (2007)
5. Li, C., Weihua, Z.: Nonline-of-sight Error Mitigation in Mobile Location. IEEE Transactions on Wireless Communications 4(2), 560–573 (2005)
6. Wann, C.D., Hsueh, C.S.: Non-line of Sight Error Mitigation in Ultra-wideband Ranging Systems Using Biased Kalman Filtering. Journal of Signal Processing Systems 10(6), 1–12 (2010)
7. Jourdan, D.B., Deyst, J.J., Win, M.Z.: Monte Carlo Localization in Dense Multipath Environments Using UWB Ranging. In: IEEE International Conference on Ultra-Wideband, Clearwater Beach, pp. 314–319 (2005)
8. Sahinoglu, Z., Gezici, S.: Ranging in The IEEE 802.15.4a Standard. In: Wireless and Microwave Technology Conference, Clearwater Beach, pp. 1–5 (2006)

Ensemble Modeling Difficult-to-Measure Process Variables Based the PLS-LSSVM Algorithm and Information Entropy

Jian Tang[1], Li-Jie ZHao[2,1], Shao-Wei Liu[3], and Dong Yan[3]

[1] Research Center of Automation, Northeast University, Shenyang, China.
[2] College of Information Engineering, Shenyang Institute of Chemical Technology, Shenyang, China.
[3] Unit 92941, PLA, Huludao, China
freeflytang@gmail.com, zlj_lunlun@163.com,
shaoweiliu888@126.com, frog_dongdong@126.com

Abstract. Many difficult-to-measure process variables of the industrial process, such as ball mill load, cannot be measured by hardware sensors directly. However, the frequency spectrum of the vibration and acoustical signals produced by the industrial mechanical devices contain information about these variables. An ensemble modeling approach based on the partial least square (PLS), least square support vector machines (LSSVM) and the information entropy is proposed to estimate the load of the wet ball mill. At first, the PLS algorithm is used to extract the latent features of the mill power, vibration and acoustical spectrum respectively. Then, the extracted latent features are used to construct the mill load sub-models. At last, the finals ensemble model is obtained based on the information entropy of the prediction errors of the sub-models. Studies based the laboratory-scale ball mill show that the proposed modeling approach has better predictive performance.

Keywords: frequency spectrum, information fusion, partial least square, least square support vector machines.

1 Introduction

Some important process variables, which give information about the final product quantity and quality, cannot sometimes be measured by a sensor directly or not reliable enough. Due to the lack of reliable on-line sensors to measure the load of the wet ball mill in the grinding process, the grinding circuits have to always run at low grinding production rate [1]. The mechanical grinding of the ball mill produces strong vibration and acoustic signals which are stable and periodic over a given time interval. The interested signals from grinding are buried in a wide-band random noise signal "white noise" in the time domain [2]. Based on the power spectral densities (PSD) of these signals, Zeng et al constructed the partial least square (PLS) and principle component regression (PCR) models between the operating parameters such as pulp density (PD), particle size etc and characteristic frequency sub-bands [2]. But the axis vibration signal is dispersed and disturbed by the transfer system of the mill. The acoustic signal has acoustical crosstalk with adjacent mill. Recently studies show that the on-contact vibration based system has at least twice the resolution of the traditional acoustical signal based system [3].Tang et al proposed a genetic algorithm-partial least square (GA-PLS) approach to model the mill load parameters inside the wet ball mill based on the shell vibration acceleration signal [4].

It is well known that the optimum number of features for classification and regression is limited by the number of training samples. Using the high-dimensional frequency spectrum data to construct model, the key problems have to avoid is "the Hughes phenomenon" and "the curse of dimensionality" [5]. Therefore, Tang et al proposed another soft sensor approach based on the multi-spectral segments principal component analysis (PCA) and support vector machines (SVM) [6]. However, the partition of the spectral segments is manually; the extracted principal components (PCs) don't take into account the correlation between inputs and outputs [7]; SVM has to solve the quadratic program (QP) problem. Studies show that different signals are mainly related to different parameters of mill load, such as shell vibration to PD [8], shell acoustical to mineral to ball volume ratio (MBVR) [9] and mill power to charge volume ratio (CVR) [10]. PLS has been proved to capture the maximal covariance between input and output data using the less latent variables (LVs) [11]. Therefore, PLS algorithm can also be used to extract the spectral features related to the parameters of the mill load. The least square-support vector machines (LS-SVM) simplifies the QP problem to solve a set of linear equations [12]. The ensemble of the sub-models can significantly improve the generalization ability of the final model [13]. The information entropy can used to ensemble the sub-models effectively [14].

Therefore, an ensemble modeling approach based on PLS, LSSVM and information entropy is proposed to estimate the mill load parameters using the mill power, the frequency spectrum of the shell vibration and acoustical signals. At first, PLS is used to extract the latent features of the frequency spectrum and mill power. Then LSSVM is used to construct the sub-models with the latent features. Finally, the sub-models are weighted to obtain the final models. The coefficients are calculated based on the information entropy.

3 Ensemble Modeling Based on PLS-LSSVM and Information Entropy

3.1 Ensemble Modeling Strategy

The mechanical grinding of ball mill produces strong vibrations and acoustic signals which are stable and periodic over a given time interval, and the vibration signal of the mill shell is more sensitive and less noise. The mill power, shell vibration and acoustic signals contains different information relate to the ML parameters. Therefore, based on the analysis of the previous sections, a new soft sensing methodology for mill load parameters is proposed, whose structure is shown in Fig. 1. Here, $j = 1,2,3$, represents vibration spectrum, acoustical spectrum and mill power respectively; $i = 1,2,3$, represents mill load parameters, namely MBVR, PD and CVR respectively; $\mathbf{x}_j$ is the input of PLS algorithm; z_{ji} is the extracted latent features using PLS; $\hat{y}_{ji}$ is the estimate output of the sub-models using the LSSVM algorithm; y_i and $\hat{y}_i$ is the real and estimate value of mill load parameters.

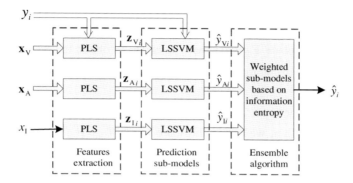

Fig. 1. Architecture of the ensemble modeling strategy

Fig. 1 shows that the proposed approach consists of the features extraction module, prediction sub-models module and ensemble algorithm module.

3.2 Features Extraction Based on Partial Least Square Algorithm

One of the challenges in mill load parameters modeling based on the mill power and spectrum is how to exact the latent features related to different parameters. Assume predictor variables $\mathbf{X} \in \Re^{k \times p}$ and response variables $\mathbf{Y} \in \Re^{k \times q}$ are normalized as $\mathbf{E}_0 = (\mathbf{E}_{01}\mathbf{E}_{02}...\mathbf{E}_{0p})_{k \times p}$ and $\mathbf{F}_0 = (\mathbf{F}_{01}\mathbf{F}_{02}...\mathbf{F}_{0q})_{k \times q}$ respectively. Let $\mathbf{t}_1$ be the first latent score vector of $\mathbf{E}_0$, $\mathbf{t}_1 = \mathbf{E}_0 \mathbf{w}_1$, and $\mathbf{w}_1$ be the first axis of the $\mathbf{E}_0$, $\|\mathbf{w}_1\| = 1$. Seemly, Let $\mathbf{u}_1$ be the first latent score vector of $\mathbf{F}_0$, $\mathbf{u}_1 = \mathbf{F}_0 \mathbf{c}_1$, and $\mathbf{c}_1$ be the first axis of the $\mathbf{F}_0$, $\|\mathbf{c}_1\| = 1$. We want to maximize the covariance between $\mathbf{t}_1$ and $\mathbf{u}_1$, thus have the following optimization problem:

$$\text{Max}(\mathbf{E}_0 \mathbf{w}_1, \mathbf{F}_0 \mathbf{c}_1)$$
$$\text{s.t.} \begin{cases} \mathbf{w}_1^T \mathbf{w}_1 = 1 \\ \mathbf{c}_1^T \mathbf{c}_1 = 1 \end{cases} \tag{1}$$

By solving (1) with Lagrange approach,

$$s = \mathbf{w}_1^T \mathbf{E}_0^T \mathbf{F}_0 \mathbf{c}_1 - \lambda_1 (\mathbf{w}_1^T \mathbf{w}_1 - 1) - \lambda_1 (\mathbf{c}_1^T \mathbf{c}_1 - 1) \tag{2}$$

where λ_1 and $\lambda_2 \geq 0$. At last, we obtain that $\mathbf{w}_1$ and $\mathbf{c}_1$ are the maximum eigenvector of matrix $\mathbf{E}_0^T \mathbf{F}_0 \mathbf{F}_0^T \mathbf{E}_0$ and $\mathbf{F}_0^T \mathbf{E}_0 \mathbf{E}_0^T \mathbf{F}_0$. So, after the $\mathbf{t}_1$ and $\mathbf{u}_1$ is obtained, we have

$$E_0 = t_1 p_1^T + E_1 \tag{3}$$

$$F_0 = u_1 q_1^T + F_1^0 \tag{4}$$

$$F_0 = t_1 r_1^T + F_1 \tag{5}$$

where, $p_1 = \dfrac{E_0^T t_1}{\|t_1\|^2}$, $q_1 = \dfrac{F_0^T u_1}{\|u_1\|^2}$, $r_1 = \dfrac{F_0^T t_1}{\|t_1\|^2}$, and E_1, F_1^0 and F_1 are the residual matrixes. Then we replace E_0 and F_0 with E_1 and F_1 to obtain the second latent score vectors t_2 and u_2. Using the same method, we get all the latent score vectors until $E_h = F_h = 0$.

Therefore, after the input data matrices X_j and output vector y_j are decomposed using the PLS algorithm, we obtain the score matrix $Z_{ji} = [t_{j1}, t_{j2}, \cdots, t_{jh_{ji}}]$, loading matrix $P_{ji} = [p_{j1}, p_{j2}, \cdots, p_{jh_{ji}}]$ and coefficients matrix $W_{ji} = [w_{j1}, w_{j2}, \cdots, w_{jh_{ji}}]$ respectively. h_{ji} is the number of the LVs, which is decided by the leave-one-out cross validation method. Therefore, Z_{ji} are the extracted latent features from the training samples. The latent features from the testing sample can be calculated by:

$$Z_{ji}^{test} = x_j^{test} W (P^T W)^{-1} + E \tag{6}$$

3.3 Prediction Sub-models Based on Least Square Support Vector Machines

As a modified version of SVM, the LS-SVM extends the application of the SVM, which has been widely used to build the soft sensor models widely [15, 16]. Given the training data set of k points $\{((z_{ji})_l, (y_i)_l)\}, l = 1, \cdots k$, with input data $(z_{ji})_l \in R^{p_j}$ and output data $(y_i)_l \in R$, one considers the following optimization problem:

$$\begin{cases} \min\limits_{\omega, b} & J_p = \dfrac{1}{2} \omega^T \omega + \dfrac{1}{2} c \sum\limits_{l=1}^{k} \xi_l^2 \\ \text{s.t:} & (y_i)_l = \omega^T \Phi((z_{ji})_l) + b + \xi_l \end{cases} \tag{7}$$

where, ω are the weighs, b is the bias, c is a regularization parameter used to decide a trade off between the training error and the margin and ξ_k is the error. Here we define RBF kernel function instead the feature map. The Lagrangian form of (7) is

$$L(\omega, b, \xi, \beta) = \dfrac{1}{2} \omega^T \omega + \dfrac{1}{2} c \sum_{l=1}^{k} \xi_k^2 - \sum_{l=1}^{k} \beta_l [((y_i)_l - (\omega^T \Phi((z_{ji})_l) + b + \xi_l)] \tag{8}$$

where β_l are Lagrange multipliers. The resulting prediction LS-SVM sub-models for different mill load parameters can be represent as

$$\hat{y}_{ji} = \sum_{l=1}^{k} (\beta_l)_{ij} k_{ij} (z_{ij}, ((z_{ij})_l)_{ij}) + b_{ij}. \tag{9}$$

3.4 Ensemble Algorithm Based on Information Entropy

Assume y_{il} is the true value of the ith mill load parameters at time l, and $\hat{y}_{jil}$ is the estimate value of the jth sub-models for the ith mill load parameters at time l. The ensemble algorithm based on the information entropy is shown as follows [14]:

Step 1: Calculate the relative prediction errors of the jth sub-model at time l:

$$e_{jil} = \begin{cases} \|(y_{il} - \hat{y}_{jil})/y_{il}\|, & 0 \le \|(y_{il} - \hat{y}_{jil})/y_{il}\| < 1 \\ 1, & \|(y_{il} - \hat{y}_{jil})/y_{il}\| \ge 1 \end{cases} \quad (10)$$

where, $l = 1, \cdots, k$, where k is the number of the training samples.

Step 2: Calculate the ratio of the relative prediction errors p_{jil}:

$$p_{jil}^k = e_{jil} / (\sum_{l=1}^{k} e_{jil}) \quad (11)$$

Step 3: Calculate the entropy value of the relative prediction errors p_{jil}:

$$E_{ji}^k = \frac{1}{\ln k} \sum_{l=1}^{k} p_{jil}^k \cdot \ln p_{jil}^k \quad (12)$$

Step 4: Calculate the weighed coefficients of the jth sub-models M_{ji}:

$$M_{ji}^k = \frac{1}{g-1}\left(1 - (1 - E_{ji}^k) / \sum_{j=1}^{z}(1 - E_{ji}^k)\right) \quad (13)$$

where $\sum_{j=1}^{g} M_{ji}^k = 1$, g is the number of the sub-models for the ith mill load parameter sub-model. In this paper, $g = 3$.

Step 5: Denote the prediction values of the sub-models for the new sample as $\hat{y}_{t,ji}^{k+1}$, then the final prediction value of the ensemble models is calculated with:

$$\hat{y}_{t,i}^{k+1} = \sum_{j=1}^{g} w_{ji}^k \cdot \hat{y}_{t,ji}^{k+1} \quad (14)$$

4 Application Study

A laboratory scale ball mill with size Φ460*460 mm is used to evaluate the soft-sensor method. The ball load, material load and water load are added to the mill, and then mill is started and run for a given period of time, the vibration, acoustical and mill power signals is measured. The other test are performed by gradually change the quantity of mineral ore and water.

Because PLS algorithm can capture the maximal covariance between the input and output data using the less latent variables (LVs), it is used to analyze the relationships between the full frequency spectrum, mill power and ML parameters. The percent variance captured with the 1th LV shows that the vibration frequency spectrum has the largest correlation coefficient with PD, 95.14% to 46.22%; the acoustical frequency spectrum has the largest correlation coefficient with MBVR, 51.48% to 57.30%; the mill power also has the largest correlation coefficient with MBVR, 45.79%. It is shows that the information fusion and features selection is necessary for construct am effective soft sensor models of mill load parameters.

In order to illustrate the effectiveness of our proposed approach, we compare PLS and PLS-LSSVM with different data set. The results are shown in Table 1, "VAI" indicates all the spectrum and mill power are combined together as one input; "V+A+I" indicates that the final model is integrated by the sub-models of different inputs. The curves of the real and predicted values with the proposed approach are shown in Fig. 2-Fig. 4.

Table 1. Error comparison of different soft sensor approach

Approach	Data set (different signals)	RMSEs (MBVR)	RMSEs (PD)	RMSEs (CVR)	RMSEs (average)
PLS	VAI	0.2546	0.1479	0.1826	0.1950
PLS-LSSVM	VAI	0.1904	0.1899	0.1948	0.1917
PLS	V	0.5163	0.2306	0.2772	0.3414
PLS	A	0.3434	0.3806	0.3599	0.3613
PLS	I	0.5247	0.3500	0.7943	0.5563
PLS-ensemble	V+A+I	0.1956	0.2909	0.3289	0.2718
PLS-LSSVM	V	0.4327	0.1213	0.1361	0.2300
PLS-LSSVM	A	0.2223	0.3602	0.3712	0.3179
PLS-LSSVM	I	0.3464	0.4792	0.3505	0.3920
This approach	V+A+I	0.1878	0.2292	0.2496	0.2222

The number of the LVs for different approach is determined with the leave-one-out cross validation approach. The result shows that for MBVR, the proposed approach has the highest prediction accuracy, which shows the information of MBVR is contained in different signals. But base the vibration spectrum, the PD and CVR has the highest prediction accuracy. These results show that ensemble all the sub-models isn't the best strategy, we maybe use the selective ensemble leaning strategy.

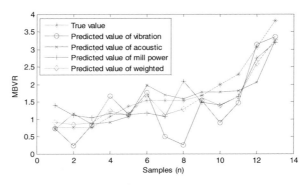

Fig. 2. Prediction result of the mineral to ball volume ratio (MBVR)

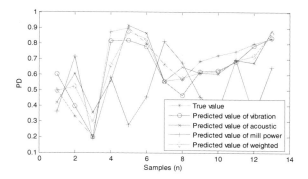

Fig. 3. Prediction results of the pulp density (PD)

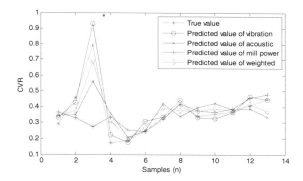

Fig. 4. Prediction results of the volume charge ratio

5 Conclusion

This paper presents an ensemble modeling approach for the parameters of wet ball mill load based on the PLS, LSSVM and the information entropy. It has made the following contributions: the PLS algorithm is used to extract the latent features of the mill power, vibration and acoustical signals, which enables the features extraction more reasonable; the extracted latent features are used to construct the nonlinear sub-models based on LSSVM algorithm; the ensemble algorithm based on information entropy is used to obtain the final models. The future research and more experiments should be done on the laboratory-scale and industrial wet ball mill.

References

1. Zhou, P., Chai, T.Y.: Intelligent Optimal-Setting Control for Grinding Circuits of Mineral Processing. IEEE Transactions on Automation Science and Engineering 6, 730–743 (2009)
2. Zeng, Y., Forssberg, E.: Monitoring Grinding Parameters by Signal Measurements for an Industrial Ball Mill. International Journal of Mineral Processing 40, 1–16 (1993)

3. Gugel, K., Palcios, G., Ramirez, J., Parra, M.: Improving Ball Mill Control with Modern Tools Based on Digital Signal Processing (DSP) Technology. In: IEEE Cement Industry Technical Conference, pp. 311–318. IEEE Press, Dalas (2003)
4. Tang, J., Zhao, L.J., Zhou, J.W., Yue, H., Chai, T.Y.: Experimental Analysis of Wet Mill Load Based on Vibration Signals ofLaboratory-scale Ball Mill Shell. Minerals Engineering 23, 720–730 (2010)
5. Fukunaga, K.: Effects of Sample Size in Classifier Design. IEEE Trans. Pattern Anal. Machine Intell. 11, 873–885 (1989)
6. Tang, J., Zhao, L.J., Yu, W., Yue, H., Chai, T.Y.: Soft Sensor Modeling of Ball Mill Load via Principal Component Analysis and Support Vector Machines. Lecture Notes in Electrical Engineering 67, 803–810 (2010)
7. Liu, J.L.: On-line Soft Sensor for Polyethylene Process with Multiple Production Grades. Control Engineering Practice 15, 769–778 (2007)
8. Spencer, S.J., Campbell, J.J., Weller, K.R., Liu, Y.: Acoustic Emissions Monitoring of SAG Mill Performance. In: Intelligent Processing and Manufacturing of Materials, Honolulu, HI, USA, pp. 936–946 (1999)
9. Wang, Z.H., Chen, B.C.: Present State and Development Trend for Ball Mill Load Measurement. Chinese Powder Science and Technology (Chinese) 7, 19–23 (2001)
10. Van Nierop, M.A., Moys, M.H.: Exploration of Mill Power Modeled as Function of Load Behavior. Minerals Engineering 14, 1267–1276 (2001)
11. Höskuldsson, A.: PLS Regression Methods. Journal of Chemometrics 2, 211–228 (1988)
12. Suykens, J.A.K., Vandewalle, J.: Least Squares Support Vector Machine Classifiers. Neural Processing Letters 99, 293–300 (1999)
13. Niu, D.P., Wang, F.L., Zhang, L.L., He, D.K., Jia, M.X.: Neural Eetwork Ensemble Modeling for Nosiheptide Fermentation Process Based on Partial Least Squares Regression. Chemometrics and Intelligent Laboratory Systems 105, 125–130 (2011)
14. Wang, C.S., Wu, M., Cao, W.H., He, Y.: Intelligent Integrated Modeling and Synthetic Optimization for Blending Process in Lead-Zinc Sintering. Acta Automatica Sinica 35, 605–612 (2009)
15. Suykens, J.A.K., Vandewalle, J.: Least Squares Support Vector Mmachine Classifiers. Neural Processing Letters 99, 293–300 (1999)
16. Cortes, C., Vapnik, V.: Support-Vector Networks. Machine Learning 20, 273–297 (1995)

E-Commerce Capabilities and Firm Performance: A Empirical Test Based on China Firm Data

Kuang Zhijun

School of Economic and Management, East China Jiaotong University,
Nanchang, Jiangxi China
kuang1130@tom.com

Abstract. In recent rapidly changing competitive environment, electric commerce plays an important role on value chain. Many firms invest IT to ensure their success on e-commerce. However, how affect the firm performance when using information systems in supporting organizational e-commerce activities is still a question. The purpose of the research is to explore how the IT infrastructure and e-commerce capability affect firm performance. The paper proposes and test a research framework to analyze the relationship among IT infrastructure, e-commerce capability and firm performance based resource-based view. At last, the paper points out some research suggestions for the future study related to the e-commerce capability and business value research.

Keywords: firm performance; IT infrastructure; e-commerce capability; resource- based view.

1 Introduction

Considering the rapidly changing competitive environment today, developing direct access to customers via IT infrastructure is the dominant e-commerce strategy among firms. The strategic value of IT infrastructure has generally been associated with its ability to allow a firm to adapt successfully to changes in the external environment (Broadbent et al., 1999; Byrdand Turner, 2001b; Weill et al., 2002), but the mechanisms by which IT infrastructure impact firm performance when using IT in supporting organizational e-commerce activities are still not well understood. A potential framework for analyzing the relationship between IT infrastructure, e-commerce capability and firm performance is the resource based view. In this paper we present a research model to analyze the relations among IT infrastructure capabilities and e-commerce capability in the aim for better firm performance. The paper is structured as follow: The background literature is briefly examined to build the rationale for our research model. The research method including data collection and data analysis approaches is discussed in the following section. This is followed by the study results and discussion in the subsequent section. These are followed by a brief set of conclusions and directions for future research.

2 Literature Review

2.1 IT Infrastructure Capability

IT infrastructure is viewed as a platform technology consisting of the processing hardware and operating system, networking and communication technologies, data, and core data processing applications(Dunkan, 1995; Weil and Broadbent, 1998). IT infrastructure is used in all e-business initiatives to connect different parts of the firm and to link to suppliers, customers, and allies(Weill et al., 2002).

IT infrastructure capability is a integrated set of reliable IT infrastructure services available to support both existing applications and new initiatives(Weill et al., 2002). IT infrastructure capability is reflected in the range and number of IT infrastructure services (Broadbent and Weill, 1997;Broadbent et al., 1999). IT infrastructure capability is a multi-dimension construct which includes reach, range, data management(Keen,1991). Reach refers to the locations that can be connected via the IT infrastructure from local workstations and computers to customers and suppliers domestically, to international locations, or to anyone, anywhere. Range determines the level of functionality (i.e., information and/or transaction processing) that can be shared automatically and seamlessly across each level of "reach", and answers the question of "what services can be offered." (Keen,1991). Another dimension is data management capability. Data has become an organizational resource shared by multiple users at different levels of management and across various functions and shared by multiple IT applications. Data resource itself is an integral part of IT infrastructure. The capability to manage the data is an important measure of IT infrastructure capability.

2.2 E-Commerce Capability

E-commerce is defined as business activities conducted over the Internet, continues to penetrate the enterprise value (Zhu 2004). The purpose of the adoption of e-commerce is predominantly to broaden their customer base by exploring new marketing channels, or to create competition for the traditional channels (Amit and Zott 2001, Chircu and Kauffman 2000). E-commerce capabilities reflect a company's strategic initiatives to use the Internet to share information, facilitate transactions, improve customer services, and strengthen supplier integration. E-commerce functionalities may range from static information to online order tracking and from digital product catalogues to integration with suppliers' databases.

Upon resource-based view, it is how firms leverage their IT infrastructure to create unique e-commerce capabilities that determines a firm's overall performance. Regardless of how commodity-like the technology components may be, the architecture that removes the barriers of system incompatibilities and makes it possible to build a corporate platform for launching e-business applications is clearly not a commodity. Moreover, e-commerce capabilities are often tightly connected to the the firm's resource base and embedded in the business processes of the firm. Therefore, e-commerce conveys to the firm a resource that cannot be substituted for or easily imitated (such as customer proprietary data and shared information). The exploitation of these resources will lead to performance advantages for net-enhanced organizations (Lederer et al. 2001).

3 Concept Framework and Hypotheses

The theoretical discussions above lead us to believe that e-commerce capabilities which combine with IT infrastructure can explain performance variance across firms. IT infrastructure capability is critical to enabling e-commerce capabilities. E-commerce capability represents a firm's ability to interact with its customers and business partners and conduct businesses over the Internet (Zhu and Kraemer 2002). The concept framework used for this study is showed in Fig.1. Following we will discuss the Hypotheses about the relationship among IT infrastructure capability, E-commerce capability and firm performance.

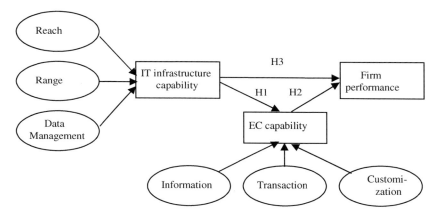

Fig. 1. Concept Model

3.1 IT Infrastructure Capability and Firm Performance

IT infrastructure capability of firms tends to be highly firm specific and evolves over long periods of time, during which gradual enhancements are made to reflect changing business needs (Scott-Morton 1991). The possibility that IT will enhance business performance and bring sustainable competitive advantage is through the development of efficacious IT infrastructure capabilities which cannot be readily assembled through markets, instead, they must be built over time(Weill et al., 2002). These arguments suggest:

Hypothesis 1 (H1): Greater IT infrastructure capability is associated with higher level of firm performance.

3.2 E-Commerce Capability and Firm Performance

E-commerce capability is the capabilities of a firm to provide information, facilitate transactions, offer customized services, and integrate the back end and fulfillment(zhou,2004). E-commerce capability can be viewed as a firm's ability to deploy and leverage e-commerce resources to support e-commerce activities. Moreover, e-commerce capabilities are often tightly connected to the the firm's resource base and embedded in the business processes of the firm. Therefore, e-commerce capability cannot be substituted for or easily imitated (such as customer proprietary data and shared information). The exploitation of these capabilities will lead to performance advantages for net-enhanced organizations (Lederer et al. 2001). These arguments suggest:

Hypothesis 2 (H2): Greater e-commerce capability is associated with higher level of firm performance.

3.3 IT Infrastructure Capability and e-Commerce Capability

IT infrastructure is used in all e-business initiatives to connect different parts of the firm and to link to suppliers, customers, and allies. The reach, range and data manangent capabity of IT infrastructure is considered to be key for initiating e-commerce activity(Weill et al., 2002). Zhu(2004) also shows that IT infrastructure capability exhibit positive relationships to e-commerce capability measures. These arguments suggest:

Hypothesis 3 (H3): Higher levels of IT infrastructure capability is associated with greater e-commerce capability.

4 Methods

4.1 Data Collection

A survey design was used for this study and questionnaires were sent to 300 top or senior IT executives (CIO, vice president of IT, director of IT) in a variety of industries randomly when a IT application meeting is held in shanghai in 2009. A total of 210 usable responses were received, resulting a response rate of about 70 percent. There were totally 138 valid questionnaires after the deduction of incomplete ones. We conducted a Chi-square test to determine whether responses varied by the industry and their revenues. No significant differences in Chi-square at the .05 level were noted, which suggests that perceptual measures are unbiased by variations in the industry and their revenues.

4.2 Operationalization of Measures

According to the literature review in the section 2, IT infrastructure capability is a second-order construct which is represented by a high-level, multidimensional construct based on several levels of reach,range and data management capability. E-commerce capability is defined as a high-level, multidimensional construct generated from a set of specific variables measuring e-commerce functionalities. Firm performance is a first-order construct which measure financial indicators of the firm.

Table 1 shows the operationalization of variables. All of the questions were asked from a scale ranging from 1 to 7, where 1 refers to the lowest score in the measure and 7 represents the highest score on the measure. The scales for various constructs were adopted from a review of the literature. If existing measures were not available, a list of items covering the domain of the variables under investigation was developed.

Table 1. Operationalization of Constructs

Constructs	Measures	Source of measures	Indicators
IT infrastructure	Reach	Keen(1991) Bharadwaj(2000) Weill et al.(2002)	IT infrastructures can be easily extended to new office
			Business interactions with suppliers are through IT
			IT facilitates good relations with our customers
	Range	Keen(1991) Bharadwaj(2000) Weill et al.(2002)	IT applications are compatible with each other
			New function(s) can be easily added to IT applications
			Different IT applications are integrated under the same architecture
	Data Management	Keen(1991) Bharadwaj(2000) Weill et al.(2002)	Data can be shared among IT applications
			Branch offices have remote access to the same data
			IT hardware/software are standardized
E-commerce capability	Information	Klein etal.(2001) McKinney (2001) Zhou(2004)	Product catalog is available online
			Web site offers quickly search capability
			Customers have real experience for the product information
	Transaction	Klein etal.(2001) McKinney (2001) Zhou(2004)	Customers can place the order online via the Web site
			Customers can track the status of the order
			Web site can protect the security of transactions and customers data
	Customization	Klein etal.(2001) McKinney (2001) Zhou(2004)	Customers can customize the content viewed on the Web site
			Customer can register online and set up a personalized account to gain access to private messages
			Customer can configure product features via the Web site
			Online technical support handled by live representatives
Firm performance		Devaraj etal.(2003) Zhou(2004)	Return on investment (ROI) is greater than other competitors
			Product or service sales profit is greater than other competitors
			Market share is greater than other competitors
			Product growth rate is greater than other competitors
			Customer retention rate is greater than other competitors
			Customer satisfaction is greater than other competitors

4.3 Reliability Analysis

Cronbach's alpha, composite reliability, and the average variance extracted were computed to assess the internal consistency of each dimension (Hair et al. 1998). The results in Table 2 show that all Cronbach's alpha and composite reliabilities exceeded Nunnally's (1978) criterion of 0.7 while the average variances extracted for these constructs were above the recommended 0.5(Hair et al. 1998).

Table 2. Assessment of consistency reliability indices of each construct

Construct/dimension	Cronbach' α	Composite Reliability	Average Variance Extracted
IT infrastructure	0.8573	0.896	0.610
Reach (3)	0.7638	0.763	
Range (3)	0.8290	0.724	
Data Management (3)	0.7895	0.735	
E-commerce capability	0.8665	0.866	0.723
Information (3)	0.7813	0.902	
Transaction (3)	0.8211	0.815	
Customization (4)	0.8404	0.772	
Firm performance (6)	0.8525	0.813	0.683

4.4 Convergent Validity

Convergent validity ensures that all items measure a single underlying construct (Bogozzi and Fornell 1982). To test the convergent validity, we performed confirmatory factor analyses of the eight multiple- items constructs using LISREL 8.7. As shown in Table 3, a high value of composite reliability, ranging from .724 to .902, suggests the convergent validity of the constructs. The model fit indices (Table 3) also provide adequate evidence of the unidimensiona- lity of the items. All indices were quite close to their criterion level.

Table 3. Assessment of Convergent Validity for the Measurement Model

Goodness of Fit Indices	Desired Levels	IT infrastructure	E-commerce capability	Firm performance
χ^2/df	<2	1.871	2.118	2.478
GFI	<0.9	0.913	0.921	0.929
AGFI	<0.9	0.927	0.932	0.932
NFI	<0.9	0.909	0.919	0.919
RMSEA	<0.05	0.056	0.054	0.051

4.5 Testing the Structural Models

Having demonstrated that our measures possessed adequate validity and reliability, we then proceeded to test hypotheses in a structural equation model. Fig.2 shows the standardized loadings of the items on each construct, as well as the path loadings between constructs. The overall model fit (RMSEA =0.049, CFI=0.962, AGF=0.912

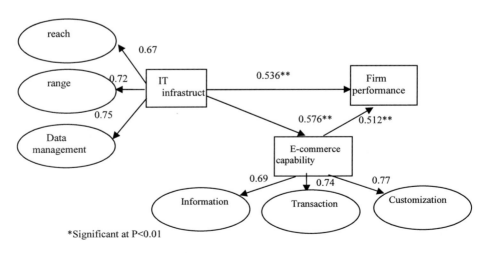

*Significant at P<0.01

Fig. 2. SEM Analysis Result

and χ^2/d.f.=2.872) shows that our structural model has good model validity according to the heuristics for statistical conclusion validity (Gefen et al 2000).Therefore, the model is acceptable. The hypotheses that we proposed have been proved for the results show that all the hypotheses are supported at the P<0.001level.

5 Discussion and Conclusions

During the past years, the rise of the Internet era prompted most corporations to reexamine their strategic logic and the role of information technologies in shaping their business strategies. Many firms begin to share a common understanding that IT play a fundamental role in their ability to enhance their business performance through e-commerce capabilities in products, services, channels, and customer segments. Therefore, the goal of this paper has been to develop a theoretical perspective for understanding the connections among IT infrastructure, e-commerce capability and firm performance. We construct a theory framework that describes IT infrastructure can enhance the e-commerce capability which can improve firm performance. But more importantly, e-commerce is a dynamic capability that requires firms to build and then dynamically reconfigure in order to align with changing technology and business environments.

Although this study provided a profound theoretical foundation for investigating the value of IT and a thoroughly empirical data analysis, there is still more work to do in the future. First, we could adopt qualitative methods such as case study to strengthen the potential for greater generalizability of the research findings. Second, since we examined the path element of IT infrastructure and e-commerce capability with cross-sectional data, we do not examine the long-term role of IT infrastructure. Longitudinal research is needed to trace the dynamics of business activities over time. In conclusion, this study provides not only a profound theoretical foundation for investigating the value of IT but also examines these empirical findings in a way which might help companies make wise decisions about IT adoption and use.

Acknowledgments. The research was supported by the National Natural Science Foundation of China under Grant 70761003.

References

1. Zhu, K.: The complementarity of Information Technology Infrastructure and E-Commerce Capability: A Resource-Based Assessment of Their Business Value. Journal of Management Information Systems 21(1), 167–202 (2004)
2. Pires, G.D., Aisbett, J.: The Relationship between Technology Adoptionand Strategy in Business-to-Business Markets: The Case of E-Commerce. Industrial Marketing Management 32(4), 291–300 (2003)
3. Melville, N., Kraemer, K., Gurbaxani, V.: Review: Information technology and organizational performance: An integrative model of IT business value. MIS Quarterly 28(2), 283–322 (2004)
4. Grant, R.M.: The resource-based theory of competitive advantage: Implications for strategy formulation. California Management Review 33(3), 114–135 (1991)

5. Bharadwaj, A.S.: A Resource-Based Perspective on Information Technology Capability and Firm Performance: An Empirical Investigation. MIS Quarterly 24(1), 169–196 (2000)
6. Sambamurthy, V., Zmud, R.W.: Arrangements for information technology governance: A theory of multiple contingencies. MIS Quarterly 23(2), 261–288 (1999)
7. Broadbent, M., Weill, P., Clair, D.: The Implications of Information Technology Infrastructure for Business Process Redesign. MIS Quarterly 23(2), 159–182 (1999)
8. Bhatt, G.D., Grover, V.: Types of Information Technology Capabilities and Their Role in Competitive Advantage: An Empirical Study. Journal of Management Information Systems 22(2), 253–278 (2005)
9. Daniel, E.M., Wilson, H.N.: The role of dynamic capability in e- business transformation. European Journal of Information Systems (12), 282–298 (2003)
10. Devaraj, S., Kohli, R.: Measuring information technology payoff: a meta-analysis of structural variables in firm-level empirical research. Information Systems Research 14(2), 127–145 (2003)
11. Zhu, K., Weyant, J.: Strategic Decisions of New Technology Adoption under Asymmetric Information. Decision Sciences 34(4), 643–675 (2003)
12. Eisenhardt, K., Martin, J.: Dynamic Capability: What are they? Strategic Management Journal (21), 1105–1121 (2000)
13. Teece, D.J., Pisano, G., Shuen, A.: Dynamic Capability and Strategic Management. Strategic Management Journal (18), 509–533 (1997)

The Development of Experimental Apparatus for Measurement of Magnetostrictive Coefficient

Lincai Gao, Bao-Jin Peng[*], Fei Xu,
Jia-Qi Hong, Wenhao Zhang, Xiao-Dong Li, and Cui-Ping Qian

[1] Institute of Information Optics Zhejiang Normal University
Jinhua, Zhejiang, China
jhpbj@zjnu.cn

Abstract. Design an experimental apparatus for measurement of magnetostrictive coefficient with simple operation and reasonable price. Use the characteristics of fiber grating whose central wavelength is sensitive to external strain to turn the length change of the magnetic material into the central wavelength change of the fiber grating by a mechanical device. Then turn the central wavelength change into light intensity change by the long period fiber grating, and the light intensity change will be turned into appropriate voltages for computer collection by an amplifier circuit with a pin tube. The computer will also collect the signals of external temperature (to reflect the change of the central wavelength of the fiber grating because of temperature change) and Magnetic changes of the device at the same time. Then draw a diagram of the magnetic field intensity and the length change of the magnetic material, thus, getting the magnetostrictive coefficient. The experiment shows that this experimental apparatus is feasible and practical with simple operation and low cost. It can not only overcome the shortages of the usual measuring method by non-balance electric bridge, but also be promoted to students experiments.

Keywords: magnetostrictive coefficient; fiber grating; long period fiber grating; computer technology.

1 Introduction

The magnetic domain of ferromagnetics will directionally arrange in the effect of external magnetic fieldl causing change of lattice spacing of the medium, thus, resulting the length change of ferromagnetic which is called magnetostrictive effect or joules effect. Magnetostriction not only has important influence on the magnetic material, but also has extensive application in practice[1,2]. Different magnetic materials have different magnetostrictive length change, we use the magnetostrictive coefficient λ ($\lambda = \Delta l / l$) to the characterize the length change. So it is very important to measure the magnetostrictive coefficient λ accurately. Since length

---

[*] Corresponding author. phone +86-579-82298849; The paper supported by the Zhejiang Jinhua science and technology project (2009-1-092) and Zhejiang New Talents Scheme Project(2011).

change of the material caused by magnetostrictive effect is relatively small, usually only 10^{-5} m to 10^{-6} m, some specific methods for measuring must be used.

The measurement of Magnetostrictive coefficient is a common experiment in general physics experiments, and is often measured by non-balance electrsc bridge. This method can turn the length change of the sample into the change of the bridge resistance for measurement by sticking a resistance strain gage to the sample as the bridge arm. However, serious drift phenomena of the circuit will appear due to the influnce of temperature, magnetoresistance effect and so on. In order to overcome the disadvantages above, we use the characteristics of fiber grating that its central wavelength is sensitive to external strain and the computer collection technology to design an experimental apparatus for the measurement of magnetostrictive coefficient with low cost, simple operation, and great practical value.

2 Experimental Principle

Below is the schematic diagram of this experiment:

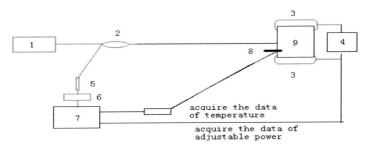

1 light source; 2 coupler; 3 coils; 4 adjustable power supply; 5 edge filter;
6 amplifier circuit; 7 computer; 8 temperature probe; 9 length chang collection device

Fig. 1. Schematic diagram of the experiment

As shown above, put the length change collection device between two electrified coils, open the power supply.The current change in coils will cause the change of magnetic field intensity, making the magnetostrictive material a small deformation. The deformation will be turned into central wavelength signal of fiber grating. through this device (3.1 shows the specific structure and principle).

The wavelength signal will be turned into signal when reaches the edge filter (the long period fiber grating) via coupler (3.2 shows specific principles).

The light intensity signals will be turned into voltage signal through the amplifier circuit (3.3 shows specific principles), and finally collected by the computer.

The computer will collect circuit voltage signal which shows the change of the central wavelength, coil voltage signal which shows the magnetic field intensity as well as eaternal temperature signal (to compensate temperature drift of the fiber grating), and then display a diagram of the magnetic field intensity and the length change of the magnetic material directly through programming, thus, getting the magnetostrictive coefficient.

3 Experimental Device and Process

3.1 The Length Change Collection Device

As the figure below shows, the main components of the length change collection device are 7 magnetostrictive material trough and 8 fiber grating trough.

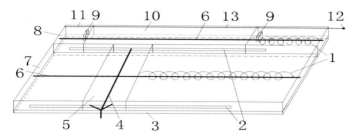

1 spring; 2 hollowed trough; 3 base; 4 rotating screw; 5 slider 1; 6 support bar; 7 magnetostrictive material trough, 8 fiber grating trough, 9 baffle; 10 slider 2; 11 fiber grating; 12 FC probe; 13 fiber

Fig. 2. The length change collection device

1) The magnetostrictive material trough
The main function of this trough is to turn the length change of the magnetic material into the sliding of slider 1. As the figure shows, the main components are slider 1, the support bar, the spring and the rotating screw. The slider and spring can only move in the direction of the support bar for they are attached to it. Put the magnetic material on the left of slider 1, and the length change will be reflected to the sliding distance change of slider 1.

Hollow one slot on each side of the trough, through which the rotating screw can go through slider 1 (the screw cannot encounter support bar when go through slider 1), the roating screw is used to withstand slider 2 in fiber grating trough.

2) Fiber grating trough
The main function of this trough is to turn the sliding distance of the slider into the change of the central wavelength of fiber grating. As the figure shows, its main components are slider 2, the support bar, the spring, the fiber grating and the baffle. The function of slider 2, the spring and the support bar are same to the corresponding components in magnetostrictive materials trough. Make slider 2 a little longer in order to make it withstood easier by the screw cap.

Since central wavelength of the fiber grating will be affected due to the drift of temperature, we need make a calibration for the drift at first. Put the temperature probe in the the length change collection device, and turn off the adjustable power supply ,and then put the device in the box with constant temperature of 0 °C, 5 °C, 10 °C... 70 °C. Collect the voltage change of each corresponding temperaturel (the way we collect the data will be talked in chapter 3.4) , and input the measurements to computer.

After measurement, stick one end of the effective part to the slider 2 and the other to the left end of the fiber grating trough. The central wavelength of the fiber grating and the change of pressure force(or pull force) have good linear relationship, repeatability and high correlation coefficient. When the axial variable of the fibe is $\Delta\varepsilon$, the moving distance change $\Delta\lambda_{BS}$ of fibe Bragg grating can be expressed as[3,4]:

$$\Delta\lambda_{BS} = \lambda_B (1 - P_e) \Delta\varepsilon$$

(P_e is the elastic-optic coefficient of the fiber)

When the length of the magnetostrictive material change, slider 1 and 2 will slide a small displacement, causing strain of the fiber grating and the drift of its central wavelength. Thus, transmitting the corresponding $\Delta\lambda_{BS}$ to external devices.

In addition, we need to take protective measures for the fiber grating is easily broken. Make a baffle on each end of slider 2 to prevent fiber grating from snapping. Moreover, seal the trough by making a sliding on its top to prevent students touching the fiber grating in operation.

3.2 The Long Period Fiber Grating

Displayed The long period fiber grating is a new type of fiber grating, whose grating period is generally bigger than 100μm. Its basic principle is that the light will attenuate after a transmission in the long period fiber grating. on each side of transmission peak of the long period fiber grating there's a period of approximatly linear range in which the LPG transmission rate and the wavelength have approximatly linear relationship[5]. Based on this characteristic, the long period fiber grating can be used as edge filter. In the linear period near the transmission peak of the long period fiber grating, the transmission loss and the light wavelength have a linear relationship[6,7]:

$$H=k*K+const$$

In this equation, k is a constant which is relevant to the characteristics of long period fiber grating, H means s transmissitvity of the long period fiber grating and K represents wavelength position. The figure of light intensity and wavelength is as follows:

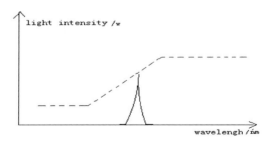

Fig. 3. the shematic diagram about linear demodulation of edge filter

In the figure, the dotted line is the transfer function curve of edge filter.

According to the characteristics of the long period fiber grating, we can turn the wavelength change from the length change device into the change of light intensity, which can be picked by pin tube and then collected by computer via the amplifier circuit.

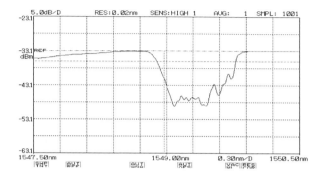

Fig. 4. Transmission spectrum of the long period fiber grating

As it shows, the center of the failling edge is at 1549.0 nm, so we can use this period of approximatly linear range to realize transition of wavelength information to light intensity information.

3.3 Amplifying Circuit

As showed above, D_7 is a pin tube, and it can turn the light intensity signal from the LPG into a current signal.

The left part of the circuit in the figure can turn the current signal into a voltage signal of point b. According to the principle of amplifying circuit, the voltage of point a is 0, so we can get the boltage of point b though the formula $U_b = I_{D7} * R_{W1}$.

The right part of the circuit in the figure can amplify the voltage in point b. Because the computer can not collect a minus signal, we design a slide rheostat so that the voltage sends to the computer will never be minus. And the voltage is decided by the formula $U = \beta * U_b$.

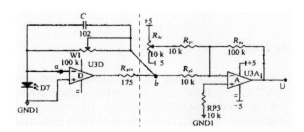

Fig. 5. Principle diagram of amplifying circuit

3.4 Collect the Temperature

In the experiment process, the changing of the open air temperature will possibly bring the disturbance when we measure the magnetostrictive coefficient .So we should gather the temperature of the installment at the same time. Just put the temperature probe in the installment, and use the computer to collect the signal through a amplifying circuit similar to amplifying circuit in 3.3.

3.5 Collect the Magnet Field

The magnet field of the device is decided by the current though the coil.If we collect the voltage of the coil,we can get the magnet field though the formula below[8]:

$$B = \mu_0 nI = \mu_0 n \frac{U}{R}$$

3.6 Authoring Software

The computer will collect the signals of external environment temperature (to show the change of the external wavelength of the fiber grating because of temperature change) ,magnetic field changes of the device and the voltage which connect to the length change of the magnetostrictive material and save them in the software.Then we can calculate the magnetostrictive coefficient with the data above and the data of the voltage reflects the temperature changes.This is the figure of the software:

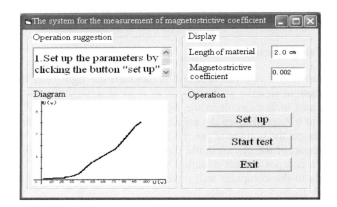

Fig. 6. The operation interface of the software

4 Experiment Result

According to the principle as shown in fig 1,we design a series of device.Firstly, we use the "300V10A"DC regulator with adjustable output power supply which is producted by Ji nan Le pu Power supply technology Company as our adjustable power supply. This kind of power supply can be controled and monitored by the computer though interface RS485.Then we put TbFe2(the length is 3 centimeters) in the magnetostrictive material trough as shown in fig 2.turn on the power,and the coil with 5000 circles will creat a magnetic field.We can set up parameters by clicking the button "set up",including the lenth of the magnetostrictive material,the frequency the magnetic field is changing and so on.Secondly,we click the "start test" button,so the computer will control the power supply changing from 0V by 2V every time every 2 seconds.And the computer will draw a diagram of voltage(reflect the voltage of the power supply)-voltage (reflect the changing length of the magnetostrictive material) as shown in fig 6.The computer will collect three signals talked above and calculate the magnetostrictive coefficient all the time though the correspondence between these data we write in the software.

The experimenta l results indicate that the magnetostrictive coefficient is about 10-3.It will be obvious when the device is in a highfield and the relatively error is less than 5% when compare to theoretical quantitative value ..But it is not so good when the magnetic field is low.

5 Conclusion

The experimental result indicated that using the fiber grating to make a device to measure magnetostrictive coefficient is feasible.it might not only overcome formerly the non-balance electrsc bridge method survey's shortcomings, but also the cost is low, the operator is simple,and may promote to the students' experiment operations like the measurement of magnetostrictive coefficient in electromagnetics and the experiment of sensors in electrooptical technology courses.

References

1. Smith, T.F., Waterman, M.S.: Identification of Common Molecular Subsequences. J. Mol. Biol. 147, 195–197 (1981)
2. Wang, S.-l.: Design and Experimental Study of the Magnetostrictive Actuator. Equipment Manufacturing Technology 8, 67–71 (2010)
3. Ma, Z.-x.: A Power Supply for Giant Magnetostrictive Actuator. Mechanical Engineering & Automation 5, 56–61 (2010)
4. Peng, B.-j., Miao, Y.-b., Min, Z.: New method of measuring fibers valid elastic-optic constant. Optical Technique 9(5), 656–660 (2005)
5. Peng, B.-j., Zhao, Y., Ying, C.-f., et al.: Micro-vibration or displacement measurement based on a novel bragg grating pair 6625, 66251Z1–66251Z6 (2008)
6. Cheng, S.-z., Jiang, Z.-y.: General physics(M) (2006)
7. Zhao, C.-l.: Optical fiber sensors based on long-period gratings in photonic crystal fibers. Journal of Optoelectronics. Laser (2011)
8. Yong, Z.: Recent Development of the Seawater Salinity Measurement Technology. Opto-Electronic Engineering (2008)
9. Mei, S.-n., Zhang, W.-h. Research of FBG Multiplexing/Demodulation on LPG(A), 89–92 (2007)

A New Driving Topology for DC Motor to Suppress Kickback Voltage

Jiang Sun[1], Bo Zhang[1], Haishi Wang[1], Ke Xiao[2], and Peng Ye[2]

[1] State Key Laboratory of Electronic Thin Films and Integrated Devices, University of Electronic Science and Technology of China, Chengdu 610054, P.R. China
[2] Chip-all Microelectronics Inc. of Chengdu
Jiang Sun is with the State Key Laboratory of Electronic Thin Films and Integrated Devices, University of Electronic Science and Technology of China, P.R. China
sun_river1980@yahoo.cn

Abstract. A new driving topology for DC motor to suppress kickback voltage is presented in this paper. By adopting a power isolation block, the proposed new driving topology for DC motor can suppress the kickback voltage of DC motor without large capacitor, power Zener diode, and it also no need high speed comparators for detecting the kickback voltage of DC motor. Furthermore, all the devices used in the new driving topology are available in the ordinary driving circuit of DC motor. Even if the power isolation block can be realized through adjusting connection of the ordinary driving circuit of DC motor. Therefore it can suppress the kickback voltage of DC motor with low-cost. The proposed new driving topology Consists of two parts: one is to generating the logic control signal, the other is power driving devices. And test results demonstrate that the new topology could suppress the kickback voltage of DC motor effectively.

Keywords: DC motor drives, DC motor protection, power isolation, regeneration.

1 introduction

A DC motor is used in various applications of electromechanical systems because of its high efficiency and good controllability over a wide range of speed. However, if the current path of driving current through the DC motor is switched off abruptly, there will be large voltage spike, kickback voltage, and it could result in the destruction of drive circuit components if not handled. Therefore, some methods were proposed to deal with the kickback voltage with additional devices or circuits. In [1] the authors proposed storing the recirculation energy into super capacitor through chopper controlling, which needs a super capacitor. In [2] the authors presented a topology of input power which could control both the power work well in motor driving mode and regenerative mode to weaken the kickback voltage. However, this will needs more additional devices in power input stage.

The authors [3], [4] proposed a regenerative circuit which allows a bidirectional power flow between load and source during the recirculation period. It can increase the efficiencies, while they also need complex logic circuit, and increase the cost of the control drive IC. Furthermore it requires strong current sink capability of power supply. Moreover, the ordinary method to suppress the kickback voltage is based on the off-chip power Zener diodes just as shown in Fig.1. The power Zener will clamp the DC motor kickback voltage to safe range, instead of exceeding the maximum voltage specification of the components in this system. However, the area of the integrated power Zener diode must be large enough to endure large re-circulation current, which will increase the cost of the chip largely. Furthermore, power Zener diode could not always be integrated in the driving IC, because many technology processes do not supply it.

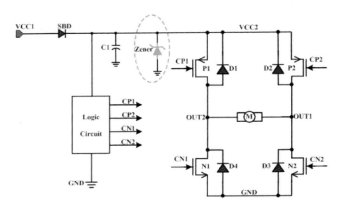

Fig. 1. Conventional circuit for suppressing the kickback voltage of DC motor

Therefore, in Fig. 2 the author [5] proposed an inductive load kickback absorption scheme which just needs high speed comparators to sense the terminal voltage of DC motor and could be built in different technology processes, even if they do not supply power Zener diodes.

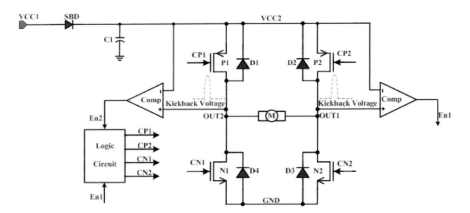

Fig. 2. The suppression of kickback voltage through detecting of high-speed comparators

In this paper, the author proposed a novel driving topology for DC motor to suppress kickback voltage without large capacitor, power Zener diode and it also do not require high-speed comparators to sense kickback voltage. Furthermore, applying the proposed novel topology, the DC motor driving circuit can be capable of auto-adjusting the kickback absorption voltage according to different power supply. The proposed driving topology could be applied in brushless DC motor driver, especially in brushless DC cooling fan motor used for cooling electronic equipment, such as power supply, CPU, VGA, LCD, LED etc. It can be used to suppress kickback voltage during power devices are switched off abruptly.

2 Proposed New Driving Toplogy for DC Motor

The DC motor kickback voltage is a large voltage spike that occurs whenever the current flowing through an inductive load is switched off abruptly [6], [7]. There is one assumption used in the single-phase brushless DC motor to show the occurrence of kickback voltage. It is that power driving transistors in the same side couldn't be turned on at the same time, for there is a dead zone between up side power transistors and low side.

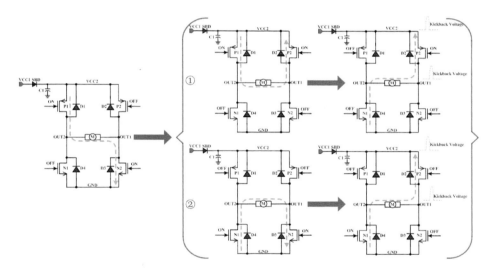

Fig. 3. Occurrence of the DC motor load kickback voltage

In Fig. 3, assume that firstly the power transistors P1, N2 are on while P2, N1 off, and the current in DC motor is flowing from OUT2 to OUT1. Secondly, as the phase change draws near, there are two paths to let the current in the DC motor flowing in the original direction:

① The power transistors P1, P2 are on and N1, N2 are off, and the re-circulation current in DC motor will keep flowing from OUT2 to OUT1 through the up side loop just as shown in Fig. 3. Then during the phase change, power transistor P2 is on while

P1, N1 and N2 off, the re-circulation current flowing in motor from OUT2 to OUT1 continues through the path D4, DC motor and P2 or P2's re-circulation diode D2.

② The power transistors N1, N2 are on and P1, P2 are off, and the re-circulation current in DC motor will keep flowing from OUT2 to OUT1 through the low side loop. Then during the phase change, power transistor N1 is on while P1, P2 and N2 off, the re-circulation current flowing in motor from OUT2 to OUT1 continues through the path N1 or N1' re-circulation diode D4, DC motor and P2's re-circulation diode D2.

Although, there are two ways for the re-circulation current to flow in DC motor in original direction, there is no path leading the re-circulation current from VCC2 to GND except capacitor C1. Usually, C1 is small and could not hold all the energy stored in DC motor. Therefore the exceeding energy will make VCC2 higher and enable the capacitor C1 hold all the energy stored in the DC motor. But the high VCC2 may exceed the maximum voltage specification of other components in the driving system. Therefore, there should be some effectively method to suppress the kickback voltage in a safe range. Furthermore, as there is a Schottky Barrier Diode (SBD), connected between system power supply VCC1 and the power supply of driving circuit VCC2, which is used to prevent reverse current from destroying the system power supply VCC1, the kickback voltage of DC motor will not affect the system power supply VCC1.

In this section, the proposed new driving topology for DC motor to suppress kickback voltage is presented just as shown in Fig.4, which adds a block named power isolation. The block to be used for separating power supply is directly connected to the system power supply terminal VCC1, power supply terminal of power driver H-bridge VCC2 and power supply terminal of the logic control circuit VCC3. When VCC1 is greater than VCC2, the power isolation circuit conduction, and the system power supply terminal VCC1 to provide the current passed to the VCC2; When the system power supply terminal VCC1 is less than VCC2, the power isolation circuit blocks and the energy from VCC2 can not pass to the system power supply terminal VCC1. This will not only prevent reversed current of DC motor from damaging the system power supply terminal VCC1, but also can keep the power supply terminal VCC2 and power supply terminal of logic control circuit VCC3 separating.

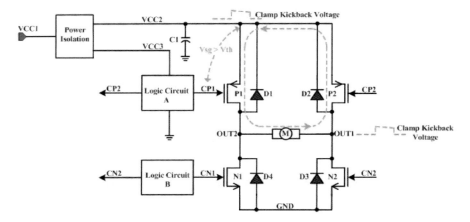

Fig. 4. Proposed new driving topology for DC motor to suppress the kickback voltage

A New Driving Topology for DC Motor to Suppress Kickback Voltage 1005

When kickback voltage of DC motor occurs at the power supply terminal VCC2, the power isolation circuit will be in blocking state to prevent the kickback high voltage of DC motor from destroying the VCC1 and VCC3. Capacitor C1 connected between VCC2 and ground is used to prevent kickback voltage from being much high in a short time when it appears.

Fig. 5 is one sequence chart of logic control signal for controlling the H-bridge. And the proposed driving circuit works as follows according to sequence chart: First, assume that the logic circuit A is use to control the power transistors P1 and P2, and the logic circuit B to control the power transistors N1 and N2. Furthermore, the logic circuit make P1 and N2 on, P2 and N1 off, therefore the driving current in DC motor load will flow from the output drive terminal OUT2 to the output drive terminal OUT1.

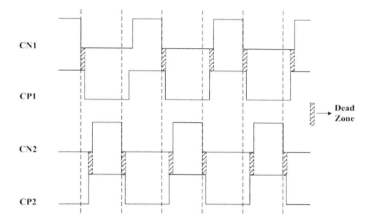

Fig. 5. Logic control signal during phase changing

Second, during the phase changing, in order to prevent the power transistors P2 and N2 are on at the same time, the power transistor P1 is on only and the power transistors P2, N1 and N2 are off. This status will remain for some time, and then power transistor P2 will be on controlled by the logic circuit A. During this time, the re-circulation current in DC motor load will flow in original direction from the path P1, DC motor load and P2 or P2 re-circulation diode D2. Then the logic circuit B will turn off the power transistor P1, the re-circulation will keep flowing in the original direction from OUT2 to OUT1 through the path N1 re-circulation diode D4, DC motor load and P2 or P2 re-circulation diode D2. But there is no other path leading the re-circulation current from VCC2 to GND except capacitor C1. Usually, C1 is small. Once the re-circulation current flowing through the capacitor to ground, the power supply terminal VCC2 raises quickly, and the DC motor kickback voltage occurs. Fortunately, power supply isolation circuit will block kickback voltage, and do not pass to the system power supply terminal VCC1 and the logic control power supply terminal VCC3. When VCC2 is greater than VCC1 and VCC3, the voltage difference between VCC3 and VCC2 will make the original shutdown power transistor P1 be on, therefore the re-circulation current in the power transistor P1, DC motor load, power transistor P2 or P2's re-circulation diode consisting of continuous current loop, just as shown in Figure.

4. Once power transistor P1 is on, the kickback voltage of DC motor is suppressed. The suppression point of kickback voltage is decided by the re-circulation current and the characteristic of power transistors P1 and P2. In addition, as power driving terminal OUT2 is zero, even if the power transistor N1 is on, there is no current. And the re-circulation current in DC motor will gradually decrease until zero. After that, kickback voltage of VCC2 disappears and the voltage difference between VCC2 and VCC3 eliminates. The power transistor P1 will be re-controlled by the logic circuit A and be off. Therefore, Only P2 and N1 are on in H-bridge and the driving current will flow from P2, DC motor to N1, to achieve the phase changing. The kickback voltage of DC motor has been suppressed effectively during the phase changing.

Power isolation circuit can be implemented in various ways. In Fig. 6, we propose three simple ways:

① Adding a Schottky Barrier Diode SBD1 between system power supply terminal VCC1 and the power supply terminal of power driver H-bridge VCC2. Furthermore, it also needs a Schottky Barrier Diode SBD2 between system power supply terminal VCC1 and the logic control power supply terminal VCC3. The diodes not only can separate the power supply terminal of logic control circuit VCC3 from the power supply terminal of power driver H-bridge VCC2, but also can prevent system power supply terminal VCC1 from damage.

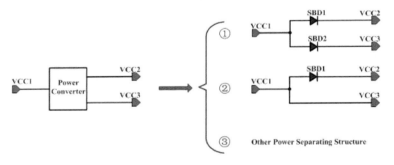

Fig. 6. Several simple ways for power isolation

② Adding just one Schottky Barrier Diode SBD1 between the system power supply terminal VCC1 and the power supply terminal of power driver H-bridge VCC2 and connecting power supply terminal of the logic control circuit VCC3 to VCC1 directly. Usually, VCC3 just needs small current, no inrush current and no kickback voltage, therefore it can be connect to VCC1 directly. As for VCC2, there is a Schottky Barrier Diode SBD1. Especially as there is no way to return re-circulation current to VCC1 and VCC3, it may prevent VCC1 and VCC3 from rising, Therefore, it can prevent system power supply terminal VCC1 from damage.

③ Using other power isolation methods, such as diode-connected power MOSFET or bipolar transistors, even other circuits can separate VCC3 from VCC2. At the same time the methods also needs to block re-circulation current to VCC1.

From the analysis above, method two is simply, and no needs other devices. Because there is a Schottky Barrier Diode SBD1 available in the ordinary DC motor driver system. We just need adjust the connection of the power supply terminal of logic control circuits VCC3 to connect to system power supply terminal VCC1 directly. Therefore, in the coming analysis, we will use the method two to show how the new driving topology for DC motor to suppress the kickback voltage.

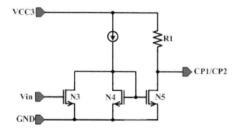

Fig. 7 (a) Fist way of generating logic signal CP1 and CP2

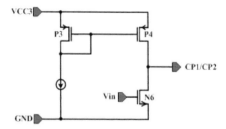

Fig. 7 (b) Second way of generating logic signal CP1 and CP2

The logic circuit A can be achieved by a variety of configurations. Fig. 7 shows two ordinary methods to implement it. The first configuration is shown in Fig.7 (a), and it consists of three transistors N3, N4, N5, a current source and a resistance R1. The power supply side is connected to power supply terminal of logic control circuit VCC3. The second configuration is shown in Fig. 7 (b), and it consists of three transistors P3, P4, N6 and a current source. he power supply side is connected to power supply terminal of logic control circuit VCC3.

3 Experimental Results and Discussion

The proposed new driving topology for DC motor to suppress the kickback voltage has been implemented through pre-driver IC EUM6867+ H-bridge driving structure just as shown in Fig. 8. The pre-driver IC EUM6867, resistors R1, R2, R3, R4, R5 and hall sensor are used for generating the logic control signal CP1, CP2, CN1 and CN2, which just works as the logic circuit A and logic circuit B shown Fig. 4. And power MOSFETS P1, P2, N1 and N2 are formed the H-bridge driving structure to driving the DC motor load.

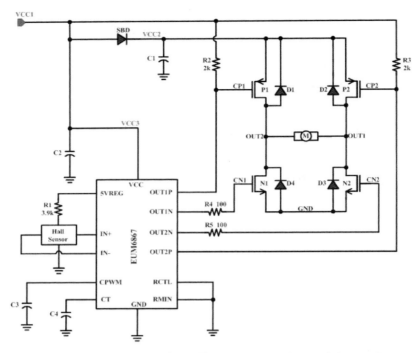

Fig. 8. The whole schematic for verifying the proposed new driving topology

The DC motor load is using a 118 mm square and 38 mm thick fan which is usually used in the applications of sever and switching power supply for the cooling system. The experimental setup is shown in Fig. 9 and the disassembly of the DC motor load is shown in Fig. 10.

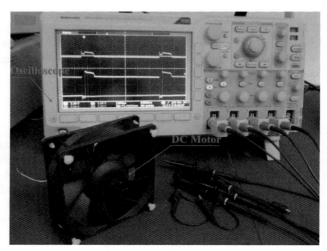

Fig. 9. The setup condition of experiment

A New Driving Topology for DC Motor to Suppress Kickback Voltage   1009

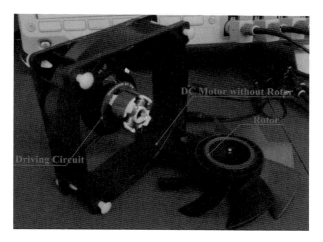

Fig. 10. The structure of the DC motor load

And measured curves are shown in Fig.11, Fig.12 and Fig.13. All the curves are under the condition of temperature=27□, C1=0.1μF, $V_{breakdown(Zener)}$=18V and DC motor equivalent to inductor: 5.2mH, resistor: 2.5Ω. In addition, in Fig.11, Fig.12 and Fig.13, Channel 1, Channel 2, Channel 3 and Channel 4 represent the voltage of OUT1, OUT2, VCC2 and VCC3 respectively.

As the kickback voltage is very large, the input power supply VCC1 is just 5.0 V in the driving system without any circuit to suppress the DC motor kickback voltage. And the measured curves are shown in Fig. 11. When the DC motor load kickback voltage occurs, the voltage of OUT1 (OUT2) will become much higher than VCC1. And the re-circulation diode D2 (D1) is turned on to let the current flowing in the DC motor in primary direction. Then for the limited absorption ability of the small capacitor C1, the voltage of VCC2 will rise to 36.0V in order to increase the C1 absorption ability, just as shown in Fig. 11.

Furthermore, as there is a Schottky Barrier Diode (SBD), the inductive kickback voltage will not affect the system power supply VCC1 and VCC3.But the increasing kickback voltage may exceed the maximum voltage specification of other components in this system and destroy them.

In the following test, as the two driving topologies have suppression ability of DC motor kickback voltage, the input power supply voltage is 12.0V. Fig.12 shows the measured curves of the system with power Zener diode. Once inductive load kickback voltage occurs, the voltage of OUT1 (OUT2) will rise to 18.0V till the Zener diode is turned on. This will re-circulate the DC motor current from VCC2 to ground and clamp the output voltage of VCC2 and OUT1 (OUT2) as shown in Fig.12. When the DC motor kickback voltage occurs, power Zener diode will clamp OUT1 (OUT2) and VCC2 to a safe range, instead of exceeding the maximum voltage specification of other components in this system and destroying the circuit system.

1010   J. Sun et al.

Fig.13 is showing the DC motor load kickback voltage's suppression with the proposed new driving topology. As DC motor load kickback voltage occurs, the output voltage OUT1 (OUT2) will go up. And VCC2 will be pulled up through P1 and P2's re-circulation diodes D1 and D2 just as shown in Fig.13. Then VCC2 is greater than VCC1 and VCC3. The voltage difference between VCC3 and VCC2 will make the original shutdown power transistor P1 (P2) open, therefore the re-circulation current in the power transistor P1, DC motor load, power transistor P2 or P2's re-circulation

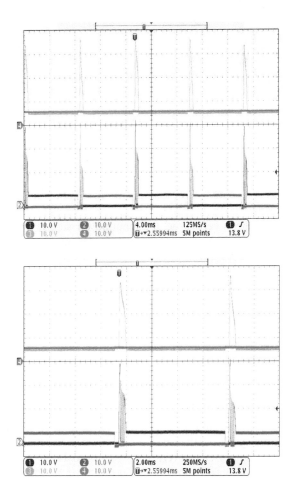

Fig. 11. Measured curves of the driving system without any kickback suppression circuit

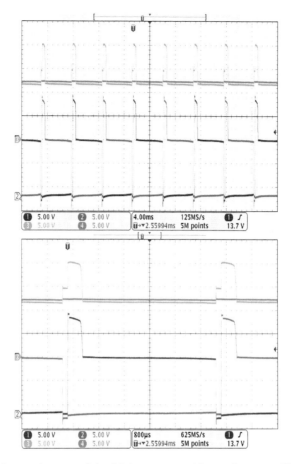

Fig. 12. Measured curves of the driving system with exterior power Zener diode

diode (P2, DC motor load, power transistor P1 or P1's re-circulation diode) consists of continuous re-circulation current loop. The kickback voltage of power supply terminal VCC2 is suppressed. As the re-circulation current in DC motor will gradually decrease until zero, the kickback voltage of VCC2 disappears and the voltage difference between VCC2 and VCC3 eliminates. The power transistor P1 (P2) will be re-controlled by EUM6867 and be off. Then the kickback voltage of DC motor has been suppressed effectively. While actually, the maximum value VCC2 is 16V just as shown in Fig.13.

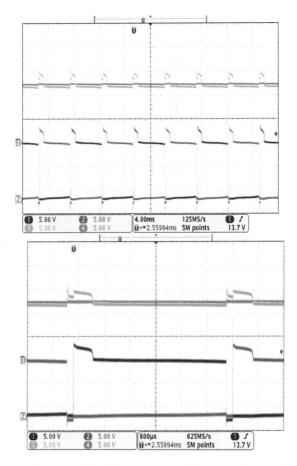

Fig. 13. Measured curves of the driving system with proposed new driving topology

4 Conclusion

A new driving topology for DC motor to suppress the kickback voltage has been proposed. All the devices used in the new driving topology are available in the ordinary driving circuit of DC motor.

Just through adjusting connection of the ordinary driving circuit of DC motor, the power isolation function can be realized. Therefore the proposed new driving topology for DC motor can suppress the kickback voltage without any additional cost compared to the ordinary the driving topology.

The proposed new driving topology consists of two parts: one is used for generating the logic control signal and the other is power driving devices.

Test results demonstrate that the new driving topology for DC motor load could suppress the kickback voltage effectively and without any additional cost.

Acknowledgment

The authors wish to thank the engineering staff at test team in Eutechmicro Semiconductor Inc. of Taiwan for their technical support to obtain the measurement data.

References

1. Koichi, Y., Takeo, T., Kouki, M.: Performance of the Inverter with the Super Capacitor for Vector Controlled Induction Motor Drives. In: IEEE Industrial Electronics, IECON, pp. 946–951 (2006)
2. Lezana, P., Rodríguez, J., Oyarzun, D.A.: Cascaded Multilevel Inverter With Regeneration Capability and Reduced Number of Switches. IEEE Trans. Ind. Electron. 55, 1059–1066 (2008)
3. Rodriguez, J., Moran, L., Pontt, J., Hernandez, J.L., Silva, L., Silva, C., Lezana, P.: High-Voltage Multilevel Converter With Regeneration Capability. IEEE Trans. Ind. Electron. 49, 839–846 (2002)
4. Yang, M., Jhou, H., Ma, B., Shyu, K.: A Cost-Effective Method of Electric Brake With Energy Regeneration for Electric Vehicles. IEEE Trans. Ind. Electron. 58, 2203–2212 (2009)
5. Sun, J., Zhang, B., Wang, H., Ming, X., Xiao, K., Ye, P., Cao, L.: An Inductive Kickback Absorption Scheme Without Power Zener and Large Capacitor. IEEE Trans. Ind. Electron. 56, 709–716 (2011)
6. Trafton, F., Ziemer, K.: A method for overvoltage protection of a motor driver IC. In: Proc. IEEE ISPSD Conference, pp. 123–126 (May 2005)
7. Haskew, T.A., Hill, E.M.: Regeneration mechanisms in a DC motor with an H-bridge inverter. In: Proc. IEEE IEMD Conference, pp. 531–533 (May 1999)

The Reduction of Facial Feature Based on Granular Computing

Runxin He[1] and Nian He[2]

[1] Electronic Engineering Department of Fudan University,
Shanghai 200433, China
[2] School of Computer Science and Telecommunication Engineering,
Jiangsu University, Zhenjiang Jiangsu 212013, China
{hnhn69,hrunx}@163.com

Abstract. Granular computing is a popular method of attribute reduction and the granularity is divided to two kinds coarse-grain and fine-grain. However, it is hard to determine the size of granularity normally. This paper presents a method to divide the original features from the actual condition of facial feature selected in Gabor. Then we use binary encoding properly to arrange the data based on the number of the samples. When the information of more coarse grain becomes simplified it can be reduced in granular computing. The experiment shows that this method can remove redundancy accurately. It will reduce the running time and improve the operation efficiency.

Keywords: Granular Computing, Attribute Reduction, Facial Feature, Granularity.

1 Introduction

It is found that it is difficult to improve the efficiency of many systems by pattern recognition with the increasing degree of pattern identification. Therefore more attention is paid to the attribute reduction. A more compendious feature set can increase the efficiency of pattern recognition easier. At present there are four kinds of methods of reduction which are respectively the attribute reduction based on differential matrix, the attribute reduction based on information representation, the attribute reduction based on positive region and the attribute reduction based on granular computing. The operation efficiency of the feature reduction based on granular computing is high, and this method is suitable for decision table of large-scale. So it is becoming a research focus. The core idea of this method is considering a set of elements as a granularity, and then deciding whether it should be deleted according to the importance of the granularity. Because of the wide range application we consider that it can be applied to the reduction of facial feature.

2 The Basic Theory and Principle

First of all, we introduce some related concepts and theorems.

2.1 Information System and Granular Amount of Knowledge

There are some important definitions.

- **Definition 1.** A system information can be seen as $S = (U, A, V, f)$ denoted as $S = (U, A)$, where $U \neq \emptyset$ is a universe. A is a set of attributes. V is the range of a: $V = \bigcup_{a \in A} V_a$. f is the information function of the system, $f : U \times A \to V$. For each set of attributes $P \subseteq A$, the related indiscernibility relation can be expressed as: $IND(P) = \{(x, y) \in U \times U \mid \forall p \in P, f(x, p) = f(y, p)\}$.

- **Definition 2.** Given knowledge base $S = (U, A, V, f)$, if there is a set of attribute R, the universe U is divided to equivalence class $U / IND(R) = \{X_1, X_2, ..., X_n\}$. The number of it is called Granular Amount of Knlwledge.

- **Definition 3.** Given knowledge base $S = (U, A, V, f)$, if there is a set of attribute R, the universe U is divided to equivalence class $U / IND(R) = \{X_1, X_2, ..., X_n\}$. Therefore $(|X_1|^2 + |X_2|^2 + ... + |X_n|^2) / (\underbrace{|U| + |U| + ... + |U|}_{n}) = \sum_{i=1}^{n} |X_i|^2 / n|U|$ means the coefficient nearly 1 of R : $1^*(R)$.

- **Property 1.** $1/|U| \leq 1^*(R) \leq |U|$ [2]

The value of $1^*(R)$ can describe the classification of R. The closer to 1, the more moderate is the classification.

- **Definition 4.** We should master another important concept that is Average Granular Particle Amount of Knowledge denoted as $AGP(R)$: $AGP(R) = |U| / GA(R)$.

- **Definition 5.** On the basis of the above, it's time to introduce the concept of Granularity of Knowledge denoted as $GK(R)$: $GK(R) = 1^*(R) / AGP(R) = (\sum_{i=1}^{n} |x_i|^2 / n|U|) / (|U| / n) = \sum_{i=1}^{n} |x_i|^2 / |U|^2$.

- **Property 2.** $1/|U| \leq GK(R) \leq 1$ [2]

2.2 The Importance of Information and the Reduction of Knowledge

According to the granularity of knowledge it is apt to analyze the importance of each set of attributes in one information system. Given a information system $S = (U, A)$. There is a attribute in A called a : $a \in A$, so we can measure the value of importance through the variable quantity of the granularity of knowledge of A when a is removed. The greater variation it will cause the more important is a, and the more important it will be to A. Given a set of attributes named C : $C \subseteq A$. If $a \in A - C$, we should measure whether this attribute is important or not through the variable quantity of the granularity of knowledge of C when a is added in. The greater variation it will cause the more important is a, and the more important to C it will be.

- **Definition. 6.** The value of significance of a to A can be denoted as $Sig_{A-\{a\}}(a)$ and it is defined as $GK(A-\{a\}) - GK(A)$.

- **Definition. 7.** Given an information system $S = (U, A)$, an attribute $a \in A$, we definite that a is redundant to A if the value of $GK(A-\{a\})$ is equality to $GK(A)$. On the opposite side, a is necessary to A. The set of all the attributes necessary is the core of A denoted as $Core(A)$.

- **Definition. 8.** There is an information system denoted as $S = (U, A)$. Given a set of attributes $P : P \subseteq A$. We define P as a reduction of A if P is independent. All the reduction of A is denoted as $\operatorname{Re}d(A)$.

- **Property. 3.** $Core(A) \subseteq \operatorname{Re}d(A)$

3 The Reduction of Facial Feature Based on Granular Computing

So the set of facial features can be deduced by granular computing.

3.1 Algorithm Description

This paper will describe the attribute reduction based on granular computing specifically. On one side, the input of this system is an information system $S = (U, A)$ where U is the universe of this information system and A is the set of features. On the other side, the output of this method is the minimum reduction of A denoted as P.

The next step is to describe the concrete step of this method.

Step1. Assume the granularity of knowledge of A denoted as $GK(A)$.

Step2. Calculate the core of the set of attributes denoted as $Core(A)$. Then calculate the value of important of each feature of A denoted as $Sig_{A-\{a\}}(a)$ and choose the set of the features whose value of importance is not equal to 0, which is $Sig_{A-\{a\}}(a) \neq 0$, then constitute the core of A that is $Core(A)$.

Step3. Denote a set of attributes as $C := Core(A)$.

Step4. Determine whether the value of $GK(C)$ is equal to the value of $GK(A)$. It is equivalent C is the minimum reduction of this information system or we should go to step5.

Step5. Calculate the value of the importance of all the feature of $A - C$ to C denoted as $Sig_C(a)$. Choose a feature a which the value of the important to C is the maximum that is $Sig_C(a) = \max_{a' \in A-C} Sig_C(a')$.

Step6. Add a to the core that is $C := C \bigcup \{a\}$.

Step7. Seed off the minimum reduction of A denoted as P.

3.2 Time Complexity

Looking for minimum reduction is a NP - hard-core problem, its complexity mainly depends on the decision table attribute combination and data quantity, so the entire time complexity mainly depends on the decision table attribute combination and data quantity.

If we want to calculate the nuclear $Core(A)$ has to calculate the $|A|$ times $Sig_{A-a}\{a\}$, we have to judge whether it greater than 0, and the time complexity of calculating $Sig_{A-a}\{a\}$ is $O(|A||U|^2)$, so the time complexity of calculating the nuclear is $O(|A|^2|U|^2)$;

When we calculating $\text{Re}\,d(A')$, we have to calculate $|A|+(|A|-1)+(|A|-2)+\ldots+1 = |A|(|A|+1)/2 = O(|A|^2)$ times $Sig_C(a)$ for the most. The time complexity of calculating $Sig_C(a)$ is $O(|A||U|^2)$, so the time complexity of $\text{Re}\,d(A')$ is $O(|A|^3|U|^2)$;

The time complexity of the calculation of $GK(A)$ and $GK(A-\{a\})$ is $O(|U|)$.

To sum up, the time complexity of the algorithm which is used by this paper is $O(|A|^3|U|^2)$.

3.3 Application Example

Table1 is the facial features of eight pictures.

Table 1. Original features.

U/A	b_1	b_2	b_3	b_4	b_5	b_6	b_7	b_8	b_9	b_{10}	b_{11}	b_{12}
x_1	1	2	2	1	1	2	3	2	1	2	3	2
x_2	1	2	3	2	1	2	2	1	0	2	3	2
x_3	1	2	3	2	1	3	3	2	1	3	3	2
x_4	1	2	2	2	0	3	3	2	1	3	3	2
x_5	1	2	3	2	1	2	3	2	0	3	3	2
x_6	1	2	2	1	1	2	2	2	0	3	3	2
x_7	1	2	2	1	0	2	2	2	0	3	3	2
x_8	1	2	2	1	0	2	3	2	1	3	3	2

First of all it should be calculated, following the steps above.
Step1: $U / IND(A) = \{\{x_1\},\{x_2\},\{x_3\},\{x_4\},\{x_5\},\{x_6\},\{x_7\},\{x_8\}\}$,

$$GK(A) = \sum_{i=1}^{n}|X_i|^2 / |U|^2 = \sum_{i=1}^{8}|X_i|^2 / |U|^2 = 1^2 \times 8 / 8^2 = 1/8.$$

Step2: $U / IND(A-\{b_1\}) = U / IND(A-\{b_2\}) = U / IND(A-\{b_3\})$
$= U / IND(A-\{b_4\}) = U / IND(A-\{b_5\}) = U / IND(A-\{b_6\})$
$= U / IND(A-\{b_7\}) = U / IND(A-\{b_8\}) = \{\{x_1\},\{x_2\},\{x_3\},\{x_4\},\{x_5\},\{x_6\},\{x_7\},\{x_8\}\}$

$GK(A-\{b_1\}) = GK(A-\{b_2\}) = GK(A-\{b_3\}) = GK(A-\{b_4\}) = GK(A-\{b_5\})$
$= GK(A-\{b_6\}) = GK(A-\{b_7\}) = GK(A-\{b_8\}) = GK(A) = 1/8$

Calculate the value of importance:

$Sig_{A-\{b_1\}}(b_1) = GK(A-\{b_1\}) - GK(A) = Sig_{A-\{b_2\}}(b_2) = Sig_{A-\{b_3\}}(b_3) = Sig_{A-\{b_4\}}(b_4)$
$= Sig_{A-\{b_5\}}(b_5) = Sig_{A-\{b_6\}}(b_6) = Sig_{A-\{b_7\}}(b_7) = Sig_{A-\{b_8\}}(b_8) = 1/8 - 1/8 = 0$

Obviously, if one feature is seen as a granularity it is too hard to reduce the attributes and it's not so helpful to the feature deduction. It can be treated by binary encoding if the set of data is 8 bits. The data been treated is shows in table2.

Table 2. Use binary encoding to treat data seeing tree features as a granularity

U/A	a_1	a_2	a_3	a_4
x_1	001	001	001	001
x_2	010	010	010	001
x_3	010	011	001	010
x_4	001	100	001	010
x_5	010	010	011	010
x_6	001	001	100	010
x_7	001	101	100	010
x_8	001	101	001	010

Calculation steps:
Step1:
$U / IND(A) = \{\{x_1\},\{x_2\},\{x_4\},\{x_5\},\{x_6\},\{x_7\},\{x_8\}\}$,

So the granularity of A is that:

$$GK(A) = \sum_{i=1}^{n}|X_i|^2 / |U|^2 = (1^2 +1^2 +1^2 +1^2 +1^2 +1^2 +1^2 +1^2)/8^2 =1/8;$$

Step2:

$U/IND(A-\{a_1\}) = \{\{x_1\},\{x_2\},\{x_3\},\{x_4\},\{x_5\},\{x_6\},\{x_7\},\{x_8\}\}$;
$U/IND(A-\{a_2\}) = \{\{x_1\},\{x_2\},\{x_3\},\{x_4,x_8\},\{x_5\},\{x_6,x_7\}\}$;
$U/IND(A-\{a_3\}) = \{\{x_1\},\{x_2\},\{x_3\},\{x_4\},\{x_5\},\{x_6\},\{x_7,x_8\}\}$;
$U/IND(A'-\{a_4\}) = \{\{x_1\},\{x_2\},\{x_3\},\{x_4\},\{x_5\},\{x_6\},\{x_7\},\{x_8\}\}$

So,

$$GK(A-\{a_1\}) = \sum_{i=1}^{n}|X_i|^2/|U|^2 = 1^2 \times 8/8^2 = 1/8;$$

$$GK(A-\{a_1\}) = \sum_{i=1}^{n}|X_i|^2/|U|^2 = (1^2+1^2+1^2+2^2+1^2+2^2)/8^2 = 3/16;$$

$$GK(A-\{a_3\}) = \sum_{i=1}^{n}|X_i|^2/|U|^2 = (1^2+1^2+1^2+1^2+1^2+1^2+2^2)/8^2 = 5/32;$$

$$GK(A-\{a_4\}) = \sum_{i=1}^{n}|X_i|^2/|U|^2 = 1^2 \times 8/8^2 = 1/8;$$

Calculate the value of importance:

$Sig_{A-\{a_1\}}(a_1) = GK(A-\{a_1\}) - GK(A) = 1/8 - 1/8 = 0$;
$Sig_{A-\{a_2\}}(a_2) = GK(A-\{a_2\}) - GK(A) = 3/16 - 1/8 = 1/16$;
$Sig_{A-\{a_3\}}(a_3) = GK(A-\{a_3\}) - GK(A) = 5/32 - 1/8 = 1/32$;
$Sig_{A-\{a_4\}}(a_4) = GK(A-\{a_4\}) - GK(A) = 1/8 - 1/8 = 0$.

So $Core(A) = \{a \in A | Sig_{A-\{a\}}(a) > 0\} = \{a_2, a_3\}$,

$GK(Core(A)) = GK(a_2, a_3) = 1/8, GK(Core(A)) = GK(A)$. So $Core(A)$ is the minimum reduction of the information system A, that is P is equal to A and the final set of features treated by granular computing is $\{a_2, a_3\}$.

4 Conclusion

Granular computing is one of the popular methods of attributes reduction. But the space of one feature is large. If the granularity we defined is the fine-grained it is too hard to deduce the attributes and it is hard to remove the redundancy accurately. Therefore it will increases the burden of operation instead of improve operating efficiency. So we can use the attribute reduction when seen a team of elements as a granularity. It will decrease the time caused by granular computing, deduce the original attributes properly and improve the efficiency of using the features reduced later.

References

1. Wu, W.-Z., Leung, Y., Mi, J.-S.: IEEE Transactions On Knowledge and Data Engineering 21(10) (October 2009)
2. Gao, L.-y., Sang, L., Hu, Y.-c., Zhou, L.-l.: Research on Granular Computing Cased on Rough Set Theory and Its Application. Control and Automation Publication Group 24(12-3), 189–191 (2008)
3. Li, H.: Research on Knowledge Reduction based on Knowledge Granularity 25(2), 16–19 (2010)
4. Xie, K., Xie, J., Du, L., Xu, X.: Granluar Computing and Neural Network Integrate Algorithm Applied in Fault Diagnosis, p. 564. IEEE, Los Alamitos (2009), doi:10.1109/FSKD
5. Hu, J., Wang, G., Zhang, Q., Liu, X.: Attribute reduction based on granular computing. In: Greco, S., Hata, Y., Hirano, S., Inuiguchi, M., Miyamoto, S., Nguyen, H.S., Słowiński, R. (eds.) RSCTC 2006. LNCS (LNAI), vol. 4259, pp. 458–466. Springer, Heidelberg (2006)
6. Zhao, Y., Wang, L., Han, Y.: Research of Image Feature Extraction Based on Morphology and Clustering. Journal of Projectiles, Rockets, Missiles and Guidance 30(2) (April 2010)

Control Methods and Simulations of Micro-grid

He-Jin Liu[1], Ke-Jun Li[1], Ying Sun[1], Zhen-Yu Zou[2], Yue Ma[2], and Lin Niu[3]

[1] School of Electrical Engineering, Shandong University, Jinan 250061, China
[2] Shandong Electric Power Engineering Consulting Institute Corporation Limited, Jinan 250013, China
[3] State Grid of China Technology College, Jinan 250002, China
sdulhj@163.com

Abstract. The micro-grid's control methods and simulations are presented in the paper. The control methods are discussed in preset conditions, which are called the grid-connected mode and islanded mode. During the grid-connected mode, the *P/Q* control is adopted. However, the control method is changed to *U/f* control when it turns to the islanded mode. To demonstrate the operation of the micro-grid in the two modes, a basic model of the micro-grid is established with Matlab/Simulink, which enables simulations for both the steady and dynamic characteristics of the three-phase micro-grid. The validity of the control methods, which are studied in the paper, is verified with simulations of the micro-grid.

Keywords: micro-grid; grid-connected mode; islanded mode; *P/Q* control; *U/f* control; simulation.

1 Introduction

New energy and renewable power generation - Distributed Generation (DG) has become a research hotspot nowadays. DG has sighted a significant development in recent years, due to its high operating efficiencies and improved reliabilities [1]. The micro-grid comprises low voltage distribution systems with distributed energy resources, such as photovoltaic power systems and wind turbines, together with storage devices.

The micro-grid concept has been proposed as a solution to the conundrum of integrating large amounts of micro-generations without disrupting the operation of the utility network [2-3]. The micro-grid concept assumes a cluster of loads and microsources operating as a single controllable system that provides both power and heat to a local area, for which some forms of energy storage equipments are usually required. It can be operated either in the grid-connected mode or in the islanded mode, depending on demands from the customers and economic reasons.

This paper emphasizes mainly on the control methodss in the two operational modes, and the micro-grid is simulated within a platform by Matlab/Simulink to account for both the transient and steady-state system characteristics. The micro-grid simulation confirms that the control methods are effective in dominating the micro-grid during either the grid-connected mode or islanded mode.

2 Structure of Micro-grid System

In a typical micro-grid system, several kinds of DGs and energy storage systems are involved. The microsources may be rotating generators or Distributed Energy Resources (DER) interfaced by power electronic inverters [4]. The installed DERs may be solar, wind, biomass, fuel cells and steam or gas turbines. The intermediate energy storage devices may be an inverter-interfaced battery bank, super-capacitors or flywheel, the storage devices ensure the balance between energy generation and consumption especially during large fluctuation in load or generation.

Fig.1 shows the general micro-grid topology architecture, which consists of a group of radial feeders: A, B and C, together with a collection of loads. The radial system is normally connected to the large power distribution system through a separate device called point of common coupling, which is usually a static switch [5]. The voltage feeders at the loads side are usually rated at 400V or less. Feeder A indicates the co-existence of several microsources with one providing both power and heat supply. Each feeder has a specific circuit breaker and a power flow controller. The controller regulates the feeder power flow at a certain level prescribed by the energy manager. As loads change randomly, the local microsources either increase or decrease their output to hold a balance of the power flow within the micro-grid.

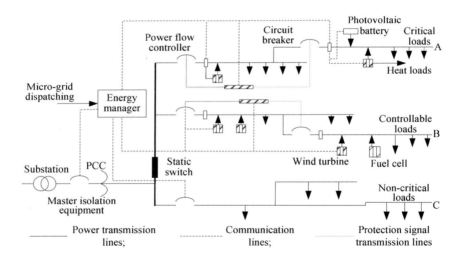

Fig. 1. Topology architecture of the micro-grid

In Fig.1, feeders A and B are attached with critical loads which require local generations (microsources), while feeder C is assumed to have non-critical loads which can be shed off if necessary. When the micro-grid is operating in the grid-connected mode, the power from the local generations can be directed to the critical and non-critical loads. If there are disturbances or malfunctions on the large distribution system, the micro-grid can be islanded for independent operation, and the non-critical feeder can also be disconnected from the micro-grid if necessary [6][7].

3 Control Methods of Micro-grid

The micro-grid can operate in the grid-connected mode and islanded mode. In the grid-connected mode, the micro-grid either draws or supplies power from or to the main grid, depending on the generation and load with suitable market policies. The micro-grid can separate itself from the main grid whenever a power quality event or malfunction occurs in the main grid [8][9].

3.1 Grid-Connected Mode

P/Q control is generally used when the micro-grid operates in the grid-connected mode. In this mode, the micro-grid is regulated by the main grid when there are undulations in the loads, and disturbances in the frequency and voltage. The microsources don't participate in the regulation of the frequency and voltage, the frequency and voltage of the main grid are adopted directly by the micro-grid [10].

Fig.2 shows the structure diagram of *P/Q* control, which is realized by dominating the inverters directly in the micro-grid. The inverters export active and reactive power according to the reference values given by the controller.

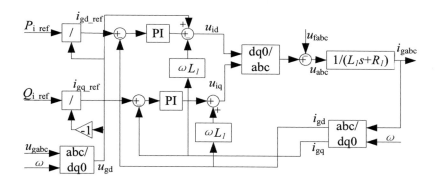

Fig. 2. Schematic diagram of *P/Q* control

Under the *P/Q* control condition, the inverters output voltage which is changed from abc coordinate to dq0 coordinate in Park transform by choosing a rational synchronous revolution axis. And the voltage of the q axis is changed to zero in the transition ($u_{gq}=0$). The output power of the inverters can be expressed as the following expressions.

$$\begin{cases} P_{ref} = u_{gd} \times i_{gd} + u_{gq} \times i_{gq} = u_{gd} \times i_{gd} \\ Q_{ref} = -u_{gd} \times i_{gq} + u_{gq} \times i_{gd} = -u_{gd} \times i_{gq} \end{cases} \quad (1)$$

Then the reference value of the current loop under the d-q axis can be given.

$$\begin{cases} i_{gd,ref} = P_{ref} / u_{gd} \\ i_{gq,ref} = -Q_{ref} / u_{gd} \end{cases} \quad (2)$$

3.2 Islanded Mode

When a power quality event or malfunction occurs in the main grid, the micro-grid can operate in the islanded mode. In this mode, the voltage and frequency of the micro-grid can be dominated appropriately. The microsources are regarded as powers with constant voltage and frequency. And the output of the inverters is detected in real time under the control of the PI adjuster, which works via setting the reference value of the voltage and frequency. Fig.3 shows the structure diagram of U/f control [10].

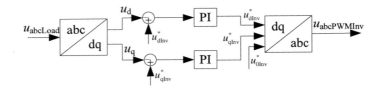

Fig. 3. Schematic diagram of U/f control

In Fig.3, only the voltage, which is the output of the inverters, is detected in the U/f control, while the frequency is a constant value. The inverters dominate the voltage by adjusting the modulation coefficient of the inverters.

Frequency control means the frequency of the micro-grid is restored to the rated value by adjusting the output of the microsources. Fig.4 (a) shows the frequency control curve. When moving the frequency curve to the left (or the right) in parallel, the frequency is kept to be a constant (such as 50Hz).

Analogously, the voltage control is implemented by adjusting the voltage curve to the left (or the right) in parallel, and the voltage, which is the output of the micro-grid, is kept to be a constant. Fig.4 (b) shows the voltage control curve.

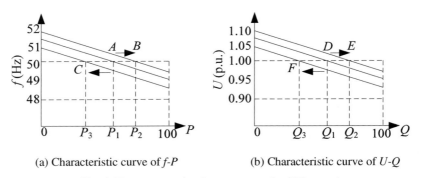

(a) Characteristic curve of f-P (b) Characteristic curve of U-Q

Fig. 4. Frequency and voltage curve under U/f control

4 Simulations of Micro-grid

The micro-grid is simulated within a platform by Matlab/Simulink, and the schematic diagram of Simulink based model is shown in Fig.5. The micro-grid model consists of DGs, controllers, critical and non-critical loads.

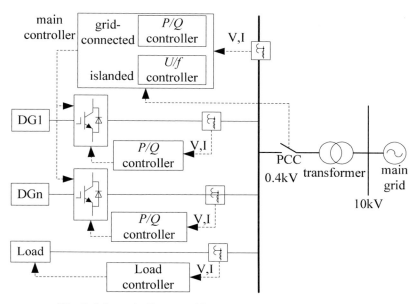

Fig. 5. Schematic diagram of Simulink based model of micro-grid

To examine the validity of the simulation platform, various operating conditions are considered, such as being interconnected with the main grid, being disconnected from the main grid and loads fluctuate in the micro-grid. The curves of the frequency and active power in the micro-grid can be obtained within the simulation.

In this model, the main grid supplies its own load of 100kW and the micro-grid supplies its critical loads of 20kW and 30kW, along with a non-critical load of 30kW. At the time of 0.4 sec, the non-critical load is increased to 50kW, and then 0.4 second later, the micro-grid is interconnected with the main grid. At the time of 1.2 sec, the micro-grid is disconnected from the main grid and the non-critical load restores to the primary value of 30kW.

The following figures demonstrate the simulation results. In the beginning, the micro-grid system experiences a starting period, but it arrives at a stable status in about 0.2 sec.

The main grid power is illustrated in Fig.6. It shows that the main grid supplies its own load up to t=0.8 sec and after that its power increases as it shares a portion of micro-grid load. Evidently, the power decreases to the initial values when the main grid is disconnected from the micro-grid.

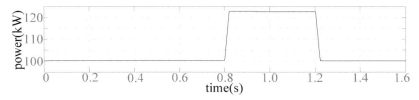

Fig. 6. The power of the main grid

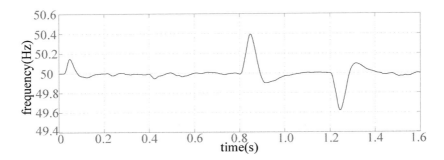

Fig. 7. The frequency of the micro-grid

Fig.7 shows the frequency curve of the micro-grid. During the grid-connected mode, the main grid and microsources provide power to the loads. At the time of 0.4 sec, the non-critical load is increased to 50kW. The micro-grid is interconnected with the main grid at the time of 0.8 sec, while it is disconnected from the main grid at t=1.2 sec. At the time of 0.4 sec, 0.8 sec and 1.2 sec, the frequency curve of the micro-grid fluctuates, but it restores to the nominal value of 50Hz in about 0.2 sec.

Fig.8 shows the active power consumed by the critical load of DG1, which is found to decrease during the grid-connected period as a portion of it is shared by the main grid. Similarly, the active power consumed by the critical load of DGn is represented in Fig.9. It also behaves in the similar way as the critical load of DG1. The active power consumed by the non-critical load is shown in Fig.10 which illustrates its increase at t=0.4 sec, and it restores to the initial value of 30kW at t=1.2 sec.

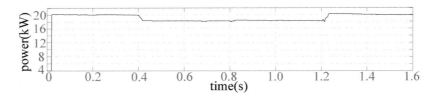

Fig. 8. Power consumed by the critical load of DG1

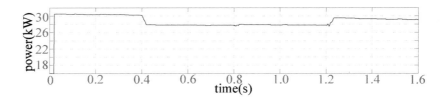

Fig. 9. Power consumed by the critical load of DGn

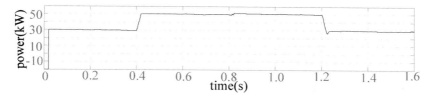

Fig. 10. Power consumed by the non-critical load

5 Conclusions

Control methods and a Simulink based model of the micro-grid are represented in the paper. During the simulation, several operational conditions are taken into account. The frequency and active power are observed when the non-critical load increases during the islanded mode, and they are also observed when the micro-grid is interconnected with the main grid and disconnected from the main grid.

The results of simulations show that, the frequency fluctuates with the undulation of the loads, and it also fluctuates at the time that the micro-grid is interconnected with the main grid and disconnected from the main grid, but it restores to the nominal value of 50Hz in a short time. The micro-grid simulation confirms that the control methods are effective in dominating the micro-grid during either the grid-connected or islanded mode.

In the future work, the microsources including diesel generator, wind turbine, fuel cells and photovoltaic generators will be studied and integrated into the micro-grid for further investigations. It is also significant to study some more advanced control schemes to further reduce the restoration time and enhance stability of the system in the following stage.

Acknowledgments. The Project is supported by Graduate Innovation Foundation of Shandong University, GIFSDU (yyx10117); in part by Shandong Provincial Natural Science Foundation, China (ZR2010EM033).

References

1. Pogaku, N., Prodanovic, M., Green, T.C.: Modeling, Analysis and Testing of Autonomous Operation of an Inverter-Based Microgrid. IEEE Transactions on Power Electronics 22(2), 613–625 (2007)
2. Lasseter, B.: Microgrids(distributed power generation). In: Power Engineering Society Winter Meeting, pp. 146–149. IEEE Press, New York (2001)
3. Salam, A.A., Mohamed, A., Hannan, M.A.: Technical Challenges on Microgrids. J. ARPN Journal of Engineering and Applied Sciences 3(6), 64–69 (2008)
4. Wei, H., Changhui, S., Ziping, W., Jianhua, Z.: A Review on Microgrid Technology Containing Distributed Generation System. J. Power System Technology 33(9), 14–18 (2009)
5. Hatziargyriou, N., Asano, H., Iravani, R., Marnay, C.: Microgrids. In: Power and Energy Magazine, pp. 78–83. IEEE Press, New York (2007)

6. Duan, Y., Gong, Y., Li, Q., Wang, H.: Modeling and Simulation of the Microsources Within a Microgrid. In: Electrical Machines and Systems (ICEMS), pp. 2667–2671. IEEE Press, New York (2008)
7. Wang, L., Li, P., Li, X., Liu, Z.: Modeling of Micro-powers and Its Application in Microgrid Simulation. Proceeding of the CSU-EPSA 22(3), 32–38 (2010)
8. Basak, P., Saha, A.K., Chowdhury, S., Chowdhury, S.P.: Microgrid: Control Techniques and Modeling. In: 44th International Proceeding on Universities Power Engineering Conference (UPEC), pp. 1–5. IEEE Press, New York (2009)
9. Kanellos, F.D., Tsouchnikas, A.I., Hatziargyriou, N.D.: Micro-Grid Simulation during Grid-Connected and Islanded Modes of Operation. In: International Conference on Power Systems Transients (IPST 2005), Montreal, pp. 113–119 (2005)
10. Engler, A.: Applicability of Droops in Low Voltage Grids. International Journal of Distributed Energy Sources 1(1), 3–15 (2004)

Harmonic Analysis of the Interconnection of Wind Farm

Jidong Wang[1], Xuhao Du[1], Guodong Li[2], and Guanqing Yang[1]

[1] Key Laboratory of Power System Simulation and Control of Ministry of Education
Tianjin University
300072 Tianjin, China
[2] Tianjin Electric Power Science & Research Institute,
300022 Tianjin, China
jidongwang@tju.edu.cn, {duxuhao97,tjlgd,ygq0530}@163.com

Abstract. With the large-scale wind farms accessing systems continuously and a wide range of variable speed wind turbines' application, the harmonic problems caused by the interconnection of wind farms have got more and more attention. This paper analyzes the harmonic mechanism caused by wind farms connected with the grid; based on Dig SILENT / Power Factory simulation platform, establishes a permanent magnet direct drive wind turbine model. Finally, the harmonic simulation of an actual example for a wind farm connected with the grid is given in this paper. Simulation results show that the second harmonic current of the wind farm accessing point and the third harmonic current of point of common coupling are exceeded, which should be suppressed; the contents of other harmonics are in the allowed range of Chinese National Standard.

Keywords: wind farm; interconnection; harmonic; simulation.

1 Introduction

Wind energy is a green renewable energy, which is safe, clean, and pollution-free. Development and application of wind energy has an important practical significance for solving the shortage of fossil energy sources and improving the ecological environment in China. Wind power generation as one of the main type of wind energy utilization is a reliable, simple, effective and practical technology. In 2006, the Global Wind Energy Council released "Wind Power Development Outlook 2050" in that, if active measures have been taken in 2030 and 2050, the world's total installed wind power capacity will reach 2.1 billion and 3.0 billion kW, generated electrical energy will reach 5 trillion kWh and 8 trillion kWh [1]. Wind power generation has the most mature technology, largest developing scale and brightly commercial prospects in the renewable energy power generation. In recent years, as China's "Renewable Energy Law" was promulgated and the supporting regulations were introduced, China's wind power generation has been got an unprecedented development.

Harmonics harm is mainly reflected in two aspects of electrical hazards and signal interference [2]. Harmonic analysis has always been the focus of power quality, with more and more wind farms interconnected with grid and increasing interconnection capacity, the harmonics problem will become increasingly prominent [3]. Variable speed wind turbine contains a number of power electronic control devices, which are the main harmonic source. If harmonic problems are so severe, the installed capacity of the interconnection wind farms will be restricted. Before the operation of the interconnection wind farms, the harmonic analysis is necessary; when the harmonic problems are serious, the viable harmonic suppression scheme shall be made in combination with the local background harmonic.

The main wind turbine types include fixed speed induction wind turbines, doubly-fed induction wind turbines, and permanent magnet direct-drive synchronous wind turbines. Permanent magnet direct-drive wind power generation system can improve the efficiency and reduce system maintenance costs, which has become one of the directions of wind power generation. This paper analyzes the harmonic mechanism caused by the interconnection wind farms; Based on Dig SILENT / Power Factory simulation platform, establishes a permanent magnet direct-drive wind turbine model. Lastly, the harmonic simulation of an actual example for one interconnection wind farm is given in this paper, which has some reference value for harmonic evaluation of the future.

2 Harmonic Mechanism Caused by Interconnection

In modern power system, harmonics is an important indicator to measure the power quality. Harmonics generated by wind farms mainly contains two aspects: one is itself of wind turbine; the other is the power electronic components in the wind power generation system.

Harmonics generated by the wind generator itself includes the inherent harmonic produced by generator structure and the harmonic generated by the excitation system. Air-gap space harmonic magnetic potential is the main performance of the inherent harmonic. Currently, wind turbines mainly use sinusoidal pulse width modulation (SPWM) inverter to provide AC excitation. SPWM inverter's output voltage is rich in harmonics; these harmonic components are amplified in the stator side by the motor air-gap magnetic field, which result in harmonic harm [4].

In the wind power generation system, there are many power electronic components, such as power electronic converters. These converters bears power transmission, load switching and other important tasks; but their operation will cause waveform distortion of grid current and voltage, which led to harmonic pollution. Any type of wind turbines, generator self-generated harmonics are very small, power electronic components in wind turbines are the main harmonic source [5]. When the unit operation, the converter will produce large amounts of harmonic current. Harmonic current's size and output power is essentially linear. In the normal state, the severity of harmonic interference depends on converter device structure, filtering device conditions, grid short-circuit capacity and other factors.

In addition, the wind turbine's frequent starts and stops also generate harmonic distortion current, thus affecting the power quality [6, 7]. There are some contents of low order harmonics in power system itself, which may interact with the wind turbines to generate new harmonic contents. If the power system is imbalanced, the power grid contains negative sequence components, which will also produce low order harmonics.

The step-up transformer of wind farm will also generate harmonics, due to core saturation, nonlinear magnetization curve, and the design consideration of economic and other reasons. The transformer magnetizing current waveform is spired, resulting in odd harmonics [8].

3 Wind Turbine Model

3.1 Permanent Magnet Direct-Drive Wind Turbine System Structure and the Mathematical Model

Permanent magnet direct-drive wind power system structure is shown in Fig.1, includes the following basic components: variable pitch wind turbine, synchronous generator, full-power converter, filter capacitor and control system.

Wind turbine changes the wind energy to mechanical energy in order to driving permanent magnet synchronous generator, while the permanent magnet synchronous generator's speed varies with wind speed, so the output power's voltage and frequency are varying. In order to obtain constant voltage and frequency, the output of the inverter must be transformed into sine wave output through AC-DC-AC converting. Full-power converter module including rectifier and inverter, wind turbine directly coupled with permanent magnet synchronous generator. The generator connected with the grid through full-power converter. Compared with induction wind turbine, directly-drive wind turbine's reliability and utilization are higher, and active and reactive power's control is more flexible [9].

Fig. 1. Permanent magnet direct-drive wind turbine system structure

Taken the generator rotor flux as reference coordinate system, the permanent magnet synchronous motor rotor flux is oriented in the d axis of the two-phase synchronous rotating d-q coordinate system; q axis is 90 degrees ahead of the d-axis. The mathematical relations of permanent magnet synchronous generator's voltage and electromagnetic torque in d-q coordinate system are as following [10]:

$$v_d = -(r + \frac{d}{dt}L_d)i_d + \omega_r L_q i_q \qquad (1)$$

$$v_q = -(r + \frac{d}{dt}L_q)i_q - \omega_r L_d i_d + \omega_r \psi_m \qquad (2)$$

$$T_e = \frac{3}{2} \times \frac{p}{2} \times [(L_d - L_q)i_q i_d - \psi_m i_q] \qquad (3)$$

In the equations, L_d, L_q are generator stator armature inductance d axis, q axis component; r is armature resistance of per phase winding of generator stator; i_d, i_d are generator stator output current d-axis, q-axis component; v_d, v_d are generator stator output voltage d-axis, q-axis component; ω_r is the angular velocity of generator rotor; ψ_m is generator q-axis winding leakage flux; p is generator pole pairs.

3.2 Dig SILENT / Power Factory Software Modeling

In this paper, taken the back to back double PWM converter as the direct-drive wind power generation system's full-power converter, system uses two PWM converter, the PWM rectifier + PWM inverter topology. Grid-side converter stables the DC bus voltage and controls the power factor of grid-side, machine-side converter control the generator's power and speed. Dual PWM system is flexible for control and can improve the operating characteristics of the system [11].

In Dig SILENT / Power Factory, the permanent magnet direct-drive generator and converter control structure are shown in Fig. 2 and Fig. 3.

Fig. 2. Permanent magnet direct-drive generator control structure

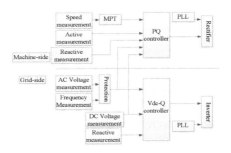

Fig. 3. Control structure of full-power converter

In Fig.2, the original motivation model consists of three modules: wind turbines, pitch angle control module and the shaft. In this paper, wind turbine uses the variable-pitch regulation strategies to achieve variable speed control [12]. In Fig.3, for the machine-side converter, the electrical values that the three modules of the top left required are measured directly, and they are the generator speed, generator active and reactive power. The measured generator speed is through the maximum power point tracking module (MPT) to get a standard reference value of generator's active power. For the grid-side converter, the electrical values that the four modules of the bottom right required are measured directly, and they are AC voltage and frequency of the point of interconnection, DC bus voltage and reactive power of the point of interconnection. Protection module controls the AC voltage and frequency of the point of interconnection within the allowable range, and participates in the control of PQ module and Vdc-Q module. After the control of PQ module and Vdc-Q module, machine-side converter's control variables and grid-side converter's control variables were obtained. The PLL module of the top right is connected to the output of synchronous motor and sends the measured phase information back to the machine-side PWM converter. The PLL module of the bottom right is connected to the AC bus of the point of interconnection and sends the measured phase information back to the grid-side PWM converter.

4 Harmonic Analysis of an Interconnection Example

This paper gives a harmonic analysis for a wind farm interconnected with the grid. The planning installed capacity of this wind farm is 49.5MW, which planned to install 33 sets of 1500kW wind turbines. Interconnection program takes single-circuit 110kV line into the nearby 110kV substation; the interconnection program of the wind farm is shown in Fig.4.

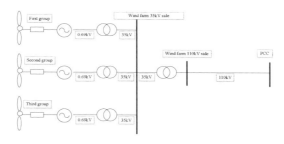

Fig. 4. The interconnection program of wind farm

The measured data of a 1.5 MW permanent magnet direct-drive wind turbine to be used in this wind farm is shown in Table 1.

Table 1. Harmonic current of a 1.5MW permanent magnet direct-drive wind turbine (690V side)

Harmonic order	Harmonic current(%)	Harmonic order	Harmonic current (%)
2	2.81	3	2.32
4	1.56	5	1.72
6	1.08	7	0.85
8	0.49	9	0.33
10	0.24	11	0.44
12	0.16	13	0.62

Note: The harmonic current is expressed as a percentage of the harmonic current and the fundamental current. The 1.5MW wind turbine's rated current is 1.255kA.

According to the wind farm's interconnection structure and the measured data of the related wind turbine's harmonic current, the simulation model is established. The harmonic currents and harmonic voltage of the wind farm 35kV side (access point) and the 110kV side (point of common coupling) are simulated.

4.1 Calculation of Harmonic Current

According to the interconnection structure and the measured data of the harmonic current, by Dig SILENT / Power Factory simulation software, we can get the harmonic currents of the wind farm 35kV side and the 110kV side (PCC), the results are shown in Table 2 and Table 3.

Table 2. Injection harmonic currents and allowed harmonic current of the wind farm 35kV side

Harmonic order	Injection harmonic current(A)	Allowed harmonic current(A)	Harmonic order	Injection harmonic current(A)	Allowed harmonic current(A)
2	19.29	16.43	3	10.36	13.15
4	5.49	8.44	5	4.75	13.15
6	2.41	5.59	7	1.68	9.64
8	0.85	4.16	9	0.49	4.49
10	0.33	3.40	11	0.55	6.14
12	0.18	2.85	13	0.66	5.15

Table 3. Injection harmonic currents and allowed harmonic current of the PCC

Harmonic order	Injection harmonic current(A)	Allowed harmonic current(A)	Harmonic order	Injection harmonic current(A)	Allowed harmonic current(A)
2	5.87	6.29	3	3.11	2.97
4	1.64	3.15	5	1.45	3.27
6	0.72	2.10	7	0.51	2.70
8	0.26	1.57	9	0.15	1.46
10	0.10	1.26	11	0.17	2.18
12	0.05	1.05	13	0.20	1.94

According to the injection harmonic current limits under corresponding voltage levels and the allowed harmonic current's conversion method in the GB / T 14549-1993 "Power quality Utility grid harmonic " [13], and combined with the minimum short-circuit capacity of the measured point, the allowed values of every order's harmonics current can be obtained.

By the simulation results in Table 2 and Table 3, we can get that the second harmonic current of the wind farm 35kV side and the third harmonic current of PCC are exceeded, which should be suppressed; the contents of other harmonics are in the allowed range of Chinese National Standard.

4.2 Calculation of Harmonic Voltage

Harmonic voltage content ratio limits under the corresponding voltage levels in the GB / T 14549-1993 "Power quality Utility grid harmonic " are shown in Table 4. Taking the harmonic power flow in the simulation model, we can obtain the various voltage harmonic content ratios in the point of wind farm 35kV side and PCC, as shown in Fig.5 and Fig.6.

By the simulation results in Fig.5 and Fig.6, we can get that the various voltage harmonic content ratios in the wind farm 35kV side are all in the allowed range of Chinese national standard, the total voltage distortion rate is 1.49% which is less than the national standard limit. In PCC, the maximum of various harmonic voltage content ratios doesn't exceed 0.2%, the total voltage distortion rate is 0.29% which is less than the national standard limit.

Table 4. Harmonic voltage limits of the utility grid

Nominal voltage(kV)	Total voltage distortion (%)	Harmonic voltage content ratio(%)	
		Even	Odd
0.38	5.0	4.0	2.0
6 10	4.0	3.2	1.6
35 66	3.0	2.4	1.2
110	2.0	1.6	0.8

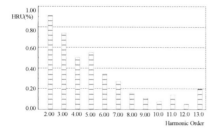 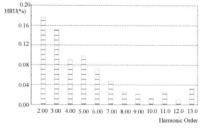

Fig. 5. Various voltage harmonic content ratios in the wind farm 35kV side

Fig. 6. Various voltage harmonic content ratios in the PCC

5 Conclusion

This paper analyzes the harmonic mechanism caused by the interconnection of the wind farms, establishes a permanent magnet direct-drive wind turbine model by Dig SILENT / Power Factory simulation platform, and applies it in the specific engineering instance, which has some reference value for the wind power interconnection harmonic evaluation of the future. Simulation results show that: Due to own structure reasons, permanent magnet direct-drive wind turbines will bring harmonics when they are interconnected with grid. The two or three low-order harmonic contents injected to the system are relatively large, which should be limited by installing the appropriate filter device. Moreover, with the increasing capacity of wind farms, the harmonic problem will be more severe, which should be given adequate attention in the actual process of building wind farms. When the harmonic contents are exceeded, a reasonable filtering scheme should be developed. If the third harmonic is too large, you can set the low-voltage side of step-up transformer of wind farm to Delta connection, which will avoid the third harmonic injected to grid.

Acknowledgments

The authors would like to thank National Natural Science Foundation of China (No.51007062; No.50837001) and Special Fund of the National Basic Research Program of China (No.2009CB219700).

References

1. Xiao, C.: Europe and the USA Wind Power Development Experience and Enlightenment to China, pp. 7–13. China Electric Power Press, Beijing (2010)(in Chinese)
2. Xiao, X., Han, M., Xu, Y., et al.: Analysis and Control of Power Quality, pp. 164–201. China Electric Power Press, Beijing (2004) (in Chinese)
3. Stavros, A.P., Michael, P.P.: Harmonic Analysis in a Power System with Wind Generation. IEEE Transactions On Power Delivery, 2006–2016 (2006)
4. Cao, Y., Jiang, X.: The Effect and Optimization of Wind farms Interconnection on Power Quality. Demand Side Management, 17 (2008) (in Chinese)
5. Gao, Y.: Power Quality Analysis with the Wind Farm Integrating in Power System. Southern Power System Technology 3(4), 68–72 (2009) (in Chinese)
6. Ding, M., Zhang, Y., Mao, M.: Key Technologies for Microgrids Being Researched. Power System Technology 33(11), 6–11 (2009) (in Chinese)
7. Zhao, H., Wu, T.: Micro Network Technology Based Distributed Generation. In: Proceedings of the CSU-EPSA, vol. 20(1), pp. 121–128 (2008) (in Chinese)
8. Guo, X., Wei, P.: Power Quality of Wind Farm. Journal of Northwest Hydroelectric Power 23(1), 23–25 (2007) (in Chinese)
9. Wu, D., Zhang, J.: Variable Speed Permanent Magnet Direct-drive Wind Turbine Control System. Auxiliary and other 6, 51–55 (2006) (in Chinese)
10. Hou, Q.: Middle-Small Permanent Magnet Direct-drive Wind Power Control System Design and Simulation. In: Master's thesis. pp. 9-15. Shenyang University of Technology Press, Shenyang (2006) (in Chinese)
11. Li, J., Xu, H.: Power Electronic Converter Technology in Wind Power, pp. 9–15. Mechanical Engineering Press, Beijing (2008) (in Chinese)
12. Chen, M., Yang, G.: Wind Power Technology and its Development Trend. Journal of Electric Power 23(4), 272–275 (2008) (in Chinese)
13. GB/T 14549-1993, Power Quality Utility Grid Harmonic (in Chinese)

Simulation of Voltage Sag Detection Method Based on DQ Transformation

Jidong Wang, Kun Liu, and Guanqing Yang

Key Laboratory of Power System Simulation and Control of Ministry of Education
Tianjin University
300072 Tianjin, China
jidongwang@tju.edu.cn, liukunjordan@yahoo.cn, ygq0530@163.com

Abstract. The rapid detection of characteristics of voltage sag is an important prerequisite for using dynamic voltage restorer (DVR) for voltage compensation. This paper introduced an improved derivation detection method based on *dq* transformation. Simulation was performed to analyze the advantages and disadvantages of both methods and compare them. Results showed that the improved detection method had shorter delay and better waveform of detection and met the requirements of accuracy and timeliness better.

Keywords: voltage sag; detection method; *dq* transformation; derivation

1 Introduction

Voltage sag is considered as the most serious problem of dynamic power quality which affects normal operation of electrical equipments. At present, the most economical and efficient equipment to control voltage sag problem is the dynamic voltage restorer (DVR) [1] which can produce the required compensation voltage by controlling circuit. DVR requires very fast response, while the time of rectifier and inverter side is usually longer. So the speed and accuracy of signal detection and control has high requirements. The actual compensation time of DVR is used to be 2ms [2]. Considering the detection, control and switching time, we must obtain the response time of millisecond level strictly. Therefore, the rapid and accurate detection of characteristics of voltage sag is an important prerequisite for achieving compensation.

There are some traditional voltage sag detection methods, such as RMS detection, missing voltage technique, peak value detection and fundamental component method and so on. Although these methods can meet the basic requirements, they can not ensure the accuracy and timeliness of detection. Now the most basic and common method is *dq* transformation on which domestic and foreign scholars proposed many detection methods based. References [3] and [4] constructed β-axis component by making single-phase instantaneous voltage as α-axis component in the stationary coordinate system and achieved detection by *dq* transformation; Reference [5-7] made double *dq* transformation to voltage signal and obtained positive, negative and zero sequence fundamental components by low-pass filtering of morphological filters, and got the characteristics of voltage sag; Reference [8] combined *abc-dq* transformation and 90°phase shift to detect the type of voltage sag effectively and reduce the delay.

In this paper, *dq* transformation and improved derivation detection method based on it are introduced and compared. And their advantages and disadvantages are analyzed by simulation. Results showed that improved derivation detection method can solve some shortcomings of existing methods and have better timeliness and accuracy.

2 The Basic Principle of DQ Transformation

The most common detection method in DVR is *dq* transformation method based on instantaneous reactive power theory. Its basic principle is to make Park transformation to three-phase vector in *abc* coordinates and convert it to *dq0* coordinates. DC component in *dq0* coordinate system is extracted by low-pass filter and the distortion component is separated from detection voltage. Ultimately, the characteristics of voltage sag can be obtained by inverse Park transformation.

From the instantaneous reactive power theory we know that the relation of three-phase voltage and *dq* coordinate is:

$$\begin{bmatrix} u_d \\ u_q \end{bmatrix} = C \begin{bmatrix} u_a \\ u_b \\ u_c \end{bmatrix} \quad (1)$$

Where

$$C = \sqrt{\frac{2}{3}} \begin{bmatrix} \sin \omega t & \sin(\omega t - 2\pi/3) & \sin(\omega t + 2\pi/3) \\ -\cos \omega t & -\cos(\omega t - 2\pi/3) & -\cos(\omega t + 2\pi/3) \end{bmatrix} \quad (2)$$

For an ideal three-phase three-wire system, assuming three-phase voltage is:

$$\begin{cases} u_a = \sqrt{2}U \sin \omega t \\ u_b = \sqrt{2}U \sin(\omega t - 2\pi/3) \\ u_c = \sqrt{2}U \sin(\omega t + 2\pi/3) \end{cases} \quad (3)$$

The results of transformation are:

$$u_d = \sqrt{3}U \quad (4)$$

$$u_q = 0 \quad (5)$$

From (4) and (5) we can know, d-axis component reflects the RMS voltage. Then instantaneous amplitude of the voltage sag can be determined. Obviously, coordinate transformation above is only used in three-phase three-wire circuit instead of asymmetric circuit. However, most of the actual voltage sag events is single-phase fault. A virtual three-phase voltage system needs to be constructed to make analysis of voltage sag through the mentioned coordinate transformation. Take A-phase as the example, considering that the waveform of two adjacent phases are same and the difference of phase angle between two is 120 degree, so we can get the other two-phase voltage through delaying u_a and extract DC component of d-axis and q-axis voltage component obtained by making transformation to constructed three-phase voltage according to (1), as follows:

$$U_{d\alpha} = \sqrt{3} U_{sag} \cos \alpha \tag{6}$$

$$U_{q\alpha} = -\sqrt{3} U_{sag} \sin \alpha \tag{7}$$

DC component $U_{d\alpha}$ and $U_{q\alpha}$ can be extracted by filtering, so we can obtain the amplitude of voltage sag

$$U_{sag} = \frac{\sqrt{3}}{3} \sqrt{U_{d\alpha}^2 + U_{q\alpha}^2} \tag{8}$$

Phase angle jump is

$$\alpha = \arcsin\left(-\frac{\sqrt{3} U_{q\alpha}}{3 U_{sag}}\right) = \arcsin\left(-\frac{U_{q\alpha}}{\sqrt{U_{d\alpha}^2 + U_{q\alpha}^2}}\right) \tag{9}$$

The *dq* transformation can achieve accurate detection to symmetrical fault. For the common asymmetric fault, new three-phase voltage must be constructed through delaying 60 °. Considering that power frequency cycle is 0.02s, delay of 60 ° can not be ignored. If voltage drop can not be detected in the initial stage of sag, this detection method can not achieve requirements of timeliness and accuracy to some sensitive electrical equipment in power system [9].

3 The Improved Derivation Detection Method

Because of time delay in constructing voltages, this paper focuses on a derivation detection algorithm which constructs three-phase voltage u_a, u_b, u_c by derivative calculation. It solves the problem of data inconsistency of the conventional methods in that delay does not occur in the process of construction. In this paper, *dq* transformation is improved.

In the improved detection method, through derivation [10, 11] of single-phase voltage, the cosine signal of this phase voltage can be obtained. According to that the waveform of two adjacent phases are same and the difference of phase angle between two is 120 degree, other two virtual phase voltage can be obtained by trigonometric formulas.

Assuming that symmetrical three-phase voltage is:

$$\begin{bmatrix} u_a \\ u_b \\ u_c \end{bmatrix} = \begin{bmatrix} U \sin \omega t \\ U \sin(\omega t - 2\pi/3) \\ U \sin(\omega t + 2\pi/3) \end{bmatrix} \tag{12}$$

Making derivation to A-phase voltage: $u_a' = U \omega \cos \omega t \tag{13}$

Expanding u_b by trigonometric functions:

$$u_b = -\frac{1}{2}U\sin\omega t - \frac{\sqrt{3}}{2}U\cos\omega t = -\frac{1}{2}u_a - \frac{\sqrt{3}}{2\omega}u_a' \qquad (14)$$

$$\text{Similarly } u_c = -\frac{1}{2}u_a + \frac{\sqrt{3}}{2\omega}u_a' \qquad (15)$$

In the new three-phase voltage system, both u_b and u_c are associated with u_a. Therefore, DC component $U_{d\alpha}$ and $U_{q\alpha}$ can be obtained by *abc-dq* coordinate transformation and low-pass filtering. Voltage sag magnitude and phase angle can be obtained through (8) and (9) (The same process is not repeated).

4 Results of Simulation

In the actual power system, since most of voltage sag is caused by single phase fault [12], this paper uses module of power system in MATLAB Simulink to create a system model and make simulation of situation that single phase fault occurs in the power system. Input is three-phase power frequency AC power supply. Ground fault occurs in the A-phase line from 0.3s to 0.6s. The waveform of A-phase fault is shown as Fig.1:

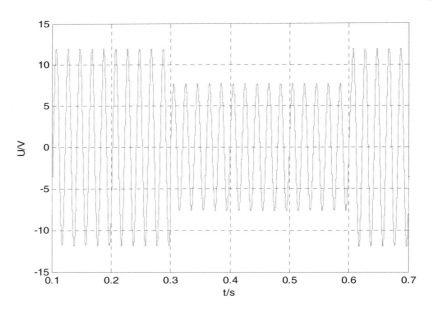

Fig. 1. The waveform of A-phase fault

For detection of voltage sag, 2-order Butterworth filter of 60Hz cutoff frequency is adopted in the simulation. Sampling frequency of system is set to 20KHZ. Time delay is simulated through module "Transport Delay", and detection results of amplitude and phase are shown as Fig.2 and Fig.3:

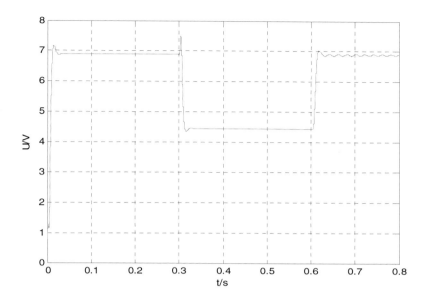

Fig. 2. The waveform of amplitude of *dq* transformation

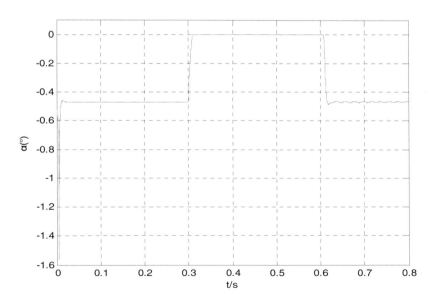

Fig. 3. The waveform of phase angle of *dq* transformation

Improved *dq* transformation method uses module "*du/dt*" to simulate derivation. Detection results of amplitude and phase through simulation with the same sampling frequency of system and cutoff frequency of filter are shown as Fig.4 and Fig.5:

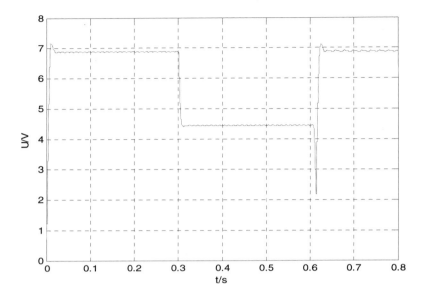

Fig. 4. The waveform of amplitude of improved method

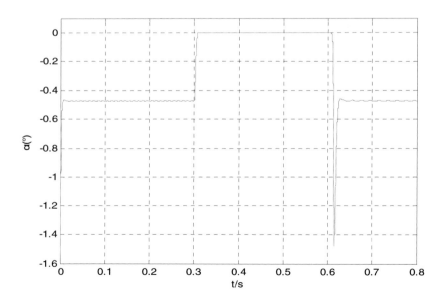

Fig. 5. The waveform of phase angle of improved method

Results of simulation showed that, both *dq* transformation and improved derivation method could detect the voltage sag and have the same characteristics of voltage sag. However, detection time of *dq* transformation was longer. Although improved detection method also had time delay, it could achieve detection more accurately within the scope of requirements. If we consider 90% of normal voltage amplitude as starting threshold of DVR devices, improved detection method had better timeliness result. Moreover, in the initial stage of voltage sag, waveform of *dq* transformation method appeared to have short-term disturbance which may cause malfunction of device, while the waveform of improved method appeared to be smooth.

Comparing the simulation results of two methods further, waveform of improved method had small fluctuation. This is because there is harmonics of different order in the system and derivation of harmonics was ignored in the calculation. Due to discontinuity of voltage in the beginning and ending time, overshoot phenomenon appeared in the waveform. Discontinuous derivation may occur to be mutation, but it does not have a significant impact on the results. In addition, low-pass filter may also bring time delay to the simulation [13]. All these are problems to be solved, but on the whole, results showed that improved detection method met the requirements of accuracy and timeliness for voltage sag compensation better.

5 Conclusion

This paper introduced two detection methods of voltage sag —*dq* transformation and improved derivation method and compared two methods through simulation. Results showed that both two detection methods can achieve detection of voltage sag, but improved derivation method had shorter delay and met the requirements of accuracy and timeliness better. However, simulation ignored the impact of harmonic problems and delay of filter which would have an impact on the results, which need further study.

Acknowledgments

The authors would like to thank National Natural Science Foundation of China (No. 51007062; No. 50837001) and Special Fund of the National Basic Research Program of China (No. 2009CB219700).

References

1. Xiao, X.: Analysis and Control of Power Quality. China Electric Power Press (2004)
2. Zhou, W., Mao, Z., Rao, Q., et al.: Research on Detection Methods of Short Duration Voltage Change. Hebei Electric Power 27(3) (2008) (in Chinese)
3. Liu, H., Xu, L., Luo, Y.: New Method to Detect Voltage Sag. Electronic Measurement Technology 31(10), 46–49 (2008) (in Chinese)
4. Fitzer, C., Barnes, M., Green, P.: Voltage Sag Detection Technique for A Dynamic Voltage Restorer. IEEE Trans on Industry Applications 40(1), 203–212 (2004)

5. Du, X., Zhou, E., Xu, K.: Classification of Short Circuit Faults Causing Voltage Sags Based on Double dq Transformation. Automation of Electric Power Systems 34(5), 86–90 (2010) (in Chinese)
6. Zhe, C., Mu, W.: A Voltage Quality Detection Method, DRPT (2008)
7. Liu, Y., Huang, C., Ou, L., et al.: Method for Unbalanced Voltage Sags Detection Based on dq Transform. Proceedings of the Chinese Society of Universities for Electric Power System and Automation 19(3), 72–76 (2007) (in Chinese)
8. Edris, P., Mudathir, F., Mojgan, H.: A Hybrid Algorithm for Fast Detection and Classification of Voltage Disturbances in Electric Power Systems. Euro. Trans. Electr. Power (2010)
9. Zhou, H., Qi, Z.: A Survey on Detection Algorithm and Restoring Strategy of Dynamic Voltage Restorer. Power System Technology 30(6), 23–29 (2006) (in Chinese)
10. Yang, S., Du, B.: A Novel Integrated Derivation Detection Algorithm Based on dq Transformation for the Dynamic Voltage Restorer. Automation of Electric Power Systems 32(2), 40–44 (2008) (in Chinese)
11. Hou, S., Liu, Z., Ji, L., et al.: A Derivation Algorithm to Detect Characteristic Quantity of Single-Phase Voltage Sag. Power System Technology 33(14), 52–56 (2009) (in Chinese)
12. Choi, S.S., Li, J.D., Vilathgamuwa, D.M.: A Generalized Voltage Compensation Strategy for Mitigation the Impact of Voltage Sags/Swells. IEEE Trans on Power Delivery 20(2), 2289–2297 (2005)
13. Feng, X., Yang, R.: A Novel Integrated Morphology-DQ Transformation Detection Algorithm for Dynamic Voltage Restorer. Proceedings of the CSEE 24(11), 193–198 (2004) (in Chinese)

Simulation of Shunt Active Power Filter Using Modified SVPWM Control Method

Jidong Wang, Guanqing Yang, and Kun Liu

Key Laboratory of Power System Simulation and Control of Ministry of Education
Tianjin University
300072 Tianjin, China
jidongwang@tju.edu.cn, ygq0530@163.com, liukunjordan@yahoo.cn

Abstract. Space vector pulse width modulation (SVPWM) was widely used in inverter control for the advantages of fixed switching frequency, full utilization of DC-link voltage and better control precision. In recent years, SVPWM is gradually applied in the control method of active power filter (APF) as the APF mains is just one invert. The voltage vectors are composed based on the α-β axis in traditional SVPWM however, in this paper one x-y composed reference axis at 60 degree angles from each other is used instead. Compared with traditional method, the principle of the method is relatively simple and easy to implement. At the same time PI regulator is adopted for the DC-link voltage control. Finally simulation results show that the APF has a better compensation effect using this control method.

Keywords: active power filter (APF), space vector PWM, PI control, simulation.

1 Introduction

Since more and more nonlinear loads especially power electronics equipments were accessed, serious harmonic problems occurred in power system. There are lots of drawbacks in traditional passive filter (PF) such as only suppressing specific harmonic, compensating fixed reactive power but no precise compensation for reactive power fluctuations. However active power filter (APF) arouses more attention which is dominant other than PF in some characteristics [1, 2]. APF is mainly composed by inverter mains, harmonic detection, controlled method and drive-pulse generator. And control method is an important part to determine compensation performance. Among those methods hysteresis control and triangle wave control are preferred methods however, the former is high precision, fast response but variable switching frequency and the latter is complete opposite [3, 4].

Space vector pulse width modulation (SVPWM) is a novel control strategy based on space voltage vector which is widely used in converter control for lots of advantages such as full utilization of DC-link voltage and better control precision [5]. Also SVPWM is gradually applied in APF control method as in fact APF main is just an inverter [6, 7]. Traditional space voltage vector [8] is composed based on α-β coordinates but the procedure division is very complicated. In [9] a novel space voltage vector composition method based on *x-y* axis at 60 degree angles from each other is proposed where the principle is relatively simple and easy to implement. In this paper the novel vector composition method proposed in [9] is applied in APF SVPWM control method. On the other hand the stability of DC-link voltage is a great impact on the compensation effect so traditional PI regulator is adopted to maintain voltage unchanged. Finally simulation results show that the APF has a better compensation effect using the proposed method.

2 Principle and Implementation of SVPWM Method Based on *x-y* Reference Axis

2.1 Principle of SVPWM Method

Before introducing the principle of SVPWM first transformation from *abc* axis to *x-y* axis at 60 degree angles from each other is discussed where the transform is referred to (1) as shown in Fig. 1.

$$\begin{bmatrix} U_x \\ U_y \end{bmatrix} = C \begin{bmatrix} U_a \\ U_b \\ U_c \end{bmatrix} = \begin{bmatrix} 1 & -1 & 0 \\ 0 & 1 & -1 \end{bmatrix} \begin{bmatrix} U_a \\ U_b \\ U_c \end{bmatrix} \quad (1)$$

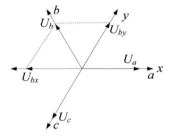

Fig. 1. Transform from *abc* to *xy* reference axis

The principle of shunt APF mains in 3-phase 3-wire system is shown in Fig. 2 and $V_1 \sim V_6$ are 6 full controlled switches where switches in each bridge work in a complementary state. So the switch function can be defined as follows.

$$S_i = \begin{cases} 1, & \text{Phase } i \text{ upper bridge close, lower bridge open} \\ 0, & \text{Phase } i \text{ upper bridge open, lower bridge close} \end{cases} \quad (i=a, b, c) \quad (2)$$

And three 3-phase voltages of inverter [8] can be represented as follows.

$$\begin{bmatrix} U_{ca} \\ U_{cb} \\ U_{cc} \end{bmatrix} = \frac{1}{3} U_{dc} \begin{bmatrix} 2 & -1 & -1 \\ -1 & 2 & -1 \\ -1 & -1 & 2 \end{bmatrix} \begin{bmatrix} S_a \\ S_b \\ S_c \end{bmatrix} \quad (3)$$

Through transforming (3) to x-y reference axis we have

$$\begin{bmatrix} U_{cx} \\ U_{cy} \end{bmatrix} = C \begin{bmatrix} U_{ca} \\ U_{cb} \\ U_{cc} \end{bmatrix} = \begin{bmatrix} 1 & -1 & 0 \\ 0 & 1 & -1 \end{bmatrix} \begin{bmatrix} S_a \\ S_b \\ S_c \end{bmatrix} U_{dc} \tag{4}$$

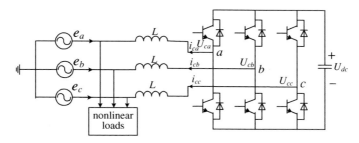

Fig. 2. Topology of 3-phase 3-wire APF

According to different combinations of switching function value components x and y of the voltage vector of the APF inverter is obtained as shown in Table 1. From Table 1 there are all 8 switch combinations together and all represent 8 basic voltage vectors as depicted by Fig. 3.

Table 1. Components of voltage vector under each switch combination in x-y axis

S_{abc}	000	100	110	010	011	001	101	111
U_{cx}	0	U_{dc}	0	$-U_{dc}$	$-U_{dc}$	0	U_{dc}	0
U_{cy}	0	0	U_{dc}	U_{dc}	0	$-U_{dc}$	$-U_{dc}$	0

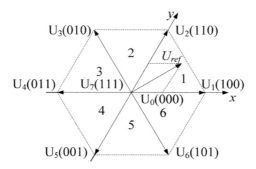

Fig. 3. Distribution of space voltage vectors in x-y coordinates

2.2 Implementation of SVPWM Method in Inverter

There is an important theory in sampling control that there is the same effect when a narrow pulse with equal impulse and different shapes is added to inertia segments.

SVPWM control method is realized just based on this theory. First set the reference voltage vector as U_{ref} and step-by-step analysis will be discussed as follows [8].
(1) Sextants criterion of reference voltage vector
   First resolve reference voltage vector U_{ref} into components of U_x and U_y under x-y reference axis, and then we have

$$U_{ref} = U_x x + U_y y \tag{5}$$

where x, y is the x-y corresponding direction unit measures.
   From Fig. 3 the criterion in each sector can be concluded in Table 2.

Table 2. Criterion in each sextant of reference voltage vector

Sextants	1	2	3	4	5	6
Con1	$U_x>0$	$U_x<0$	$U_x<0$	$U_x<0$	$U_x>0$	$U_x>0$
Con2	$U_y>0$	$U_y>0$	$U_y>0$	$U_y<0$	$U_y<0$	$U_y<0$
Con3	--	$|U_x|<|U_y|$	$|U_x|>|U_y|$	--	$|U_x|<|U_y|$	$|U_x|>|U_y|$

(2) The pulse duration of reference voltage vector in each of the adjacent states
   Assuming the vector in sextant 1 once the component vector U_x and U_y have been obtained, the pulse duration in each of the adjacent states T_1 and T_2 may be determined by (6) and (7), where T_0 in the corresponding adjacent null state will be the remaining up to the switching time T.

$$\begin{cases} U_1 T_1 + U_2 T_2 = U_{ref} T \\ T = T_1 + T_2 + T_0 \end{cases} \tag{6}$$

$$\begin{cases} T_1 = T U_x / U_{dc} \\ T_2 = T U_y / U_{dc} \end{cases} \tag{7}$$

Similarly the pulse duration in other sextants will be obtained. Set the time variable XYZ determined by (8) and Table 3 gives a summary of T_i and T_{i+1} values in different sextants. We set $T_i = T_i * T/(T_i+T_{i+1})$, $T_{i+1} = T_{i+1} * T/(T_i+T_{i+1})$ when $T<T_i+T_{i+1}$.

$$\begin{cases} X = T U_x / U_{dc} \\ Y = T U_y / U_{dc} \\ Z = T(U_x + U_y) / U_{dc} \end{cases} \tag{8}$$

Table 3. T_i, T_{i+1} values in different sextants

Sextants	1	2	3	4	5	6
T_i	X	Z	Y	-X	-Z	-Y
T_{i+1}	Y	-X	-Z	-Y	X	Z

Taking the origin at the beginning of commutation period, the instants of commutation T_{aon}, T_{bon} and T_{con} in each switching period will be given by (9), and a summary in different sextants is shown in Table 4.

$$\begin{cases} T_{aon} = (T - T_i - T_{i+1})/4 \\ T_{bon} = T_{aon} + T_i/2 \\ T_{con} = T_{bon} + T_{i+1}/2 \end{cases} \quad (9)$$

Table 4. T_{aon}, T_{bon} and T_{con} values in different sextants

Sextants	1	2	3	4	5	6
Phase A	T_{aon}	T_{bon}	T_{con}	T_{con}	T_{bon}	T_{aon}
Phase B	T_{bon}	T_{aon}	T_{aon}	T_{bon}	T_{con}	T_{con}
Phase C	T_{con}	T_{con}	T_{bon}	T_{aon}	T_{aon}	T_{bon}

3 SVPWM Control Method in Shunt APF and PI Regulator in DC-Link Voltage

As shown in Fig. 2 ignoring circuit resistance, differential equations of APF mains can be described as follows[1]

$$\begin{cases} U_{ca} = L\dfrac{di_{ca}}{dt} + e_a \\ U_{cb} = L\dfrac{di_{cb}}{dt} + e_b \\ U_{cc} = L\dfrac{di_{cc}}{dt} + e_c \end{cases} \quad (10)$$

Transform (10) into x-y reference axis and set $\vec{U}_c = U_{cx}\vec{x} + U_{cy}\vec{y}$, $\vec{i}_c = i_{cx}\vec{x} + i_{cy}\vec{y}$, $\vec{e}_s = e_x\vec{x} + e_y\vec{y}$ then equation (10) can be turned into

$$\vec{U}_c(t) = L\frac{d\vec{i}_c(t)}{dt} + \vec{e}_s(t) \quad (11)$$

Under normal circumstances the reference signal obtained by harmonic detection is just current signal, however the reference vector in SVPWM is obvious voltage signal. So in this control strategy one important step is that turning the reference current vectors i_{abc}^* into voltage vectors U_{abc}^*.

For equation (11) disperse the differential part then we get [8]

$$\vec{U}_c(k) = \vec{e}_s(k) + L\frac{\vec{i}_c(k+1) - \vec{i}_c(k)}{\Delta T} \quad (12)$$

where ΔT is sampling period, $\vec{i}_c(k+1)$ and $\vec{i}_c(k)$ is actual compensation current from APF mains in the next and present sampling period respectively. In order to make a close loop for better compensation just replace $\vec{i}_c(k+1)$ by reference current vector $\vec{i}_{ref}(k)$ then (12) will be turned into

$$\vec{U}_c(k) = \vec{e}_s(k) + L\frac{\vec{i}_{ref}(k) - \vec{i}_c(k)}{\Delta T} \quad (13)$$

In this way through differential operation reference voltage vector is directly obtained as shown in Fig. 4. Then trigger pulse signal obtained through SVPWM modulation will control switches open or close and so the output current from APF would track the reference current signal.

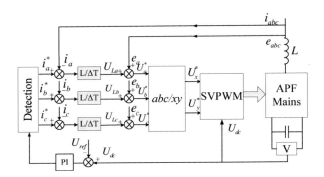

Fig. 4. Flow chart of APF system

In order to achieve required compensation effect, to maintain DC-link voltage as a constant is completely necessary. Also for SVPWM modulation in every switching period the DC-link voltage is involved in calculations, to keep the voltage stable is apparently important. In the lower part of Fig. 4 is the diagram of DC-link voltage control circuit. In order to maintain the DC bus voltage stable the detected DC bus voltage is compared with a reference voltage and the result is fed to the reference current through a limited PI controller. This negative feedback system would realize the control goals.

How to adjust the parameters of PI regulator for DC-link voltage quickly responding to steady-state values in several periods is a complicated task. Some experiences from simulation procedure can be concluded as follows: First try proportion factor to make voltage rise speed meet the qualification but the overshoot is in certain scope; then regulate the integral factor and simultaneously give dual attention to the proportion factor to reduce the error until satisfying the request gradually.

4 Simulation Results

In order to verify the correctness and feasibility of the control method proposed in this paper a simulate model for 3-phase 3-wire shunt APF is presented with MATLAB. Artificial systematic parameters: Supply line voltage 380V; RL load R=10Ω L=0.01H; Frequency 50Hz; DC-link voltage 800V; AC side line inductance 3mH; DC capacitor 1800μF.

According to the described principles parameters of PI regulator can be obtained as K_P=0.28, K_I=6.9. Simulation results show that through PI regulator adjusting the DC-link voltage will be stable in 4 line-frequency periods as shown in Fig. 5.

The simulation model is a 3-phase balanced system. Fig. 6(a) is the current waveform of phase A before compensating and spectrum by FFT depicts that the total distortion factor THD is over 24.04% as shown in Fig. 7(a) which exceeds the national standards greatly. Fig. 6(b) is the current waveform after compensating and FFT results represent that THD is decreased to 6.15% as shown in Fig. 7(b) which meets the national standards for power quality of public harmonics.

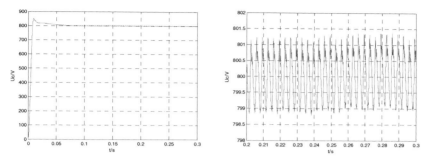

(a) The whole diagram of DC-link voltage   (b) Partial enlargement of DC-link voltage

Fig. 5. DC-link voltage response with APF

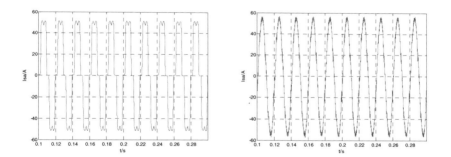

(a) Phase A supply current without APF   (b) Phase A supply current with APF

Fig. 6. Supply current waveform without and with APF

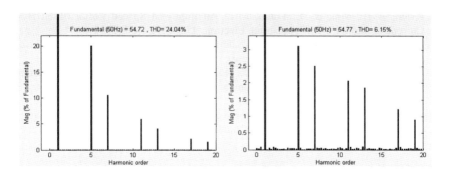

(a) Spectrum without APF   (b) Spectrum with APF

Fig. 7. Spectrum of supply current without and with APF

5 Conclusions

In this paper SVPWM control strategy for APF based on *x-y* reference axis is proposed and theoretical analysis shows that compared with the traditional SVPWM strategy the proposed method is simple and easy to implement. For DC-link voltage control the PI regulator is adopted. The results of simulation show that the APF offers a good quality of filtering of harmonic currents with the correction of the power factor by using the proposed SVPWM method, where the THD of the load current has fell from 24.04% to 6.15%. And the DC-link voltage has also been maintained stable in 4 line-frequency periods.

Acknowledgments

The authors would like to thank National Natural Science Foundation of China (No.51007062; No.50837001) and Special Fund of the National Basic Research Program of China (No.2009CB219700).

References

1. Wang, Z., Yang, j., Liu, J., et al.: Harmonic suppression and reactive power compensation. China Machine Press, Beijing (2006) (in Chinese)
2. Chen, G., Lv, Z., Qian, Z.: The general principle of active filter and its application. Proceeding of the CSEE 20(9), 17–21 (2000) (in Chinese)
3. Wang, W., Zhou, L., Xu, M.: Control method of the active power filter. Power system protection and control 34(20), 81–86 (2006) (in Chinese)
4. Buso, S., Malesai, L., Mattavelli, P.: Comparison of current control techniques for active filter applications. IEEE Trans. on Industrial Electronics 45(5), 722–729 (1998)
5. Xiong, J., Kang, Y., Zhang, K., Chen, J.: Comparison study of voltage space vector PWM and conventional SPWM. Power Electronics (1), 25–28 (1999) (in Chinese)
6. Jiang, J., Liu, H., Chen, Y., Sun, J.: A novel double hysteresis current control method for active power filter with voltage space vector. Proceedings of the CSEE 24(10), 82–86 (2004) (in Chinese)
7. Kwon, B.-H., Min, B.-D., Youm, J.-H.: An Improved Spaced-Vector-Based Hysteresis Current Controller. IEEE Trans. Ind. Electron. 45(5), 752–760 (1998)
8. Qu, X., Qu, J., Yu, M.: Power electronic filter technology and application. Electronic Industry Press, Beijing (2008) (in Chinese)
9. Lamich, M., Balcells, J., Gonzalez, D.: New Method for Obtaining SV-PWM Patterns Following an Arbitrary Reference. In: Annual Conference of the Industrial Electronics Society IECON 2002, vol. 1, pp. 18–22 (2002)
10. Zhang, D., Lv, Z., Chen, G.: Capacitor voltage control of shunt active power filter. Power electronics 41(10), 77–79 (2007) (in Chinese)

Author Index

Badie, Kambiz   289
Baokun, Li   461
Baoyuan, Wu   403
Bo, Jin   671
Bu, Leping   903

Cai, Di   281
Cai, Xu   601
Cao, Baoxiang   817
Chai, Yanli   687, 695
Chan-Juan, Shang   657
Chauhan, Durg Singh   735
Chen, Guodong   601
Chen, Hongyun   77
Chen, Huipeng   719
Chen, Jun-Jie   355
Chen, Jun-jie   363
Chen, Leigang   273
Chen, Shujian   719
Chen, Tianhua   953
Cheng, Shao   235
Cheng, Shuying   633
Cheng, Siyang   371
Chiang, Chiang Ming   203
Chun-Ling, He   663
Cui, Leilei   871

Dai, Wenjin   803, 811
Depeng, Song   395
Ding, Jia-Hui   443, 453
Ding, Shoucheng   477
Dong, Aihua   711
Dong-Ming, Li   63
Dong, Yangmei   711

Du, Xuhao   1031
Duan, Xinyu   547

Faieghi, Mohammadreza   29
Fan, Li   937
Fan, Liping   503
Fang, Xinbing   839
Fei, Tang   743
Florino, José Augusto Coeve   153

Gao, Guowei   547
Gao, Lincai   993
Gao, Minguang   371
Gao, MingXing   771
Gao, Tie-Gang   937
Gasmelseed, Akram   109
Genwang, Mao   403
Gou, Lin-Feng   315, 493
Gui-Fen, Chen   63
Guo, Chuangxin   165
Guo, Maopai   897
Guo, XiaoRong   85
Guo, Yuanshan   485
Guo-feng, Jin   411
Guo-wei, Wang   411

Hang, Chen   63
He, Nian   1015
He, Runxin   1015
He, You   443, 453
Hong, Jia-Qi   993
Hong, Man   235
Hong-Bin, Li   9
Hong-Ling, Wang   55
Honglin, He   145

Author Index

Hou, Yunhai 23
Hu, Baoan 825
Hu, Huiyong 349
Hu, Ran 755, 763
Hu, Wenjing 831
Huang, Lei 177
Huang, Yanqiong 791, 797
Huang, Yushui 185
Huangwei 847
Hui-Ru, Chen 657, 663

Ji, Aiming 299, 307
Ji, Hong-Bing 215
Jia, Jitao 435
Jiang, Liao 749
Jiang, Xiangyuan 969
Jiang, Yong 343
Jin, Ling 371
Junwei-Ma 541

Khademzadeh, Ahmad 289
Kodama, Toshio 135
Kunii, Tosiyasu L. 135

Lan, Bing 509
Lei, Shuai 343, 349
Lei, Zhang 39
Li, Ershuai 23
Li, Xiao-Dong 993
Li, An 85
Li, Bin 195
Li, FuFang 1
Li, Guangzheng 625
Li, Guochang 379, 387
Li, Guodong 1031
Li, Hongru 649
Li, Hua 469
Li, Jinfeng 257
Li, Jingxian 953
Li, Jufang 177
Li, Ke-Jun 1023
Li, Liang 711
Li, Sheng 371
Li, Ting 251
Li, Xiao bin 871
Li, Yibin 195
Li, Yi-Jhen 571
Li, Zhen 777
Li, Zhihui 831
Li-Juan, Zhang 63

Li-Wan, Chen 9
Liang, Lingmin 811
Lin, Hong 525, 533
Lin, Ruilin 859, 865
Liu, Chong 185
Liu, He-Jin 1023
Liu, Jianbao 903, 945
Liu, Jiaxu 703
Liu, Junyin 727
Liu, Kun 1039, 1047
Liu, Lan-Juan 15
Liu, Qiankun 349
Liu, Shao-wei 977
Liu, Tongtong 251
Liu, Xianlin 485
Liu, Yuxuan 91
Liu, Zhizhen 831
Lou, Yuansheng 47
Lu, Huanda 243
Lu, Yinchao 727
Lu, Zhu 411
Luo, Dongyun 159
Luo, Fei 1
Lutao, Wang 39
Lv, Youxin 929

Ma, Jianming 71
Ma, Taotao 165, 177
Ma, Wenbo 785
Ma, Yue 1023
Maheshwari, Sudhanshu 735
Mahmood, Nasrul Humaimi 109
Mao, Lingfeng 299
Melo, Leonimer Flávio de 153
Meng, Jian 195
Min, Zunnan 847

Nejad, Ebrahim Behrouzian 289
Nejad, Mohammad Behrouzian 289
Nemati, Abbas 29
Niu, Lin 1023

Ouyang, Cheng 215
Ouyang, Hua 945

Peijie, Lin 959
Peng, Bao-Jin 993
Peng, Wenbin 185
Pu, Wang 617

Author Index

Qi, DeYu   1
Qian, Cui-Ping   993
Qiang, Chen   9
Qian, Ping   889, 897
Qiang, Shen   395
Qing, Wang   117
Qu, Baozeng   257
Qu, Bo   525, 533

Rahmani, Amir Masoud   289
Ran, Feng   541
Raobin   847
Rao, Min   281
Ren, Gu   587, 593
Ren, Li   679
Ren, Lixin   679
Rong, Xuewen   195
Ruan, Jing   263

SanFeng, Chen   743
Sanxiu, Wang   127
Seki, Yoichi   135
Sen, Guo   749
Shan, Weiwei   727
Shan, Yong   101
Shao, Yiou   243
Shen, Qiang   315, 493
Shengtao, Jiang   127
Shi, Ji-Hui   15
Shi-Wei, Wang   191
Shu, Bin   349
Shuying, Cheng   959
Singh, Sajai Vir   735
Song, Jianjun   343
Song, Jianjun   349
Su, Xiangjing   485
Sun, Chuanjiao   77
Sun, Haiyan   871
Sun, Huafang   727
Sun, Jiang   1001
Sun, Jingru   85, 227
Sun, Shihua   23
Sun, Xiaodong   273
Sun, Ying   1023
SunDong, Liu   743

Tang, Jian   977
Tang, Xiao-Ming   443, 453
Tian, Cunjian   803

Tong, JingJing   371
Tseng, Hwai-En   881

Wan, Lei   325
Wang, Shilong   325
Wang, Chunhua   85, 227
Wang, Haishi   1001
Wang, Jianhua   871
Wang, Jidong   1031, 1039, 1047
Wang, Jun   15
Wang, Liqiang   421
Wang, Lujun   913
Wang, Mu-Chun   553, 561, 571
Wang, Xiaoyan   921
Wang, Xinzhi   903
Wang, Yiming   307
Wang, Yuanzheng   703
Wang, Zhijian   47
Wang, Zhuo   755, 763
Wang, Ziqi   485
WanMing, Chen   743
Wei, Wei   791, 797
Wei, XiuLi   371
Wei, Zhang   411
Wenjun, Chen   145
Wu, Lei   101
Wu, Liaolan   641
Wu, Min   825
Wu, Minhua   281
Wu, Wang   55

Xia, Li   101
Xia, Naijie   871
Xia, Xiaona   817
Xiangning, Xiao   587, 593
Xian-Yi, Cheng   517
Xiao, Jinxing   165
Xiao, Ke   1001
Xiao, Liangfen   903
Xiao, Limei   477
Xiaoe, Lin   395
Xiaofeng, Wan   395
Xiaolong, Duan   403
Xie, DongQing   1
Xie, Guowen   1
Xie, Jun   281
Xie, Youhui   335
Xing, Suxia   953
Xi, Sale   435
Xu, Fei   993

Xu, Jianfeng   755, 763
Xu, Jingqi   165
Xu, Liang   371
Xu, Yuru   325
Xuan, Rongxi   349
Xuan, Xiao   617
Xue, Jun   839
Xueren, Gan   395

Yan, Dong   977
Yan, Pan   517
Yan, Xiang'an   421
Yan, Zhengguo   71
Yang, Guanqing   1031, 1039, 1047
Yang, Hsin-Chia   553, 561, 571
Yang, Jin-Long   215
Yang, Qun-Ting   937
Yang, Ren-Hau   561
Yang, Xuecheng   509
Yang, Xue-Min   315, 493
Yang, Yang   395
Yang, Zhonglin   903, 945
Ye, Peng   1001
Ye, Yunqing   633
Yecai, Guo   461
Yeh, Jih Pin   203
Yin, Chengqiang   601
Yingge, Han   461
Yongsheng, Lian   959
Yu, Bin   165
Yu, Chengbo   91
Yu, Jiguo   817
Yu, Qiongfang   711
Yu, Tao   609
Yu, Xin   243
Yu, Yazhou   503
Yu, Yongyan   47
Yuan, Chao   227
Yuan, Guici   477
Yuan-jia, Song   411
Yuan, Yanyi   913
YuFan, Long   145

Zadeali, Ahmad   289
Zhang, Ailian   719
Zhang, Bo   1001
Zhang, Cai-Sheng   443, 453

Zhang, Changsheng   263
Zhang, Heming   343, 349
Zhang, Huanshui   969
Zhang, Huijie   579
Zhang, Jiatian   71
Zhang, Jie   77
Zhang, Jin   91
Zhang, Jun-Chao   355
Zhang, Jun-Chao   363
Zhang, Kaisheng   771, 777, 785
Zhang, Lijun   299, 307
Zhang, Luhua   601
Zhang, Wei   579, 601
Zhang, Wenhao   993
Zhang, Xian   641
Zhang, Xiao-Song   469
Zhang, Yimeng   91
Zhang, Yong   889
Zhang, Yongliang   839
Zhang, Youmu   159
Zhang, Yuanyuan   427, 435
Zhang, Yujie   427, 435
Zhang, Yujun   371
ZHao, Li-jie   977
Zhao, Rong-Xiang   363
Zhao, Wenhua   839
Zhao, Zhanxi   921
Zhao, Zhiqin   929
Zheng, Boren   929
Zheng, Jiong   469
Zheng-Min, Bai   55
Zheng-wei, Yang   411
Zhijun, Kuang   985
Zhongjie, Xu   403
Zhong, Tingjian   847
Zhong, Yantao   921
Zhou, Yanming   335
Zhou, Zhongyuan   641
Zhu, Canyan   299, 307
Zhu, Haoyang   703
Zhu, Huangqiu   273
Zhu, Mingyong   913
Zhu, Shaohua   165
Zhu, Weibing   579
Zhu, Yong   649
Zou, Qingdong   641
Zou, Zhen-Yu   1023